Abscisic Acid: Metabolism, Transport and Signaling

Da-Peng Zhang
Editor

Abscisic Acid: Metabolism, Transport and Signaling

Springer

Editor
Da-Peng Zhang
School of Life Sciences
Tsinghua University
Beijing
China

ISBN 978-94-017-9423-7 ISBN 978-94-017-9424-4 (eBook)
DOI 10.1007/978-94-017-9424-4

Library of Congress Control Number: 2014950346

Springer Dordrecht Heidelberg New York London

Printed on acid-free paper

Springer is part of Springer Science+Business Media (www.springer.com)

Preface

Abscisic acid (ABA) is one of the five classic phytohormones, which regulates many aspects of plant growth and development, and is widely known as a 'stress hormone' that modulates plant response to a variety of environmental cues, including both abiotic and biotic challenges. Over the past 30 years, extensive studies with biochemical, molecular genetic and genomic approaches have revealed the main framework of ABA biosynthesis and catabolism pathways, and identified ABA transporters and more than a hundred signaling components of ABA signaling pathways. Especially, recent identification of ABA receptors allows to elucidate ABA signal transduction from the primary signaling events to downstream gene expression or effector activation, which deepens considerably our understanding of functional mechanisms of ABA. Additionally and importantly, numbers of the ABA-related genes have been engineered in crops for stress resistance improvement, which provides opportunities for the development of new crop varieties with enhanced stress resistance.

The objective of this book is to provide a comprehensive review of all aspects of the mechanisms of ABA metabolism, transport and signal transduction, covering the current state of knowledge of ABA and recent advances in this field. Chapters 1–5 deal with ABA metabolism and transport, which firstly introduce basic knowledge of chemical structure of ABA and its derivatives in relation to their physiological functions (Chap. 1), followed by critical reviews of ABA metabolic pathways (Chaps. 2 and 5) and ABA transport between cells and at the whole plant level in relation to its function (Chaps. 3 and 4). Chapters 6 and 7 review ABA signal perception by three classes of ABA receptors or candidate receptors with structural insights into how a family of PYR/PYL/RCAR soluble receptors function to perceive ABA, and signaling pathways downstream of the PYR/PYL/RCAR receptors and a chloroplast CHLH/ABAR candidate receptor. Protein kinases and phosphatases involved in ABA signaling are reviewed in Chap. 8, which helps to understand the critical roles of reversible protein phosphorylation in ABA signaling and especially in the PYR/PYL/RCAR-mediated signaling pathway. Some other key processes of ABA signaling, mediated by protein ubiquitination and sumoylation, reactive oxygen species (ROS) and transcription factors, are reviewed in Chaps. 9–11. Crosstalk of signaling pathways between ABA and other phytohormones as well as between ABA and light attracts much attention in recent years, of which the advances are reviewed in Chaps. 12 and 13. Additionally, ABA has been shown to be a signal to promote ripening of fleshy fruits, and the advances in its metabolism and signaling in fleshy fruits are summarized in Chap. 14. The topics of Chaps. 15–20 are focused on advances in ABA regulation of a diversity of physiological responses, including stomatal movement (Chap. 15), plant responses to drought, salt (Chap. 16) and cold stresses (Chap. 17), the floral transition (Chap. 18), circadian clock (Chap. 19), and plant response to biotic stresses (Chap. 20). The principle and practice of ABA analysis are also addressed (Chap. 21). The last chapter discusses agricultural significance of the information on ABA metabolism and signaling and reviews actuality of improvement of stress tolerance in crops by genetic manipulation of ABA metabolism and signaling (Chap. 22).

I am very grateful to the authors for their excellent contributions to this book, which are of great value to the readers with interests in plant biology and agriculture.

Da-Peng Zhang

Contents

Contributors

Ling Bai Institute of Plant Stress Biology, State Key Laboratory of Cotton Biology, Department of Biology, Henan University, Kaifeng, China

Ligang Chen Key Laboratory of Tropical Forest Ecology, Xishuangbanna Tropical Botanical Garden, Chinese Academy of Sciences, Kunming, Yunnan, China

Zhongzhou Chen State Key Laboratory of Agrobiotechnology, China Agricultural University, Beijing, China

Lucio Conti Department of Biosciences, University of Milan, Milan, Italy

Hao Du National Key Laboratory of Crop Genetic Improvement and National Center of Plant Gene Research (Wuhan), Huazhong Agricultural University, Wuhan, China

Akira Endo Department of Cell and Systems Biology, University of Toronto, Toronto, Canada; Crop Breeding Research Division, NARO Hokkaido Agricultural Research Center, Sapporo, Japan

Massimo Galbiati Department of Biosciences, University of Milan, Milan, Italy; Fondazione Filarete, Milan, Italy

Zhigang Huang Hunan Provincial Key Laboratory of Phytohormones and Growth Development, Hunan Provincial Key Laboratory for Crop Germplasm Innovation and Utilization, Hunan Agricultural University, Changsha, Hunan, China

Inhwan Hwang Division of Molecular and Life Sciences, Pohang University of Science and Technology, Pohang, South Korea; Division of Integrative Biosciences and Biotechnology, Pohang University of Science and Technology, Pohang, South Korea

Wensuo Jia College of Agronomy and Biotechnology, China Agricultural University, Beijing, China

Lun Jiang State Key Laboratory of Agrobiotechnology, China Agricultural University, Beijing, China

Mohammed Humayun Kabir Hunan Provincial Key Laboratory of Phytohormones and Growth Development, Hunan Provincial Key Laboratory for Crop Germplasm Innovation and Utilization, Hunan Agricultural University, Changsha, Hunan, China

Soo Young Kim Department of Molecular Biotechnology and Kumho Life Science Laboratory, College of Agriculture and Life Sciences, Chonnam National University, Gwangju, South Korea

Tomokazu Koshiba Department of Biological Sciences, Tokyo Metropolitan University, Tokyo, Japan

Bingbing Li College of Agronomy and Biotechnology, China Agricultural University, Beijing, China

Chuanyou Li State Key Laboratory of Plant Genomics, National Centre for Plant Gene Research, Institute of Genetics and Developmental Biology, Chinese Academy of Sciences, Beijing, China

Shan Liang Center for Plant Biology, School of Life Sciences, Tsinghua University, Beijing, China

Rongcheng Lin Key Laboratory of Photobiology, Institute of Botany, The Chinese Academy of Sciences, Beijing, China

Qing Liu Hunan Provincial Key Laboratory of Phytohormones and Growth Development, Hunan Provincial Key Laboratory for Crop Germplasm Innovation and Utilization, Hunan Agricultural University, Changsha, Hunan, China

Yun Ma Key Laboratory of Plant Molecular Physiology, Institute of Botany, Chinese Academy of Sciences, Beijing, China

Masanori Okamoto Arid Land Research Center, Tottori University, Tottori, Japan

Sergi Portolés School of Life Sciences, Center for Plant Biology, Tsinghua University, Beijing, China

Feng Qin Key Laboratory of Plant Molecular Physiology, Institute of Botany, Chinese Academy of Sciences, Beijing, China

Jocelyn K.C. Rose Department of Plant Biology, Cornell University, Ithaca, NY, USA

Mitsunori Seo RIKEN Center for Sustainable Resource Science, Yokohama, Kanagawa, Japan

Yuan-Yue Shen Beijing Key Laboratory for Agricultural Application and New Technique, Beijing University of Agriculture, Beijing, China

Yiting Shi State Key Laboratory of Plant Physiology and Biochemistry, College of Biological Sciences, China Agricultural University, Beijing, China

Chun-Peng Song Institute of Plant Stress Biology, State Key Laboratory of Cotton Biology, Department of Biology, Henan University, Kaifeng, China

Jiaqiang Sun National Key Facility for Crop Gene Resources and Genetic Improvement, Institute of Crop Science, Chinese Academy of Agricultural Sciences, Haidian District, Beijing, China

Weijiang Tang Key Laboratory of Photobiology, Institute of Botany, The Chinese Academy of Sciences, Beijing, China

Yasushi Todoroki Graduate School of Agriculture, Shizuoka University, Shizuoka, Japan

Chiara Tonelli Department of Biosciences, University of Milan, Milan, Italy

Jianhua Tong Hunan Provincial Key Laboratory of Phytohormones and Growth Development, Hunan Provincial Key Laboratory for Crop Germplasm Innovation and Utilization, Hunan Agricultural University, Changsha, Hunan, China

Pengtao Wang Institute of Plant Stress Biology, State Key Laboratory of Cotton Biology, Department of Biology, Henan University, Kaifeng, China

Xiao-Fang Wang School of Life Sciences, Center for Plant Biology, Tsinghua University, Beijing, China

Yong-Fei Wang National Key Laboratory of Plant Molecular Genetics, Institute of Plant Physiology and Ecology, Shanghai Institutes for Biological Sciences, Chinese Academy of Sciences, Shanghai, China

Yaorong Wu State Key Laboratory of Plant Genomics, National Center for Plant Gene Research, Institute of Genetics and Developmental Biology, Chinese Academy of Sciences, Chaoyang, Beijing, China

Langtao Xiao Hunan Provincial Key Laboratory of Phytohormones and Growth Development, Hunan Provincial Key Laboratory for Crop Germplasm Innovation and Utilization, Hunan Agricultural University, Changsha, Hunan, China

Yuan Xiao Hunan Provincial Key Laboratory of Phytohormones and Growth Development, Hunan Provincial Key Laboratory for Crop Germplasm Innovation and Utilization, Hunan Agricultural University, Changsha, Hunan, China

Qi Xie State Key Laboratory of Plant Genomics, National Center for Plant Gene Research, Institute of Genetics and Developmental Biology, Chinese Academy of Sciences, Chaoyang, Beijing, China

Lizhong Xiong National Key Laboratory of Crop Genetic Improvement and National Center of Plant Gene Research (Wuhan), Huazhong Agricultural University, Wuhan, China

Zheng-Yi Xu Division of Molecular and Life Sciences, Pohang University of Science and Technology, Pohang, South Korea

Shuhua Yang State Key Laboratory of Plant Physiology and Biochemistry, College of Biological Sciences, China Agricultural University, Beijing, China

Yun-Joo Yoo School of Interdisciplinary Bioscience and Bioengineering, Pohang University of Science and Technology, Pohang, South Korea

Diqiu Yu Key Laboratory of Tropical Forest Ecology, Xishuangbanna Tropical Botanical Garden, Chinese Academy of Sciences, Kunming, Yunnan, China

Feifei Yu State Key Laboratory of Plant Genomics, National Center for Plant Gene Research, Institute of Genetics and Developmental Biology, Chinese Academy of Sciences, Chaoyang, Beijing, China

Da-Peng Zhang School of Life Sciences, Center for Plant Biology, Tsinghua University, Beijing, China

Xingliang Zhang Department of Pediatrics, The Affiliated Hospital of Guangdong Medical College, Zhanjiang, Guangdong, China

Chapter 1
ABA and Its Derivatives: Chemistry and Physiological Functions

Yasushi Todoroki

Abstract ABA, a sesquiterpene, is a small molecule that has a non-planar shape and contains various functional groups. The C_{15} ABA-skeleton is common in biosynthetic precursors (xanthoxin, abscisic alcohol, and abscisic aldehyde) and oxidised catabolites (8′-hydroxy-ABA, phaseic acid, and dihydrophaseic acid). Thus, the conformation of the molecule and its functional groups are the key factors governing whether these ABA derivatives function as ABA mimics that can activate ABA receptors. Recent reports of the crystal structures of the ABA receptor proteins, PYR/PYL/RCAR (PYL), give significant clues regarding the structural requirements for eliciting ABA responses. The findings from structural studies of PYL are generally consistent with those from structure–activity studies based on bioassays using ABA analogues. The structural requirements for biosynthetic and catabolic enzymes and for ABA transporters remain unknown, as these structures have not yet been solved. However, ABA 8′-hydroxylases have been well investigated using in vitro enzyme assays. These studies show that different ABA-binding proteins have somewhat different structural requirements. Based on this knowledge, we can design a selective ligand that targets a specific ABA-binding protein.

Keywords ABA · Derivatives · Functions · Structure requirements

---

Y. Todoroki (✉)
Graduate School of Agriculture, Shizuoka University, 836 Ohya,
Suruga-Ku, Shizuoka 422-8529, Japan
e-mail: aytodor@ipc.shizuoka.ac.jp

D.-P. Zhang (ed.), *Abscisic Acid: Metabolism, Transport and Signaling*,
DOI 10.1007/978-94-017-9424-4_1

1.1 Structures, Physicochemical Properties, and Physiological Functions

1.1.1 ABA

ABA(**1**) has a chiral centre (C-1′). Consequently, two enantiomers, *S*-(+)-ABA and *R*-(−)-ABA (Fig. 1.1), are formed during chemical synthesis. All naturally occurring ABAs have the *S*-configuration, and its *R* counterpart has never been detected in plants. The physicochemical properties of both isomers of ABA are identical except for their optical characteristics, such as optical rotation and circular dichroism. ABA is sufficiently heat-stable to be dissolved in boiling water without decomposition occurring. Although ABA is relatively stable towards a broad range of pH values, it is converted to γ-lactone (**2**) under strong acidic conditions such as formic acid–hydrochloric acid (Mallaby and Ryback 1972) (Fig. 1.2). Under alkaline conditions, the hydrogens at C-3′, -5′, -7′, -2, and -6 exchange with deuterium when dissolved in deuterated water (Milborrow 1984; Willows et al. 1991; Gray et al. 1974) (Fig. 1.2). The six hydrogens at C-3′, -5′, and -7′ are particularly easily exchanged with deuterium, making ABA-d_6 (**3**) a useful internal standard for LC–MS and GC–MS analysis.

ABA is moderately photosensitive due to the dienoic acid side chain and the ring enone. Irradiation with UV at 365 nm causes photoisomerization at the C-2 double bond to give an equimolar mixture of ABA and 2*E*-ABA (2-*trans*-ABA, **5**) (Fig. 1.2), which is biologically inactive (Todoroki et al. 2001). This isomerization occurs even upon irradiation using fluorescent and tungsten lamps. UV light at wavelengths shorter than 305 nm decomposes the ABA molecule into unidentified compounds, in addition to causing photoisomerization (Cornelussen et al. 1995). This photolabile property of ABA limits its applications in agriculture. Consequently, several ABA analogues have been designed and synthesised to increase the photostability of ABA. For example, compounds **6–8** have a rigid side chain with an aromatic ring (Chen and Mctaggart 1986; Kim et al. 1992; Asami and Yoshida 1999) (Fig. 1.3). Recently, Wenjian et al. (2013) reported that 2,3-cyclopropanated ABA (**9**) is more photostable than ABA (Fig. 1.3).

ABA behaves as a weak acid due to the side chain carboxy group. Thus, the lipophilicity of ABA depends greatly on pH and is more lipophilic at lower pH. Because

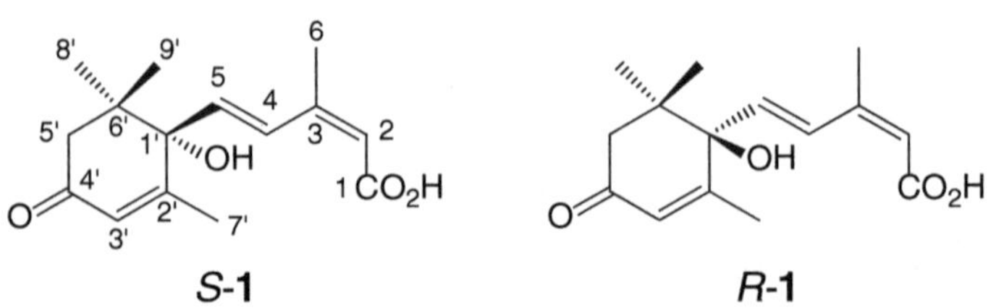

Fig. 1.1 Chemical formula of *S*-(+)-ABA (*S*-**1**) and *R*-(−)-ABA (*R*-**1**)

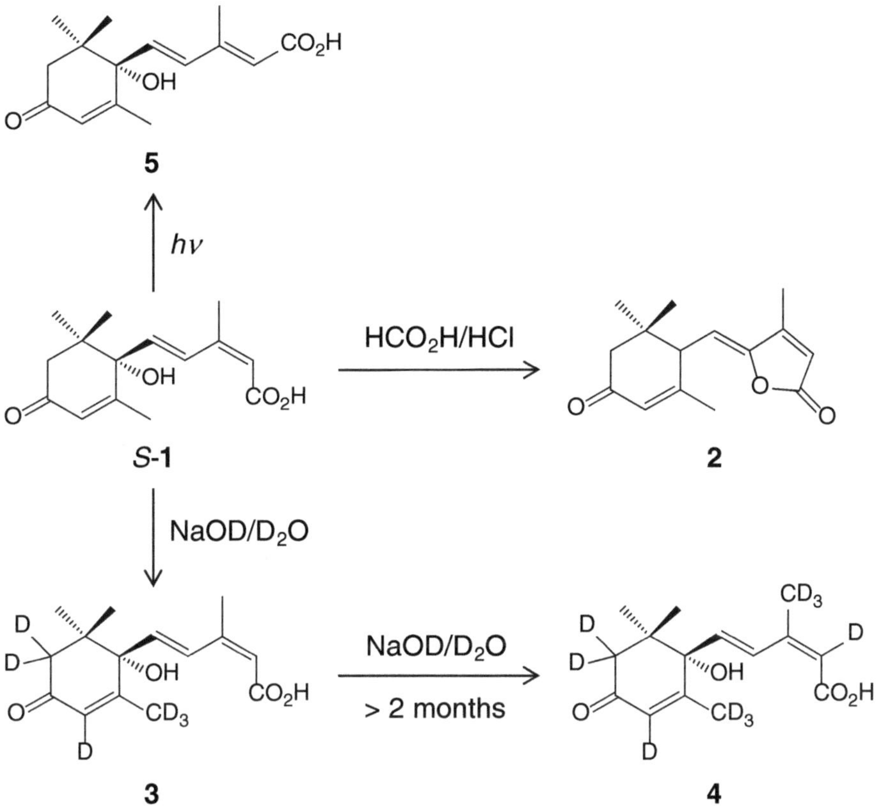

Fig. 1.2 Acidic and alkaline reactions and photoisomerization of ABA

the undissociated form of ABA is cell-membrane-permeable, the in vivo distribution of ABA changes depending on the pH conditions (Wilkinson et al. 1998; Jiang and Hartung 2008). This means that the movement of ABA within plants does not always require specific transport proteins; nevertheless, some ABA transporters seem to be involved in physiological processes mediated by ABA (Kang et al. 2010; Kuromori et al. 2010; Seo and Koshiba 2011; Kanno et al. 2012).

The shape of the ABA molecule depends largely on the conformation of the cyclohexenone ring. The crystal structure of ABA shows that the ring adopts a slightly distorted sofa conformation with the pseudoaxial side chain (Ueda and Tanaka 1977; Schmalle et al. 1977) (Fig. 1.4). The preferred form in solution, revealed by NMR and CD analyses, is a half-chair with the pseudoaxial side chain (Milborrow Milborrow 1984; Willows and Milborrow 1993; Koreeda et al. 1973; Harada 1973). Theoretical conformational analyses revealed that the minimum-energy conformer is a half-chair with the pseudoaxial side chain (Todoroki et al. 1996) (Fig. 1.4). Because the cyclohexenone ring can invert into another half-chair with the pseudoequatorial side chain by paying only a small energy penalty

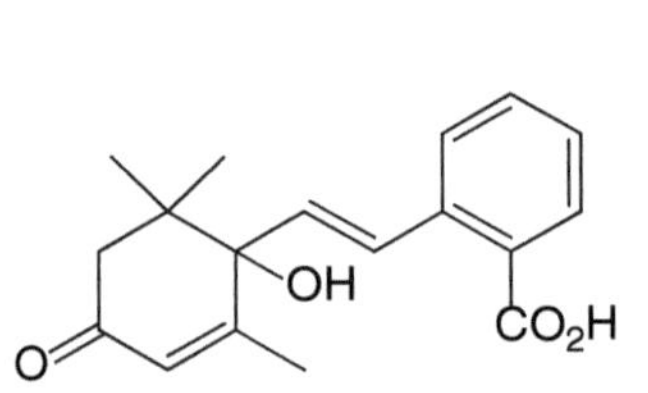

6

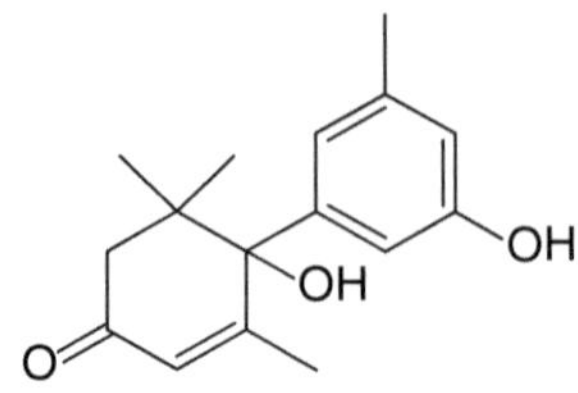

7

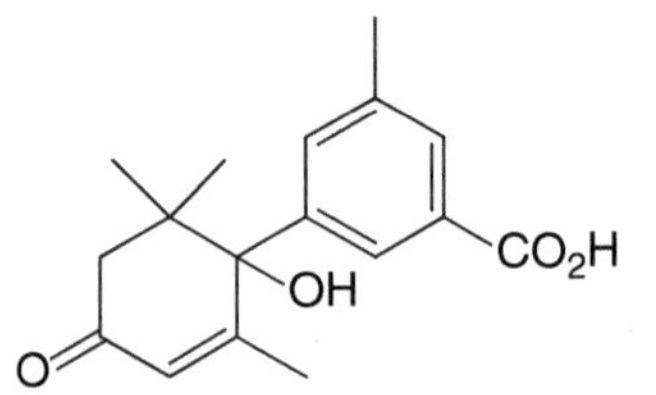

8

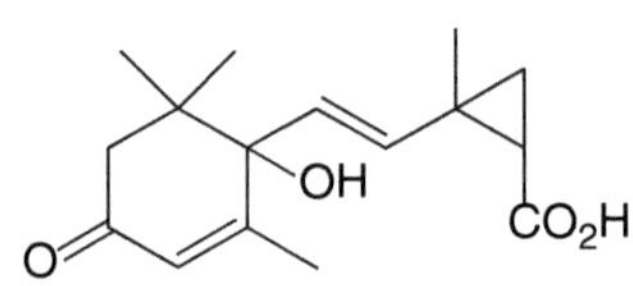

9

Fig. 1.3 Photoisomerization-resistant analogues of ABA

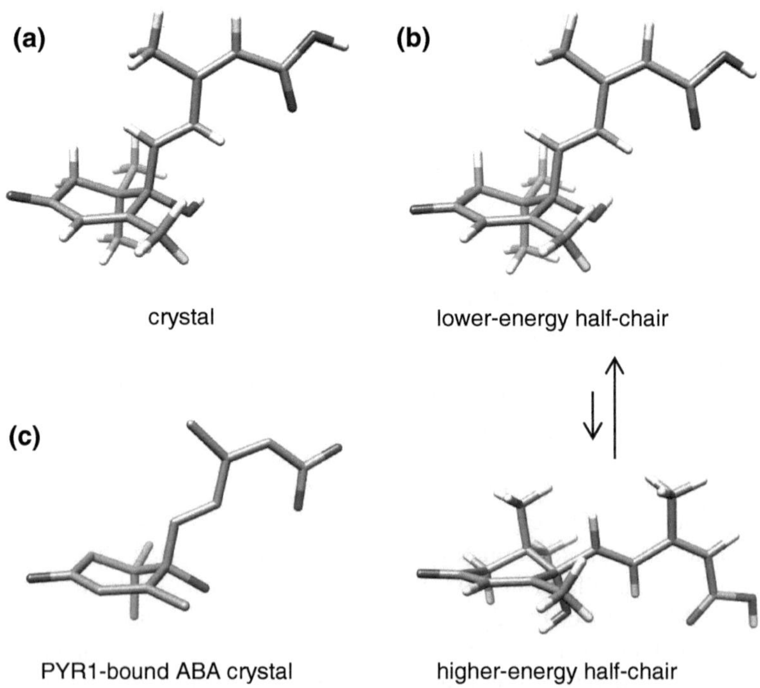

Fig. 1.4 Conformation of ABA. **a** Crystal structure determined from crystals grown in ethanol solution, **b** preferred and non-preferred half-chairs calculated in B3LYP/6-31G(d) level of theory, and **c** the extracted structure from a PYR1-ABA crystal (PDB code: 3K90)

(~1.4 kcal mol^{-1}) (Todoroki and Hirai 2000a, b) (Fig. 1.4), the shape of the ABA molecule when bound to proteins such as receptors, transporters, and catabolic enzymes cannot be estimated based on the preferred form of the uncomplexed ABA molecule. Todoroki et al. (1996) demonstrated that the biologically active conformation of ABA is a half-chair with the pseudoaxial side chain, based on the biological activity of cyclopropane analogues of ABA. Furthermore, the crystal structures of the ABA receptor proteins PYR/PYL/RCAR (PYL) bound to ABA revealed that ABA adopts a half-chair with the pseudoaxial side chain in the binding pocket (Melcher et al. 2009; Miyazono et al. 2009; Nishimura et al. 2009; Santiago et al. 2009; Yin et al. 2009) (Fig. 1.4). However, this does not guarantee that the conformation of ABA in other binding proteins, including transporters and catabolic enzymes, is also a half-chair with the pseudoaxial side chain.

Many reports have demonstrated that both the *S*- and *R*-isomers of ABA show similar hormonal activities in several assay systems (Lin et al. 2005), suggesting that plants have a mechanism which permits the *R*-isomer, which is not naturally occurring, to mimic the endogenous hormone. The ABA molecule, although a chiral molecule, is relatively symmetric about the plane constructed by C-5, C-1′, and C-4′ (Fig. 1.5). When the structural formula of *R*-ABA is drawn on paper, the orientation of the side chain and the hydroxy group at C-1′ of *S*-ABA is usually exchanged. However, this method of drawing does not accurately represent the pseudo-symmetric property of the ABA molecule. Instead, the molecule should be flipped to have the side chain and hydroxy group in the same orientation, as in the case of *S*-ABA. By doing so, a large structural difference between the enantiomers

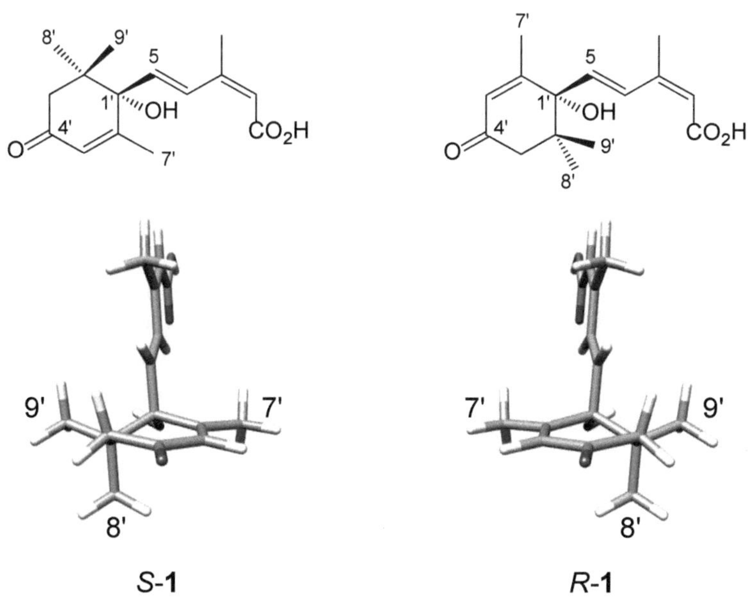

S-**1** *R*-**1**

Fig. 1.5 Structural comparison of enantiomers of ABA

becomes evident at the α-oriented, axial methyl group (C-8′). This means that the enantiomeric recognition of ABA by ABA-binding proteins depends on their sensitivity to the location of this methyl group (Fig. 1.5). The crystal structure of the PYL3-*R*-ABA complex shows that *R*-ABA is oriented in the pocket in a manner similar to *S*-ABA (Zhang et al. 2013), indicating that the ability to accommodate C-8′ provides PYL with a relatively high affinity for *R*-ABA. However, although some PYL proteins have measurable affinities for *R*-ABA, generally this affinity is weaker than that for *S*-ABA (Park et al. 2009; Zhang et al. 2013). Of the Arabidopsis PYL proteins, PYL5 shows the strongest binding affinity for *R*-ABA, whereas PYL9 does not function in the presence of *R*-ABA (Zhang et al. 2013). On the other hand, recombinant ABA 8′-hydroxylases (CYP707A enzymes), which oxidise the C-8′ of ABA, do not bind *R*-ABA (Kushiro et al. 2004). This is as expected, since CYP707A must recognise C-8′ in order to conduct oxidation, so stereospecific ligand binding is critical for this enzyme. Interestingly, C-8′ recognition by CYP707A seems to depend on the ring enone of ABA (Ueno et al. 2007). The difference in sensitivity to the orientation of C-8′ between CYP707A and PYL proteins may explain the comparable biological activity of *R*-ABA and *S*-ABA in some bioassays. *R*-ABA is metabolized to (−)-7′-hydroxy-ABA (**21**) (Cowan and Railton 1987; Boyer and Zeevaart 1986; Balsevich et al. 1994a) and (+)-PA (**23**) (Gillard and Walton 1976; Balsevich et al. 1994b; Dashek et al. 1979; Okamoto and Nakazawa 1993) (Fig. 1.6). Since the recombinant CYP707A enzymes do not bind *R*-ABA (Kushiro et al. 2004), these metabolites may be produced by the nonspecific oxidation of *R*-ABA.

1.1.2 Biosynthetic Precursors

The biosynthetic precursors that have an ABA-skeleton are xanthoxin (**10**), abscisic alcohol (ABAlc, **16**), and abscisic aldehyde (ABAld, **11**) (Fig. 1.6). The side chain of xanthoxin cannot be axial owing to the 1′,2′-epoxide, so the conformation of xanthoxin must be quite different from that of ABAlc, ABAld, and ABA. Crystal structures of the PYL–ABA complex suggest that these precursors, which have no carboxylic acid moiety, generally have poor affinity to PYL proteins because the salt bridge between the C-1 carboxylate of ABA and the ε-ammonium group of Lys of PYL is critical for forming a stable PYL–ABA complex. In fact, these precursors induce a slight reduction in PP2C activity in the presence of PYL proteins (Kepka et al. 2011).

1.1.3 Catabolites

Common catabolites of ABA in many plants include phaseic acid (PA, **13**), dihydrophaseic acid (DPA, **14**), *epi*-dihydrophaseic acid (*epi*-DPA, **15**),

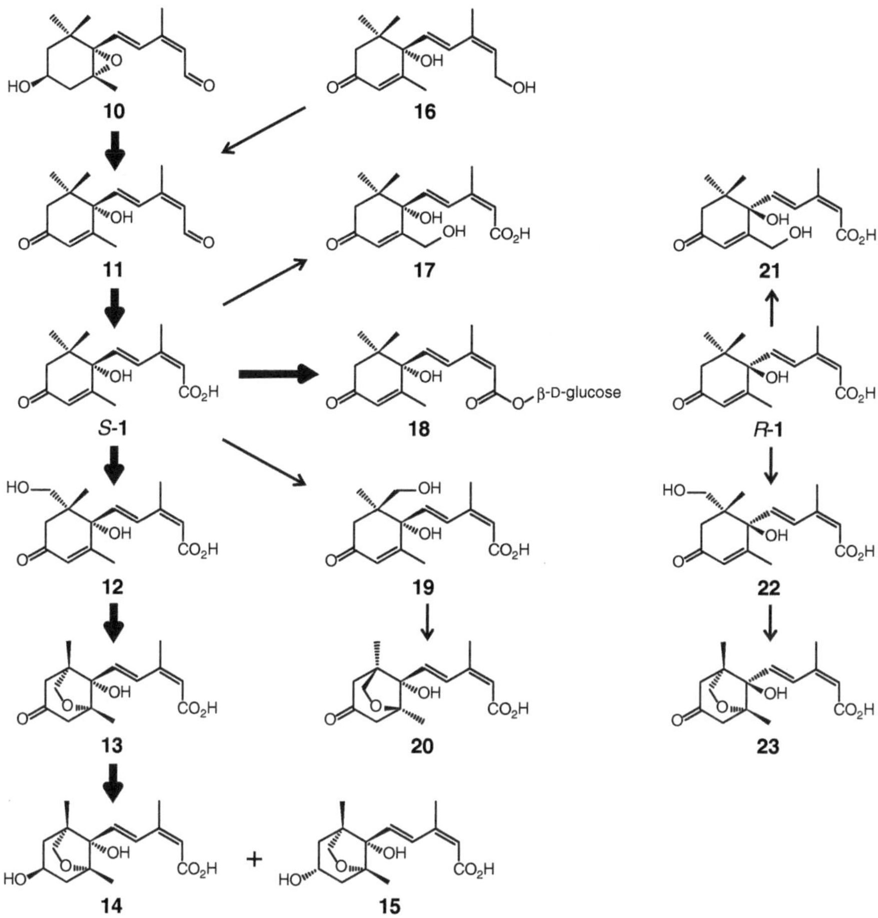

Fig. 1.6 Biosynthetic and catabolic pathway of ABA from xanthoxin. The pathway shown with *bold arrows* is the main metabolic pathway found commonly in many higher plants

7′-hydroxy-ABA (**17**), and ABA glucose ester (ABA-GE, **18**) (Ref) (Fig. 1.6). 9′-Hydroxy-ABA (**19**) and neoPA (**20**) are found in some plants, including Arabidopsis (Zhou et al. 2004; Okamoto et al. 2011). 8′-Hydroxy-ABA (**12**), which is produced by the hydroxylation of C-8′ by CYP707A enzymes, is thermodynamically unstable and spontaneously isomerizes to the more stable tautomer, PA. Thus, 8′-hydroxy-ABA is not stably maintained in the absence of 8′-*O*-protection following isolation. Zou et al. (1995) isolated 8′-hydroxy-ABA as a borate complex by heating PA and boric acid in glacial acetic acid. The isomerization of 8′-hydroxy-ABA to PA is an intramolecular Michael-type reaction which is accelerated under basic conditions. At 25 °C, the half-life of 8′-hydroxy-ABA is 30 h at pH 3, 4 h at pH 7, and shorter than 1 min at pH 10 (Todoroki and Hirai 2000a, b). 8′-Hydroxy-ABA is observed upon HPLC analysis of enzyme assay mixtures using recombinant CYP707A (Saito et al. 2004). This suggests that CYP707A catalyses only the hydroxylation reaction, and not the isomerization reaction to PA.

Is 8′-hydroxy-ABA biologically active? It is difficult to answer this question because its thermodynamic instability prevents us from determining its true activity. Zou et al. (1995) reported that a borate complex of 8′-hydroxy-ABA is active in lipid and oleosin biosynthesis in microspore-derived embryos of *B. napus*. Kepka et al. (2011) extrapolated the activity of 8′-hydroxy-ABA from that of the tetralone analogue of 8′-hydroxy-ABA (**24**). This analogue cannot isomerize to a PA-type derivative because the C-2′ is in an aromatic ring (Fig. 1.7). Tetralone-ABA is as effective as ABA in some bioassays using Arabidopsis, whereas the 8′-hydroxylated analogue is much less effective than tetralone-ABA. In PP2C assays using the Arabidopsis ABA receptors, PYL9 and PYR1, 8′-hydroxylated tetralone-ABA is less effective than tetralone-ABA and ABA by a factor of 100–1000. This suggests that 8′-hydroxy-ABA is essentially inactive, at least in Arabidopsis. Because some space is evident around the C-8′ of ABA in both the PYR1-ABA and PYL9-ABA complexes (PDB code: 3K90 and 3OQU, respectively), the 8′-hydroxy group is accommodated sterically. Furthermore, since ABA stabilizes the gate-closed conformation of PYL proteins by producing a hydrophobic network, a protic polar moiety around C-8′ may impair the critical function of the proteins. In any case, currently there is no experimental evidence to help resolve the discrepancy in ABA activity between *Brassica napus* and Arabidopsis.

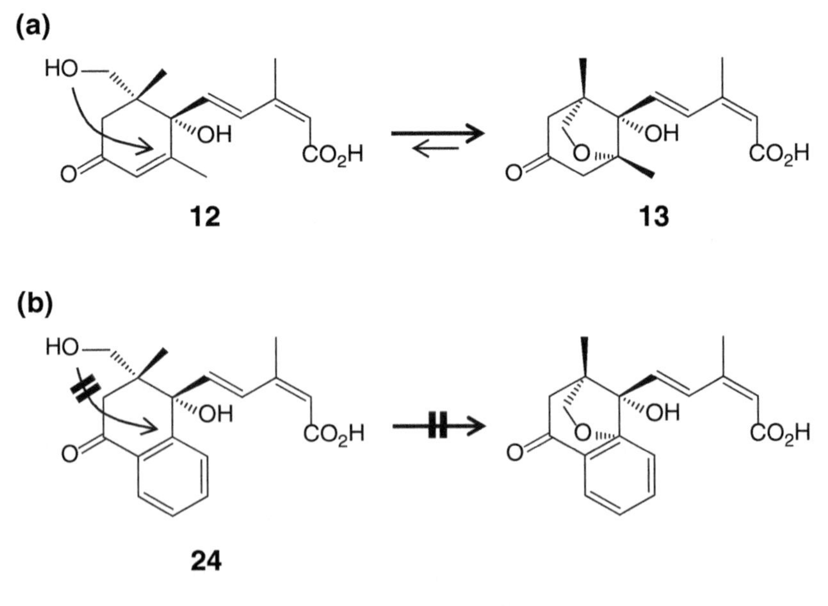

Fig. 1.7 Intramolecular Michael-type reaction of 8′-hydroxylated compounds. **a** 8′-Hydroxy-ABA (**12**) is isomerized to PA (**13**), whereas **b** the 8′-hydroxylated tetralone analogue of ABA is not converted to the bicyclized compound owing to the stable aromatic ring

PA is a more rigid molecule than ABA. The conformation of the six-membered ring of PA is a rigid chair with an axial side chain. This conformation is similar to that of ABA bound to PYL proteins; however, PA has much less affinity with PYL compared to ABA (Kepka et al. 2011). PA has an α-axial substituent at C-2′, like the substituent at C-6′ of *R*-ABA. However, the C-2′ substituent of PA is not a methyl group, but an ether oxygen. This is a significant disadvantage for interacting with PYL because the ether oxygen reduces the affinity of PA with PYL due to steric hindrance and hydrophilicity. In fact, PA is inactive in most bioassays. Although rarely mentioned, solutions of PA inevitably contain a trace amount of 8′-hydroxy-ABA; the equilibrium ratio of 8′-hydroxy-ABA/PA is 2:98 at pH 3–10 (Todoroki and Hirai 2000a, b). Consequently, a peak and signals due to 8′-hydroxy-ABA are evident in the HPLC and ^{1}H NMR analyses of PA, respectively.

1.1.4 ABA Mimics Lacking an ABA-skeleton

Although ABA is produced by fermentation using ABA-producing fungi at relatively low cost, ABA mimics with a simpler structure than ABA would be practical and advantageous if their production costs were lower than that of ABA (Fig. 1.8). In early studies, lunularic acid (**25**) was considered to play an ABA-like physiological role in liverworts (Pryce 1972), although currently this hypothesis is in question because ABA has been detected in two species of liverworts (Nakayama et al. 1993). Some synthetic ABA mimics containing an aromatic ring instead of the cyclohexenone ring of ABA (**26**–**28**) have been reported, although their activities are lower than that of ABA (Bittner et al. 1977; Ladyman et al. 1988; Yoshikawa et al. 1992). The Lewis formulae of these compounds somewhat resemble ABA. However, these molecules are relatively flat, so they bear little resemblance to ABA sterically.

Pyrabactin (**29**), a key compound for identifying ABA receptors amongst PYL proteins, exhibits weak ABA-like activity (Park et al. 2009). Although this compound does not superficially resemble ABA, the bent conformation functions as an effective gate-closing promoter in PYL1 (Melcher et al. 2010) (Fig. 1.8). Interestingly, pyrabactin binds to PYL2 in a manner different from its binding to PYL1 and does not induce closure of the gate. The ABA agonistic effect of pyrabactin is observed in in vitro experiments only for PYR1 and PYL1 (Okamoto et al. 2013). Quinabactin (**30**), which contains aromatic rings and a sulfonamide similar to pyrabactin, is a stronger ABA mimic than is pyrabactin (Okamoto et al. 2013). In vitro analysis revealed that quinabactin activates mainly dimeric receptors to elicit stomatal closure, ABA-regulated gene expression, and drought tolerance. Quinabactin also adopts a bent conformation in PYL2 (Fig. 1.8). The dimeric receptor-selective binding of quinabactin may depend on its slightly larger molecular size compared to ABA.

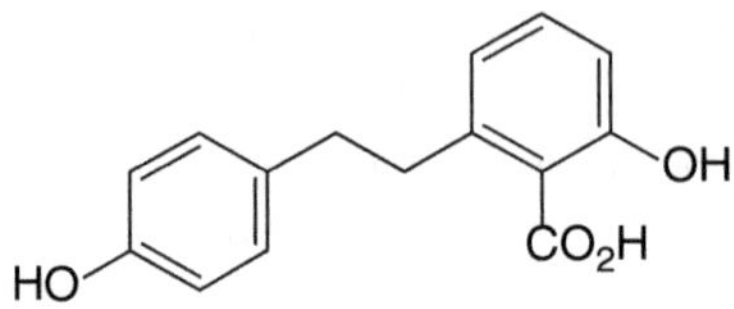

25

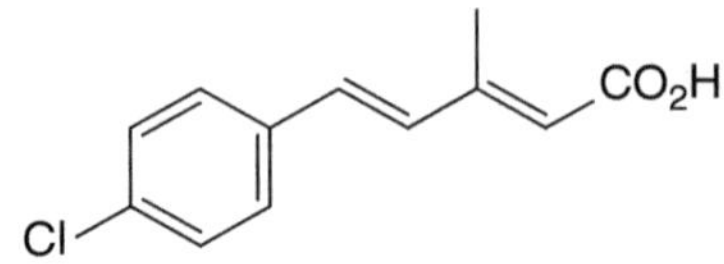

27

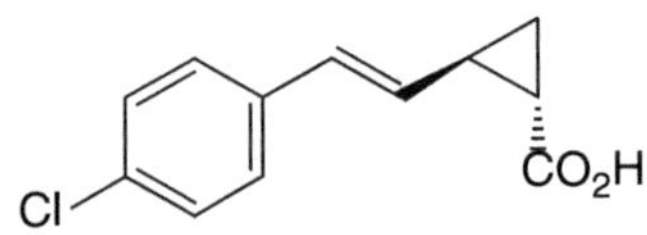

26

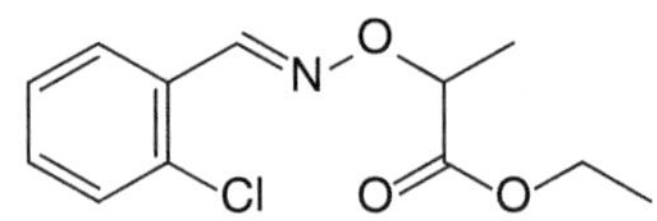

28

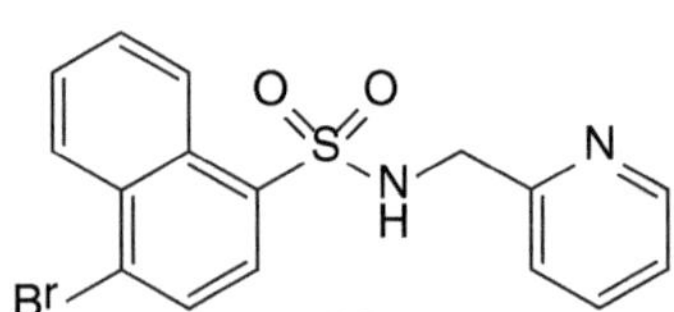

29

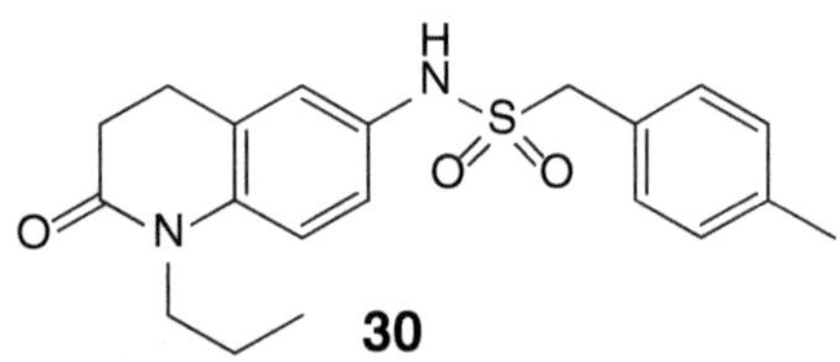

30

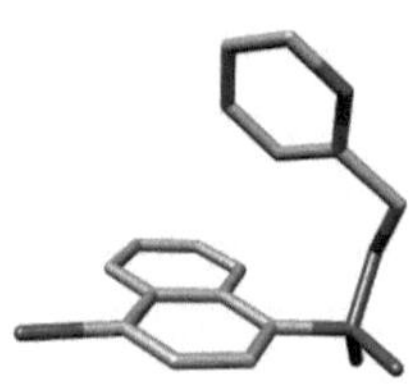

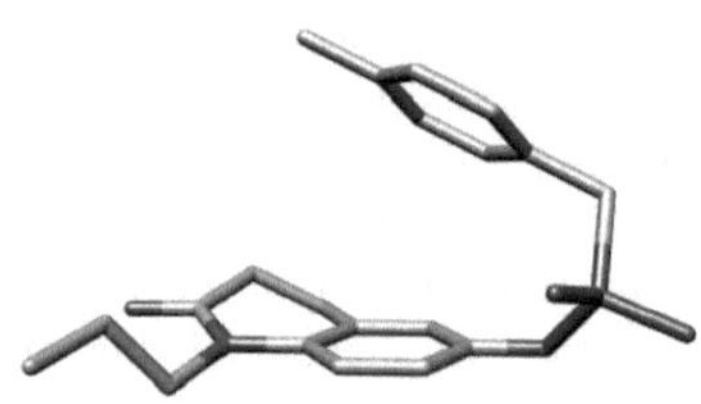

Fig. 1.8 ABA mimics lacking an ABA-skeleton. Two molecular models extracted from the PYL1-pyrabactin (**29**) and PYL2-quinabactin (**30**) crystal structures (PDB code: 3NEF and 4LA7, respectively)

1.2 Structural Requirements

1.2.1 For Exhibiting Biological Activity

The biological activities of many ABA analogues have been examined using various plants and tissues under various conditions (Fig. 1.9). Although there are some exceptions, the absence of an existing functional group, or the addition of a new functional group, generally reduces the ABA activity. The most critical functional group is the C-1 carboxy group, although ABA analogues that have a C-1 hydroxy (ABAlc), aldehyde (ABAld), or methyl ester (Me ABA, **31**) are relatively active under some assay conditions. For example, the *aba3* mutant, which is deficient in molybdenum cofactor (MoCo) sulfurase necessary for ABAld oxidase (AAO3), is insensitive to ABAlc and ABAld (Kepka et al. 2011). The relatively high activity

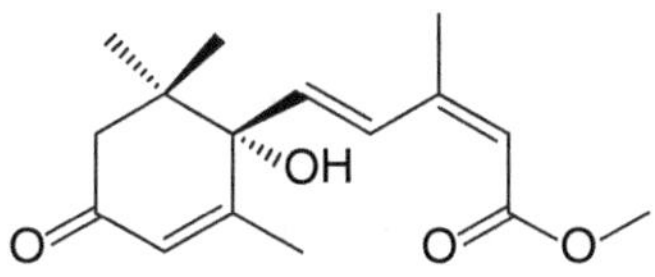

31

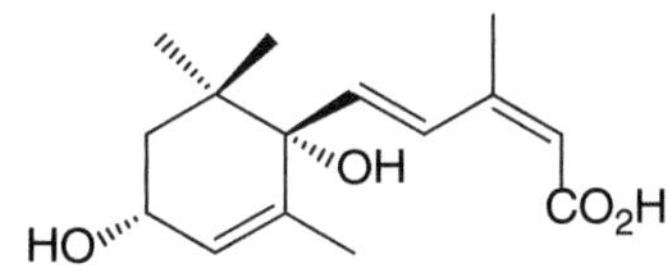

36

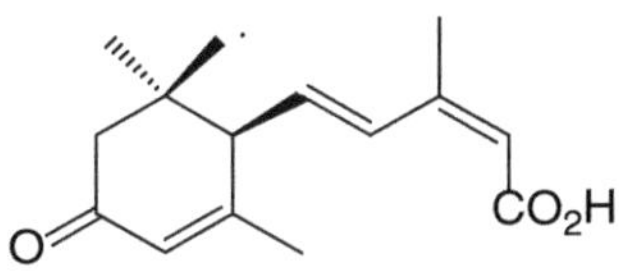

32

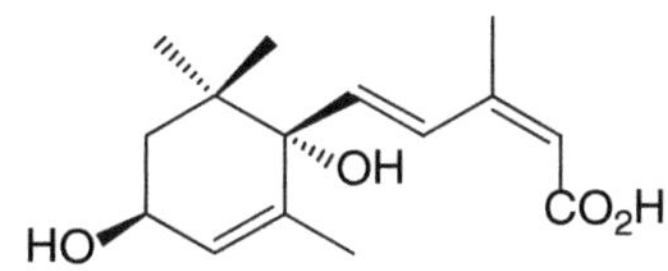

37

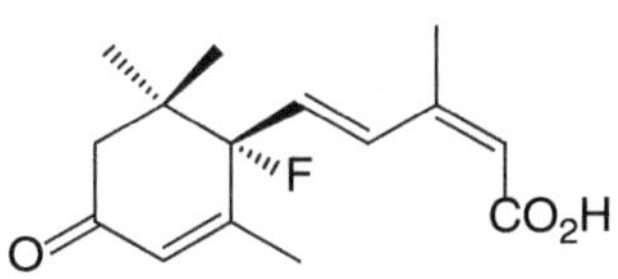

33

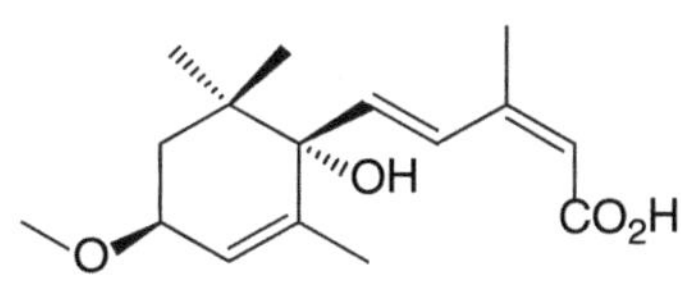

38

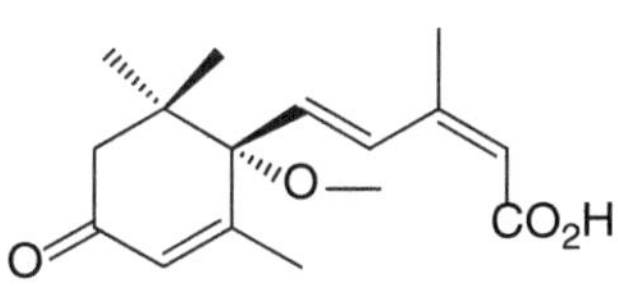

34

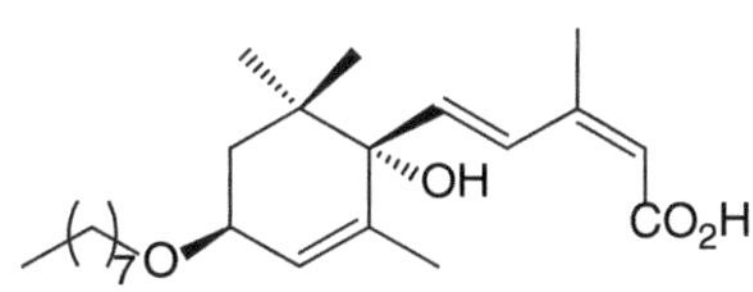

39

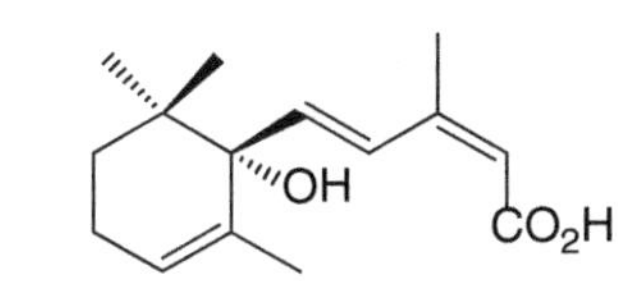

35

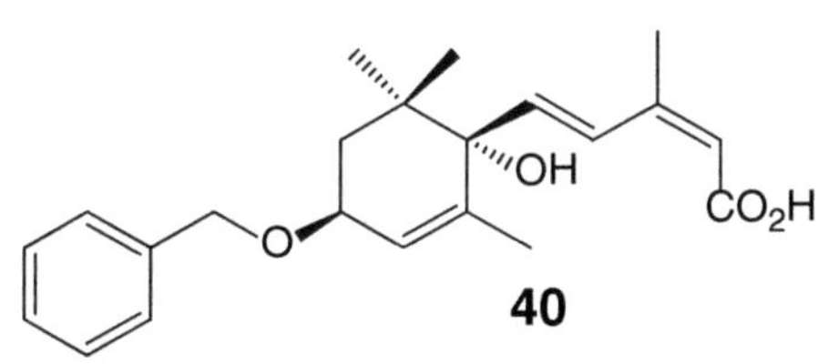

40

Fig. 1.9 The C-1, C-1′, and C-4′ modified ABA analogues

of these ABA precursors results from ABA being generated from them in plants. Since plant cells generally exhibit high esterase activity and rapidly hydrolyse esters (Kusaka et al. 2009), the activity of Me ABA must also result from its conversion to ABA by enzymatic hydrolysis.

The other two polar groups in ABA are also important for ABA activity, although they are less crucial than the C-1 carboxy group: The reported activities of

1′-deoxy-ABA (**32**), 1′-fluoro-ABA (**33**) (Todoroki et al. 1995a, b, c), 1′-*O*-methyl-ABA (**34**) (Rose et al. 1996), 4′-deoxo-ABA (**35**) (Oritani and Yamashita 1974), 1′,4′-*cis*/*trans*-diol-ABAs (**36** and **37**) (Walton and Sondheimer 1972), and *trans*-4′-alcoxy-ABA (**38–40**) (Asami et al. 2000; 2002) are weaker than that of ABA by a factor of 10–100. The *cis* isomer of 4′-alcoxy-ABA is much less potent than the *trans* isomer. Interestingly, some ABA analogues containing a larger functional group than ABA's functional group at C-4′ (**41–44**) (Fig. 1.10) retain ABA activity, albeit weaker than ABA's by a factor of 10–100 (Kohler et al. 1997; Asami et al. 1997; Kitahata et al. 2005). Because these compounds are substituted using a hydrozone linker at C-4′, it is possible that ABA is released by hydrolysis of the analogue in plant cells. On the other hand, the activity of 4′-octoxy-ABA (**39**) and 4′-benzyl-ABA (**40**) (Asami et al. 2002) cannot be derived from released ABA and must be due to the analogue, since an ether bond is more resistant to hydrolysis than a hydrazone. The significance of this will be discussed later.

The two methyl groups at C-3 in the side chain and at C-2′ in the ring (C-6 and C-7′, respectively) are more important for activity than the geminal methyl groups at C-6′ in the ring (C-8′ and C-9′). The C-6 and C-7′ methyl groups may contribute to specific recognition rather than to nonspecific hydrophobic interactions. The modification of C-8′ or C-9′ sometimes results in an increase in activity. In particular, ABA analogues, in which C-8′ is replaced by a trifluoromethyl (**45**) or an acetylenyl (**46**), and where C-9′ is replaced by a propagyl (**47**), are the most potent analogues tested to date in bioassays (Todoroki et al. 1995a, b, c; Cutler et al. 2000) (Fig. 1.11). 5′α,8′-Cyclo-ABA (**48**), which contains a direct linkage between C-5′ and C-8′, is also remarkably effective in some bioassays (Todoroki et al. 1996). The amplified activity of these analogues may depend partially on their long lifetime in plant cells, since modification of C-8′ or its neighbouring groups makes these compounds resistant to CYP707A enzymes (Cutler et al. 2000). Nevertheless, since recombinant Arabidopsis ABA 8′-hydroxylase CYP707A3 is not inhibited by the 8′-acetylenic analogue of ABA (Ueno et al. 2005), another factor may be involved. 2*E*-ABA (**5**) is also inactive, probably because the 2*E*-configuration greatly affects the orientation of the C-1 carboxy group. The C-2′ double bond in the ring can be changed to a single bond with only a slight loss of activity if C-7′ is *cis* to the side chain (**49**), that is, when the ring prefers a half-chair conformation with the pseudoaxial side chain (Fig. 1.11). This agrees with the finding that the ring conformation is a significant factor for activity. As described above, the active conformation of ABA is a half-chair with the pseudoaxial side chain.

1.2.2 For Binding to PYL Proteins

The affinity of ABA analogues to PYL proteins has rarely been measured directly or indirectly. We can only speculate on their affinity on the basis of the crystal structure of the PYL–ABA complex, where ABA is almost entirely

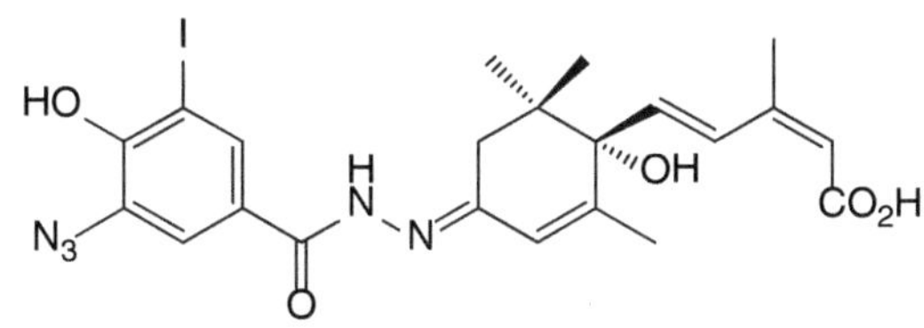

41

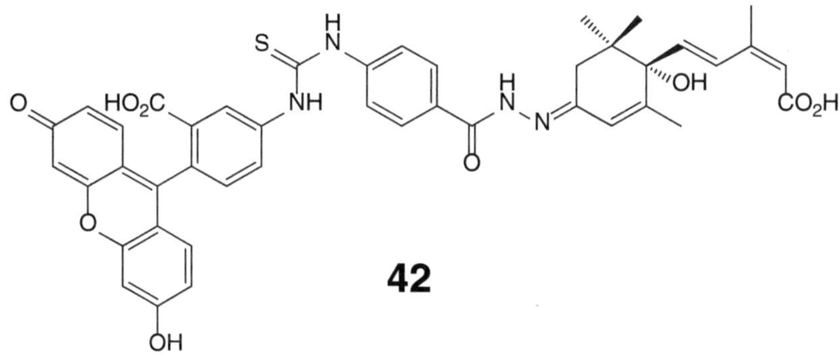

42

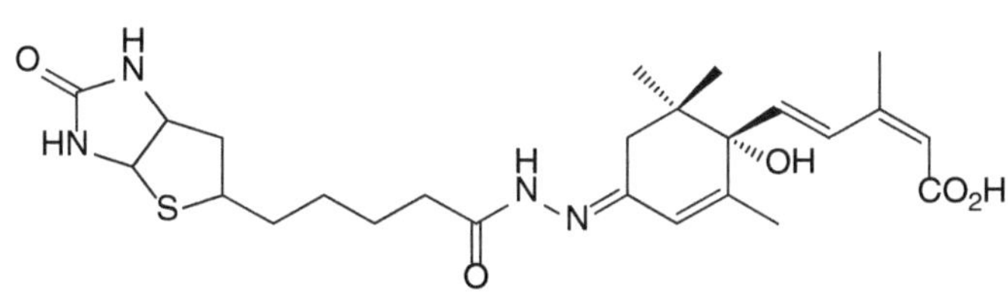

43

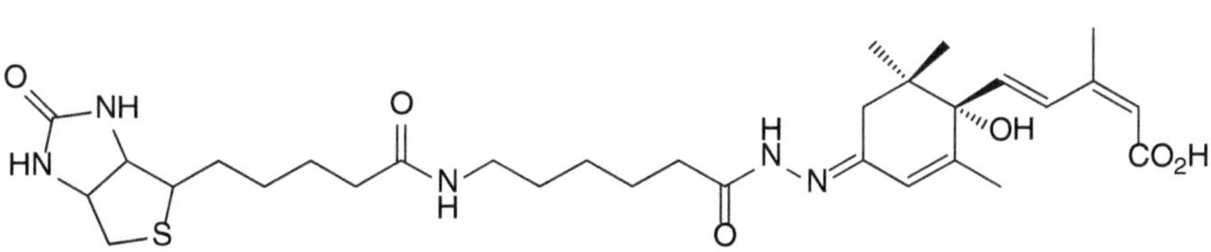

44

Fig. 1.10 ABA analogues containing a larger functional group than found at C-4′ of ABA

surrounded by the PYL protein. This binding mode suggests that the addition of a large substituent anywhere in the structure reduces the affinity because of interference with gate-closing or PP2C binding. This is generally consistent

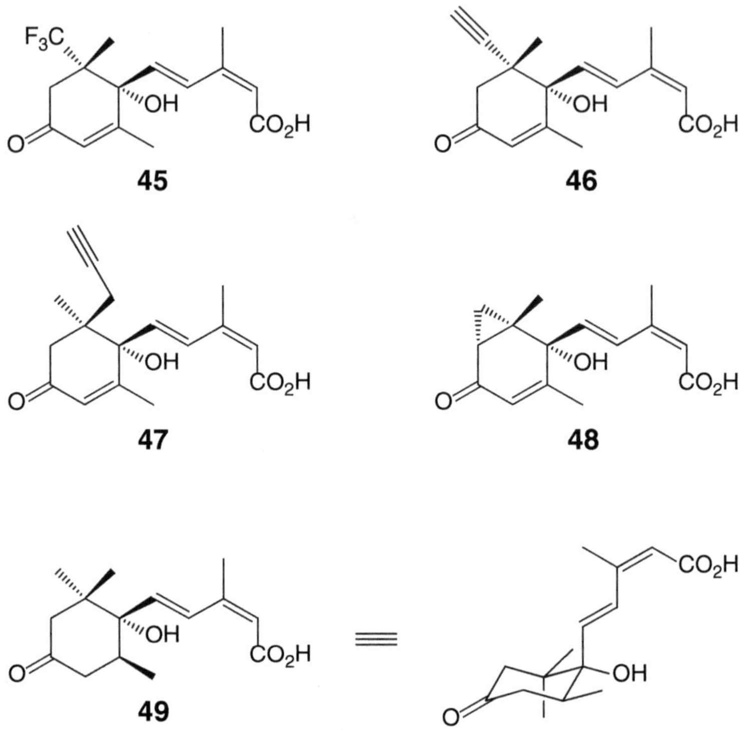

Fig. 1.11 The C-8′ and C-9′ modified ABA analogues and (1′*S*, 2′*S*)-2′,3′-dihydro-ABA

with the structural requirements for ABA activity in bioassays, discussed above. Nevertheless, some exceptions exist. The most significant inconsistency is observed in C-4′. As described above, 4′-octyl-ABA and 4′-benzyl-ABA (**39** and **40**) retain ABA activity. In the crystal structure of the PYL–ABA–PP2C complex, the C-4′ ketone of ABA forms an indirect hydrogen bond via a water molecule with the indole moiety of Trp in PP2C. In addition, the gate Pro in PYL is close to the C-4′ ketone. Thus, the introduction of a large substituent at C-4′ must impair the function of PYL as an inhibitor of PP2C. *n*-Octoxy and benzyloxy groups, with lengths of ∼10 and 5 Å, respectively, seem to be too large to fit into the space around C-4′ of ABA in the PYL–ABA–PP2C complex. In this complex, the distance between the 4′-carbonyl oxygen of ABA and the indole nitrogen of Trp in PP2C is calculated to be approximately 4.5–5 Å, based on the crystal structures (Fig. 1.12).

Recently, Takeuchi et al. (2014) reported an ABA analogue, AS6, that acts as an inhibitor of PYL proteins. This molecule has an *S*-hexyl group at C-3′ and is an effective gate-closing promoter of PYL. Crystallographic data of PYR1 bound to AS6 demonstrate that the long *S*-hexyl chain of AS6 protrudes through a solvent-exposed tunnel to prevent PYL–PP2C interactions.

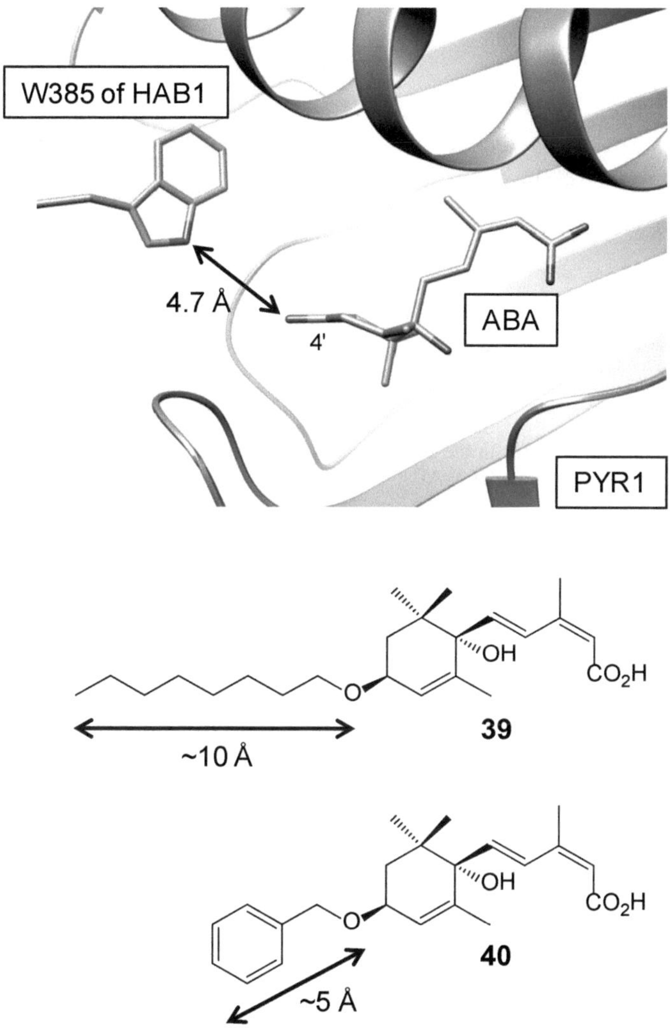

Fig. 1.12 The space around C-4′ of ABA in the PYR1–ABA–HAB1 crystal structure (PDB: 3QN1) and the C-4′ modified ABA analogues

1.2.3 For Inhibition of CYP707A Enzymes

ABA 8′-hydroxylases, which are CYP707A enzymes, are not inhibited by *R*-ABA (Kushiro et al. 2004). This strict stereoselective recognition is in marked contrast to the loose recognition requirements of PYL proteins. Ueno et al. (2007) revealed that the C-1 carboxy and C-4′ carbonyl groups play a significant role in stereoselective binding. Based on the hypothesis that ABA is anchored via these two polar groups on the helix I residues in the substrate-binding pocket of CYP707A,

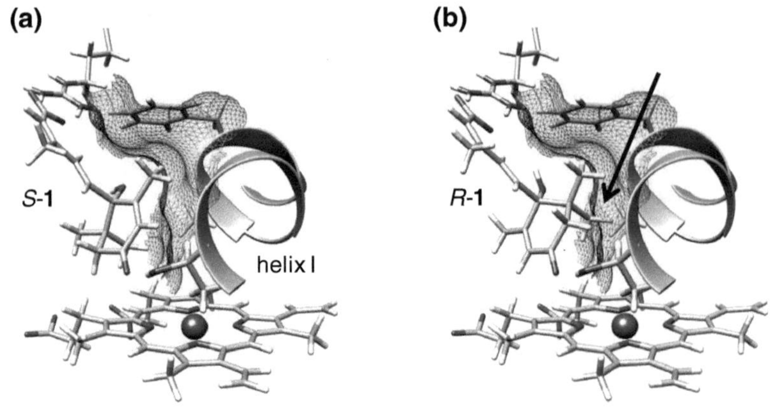

Fig. 1.13 Molecular surfaces in the active site of the 3D structural model of CYP707A3. **a** The *S*-ABA bound active site, and **b** *R*-ABA superimposed and overlaid onto *S*-ABA (omitted for simplicity)

the C-2′ side of the ring must be close to helix I (SRS 4) to keep C-8′ in close proximity to the heme iron. If *R*-ABA binds in the same manner as *S*-ABA, C-8′ should contact helix I (Fig. 1.13). Determination of the exact mechanism requires the crystal structure of CYP707A, which has yet to be obtained. Another large difference is that the side chain methyl (C-6) and C-1′ hydroxy groups which are required for the high ABA activity is dispensable for binding to CYP707A enzymes (Ueno et al. 2005) (Fig. 1.14). Focusing on this difference, Ueno et al. (2005) developed AHI1 (**50**), an ABA analogue which functions as a competitive inhibitor of CYP707A without causing an ABA response.

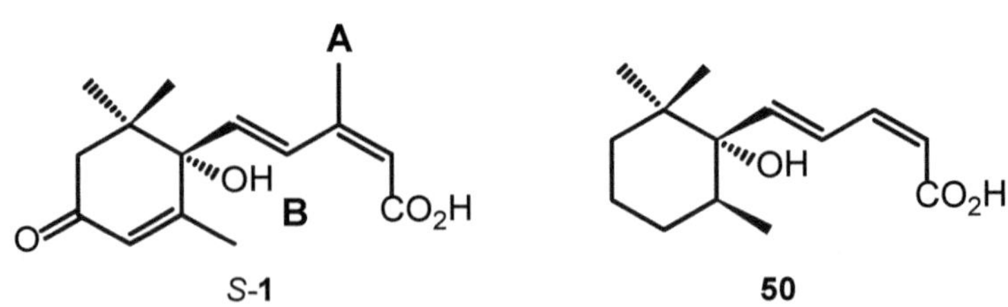

	Required for	
	CYP707A3 inhibition	ABA acitivty
A	CH_3, **H**	CH_3
B	OH, **H**, **F**	OH

Fig. 1.14 Differences in structural requirements for CYP707A3 inhibition and for ABA activity, and AHI1, a CYP707A3 inhibitor lacking ABA activity

References

Asami T, Yoshida S. Abscisic acid analogues possessing hetero five-membered ring: design, synthesis and activity. RIKEN Rev. 1999;21:17–9.

Asami T, Lin S, Yamamoto M, Robertson M, Min YK, Murofushi N, Yoshida S. Fluorescence-labeled abscisic acid possessing abscisic acid-like activity in barley aleurone protoplasts. Biosci Biotechnol Biochem. 1997;61:1198–9.

Asami T, Min YK, Han SY, Kitahata N, Oh K, Murofushi N, Yoshida S. 4′-Alkoxy derivatives of abscisic acid: a promising new probe for mechanisms of abscisic acid perceptions. Plant Growth Regul. 2002;38:237–41.

Asami T, Min YK, Matsuyama T, Murofushi N, Yamaguchi I, Yoshida S. Synthesis and biological activity of 4′-methoxy derivatives of abscisic acid. Bioorg Med Chem Lett. 2000;10:1571–4.

Balsevich J, Abrams SR, Lamb N, Konig W. Identification of unnatural phaseic acid as a metabolite derived from exogenouslyadded (−)-abscisic acid in a maize cell suspension culture. Phytochemistry. 1994a;36:647–50.

Balsevich J, Cutler A, Lamb N, Friesen L, Kurz E, Perras M, Abrams SR. Response of cultured maize cells to (+)-abscisic acid, (−)-abscisic acid, and their metabolites. PlantPhysiol. 1994b;106:135–42.

Bittner S, Gorodetsky M, Har-Paz I, MIzrahi Y, Richmond AE (1977). Synthesis and biological effect of aromatic analogues of abscisic acid. Phytochemistry 16: 1143–51.

Boyer GL, Zeevaart JAD. 7′-Hydroxy (−)-R-abscisic acid: ametabolite of feeding (−)-R-abscisic acid to *Xanthium strumarium*. Phytochemistry. 1986;25:1103–5.

Chen SC, Mctaggart JM. Abscisic acid analogueues with a geometrically rigid conjugated acid side-chain. Agric Biol Chem. 1986;50:1097–100.

Cornelussen MHM, Karssen CM, van Loon LC. UV-induced cross-linking of abscisic acid to binding proteins. Phytochemistry. 1995;39:959–68.

Cowan AK, Railton ID. The catabolism of (±)-abscisic acid by excised leaves of *Hordeum vulgare* L. cv Dyan and its modification by chemical and environmental factors. Plant Physiol. 1987;84:157–63.

Cutler AJ, Rose PA, Squires TM, Loewen MK, Shaw AC, Quail JW, Krochko JE, Abrams SR. Inhibitors of abscisic acid 8′-hydroxylase. Biochemistry. 2000;39:13614–24.

Dashek WV, Singh BN, Walton DC. Abscisic acid localization and metabolism in barley aleurone layers. Plant Physiol. 1979;64:43–8.

Gillard DF, Walton DC. Abscisic acid metabolism by a cell-free preparation from *Echinocystis lobata* liquid endoserum. Plant Physiol. 1976;58:790–5.

Gray RT, Mallaby R, Ryback G, Williams VP. Mass spectra of methyl abscisate and isotopically labelled analogueues. J Chem Soc Perkin Trans II. 1974;1974:919–24.

Harada N. Absolute configuration of (+)-trans-abscisic acid as determined by a quantitative application of the exciton chirality method. J Am Chem Soc. 1973;95:240–2.

Jiang F, Hartung W. Long-distance signalling of abscisic acid (ABA): the factors regulating the intensity of the ABA signal. J Exp Bot. 2008;59:37–43.

Kang J, Hwang J, Lee M, Kim Y, Assmann SM, Martinoia E, Lee Y. PDR-type ABC transporter mediates cellular uptake of the phytohormone abscisic acid. Proc Natl Acad Sci USA. 2010;107:2355–60.

Kanno Y, Hanada A, Chiba Y, Ichikawa T, Nakazawa M, Matsui M, Koshiba T, Kamiya Y, Seo M (2012). Identification of an abscisic acid transporter by functional screening using the receptor complex as a sensor. Proc Natl Acad Sci USA 109: 9653–8.

Kepka M, Benson CL, Gonugunta VK, Nelson KM, Christmann A, Grill E, Abrams SR. Action of natural abscisic acid precursors and catabolites on abscisic acid receptor complexes. Plant Physiol. 2011;157:2108–19.

Kim BT, Asami T, Morita K, Soh CH, Murofushi N, Yoshida S. Synthesis of new abscisic acid (ABA) analogues possessing a geometrically rigid cyclized side chain. Biosci Biotechnol Biochem. 1992;56:624–9.

Kitahata N, Nakano T, Kuchitsu K, Yoshida S, Asami T. Biotin-labeled abscisic acid as a probe for investigating abscisic acid binding sites on plasma membranes of barley aleurone protoplasts. Bioorg Med Chem. 2005;13:3351–8.

Kohler AD, Beale MH, Rollason R, Barratt DHP, Lewis MJ, Van der Meulen RM, Wang M. A radio-iodinated abscisic acid photoaffinity probe. J Chem Soc Perkin Trans. 1997;1:1543–7.

Koreeda M, Weiss G, Nakanishi K. The absolute configuration of natural (+)-abscisic acid. J Am Chem Soc. 1973;95:239–40.

Kuromori T, Miyaji T, Yabuuchi H, Sugimoto E, Kamiya A, Moriyama Y, Shinozaki K. ABC transporter AtABCG25 is involved in abscisic acid transport and responses. Proc Natl Acad Sci USA. 2010;107:2361–6.

Kusaka N, Maisch J, Nick P, Hayashi KI, Nozaki H. Manipulation of intercellular auxin in a single cell by light with esterase-resistant caged auxins. ChemBioChem. 2009;10:2195–202.

Kushiro T, Okamoto M, Nakabayashi K, Yamagishi K, Kitamura S, Asami T, Hirai N, Koshiba T, Kamiya Y, Nambara E. The Arabidopsis cytochrome P450 CYP707A encodes ABA 8′-hydroxylases: key enzymes in ABA catabolism. EMBO J. 2004;23:1647–56.

Ladyman JA, Sanborn JR, Eelkema EE. Biological activity of an aromatic analogueue of abscisic acid. Phytochemistry. 1988;27:3751–5.

Lin BL, Wang HJ, Wang JS, Zaharia LI, Abrams SR. Abscisic acid regulation of heterophylly in *Marsilea quadrifolia* L.: effects of R-(−) and S-(+) isomers. Plant Physiol. 2005;106:135–42.

Mallaby R, Ryback G. Chemistry of a color test for abscisic acid. J Chem Soc. 1972;8:919–21.

Melcher K, Ng LM, Zhou XE, Soon F-F, Xu Y, Suino-Powell KM, Park S-Y, Weiner JJ, Fujii H, Chinnusamy V, Kovach A, Li J, Wang Y, Li J, Peterson FC, Jensen DR, Yong E-L, Volkman BF, Cutler SR, Zhu J-K, Xu HE. A gate–latch-lock mechanism for hormone signalling by abscisic acid receptors. Nature. 2009;462:602–8.

Melcher K, Xu Y, Ng LM, Zhou XE, Soon FF, Chinnusamy V, Suino-Powell KM, Kovach A, Tham FS, Cutler SR, Li J, Yong EL, Zhu JK, Xu HE. Identification and mechanism of ABA receptor antagonism. Nat Struct Mol Biol. 2010;17:1102–8.

Milborrow BV. The conformation of abscisic acid by N.M.R. and a revision of the proposed mechanism for cyclization during its biosynthesis. Biochem J. 1984;220:325–32.

Miyazono K, Miyakawa T, Sawano Y, Kubota K, Kang HJ, Asano A, Miyauchi Y, Takahashi M, Zhi Y, Fujita Y, Yoshida T, Kodaira KS, Yamaguchi-Shinozaki K, Tanokura M. Structural basis of abscisic acid signalling. Nature. 2009;462:609–14.

Nakayama M, Takase T, Yokota T. Abstract of the XV international botanical congress. Yokohama: Japan; 1993. p. 388.

Nishimura N, Hitomi K, Arvai AS, Rambo RP, Hitomi C, Cutler SR, Schroeder JI, Getzoff ED. Structural mechanism of abscisic acid binding and signaling by dimeric PYR1. Science. 2009;326:1373–9.

Okamoto M, Nakazawa H. Stereoselectivity of abscisic acid oxygenase in avocado fruits. Biosci Biotechnol Biochem. 1993;57:1768–9.

Okamoto M, Kushiro T, Jikumaru Y, Abrams SR, Kamiya Y, Seki M, Nambara E. ABA 9′-hydroxylation is catalyzed by CYP707A in Arabidopsis. Phytochemistry. 2011;72:717–22.

Okamoto M, Peterson FC, Defries A, Park SY, Endo A, Nambara E, Volkman BF, Cutler SR. Activation of dimeric ABA receptors elicits guard cell closure, ABA-regulated gene expression, and drought tolerance. Proc Natl Acad Sci USA. 2013;110:12132–7.

Oritani T, Yamashita K. Syntheses and biological activities of analogues of abscisic acid. Agric Bioi Chem. 1974;38:801–8.

Park SY, Fung P, Nishimura N, Jensen DR, Fujii H, Zhao Y, Lumba S, Santiago J, Rodrigues A, Chow TF, Alfred SE, Bonetta D, Finkelstein R, Provart NJ, Desveaux D, Rodriguez PL, McCourt P, Zhu JK, Schroeder JI, Volkman BF, Cutler SR. Abscisic acid inhibits type 2C protein phosphatases via the PYR/PYL family of START proteins. Science. 2009;324:1068–71.

Pryce RJ. The occurrence of lunularic and abscisic acids in plants. Phytochemistry. 1972;11:1759–61.

Rose PA, Lei B, Shaw AC, Barton DL, Walker-Simmons MK, Abrams SR. Probing the role of the hydroxyl group of ABA: analogues with methyl ether at C-1′. Phytochemistry. 1996;41:1251–8.

Saito S, Hirai N, Matsumoto C, Ohigashi H, Ohta D, Sakata K, Mizutani M. Arabidopsis CYP707As encode (+)-abscisic acid 8′-hydroxylase, a key enzyme in the oxidative catabolism of abscisic acid. Plant Physiol. 2004;134:1439–49.

Santiago J, Dupeux F, Round A, Antoni R, Park SY, Jamin M, Cutler SR, Rodriguez PL, Marquez JA. The abscisic acid receptor PYR1 in complex with abscisic acid. Nature. 2009;462:665–8.

Schmalle VHW, Klaska KH, Jarchow O. Die Kristall- und Molekülstruktur von (R, S)-cis, trans-Abscisinsäure:5-(1-Hydroxy-2,6,6-trimethyl-4-oxo-2-cyclohexen-1-yl)-3-methyl-2,4-pentadiensäure. Acta Cryst. 1977;B33:2218–24.

Seo M, Koshiba T. Transport of ABA from the site of biosynthesis to the site of action. J Plant Res. 2011;124:501–7.

Takeuchi J, Okamoto M, Akiyama T, Muto T, Yajima S, Sue M, Seo M, Kanno Y, Kamo T, Endo A, Nambara E, Hirai N, Ohnishi T, Cutler SR, Todoroki Y. Designed abscisic acid analogues as antagonists of PYL-PP2C receptor interactions. Nat Chem Biol. 2014;10:477–82.

Todoroki Y, Hirai N. Analysis of isomerization process of 8′-hydroxyabscisic acid and its 3′-fluorinated analogue in aqueous solutions. Tetrahedron. 2000a;56:1649–53.

Todoroki Y, Hirai N. Conformational analysis of the cyclohexenone ring in abscisic acid and its analogues with a fused cyclopropyl ring. Tetrahedron. 2000b;56:8095–100.

Todoroki Y, Hirai N, Koshimizu K. 8′,8′,8′-Trifluoroabscisic acids as highly potent, long-lasting analogueues of abscisic acid. Phytochemistry. 1995a;38:561–8.

Todoroki Y, Hirai N, Koshimizu K. Synthesis and biological activity of 1′-deoxy-1′-fluoro- and 8′-fluoroabscisic acids. Phytochemistry. 1995b;40:633–41.

Todoroki Y, Nakano S, Hirai N, Ohigashi H. Synthesis, biological activity and metabolism of (S)-(+)-3′-Fluoroabscisic acid. Tetrahedron. 1995c;51:6911–26.

Todoroki Y, Nakano S, Hirai N, Ohigashi H. Ring conformational requirement for biological activity of abscisic acid probed by the cyclopropane analogueues. Tetrahedron. 1996;52:8081–98.

Todoroki Y, Tanaka T, Kisamori M, Hirai N. 3′-Azidoabscisic acid as a photoaffinity reagent for abscisic acid binding proteins. Bioorg Med Chem Lett. 2001;11:2381–4.

Ueda H, Tanaka J. The crystal and molecular structures of dl-2-cis-4-trans-abscisic acid. Bull Chem Soc Jpn. 1977;50:1506–9.

Ueno K, Araki Y, Hirai N, Saito S, Mizutani M, Sakata K, Todoroki Y. Differences between the structural requirements for ABA 8′-hydroxylase inhibition and for ABA activity. Bioorg Med Chem. 2007;13:3359–70.

Ueno K, Araki Y, Hirai N, Saito S, Mizutani M, Sakata K, Todoroki Y. Differences between the structural requirements for ABA 8′-hydroxylase inhibition and for ABA activity. Bioorg Med Chem. 2005;13:3359–70.

Walton DC, Sondheimer E. Activity and metabolism of C-(±)-abscisic acid derivatives. Plant Physiol. 1972;49:290–2.

Wenjian L, Xiaoqiang H, Yumei X Jinlong F, Yuanzhi Z, Huizhe L, Mingan W, Zhaohai Q (2013). Synthesis, photostability and bioactivity of 2,3-cyclopropanated abscisic acid. Phytochemistry. 2013;96:72–80.

Wilkinson S, Corlett JE, Oger L, Davies WJ. Effects of xylem pH on transpiration from wild-type and flacca tomato leaves. Plant Physiol. 1998;117:703–9.

Willows RD, Milborrow BV. Configurations and conformations of abscisic acid. Phytochemistry. 1993;34:233–7.

Willows RD, Netting AG, Milborrow BV. Synthesis of stably deuteriated abscisic acid, phaseic acid and related compounds. Phytochemistry. 1991;30:1483–5.

Yin P, Fan H, Hao Q, Yuan X, Wu D, Pang Y, Yan C, Li W, Wang J, Yan N. Structural insights into the mechanism of abscisic acid signaling by PYL proteins. Nat Struct Mol Biol. 2009;16:1230–6.

Yoshikawa H, Fujimoto E, Doi K (1992). Synthesis and biological activity of benzaldehyde O-alkyloximes as abscisic acid mimics (Part 1). Biosci Biotechnol Biochem 56: 256–60.

Zhang X, Jiang L, Wang G, Yu L, Zhang Q, Xin Q, Wu W, Gong Z, Chen Z. Structural insights into the abscisic acid stereospecificity by the ABA receptors PYR/PYL/RCAR. PLoS ONE. 2013;8:e67477.

Zhou R, Cutler AJ, Ambrose SJ, Galka MM, Nelson KM. A new abscisic acid catabolic pathway. Plant Physiol. 2004;134:361–9.

Zou J, Abrams GD, Barton DL, Taylor DC, Pomeroy MK, Abrams SR. Induction of lipid and oleosin biosynthesis by (+)–abscisic acid and its metabolites in microspore-derived embryos of Brassica napus L. cv. Reston. Biological responses in the presence of 8′-hydroxy abscisic acid. Plant Physiol. 1995;108:563–71.

Chapter 2
ABA Biosynthetic and Catabolic Pathways

Akira Endo, Masanori Okamoto and Tomokazu Koshiba

Abstract Abscisic acid (ABA) is a phytohormone that regulates physiological processes such as seed maturation, seed dormancy, and stress adaptation. These physiological responses are triggered by the fluctuation of endogenous ABA levels in accordance with changing surroundings or developmental stimuli. Endogenous ABA levels are largely controlled by the balance between biosynthesis and catabolism. ABA is also synthesized in various kinds of organisms other than plants. To manipulate ABA levels, we first need to understand the pathways for ABA biosynthesis and catabolism in each organism. The biosynthetic pathway has been extensively studied in plants and phytopathogenic fungi. The catabolic pathway has been mostly established in plants. Extensive investigations of mutants defective in ABA metabolism using biochemical, molecular genetic, and genomic approaches have helped to reveal the main framework of these pathways. This chapter reviews our current understanding of the pathways of ABA biosynthesis and catabolism. In addition, inhibitors of ABA biosynthesis and catabolism are introduced. These inhibitors can be used to manipulate endogenous ABA levels and are useful tools to investigate ABA action in plants.

Keywords ABA · Biosynthesis · Catabolism

A. Endo
Department of Cell and Systems Biology,
University of Toronto, Toronto, Canada
e-mail: akiraendo@affrc.go.jp

Present Address
Crop Breeding Research Division,
NARO Hokkaido Agricultural Research Center, Sapporo, Japan

M. Okamoto
Arid Land Research Center, Tottori University, Tottori, Japan
e-mail: okamo@alrc.tottori-u.ac.jp

T. Koshiba (✉)
Department of Biological Sciences,
Tokyo Metropolitan University, Tokyo, Japan
e-mail: koshiba-tomokazu@tmu.ac.jp

D.-P. Zhang (ed.), *Abscisic Acid: Metabolism, Transport and Signaling*,
DOI 10.1007/978-94-017-9424-4_2

2.1 ABA Biosynthesis—Two Distinct Routes of ABA Biosynthesis

Abscisic acid (ABA) is a C15 sesquiterpene containing fifteen carbon atoms in its structure that originates from isoprene known as isopentenyl pyrophosphate (IPP). Now, ABA is known to be synthesized via two distinct pathways (Nambara and Marion-Poll 2005; Oritani and Kiyota 2003; Schwartz and Zeevaart 2010). One is the direct pathway that occurs in phytopathogenic fungi. The other is the indirect pathway that operates in plants. IPP for the direct pathway is synthesized from the mevalonate (MVA) pathway that exists in prokaryotes and almost all eukaryotes (Newman and Chappell 1999). On the other hand, the indirect pathway uses the methylerythritol phosphate (MEP) pathway as a source of IPP. The MEP pathway seems to exist in cyanobacteria and all photosynthetic eukaryotes (Lichtenthaler 1999).

2.2 Direct Pathway in Fungi

In phytopathogenic fungi such as *Botrytis cinerea* and *Cercospora cruenta*, ABA is synthesized from MVA. When radiolabeled MVA or farnesyl diphosphate (FDP) was fed to ABA-producing fungi, the labels were effectively incorporated into ABA (Neill et al. 1982; Norman et al. 1983). In addition, when [1-^{13}C] glucose was fed to ABA-producing fungi or plants, the positions of the labeled carbons in ABA were different between plants and fungi, because the labeled carbon was differently incorporated into IPP depending on its source, the MVA or MEP pathway (Hirai et al. 2000). These results strongly indicate that fungal ABA is synthesized from IPP produced in the MVA pathway (Hirai et al. 2000). Since all intermediates between FDP and ABA are sesquiterpenes, the ABA biosynthetic pathway in fungi has been referred to as the direct pathway (Nambara and Marion-Poll 2005; Oritani and Kiyota 2003; Zeevaart and Creelman 1988).

The direct pathway involves several modifications of FDP to generate ABA (Fig. 2.1). Isomers of ionylideneethanol and/or ionylideneacetic acid have been identified from several fungi and are supposed to be the endogenous precursors of ABA in fungi (Oritani and Kiyota 2003). As shown in Fig. 2.1, similar intermediates are utilized among diverse fungal genera, suggesting that the ABA biosynthetic pathway might be mostly similar but distinct among these fungi. Recently, allofarnesenes and ionylideneethanes were shown to be the endogenous precursors in *Botrytis cinerea* and *Cercospora cruenta* (Inomata et al. 2004a, b). The allofarnesenes were supposed to be converted from FDP to allofarnesenes via 4,5-didehydrofarnesyl diphosphate or 6*E*,10*E*-2,6,10-trimethyl-2,6,10-dodecatriene (Inomata et al. 2004a, b). Then, cyclization and isomerization of allofarnesenes give ionylideneethanes (Fig. 2.1) (Inomata et al. 2004a, b). In addition, possible carotenoid precursors having a γ-ring to produce ionylideneacetoaldehyde were not detected in *C. cruenta* (Inomata et al. 2004b). These results

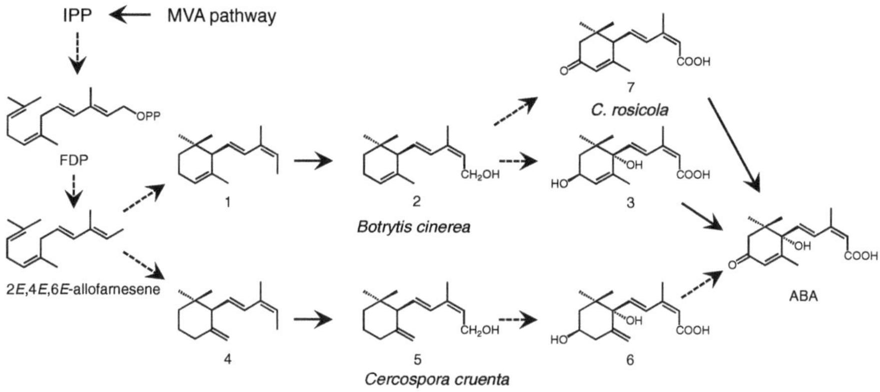

Fig. 2.1 Proposed biosynthetic pathways of ABA in fungi. The direct pathway starts with FDP produced from IPP via the MVA pathway. *Solid arrows* indicate one-step modification of an intermediate and *dashed arrows* represent multistep modifications of an intermediate. Abbreviations: *IPP* isopentenyl pyrophosphate, *FDP* Farnesyl diphosphate, *1* 2*Z*,4*E*-α-ionylideneethane, *2* 2*Z*,4*E*-α-ionylideneethanol, *3* 1′,4′-*t*-diol-ABA, *4* 2*Z*,4*E*-γ-ionylideneethane, *5* 2*Z*,4*E*-γ-ionylideneethanol, *6* 2*Z*,4*E*-1′,4′-Dihyrdoxy-γ-ionylideneacetic acid, *7* 1′-deoxy ABA

indicate that ABA is synthesized by the direct pathway via the isomerization and cyclization of allofarnesene and several oxidation steps of ionylideneethanes in these fungi (Fig. 2.1). After the synthesis of ionylideneethanes, ionylideneethanol is supposed to be generated by their oxidation (Inomata et al. 2004a, b). In *C. rosicola*, α-ionylideneethanol and α-ionylideneacetic acid were converted to ABA and 1′-deoxy ABA (Fig. 2.1) (Neill and Horgan 1983). 1′-deoxy ABA is thought to be the precursor of ABA in this fungus, because it is also oxidized to ABA (Fig. 2.1) (Neill et al. 1982). On the one hand, in *B. cinerea* and *C. pinidensiflorae*, 1′,4′-trans-diol ABA is likely the predominant precursor whose endogenous levels are correlated with ABA synthesis (Fig. 2.1) (Hirai et al. 1986; Okamoto et al. 1988). On the other hand, 1′,4′-trans-dihydro-γ-ionylideneacetic acid is thought to be the intermediate of ABA biosynthesis in *C. cruenta* (Fig. 2.1) (Oritani et al. 1985). During the conversion from ionylideneethanes to ABA, atmospheric oxygen is incorporated into ABA at C-1, -1, C-1′, and C-4′ in *C. cruenta* and *B. cinerea* (Inomata et al. 2004a, b). P450 was supposed to be involved in these oxidation steps since P450 inhibitors could effectively block ABA production (Norman et al. 1986). Based on this notion, Siewers et al. (2004) tried to knock out two genes in *Botrytis cinerea, BcCPR1* and *BcABA1*, encoding P450 oxidoreductase and P450, respectively. The ABA level in the *Bccpr1* mutant was reduced compared with the wild type. ABA production was completely abolished in the *Bcaba1* mutant (Siewers et al. 2004). The same group also found three additional genes located around the *BcABA1* gene on the genome (Siewers et al. 2006). These genes were named *BcABA2*, *BcABA3,* and *BcABA4* and encoded P450, an unknown protein and a short-chain dehydrogenase/reductase, respectively (Siewers et al. 2006). ABA was undetectable in the *Bcaba3* mutant but not in

Bcaba2 and *Bcaba4*. The authors speculated that the *Bcaba2* and *Bcaba4* mutants likely accumulated ABA precursors that could not be distinguished from ABA by their immunological ABA detection system (Siewers et al. 2006). Biochemical analysis of these proteins or feeding experiments combining several known precursors with these mutants will be powerful tools to investigate the direct pathway in detail.

2.3 Indirect Pathway in Higher Plants

ABA was discovered in the 1960s (Addicott et al. 1968; Ohkuma et al. 1965; Schwartz and Zeevaart 2010) and subsequently xanthoxin (Xan) was isolated as a plant growth inhibitor like ABA (Taylor and Burden 1970b; Taylor and Smith 1967). Xan had structural similarity to part of an epoxycarotenoid (Fig. 2.2). Indeed, the cleavage of all-*trans*- or 9-*cis*-epoxycarotenoids by light or lipoxygenase successfully generated 2-*trans*,4-*trans*- or 2-*cis*,4-*trans*-Xan (Firn and Friend 1972; Taylor and Burden 1970a). A conversion experiment from Xan to ABA using cell-free extracts from various plant species indicated that 2-*cis*,4-*trans*-Xan was the possible precursor of ABA that originates from 9-*cis*-epoxycarotenoids (Sindhu and Walton 1987). In addition, most carotenoid deficient mutants were also ABA deficient (Moore and Smith 1985; Neill et al. 1986). These facts suggested that epoxycarotenoids could be the precursors of ABA in higher plants. On the other hand, terpenoids such as carotenoids, plastoquinone, sterol, and phytol are synthesized from IPP produced in the MEP pathway in plastids (Lichtenthaler 1999). The *Arabidopsis chloroplasts altered 1* (*cla1*) mutant has a defect in the synthesis of 1-deoxy-D-xylulose-5-phosphate in the MEP pathway and presents carotenoid deficiency resulting in decreased levels of ABA (Estevez et al. 2001). These facts support that C15 ABA is synthesized via C40 epoxycarotenoids that are composed of IPP from the MEP pathway in plants. Therefore, this ABA biosynthetic pathway is referred to as the indirect pathway since C40 epoxycarotenoids are the intermediates of ABA in contrast to the direct pathway in fungi (Figs. 2.1 and 2.2). The indirect pathway has been revealed by biochemical and molecular genetic studies of ABA-deficient mutants as described below. Mutants impaired in carotenoid biosynthesis or molybdenum cofactor (MoCo) synthesis present ABA deficiency as part of a pleiotropic phenotype. We will not deal with these biosynthetic pathways or their corresponding mutants in this chapter since they are not specifically involved in ABA biosynthesis. The epoxidation steps of zeaxanthin are set as the starting point of the indirect pathway.

Zeaxanthin epoxidase (ZEP) converts zeaxanthin into violaxanthin via antheraxanthin by a two-step epoxidation (Fig. 2.2). Zeaxanthin can be also produced from violaxanthin by violaxanthin de-epoxidase (VDE) (Fig. 2.2). This cyclic reaction is called the xanthophyll cycle and is involved in nonphotochemical quenching for photoprotection (Li et al. 2009). The *ZEP* gene was first identified in the study of a tobacco ABA-deficient mutant, *Npaba2*, in *Nicotiana*

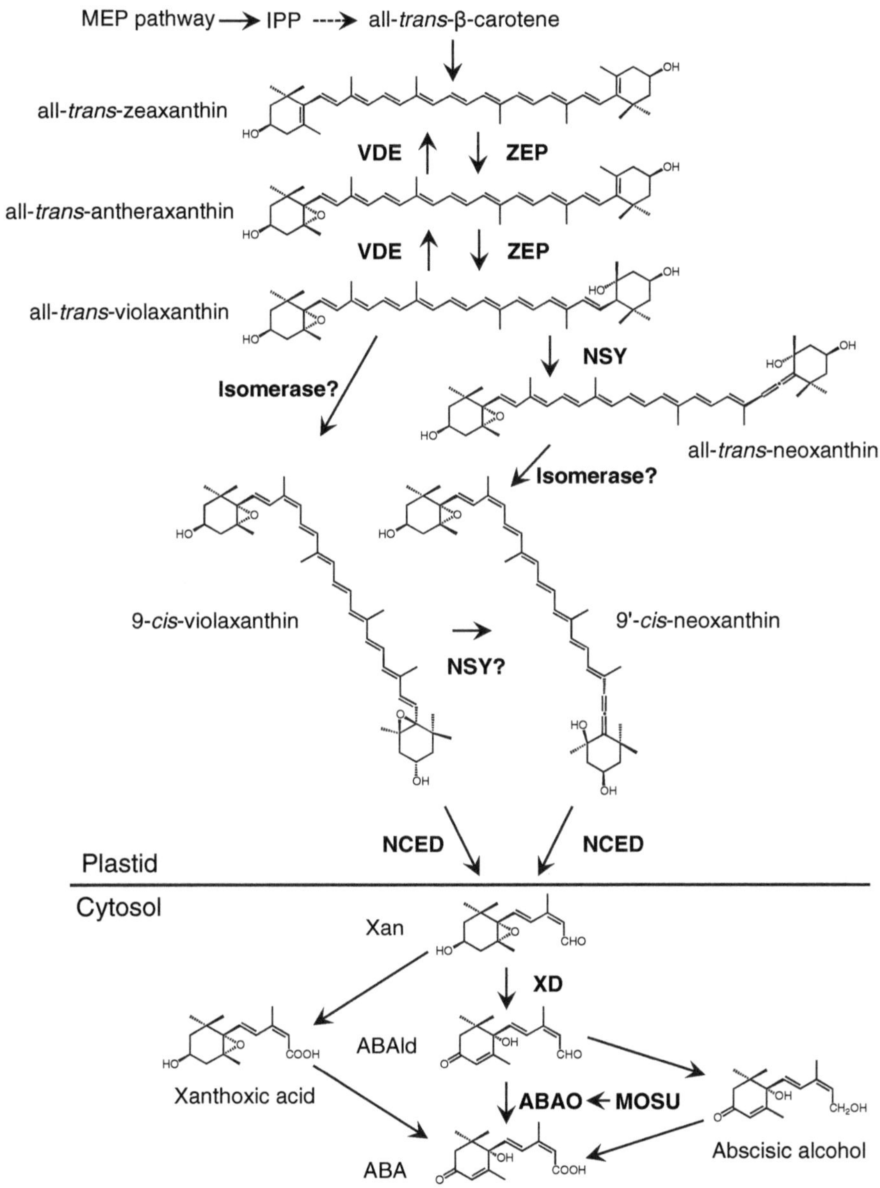

Fig. 2.2 ABA biosynthetic pathway in higher plants. The ABA precursor, a C40 carotenoid, is synthesized from IPP originating from the MEP pathway. *Solid arrows* indicate one-step modification of an intermediate and *dashed arrows* represent multistep modifications of an intermediate. Enzyme names are given in *bold*. Abbreviations: *IPP* isopentenyl pyrophosphate, *ZEP* zeaxanthin epoxidase, *VDE* violaxanthin de-epoxidase, *NSY* neoxanthin synthase, *NCED* 9-*cis*-epoxycarotenoid dioxygenase, *XD* xanthoxin dehydrogenase, *ABAO* abscisic aldehyde oxidase, *MOSU* MoCo sulfurase, *Xan* xanthoxin, *ABAld* abscisic aldehyde

plumbaginifolia (Marin et al. 1996). The *ZEP* gene encodes a monooxygenase with a FAD binding domain. Recombinant ZEP protein required additives existing in the stroma fraction for its conversion activity from zeaxanthin to violaxanthin (Marin et al. 1996). Then, it was revealed that the activity of ZEP depended on a reduced ferredoxin (Bouvier et al. 1996). Tobacco *Npaba2*, *Arabidopsis Ataba1*, rice *Osaba1*, and tomato *hp3* mutants impaired in *ZEP* genes show wilty and non-dormant seed phenotypes, and accumulate high levels of zeaxanthin but not epoxycarotenoids such as violaxanthin and neoxanthin (Table 2.1) (Agrawal et al. 2001; Duckham et al. 1991; Galpaz et al. 2008; Marin et al. 1996; Rock and Zeevaart 1991). VDE may not be responsible for ABA biosynthesis, as the *Arabidopsis npq1* mutant defective in the *VDE* gene accumulated significant amounts of epoxycarotenoids such as violaxanthin and neoxanthin that are precursors of ABA (Niyogi et al. 1998).

After the epoxidation step, all-*trans*-violaxanthin is converted to the 9-*cis* isomer prior to oxidative cleavage of the epoxycarotenoids to form xanthoxin by 9-*cis*-epoxycarotenoid dioxygenase (NCED) (Fig. 2.2). There are two possible substrates for NCED in plants, the 9-*cis* isomers of violaxanthin and neoxanthin (Fig. 2.2). In *Arabidopsis* leaves, all-*trans*-violaxanthin and 9′-*cis*-neoxanthin were found to be the two major epoxycarotenoids (40.2 and 51.8 %, respectively) of the total epoxyxanthophyll content. All-*trans*-neoxanthin and 9-*cis*-violaxanthin represented only 5.4 and 2.6 %, respectively (North et al. 2007). In tomato leaves, a similar composition of epoxyxanthophylls was observed (Parry and Horgan 1991). Therefore, 9′-*cis*-neoxanthin was supposed to be the major substrate of NCED in leaves. Two enzymatic steps are thought to be involved in the synthesis of 9′-*cis*-neoxanthin from all-*trans*-violaxanthin, involving neoxanthin synthase (NSY) and an unknown isomerase (Fig. 2.2). Three different types of NSY have been reported to date (Table 2.1). The first NSYs were biochemically isolated from tomato and potato as homologs of capsanthin capsorubin synthase or lycopene β-cyclase (Al-Babili et al. 2000; Bouvier et al. 2000). The tomato NSY enzyme was then found to be identical to the *B* gene encoding a fruit-specific lycopene β-cyclase isoform (Ronen et al. 2000), suggesting that NSY might be capable of converting both lycopene to β-carotene and violaxanthin to neoxanthin in tomato. However, neoxanthin was still produced in the *old-gold* mutant, a loss of function allele of the *B* gene (Hirschberg 2001; Ronen et al. 2000). Other genes responsible for neoxanthin synthesis were identified in the *Arabidopsis aba4* and tomato *neoxanthin-deficient 1* (*nxd1*) mutants lacking both isomers of neoxanthin (Neuman et al. 2014; North et al. 2007). The mutated genes in each mutant encoded different types of unknown proteins. The *Arabidopsis aba4* mutant had a defect in a gene encoding a functionally unknown chloroplastic protein with four transmembrane domains. The loss of neoxanthin in the mutant resulted in reduced levels of ABA, and the phenotype was obvious under dehydration conditions. The ABA-deficient phenotypes of the *aba4* mutant were milder than those of *aba1* since *aba4* was able to produce a small amount of ABA. The existence of 9-*cis*-violaxanthin in the *aba4* mutant could account for the mild phenotypes (North et al. 2007). Biochemical analysis of recombinant ABA4 protein had

Table 2.1 Characterized enzyme list for ABA biosynthesis and catabolism

	Gene name	Species	Gene ID	References
ABA biosynthetic enzyme				
Zeaxanthin epoxidase	*NpABA2*	*Nicotiana plumbaginifolia*	X95732	Marin et al. (1996)
	AtABA1/LOS6	*Arabidopsis thaliana*	At5g67030	Duckham et al. (1991), Xiong et al. (2002)
	HP3	*Solanum lycopersicum*	EU004202	Galpaz et al. (2008)
	OsABA1	*Oryza sativa*	Os04g0448900	Agrawal et al. (2001)
	VuABA1	*Vigna unguiculata*	AB030295	Iuchi et al. (2000)
Neoxanthin synthase (1)	*NSY*	*Solanum lycopersicum*	Y18297	Bouvier et al. (2000), Ronen et al. (2000)
	NXS	*Solanum tuberosum*	AJ272136	Al-Babili et al. (2000)
Neoxanthin synthase (2)	*AtABA4*	*Arabidopsis thaliana*	At1g67080	North et al. (2007)
Neoxanthin synthase (3)	*Nxd1*	*Solanum lycopersicum*	Solyc12g041880	Neuman et al. (2014)
	AtNxd1	*Arabidopsis thaliana*	At1g28100	Neuman et al. (2014)
9-*cis*-epoxycarotenoid dioxygenase	*VP14*	*Zea mays*	U95953	Schwartz et al. (1997b)
	NOTABILIS	*Solanum lycopersicum*	Z97215	Burbidge et al. (1999)
	PvNCED1	*Phaseolus vulgaris*	AF190462	Qin and Zeevaart (1999)
	PaNCED1	*Persea americana*	AF224672	Chernys and Zeevaart (2000)
	PaNCED3	*Persea americana*	AF224671	Chernys and Zeevaart (2000)
	VuNCED1	*Vigna unguiculata*	AB030293	Iuchi et al. (2000)
	AtNCED2	*Arabidopsis thaliana*	At4g18350	Iuchi et al. (2001)
	AtNCED3	*Arabidopsis thaliana*	At3g14440	Iuchi et al. (2001)
	AtNCED5	*Arabidopsis thaliana*	At1g30100	Iuchi et al. (2001), Toh et al. (2008)
	AtNCED6	*Arabidopsis thaliana*	At3g24220	Iuchi et al. (2001)
	AtNCED9	*Arabidopsis thaliana*	At1g78390	Iuchi et al. (2001)
	CrNCED1	*Cuscuta reflexa*	AY974807	Qin et al. (2008)
	CrNCED2	*Cuscuta reflexa*	AY974808	Qin et al. (2008)

(continued)

Table 2.1 (continued)

	Gene name	Species	Gene ID	References
Xanthoxin dehydrogenase	*AtABA2*	*Arabidopsis thaliana*	At1g52340	Cheng et al. (2002), Gonzalez-Guzman et al. (2002)
Abscisic aldehyde oxidase	*AAO3*	*Arabidopsis thaliana*	At2g27150	Sekimoto et al. (1998), Seo et al. (2000b)
	SITIENS	*Solanum lycopersicum*	HQ317906	Harrison et al. (2011)
	PsAO3	*Pisum sativum*	EF491600	Zdunek-Zastocka and Sobczak (2013)
MoCo sulfurase	*AtABA3/LOS5*	*Arabidopsis thaliana*	At1g16540	Bittner et al. (2001), Xiong et al. (2001)
	FLACCA	*Solanum lycopersicum*	Solyc07g066480	Sagi et al. (2002)
	NpABA1/CKR1	*Nicotiana plumbaginifolia*		Leydecker et al. (1995), Rousselin et al. (1992)
ABA catabolic enzyme				
ABA 8′-hydroxylase	*AtCYP707A1*	*Arabidopsis thaliana*	At4g19230	Kushiro et al. (2004), Saito et al. (2004)
	AtCYP707A2	*Arabidopsis thaliana*	At2g29090	Kushiro et al. (2004), Saito et al. (2004)
	AtCYP707A3	*Arabidopsis thaliana*	At5g45340	Kushiro et al. (2004), Saito et al. (2004)
	AtCYP707A4	*Arabidopsis thaliana*	At3g19270	Kushiro et al. (2004) Saito et al. (2004)
	HvABA8′OH1	*Hordeum vulgare*	DQ145932	Chono et al. (2006), Millar et al. (2006)
	HvABA8′OH2	*Hordeum vulgare*	DQ145933	Chono et al. (2006), Millar et al. (2006)
	LsABA8ox1	*Lactuca sativa*	AB235917	Sawada et al. (2008)
	LsABA8ox2	*Lactuca sativa*	AB235918	Sawada et al. (2008)
	LsABA8ox3	*Lactuca sativa*	AB235919	Sawada et al. (2008)
	LsABA8ox4	*Lactuca sativa*	AB235920	Sawada et al. (2008)
	OsABA8ox1	*Oryza sativa*	Os02g0703600	Saika et al. (2007), Yang and Choi (2006)
	OsABA8ox2	*Oryza sativa*	Os08g0472800	Saika et al. (2007), Yang and Choi (2006)
	OsABA8ox3	*Oryza sativa*	Os09g0457100	Saika et al. (2007)
	PvCYP707A1	*Phaseolus vulgaris*	DQ352541	Yang and Zeevaart (2006)
	PvCYP707A2	*Phaseolus vulgaris*	DQ352542	Yang and Zeevaart (2006)
	PvCYP707A3	*Phaseolus vulgaris*	DQ352543	Yang and Zeevaart (2006)

(continued)

Table 2.1 (continued)

	Gene name	Species	Gene ID	References
ABA glucosyltransferases	*AOG*	*Vigna angularis*	AB065190	Xu et al. (2002)
	UGT71B6	*Arabidopsis thaliana*	At3g21780	Lim et al. (2005), Priest et al. (2006)
ABA specific β-glucosidases	*AtBG1*	*Arabidopsis thaliana*	At1g52400	Lee et al. (2006)
	AtBG2	*Arabidopsis thaliana*	At2g32860	Xu et al. (2012)

Well-recognized mutant or biochemically characterized enzymes are shown in this list difficulty showing the NSY activity since the protein tended to be insoluble (North et al. 2007). The tomato *nxd1* mutant had a defect in an unknown protein without an obvious chloroplast targeting sequence. Although the tomato *nxd1* mutant did not produce either isomer of neoxanthin, it did not show any ABA-deficient phenotypes. The existence of 9-*cis*-violaxanthin could compensate for the loss of neoxanthin in the *nxd1* mutant (Neuman et al. 2014). *Arabidopsis* NXD1 protein expressed in *E. coli* also failed to show NSY activity even when it was coexpressed with ABA4 (Neuman et al. 2014). Further analysis will be necessary to reveal how these genes contribute to neoxanthin synthesis. The results from the analyses of neoxanthin-deficient mutants indicate that the preferred substrate of NCED could differ according to plant species and both 9-*cis*-violaxanthin and 9′-*cis*-neoxanthin could be the in vivo substrates of NCED. In fact, the parasitic plant *Cuscuta reflexa* does not have neoxanthin but still accumulates ABA under dehydration (Qin et al. 2008).

All-*trans*-epoxycarotenoids, violaxanthin and neoxanthin, are then converted into 9-*cis* isomers by an unknown isomerase (Fig. 2.2). It has been a long-standing question that gene encodes the 9-*cis*-epoxycarotenoid-forming isomerase in ABA biosynthesis. Recently, Alder et al. (2012) (Alder et al. 2012) found that the rice strigolactone biosynthetic enzyme D27 had isomerase activity converting all-*trans*-β-carotene into 9-*cis*-β-carotene and that the conversion activity was reversible. The structural similarity between β-carotene and epoxycarotenoids evokes the possible involvement of D27 in ABA biosynthesis. Rice has two additional genes homologous to *D27,* and *Arabidopsis* has three genes showing similarity to *D27.* Further analyzes of these genes and the corresponding mutants will help to discover the isomerase. Identification of the isomerase is the final piece for establishing the main framework of the ABA biosynthetic pathway in plants.

9-*cis* isomers of violaxanthin and neoxanthin are then cleaved into C15 and C25 compounds to produce the C15 xanthoxin, the direct precursor of ABA (Fig. 2.2). The first discovery of 9-*cis*-epoxycarotenoid dioxygenase came in the study of the viviparous, ABA-deficient, maize *vp14* mutant (Schwartz et al. 1997b; Tan et al. 1997). *VP14* encodes a non-heme iron (II)-dependent dioxygenase. Recombinant VP14 recognized the 9-*cis* configuration of C40 carotenoids and was able to cleave 9-*cis*-violaxanthin, 9′-*cis*-neoxanthin, and 9-*cis*-zeaxanthin at the 11–12 position but not all-*trans* isomers (Schwartz et al. 1997b). After the cleavage reaction in vitro, C25 compounds such as epoxy-apo-aldehyde or allenic-apo-aldehyde were generated as by-products with C15 xanthoxin. Whereas xanthoxin was detectable in plants, neither of these two C25 by-products could be detected (Parry and Horgan 1991; Schwartz and Zeevaart 2010). After the cloning of *VP14*, *NCED* genes were isolated from various plant species (Table 2.1). Overexpression of *NCED* genes resulted in higher accumulation of ABA in the transgenic plants, indicating that NCED catalyzes the rate-limiting step of ABA biosynthesis. *NCED* genes form a small gene family in various plant species. Most plants have a drought-inducible *NCED* gene whose expression levels are correlated with the accumulation of ABA in response to drought. Carotenoids including the substrate of NCED are abundantly localized at the thylakoid membrane in the chloroplast (Parry and Horgan 1991). The PvNCED1 protein is also predominantly localized at the thylakoid membrane in the chloroplast (Qin and Zeevaart 1999). The N-terminal amphipathic sequence of VP14 was necessary to bind to the thylakoid membrane (Tan et al. 2001). The crystal structure of VP14 revealed how it accesses the substrate in the thylakoid membrane via its N-terminal amphipathic region (Messing et al. 2010). In *Arabidopsis*, drought-inducible AtNCED3 was detected in two forms with different molecular weights in the chloroplast (Endo et al. 2008). The larger form was localized in the thylakoid membrane and the smaller existed in the stroma. A similar distribution was observed in a chloroplast import assay of AtNCED proteins (Tan et al. 2003). The meaning of the differential localization of NCED in the chloroplast remains to be resolved.

After the oxidative cleavage of 9-*cis*-epoxycarotenoids by NCED, xanthoxin (Xan) is translocated from the plastid to the cytosol and then converted into abscisic aldehyde (ABAld) (Fig. 2.2). The *Arabidopsis aba2* mutant was supposed to be impaired in this step since a cell-free extract from the *aba2* mutant did not show conversion activity from Xan to ABAld (Schwartz et al. 1997a). The *aba2* mutants have a defect in a member of the short-chain dehydrogenase/reductase (SDR) gene family, which includes 56 members in the *Arabidopsis* genome, and ABA2 was identified as a cytosolic NAD-dependent oxidoreductase with xanthoxin dehydrogenase (XD) activity (Cheng et al. 2002; Gonzalez-Guzman et al. 2002). Intragenic complementation of *aba2* mutant alleles suggested that the activity of mutated ABA2 enzymes could be recovered through multimer formation (Merlot et al. 2002; Rook et al. 2001). *Arabidopsis aba2* mutants were originally isolated based on their ability to germinate in the presence of inhibitors of gibberellin biosynthesis (LeonKloosterziel et al. 1996; Nambara et al. 1998). Different alleles of *aba2* such as *gin1, sis4, isi4, sre1,* and *san3* were isolated by various kinds of screening that utilized their ability to germinate and grow in toxic concentrations of glucose, sucrose, or NaCl (Cheng et al. 2002; Gonzalez-Guzman et al. 2002; Laby et al. 2000; Rook et al. 2001), or by using thermal imaging screening (Merlot et al. 2002). These facts indicate that *aba2* mutants always have a short supply of ABA during their life cycle. In other words, there are no functionally redundant genes for *ABA2* in the *Arabidopsis* genome. Hwang et al. (2012) reported that neither AtSDR2 nor AtSDR3, the closest homologs of ABA2, was likely to substitute the function of ABA2, supporting the uniqueness of ABA2 in the *Arabidopsis* genome. Although orthologs of *Arabidopsis* ABA2 are likely to be conserved in various plant species (Hanada et al. 2011), there have been no reports of additional XDs other than *Arabidopsis* ABA2.

After the conversion of Xan to ABAld, ABA is produced by the oxidation of ABAld. Two enzymes are involved in this oxidation step, abscisic aldehyde oxidase (ABAO) and MoCo sulfurase (Fig. 2.2). The conversion activity from ABAld to ABA was diminished in cell-free extracts of two wilty tomato mutants, *flacca* and *sitiens* (Sindhu and Walton 1988; Taylor et al. 1988). The enzyme that oxidizes ABAld to ABA was supposed to be an aldehyde oxidase (AO) (Leydecker et al. 1995; Walkersimmons et al. 1989) and four *Arabidopsis aldehyde oxidase* genes (*AAO1-4*) were isolated (Table 2.1) (Sekimoto et al. 1998). In-gel activity staining of AAOs revealed that the homodimer of AAO3, could utilize ABAld as a substrate, and recombinant AAO3 effectively converted ABAld into ABA (Seo et al. 2000a). In addition, the transcript level of AAO3 was increased by dehydration but transcripts of the other AAOs were not (Seo et al. 2000a). The *Arabidopsis v11* mutant impaired in AAO3 could not synthesize ABA under drought stress, resulting in a wilty phenotype (Seo et al. 2000b). While AAO3 is the major ABAO it was reported that, the other members of the AAO family have minor contributions to ABA biosynthesis in seeds (Seo et al. 2004). In addition, ectopic expression of AAO3 successfully rescued the ABA-deficient tomato mutant *sitiens* (Okamoto et al. 2002). *sitiens* was supposed to impaired in newly identified *AO* gene other than previously reported *TAOs* (Table 2.1) (Harrison et al. 2011; Min et al. 2000). Recently, three *AO* genes were isolated from *Pisum sativum* (Zdunek-Zastocka 2008). Recombinant PsAO3 protein showed abscisic aldehyde oxidase activity and the levels of *PsAO3* transcript were increased by progressive drought stress in leaves and roots (Zdunek-Zastocka and Sobczak 2013).

Aldehyde oxidase (AO), xanthine dehydrogenase (XDH), sulfite oxidase (SO), and nitrate reductase (NR) are MoCo-containing enzymes in plants (Mendel and Hänsch 2002). Whereas the activities of SO and NR need a dioxo Mo-center in the MoCo, AO and XDH require the sulfuration of MoCo to generate a monooxo Mo-center for their activity. The tomato *flacca,* tobacco *Npaba1/CKR1,* and *Arabidopsis aba3/los5* mutants were isolated as ABA-deficient phenotypes because of the loss of activities in AO and XDH but not NR (Leydecker et al. 1995; Rousselin et al. 1992; Sagi et al. 1999; Schwartz et al. 1997a). The *aba3/los5* and *flacca* mutants are impaired in the gene encoding the MoCo sulfurase responsible for sulfuration of MoCo (Table 2.1) (Bittner et al. 2001; Sagi et al. 2002; Xiong et al. 2001). This is the reason why the biochemical defects of *flacca* and *aba3* mutants are similar to those of *sitiens* and *aao3* with respect to their inability to convert ABAld into ABA.

2.4 Indirect Pathway—Possible Minor Routes After the Production of Xanthoxin

As mentioned above, Xan is first converted to ABAld, which is then converted to ABA. On the other hand, Xan has been supposed to be converted to ABA via two possible minor routes in plants (Cowan 2000; Nambara and Marion-Poll 2005; Seo and Koshiba 2002).

One pathway might operate via xanthoxic acid (Fig. 2.2). In ripening avocado mesocarp, inhibition of AO activity by tungstate, a potent inhibitor of the molybdo-enzymes in plants, results in the accumulation of xanthoxin, suggesting that xanthoxin is a substrate of AO (Lee and Milborrow 1997). When Xan was utilized as a substrate of AO, xanthoxic acid might be produced by the oxidation of xan by AO. Although xanthoxic acid is thought to then be converted into ABA by an enzyme having activity like *Arabidopsis* ABA2, AtABA2 did not show conversion activity from xanthoxic acid to ABA (Cheng et al. 2002). While xanthoxic acid was not effectively converted to ABA in a cell-free conversion assay, it could be converted to ABA (Sindhu and Walton 1988). These results suggest that xanthoxic acid could be a precursor in ABA biosynthesis.

The other pathway, via abscisic alcohol, might be activated in some mutants (Fig. 2.2). When ABAld was supplied to *flacca* or *sitiens* mutants, it was converted to abscisic alcohol, showing that ABAld is reduced to abscisic alcohol and then oxidized to ABA via a shunt pathway (Rock et al. 1991; Taylor et al. 1988). This phenomenon was also observed in the *Npaba1/CKR1* mutant in *Nicotiana plumbaginifolia* (Parry et al. 1991). The shunt pathway appears to be a minor source of ABA in wild-type plants but might play a significant role in mutants defective in the oxidation of ABAld to generate ABA directly. One might speculate on the possibility of the minor ABA biosynthetic pathways operating in an organ and/or developmental stage-dependent manner.

2.5 ABA Catabolism

In contrast to biosynthesis, ABA is catabolized through several pathways in plants. ABA catabolism is largely categorized into two types of reaction, hydroxylation and conjugation (Fig. 2.3). Among them, ABA 8′-hydroxylation is a key step in the major ABA catabolic route in several plant species (Nambara and Marion-Poll 2005). Hydroxylation at C-8′ of ABA is catalyzed by cytochrome P-450 type mono-oxygenases, and unstable 8′-hydroxy-ABA is then isomerized spontaneously to phaseic acid (PA). Although PA has faint ABA-like activity (Kepka et al. 2011; Walton 1983), substantial PA activity is observed in specific tissues such as barley aleurone layers (Hill et al. 1995; Todoroki et al. 1995). PA is further metabolized by an unidentified reductase to form dihydrophaseic acid (DPA) or *epi*-DPA, which have almost no biological activity (Walton 1983). ABA 8′-hydroxylases are encoded by the CYP707A family (Kushiro et al. 2004; Saito et al. 2004) (Table 2.1). The ABA 8′-hydroxylation reaction catalyzed by CYP707A requires both NADPH and P450 reductase (Kushiro et al. 2004; Saito et al. 2004). CYP707A selectively catalyzes the naturally occurring (+)-S-ABA enantiomer, but not the unnatural type (−)-R-ABA (Kushiro et al. 2004; Saito et al. 2004). Since ABA 8′-hydroxylase activity was observed in the microsomal fraction of suspension-cultured corn cells and CYP707A-green fluorescent protein (GFP) fusion protein localizes to the endoplasmic reticulum (ER), the ABA 8′-hydroxylation reaction is thought to take place in the ER (Krochko et al. 1998; Saika et al. 2007). Multiple mutants of CYP707A in *Arabidopsis* accumulate a large amount of ABA, whereas overexpression of CYP707A effectively reduces endogenous ABA (Millar et al. 2006; Okamoto et al. 2006, 2010, 2011; Umezawa et al. 2006). Thus, CYP707A plays a major regulatory role in controlling the level

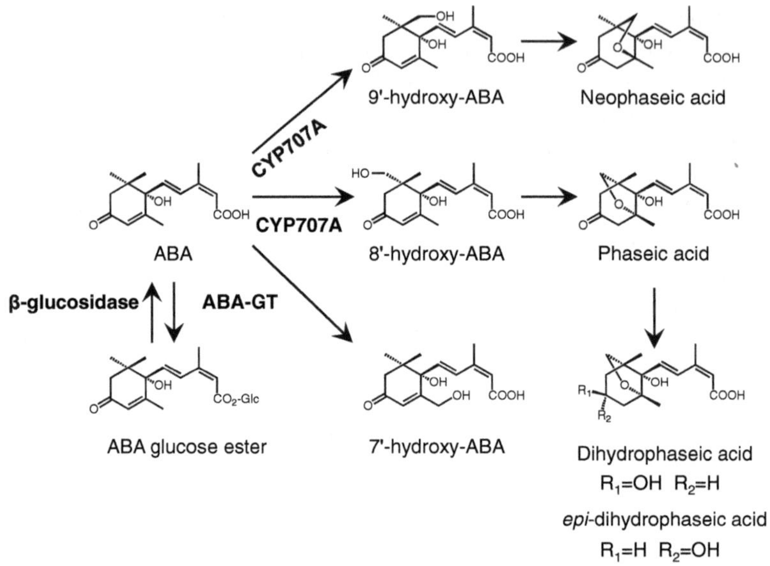

Fig. 2.3 ABA catabolic pathways in higher plants. In hydroxylation pathways, ABA 8′-hydroxylation is thought to be a major ABA catabolic route. ABA is also inactivated to ABA glucosyl ester (ABA-GE) by ABA-glucosyltransferase (ABA-GT), but ABA-GE is converted to free ABA by β-glucosidase

of ABA. In addition to the ABA 8′-hydroxylation pathway, ABA is also hydroxylated at C-7′ and C-9′ to form 7′- and 9′-hydroxy-ABA, respectively (Hampson et al. 1992; Zhou et al. 2004). These hydroxylated ABA catabolites have substantial biological activity (Hill et al. 1995; Zhou et al. 2004). Unstable 9′-hydroxy-ABA is spontaneously isomerized to form neophaseic acid (NeoPA) like PA (Zhou et al. 2004). NeoPA is a minor catabolite, but exists in a wide range of plant species (Zhou et al. 2004). The levels of neoPA accumulate substantially in restricted tissues such as developing seeds of *Arabidopsis thaliana* and *Brassica napus* (Kanno et al. 2010; Zhou et al. 2004). *Arabidopsis* CYP707As have minor activity for the 9′-hydroxylation of ABA in vitro (Okamoto et al. 2011). Additionally, endogenous neoPA levels in the *Arabidopsis cyp707a* mutant are lower than in the wild type, as with PA (Okamoto et al. 2011). Therefore, neoPA is thought to be produced by CYP707A as a side reaction. On the other hand, the gene encoding 7′-hydroxylase has not been identified yet. Hydroxylated catabolites of ABA are further conjugated with glucose (Cutler and Krochko 1999; Nambara and Marion-Poll 2005).

Among several conjugated catabolites, ABA glucosyl ester (ABA-GE) is the most widespread (Hartung et al. 2002). ABA-GE has been thought to be an inactive pool of ABA and accumulate in the vacuole or apoplast. However, ABA-GE is considered a storage or long-distance transport form of ABA that functions as a mobile stress signal from roots to shoots (Hartung et al. 2002; Jiang and Hartung

2008; Sauter et al. 2002). Glucosylation at the carboxyl group of ABA is catalyzed by glucosyltransferase (Lim et al. 2005; Xu et al. 2002). Eight ABA glucosyltransferases have been reported in *Arabidopsis*, although only one (UGT71B6) selectively recognizes (+)-ABA, but not (−)-R-ABA (Lim et al. 2005) (Table 2.1). The single knockout mutant of UGT71B6 does not show a significant difference in the levels of ABA and its metabolites compared with the wild type (Priest et al. 2006). In contrast, overexpression of UGT71B6 (UGT71B6OX) causes the accumulation of a large amount of ABA-GE (Priest et al. 2006). Nevertheless, UGT71B6OX displays only minor changes in ABA levels (Priest et al. 2006). This might be because the ABA 8′-hydroxylation pathway is predominantly involved in homeostatic regulation of ABA.

β-glucosidase hydrolyzes ABA-GE to form active ABA. There are two homologous β-glucosidases in *Arabidopsis*, AtBG1 and AtBG2, which localize to ER and vacuole, respectively (Lee et al. 2006; Xu et al. 2012) (Table 2.1). The enzymatic activity of AtBG1 is enhanced by dehydration-induced polymerization (Lee et al. 2006). It was reported that the purine metabolite allantoin promotes polymerization of AtBG1 and increases endogenous ABA levels (Watanabe et al. 2014). In contrast, the AtBG2 protein exists in high molecular weight complexes through polymerization under non-stress conditions and is protected from degradation under dehydration stress (Xu et al. 2012). However, the details of how allantoin regulates the polymerization of AtBG1 and how protein degradation of vacuole-localized AtBG2 is inhibited under stress remain unknown. The *atbg2* mutant does not affect endogenous ABA levels, whereas the *atbg1* mutant shows markedly reduced ABA levels in the extracellular space, but not in the intracellular space or xylem sap (Lee et al. 2006; Xu et al. 2012). Additionally, the *atbg1 atbg2* double mutant shows a more sensitive phenotype to drought stress (Xu et al. 2012). Therefore, β-glucosidase-mediated ABA production from ABA-GE is also considered a key pathway for regulating the local ABA concentration in response to environmental stimuli.

2.6 Chemical Inhibitors of ABA Metabolism

Inhibitors of ABA metabolism have been utilized in studies of the physiological roles of ABA in plants. They are especially useful for analyzing ABA physiology in plant species that are difficult to perform genetic analyses on.

Many researchers have used fluridone and norflurazon as inhibitors of ABA biosynthesis (Fig. 2.4). These inhibitors target phytoene desaturase, which functions in carotenoid biosynthesis, and cause concurrent bleaching of plants due to the destruction of chlorophyll resulting in loss of ABA biosynthesis (Gamble and Mullet 1986). This approach cannot distinguish the effects of ABA from the consequences of depleting most carotenoids and their derivatives (Taylor et al. 2005). As mentioned above, carotenoid deficiency causes a pleiotropic phenotype including ABA deficiency. Therefore, we need to use these inhibitors carefully.

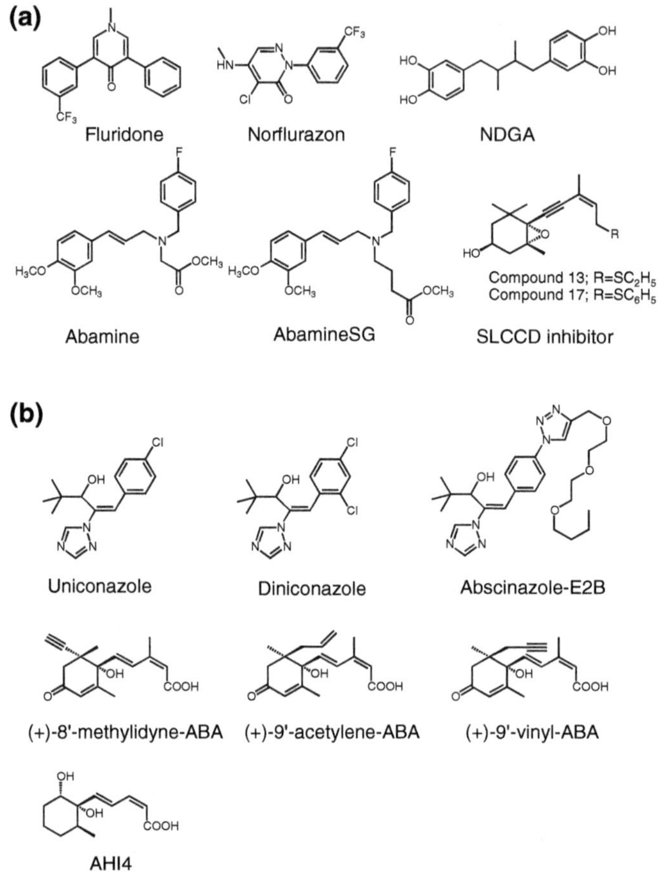

Fig. 2.4 Inhibitors of ABA biosynthetic and catabolic pathways. **a** ABA biosynthetic inhibitors. These chemicals inhibit phytoene desaturase or NCED activity. Abbreviation: *NDGA* nordihydroguaiaretic acid. **b** ABA catabolic inhibitors. Both the azole types (uniconazole, diniconazole, and abscinazole-E2B) and ABA analogues shown in this figure inhibit CYP707A activity

Nordihydroguaiaretic acid (NDGA) is an inhibitor of lipoxygenase, which catalyzes dioxygenation of polyunsaturated fatty acids (Fig. 2.4). NDGA is supposed to inhibit carotenoid cleavage dioxygenases including NCED (Creelman et al. 1992). The application of NDGA to osmotically stressed soybean suspension cells inhibited the accumulation of ABA (Creelman et al. 1992). Abamine was first developed as an NCED-specific inhibitor based on the structure of NDGA (Han et al. 2004). Abamine inhibited ABA production in osmotically stressed spinach leaf disks more effectively than NDGA (Han et al. 2004). Kitahata et al. (2006) further improved abamine to generate abamineSG, which has increased specificity for NCED without the growth retardation effect seen in abamine. Abamine and abamineSG inhibited ABA accumulation by 35 and 77 %, respectively, and acted as inhibitors of cowpea NCED in vitro with K_i values of 38.8 and 18.5 μM, respectively (Kitahata et al. 2006). Another group developed several kinds of sesquiterpene-like carotenoid cleavage dioxygenase (SLCCD) inhibitors against *Arabidopsis* NCED3, based on the structures of 9-*cis*-epoxycarotenoid and xanthoxin (Boyd et al. 2009). Compounds 13 and 17 inhibited the activity of AtNCED3 in vitro with K_i values of 93 and 57 μM, respectively (Fig. 2.4) (Boyd et al. 2009). Compound 13 had a greater inhibitory effect on ABA biosynthesis *in planta* than compound 17 and abamineSG (Boyd et al. 2009).

Most reported ABA catabolic inhibitors are chemicals targeting CYP707A. The structural features of CYP707A inhibitors are categorized into two types, azole types and ABA analogues (Fig. 2.4). Since CYP707A is a P-450 monooxygenase, several P450 inhibitors containing an azole group are considered candidate inhibitors of ABA catabolism. Uniconazole impairs the conversion of ABA to PA and inhibits the activity of the CYP707A enzyme (Kitahata et al. 2005; Saito et al. 2006). Although uniconazole effectively confers drought stress tolerance to plants, it was originally developed as a gibberellin biosynthesis inhibitor and consequently induces dwarfism (Izumi et al. 1985; Saito et al. 2006). Diniconazole functions as a fungicide, and its chemical structure is similar to that of S-uniconazole (Fletcher et al. 1986). Therefore, diniconazole also inhibits the activity of CYP707A (Kitahata et al. 2005). However, these azole P450 inhibitors arrest plant growth because of the broad inhibition spectrum of azole compounds for P450 enzymes (Rademacher 2000; Yokota et al. 1991). The structure of uniconazole was modified, and several inhibitors were synthesized to develop a more specific inhibitor against CYP707A (Okazaki et al. 2012; Todoroki et al. 2009a, b, 2010). Unlike uniconazole, abscinazole-E2B selectively inhibits CYP707A, but not CYP701A, which is involved in gibberellin biosynthesis. Application of abscinazole-E2B increases endogenous ABA levels during dehydration and confers drought tolerance with less growth arrest.

Among non-azole-type ABA catabolic inhibitors, ABA analogues designed based on the structure of ABA have the potential to function as specific inhibitors. Several ABA 8′- or 9′-derivatives have been synthesized and tested for their effect on the activity of the CYP707A enzyme, because C-8′ and its neighboring position in ABA are thought to be the possible influence groups for the hydroxylation reaction by CYP707A. (+)-8′-methylidyne-ABA, (+)-9′-acetylene-ABA, (−)-9′-acetylene-ABA, and (+)-9′-vinyl-ABA have been reported as suicide substrates that irreversibly inhibit the activity of ABA 8′-hydroxylase (Cutler et al. 2000). Among these, (+)-9′-acetylene-ABA functions as the most effective inhibitor of ABA 8′-hydroxylase (Cutler et al. 2000). There is no clear explanation why 9′-derivatives show potent inhibition effects despite the fact that CYP707A mainly catalyzes hydroxylation at the C-8′ of ABA. It is possible that the catalytic site of CYP707A might recognize both C-8′ and -9′ methyl groups for the hydroxylation reaction, since CYP707A catalyzes both 8′- and 9′-hydroxylation of ABA. In addition to the suicide effect, ABA 8′- or 9′-derivatives are thought to possess ABA agonist activity. Indeed, ABA-responsive genes regulated by (+)-8′-methylidyne-ABA widely overlap with ABA (Huang et al. 2007). To overcome the ABA agonist activity of ABA analogues, several specific inhibitors have been synthesized and developed by integrating designs based on the structural features of ABA analogues and the biochemical activity of CYP707A (Araki et al. 2006; Ueno et al. 2005). AHI4 does not exhibit ABA activity such as growth arrest and germination inhibition, but it strongly inhibits the activity of ABA 8′-hydroxylase (Araki et al. 2006). Therefore, co-treatment of ABA and AHI4 enhances ABA's activity, and application of AHI4 confers drought stress tolerance (Araki et al. 2006). Specific inhibitors are useful not only for understanding ABA metabolic regulation, but also as agrochemicals that can control the action of ABA in crops.

References

Addicott FT, Lyon JL, Ohkuma K, Thiessen WE, Carns HR, Smith OE, Cornforth JW, Milborrow BV, Ryback G, Wareing PF. Abscisic acid: a new name for abscisin II (dormin). Science. 1968;159:1493.

Agrawal GK, Yamazaki M, Kobayashi M, Hirochika R, Miyao A, Hirochika H. Screening of the rice viviparous mutants generated by endogenous retrotransposon Tos17 insertion. Tagging of a zeaxanthin epoxidase gene and a novel ostatc gene. Plant Physiol. 2001;125:1248–57.

Al-Babili S, Hugueney P, Schledz M, Welsch R, Frohnmeyer H, Laule O, Beyer P. Identification of a novel gene coding for neoxanthin synthase from Solanum tuberosum. FEBS Lett. 2000;485:168–72.

Alder A, Jamil M, Marzorati M, Bruno M, Vermathen M, Bigler P, Ghisla S, Bouwmeester H, Beyer P, Al-Babili S. The path from beta-carotene to carlactone, a strigolactone-like plant hormone. Science. 2012;335:1348–51.

Araki Y, Miyawaki A, Miyashita T, Mizutani M, Hirai N, Todoroki Y. A new non-azole inhibitor of ABA 8′-hydroxylase: effect of the hydroxyl group substituted for geminal methyl groups in the six-membered ring. Bioorg Med Chem Lett. 2006;16:3302–5.

Bittner F, Oreb M, Mendel RR. ABA3 is a molybdenum cofactor sulfurase required for activation of aldehyde oxidase and xanthine dehydrogenase in *Arabidopsis thaliana*. J Biol Chem. 2001;276:40381–4.

Bouvier F, d'Harlingue A, Hugueney P, Marin E, Marion-Poll A, Camara B. Xanthophyll biosynthesis. Cloning, expression, functional reconstitution, and regulation of beta-cyclohexenyl carotenoid epoxidase from pepper (*Capsicum annuum*). J Biol Chem. 1996;271:28861–7.

Bouvier F, D'Harlingue A, Backhaus RA, Kumagai MH, Camara B. Identification of neoxanthin synthase as a carotenoid cyclase paralog. Eur J Biochem. 2000;267:6346–52.

Boyd J, Gai Y, Nelson KM, Lukiwski E, Talbot J, Loewen MK, Owen S, Zaharia LI, Cutler AJ, Abrams SR, Loewen MC. Sesquiterpene-like inhibitors of a 9-cis-epoxycarotenoid dioxygenase regulating abscisic acid biosynthesis in higher plants. Bioorg Med Chem. 2009;17:2902–12.

Burbidge A, Grieve TM, Jackson A, Thompson A, McCarty DR, Taylor IB. Characterization of the ABA-deficient tomato mutant notabilis and its relationship with maize Vp14. Plant J. 1999;17:427–31.

Cheng WH, Endo A, Zhou L, Penney J, Chen HC, Arroyo A, Leon P, Nambara E, Asami T, Seo M, Koshiba T, Sheen J. A unique short-chain dehydrogenase/reductase in *Arabidopsis* glucose signaling and abscisic acid biosynthesis and functions. Plant Cell. 2002;14:2723–43.

Chernys JT, Zeevaart JA. Characterization of the 9-cis-epoxycarotenoid dioxygenase gene family and the regulation of abscisic acid biosynthesis in avocado. Plant Physiol. 2000;124:343–53.

Chono M, Honda I, Shinoda S, Kushiro T, Kamiya Y, Nambara E, Kawakami N, Kaneko S, Watanabe Y. Field studies on the regulation of abscisic acid content and germinability during grain development of barley: molecular and chemical analysis of pre-harvest sprouting. J Exp Bot. 2006;57:2421–34.

Cowan AK. Is abscisic aldehyde really the immediate precursor to stress-induced ABA? Trends Plant Sci. 2000;5:191–2.

Creelman RA, Bell E, Mullet JE. Involvement of a lipoxygenase-like enzyme in abscisic acid biosynthesis. Plant Physiol. 1992;99:1258–60.

Cutler AJ, Krochko JE. Formation and breakdown of ABA. Trends Plant Sci. 1999;4:472–8.

Cutler AJ, Rose PA, Squires TM, Loewen MK, Shaw AC, Quail JW, Krochko JE, Abrams SR. Inhibitors of abscisic acid 8′-hydroxylase. Biochemistry-Us. 2000;39:13614–24.

Duckham SC, Linforth RST, Taylor IB. Abscisic-acid-deficient mutants at the aba gene locus of *Arabidopsis thaliana* are impaired in the epoxidation of zeaxanthin. Plant, Cell Environ. 1991;14:601–6.

Endo A, Sawada Y, Takahashi H, Okamoto M, Ikegami K, Koiwai H, Seo M, Toyomasu T, Mitsuhashi W, Shinozaki K, Nakazono M, Kamiya Y, Koshiba T, Nambara E. Drought induction of Arabidopsis 9-cis-epoxycarotenoid dioxygenase occurs in vascular parenchyma cells. Plant Physiol. 2008;147:1984–93.

Estevez JM, Cantero A, Reindl A, Reichler S, Leon P. 1-Deoxy-D-xylulose-5-phosphate synthase, a limiting enzyme for plastidic isoprenoid biosynthesis in plants. J Biol Chem. 2001;276:22901–9.

Firn RD, Friend J. Enzymatic production of the plant growth inhibitor, xanthoxin. Planta. 1972;103:263–6.

Fletcher RA, Hofstra G, Gao JG. Comparative fungitoxic and plant-growth regulating properties of triazole derivatives. Plant and Cell Physiol. 1986;27:367–71.

Galpaz N, Wang Q, Menda N, Zamir D, Hirschberg J. Abscisic acid deficiency in the tomato mutant high-pigment 3 leading to increased plastid number and higher fruit lycopene content. Plant J. 2008;53:717–30.

Gamble PE, Mullet JE. Inhibition of carotenoid accumulation and abscisic-acid biosynthesis in fluridone-treated dark-grown barley. Eur J Biochem. 1986;160:117–21.

Gonzalez-Guzman M, Apostolova N, Belles JM, Barrero JM, Piqueras P, Ponce MR, Micol JL, Serrano R, Rodriguez PL. The short-chain alcohol dehydrogenase ABA2 catalyzes the conversion of xanthoxin to abscisic aldehyde. Plant Cell. 2002;14:1833–46.

Hampson CR, Reaney MJT, Abrams GD, Abrams SR, Gusta LV. Metabolism of (+)-abscisic acid to (+)-7′-hydroxyabscisic acid by bromegrass cell-cultures. Phytochemistry. 1992;31:2645–8.

Han SY, Kitahata N, Sekimata K, Saito T, Kobayashi M, Nakashima K, Yamaguchi-Shinozaki K, Shinozaki K, Yoshida S, Asami T. A novel inhibitor of 9-cis-epoxycarotenoid dioxygenase in abscisic acid biosynthesis in higher plants. Plant Physiol. 2004;135:1574–82.

Hanada K, Hase T, Toyoda T, Shinozaki K, Okamoto M. Origin and evolution of genes related to ABA metabolism and its signaling pathways. J Plant Res. 2011;124:455–65.

Harrison E, Burbidge A, Okyere JP, Thompson AJ, Taylor IB. Identification of the tomato ABA-deficient mutant sitiens as a member of the ABA-aldehyde oxidase gene family using genetic and genomic analysis. Plant Growth Regul. 2011;64:301–9.

Hartung W, Sauter A, Hose E. Abscisic acid in the xylem: where does it come from, where does it go to? J Exp Bot. 2002;53:27–32.

Hill RD, Liu JH, Durnin D, Lamb N, Shaw A, Abrams SR. Abscisic acid structure-activity relationships in barley aleurone layers and protoplasts (biological activity of optically active, oxygenated abscisic acid analogs). Plant Physiol. 1995;108:573–9.

Hirai N, Okamoto M, Koshimizu K. The 1',4'-trans-diol of abscisic-acid, a possible precursor of abscisic-acid in botrytis-cinerea. Phytochemistry. 1986;25:1865–8.

Hirai N, Yoshida R, Todoroki Y, Ohigashi H. Biosynthesis of abscisic acid by the non-mevalonate pathway in plants, and by the mevalonate pathway in fungi. Biosci Biotechnol Biochem. 2000;64:1448–58.

Hirschberg J. Carotenoid biosynthesis in flowering plants. Curr Opin Plant Biol. 2001;4:210–8.

Huang DQ, Jaradat MR, Wu WR, Ambrose SJ, Ross AR, Abrams SR, Cutler AJ. Structural analogs of ABA reveal novel features of ABA perception and signaling in *Arabidopsis*. Plant J. 2007;50:414–28.

Hwang SG, Lin NC, Hsiao YY, Kuo CH, Chang PF, Deng WL, Chiang MH, Shen HL, Chen CY, Cheng WH. The *Arabidopsis* short-chain dehydrogenase/reductase 3, an abscisic acid deficient 2 homolog, is involved in plant defense responses but not in ABA biosynthesis. Plant Physiol Biochem. 2012;51:63–73.

Inomata M, Hirai N, Yoshida R, Ohigashi H. Biosynthesis of abscisic acid by the direct pathway via ionylideneethane in a fungus, *Cercospora cruenta*. Biosci Biotechnol Biochem. 2004a;68:2571–80.

Inomata M, Hirai N, Yoshida R, Ohigashi H. The biosynthetic pathway to abscisic acid via ionylideneethane in the fungus Botrytis cinerea. Phytochemistry. 2004b;65:2667–78.

Iuchi S, Kobayashi M, Yamaguchi-Shinozaki K, Shinozaki K. A stress-inducible gene for 9-cis-epoxycarotenoid dioxygenase involved in abscisic acid biosynthesis under water stress in drought-tolerant cowpea. Plant Physiol. 2000;123:553–62.

Iuchi S, Kobayashi M, Taji T, Naramoto M, Seki M, Kato T, Tabata S, Kakubari Y, Yamaguchi-Shinozaki K, Shinozaki K. Regulation of drought tolerance by gene manipulation of 9-cis-epoxycarotenoid dioxygenase, a key enzyme in abscisic acid biosynthesis in *Arabidopsis*. Plant J. 2001;27:325–33.

Izumi K, Kamiya Y, Sakurai A, Oshio H, Takahashi N. Studies of sites of action of a new plant-growth retardant (E)-1-(4-chlorophenyl)-4,4-dimethyl-2-(1,2,4-triazol-1-Yl)-1-penten-3-Ol (S-3307) and comparative effects of its stereoisomers in a cell-free system from cucurbita-maxima. Plant and Cell Physiol. 1985;26:821–7.

Jiang F, Hartung W. Long-distance signalling of abscisic acid (ABA): the factors regulating the intensity of the ABA signal. J Exp Bot. 2008;59:37–43.

Kanno Y, Jikumaru Y, Hanada A, Nambara E, Abrams SR, Kamiya Y, Seo M. Comprehensive hormone profiling in developing *Arabidopsis* seeds: examination of the site of ABA biosynthesis, ABA transport and hormone interactions. Plant and Cell Physiol. 2010;51:1988–2001.

Kepka M, Benson CL, Gonugunta VK, Nelson KM, Christmann A, Grill E, Abrams SR. Action of natural abscisic acid precursors and catabolites on abscisic acid receptor complexes. Plant Physiol. 2011;157:2108–19.

Kitahata N, Saito S, Miyazawa Y, Umezawa T, Shimada Y, Min YK, Mizutani M, Hirai N, Shinozaki K, Yoshida S, Asami T. Chemical regulation of abscisic acid catabolism in plants by cytochrome P450 inhibitors. Bioorgan Med Chem. 2005;13:4491–8.

Kitahata N, Han SY, Noji N, Saito T, Kobayashi M, Nakano T, Kuchitsu K, Shinozaki K, Yoshida S, Matsumoto S, Tsujimoto M, Asami T. A 9-cis-epoxycarotenoid dioxygenase inhibitor for use in the elucidation of abscisic acid action mechanisms. Bioorg Med Chem. 2006;14:5555–61.

Krochko JE, Abrams GD, Loewen MK, Abrams SR, Cutler AJ. (+)-abscisic acid 8′-hydroxylase is a cytochrome P450 monooxygenase. Plant Physiol. 1998;118:849–60.

Kushiro T, Okamoto M, Nakabayashi K, Yamagishi K, Kitamura S, Asami T, Hirai N, Koshiba T, Kamiya Y, Nambara E. The *Arabidopsis* cytochrome P450 CYP707A encodes ABA 8′-hydroxylases: key enzymes in ABA catabolism. EMBO J. 2004;23:1647–56.

Laby RJ, Kincaid MS, Kim D, Gibson SI. The *Arabidopsis* sugar-insensitive mutants sis4 and sis5 are defective in abscisic acid synthesis and response. Plant J. 2000;23:587–96.

Lee HS, Milborrow BV. Endogenous biosynthetic precursors of (+)-abscisic acid. V. Inhibition by tungstate and its removal by cinchonine shows that xanthoxal is oxidised by a molybdo-aldehyde oxidase. Aust J Plant Physiol. 1997;24:727–32.

Lee KH, Piao HL, Kim HY, Choi SM, Jiang F, Hartung W, Hwang I, Kwak JM, Lee IJ, Hwang I. Activation of glucosidase via stress-induced polymerization rapidly increases active pools of abscisic acid. Cell. 2006;126:1109–20.

LeonKloosterziel KM, Gil MA, Ruijs GJ, Jacobsen SE, Olszewski NE, Schwartz SH, Zeevaart JAD, Koornneef M. Isolation and characterization of abscisic acid-deficient *Arabidopsis* mutants at two new loci. Plant J. 1996;10:655–61.

Leydecker MT, Moureaux T, Kraepiel Y, Schnorr K, Caboche M. Molybdenum cofactor mutants, specifically impaired in xanthine dehydrogenase activity and abscisic acid biosynthesis, simultaneously overexpress nitrate reductase. Plant Physiol. 1995;107:1427–31.

Li Z, Wakao S, Fischer BB, Niyogi KK. Sensing and responding to excess light. Annu Rev Plant Biol. 2009;60:239–60.

Lichtenthaler HK. The 1-deoxy-D-xylulose-5-phosphate pathway of isoprenoid biosynthesis in plants. Annu Rev Plant Phys. 1999;50:47–65.

Lim EK, Doucet CJ, Hou B, Jackson RG, Abrams SR, Bowles DJ. Resolution of (+)-abscisic acid using an *Arabidopsis* glycosyltransferase. Tetrahedron-Asymmetr. 2005;16:143–7.

Marin E, Nussaume L, Quesada A, Gonneau M, Sotta B, Hugueney AF, Marion-Poll A. Molecular identification of zeaxanthin epoxidase of nicotiana plumbaginifolia, a gene involved in abscisic acid biosynthesis and corresponding to the ABA locus of *Arabidopsis thaliana*. EMBO J. 1996;15:2331–42.

Mendel RR, Hänsch R. Molybdoenzymes and molybdenum cofactor in plants. J Exp Bot. 2002;53:1689–98.

Merlot S, Mustilli AC, Genty B, North H, Lefebvre V, Sotta B, Vavasseur A, Giraudat J. Use of infrared thermal imaging to isolate *Arabidopsis* mutants defective in stomatal regulation. Plant J. 2002;30:601–9.

Messing SA, Gabelli SB, Echeverria I, Vogel JT, Guan JC, Tan BC, Klee HJ, McCarty DR, Amzel LM. Structural insights into maize viviparous14, a key enzyme in the biosynthesis of the phytohormone abscisic acid. Plant Cell. 2010;22:2970–80.

Millar AA, Jacobsen JV, Ross JJ, Helliwell CA, Poole AT, Scofield G, Reid JB, Gubler F. Seed dormancy and ABA metabolism in Arabidopsis and barley: the role of ABA 8′-hydroxylase. Plant J. 2006;45:942–54.

Min XJ, Okada K, Brockmann B, Koshiba T, Kamiya Y. Molecular cloning and expression patterns of three putative functional aldehyde oxidase genes and isolation of two aldehyde oxidase pseudogenes in tomato. Biochim Biophys Acta. 2000;1493:337–41.

Moore R, Smith JD. Graviresponsiveness and abscisic-acid content of roots of carotenoid-deficient mutants of zea-mays-l. Planta. 1985;164:126–8.

Nambara E, Marion-Poll A. Abscisic acid biosynthesis and catabolism. Annu Rev Plant Biol. 2005;56:165–85.

Nambara E, Kawaide H, Kamiya Y, Naito S. Characterization of an *Arabidopsis thaliana* mutant that has a defect in ABA accumulation: ABA-dependent and ABA-independent accumulation of free amino acids during dehydration. Plant and Cell Physiol. 1998;39:853–8.

Neill SJ, Horgan R. Biosynthesis of Aba in C-Rosicola. 2. Incorporation of alpha-ionylidene ethanol and alpha-ionylidene acetic-acid into abscisic-acid by cercospora-rosicola. Phytochemistry. 1983;22:2469–72.

Neill SJ, Horgan R, Walton DC, Lee TS. The biosynthesis of abscisic-acid in cercospora-rosicola. Phytochemistry. 1982;21:61–5.

Neill SJ, Horgan R, Parry AD. The carotenoid and abscisic-acid content of viviparous kernels and seedlings of zea-mays-l. Planta. 1986;169:87–96.

Neuman H, Galpaz N, Cunningham FX Jr, Zamir D, Hirschberg J. The tomato mutation nxd1 reveals a gene necessary for neoxanthin biosynthesis and demonstrates that violaxanthin is a sufficient precursor for abscisic acid biosynthesis. Plant J. 2014;78:80–93.

Newman JD, Chappell J. Isoprenoid biosynthesis in plants: carbon partitioning within the cytoplasmic pathway. Crit Rev Biochem Mol Biol. 1999;34:95–106.

Niyogi KK, Grossman AR, Bjorkman O. *Arabidopsis* mutants define a central role for the xanthophyll cycle in the regulation of photosynthetic energy conversion. Plant Cell. 1998;10:1121–34.

Norman SM, Bennett RD, Maier VP, Poling SM. Cytokinins inhibit abscisic-acid biosynthesis in cercospora-rosicola. Plant Sci Lett. 1983;28:255–63.

Norman SM, Bennett RD, Poling SM, Maier VP, Nelson MD. Paclobutrazol inhibits abscisic acid biosynthesis in cercospora rosicola. Plant Physiol. 1986;80:122–5.

North HM, De Almeida A, Boutin JP, Frey A, To A, Botran L, Sotta B, Marion-Poll A. The *Arabidopsis* ABA-deficient mutant aba4 demonstrates that the major route for stress-induced ABA accumulation is via neoxanthin isomers. Plant J. 2007;50:810–24.

Ohkuma K, Addicott FT, Smith OE, Thiessen WE. The structure of abscisin II. Tetrahedron Lett. 1965;6:2529–35.

Okamoto M, Hirai N, Koshimizu K. Biosynthesis of abscisic-acid from alpha-ionylideneethanol in cercospora-pini-densiflorae. Phytochemistry. 1988;27:3465–9.

Okamoto M, Min X, Seo M, Nakabayashi K, Kamiya Y, Nambara E, Koshiba T. Complementation of a tomato ABA-deficient sitiens mutant by an *Arabidopsis* aldehyde oxidase gene, AAO3. Plant and Cell Physiol. 2002;43:S42–S42.

Okamoto M, Kuwahara A, Seo M, Kushiro T, Asami T, Hirai N, Kamiya Y, Koshiba T, Nambara E. CYP707A1 and CYP707A2, which encode abscisic acid 8′-hydroxylases, are indispensable for proper control of seed dormancy and germination in *Arabidopsis*. Plant Physiol. 2006;141:97–107.

Okamoto M, Tatematsu K, Matsui A, Morosawa T, Ishida J, Tanaka M, Endo TA, Mochizuki Y, Toyoda T, Kamiya Y, Shinozaki K, Nambara E, Seki M. Genome-wide analysis of endogenous abscisic acid-mediated transcription in dry and imbibed seeds of *Arabidopsis* using tiling arrays. Plant J. 2010;62:39–51.

Okamoto M, Kushiro T, Jikumaru Y, Abrams SR, Kamiya Y, Seki M, Nambara E. ABA 9'-hydroxylation is catalyzed by CYP707A *in Arabidopsis*. Phytochemistry. 2011;72:717–22.

Okazaki M, Kittikorn M, Ueno K, Mizutani M, Hirai N, Kondo S, Ohnishi T, Todoroki Y. Abscinazole-E2B, a practical and selective inhibitor of ABA 8′-hydroxylase CYP707A. Bioorgan Med Chem. 2012;20:3162–72.

Oritani T, Kiyota H. Biosynthesis and metabolism of abscisic acid and related compounds. Nat Prod Rep. 2003;20:414–25.

Oritani T, Niitsu M, Kato T, Yamashita K. Isolation of (2z,4e)-gamma-ionylideneethanol from cercospora-cruenta, a fungus producing (+)-abscisic acid. Agr Biol Chem Tokyo. 1985;49:2819–22.

Parry AD, Horgan R. Carotenoids and Abscisic-Acid (Aba) biosynthesis in higher-plants. Physiol Plantarum. 1991;82:320–6.

Parry AD, Blonstein AD, Babiano MJ, King PJ, Horgan R. Abscisic-acid metabolism in a wilty mutant of nicotiana plumbaginifolia. Planta. 1991;183:237–43.

Priest DM, Ambrose SJ, Vaistij FE, Elias L, Higgins GS, Ross ARS, Abrams SR, Bowles DJ. Use of the glucosyltransferase UGT71B6 to disturb abscisic acid homeostasis in *Arabidopsis thaliana*. Plant J. 2006;46:492–502.

Qin X, Zeevaart JA. The 9-cis-epoxycarotenoid cleavage reaction is the key regulatory step of abscisic acid biosynthesis in water-stressed bean. Proc Natl Acad Sci USA. 1999;96:15354–61.

Qin X, Yang SH, Kepsel AC, Schwartz SH, Zeevaart JA. Evidence for abscisic acid biosynthesis in Cuscuta reflexa, a parasitic plant lacking neoxanthin. Plant Physiol. 2008;147:816–22.

Rademacher W. Growth retardants: effects on gibberellin biosynthesis and other metabolic pathways. Annu Rev Plant Phys. 2000;51:501–31.

Rock CD, Zeevaart JA. The aba mutant of *Arabidopsis thaliana* is impaired in epoxy-carotenoid biosynthesis. Proc Natl Acad Sci USA. 1991;88:7496–9.

Rock CD, Heath TG, Gage DA, Zeevaart JA. Abscisic alcohol is an intermediate in abscisic acid biosynthesis in a shunt pathway from abscisic aldehyde. Plant Physiol. 1991;97:670–6.

Ronen G, Carmel-Goren L, Zamir D, Hirschberg J. An alternative pathway to beta-carotene formation in plant chromoplasts discovered by map-based cloning of beta and old-gold color mutations in tomato. Proc Natl Acad Sci USA. 2000;97:11102–7.

Rook F, Corke F, Card R, Munz G, Smith C, Bevan MW. Impaired sucrose-induction mutants reveal the modulation of sugar-induced starch biosynthetic gene expression by abscisic acid signalling. Plant J. 2001;26:421–33.

Rousselin P, Kraepiel Y, Maldiney R, Miginiac E, Caboche M. Characterization of 3 hormone mutants of *Nicotiana plumbaginifolia*—evidence for a common Aba deficiency. Theor Appl Genet. 1992;85:213–21.

Sagi M, Fluhr R, Lips SH. Aldehyde oxidase and xanthine dehydrogenase in a flacca tomato mutant with deficient abscisic acid and wilty phenotype. Plant Physiol. 1999;120:571–8.

Sagi M, Scazzocchio C, Fluhr R. The absence of molybdenum cofactor sulfuration is the primary cause of the flacca phenotype in tomato plants. Plant J. 2002;31:305–17.

Saika H, Okamoto M, Miyoshi K, Kushiro T, Shinoda S, Jikumaru Y, Fujimoto M, Arikawa T, Takahashi H, Ando M, Arimura S, Miyao A, Hirochika H, Kamiya Y, Tsutsumi N, Nambara E, Nakazono M. Ethylene promotes submergence-induced expression of OsABA8ox1, a gene that encodes ABA 8′-hydroxylase in rice. Plant and Cell Physiol. 2007;48:287–98.

Saito S, Hirai N, Matsumoto C, Ohigashi H, Ohta D, Sakata K, Mizutani M. *Arabidopsis CYP707A*s encode (+)-abscisic acid 8′-hydroxylase, a key enzyme in the oxidative catabolism of abscisic acid. Plant Physiol. 2004;134:1439–49.

Saito S, Okamoto M, Shinoda S, Kushiro T, Koshiba T, Kamiya Y, Hirai N, Todoroki Y, Sakata K, Nambara E, Mizutani M. A plant growth retardant, uniconazole, is a potent inhibitor of ABA catabolism in *Arabidopsis*. Biosci Biotech Bioch. 2006;70:1731–9.

Sauter A, Dietz KJ, Hartung W. A possible stress physiological role of abscisic acid conjugates in root-to-shoot signalling. Plant, Cell Environ. 2002;25:223–8.

Sawada Y, Aoki M, Nakaminami K, Mitsuhashi W, Tatematsu K, Kushiro T, Koshiba T, Kamiya Y, Inoue Y, Nambara E, Toyomasu T. Phytochrome- and gibberellin-mediated regulation of abscisic acid metabolism during germination of photoblastic lettuce seeds. Plant Physiol. 2008;146:1386–96.

Schwartz SH, Zeevaart JAD. Abscisic acid biosynthesis and metabolism. In: Davies PJ, editor. Plant hormones. 3rd ed. New York: Springer; 2010. p. 137–55.

Schwartz SH, Leon-Kloosterziel KM, Koornneef M, Zeevaart JA. Biochemical characterization of the aba2 and aba3 mutants in *Arabidopsis thaliana*. Plant Physiol. 1997a;114:161–6.

Schwartz SH, Tan BC, Gage DA, Zeevaart JA, McCarty DR. Specific oxidative cleavage of carotenoids by VP14 of maize. Science. 1997b;276:1872–4.

Sekimoto H, Seo M, Kawakami N, Komano T, Desloire S, Liotenberg S, Marion-Poll A, Caboche M, Kamiya Y, Koshiba T. Molecular cloning and characterization of aldehyde oxidases in *Arabidopsis thaliana*. Plant Cell Physiol. 1998;39:433–42.

Seo M, Koshiba T. Complex regulation of ABA biosynthesis in plants. Trends Plant Sci. 2002;7:41–8.

Seo M, Koiwai H, Akaba S, Komano T, Oritani T, Kamiya Y, Koshiba T. Abscisic aldehyde oxidase in leaves of *Arabidopsis thaliana*. Plant J. 2000a;23:481–8.

Seo M, Peeters AJ, Koiwai H, Oritani T, Marion-Poll A, Zeevaart JA, Koornneef M, Kamiya Y, Koshiba T. The *Arabidopsis* aldehyde oxidase 3 (AAO3) gene product catalyzes the final step in abscisic acid biosynthesis in leaves. Proc Natl Acad Sci USA. 2000b;97:12908–13.

Seo M, Aoki H, Koiwai H, Kamiya Y, Nambara E, Koshiba T. Comparative studies on the *Arabidopsis* aldehyde oxidase (AAO) gene family revealed a major role of AAO3 in ABA biosynthesis in seeds. Plant Cell Physiol. 2004;45:1694–703.

Siewers V, Smedsgaard J, Tudzynski P. The P450 monooxygenase BcABA1 is essential for abscisic acid biosynthesis in Botrytis cinerea. Appl Environ Microbiol. 2004;70:3868–76.

Siewers V, Kokkelink L, Smedsgaard J, Tudzynski P. Identification of an abscisic acid gene cluster in the grey mold *Botrytis cinerea*. Appl Environ Microbiol. 2006;72:4619–26.

Sindhu RK, Walton DC. Conversion of xanthoxin to abscisic acid by cell-free preparations from bean leaves. Plant Physiol. 1987;85:916–21.

Sindhu RK, Walton DC. Xanthoxin metabolism in cell-free preparations from wild type and wilty mutants of tomato. Plant Physiol. 1988;88:178–82.

Tan BC, Schwartz SH, Zeevaart JA, McCarty DR. Genetic control of abscisic acid biosynthesis in maize. Proc Natl Acad Sci USA. 1997;94:12235–40.

Tan BC, Cline K, McCarty DR. Localization and targeting of the VP14 epoxy-carotenoid dioxygenase to chloroplast membranes. Plant J. 2001;27:373–82.

Tan BC, Joseph LM, Deng WT, Liu LJ, Li QB, Cline K, McCarty DR. Molecular characterization of the *Arabidopsis* 9-cis epoxycarotenoid dioxygenase gene family. Plant J. 2003;35:44–56.

Taylor HF, Burden RS. Identification of plant growth inhibitors produced by photolysis of violaxanthin. Phytochemistry. 1970a;9:2217–23.

Taylor HF, Burden RS. Xanthoxin, a new naturally occurring plant growth inhibitor. Nature. 1970b;227:302–4.

Taylor HF, Smith TA. Production of plant growth inhibitors from xanthophylls: a possible source of dormin. Nature. 1967;215:1513–4.

Taylor IB, Linforth RST, Alnaieb RJ, Bowman WR, Marples BA. The wilty tomato mutants Flacca and Sitiens are impaired in the oxidation of Aba-Aldehyde to Aba. Plant, Cell Environ. 1988;11:739–45.

Taylor IB, Sonneveld T, Bugg TDH, Thompson AJ. Regulation and manipulation of the biosynthesis of abscisic acid, including the supply of xanthophyll precursors. J Plant Growth Regul. 2005;24:253–73.

Todoroki Y, Hirai N, Koshimizu K. 8′,8′-Difluoroabscisic acid and 8′,8′,8′-trifluoroabscisic acid as highly potent, long-lasting analogs of abscisic-acid. Phytochemistry. 1995;38:561–8.

Todoroki Y, Aoyama H, Hiramatsu S, Shirakura M, Nimitkeatkai H, Kondo S, Ueno K, Mizutani M, Hirai N. Enlarged analogues of uniconazole, new azole containing inhibitors of ABA 8′-hydroxylase CYP707A. Bioorg Med Chem Lett. 2009a;19:5782–6.

Todoroki Y, Kobayashi K, Shirakura M, Aoyama H, Takatori K, Nimitkeatkai H, Jin MH, Hiramatsu S, Ueno K, Kondo S, Mizutani M, Hirai N. Abscinazole-F1, a conformationally restricted analogue of the plant growth retardant uniconazole and an inhibitor of ABA 8′-hydroxylase CYP707A with no growth-retardant effect. Bioorg Med Chem. 2009b;17:6620–30.

Todoroki Y, Naiki K, Aoyama H, Shirakura M, Ueno K, Mizutani M, Hirai N. Selectivity improvement of an azole inhibitor of CYP707A by replacing the monosubstituted azole with a disubstituted azole. Bioorg Med Chem Lett. 2010;20:5506–9.

Toh S, Imamura A, Watanabe A, Nakabayashi K, Okamoto M, Jikumaru Y, Hanada A, Aso Y, Ishiyama K, Tamura N, Iuchi S, Kobayashi M, Yamaguchi S, Kamiya Y, Nambara E, Kawakami N. High temperature-induced abscisic acid biosynthesis and its role in the inhibition of gibberellin action in *Arabidopsis* seeds. Plant Physiol. 2008;146:1368–85.

Ueno K, Araki Y, Hirai N, Saito S, Mizutani M, Sakata K, Todoroki Y. Differences between the structural requirements for ABA 8′-hydroxylase inhibition and for ABA activity. Bioorgan Med Chem. 2005;13:3359–70.

Umezawa T, Okamoto M, Kushiro T, Nambara E, Oono Y, Seki M, Kobayashi M, Koshiba T, Kamiya Y, Shinozaki K. CYP707A3, a major ABA 8′-hydroxylase involved in dehydration and rehydration response in *Arabidopsis thaliana*. Plant J. 2006;46:171–82.

Walkersimmons M, Kudrna DA, Warner RL. Reduced accumulation of aba during water-stress in a molybdenum cofactor mutant of barley. Plant Physiol. 1989;90:728–33.

Walton DC. Structure-activity relationships of abscisic acid analogs and metabolites. In: Addicott FT, editor. Abscisic acid. New York: Praeger; 1983. p. 113–46.

Watanabe S, Matsumoto M, Hakomori Y, Takagi H, Shimada H, Sakamoto A. The purine metabolite allantoin enhances abiotic stress tolerance through synergistic activation of abscisic acid metabolism. Plant, Cell Environ. 2014;37:1022–36.

Xiong L, Ishitani M, Lee H, Zhu JK. The *Arabidopsis* LOS5/ABA3 locus encodes a molybdenum cofactor sulfurase and modulates cold stress- and osmotic stress-responsive gene expression. Plant Cell. 2001;13:2063–83.

Xiong L, Lee H, Ishitani M, Zhu JK. Regulation of osmotic stress-responsive gene expression by the LOS6/ABA1 locus in *Arabidopsis*. J Biol Chem. 2002;277:8588–96.

Xu ZJ, Nakajima M, Suzuki Y, Yamaguchi I. Cloning and characterization of the abscisic acid-specific glucosyltransferase gene from adzuki bean seedlings. Plant Physiol. 2002;129:1285–95.

Xu ZY, Lee KH, Dong T, Jeong JC, Jin JB, Kanno Y, Kim DH, Kim SY, Seo M, Bressan RA, Yun DJ, Hwang I. A vacuolar beta-glucosidase homolog that possesses glucose-conjugated abscisic acid hydrolyzing activity plays an important role in osmotic stress responses in *Arabidopsis*. Plant Cell. 2012;24:2184–99.

Yang SH, Choi DS. Characterization of genes encoding ABA 8′-hydroxylase in ethylene-induced stem growth of deepwater rice (*Oryza sativa* L.). Biochem Bioph Res Co. 2006;350:685–90.

Yang SH, Zeevaart JAD. Expression of ABA 8′-hydroxylases in relation to leaf water relations and seed development in bean. Plant J. 2006;47:675–86.

Yokota T, Nakamura Y, Takahashi N, Nonaka H, Oshio H, Takatsuo S. Inconsistency between growth and endogenous levels of gibberellins, brassinosteroids, and sterols in *Pisum sativum* treated with uniconazole antipodes. In: Takahashi N, Phinney BO, MacMillan J, editors. Gibberellins. New York: Springer; 1991. pp. 339–49.

Zdunek-Zastocka E. Molecular cloning, characterization and expression analysis of three aldehyde oxidase genes from *Pisum sativum* L. Plant Physiol Biochem. 2008;46:19–28.

Zdunek-Zastocka E, Sobczak M. Expression of *Pisum sativum* PsAO3 gene, which encodes an aldehyde oxidase utilizing abscisic aldehyde, is induced under progressively but not rapidly imposed drought stress. Plant Physiol Biochem. 2013;71:57–66.

Zeevaart JAD, Creelman RA. Metabolism and physiology of abscisic-acid. Annu Rev Plant Phys. 1988;39:439–73.

Zhou R, Cutler AJ, Ambrose SJ, Galka MM, Nelson KM, Squires TM, Loewen MK, Jadhav AS, Ross ARS, Taylor DC, Abrams SR. A new abscisic acid catabolic pathway. Plant Physiol. 2004;134:361–9.

Chapter 3
ABA Transmembrane Transport and Transporters

Mitsunori Seo

Abstract Two different types of abscisic acid (ABA) transmembrane transporters, the ATP-binding cassette (ABC) type and the NRT1/PTR FAMILY (NPF) member proteins, have been identified in Arabidopsis (*Arabidopsis thaliana*). Previous studies indicated that ABA is actively synthesized in cells that are located around the vasculature in response to water deficit. This observation suggests that ABA has to be translocated to the guard cells to induce stomatal closure. In accordance with this hypothesis, the gene encoding an ABA exporter (AtABCG25) is expressed in vascular tissues, whereas the gene encoding an ABA importer (AtABCG40) is expressed in guard cells. The expression of another ABA importer gene (*AtNPF4.6/NRT1.2/AIT1*) in the vascular tissues implies that the balance between ABA export from and ABA import back into the vascular tissues determines the amount of mobile ABA present in the apoplastic space. The existence of closely related homologues in the AtABCG and AtNPF members that might function as ABA transporters suggests that ABA transport is regulated in a complex manner.

Keywords ABA · Transmembrane transport · Transporters · Regulation

3.1 Passive and Active ABA Transport Across Biological Membranes

Movement of abscisic acid (ABA) within plants has been shown in several ways, for example, in experiments using radioisotope-labeled ABA as a tracer and by grafting between ABA-deficient mutants and wild-type plants (Ikegami et al. 2009; Frey

M. Seo (✉)
RIKEN Center for Sustainable Resource Science, 1-7-22 Suehiro-cho, Tsurumi-ku, Yokohama, Kanagawa 230-0045, Japan
e-mail: mitsunori.seo@riken.jp

D.-P. Zhang (ed.), *Abscisic Acid: Metabolism, Transport and Signaling*,
DOI 10.1007/978-94-017-9424-4_3

et al. 2004; Holbrook et al. 2002; Everatbourbouloux 1982; Zeevaart and Boyer 1984). ABA is synthesized inside cells. Thus, transmembrane transport of ABA from the inside to the outside of ABA-synthesizing cells is the first step required for movement unless ABA moves through plasmodesmata. ABA has been detected in xylem sap (Tardieu et al. 2010; Hartung et al. 2002), indicating that ABA is exported to the outside of cells after biosynthesis. Once exported from the site of biosynthesis, ABA can be delivered to neighboring or distant cells and can induce physiological responses. Sites for ABA recognition by receptors that occur both inside and on the surface (plasma membrane) of cells have been proposed (Allan et al. 1994; Anderson et al. 1994; Gilroy and Jones 1994; Schwartz et al. 1994; Yamazaki et al. 2003). Studies in Arabidopsis demonstrated that at least two different types of proteins, the plasma membrane-localized GPCR-type G proteins and the soluble START proteins (PYR/PYL/RCAR), function as ABA receptors (Ma et al. 2009; Pandey et al. 2009; Park et al. 2009). The soluble receptor PYR/PYL/RCAR family plays a central role in most of the physiological responses mediated by ABA since multiple knockout mutants in this family display severe ABA-insensitive phenotypes (Gonzalez-Guzman et al. 2012; Park et al. 2009). Therefore, ABA present in the apoplast needs to be taken into cells again across a biological membrane to induce physiological responses through the PYR/PYL/RCAR receptors.

ABA is a weak acid with a p*K*a of 4.7 and is ionized at its carboxyl group to form ABA^- depending on the pH. Compared to ionized ABA^-, non-ionized ABA (protonated ABA; ABA-H) can pass relatively easily through biological membrane by passive diffusion because of its hydrophobic nature. Therefore, pH conditions can affect the distribution of ABA between the inside and the outside of cells. The last step of ABA biosynthesis takes place in the cytosol where the pH is estimated to be 7.2–7.4. Thus, most of ABA in the cytosol exists as ABA^- that requires specific transporters for extracellular ABA export. Once exported, ABA can move through the apoplastic space unless it is taken into cells (Kramer 2006). The pH in the apoplast (estimated to be 5.0–6.0) is relatively lower than that in the cytosol; however, the pH of the apoplast is still higher than the p*K*a of ABA, suggesting that active transport mechanisms are also required for efficient uptake of apoplastic ABA into cells. Several studies have indicated the presence of saturable components (carriers or transporters) that mediate cellular ABA uptake (Bianco-Colomas et al. 1991; Perras et al. 1994; Windsor et al. 1994; Windsor et al. 1992). On the other hand, apoplastic pH increases up to approximately pH 7.0 during water stress, enabling a relatively large proportion of ABA-H to move long distances through the apoplast (Wilkinson 1999; Wilkinson and Davies 2002; Hartung et al. 2002; Sauter et al. 2001).

3.2 The Sites of ABA Biosynthesis and Action

As described above, transmembrane ABA transport could be associated with local and/or long-distance cell-to-cell movement of ABA. Despite many investigations, the significance of transmembrane transport of ABA and, hence, local and/or long-distant transport of ABA between cells in the regulation of physiological responses is largely unknown. ABA has long been considered as a mobile signal that is produced in roots upon soil drying and is transported to shoots (leaves) to regulate water loss by reducing stomatal conductance (Davies and Zhang 1991; Schachtman and Goodger 2008). In contrast, root-derived ABA is not always required for stomatal closure because leaves can potentially synthesize ABA and induce stomatal closure in response to water deficit (Ikegami et al. 2009; Holbrook et al. 2002; Christmann et al. 2007). Also, ABA moves from leaves to leaves, leaves to roots, vegetative organs to seeds, and maternal tissues to zygotic tissues (embryo and/or endosperm) in seeds (Everatbourbouloux 1982; Zeevaart and Boyer 1984; Rodriguez-Gamir et al. 2011; Kudoyarova et al. 2011; Ernst et al. 2010; Goodger and Schachtman 2010; Ikegami et al. 2009; Frey et al. 2004; Karssen et al. 1983; Kanno et al. 2010).

To understand the physiological responses mediated through ABA transport, it is necessary to define the site of ABA biosynthesis and the sites of ABA action. Earlier studies deduced the site of ABA biosynthesis and subsequent movement by measuring endogenous concentrations of ABA from different parts of a plant as a function of time. For example, application of water stress to a limited part of the root system without changing the leaf water status resulted in reduced stomatal conductance, and this phenomenon was associated with an increase in xylem sap ABA levels (Davies and Zhang 1991). This finding suggests that the ABA synthesized in roots moved to the leaves (guard cells) through the xylem; however, measurements of ABA concentrations from plant tissues at a specific time point do not simply reveal the sites of ABA biosynthesis and movement, especially if ABA synthesis takes place ubiquitously within plants and if the direction of ABA movement is nonpolar or multi-directional.

Another strategy to identify the site for ABA biosynthesis is to assume that ABA is synthesized where ABA biosynthetic enzymes are present. Almost all of the genes and enzymes involved in ABA biosynthesis have been identified (Nambara and Marion-Poll 2005; Seo and Koshiba 2011), thereby facilitating our investigation of the site of ABA biosynthesis. Immunohistochemical staining was used to determine the localization of three ABA biosynthesis enzymes, AtNCED3, AtABA2, and AAO3, in Arabidopsis by using antibodies raised against the respective proteins (Koiwai et al. 2004; Endo et al. 2008). AtABA2 and AAO3 were consistently present in leaf vascular tissues, irrespective of water stress conditions. In contrast, AtNCED3 was detected in the same region as AtABA2 and AAO3 only after, but not before, the stress treatment. These observations indicate that ABA biosynthesis is activated in vascular tissues in response to water stress. This, in turn, indicates that ABA synthesized in vascular tissues has to be translocated to the guard cells to close stomata in response to water stress; however, we cannot exclude the possibility that ABA is synthesized not only in vascular tissues but also other tissues. It is still possible that ABA biosynthetic enzymes existing at undetectable levels could contribute to physiological responses to some extent. Transient expression of two ABA biosynthesis genes, *AtNCED3* and *AAO3*, in the guard cells of broad beans (*Vicia faba*) after particle bombardment induced stomatal closure (Melhorn et al. 2008). Similarly, expression of the functional *AtABA3* gene in the Arabidopsis *aba3* mutant background under the control of a guard cell-specific promoter complemented the defective stomatal closure at least in part (Bauer et al. 2013). These studies suggest that guard cells can potentially synthesize ABA and induce stomatal closure. Also, expression of some ABA-inducible genes are up-regulated in vascular tissues, whereas others are activated in non-vascular cells (Endo et al. 2008), indicating that ABA acts at the site of its synthesis and also acts in surrounding and/or distant cells. Therefore, the contribution of ABA derived from a different source to a particular physiological response at a particular site needs to be considered very carefully.

3.3 Transmembrane ABA Transporters

Several auxin transporters have been identified by characterizing mutants isolated based on their defects in organ development or based on their altered responses to exogenously applied auxins or environmental stimuli (Benjamins and Scheres 2008; Petrasek and Friml 2009). Such mutants contributed significantly toward understanding the physiological processes mediated by auxin transport. In contrast, forward genetics approaches based on physiological or hormone responses have not succeeded in isolating mutants defective in the transport of other hormones including ABA. These findings suggest that hormone transport systems are highly redundant or that hormone transport is not required for physiological functions. The latter hypothesis seems unlikely since two different types of ABA transporters, the ATP-binding cassette (ABC) type and the NRT1/PTR FAMILY (NPF) member proteins, were identified recently in Arabidopsis (Kuromori et al. 2010; Kang et al. 2010; Kanno et al. 2012). These studies demonstrated the importance of ABA transmembrane transport in regulating physiological responses.

3.3.1 ABCG

Two independent studies identified ABC proteins belonging to subfamily G in Arabidopsis, namely ABCG25 and ABCG40, as an ABA exporter and ABA importer, respectively. ABC proteins are defined as proteins that contain an ATP-binding cassette (Rea 2007; Verrier et al. 2008; Kang et al. 2011). A full-size ABC transporter is composed of two ATP-binding cassettes and two transmembrane domains, whereas a half-size ABC transporter contains one ATP-binding cassette and two transmembrane domains. Half-size ABC proteins are proposed to function as a homo- and/or a heterodimer. ABC proteins exist in prokaryotes and eukaryotes, including bacteria, fungi, yeast, insects, animals, and plants. ABC proteins mediate the export or import of various compounds by using the energy generated by ATP hydrolysis. There are about 130 ABC genes in Arabidopsis. ABC proteins from eukaryotes are grouped into 8 subfamilies (A–H), but plants do not have the H subfamily. Instead, plants contain prokaryote types (subfamily I).

An Arabidopsis mutant defective in *AtABCG25*, encoding a half-size ABC transporter, was isolated from a population of Ac/Ds insertion lines based on its altered response to exogenously applied ABA during seed germination and/or early seedling growth (Kuromori et al. 2010). The *abcg25* mutant was more sensitive to exogenously applied ABA than wild type, suggesting that the function of AtABCG25 is related to ABA transport. Using isolated membrane vesicles from Sf9 insect cells, the AtABCG25 protein expressed with the baculovirus expression system was shown to have sufficient activity to transport ABA (Kuromori et al. 2010). The K_m value of AtABCG25 for ABA [racemic mixture of (+) and (−) enantiomers] is 260 nM, and the ABA transport activity of AtABCG25 was inhibited by (+)-ABA but not by (−)-ABA, gibberellin (GA), or indole-3-acetic acid (IAA). These results suggest that AtABCG25 has a relatively high affinity and specificity for the naturally occurring (+) enantiomer of ABA. Promoter activity of *AtABCG25* was induced by ABA in vascular tissues, suggesting that AtABCG25 regulates ABA export from ABA-synthesizing cells in response to elevated endogenous ABA concentrations. Transgenic plants overexpressing *AtABCG25* had higher leaf surface temperatures with reduced water loss compared to wild type. This result suggests that elevated apoplastic ABA levels caused by constitutive ABA export activities of AtABCG25 resulted in an increased amount of ABA to be transported from vascular tissues to the guard cells. However, the guard cells of overexpressors should have more ABA exporting activity that could inhibit stomatal closure even if increased levels of ABA were present in the apoplast. The use of tissue/cell-specific promoters to locally regulate ABA transport activities should help elucidating the roles of the ABA transporter at specific sites.

AtABCG40 (PDR12), a full-size ABC transporter, was originally identified based on its involvement in lead (PbII) resistance (Kang et al. 2010). Expression of *AtABCG40* was induced by PbII treatment. Mutants defective in *AtABCG40* were sensitive whereas *AtABCG40* overexpressors were resistant to PbII. Some members of the full-size ABCG subfamily have been implicated the transport of terpenoids and are also involved in responses to pathogens, salinity, cold, and heavy metal stresses (Lee et al. 2005; Moons 2008; Stein et al. 2006; Campbell et al. 2003; Rea 2007). These observations motivated Kang et al. (2010) to examine the involvement of this family in ABA transport, because ABA is a terpenoid (sesquiterpene) and is involved in responses to biotic and abiotic stresses. Knockout mutants of 13 out of 15 Arabidopsis ABCG/PDR family members were investigated for their phenotypes. One mutant, *abcg40*, was found to be less sensitive to exogenously applied ABA than wild type in terms of inhibiting seed germination and inducing stomatal closure. These results indicate that AtABCG40 is involved in ABA transport. Yeast and tobacco BY2 cells expressing AtABCG40 incorporated more ABA than the corresponding control cells. The K_m value of AtABCG40 for ABA is 1 μM when determined in the tobacco BY2 system. Also, isolated mesophyll cells from *atabcg40* were less active in taking up ABA compared to wild-type cells. All these data suggest that AtABCG40 is an ABA importer. In leaves, *AtABCG40* promoter activity was abundant in guard cells, suggesting that this protein mediates ABA uptake into guard cells. *AtABCG40* is expressed also in other tissues including roots; however, the roles of this protein in these tissues remain to be identified.

A mutant defective in AtABCG22, a half-size ABC transporter related to AtABCG25, was found to be defective in stomatal closure (Kuromori et al. 2011). It is possible that ABCG22 is involved in ABA transport; however, no ABA transport activity has been detected for AtABCG22. Alternatively, AtABCG22 may regulate stomatal aperture independent of ABA transport as reported for AtABCC5/MRP5 and AtABCB14. ABCC5/MRP5 regulates ABA and $Ca2^{+}$ signaling through the modulation of anion-channel activities; however, the substrate of this protein has not been identified (Klein et al. 2003; Suh et al. 2007). Moreover, ABCB14 is a malate importer that is highly expressed in guard cells and regulates stomatal responses to CO_2 (Lee et al. 2008).

3.3.2 NPF

A yeast two-hybrid system that can detect the ABA-dependent interactions between the ABA receptor PYR/PYL/RCAR and PP2C protein phosphatases was used as a sensor to monitor ABA concentrations in yeast cells and, hence, the transport activities of proteins expressed by the yeast cells (Kanno et al. 2012). A screen using the yeast two-hybrid system against an Arabidopsis cDNA library identified some NPF proteins as candidates for ABA importers; the candidate NPF proteins induced interactions between PYR/PYL/RCAR and PP2C in yeast cells at low ABA concentrations. Mutants defective in one of the candidate genes, *AtNPF4.6/AIT1*, were less sensitive to exogenously applied ABA, whereas overexpression of *AtNPF4.6* resulted in an enhanced response to ABA during seed germination and/or early seedling growth. These data, together with the plasma membrane localization of AtNPF4.6, suggest that the protein mediates cellular ABA uptake. The inflorescence stems of *npf4.6* mutants had lower surface temperatures with more open stomata compared to wild type. Considering that AtNPF4.6 is an ABA importer and that the *npf4.6* mutant is defective in stomatal closure, it was speculated that AtNPF4.6 mediated ABA uptake into guard cells; however, *AtNPF4.6* was expressed in vascular tissues. Thus, AtNPF4.6 might regulate the amount of ABA transported out from vascular tissues, where ABA is actively synthesized, in association with the function of AtABCG25, an ABA exporter. Interestingly, overexpression of *AtNPF4.6* under the control of the CaMV35S promoter reduced the surface temperature of leaves and stems. This result implies that constitutive ABA uptake activity reduced the amount of mobile ABA present in the apoplastic space.

NPF proteins were originally identified as low-affinity nitrate transporters (NRT1) or di/tri-peptide transporters (PTR) (Tsay et al. 2007; Leran et al. 2014).

AtNPF4.6 is identical to the previously characterized nitrate transporter NRT1.2 (Huang et al. 1999). The K_m values of AtNPF4.6 for nitrate and ABA were about 5.9 mM and 5 μM, respectively (Kanno et al. 2012; Huang et al. 1999). A competition assay using a modified yeast two-hybrid system with the receptor complex showed that nitrate did not compete with ABA as a substrate for AtNPF4.6 (Kanno et al. 2013), suggesting that ABA is the better substrate of this protein. In addition, AtNPF4.6 preferred the naturally occurring (+) enantiomer of ABA to the synthetic (−) enantiomer and did not transport GA, IAA, or jasmonic acid. These results indicate that AtNPF4.6 is a specific ABA transporter. AtNPF4.1, a closely related homologue of AtNPF4.6 that was identified also as an ABA importer candidate, transported GA in addition to ABA (Kanno et al. 2012). The physiological functions of AtNPF4.1 remain to be determined.

Several other NPF members are reported to transport IAA or glucosinolates (Nour-Eldin et al. 2012; Krouk et al. 2010). These findings indicate that NPF proteins transport a variety of compounds. Thus, although several NPF proteins have been characterized as nitrate or peptide transporters, careful investigation including detailed analysis of their substrate specificities will be required to understand their in vivo functions. The *nip/lated* mutants that are defective in an NPF protein in *Medicago truncatula* (MtNPF1.7) have defects in lateral root development and root nodule formation (Veereshlingam et al. 2004; Yendrek et al. 2010). The defect in lateral root development of *nip/lated* is partially rescued by exogenously applied ABA (Liang et al. 2007). Also, the mutants are less sensitive to exogenously applied ABA in terms of stomatal closure and seed germination (Liang et al. 2007). These observations indicate that MtNPF1.7/NIP/LATED is somehow involved in the physiological responses mediated by ABA; however, ABA transport activities of MtNPF1.7/NIP/LATED have not been reported and this protein may function as a high-affinity nitrate transporter (Bagchi et al. 2012).

3.4 Further Prospects

Endogenous ABA levels are regulated by several steps during biosynthesis and catabolism. Thus, transmembrane transport of ABA precursors and metabolites as well as bioactive ABA will all contribute to define the ABA concentration at a specific site. ABA is derived from carotenoids, and early stages of ABA biosynthesis take place in chloroplasts (plastids). After the cleavage of epoxycarotenoids by 9-*cis* epoxycarotenoid dioxygenase (NCED), the C_{15} product xanthoxin is converted to ABA via two steps in the cytosol. Since NCEDs are localized to the stroma and/or thylakoids (Endo et al. 2008; Tan et al. 2003), xanthoxin has to be translocated from the inside to the outside of chloroplasts. The first step in the major ABA catabolism pathway is the hydroxylation of the 8′ position by the CYP707A subfamily of cytochrome P450 (P450) enzymes (Cutler and Krochko 1999; Kushiro et al. 2004; Saito et al. 2004). P450s are generally localized in the endoplasmic reticulum (ER). Thus, ABA synthesized in the cytosol needs to be

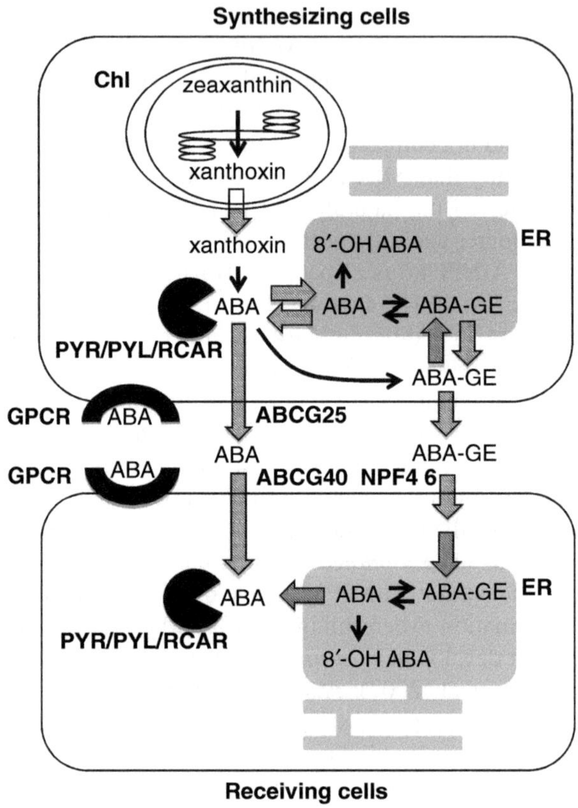

Fig. 3.1 A simplified model for the transport of ABA. For simplification, the transport of ABA, ABA precursors and metabolites into ABA-synthesizing cells (*upper*), and ABA biosynthesis in the receiving cells (*lower*) are not considered. *Simple arrows* and *bold outlined arrows* indicate metabolism (biosynthesis and catabolism) of ABA and transmembrane transport of ABA, respectively. The early steps of ABA biosynthesis take place in chloroplasts (*Chl*). Xanthoxin, a C_{15} product, is produced by the reaction mediated by 9-*cis* epoxycarotenoid dioxygenase (*NCED*) in chloroplasts. Specific transporters are required for the transmembrane transport of xanthoxin from the inside to the outside of chloroplasts. The final reactions of ABA biosynthesis proceed in the cytosol. ABA is recognized by the soluble PYR/PYL/RCAR receptors and induces physiological responses. ABA catabolism mediated by CYP707A to produce 8′-hydroxyl ABA (*8′-OH ABA*) takes place in the endoplasmic reticulum (*ER*), indicating that transmembrane transport of ABA into the ER is required. It is possible that ABA is inactivated also by glucosidation to produce an ABA-glucosyl ester (*ABA-GE*). Because the ABA-GE hydrolyzing enzyme BG1 is localized in the ER, ABA, or ABA-GE has to be transported into the ER. ABA synthesized from ABA-GE could induce physiological responses via PYR/PYL/RCAR. ABA in the cytosol is exported extracellularly by ABCG25 and moves through the apoplastic space. ABA in the apoplast could induce physiological responses via plasma membrane-localized GPCR receptors. After local or long-distance transport through the apoplast, ABA is imported into cells by ABCG40 and/or NPF4.6 and induces physiological responses via PYR/PYL/RCAR. It is also possible that ABA-GE is exported from cells, moves through the apoplast, is taken again into cells, and induces physiological responses

translocated to the ER to be inactivated. ABA may also be inactivated by glucosylation; glucosylated ABA (ABA-glucosyl ester; ABA-GE) is hydrolyzed to form bioactive free ABA (Nambara and Marion-Poll 2005). The β-glucosidase encoded by *BG1* in Arabidopsis hydrolyzes ABA-GE (Lee et al. 2006). Since BG1 is localized in the ER, ABA generated by BG1 in the ER has to be translocated to the cytosol to induce physiological responses. Also ABA or ABA-GE has to be translocated to the ER before or after glucosylation. Several glucosyltransferases in the UGT family that have activities to synthesize ABA-GE have been identified (Priest et al. 2006; Xu et al. 2002); however, the physiological functions of these glycosyltransferases are still not well established. ABA-GE has been detected in the apoplast, suggesting that ABA-GE moves between cells (Wilkinson and Davies 2002; Hartung et al. 2002).

Figure 3.1 illustrates a model for the possible transport or movement of ABA. Note that this is a simplified model in which ABA transport into ABA-synthesizing cells and ABA biosynthesis in the receiving cells are not taken into account. It is speculated that the transmembrane transport of ABA, ABA precursors, and ABA metabolites regulate physiological responses in a highly complex manner. As described above, three ABA transporters have been identified; however, relatively mild phenotypes of the mutants defective in the possible ABA transporters AtABCG25, AtABCG40 and AtNPF4.6 compared to typical ABA-deficient mutants such as Arabidospsi *aba1*, *aba2,* and *aba3*, indicate that the transport mechanisms are highly redundant. Therefore, further identification of ABA transporters will be required for a total understanding of ABA transport. Also, to identify how these transporters regulate ABA transport within plants, development of techniques to visualize local concentrations of ABA will be required.

References

Allan AC, Fricker MD, Ward JL, Beale MH, Trewavas AJ. Two transduction pathways mediate rapid effects of abscisic acid in *Commelina* guard cells. Plant Cell. 1994;6:1319–28.

Anderson BE, Ward JM, Schroeder JI. Evidence for an extracellular reception site for abscisic acid in *Commelina* guard cells. Plant Physiol. 1994;104:1177–83.

Bagchi R, Salehin M, Adeyemo OS, Salazar C, Shulaev V, Sherrier DJ, Dickstein R. Functional assessment of the *Medicago truncatula* NIP/LATD protein demonstrates that it is a high-affinity nitrate transporter. Plant Physiol. 2012;160:906–16.

Bauer H, Ache P, Lautner S, Fromm J, Hartung W, Al-Rasheid KA, Sonnewald S, Sonnewald U, Kneitz S, Lachmann N, Mendel RR, Bittner F, Hetherington AM, Hedrich R. The stomatal response to reduced relative humidity requires guard cell-autonomous ABA synthesis. Curr Biol. 2013;23:53–7.

Benjamins R, Scheres B. Auxin: the looping star in plant development. Annu Rev Plant Biol. 2008;59:443–65.

Bianco-Colomas J, Barthe P, Orlandini M, Le Page-Degivry MT. Carrier-mediated uptake of abscisic acid by suspension-cultured *Amaranthus tricolor* cells. Plant Physiol. 1991;95:990–6.

Campbell EJ, Schenk PM, Kazan K, Penninckx IA, Anderson JP, Maclean DJ, Cammue BP, Ebert PR, Manners JM. Pathogen-responsive expression of a putative ATP-binding cassette transporter gene conferring resistance to the diterpenoid sclareol is regulated by multiple defense signaling pathways in Arabidopsis. Plant Physiol. 2003;133:1272–84.

Christmann A, Weiler EW, Steudle E, Grill E. A hydraulic signal in root-to-shoot signalling of water shortage. Plant J. 2007;52:167–74.

Cutler AJ, Krochko JE. Formation and breakdown of ABA. Trends Plant Sci. 1999;4:472–8.

Davies WJ, Zhang JH. Root signals and the regulation of growth and development of plants in drying soil. Annu Rev Plant Physiol Plant Mol Biol. 1991;42:55–76.

Endo A, Sawada Y, Takahashi H, Okamoto M, Ikegami K, Koiwai H, Seo M, Toyomasu T, Mitsuhashi W, Shinozaki K, Nakazono M, Kamiya Y, Koshiba T, Nambara E. Drought induction of Arabidopsis 9-cis-epoxycarotenoid dioxygenase occurs in vascular parenchyma cells. Plant Physiol. 2008;147:1984–93.

Ernst L, Goodger JQ, Alvarez S, Marsh EL, Berla B, Lockhart E, Jung J, Li P, Bohnert HJ, Schachtman DP. Sulphate as a xylem-borne chemical signal precedes the expression of ABA biosynthetic genes in maize roots. J Exp Bot. 2010;61:3395–405.

Everatbourbouloux A. Transport and metabolism of labeled abscisic acid in broad bean plants (*Vicia faba* L.). Physiol Plant. 1982;54:431–9.

Frey A, Godin B, Bonnet M, Sotta B, Marion-Poll A. Maternal synthesis of abscisic acid controls seed development and yield in *Nicotiana plumbaginifolia*. Planta. 2004;218:958–64.

Gilroy S, Jones RL. Perception of gibberellin and abscisic acid at the external face of the plasma membrane of barley (*Hordeum vulgare* L.) aleurone protoplasts. Plant Physiol. 1994;104:1185–92.

Gonzalez-Guzman M, Pizzio GA, Antoni R, Vera-Sirera F, Merilo E, Bassel GW, Fernandez MA, Holdsworth MJ, Perez-Amador MA, Kollist H, Rodriguez PL. *Arabidopsis* PYR/PYL/RCAR receptors play a major role in quantitative regulation of stomatal aperture and transcriptional response to abscisic acid. Plant Cell. 2012;24:2483–96.

Goodger JQ, Schachtman DP. Re-examining the role of ABA as the primary long-distance signal produced by water-stressed roots. Plant Signal Behav. 2010;5:1298–301.

Hartung W, Sauter A, Hose E. Abscisic acid in the xylem: where does it come from, where does it go to? J Exp Bot. 2002;53:27–32.

Holbrook NM, Shashidhar VR, James RA, Munns R. Stomatal control in tomato with ABA-deficient roots: response of grafted plants to soil drying. J Exp Bot. 2002;53:1503–14.

Huang NC, Liu KH, Lo HJ, Tsay YF. Cloning and functional characterization of an Arabidopsis nitrate transporter gene that encodes a constitutive component of low-affinity uptake. Plant Cell. 1999;11:1381–92.

Ikegami K, Okamoto M, Seo M, Koshiba T. Activation of abscisic acid biosynthesis in the leaves of *Arabidopsis thaliana* in response to water deficit. J Plant Res. 2009;122:235–43.

Kang J, Hwang JU, Lee M, Kim YY, Assmann SM, Martinoia E, Lee Y. PDR-type ABC transporter mediates cellular uptake of the phytohormone abscisic acid. Proc Natl Acad Sci USA. 2010;107:2355–60.

Kang J, Park J, Choi H, Burla B, Kretzschmar T, Lee Y, Martinoia E. Plant ABC transporters. Arabidopsis Book. 2011;9:e0153.

Kanno Y, Jikumaru Y, Hanada A, Nambara E, Abrams SR, Kamiya Y, Seo M. Comprehensive hormone profiling in developing Arabidopsis seeds: examination of the site of ABA biosynthesis, ABA transport and hormone interactions. Plant Cell Physiol. 2010;51:1988–2001.

Kanno Y, Hanada A, Chiba Y, Ichikawa T, Nakazawa M, Matsui M, Koshiba T, Kamiya Y, Seo M. Identification of an abscisic acid transporter by functional screening using the receptor complex as a sensor. Proc Natl Acad Sci USA. 2012;109:9653–8.

Kanno Y, Kamiya Y, Seo M. Nitrate does not compete with abscisic acid as a substrate of AtNPF4.6/NRT1.2/AIT1 in Arabidopsis. Plant Signal Behav. 2013;8:e26624.

Karssen CM, Brinkhorstvanderswan DLC, Breekland AE, Koornneef M. Induction of dormancy during seed development by endogenous abscisic acid: studies on abscisic acid deficient genotypes of *Arabidopsis thaliana* (L.) Heynh. Planta. 1983;157:158–65.

Klein M, Perfus-Barbeoch L, Frelet A, Gaedeke N, Reinhardt D, Mueller-Roeber B, Martinoia E, Forestier C. The plant multidrug resistance ABC transporter AtMRP5 is involved in guard cell hormonal signalling and water use. Plant J. 2003;33:119–29.

Koiwai H, Nakaminami K, Seo M, Mitsuhashi W, Toyomasu T, Koshiba T. Tissue-specific localization of an abscisic acid biosynthetic enzyme, AAO3, in Arabidopsis. Plant Physiol. 2004;134:1697–707.

Kramer EM. How far can a molecule of weak acid travel in the apoplast or xylem? Plant Physiol. 2006;141:1233–6.

Krouk G, Lacombe B, Bielach A, Perrine-Walker F, Malinska K, Mounier E, Hoyerova K, Tillard P, Leon S, Ljung K, Zazimalova E, Benkova E, Nacry P, Gojon A. Nitrate-regulated auxin transport by NRT1.1 defines a mechanism for nutrient sensing in plants. Dev Cell. 2010;18:927–37.

Kudoyarova G, Veselova S, Hartung W, Farhutdinov R, Veselov D, Sharipova G. Involvement of root ABA and hydraulic conductivity in the control of water relations in wheat plants exposed to increased evaporative demand. Planta. 2011;233:87–94.

Kuromori T, Miyaji T, Yabuuchi H, Shimizu H, Sugimoto E, Kamiya A, Moriyama Y, Shinozaki K. ABC transporter AtABCG25 is involved in abscisic acid transport and responses. Proc Natl Acad Sci USA. 2010;107:2361–6.

Kuromori T, Sugimoto E, Shinozaki K. Arabidopsis mutants of *AtABCG22*, an ABC transporter gene, increase water transpiration and drought susceptibility. Plant J. 2011;67:885–94.

Kushiro T, Okamoto M, Nakabayashi K, Yamagishi K, Kitamura S, Asami T, Hirai N, Koshiba T, Kamiya Y, Nambara E. The *Arabidopsis* cytochrome P450 CYP707A encodes ABA 8′-hydroxylases: key enzymes in ABA catabolism. EMBO J. 2004;23:1647–56.

Lee M, Lee K, Lee J, Noh EW, Lee Y. AtPDR12 contributes to lead resistance in Arabidopsis. Plant Physiol. 2005;138:827–36.

Lee KH, Piao HL, Kim HY, Choi SM, Jiang F, Hartung W, Hwang I, Kwak JM, Lee IJ. Activation of glucosidase via stress-induced polymerization rapidly increases active pools of abscisic acid. Cell. 2006;126:1109–20.

Lee M, Choi Y, Burla B, Kim YY, Jeon B, Maeshima M, Yoo JY, Martinoia E, Lee Y. The ABC transporter AtABCB14 is a malate importer and modulates stomatal response to CO_2. Nat Cell Biol. 2008;10:1217–23.

Leran S, Varala K, Boyer JC, Chiurazzi M, Crawford N, Daniel-Vedele F, David L, Dickstein R, Fernandez E, Forde B, Gassmann W, Geiger D, Gojon A, Gong JM, Halkier BA, Harris JM, Hedrich R, Limami AM, Rentsch D, Seo M, Tsay YF, Zhang M, Coruzzi G, Lacombe B. A unified nomenclature of NITRATE TRANSPORTER 1/PEPTIDE TRANSPORTER family members in plants. Trends Plant Sci. 2014;19:5–9.

Liang Y, Mitchell DM, Harris JM. Abscisic acid rescues the root meristem defects of the *Medicago truncatula latd* mutant. Dev Biol. 2007;304:297–307.

Ma Y, Szostkiewicz I, Korte A, Moes D, Yang Y, Christmann A, Grill E. Regulators of PP2C phosphatase activity function as abscisic acid sensors. Science. 2009;324:1064–8.

Melhorn V, Matsumi K, Koiwai H, Ikegami K, Okamoto M, Nambara E, Bittner F, Koshiba T. Transient expression of *AtNCED3* and *AAO3* genes in guard cells causes stomatal closure in *Vicia faba*. J Plant Res. 2008;121:125–31.

Moons A. Transcriptional profiling of the PDR gene family in rice roots in response to plant growth regulators, redox perturbations and weak organic acid stresses. Planta. 2008;229:53–71.

Nambara E, Marion-Poll A. Abscisic acid biosynthesis and catabolism. Annu Rev Plant Biol. 2005;56:165–85.

Nour-Eldin HH, Andersen TG, Burow M, Madsen SR, Jorgensen ME, Olsen CE, Dreyer I, Hedrich R, Geiger D, Halkier BA. NRT/PTR transporters are essential for translocation of glucosinolate defence compounds to seeds. Nature. 2012;488:531–4.

Pandey S, Nelson DC, Assmann SM. Two novel GPCR-type G proteins are abscisic acid receptors in *Arabidopsis*. Cell. 2009;136:136–48.

Park SY, Fung P, Nishimura N, Jensen DR, Fujii H, Zhao Y, Lumba S, Santiago J, Rodrigues A, Chow TF, Alfred SE, Bonetta D, Finkelstein R, Provart NJ, Desveaux D, Rodriguez

PL, McCourt P, Zhu JK, Schroeder JI, Volkman BF, Cutler SR. Abscisic acid inhibits type 2C protein phosphatases via the PYR/PYL family of START proteins. Science. 2009;324:1068–71.

Perras MR, Abrams SR, Balsevich JJ. Characterization of an abscisic acid carrier in suspension-cultured barley cells. J Exp Bot. 1994;45:1565–73.

Petrasek J, Friml J. Auxin transport routes in plant development. Development. 2009;136:2675–88.

Priest DM, Ambrose SJ, Vaistij FE, Elias L, Higgins GS, Ross AR, Abrams SR, Bowles DJ. Use of the glucosyltransferase UGT71B6 to disturb abscisic acid homeostasis in *Arabidopsis thaliana*. Plant J. 2006;46:492–502.

Rea PA. Plant ATP-binding cassette transporters. Annu Rev Plant Biol. 2007;58:347–75.

Rodriguez-Gamir J, Ancillo G, Gonzalez-Mas MC, Primo-Millo E, Iglesias DJ, Forner-Giner MA. Root signalling and modulation of stomatal closure in flooded citrus seedlings. Plant Physiol Biochem. 2011;49:636–45.

Saito S, Hirai N, Matsumoto C, Ohigashi H, Ohta D, Sakata K, Mizutani M. Arabidopsis CYP707As encode (+)-abscisic acid 8′-hydroxylase, a key enzyme in the oxidative catabolism of abscisic acid. Plant Physiol. 2004;134:1439–49.

Sauter A, Davies WJ, Hartung W. The long-distance abscisic acid signal in the droughted plant: the fate of the hormone on its way from root to shoot. J Exp Bot. 2001;52(363):1991–7.

Schachtman DP, Goodger JQ. Chemical root to shoot signaling under drought. Trends Plant Sci. 2008;13(6):281–7.

Schwartz A, Wu WH, Tucker EB, Assmann SM. Inhibition of inward K^+ channels and stomatal response by abscisic acid: an intracellular locus of phytohormone action. Proc Natl Acad Sci USA. 1994;91:4019–23.

Seo M, Koshiba T. Transport of ABA from the site of biosynthesis to the site of action. J Plant Res. 2011;124:501–7.

Stein M, Dittgen J, Sanchez-Rodriguez C, Hou BH, Molina A, Schulze-Lefert P, Lipka V, Somerville S. *Arabidopsis* PEN3/PDR8, an ATP binding cassette transporter, contributes to nonhost resistance to inappropriate pathogens that enter by direct penetration. Plant Cell. 2006;18:731–46.

Suh SJ, Wang YF, Frelet A, Leonhardt N, Klein M, Forestier C, Mueller-Roeber B, Cho MH, Martinoia E, Schroeder JI. The ATP binding cassette transporter AtMRP5 modulates anion and calcium channel activities in *Arabidopsis* guard cells. J Biol Chem. 2007;282:1916–24.

Tan BC, Joseph LM, Deng WT, Liu L, Li QB, Cline K, McCarty DR. Molecular characterization of the *Arabidopsis* 9-*cis* epoxycarotenoid dioxygenase gene family. Plant J. 2003;35:44–56.

Tardieu F, Parent B, Simonneau T. Control of leaf growth by abscisic acid: hydraulic or non-hydraulic processes? Plant Cell Environ. 2010;33:636–47.

Tsay YF, Chiu CC, Tsai CB, Ho CH, Hsu PK. Nitrate transporters and peptide transporters. FEBS Lett. 2007;581:2290–300.

Veereshlingam H, Haynes JG, Penmetsa RV, Cook DR, Sherrier DJ, Dickstein R. *nip*, a symbiotic *Medicago truncatula* mutant that forms root nodules with aberrant infection threads and plant defense-like response. Plant Physiol. 2004;136:3692–702.

Verrier PJ, Bird D, Burla B, Dassa E, Forestier C, Geisler M, Klein M, Kolukisaoglu U, Lee Y, Martinoia E, Murphy A, Rea PA, Samuels L, Schulz B, Spalding EJ, Yazaki K, Theodoulou FL. Plant ABC proteins: a unified nomenclature and updated inventory. Trends Plant Sci. 2008;13:151–9.

Wilkinson S. pH as a stress signal. Plant Growth Regul. 1999;29:87–99.

Wilkinson S, Davies WJ. ABA-based chemical signalling: the co-ordination of responses to stress in plants. Plant, Cell Environ. 2002;25:195–210.

Windsor ML, Milborrow BV, McFarlane IJ. The uptake of (+)-*S*- and (−)-*R*-abscisic acid by suspension culture cells of hopbush (*Dodonaea viscosa*). Plant Physiol. 1992;100:54–62.

Windsor ML, Milborrow BV, Abrams SR. Stereochemical requirements of the saturable uptake carrier for abscisic acid in carrot suspension culture cells. J Exp Bot. 1994;45:227–33.

Xu ZJ, Nakajima M, Suzuki Y, Yamaguchi I. Cloning and characterization of the abscisic acid-specific glucosyltransferase gene from adzuki bean seedlings. Plant Physiol. 2002;129:1285–95.

Yamazaki D, Yoshida S, Asami T, Kuchitsu K. Visualization of abscisic acid-perception sites on the plasma membrane of stomatal guard cells. Plant J. 2003;35:129–39.

Yendrek CR, Lee YC, Morris V, Liang Y, Pislariu CI, Burkart G, Meckfessel MH, Salehin M, Kessler H, Wessler H, Lloyd M, Lutton H, Teillet A, Sherrier DJ, Journet EP, Harris JM, Dickstein R. A putative transporter is essential for integrating nutrient and hormone signaling with lateral root growth and nodule development in *Medicago truncatula*. Plant J. 2010;62:100–12.

Zeevaart JA, Boyer GL. Accumulation and transport of abscisic acid and its metabolites in Ricinus and Xanthium. Plant Physiol. 1984;74:934–9.

Chapter 4
ABA Transport and Distribution in Relation to Its Function in Plants

Bingbing Li and Wensuo Jia

Abstract Abscisic acid (ABA) is a phytohormone with key roles in various physiological processes, including germination, stomatal movement, and biotic and abiotic stress responses. Given that the sites of ABA biosynthesis and action are separated in plant cells, ABA transport is an important step in ABA signaling. ABA exhibits distinct patterns of compartmentalization, and its level in a specific compartment is not only determined by its rate of biosynthesis and catabolism, but also by the rates of ABA export and import. Compartmentalization ensures that the physiology of the plant is not affected by ABA under normal conditions and that redistribution of ABA in response to stress stimuli triggers ABA stress signaling. ABA is transported between cells or tissues and also systemically in whole plants. The expression of genes encoding key enzymes in the ABA biosynthesis pathway in vascular tissues and the overlap in the expression patterns of genes encoding enzymes involved in ABA biosynthesis and ABA transport implies the importance of ABA transport, particularly ABA systemic transport, in ABA functioning. The root-to-shoot transport of ABA further shows that ABA transport has critical roles in plant systemic signaling, and the molecular identification and functional characterization of ABA transporters provided direct evidence that ABA transport is required for ABA signaling. The ion-trap mechanism by pH and ABA transporters may operate synergistically, but in different manners, to control ABA transport and distribution. Further investigations on ABA transport and localization will contribute to our understanding of mechanisms that modulate ABA signaling in plants.

Keywords ABA long-distance transport · ABA distribution · ABA function

B. Li · W. Jia (✉)
College of Agronomy and Biotechnology, China Agricultural University,
Beijing 100094, China
e-mail: jiaws@cau.edu.cn

D.-P. Zhang (ed.), *Abscisic Acid: Metabolism, Transport and Signaling*,
DOI 10.1007/978-94-017-9424-4_4

4.1 ABA Content in Relation to ABA Biosynthesis and Catabolism

ABA is ubiquitous in plant cells. The distribution and abundance of ABA change greatly during development and under different environmental conditions. Under non-stressed conditions, the ABA level in leaves is generally much higher than in whole roots and stems. For example, the ABA content in the leaves of Zea may (maize) plants is several fold higher than that of the roots and stems (200–700 ng g^{-1} DW vs. below 100 ng g^{-1} DW, respectively) (Bahrun et al. 2002; Pekic et al. 1995). Although the level of ABA varies in different leaves of an individual plant, it does not differ much in different regions of the root. However, the level of ABA in brace roots is several folds higher than that in the upper and lower roots (Pekic et al. 1995). Furthermore, Pilet and Rivier (1981) found that the distribution of ABA was asymmetrical in the elongation zone of horizontal maize roots, with higher levels in the lower half and lower levels in the upper half of such roots. In wheat (*Triticum aestivum*) plants, the ABA concentration in whole roots is about 45 pmol g-1 fresh weight (FW), whereas it is 98 pmol g^{-1} FW in the apical parts of roots (Vysotskaya et al. 2008). Rinne et al. (1994) reported that the ABA content in the dormant buds of mountain birch (*Betzda pubescens*) is over 3,000 ng g^{-1} FW, which is more than 15 times higher than that in maize leaves. Furthermore, the ABA concentration in the terminal buds of short shoots is about four times higher than in the basal buds. In the floral organs of navel orange (*Citrus sinensis* Osbeck cv. Washington), the highest ABA levels were observed in the stigma/style shortly after anthesis (i.e., 11.5 or 3,036 ng g^{-1} FW), and these levels were several times higher than those measured in leaves (Harris and Dugoer 1986).

Relatively high ABA concentrations have been reported in fruits. For example, in ripening grape berries (*Vitis vinifera* cv. Doradillo) and strawberry (*Fragaria ananassa*), the ABA content may reach 212 ng g^{-1} FW Coombe and Hale (1973) and 150 ng g^{-1} FW (Jia et al. 2011), respectively. Much higher ABA concentrations were found in some other fruit species. For example, in bilberry (*Vaccinium myrtillus)* fruits, the ABA content may reach over 30,000 ng g^{-1} DW, which is more than 70 times higher than that in leaves (Karppinen et al. 2013). Furthermore, the ABA concentration in mangosteens (*Garcinia mangostana L.*) might be as high as 6,600 ng g^{-1} DW in some developmental stages (Kondo et al. 2002).

The ABA concentration in specific tissues is generally thought to be determined by the rate of ABA biosynthesis. However, the rate of ABA catabolism may also have an effect. The absolute rate of ABA catabolism was found to be proportional to the level of ABA, i.e., the higher the level of ABA, the faster the catabolism of this phytohormone (Jia and Zhang 1997). The rate of ABA catabolism has been demonstrated to be quite high. In maize leaves, for example, the half-life of ABA is less than an hour, and this implies that over 90 % of the ABA in a specific tissue would vanish as a result of ABA catabolism if ABA biosynthesis was to cease completely. Given the rapid rate of ABA catabolism, stable levels of ABA in specific tissues are due to a dynamic equilibrium between ABA biosynthesis and catabolism. It has been suggested that the level of xanthophyll precursors is several hundred times higher than the ABA level under non-stressed conditions (Zeevaart and Creelman 1988). Under stressed conditions, the accelerated rate of ABA biosynthesis would likely lead to the depletion of ABA precursors, such that content of ABA precursors may become a limiting factor in ABA accumulation. In support of this notion, a study by Ren et al. (2007) demonstrated that the water stress-induced ABA in maize roots depends on the delivery of ABA precursors from the leaves to the roots (Ren et al. 2007).

From a molecular perspective, the changes in ABA content at specific sites are generally thought to be determined by key enzymes in the ABA biosynthesis and catabolism pathway. ABA arises from the xanthophyll cleavage product. The first step of ABA biosynthesis, which is catalyzed by zeaxanthin epoxidase (ZEP), is the conversion of zeaxanthin to all-trans-violaxanthin, and the second step, which is catalyzed by 9-cis-epoxycarotenoid dioxygenase (NCED), is the production of xanthoxin through the oxidative cleavage of 9-cis-violaxanthin and/or 9′-cisneoxanthin. The later events of ABA biosynthesis are dominated by a major pathway that includes the conversion of xanthoxin to abscisic aldehyde, in a reaction that is catalyzed by a short-chain dehydrogenase/reductase (ABA2), and the conversion of abscisic aldehyde to ABA, which is catalyzed by ABA aldehyde oxidase (AAO3). ABA catabolism includes two major pathways, i.e., the catabolic pathway of ABA hydroxylation and the conjugation of ABA (Nambara and Marion-Poll 2005). In the ABA catabolic pathway, 8′-hydroxylation is believed to be the key regulator of ABA levels. ABA 8′-hydroxylation is catalyzed by a cytochrome P450 monooxygenase (P450). In Arabidopsis, P450 CYP707A genes that encode ABA 8′-hydroxylases have been identified (Kushiro et al. 2004; Saito et al. 2004). ABA conjugation is largely through the conjugation of the carboxyl and hydroxyl groups of ABA with glucose to produce ABA glucosyl ester (ABA-GE). The release of ABA from ABA-GE, which is catalyzed by beta-glucosidase, has been reported to play an important role in the regulation of ABA accumulation.

The first two steps of the ABA biosynthesis pathway are believed to occur in plastids and the later steps occur in the cytosol (Seo and Koshiba 2002). However, evidence suggests that ABA is largely distributed in chloroplasts rather than in the cytosol. This indicates that the sites of ABA biosynthesis may differ from those of the cellular distribution of ABA. Tan et al. (2003) have identified nine NCED genes in the complete genome sequence. Spatial expression analysis using promoter::GUS fusions in transgenic Arabidopsis plants revealed that different AtNCED proteins have different expression patterns, but that localization to chloroplasts is a distinguishing characteristic of all AtNCEDs. Furthermore, several AtNCEDs localized largely to vascular tissues. Using a similar approach, Cheng et al. (2002) reported that ABA2 is largely localized to vascular tissues in both roots and leaves in Arabidopsis seedlings. Similar to the observations for the localization of NCED and ABA2, Koiwai et al. (2004) reported that AAO3 is abundant in vascular tissues, especially in phloem companion cells and xylem parenchyma cells. These studies suggest that chloroplasts in leaves and vascular tissues play important roles in ABA biosynthesis under unstressed conditions.

4.2 Cellular Compartmentalization and Redistribution

Since ABA is a weak acid, it may exist in undissociated or protonated forms and in a dissociated anion form. The protonated form permeates freely across membranes, whereas the dissociated anion does not. Consequently, the amount of ABA distributed in different compartments is largely determined by pH; the greater the difference in pH between compartments, the greater the amount of ABA accumulates in the more alkaline compartment (Hartung 1983; Baier and Hartung 1988; Hartung and Slovik 1991). Based on related parameters, such as membrane conductance, anion permeability, pKa value, and so on, Cowan and Railton (1986) developed a mathematic model of ABA distribution, which showed that 4.4, 68.4, 17.2, and 10 % of the total cell ABA occurred in apoplasts, chloroplasts, the cytoplasm, and vacuoles, respectively. Based on a mathematical model developed using morphological and physiological data experimentally determined in *Valerlanella locusta*, Slovik and Hartung (1992) described the pattern of ABA distribution in mesophyll cells, epidermal cells, guard cells, phloem, and the apoplasm. According to their analysis (Slovik and Hartung 1992), the highest concentration of ABA occurs in guard cells (i.e., 426.4, 467.6, and 195.7 nM of ABA in the cytosol, chloroplast, and vacuole of guard cells, respectively). By contrast, the ABA content in mesophyll and epidermal cells was much lower; the ABA content in the cytosol and vacuole was only 9 and 2.8 nM for mesophyll cells and 55.1 and 5.6 nM for epidermal cells, respectively. Regardless of the cell type, chloroplasts always had the greatest ABA content.

Immunolocalization techniques using anti-ABA serum or monoclonal antibodies allow ABA to be monitored directly at the cellular and subcellular levels. Using rabbit anti-ABA serum, Sotta et al. (1985) developed an indirect immunohistochemical technique involving staining of the peroxidase–antiperoxidase (PAP) complex to indicate ABA localization in the buds and leaves of *Chenopodium* plants. The staining patterns showed an apicobasal gradient of ABA. Dense staining appeared in buds and vascular tissues. Phloem cells of the main axis and chloroplasts of the leaves were specifically labeled. In older buds, ABA was mainly distributed in quiescent meristematic cells. Using a combination of ABA monoclonal antibody and 568 goat anti-mouse IgG antibodies, Schraut et al. (2004) visualized ABA localization and transport in maize roots using confocal laser scanning microscopy and showed that ABA is mainly concentrated in the root cap and meristematic root tip and that the exodermis may act as a barrier to control the lateral transport of ABA in maize roots. Similarly, immunolocalization of ABA in roots and needles of radiate pine *(Pinus radiate)* showed that ABA is mainly located in the exodermis of roots and guard cells of needless (Diego et al. 2013).

ABA can be observed at the subcellular and ultrastructural levels using immunogold electron microscopy localization. Sossountzov et al. (1986) used this technique to show that ABA is mainly localized to two sites in the axillary-bud-bearing nodes of *Chenopodium polyspermum L*, i.e., in the plastids of cortical cells and the vascular parenchyma cells associated with sieve elements and xylem vessels, and in the cytoplasm and nucleus of cells in the axillary bud tip and procambial strands. Immunogold analysis of tomato showed that free ABA mainly localized in the apoplast, cytoplasmic vesicles, and amyloplasts, around starch grains, of root cap cells, and in the columella and meristematic cells, mainly at the junction area with root cap cells and in the wall (Bertrand et al. 1992). Water stress induced a significant increase in ABA labeling in the chloroplast, nucleus, and cell walls, with the increase being greatest in the cell wall. Based on this observation, the authors' proposed that the increase in apoplastic ABA in lavender exposed to water stress was not attributable to ABA release from the chloroplast.

The information provided by immunolocalization studies of ABA at various levels largely agrees with the data of ABA quantification in specific tissues or cells, and the spatial localization of enzymes involved in ABA biosynthesis, such as NCED, ABA2, and AAO3. All of these studies suggest that ABA is particularly concentrated in buds, root tips, vascular tissues, and stomata. However, at the cellular and subcellular levels, there is some disagreement between the data. As described above, both histochemical staining of promoter::GUS activity and immunofluorescence localization demonstrated that AAO3, the last enzyme in the ABA biosynthesis pathway, is mainly present in phloem companion cells and xylem parenchyma cells of Arabidopsis (Koiwai et al. 2004). In contrast, immunolocalization of ABA in the vascular tissues of *Chenopodiu* plants did not reveal any ABA in the parenchyma cells or the cambial zone (Sotta et al. 1985). This may be due to the different plant materials used; however, the ABA distribution pattern is thought to be basically the same among common plant species. Furthermore, it should be noted that immunolocalization analysis does not exactly reflect the distribution of ABA in plants. Because ABA is a small molecule, ABA will only be detected by immunolocalization when it is coupled to a cellular protein, which implies that the distribution of cellular proteins is an important factor affecting ABA localization. Given that the protein content in companion cells is not less than in other cells, the reduction or lack of ABA in the companion cells of vascular tissues implies that the ABA synthesized in companion cells is transported through vascular strands to various target tissues and cells.

4.3 Intercellular Transport in Relation to ABA Function

4.3.1 ABA Transport and Distribution in Relation to Stomatal Movement

ABA is a major regulator of stomatal movement and is generally thought to be a critical mediator of water deficit-induced stomatal closure. Although the signal transduction pathways underlying ABA-regulated stomatal movement have been extensively studied, there is still much debate about the origin of the ABA signal. As described above, in leaves, ABA is mainly localized to chloroplasts and vascular tissues; therefore, for ABA to mediate stomatal closure, ABA must be transported from the chloroplast or vascular tissues to their sites of action in guard cells. To reveal the mechanisms by which ABA mediates drought-induced stomatal closure, it is imperative that we identify the sites of ABA action at the guard cells. In the early 1980s, Hartung (1983) demonstrated that the primary site of ABA action at the guard cell plasmalemma is either at the outer surface of the plasmalemma or at another position that is readily accessible from the outside, because exogenous ABA is able to induce stomatal closure at high pH (i.e., 8.0), in the absence of ABA uptake. Given that Arabidopsis NCEDs and AAO3 are primarily localized in vascular tissues, Seo and Koshiba (2011) proposed that water shortages are first perceived by leaf vascular tissues and that the ABA signal originates from the vascular tissues that induce stomatal closure.

However, this hypothesis does not consider the roles of ABA in the chloroplasts of mesophyll and guard cells. As mentioned above, ABA is abundant in both the vascular tissues and chloroplasts. Furthermore, guard cells are not only directly attached to mesophyll cells, but also themselves contain chloroplasts. Therefore, the ABA in mesophyll and guard cells may also regulate stomatal movement. Actually, most of the earlier works suggested that ABA in the mesophyll and guard cells plays important roles in stomatal closure. Based on the anion-trap mechanism for ABA, Hartung and Slovik (1991) developed a mathematical model with which they analyzed the redistribution of ABA in relation to stomatal closure in leaves in response to drought stress. Based on their analysis, Hartung and Slovik (1991) proposed that drought stress causes a pH shift in different compartments of leaves and this pH shift induces a complicated redistribution of ABA among compartments. The redistribution of ABA may cause ABA to accumulate in guard cell walls to levels that are more than 16-fold over the initial value, which would be sufficient to induce stomatal closure. Furthermore, these authors proposed that it is the plasmalemma of the epidermal cells that initially senses drought stress, such that the epidermal cells play an important role in triggering stomatal closure, and the mesophyll cells function to support stomatal closure only synergistically. By quantifying the amount of (±)-3H-ABA released from (±)-3H-ABA-loaded Xanthium leaves, Bray and Zeevaart (1985) demonstrated that the efflux of endogenous ABA from cells was indeed closely correlated with the pH gradients among the various cellular compartments, and moreover that this efflux could be promoted by drought stress. In agreement with Hartung and Slovik's proposal described above, immunogold localization analysis in lavender leaves showed a fourfold increase in ABA, as reflected by the density of immunolabeling in the cell wall under drought (Pastor et al. 1998). Furthermore, Wang and Jia (1995) used immunogold electron microscopy localization to show that drought stress could lead to an evident increase in ABA content in the apoplast of the epidermis, while ABA content in mesophyll cells remained constant during the early stages of water stress. Interestingly, in response to drought stress, ABA was found to preferentially accumulate in the dorsal wall of guard cells.

While drought stress-induced stomatal closure is generally thought to be mediated by ABA coming from mesophyll and vascular cells, it is difficult to understand why stomatal movement is not affected by the ABA in the mesophyll and vascular cells under non-stressed conditions, because under non-stressed conditions both mesophyll and vascular cells contain much ABA. A reasonable explanation may be that drought stress not only results in an increase in the ABA level in mesophyll and vascular cells, but also increases the release of ABA from the mesophyll and vascular cells. Therefore, the ABA in guard cells themselves may well have a function in stomatal movement, and this possibility merits further investigation. There is strong evidence that guard cells contain a large amount of ABA in guard cells. For example, in *Valerianella locusta*, the ABA content in guard cells was estimated to be 600 fg per guard cell, corresponding to an even concentration of 0.65 mM in the cells (Behl and Hartung 1986), and in isolated protoplasts from *Vicia faba,* it was estimated to be 24–150 fg per guard cell protoplast (Lahr and Raschke 1988). More importantly, ABA accumulation in guard cells appears to increase in response to drought stress. For example, Cornish and Zeevaart (1986) found that the guard cells in *V. faba L.* plants contained 18 times more ABA when isolated from stressed leaves than when isolated from turgid leaves. Given that some ABA receptors were localized within cells, a great increase in the ABA content within the guard cells is likely to contribute to stomatal closure.

4.3.2 ABA Transport and Distribution in Relation to Seed and Fruit Development

ABA plays important roles in seed development and germination. During seed development, ABA levels changed to different degrees in different tissues. In Arabidopsis, ABA levels in whole siliques slowly increased during late development, whereas they concomitantly sharply decreased in the seeds and seed envelopes (i.e., the pedicles, receptacles, valves, replums, septa, and funiculi) during the middle to final stages of development (Kanno et al. 2010), implying that ABA may be transported among different tissues. Given that the seed coat originates from the mother plant, whereas the endosperm and embryo originate from zygotes, ABA is expected to be synthesized only in the zygotic tissues of the seeds obtained by crossing an ABA-deficient female and a wild-type male. By quantifying the ABA in the F1 and F2 populations derived from crosses between the wild type and an ABA-deficient mutant aba2-2, Kanno et al. (2010) demonstrated that in the absence of zygotic ABA, ABA synthesized in maternal tissues could be translocated into the embryos to induce seed dormancy. Grafting analyses using wild-type tobacco (*Nicotiana plumbaginifolia*) and its corresponding ABA-deficient mutants revealed that ABA synthesized in vegetative tissues could be transported to the seeds to promote seed development and growth (Frey et al. 2004). Generating crosses or grafts between the wild type and an ABA-deficient mutant is a good approach for investigating ABA transport between tissues. However, the observed ABA translocation from wild-type tissues to those of the ABA-deficient tissues may not reflect the cases in non-crossed or non-grafted plants, because it has been proposed that ABA may be transported within seed tissues as a result of passive diffusion (Bruggeman et al. 2001). Supposing that the observed translocation of ABA from wild-type tissues to ABA-deficient tissues is the result of passive ABA diffusion, one would expect ABA transport from high to low ABA-contained tissues, such that the ABA transport among tissues should be a universal and important phenomenon in regulating seed development and dormancy.

Many plant species utilize fruit to disperse seeds, and seed development is tightly coupled with the development of the whole fruit. For example, once a seed is fully developed, the fruit is also fully ripened, therefore promoting the dissemination of the seeds. The tight relationship between seed and fruit development implies that materials and messages are transferred between seeds and other tissues of the fruit (i.e., the pericarp and receptacle). Phytohormones are central regulators of both seed and fruit development and ripening. Ripening of the respiration climatic fruits is determined by ethylene. Recently, ABA has been increasingly reported to play important roles in development and ripening in many fruit species, especially in the non-respiration climatic species (Giribaldi et al. 2010; Jia et al. 2011). Deytieux et al. (2005) reported that, in grape berry (*V. vinifera* L), a non-climacteric fruit, ABA levels greatly increased in both whole berries and separated pericarps when the berries changed color and, interestingly, the ABA level in pulp rapidly declined during the fruit ripening and dramatically increased in the skin tissue. This study suggests that a specific repartition of ABA occurs between the different tissues of the berry. Knowledge of the origin and movement of the ABA signal among fruit tissues is particularly important for understanding the mechanisms underlying fruit development and ripening, but unfortunately, little is known about these topics.

4.4 Long-Distance Transport in Relation to Plant Systemic Signaling

Higher organisms not only rely on the communication of materials and signals among cells, but also among organs. The communication between organs generally requires long-distance transport of a signal, and this enables higher organisms to respond more rapidly and more positively to environmental stimuli. This is best demonstrated in animals, where nerve-mediated signaling swift and purposeful reactions, thus protecting the animal from possible harm. As in nerve-mediated signaling in animals, long-distance signaling in plants is a prerequisite for plants being capable of perceiving environmental stimuli. Recently, numerous studies have suggested that ABA may act as a long-distance transporter signal to mediate root-to-shoot communication under drought stress conditions.

Drought stress is a major factor affecting plant growth and development. As more than 90 % of water is lost through stomatal transpiration, the regulation of stomatal movement is particularly important for plants' adaptation to drought stress. Leaf stomatal conductance was initially thought to be regulated by a hydraulic signal. However, Blackman and Davies (1985) conducted a root-split experiment that demonstrated that stomatal closure could occur when no perceptible changes occurred in leaf water potential under soil-drying conditions, implying the presence of chemical signals involved in root-to-shoot communication. As ABA is known to play an important role in the regulation of stomatal movements, there has been much discussion as to whether ABA is a root-to-shoot signal. Numerous studies have found that drought stress induces a dramatic increase in ABA content in both the root and xylem sap, and the increase in ABA is closely correlated with a decrease in leaf stomatal conductance (Zhang and Davies 1989, 1990, 1991). Feeding xylem sap collected from drought plants to detached leaves was able to inhibit stomatal movement, and this inhibitory effect was relieved when ABA was removed from the xylem sap by an ABA-affinity column (Zhang and Davies 1991; Davies et al. 1994). These findings strongly suggest that ABA is capable of acting as a long-distance signal to regulate stomatal movement under drought stress conditions (Davies and Zhang 1991).

While ABA is a well-known mediator of stomatal closure through long-distance transport, numerous reports suggest that root-sourced ABA cannot mediate stomatal closure under some conditions. For instance, in some plant species, root-sourced ABA was found to have no significant effect on stomatal movement (Holbrook et al. 2002; Munns and King 1988). Supposing that drought stress is able to increase the xylem ABA concentration significantly, a reasonable explanation for the failure of root-sourced ABA to regulate stomatal movement is that the root-sourced ABA cannot be effectively accumulated at its target sites of the guard cells. As discussed above, pH is a major factor affecting the distribution and accumulation of ABA, and hence, it is of particular interest to establish whether pH modulates the root-to-shoot signaling of ABA. In agreement with this proposal, it has repeatedly been shown that drought stress is able to induce an increase in xylem sap pH in many plant species (Jia and Davies 2007; Wilkinson 1999), and the pH increase in the xylem is proposed to render the apoplast of the leaf more alkaline, and this would contribute to the sequestration of more ABA in the apoplast of guard cells, thereby promoting stomatal closure (Wilkinson 1999; Jia and Davies 2007; Wikinson and Davies 2002). By feeding artificial xylem sap buffered to different pH to detached leaves of *Commelina communis*, Wilkinson and Davies (2002) found that an increase in pH could indeed lead to a great reduction in transpiration rate in the presence of low concentrations of ABA. These studies strongly suggest that pH can function as a signal that coordinately regulates stomatal movement with the ABA signal under drought stress conditions. Given that the regulatory effect of root-sourced ABA on stomatal movement is strongly affected by apoplasitc pH, any factors that affect apoplastic pH would be expected to have an impact on stomatal movement in relation to root-to-shoot ABA signaling. This has been demonstrated by a study of Jia and Davies (2007), which showed that different forms of nitrite nutrition could modify stomatal sensitivity to root-sourced ABA, owing to modifications of the apoplastic pH in the leaves of some species.

Water deficit occurs initially from the top of roots and gradually extends toward the tip. Once the rate of water transport from soil into roots can no longer balance the rate of water loss from leaves, a decrease in leaf water potential must occur. The top region of roots in drying soil produces ABA, while the region deep within wet soil helps sustain the leaves in a state of water balance, such that the control of stomatal movement by root-sourced ABA can be achieved in the absence of a decreased leaf water potential under drought stress conditions. As with nerve-mediated signaling in animals, the root-to-shoot transport of ABA enables plants to reduce the transpiration rate soon after exposure to drought stress, thereby minimizing the effects of drought on plant growth and development.

4.5 ABA Transporters in Relation to ABA Transport and Distribution

Chemical transport across cellular membranes generally includes two major modes, i.e., passive diffusion and transporter-mediated positive transport. Investigations on ABA transport have largely focused on passive transport, while the contribution of ABA transporters was largely overlooked, owing to the fact that ABA transporters were only recently identified. The first evidence for the presence of ABA transporters came from the observations that ABA transport showed a saturable pattern in specific regions of the root (Astle and Rubery 1980; Astle and Rubery 1983). However, it was not until 2010 that two ABA transporters, i.e., ABCG25 and ABCG40, were successfully identified (Kuromori et al. 2010; Kang et al. 2010). The ABA transporters belong to the large family of ABC (ATP Binding Cassette) transporter genes in Arabidopsis (Verrier et al. 2008). Besides the two ABA transporters, another ABA transporter, AIT1, was recently been identified as a nitrate transporter, NRT1.2 (Kanno et al. 2012).

The ABA transport capability of AIT1 was demonstrated using a two-hybrid system comprised of a PYR1 (a PYR ABA-receptor)-fused GAL4 activation domain and PP2C (type 2C protein phosphatase)-fused DNA-binding domain. In the presence of external ABA, the expression of AIT1 promoted a physical interaction between PYR1 and PP2C, demonstrating entry of external ABA into the cells (Kanno et al. 2012). The transport activity of AIT1 was determined using a heterologous expression system in yeast and Sf9 cells (Kanno et al. 2012), and it was found that AIT1 showed a higher influx transport activity for (+)-ABA than for (−)-ABA. A close homologue of AIT1 and AIT3 was also demonstrated to have ABA transport activity, but this transporter did not discriminate between (+) and (−)-ABA in yeast cells. The transport activity of AtABCG25 was tested in membrane vesicles derived from AtABCG25-expressing insect cells. AtABCG25 showed a relatively high affinity for (+)-ABA, with a Km of 260 nM, and was found to be ATP-dependent. The ABA transport activity of AtABCG40 was evaluated by expression in yeast and bright yellow 2 (BY2) cells (Kang et al., 2010). AtABCG40 also showed a high affinity for (+)-ABA with a Km of 1 μM. Therefore, compared with AtABCG25, the affinity of AtABCG40 for ABA is relatively lower (Kang et al. 2010).

pAtABCG25::GUS expression analysis indicated that AtABCG25 is expressed mainly in vascular tissues. Overexpression of AtABCG25 resulted in a decrease in stomatal aperture, as reflected by a higher leaf temperature (Kuromori et al. 2010). Also, promoter–GUS analysis showed that AtABCG40 was broadly expressed in both leaves and roots, but the expression of AtABCG40 was much higher in guard cells than in mesophyll cells. Microarray analysis indicated that the expression of AtABCG40 may be eightfold higher in guard cells than in mesophyll cells. Loss-of-function atabcg40 mutants exhibited a decrease in stomatal sensitivity to ABA and drought tolerance, and moreover, the expression of ABA-responsive genes was strongly delayed compared with that in wild-type plants. Similarly, AtAIT1 was also found to be mainly expressed in vascular tissues and guard cells. Compared with the wild type, the stomatal aperture on the stem surface was larger in the ait1 mutants, as reflected by a lower surface temperature. Additionally, compared with the wild type, the AIT1 mutants were less sensitive to exogenous ABA in terms of seed germination and post-germinative growth, and conversely, overexpression of AIT1 resulted in ABA hypersensitivity under the same conditions (Kanno et al. 2012).

The identification of ABA transporters was an important breakthrough in our understanding of ABA transport and distribution. The identification of ABA transporters not only provided a more reasonable explanation for the precise control of ABA transport and localization, but also demonstrated that ABA transport is an absolutely important step in ABA functioning in plants, because the loss and gain of function of the ABA transporters altered the plants' responses to exogenous ABA, at least in terms of stomatal movement, seed germination, and gene expression. Theoretically, the ABA transporters function mainly in two aspects: (1) ABA export from its sites of synthesis and (2) ABA import to its sites of action. The large overlap in the localization of ABA transporters and key enzymes involved in ABA biosynthesis pathway suggests that ABA transporters mainly transport ABA from its sites of synthesis. However, given that ABA receptors are proposed to be mainly localized within cells (Shen et al. 2006; Santiago et al. 2009), the modulation of the plant's response to exogenous ABA by ABA transporters demonstrates that the ABA transporters function in ABA import from the apoplastic space into cells. This is a paradox that is difficult to reconcile. Therefore, although several ABA transporters have been identified to date, little is known about the mechanisms that control ABA transport and localization in relation to the function of ABA transporters. As the specific expression of ABA transporters in different tissues and at developmental stages is capable of directly governing ABA transport and localization, compared with the ion-trap mechanism, ABA transporter-mediated ABA transport appears to be a more powerful mechanism for controlling ABA transport and distribution. Particularly, the specific expression of ABA transporters in vascular tissues suggests that the ABA transporters contribute to the loading of ABA into xylem or phloem for intercellular or long-distance transport. It is likely that these two mechanisms play different roles, but operate synergistically in ABA transport and distribution. In other words, ABA transporters may play a major role in the intercellular or long-distance transport of ABA, whereas the ion-trap mechanism involving pH may play a major role in cellular

ABA compartmentalization and redistribution. Nevertheless, the existence of other mechanisms involved in ABA transport and localization in relation to the synergistic action of ABA transporters and pH needs to be investigated.

4.6 Concluding Remarks

As for any signaling molecule, the sites of ABA metabolism must be separated from its sites of perception, and the localization and transport of ABA are important factors in ABA signaling. ABA is a weak acid that is ubiquitous in plant tissues. Both the ion-trap theory of pH and experimental evidence suggest that ABA is compartmentalized in plant cells, rather than being evenly distributed. On one hand, this compartmentalization ensures that ABA does not interfere with other physiological actions in plant tissues or cells containing a high level of ABA under normal conditions. On the other hand, the redistribution of ABA following changes in pH in cellular compartments under specific conditions ensures that ABA may possibly function even if its content remains unchanged. Besides compartmentalization and redistribution, intercellular or long-distance transport is an important basis for ABA signaling in plants. Particularly, the expressional pattern of genes encoding the key enzymes in ABA biosynthesis was found to be highly similar to that of ABA transporters, and moreover, loss or gain of the ABA transporters was found to alter the response sensitivity to ABA, strongly demonstrating the importance of ABA transport in ABA functioning. The root-to-shoot signaling mediated by the long-distance transport of ABA is a good example of the critical roles of ABA transport in ABA functioning. Although the recent identification of some ABA transporters has revealed the molecular basis of ABA transport, little is known about the mechanisms that control ABA transport and localization in relation to ABA accumulation at action sites. A major reason for this is the technical difficulty involved in localizing ABA. The commonly adopted immunolocalization method is not actually an ideal indicator of ABA localization and distribution, because it depends on the conjugation of ABA to cellular proteins. Immunolocalization in combination with ultracryotomy may be a good technical system for this purpose; however, this technique needs to be further developed. The development of new technologies to observe ABA localization as well as thorough investigations of ABA transporters is required to fully understand ABA transport and localization in relation to ABA functioning.

References

Astle MC, Rubery PH. A study of abscisic acid uptake by apical and proximal root segments of *Phaseolus coccineus* L. Planta. 1980;150:312–20.

Astle MC, Rubery PH. Carriers for abscisic acid and indole-3-acetic acid in primary roots: their regional localization and thermodynamic driving forces. Planta. 1983;157:53–63.

Bahrun A, Jensen CR, Asch F, Mogensen VO. Drought-induced changes in xylem pH, ionic composition, and ABA concentration act as early signals in field-grown maize (*Zea mays L.*). J Exp Bot. 2002;53:251–63.

Baier M, Gimmler H, Hartung W. Permeability of guard cell plasma membrane and tonoplast. J Exp Bot. 1998;41:351–8.

Behl R, Hartung W. Movement and compartmentation of abscisic acid in guard cells of *Valerianella locusta*: Effects of osmotic stress, external H(+)-concentration and fusicoccin. Planta. 1986;168:360–8.

Bertrand S, Benhamou N, Nadeau P, Dostaler D, Gosselin A. Immunogold localization of free abscisic acid in tomato root cells. Can J Bot. 1992;70(5):1001–11.

Blackman PG, Davies WJ. Root to shoot communication in maize plants of the effects of soil drying. J Exp Bot. 1985;36:39–48.

Bray EA, Zeevaart JA. The compartmentation of abscisic acid and beta-d-glucopyranosyl abscisate in mesophyll cells. Plant Physiol. 1985;79(3):719–22.

Bruggeman FJ, Libbenga KR, Duijn BV. The diffusive transport of gibberellins and abscisic acid through the aleurone layer of germinating barley grain: A mathematical model. Planta. 2001;214(1):89–96.

Cheng WH, Endo A, Zhou L, Penney J, Chen HC, et al. A unique short-chaindehydrogenase/reductase in Arabidopsis glucose signaling and abscisic acid biosynthesis and functions. Plant Cell. 2002;14:2723–43.

CornishJan Katrina, Zeevaart. AD. Abscisic Acid Accumulation by in Situ and Isolated Guard Cells of *Pisum sativum L.* and *Vicia faba L.* in relation to water stress. Am Soc Plant Biol. 1986;81:1017–21.

Cowan AK, Railton ID. Chloroplasts and the biosynthesis and catabolism of abscisic acid. J Plant Growth Regul. 1986;4(21):1–24.

Davies WJ, Zhang J. Root signals and the regulation of growth and development of plants in drying soil. Annu Rev Plant Physinl Plant Mol Biol. 1991;42:55–76.

Davies WJ, Tardieu F, Trejo CL. How do chemical signals work in plants that grow in drying soil. Plant Physiol. 1994;104:309–14.

De Diego N, Rodríguez JL, Dodd IC, Pérez-Alfocea F, Moncaleán P, Lacuesta M. Immunolocalization of IAA and ABA in roots and needles of radiata pine (*Pinus radiata*) during drought and re-watering. Tree Physiol. 2013;33(5):537–49.

Deytieux C, Geny L, Doneche B. Relation between hormonal balance and polygalacturonase activity in grape berry. Acta Hortic. 2005;682:163–70.

Frey A, Godin B, Bonnet M, Sotta B, Marion-Poll A. Maternal synthesis of abscisic acid controls seed development and yield in *Nicotiana plumbaginifolia*. Planta. 2004;218:958–64.

Giribaldi M, Gény L, Delrot S, Schubert A. Proteomic analysis of the effects of ABA treatments on ripening *Vitis vinifera* berries. J Exp Bot. 2010;61:2447–58.

Harris MJ, Dugoer WM. Levels of free and conjugated abscisic acid in developing floral organs of the navel orange. Plant Physiol. 1986;82:1164–6.

Hartung W. The site of action of abscisic acid at the guard cell plasmalemma of *Valerianella locusta*. Plant Cell Environ. 1983;6:27–428.

Hartung and Slovik. Physicochemical properties of plant growth regulators and plant tissues determine their distribution and redistribution. New Phytol. 1991;119:361–82.

Holbrook NM, Shashidhar VR, James RA, Munns R. Stomatal control in tomato with ABA-deficient response of grafted plants to soil drying. J Exp Bot. 2002;53:1503–14.

Jia WS, Davies WJ. Modification of leaf apoplastic pH in relation to stomatal sensitivity to root-sourced abscisic acid signals. Plant Physiol. 2007;143:68–77.

Jia W, Zhang J. Comparison of exportation and metabolism of xylem-delivered ABA in maize leaves at different water status and xylem sap pH. Plant Growth Regul. 1997;21:43–9.

Jia HF, Chai YM, Li CL, Lu D, Luo JJ, Qin L, Shen YY. Abscisic acid plays an important role in the regulation of strawberry fruit ripening. Plant Physiol. 2011;157:188–99.

Kang J, et al. PDR-type ABC transporter mediates cellular uptake of the phytohormone abscisic acid. Proc Natl Acad Sci USA. 2010;107:2355–60.

Kanno Y, Jikumaru Y, Hanada A, Nambara E, Abrams SR, Kamiya Y, et al. Comprehensive hormone profiling in developing *Arabidopsis* seeds: examination of the site of ABA biosynthesis ABA transport and hormone interactions. Plant Cell Physiol. 2010;51:1988–2001.

Kanno Y, et al. Identification of an abscisic acid transporter by functional screening using the receptor complex as a sensor. Proc Natl Acad Sci USA. 2012;109:9653–8.

Karppinen K, Hirvelä E, Nevala T, Sipari N, Suokas M, Jaakola L. Changes in the abscisic acid levels and related gene expression during fruit development and ripening in bilberry (*Vaccinium myrtillus L.*). Phytochemistry. 2013;95:127–34.

Koiwai H, Nakaminami K, Seo M, Mitsuhashi W, Toyomasu T, et al. Tissue specific localization of an abscisic acid biosynthetic enzyme, AAO3, in Arabidopsis. Plant Physiol. 2004;134:1697–707.

Kondo Satoru, Ponrod Wanvisa, Kanlayanarat Sirichai, Hirai Nobuhiro. Abscisic acid metabolism during fruit development and maturation of mangosteens. J Am Soc Hortic Sci. 2002;127:737–41.

Kuromori T, et al. ABC transporter AtABCG25 is involved in abscisic acid transport and responses. Proc Natl Acad Sci USA. 2010;107:2361–6.

Kushiro T, Okamoto M, Nakabayashi K, Yamagishi K, Kitamura S, Asami T, Hirai N, Koshiba T, Kamiya Y, Nambara E. The Arabidopsis cytochrome P450 CYP707A encodes ABA 8′-hydroxylases: key enzymes in ABA catabolism. EMBO J. 2004;23:1647–56.

Lahr W, Raschke K. Abscisic-acid contents and concentrations in protoplasts from guard cells and mesophyll cells of *Vicia faba L.* Planta. 1988;173:528–31.

Munns R, King RW. ABA is not the only stomatal inhibitor in transpiration stream of wheat plants. Plant Physiol. 1988;88:703–8.

Nambara E, Marion-Poll A. Abscisic acid biosynthesis and catabolism. Annu Rev Plant Biol. 2005;56:165–85.

Pekic S, Stikic R, Tomljanoviv L, Andjelkovic V, Anovic MI, Quarrie SA. Characterization of maize lines differing in leaf abscisic acid content in the field. 1 Abscisic acid physiology. Ann Bot. 1995;75:67–73.

Pilet PE, Rivier L. Abscisic acid distribution in horizontal maize root segments. Planta. 1981;153:453–8.

Ren HB, Gao ZH, Chen L, Wei KF, Liu J, Fan YJ, Jia W, Zhang J. Dynamic analysis of ABA accumulation in relation to the rate of ABA catabolism in maize tissues under water deficit. J Exp Bot. 2007;58:211–9.

Rinne P, Tuominen H, Junttila O. Seasonal changes in bud dormancy in relation to bud morphology, water and starch content, and abscisic acid concentration in adult trees of *Betula pubescens*. Tree Physiol. 1994;14(6):549–61.

Saito S, Hirai N, Matsumoto C, Ohigashi H, Ohta D, Sakata K, Mizutani M. Arabidopsis CYP707As encode (+)-abscisic acid 8'-hydroxylase, a key enzymein the oxidative catabolism of abscisic acid. Plant Physiol 2004;134:1439–49.

Santiago J, Dupeux F, Round A, Antoni R, Park SY, Jamin M, Cutler SR, Rodriguez PL, Marquez JA. The abscisic acid receptor PYR1 in complex with abscisic acid. Nature. 2009;462:665–8.

Schraut D, Ullrich CI, Hartung W. Lateral ABA-transport in maize roots (*Zea mays*): visualization by immunolocalization. J Exp Bot. 2004;55:1635–41.

Seo M, Koshiba T. Complex regulation of ABA biosynthesis in plants. Trends Plant Sci. 2002;7(1):41–8.

Seo M, Koshiba T. Transport of ABA from the site of biosynthesis to the site of action. J Plant Res. 2011;124(4):501–7.

Shen YY, Wang XF, Wu FQ, Du SY, Cao Z, Shang Y, Wang XL, Peng CC, Yu XC, Zhu SY, Fan RC, Xu YH, Zhang DP. The Mg-chelatase H subunit is an abscisic acid receptor. Nature. 2006;443:823–6.

Slovik S, Hartung W. Compartmental distribution and redistribution of abscisic acid in intact leaves: III. Analysis of the stress-signal chain. Planta. 1992;187(1):37–47.

Sossountzov L, Sotta B, Maldiney R, Sabbagh I, Migniniac E. Immunoelectron—microscopy localization of abscisic acid with colloidal gold on Lowicryl-embedded tissues of *Chenopodium polyspermum* L. Planta. 1986;168:471–81.

Sotta B, Sossountzov L, Maldiney R, Sabbagh I, Tachon P, Miginiac E. Abscisic acid localization by light microscopic immunohistochemistry in *Chenopodium polyspermum L.* Effect of water stress. J Histochem Cytochem. 1985;33:201–8.

Tan BC, Joseph LM, Deng WT, Liu LJ, Li QB, et al. Molecular characterization of the Arabidopsis 9-cis epoxycarotenoid dioxygenase gene family. Plant J. 2003;35:44–56.

Verrier PJ, et al. Plant ABC proteins-a unified nomenclature and updated inventory. Trends Plant Sci. 2008;13:151–9.

Vysotskaya LB, Korobova AV, Kudoyarova GR. Abscisic acid accumulation in the roots of nutrient-limited plants: Its impact on the differential growth of roots and shoot. J Plant Physiol. 2008;165:1274–9.

Wang XC, Jia W. Stress-induced stomatal closure in relation to compartmentation and redistribution of ABA in intact leaves of *Vicia faba L.* Acta Phytophysiologica Sinica. 1995;21:324–8.

Wikinson S, Davies WJ. ABA-based chemical signaling: the co-ordination of responses to stress in plants. Plant Cell Environ. 2002;25:195–210.

Zeevaart JAD. Metabolism of abscisic acid and its regulation in Xanthium leaves during and after water stress. Plant Physiol. 1983;7(1):477–81.

Zhang J, Davies WJ. Abscisic acid produced in dehydrating roots may enable the plant to measure the water status of the soil. Plant Cell Environ. 1989;12:73–81.

Zhang J, Davies WJ. Changes in the concentration of ABA in xylem sap as a function of changing soil water status com account for changes in leaf conductance and growth. Plant Cell Environ. 1990;13:271–85.

Zhang J, Davies WJ. Anti-transpirant activity in the xylem sap of maize plants. J Exp Bot. 1991;42:317–21.

Chapter 5
ABA Conjugates and Their Physiological Roles in Plant Cells

Zheng-Yi Xu, Yun-Joo Yoo and Inhwan Hwang

Abstract The phytohormone abscisic acid (ABA) is involved in many important physiological processes such as stomatal closure, seed dormancy, plant growth and development, and responses to various environmental stresses. Along with ABA, various conjugated forms of ABA, which are inactive in ABA-related physiology, have been identified in plant cells. One of the most abundant forms of ABA conjugates is an ABA-glucosyl ester named as ABA-GE. The conjugate forms have long been thought to be by-products that are generated by one of the catabolic processes involved in lowering cellular ABA levels. However, recent studies provide evidence for the role of ABA-GE as a reservoir for the rapid production of active ABA in a compartmentalized fashion. ABA-GE can be converted to active ABA through a one-step hydrolysis by two β-glucosidases: AtBG1 and AtBG2, that localize to the ER and vacuole, respectively. Moreover, ABA produced from ABA-GE by these enzymes is crucial for proper adaptation to abiotic stresses. Thus, ABA-GE plays a crucial role in ABA-related plant physiology, which is exerted through the hydrolysis by β-glucosidases. In addition, the physiological role of ABA-GE is critically dependent on additional activities including de novo biosynthesis of ABA, conjugation of ABA with glucose by ABA UDP-glucosyltransferases, compartmentalization into subcellular organelles and long-distance transportation between tissues.

Keywords ABA-glucosyl ester · Physiological functions · Metabolism · Transport

---

Z.-Y. Xu · I. Hwang
Division of Molecular and Life Sciences,
Pohang University of Science and Technology, Pohang 790-784, South Korea

I. Hwang (✉)
Division of Integrative Biosciences and Biotechnology,
Pohang University of Science and Technology, Pohang 790-784, South Korea
e-mail: ihhwang@postech.ac.kr

Y.-J. Yoo
School of Interdisciplinary Bioscience and Bioengineering,
Pohang University of Science and Technology, Pohang 790-784, South Korea

D.-P. Zhang (ed.), *Abscisic Acid: Metabolism, Transport and Signaling*,
DOI 10.1007/978-94-017-9424-4_5

5.1 Introduction

The phytohormone ABA is a 15-carbon sesquiterpenoid which functions in series of physiological processes including seed dormancy, stomatal conductance, and early embryo morphogenesis as well as in the synthesis of seed storage proteins. In addition, ABA also is involved in various abiotic and biotic stress responses. Under different environmental conditions, ABA modulates expression of a large number of genes so as to optimize plant growth and developmental strategies to these conditions (Thomashow 1999; Shinozaki and Yamaguchi-Shinozaki 2000; Zhu 2001; Bray 2002).

For these physiological responses, alteration of cellular ABA levels is essential and this is achieved through two opposing processes: biosynthesis and catabolism of ABA. De novo ABA biosynthesis starts from carotenoids and includes many intermediates (Nambara and Marion-Poll 2005). In the lengthy biosynthetic process, the intermediate steps leading to xanthoxin occur in chloroplasts, whereas the final two steps for conversion of xanthoxin to ABA occur in the cytoplasm (Cheng et al. 2002; Gonzalez-Guzman et al. 2002; Seo et al. 2004). On the other hand, the catabolic processes of ABA consist of two different pathways: hydroxylation and conjugation. In Arabidopsis, enzymes encoded by the *AtCYP707A* gene family (*CYP707A1, 2, 3* and *4*) mediate hydroxylation of ABA to produce hydroxylated ABA which is first metabolized to PA (Saito et al. 2004), then further converted to DPA by isomerization. In the conjugation pathway, glucose is added to ABA to produce ABA-glucosyl ester which exhibits little or no biological activity and is stored in the vacuole and ER (Vilaró et al. 2006). However, recent studies showed that inactive ABA-GE can be converted to active ABA by a one-step hydrolysis carried out by certain β-glucosidase (Dietz et al. 2000; Sauter et al. 2002; Lee et al. 2006; Xu et al. 2012, 2013). This indicates that ABA-GE is not a simple by-product produced in the inactivation process of ABA but rather plays a crucial role as the stored form of ABA.

5.2 Generation of ABA-GE

Until now, the most well-known conjugated form of ABA is ABA-GE which exhibits little or no biological activity (Vilaró et al. 2006). ABA-GE is generated from ABA in the course of the inactivation. Consistent with this notion, the concentration of ABA-GE increases substantially under stress conditions (Sauter et al. 2002). Moreover, repeated water stress treatment caused increased ABA-GE levels in *Xanthium strumarium* leaf (Zeevaart 1983).

In the cell, ABA-GE is produced by ABA UDP-glucosyltransferase (UGT) through esterification of ABA with glucose (Zeevaart and Creelman 1998; Lim et al. 2005; Priest et al. 2006; Piotrowska and Bajguz 2011). In *Arabidopsis*, the UGT family consists of more than 100 homologous proteins which can be divided into 12 subfamilies (Lorenc-Kukula et al. 2004; Ross et al. 2004;

Yonekura-Sakakibara and Hanada 2011). Among these UGTs, UGT71B6 which belongs to the subfamily E displays an activity toward naturally occurring (+)-ABA for glucose conjugation in vitro with an optimum pH between 6.5 and 7.0. Indeed, *UGT71B6* overexpression in plants leads to high-level accumulation of ABA-GE. Interestingly, in these transgenic plants, the levels of ABA catabolites, such as PA, DPA and neo-PA, were significantly reduced, which can be explained by the fact that these two pathways are involved in the inactivation of ABA in a competitive manner. Moreover, these results suggest a coordination of the two catabolic pathways in lowering cellular ABA levels. Interestingly, the free ABA level was not affected in these transgenic plants, suggesting that there is a compensation mechanism to adjust the cellular ABA levels in plant cells. However, transgenic plants overexpressing *UGT71B6* were slightly less sensitive to exogenous ABA during germination and showed a slightly higher water loss rate from detached leaves. In addition, *UGT71B6* overexpressers displayed hyposensitivity to 6 % glucose at post-germination-growth stages. In contrast, *UGT71B6* loss-of-function mutants did not display any noticeable phenotypes, which might be due to the functional redundancy (Priest et al. 2006). In fact, in *Arabidopsis*, the subfamily E contains additional UGTs: *At3g21790* and *At3g21800* that are highly homologous to *UGT71B6* and are located immediately upstream and downstream of *UGT71B6*, respectively (Fig. 5.1a). However, their physiological role and biochemical activities have not been addressed at the molecular and biochemical levels.

An expression study revealed that *UGT71B6* is strongly induced under various abiotic stress conditions (Priest et al. 2006). Interestingly, *UGT71B6* was induced as early as 30 min after the start of the abiotic stress treatment. Thus, under abiotic stress conditions, the induction of *UGT71B6* is as rapid as *NCED3,* which is involved in the de novo ABA biosynthetic pathway (Endo et al. 2008). It is expected that *UGT71B6* is induced at the later time points because it is involved in lowering ABA levels after abiotic stress responses. However, the rapid induction of *UGT71B6* upon abiotic stress treatments raises the intriguing possibility that *UGT71B6* might also be involved in abiotic stress responses. One possible role of *UGT71B6* under abiotic stress conditions is to fine-tune cellular ABA levels to prevent the excessive accumulation of ABA under abiotic stress conditions. However, this idea needs to be further tested in future studies.

5.3 β-Glucosidase Homologs in the Hydrolysis of ABA-GE

Whether ABA-GE has any physiological role has been controversial for a long time. Considering that (i) the amount of free ABA under water stress conditions is greater than that of endogenous ABA-GE under normal conditions (Neil et al. 1983); (ii) ABA-GE was not able to be hydrolyzed under water stress conditions (Milborrow 1978); and (iii) the activity of ABA-GE splitting enzymes was not increased (Lehmann and Vlasov 1988), it has been generally considered that ABA-GE is a simple by-product, rather than a stored form of ABA produced

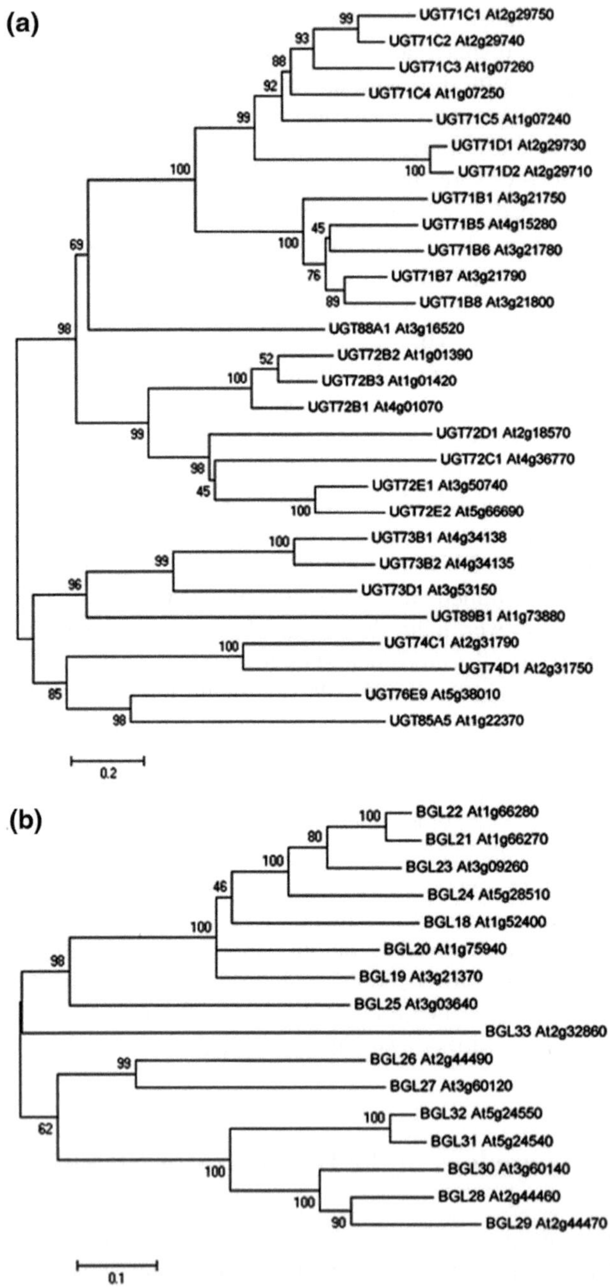

Fig. 5.1 Phylogenetic trees of *Arabidopsis* β-glucosidase homologs and UDP-glucosyltransferase homologs. The *Arabidopsis* β-glucosidase homologs and UDP-glucosyltransferase homologs were aligned by MUSCLE, using the default settings for multiple sequence alignments. The aligned sequences were used to construct a phylogenetic tree with Molecular Evolutionary Genetics Analysis version 5(MEGA5). Distance trees were created by a maximum likely-hood method. To estimate the reliability of this tree, a bootstrap analysis was performed with 1000 replicates

during ABA inactivation (Zeevart 1983; Piotrowska and Bajguz 2011). However, there have been multiple lines of evidence that ABA-GE may have a physiological role such as being a stored form of ABA. Johnson and Ferrell (1982) reported that in needles of *Pseudotsugamenziesii,* the level of conjugated ABA decreases and concomitantly the level of free ABA increases. In addition, it has been suggested that β-glucosidase homologs located at either intracellular or extracellular spaces harbor the activity that hydrolyzes glucose conjugated, biologically inactive ABA-GE to produce active ABA in plants. Indeed, in *Arabidopsis*, a large number of genes encode the putative β-glucosidases (Fig. 5.1b) and more than 30 β-glucosidases were predicted to be secreted to the apoplasmic space (Sauter et al. 2002). Consistent with this notion, Dietz et al. (2000) showed that intercellular washing fluid of barely primary leaves harbors the activity to release active ABA from physiologically inactive ABA-GE. The ABA liberated by this β-glucosidase can be taken up directly by the cells or loaded into the xylem for long-distance transportation. Based on this observation, they suggested that ABA-GE is a stored form of ABA. Moreover, considering the impermeability of lipid membranes toward ABA-GE, it has been suggested that ABA-GE is an ideal compound for long-distance transportation of ABA. Furthermore, because of this property, it has been postulated that ABA-GE may function as a long-distance signal transmitted from roots to shoots under dehydration stress conditions.

Following up on this long speculation, the most compelling evidence on the possible role of ABA-GE as a stored form of ABA was provided by recent publications on two β-glucosidases in Arabidopsis, AtBG1 and AtBG2, which can hydrolyze ABA-GE to ABA (Lee et al. 2006; Xu et al. 2012). A genetic approach showed that *atbg1* and *atbg2* mutants display increased sensitivity to dehydration and salinity stresses. Moreover, *atbg1 atbg2* double-knockout mutants showed the additive effect on dehydration stress sensitivity, suggesting that they act redundantly. By contrast, the ectopic expression of *AtBG1* and *AtBG2* displayed an enhanced resistance to osmotic stress. Moreover, *atbg1* mutants showed many ABA-deficient phenotypes. Consistent with these phenotypes, *atbg1* mutants have lower ABA levels, in particular, greatly reduced extracellular pools of ABA, compared to the wild-type plants. Intriguingly, AtBG1 and AtBG2 localize to the ER and vacuole, respectively, indicating that ABA is produced in the ER and vacuole in plant cells in addition to the cytosol through the de novo biosynthetic pathway. These findings raise fundamentally important questions such as what is the physiological meaning of ABA production in multiple compartments and how these multiple ABA biosynthetic pathways in different organelles are coordinated to maintain the active ABA pool at a specific level depending on cellular and environmental conditions. Another interesting aspect of these proteins is the regulation of their activity according to the environmental conditions. In the case of AtBG1, it undergoes polymerization into high-molecular weight forms to increase the enzymatic activity under dehydration stress conditions (Lee et al. 2006; Watanabe et al. 2014), whereas AtBG2 existing as high-molecular weight forms at all conditions accumulates to higher levels under dehydration stress to increase ABA production (Xu et al. 2012).

In regulating ABA production at the whole plant level, one mechanism would be to differentially regulate expression of genes involved in ABA production.

AtBG1 and *AtBG2* sharing a high-sequence similarity show both overlapping and also specific expression patterns. The *AtBG1* expression is specifically restricted to seed cells, hydathodes, and vascular tissues in leaf tissues (Lee et al. 2006). In contrast, *AtBG2* showed a more broad-expression pattern, such as veins of cotyledons and leaf tissues, the vascular bundles of the hypocotyl, and primary and secondary root tissues (Xu et al. 2012). Moreover, both of them are induced under osmotic stress conditions.

5.4 Compartmentalization of ABA-GE

Location of AtBG1 and AtBG2 in the ER and vacuole, respectively, implies that ABA-GE is stored in the ER and vacuole. Previous studies have shown that nearly 20 % of total ABA is stored in the vacuole as glucosyl ester forms (Harris and Dugger 1986; Xiong and Zhu 2003; Verslues and Zhu 2007; Piotrowska and Bajguz 2011). In the cytosol, high levels of ABA are lowered through a process of glucose conjugation to ABA-GE by the cytosolic ABA-specific UGTs, and subsequently, ABA-GE is stored in the vacuole (Fig. 5.2). Vacuolar storage of ABA-GE requires transportation of ABA-GE from the cytosol to the vacuole across the tonoplast, which most likely occurs via energy-consuming transporters because ABA-GE is highly impermeable to biomembranes. Recently, it has been shown that vacuolar ABCC-type ABC transporters, AtABCC1 and AtABCC2, have the ABA-GE transport activity when expressed in yeast although in vivo evidence is still lacking in plants (Fig. 5.2) (Burla et al. 2013).

In contrast to vacuolar ABA-GE, there is little information on the storage of ABA-GE in the ER. However, the fact that AtBG1 localized to the ER hydrolyzes ABA-GE to ABA, which is critical for plant development and stress responses, strongly suggests that ABA-GE is also stored in the ER. In addition, this finding implies the existence of an ABA-GE transporter at the ER membrane. Unfortunately, most of the aspects regarding ABA-GE in the ER remain elusive. For example, how much of ABA-GE is stored in the ER and which type of transporters is involved in the ABA-GE transportation across the ER membrane remains elusive (Fig. 5.2). Despite the lack of information on these aspects, it is clear that ABA-GE stored in the ER is crucial for cellular homeostasis of ABA not only for abiotic stress conditions but also under normal physiological conditions as evidenced by the fact that AtBG1 plays a crucial role in various aspects of ABA-related plant physiology.

5.5 Long-distance Transport of ABA-GE

In ABA-related plant physiology, an important unanswered question is whether ABA is transported from one tissue to another via long-distance transportation and whether such transport is necessary in plants. The fact that genes involved

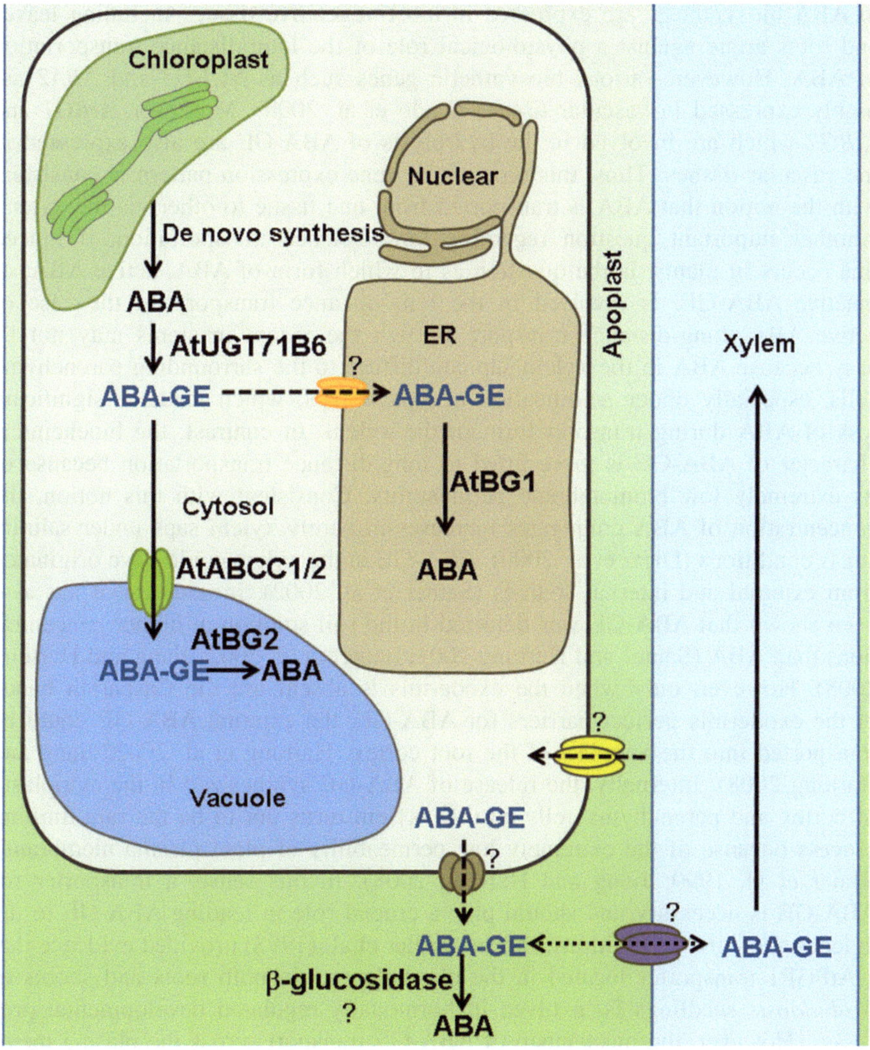

Fig. 5.2 Production and hydrolysis of ABA-GE and its transport across various subcellular membranes. ABA-GE was produced through the conjugation of ABA with glucose UGT71B6 in the cytosol. AtABCC1/2 located at the tonoplast direct translocation of ABA-GE from the cytosol to the vacuole. After translocating into the ER or vacuoles, it could be hydrolyzed to produce ABA by AtBG1 and AtBG2, respectively. Apoplastic ABA-GE could be also by hydrolyzed to ABA by β-glucosidase homologs whose identity is not known, or loaded into the xylem for long-distance transportation to other tissues. The processes which have been experimentally demonstrated are depicted in solid lines whereas those unknown are shown by broken lines. Putative ABA-GE transporters at various subcellular organelles are marked by a question mark. ER, endoplasmic reticulum

in ABA biosynthesis are expressed in most vegetative tissues including leaves and roots argue against a physiological role of the long-distance transportation of ABA. However, various biosynthetic genes such as *NCED3* and *ABA2* are highly expressed in vascular tissues (Endo et al. 2008). Moreover, *AtBG1* and *AtBG2* which are involved in the hydrolysis of ABA-GE are also expressed in the vascular tissues. Thus, this type of the gene expression pattern is consistent with the notion that ABA is transported from one tissue to other via the xylem. Another important question regarding long-distance transportation, if indeed this occurs in plants, is the question as to which form of ABA, active ABA, or inactive ABA-GE, is involved in the long-distance transport. In the case of active ABA, long-distance transport through the xylem in stems may not be easy because ABA in the xylem sap can diffuse to the surrounding parenchyma cells, especially under acidification of xylem saps, which leads to significant loss of ABA during transport through the xylem. In contrast, the biochemical character of ABA-GE is more fitted to long-distance transportation because of its extremely low biomembrane permeability. Consistent with this notion, the concentration of ABA conjugates increases in barely xylem saps under salinity stress conditions (Dietz et al. 2000). ABA-GE in the xylem could have originated from external and internal sources (Sauter et al. 2002). In addition, it has also been shown that ABA-GE was detected in the soil solution at higher concentrations than ABA (Sauter and Hartung 2002; Sauter et al. 2002; Jiang and Hartung 2008). However, only when the exodermis is absent are the Casparian bands of the exodermis perfect barriers for ABA-GE, the external ABA-GE could be transported into the apoplast of the root cortex (Hartung et al. 2002; Jiang and Hartung 2008). Internally, the release of ABA-GE synthesized in the cytoplasm of cortex and parenchyma cells into the xylem turns out to be the rate-limiting process because of the extremely low permeability of plant plasma membranes (Baier et al. 1990; Jiang and Hartung 2008). In this sense, a transporter for ABA-GE is necessary and should play a crucial role in loading ABA-GE to the xylem under stress conditions. Indeed, Sidler et al. (1998) provided evidence that a AtPGP1 transporter located in the plasmalemma in both roots and shoots of *Arabidopsis* seedlings is involved in hormonally regulated developmental processes. However, the mechanism of ABA-GE transport across the plasma membrane remains unclear.

It is generally believed that ABA-GE is transported from the root tissues to leaf tissues. At the leaf tissues, ABA-GE is subject to two different pathways. One of them involves apoplastic glucosidases that could directly hydrolyze ABA-GE to release active ABA, which is then involved in regulating stomatal opening and closing, or plant development. In another pathway, putative ABA-GE transporters might conduct the active transport of ABA-GE into the cytosol of plant cells and then subsequently from the cytosol to subcellular organelles such as the ER and vacuole where ABA-GE is stored, or hydrolyzed to active ABA by ABA-GE-specific β-glucosidases depending on cellular conditions.

5.6 Perspective and Concluding Remarks

The physiological role of ABA-GE has long been controversial. Now, it is clear that ABA-GE, one type of ABA conjugate, is crucial in ABA-related physiology by functioning as a stored form of ABA because ABA-GE can be rapidly converted to active ABA through a one-step hydrolysis by AtBG1 and AtBG2. However, there are many unanswered questions regarding the physiological meaning of ABA. First of all, the identity of the ABA-GE transporter localized at the ER and possibly plasma membrane remains elusive, not to mention their action mechanisms. In addition, how much ABA-GE is stored in different subcellular organelles and how much of the active ABA pool is contributed by the ABA-GE pathways remains unclear. Furthermore, how are these different pathways coordinated to maintain the active ABA pool at the cellular and whole plant levels? Therefore, future work should be directed to investigate: (i) the identity of ABA-GE transporters at various subcellular organellar membranes and their mechanisms of action; (ii) how these transporters are coordinated for homeostasis of ABA levels at the cellular level; (iii) what is(are) the signaling pathway(s) involved in the coordination of these multiple biosynthetic and catabolic pathways involving multiple organelles under normal growth and abiotic stress conditions at the cell level; (iv) how the long-distance transport, if there is any, contributes to the homeostasis of ABA at the whole plant level.

References

Baier M, Gimmler H, Hartung W. Permeability of the guard cell plasma-membrane and tonoplast. J Exp Bot. 1990;41:351–8.

Bray EA. Abscisic acid regulation of gene expression during water-deficit stress in the era of the *Arabidopsis* genome. Plant Cell Environ. 2002;25:153–61.

Burla B, Pfrunder S, Nagy R, Francisco RM, Lee Y, Martinoia E. Vacuolar transport of abscisic acid glucosyl ester is mediated by ATP-binding cassette and proton-antiport mechanisms in Arabidopsis. Plant Physiol. 2013;163:1446–58.

Cheng WH, Endo A, Zhou L, Penney J, Chen HC, Arroyo A, Leon P, Nambara E, Asami T, Seo M, Koshiba T, Sheen J. A unique short-chain dehydrogenase/reductase in Arabidopsis glucose signaling and abscisic acid biosynthesis and functions. Plant Cell. 2002;14:2723–43.

Dietz KJ, Sauter A, Wichert K, Messdaghi D, Hartung W. Extracellular b-glucosidase activity in barley involved in the hydrolysis of ABA glucose conjugate in leaves. J Exp Bot. 2000;51:937–44.

Endo A, Sawada Y, Takahashi H, Okamoto M, Ikegami K, Koiwai H, Seo M, Toyomasu T, Mitsuhashi W, Shinozaki K, Nakazono M, Kamiya Y, Koshiba T, Nambara E. Drought induction of Arabidopsis 9-cis-epoxycarotenoid dioxygenase occurs in vascular parenchyma cells. Plant Physiol. 2008;147:1984–93.

Gonzalez-Guzman M, Apostolova N, Belles JM, Barrero JM, Piqueras P, Ponce MR, Micol JL, Serrano R, Rodriquez PL. The short-chain alcohol dehydrogenase ABA2 catalyzes the conversion of xanthoxin to abscisic aldehyde. Plant Cell. 2002;14:1833–46.

Harris MJ, Dugger WM. The occurrence of abscisic acid and abscisyl-beta-d-glucopyranoside in developing and mature citrus fruit as determined by enzyme immunoassay. Plant Physiol. 1986;82:339–45.

Hartung W, Sauter A, Hose E. Abscisic acid in the xylem: where does it come from, where does it go to? J Exp Bot. 2002;53:27–32.

Jiang F, Hartung W. Long-distance signaling of abscisic acid (ABA): the factors regulating the intensity of the ABA signal. J Exp Bot. 2008;59:937–44.

Johnson JD, Ferrell WK. The relationship of abscisic acid metabolism to stomatal conductance in Douglas fir during water stress. Plant Physiol. 1982;55:431–7.

Lee KH, Piao HL, Kim HY, Choi SM, Jiang F, Hartung W, Hwang I, Kwak JM, Lee IJ, Hwang I. Activation of glucosidase via stress-induced polymerization rapidly increases active pools of abscisic acid. Cell. 2006;126:1109–20.

Lehmann H, Vlasov P. Plant growth and stress: the enzymic hydrolysis of an abscisic acid conjugate. J Plant Physiol. 1988;132:98–101.

Lim EK, Doucet CJ, Hou B, Jackson RG, Abrams SR, Bowles DJ. Resolution of (+)-abscisic acid using an Arabidopsis glycosyltransferase. Asymmetry. 2005;16:143–7.

Lorenc-Kukuła K, Korobczak A, Aksamit-Stachurska A, Kostyń K, Lukaszewicz M, Szopa J. Glucosyltransferase: the gene arrangement and enzyme function. Cell Mol Biol Lett. 2004;9:935–46.

Milborrow BV. The stability of conjugated abscisic acid during wilting. J Exp Bot. 1978;29:1059–66.

Nambara E, Marion-Poll A. Abscisic acid biosynthesis and catabolism. Annu Rev Plant Biol. 2005;56:165–85.

Neil SJ, Horgan R, Heald JK. Determination of the levels of abscisic acid-glucose ester in plants. Planta. 1983;157:371–5.

Piotrowska A, Bajguz A. Conjugates of abscisic acid, brassinosteroids, ethylene, gibberellins, and jasmonates. Phytochemistry. 2011;72:2097–112.

Priest DM, Ambrose SJ, Vaistij FE, Elias L, Higgins GS, Ross ARS, Abrams S, Bowles DJ. Use of the glucosyltransferase UGT71B6 to disturb abscisic acid homeostasis in *Arabidopsis thaliana*. Plant J. 2006;46:492–502.

Ross ARS, Ambrose SJ, Cutler AJ, Feurtado JA, Kermode AR, Nelson K, Zhou R, Abrams SR. Determination of endogenous and supplied deuterated abscisic acid in plant tissues by high-performance liquid chromatography-electrospray ionization tandem mass spectrometry with multiple reaction monitoring. Anal Biochem. 2004;329:324–33.

Saito S, Hirai N, Matsumoto C, Ohigashi H, Ohta D, Sakata K, Mizutani M. Arabidopsis CYP707As encode (+)-abscisic acid 8′-hydroxylase, a key enzyme in the oxidative catabolism of abscisic acid. Plant Physiol. 2004;134:1439–49.

Sauter A, Abrams SR, Hartung W. Structural requirements of abscisic acid (ABA) and its impact on water flow during radial transport of ABA analogues through maize roots. J Plant Growth Regul. 2002;21:50–9.

Sauter A, Hartung W. The contribution of internode and mesocotyl tissues to root-to-shoot signaling of abscisic acid. J Exp Bot. 2002;53:297–302.

Seo M, Aoki H, Koiwai H, Kamiya Y, Nambara E, Koshiba T. Comparative studies on the Arabidopsis aldehyde oxidase (AAO) gene family revealed a major role of AAO3 in ABA biosynthesis in seeds. Plant Cell Physiol. 2004;45:1694–703.

Shinozaki K, Yamaguchi-Shinozaki K. Molecular responses to dehydration and low temperature: differences and cross-talk between two stress signaling pathways. Curr Opin Plant Biol. 2000;3:217–23.

Sidler M, Hassa P, Hasan S, Ringli C, Dudler P. Involvement of an ABC transporter in a developmental pathway regulating hypocotyl cell elongation in the light. Plant Cell. 1998;10:1623–36.

Thomashow MF. Plant cold acclimation: Freezing tolerance genes and regulatory mechanisms. Annu. Rev. Plant Physiol. Plant Mol. Biol. 1999;50:571–99.

Verslues PE, Zhu JK. New developments in abscisic acid perception and metabolism. Curr. Opion. Plant. Biol. 2007;10:447–52.

Vilaró F, Canela-Xandri A, Canela R. Quantification of abscisic acid in grapevine leaf (*Vitisvinifera*) by isotope-dilution liquid chromatography-mass spectrometry. Anal Bioanal Chem. 2006;386:306–12.

Watanabe S, Matsumoto M, Hakomori Y, Takagi H, Shimada H, Sakamoto A. The purine metabolite allantoin enhances abiotic stress tolerance through synergistic activation of abscisic acid metabolism. Plant Cell Environ. 2014;37(4):1022–36. doi:10.1111/pce.12218 [Epub ahead of print].

Xiong L, Zhu JK. Regulation of abscisic acid biosynthesis. Plant Physiol. 2003;133:29–36.

Xu ZY, Lee KH, Dong T, Jeong JC, Jin JB, Kanno Y, Kim DH, Kim SY, Seo M, Bressan RA, Yun DJ, Hwang I. A vacuolar β-glucosidase homolog that possesses glucose-conjugated abscisic acid hydrolyzing activity plays an important role in osmotic stress responses in Arabidopsis. Plant Cell. 2012;24:2184–99.

Xu ZY, Kim DH, Hwang I. ABA homeostasis and signaling involving multiple subcellular compartments and multiple receptors. Plant Cell Rep. 2013;32:807–13.

Yonekura-Sakakibara K, Hanada K. An evolutionary view of functional diversity in family 1 glycosyltransferases. Plant J. 2011;66:182–93.

Zeevaart JAD. Metabolism of abscisic acid and its regulation in Xanthium leaves during and after water stress. Plant Physiol. 1983;71:477–81.

Zeevaart JAD, Creelman RA. Metabolism and physiology of abscisic acid. Annu. Rev. Plant Physiol. Biol. 1998;39:439–73.

Zhu JK. Cell signaling under salt, water and cold stresses. Curr Opin Plant Biol. 2001;4:401–6.

Chapter 6
ABA Signal Perception and ABA Receptors

Xiao-Fang Wang and Da-Peng Zhang

Abstract Searches for ABA receptors started from earlier analysis of ABA-binding proteins or perception sites, which suggest that ABA signal is perceived both inter- and intracellularly. The receptors or candidate receptors for ABA, localized to plasma membrane, cytosol, and chloroplast, have been identified. Mg-chelatase H subunit (CHLH)/putative ABA receptor (ABAR) is a chloroplast-membrane protein, binds ABA, and inhibits a group of WRKY transcription repressors to relieve ABA-responsive genes of inhibition. A chloroplast co-chaperonin CPN20 interacts with ABAR to antagonize ABAR–WRKY40 coupled signaling. Recent discovery of a crucial ABA signaling component SOAR1 working downstream of ABAR suggests that ABAR regulates a central ABA signaling network. GTG1 and GTG2 are a novel class of GPCR-type G proteins, which localize to plasma membrane and may perceive intercellular ABA signal. GDP-bound GTGs bind ABA, which initiates the ABA signaling cascade; GTP-bound Gα subunit GPA1 inhibits formation of the GTG–GDP, downregulates ABA binding to the GTGs and represses ABA signaling. The START-domain PYR/PYL/RCAR receptors are cytosolic/nuclear proteins and mediate a core ABA signaling pathway. ABA binding to the PYR/PYL ABA receptors inactivates negative signaling regulators type 2C protein phosphatases, leading to the activation of the SNF1-related protein kinases SnRK2s through phosphorylation by the GSK3-like BIN2 kinase, which subsequently phosphorylate downstream targets to induce a diversity of ABA responses. How to integrate known ABA signal components, especially the ABAR- and GTGs-mediated signaling pathways, into the PYR/PYL/RCAR-mediated signaling pathway, is a challenging question in the future.

Keywords ABA-binding protein · Receptors · ABA signaling

X.-F. Wang (✉) · D.-P. Zhang
School of Life Sciences, Center for Plant Biology,
Tsinghua University, Beijing 100084, China
e-mail: wangxf@biomed.tsinghua.edu.cn

D.-P Zhang
e-mail: zhangdp@tsinghua.edu.cn

D.-P. Zhang (ed.), *Abscisic Acid: Metabolism, Transport and Signaling*,
DOI 10.1007/978-94-017-9424-4_6

6.1 Introduction

Abscisic acid (ABA) regulates various aspects of plant growth and development, such as seed development, dormancy and germination, seedling growth, and stomatal movement, and plays vital roles in the adaptive responses of plants to biotic and abiotic stresses, such as drought, high salinity, chilling, and pathogen (Leung and Giraudat 1998; Finkelstein et al. 2002; Himmelbach et al. 2003; Cutler et al. 2010). It is well known that signal transduction of a hormone in cells begins with the primary perception process mediated by a hormone receptor, which triggers an intracellular, downstream signaling cascade to finally activate physiological responses (Venis 1985). Since ABA was identified and characterized by Frederick Addicott and his associates in 1963 (Ohkuma et al. 1963; Addicott and Lyon 1969), extensive research efforts with genetic and biochemical approaches have been made, which leads to identification of numerous components of ABA signaling pathways (Giraudat et al. 1992, 1994; Leung et al. 1994, 1997; Leung and Giraudat 1998; Cutler et al. 2010). Especially, recent studies have identified the receptors or candidate receptors for ABA (Cutler et al. 2010), which allows to understand the primary events of ABA signaling. In this chapter, we review the advances in the researches of ABA receptors and the insights of these discoveries into the mechanisms of the perception processes during ABA signaling.

6.2 ABA-binding Proteins and Perception Sites

6.2.1 Both Intra- and Extracellular Presence of ABA-binding or Perception Sites

A hormone receptor should meet two essential criteria: One is to perceive the signal through binding to the hormone, and another is, after binding, to be able to activate physiological response by triggering a series of intracellular signaling steps (Venis 1985). Search for ABA receptors with the approaches of forward genetics through screening mutants defective in putative ABA receptor-encoding genes was not successful for a long time, though a lot of important components of ABA signalling have been identified using these approaches. This is likely because loss-of-function mutants of ABA receptors may be lethal or show no altered phenotypes due to genetic redundancy (McCourt and Creelman 2008). Biochemical approaches through analysis and isolation of ABA-binding proteins were considered as an alternative path to lead to identification of ABA receptors with reverse genetics.

Studies to test ABA-binding proteins or perception sites used biochemical, cellular, and even molecular approaches and suggest that there exist ABA-binding or perception sites in the plasma membrane (Horton and Bruce 1972; Meidner and Willmer 1975; Milborrow 1974; Hocking et al. 1978; Hartung 1983; Hornberg and Weiler 1984; Anderson et al. 1994; Gilroy and Jones 1992, 1994; MacRobbie 1995; Ritchie and Gilroy 1998, 2000; Jeannette et al. 1999; Jacob et al. 1999; Hallouin et al. 2002; Yamazaki et al. 2003). Noteworthily, Hocking et al. (1978) isolated the membrane factions from *Vicia faba* leaves and detected the binding of [^{3}H] (±) ABA to membrane-rich fractions. They showed that the binding bears the properties of binding saturation, of which the dissociation constant is 3.5×10^{-8} M with more than one binding sites in the membrane factions. Using guard cells isolated from the epidermis of *Valerianella locusta*, Hartung (1983) found that ABA could induce stomatal closure at pH 8.0, where nearly all ABA molecules are present as the anion of ABA (ABA^-), and could not penetrate plasmalemma, which suggests that ABA could be perceived by binding sites or receptors outside the plasmalemma and induce stomatal closure. Hornberg and Weiler (1984), using guard cell protoplast of *Vicia faba* leaves, provided new, compelling evidence for high-affinity [^{3}H]-*cis*(+)ABA-binding sites, which are located at the plasmalemma with high density and potentially mediate the guard cell signaling in response to the extracellular ABA. The possible extracellular perception site for ABA in guard cells was repeatedly confirmed by several independent groups in *Commelina*, *Vicia faba,* and *Arabidopsis* (Anderson et al. 1994; MacRobbie 1995; Jacob et al. 1999; Yamazaki et al. 2003). Additionally, the perception sites of ABA in the plasma membrane were detected in the barley *(Hordeum vulgare)* aleurone protoplasts and *Arabidopsis* suspension cells (Gilroy and Jones 1992, 1994; Ritchie and Gilroy 1998, 2000; Jeannette et al. 1999; Hallouin et al. 2002).

In addition to these extracellular binding or perception sites at the cell surface, the intracellular binding or perception sites were also detected, which may potentially contribute to ABA signal perception within cells. Allan et al. (1994) synthesized and microinjected caged ABA directly into *Commelina* guard cells, where ABA was released internally and subsequently caused stomatal closure, suggesting the presence of the intracellular perception sites. Meanwhile, Schwartz et al. (1994) demonstrated that ABA could act from internal of guard cells to regulate *Commelina* stomatal closure, implying that there is an intracellular locus of ABA perception. The coexistence of the intracellular and extracellular perception sites for ABA in guard cells was consistently reported in *Commelina* by other groups (Anderson et al. 1994; MacRobbie 1995).

ABA-binding sites were also studied in other plants with various tissues and were detected at different subcellular fractions of different tissues, such as cytosolic fraction of wheat (Veliev 1991), membrane fractions of rice seedling (Chen et al. 1992), maize roots (Chen and Zhu 1996), the plasma membrane fraction of *Arabidopsis* cells cultured in vitro (Pedron et al. 1998), the mesocarp of grape berry (Zhang et al. 1999), and developing apple fruits (Zhang et al. 2001). All these studies support the presence of the ABA-binding proteins or receptors for perceiving ABA signal in both the inside and outside of plant cells (Assmann 1994; Finkelstein et al. 2002; Verslues and Zhu 2007).

6.2.2 Isolation of ABA-binding Proteins

Isolation of the ABA-binding proteins may be a major step toward identification of ABA receptors by reverse genetic approaches. Based on the information from the studies of the ABA-binding proteins and using affinity chromatography combined with carefully designed affinity probe, two types of ABA-binding proteins have been successfully purified from the epidermis of broad bean (*Vicia faba*) and embryos of *Brassica napus*, respectively (Zhang et al. 2002; Nyangulu et al. 2005). The first ABA-binding protein was purified from the epidermis of broad bean (*Vicia faba*) leaves via the ABA-affinity chromatography where ABA molecules were coupled through their carboxyl groups at C-1 to the free amino groups located in the spacer arm of EAH-Sepharose 4B, which was used to 'catch' ABA-binding protein from total proteins (Zhang et al. 2002). An ABA-binding protein with a molecular mass of 42 kD and an isoelectric point of 4.86 was isolated, which was shown to bind ABA with saturability, reversibility, high affinity, stereo-specificity for the physiologically active (+)ABA. In addition, evidence for the potential receptor nature of the purified ABA-binding protein was provided, in which the phospholipase D activity induced by ABA in guard cell protoplasts (Wang 1999; Ritchie and Gilroy 1998, 2000; Jacob et al. 1999) of broad bean leaves in relation to the ABA-binding protein was measured (Zhang et al. 2002). The 42-kD ABA-binding protein was further used to identify possible ABA receptor subsequently (see following description). The second ABA-binding protein was isolated with a different strategy by considering that the alterations of ABA molecule at the C-1 or C-4' site may have effect on the affinity for ABA receptors (Nangulu et al. 2005). A bioactive and biotinylated ABA probe (+)-17, in which the C-1 or C-4' site of ABA remains unmodified, was synthesized. A membrane-associated ABA-binding protein was purified successfully using this biotinylated ABA as an affinity probe (Nangulu et al. 2005).

6.3 Mg-chelatase H Subunit (CHLH): A Receptor for ABA in Chloroplast Membrane

6.3.1 CHLH/ABAR Is an ABA-binding Protein

As described above, Zhang et al. (2002) purified an ABA-binding protein (ABAR, for putative ABA receptor) via affinity chromatography from broad bean, according to which a reverse genetic approach was made to identify a possible receptor for ABA. The sequencing results of the ABAR protein led to cloning, from broad bean leaves, of a complementary DNA fragment encoding the carboxy-terminal half of about 770 amino acids of the putative H subunit of the magnesium protoporphyrin-IX chelatase (Mg-chelatase). The yeast-expressed products of both the cDNA fragment of broad bean and the full-length cDNA of the *Arabidopsis CHLH* were shown to bind ABA specifically with the [^{3}H]-ABA-binding assay, and the ABA-binding properties of the *Arabidopsis* CHLH meet all criteria of the ligand–receptor binding (Shen et al. 2006; Wang and Zhang 2008; Wang et al. 2011).

Further, the second line of evidence was provided for the ABA-binding ability of the *Arabidopsis* CHLH/ABAR with a newly developed ABA-affinity chromatography technique, which is based on the same principle as that of the technique used for purification of the ABA-binding protein (Zhang et al. 2002; Wu et al. 2009;

Wang et al. 2011). The C-terminal half, but not N-terminal half of ABAR/CHLH, was shown to bind ABA with this system (Wu et al. 2009). However, the ABA-affinity technique was questioned because the carboxyl group at the C-1 locus of ABA, which is required for the activity of ABA, was altered (Cutler et al. 2010). This may possibly induce some errors in purification and analysis of ABA-binding proteins. In regard to this question, Wu et al. (2009) suggested that the specific binding ability to a protein may depend on the whole structure of ABA, and the ABA molecule with the C-1-carboxyl group bound may interact with ABA-binding protein, though the binding intensity may be reduced. This postulation was supported by the multiple controls that showed specificity and reliability of ABA binding to this affinity chromatography column (Wu et al. 2009). Also, we tested the functionality of the ABA molecule with the fixed C-1-carboxyl group. Given that ABA-Sepharose-4B is too big to permeate into cells and its bioactivity could not be tested, we coupled ABA to a small molecule methotrexate to fix the C-1-carboxyl group of ABA, forming an ABA derivate ABA-methotrexate (ABA-MTX), which may enter cells to test the bioactivity of the C-1-carboxyl group-blocked ABA. The results showed that ABA-MTX inhibits seed germination, seedling growth, and promote the interaction between an ABA receptor member PYL5 and a type-2C protein phosphatase member HAB1 (Ma et al. 2009; Park et al. 2009; Santiago et al. 2009b) as ABA does, though the activity is decreased (Q. Xin, X.F. Wang, and D.P. Zhang, unpublished data), which strongly suggests that blocking the C-1-carboxyl group of ABA does not result in loss of ABA bioactivity, supporting thus the reliability of the ABA affinity chromatography based on ABA-Sepharose-4B (Zhang et al. 2002; Wu et al. 2009; Wang et al. 2011).

Several lines of genetic evidence through transgenic manipulation consistently support the ABA-binding data (Wu et al. 2009), in which expression of the ABA-binding domains, C-terminal-truncated proteins of the *Arabidopsis* ABAR/CHLH, results in ABA hypersensitivity in all the major ABA responses including seed germination, seedling growth, and stomatal movement, but expression of the N-terminal-truncated proteins of ABAR/CHLH, non-ABA-binding domains, induces only limited ABA hypersensitivity in either seed germination or seedling growth (Wu et al. 2009). These data reveal that the C-terminal half not only covers the core ABA-binding domains, but also is a core functional region to mediate ABA signaling.

The third line of evidence for the binding interaction of the *Arabidopsis* ABAR/CHLH with ABA comes from a test with the surface plasmon resonance (SPR) system, by which Du et al. (2012) showed that ABAR/CHLH binds ABA with a saturation curve typical for receptor–ligand binding. The ABA-binding data with the SPR system are easily reproduced, but the detected ABA affinity to ABAR/CHLH is low (disassociation constant Kd $\approx$ 20 μM) in comparison with our previously detected high ABA affinity (about 30 nM) with the ^{3}H-labeled ABA-binding assay (Shen et al. 2006; Wu et al. 2009; Wang et al. 2011). This is likely due to technical limitations of the SPR system to test the interaction between a huge (about 150 kDa), hydrophobic CHLH protein and a small ligand ABA (Du et al. 2012).

In contrast to the *Arabidopsis* PYR/PYL/RCAR receptor for ABA, which is a cytosolic protein with a low molecular mass (about 20 kDa) and a highly hydrophilic nature (Ma et al. 2009; Park et al. 2009; Santiago et al. 2009a; and also see description below), the properties of a chloroplast-membrane ABAR/CHLH protein (Shang et al. 2010) with a high molecular mass and a slightly hydrophobic nature make ABA-binding assay difficult (Wang et al. 2011). This may likely be a reason why ABA binding of CHLH was not detected by other groups with the ^{3}H-labeled ABA-binding assay (Müller and Hansson 2009; Tsuzuki et al. 2011).

6.3.2 CHLH/ABAR Mediates ABA Signaling Positively

CHLH/ABAR is a multi-functional protein, which participates in chlorophyll biosynthesis (Walker and Willows 1997), and plastid-to-nucleus signaling (Mochizuki et al. 2001; Surpin et al. 2002; Strand et al. 2003; Nott et al. 2006). We observed that down- and up-expressions of the *Arabidopsis ABAR* gene result in ABA insensitivity and hypersensitivity, respectively, in the major ABA responses including ABA-induced inhibition of seed germination, post-germination growth arrest, and promotion of stomatal closure and inhibition of stomatal opening, revealing that ABAR/CHLH is a positive regulator of ABA signaling (Shen et al. 2006; Wu et al. 2009). There are five *abar* mutant alleles in *Arabidopsis*, *abar-1*, *abar-2*, *abar-3*, *cch*, and *rtl1*. The *abar-1* mutant, a knockout mutant resulting from a T-DNA insertion in *ABAR* gene, is lethal (Shen et al. 2006). The ABA responses are significantly altered in all the rest four mutants. The *abar-2*, *abar-3*, *cch*, and *rtl1* mutants all harbor a single-nucleotide substitution in *ABAR* gene; the mutations of *abar-2* and *abar-3* result in a single amino acid mutation Leu348 → Phe and Ser183 → Phe, respectively, in the N-terminal half of ABAR protein (Wu et al. 2009), and the *cch* and *rtl1* mutations result also in a single amino acid mutation Pro642 → Leu and Leu690 → Phe, respectively, in the middle of ABAR protein (Mochizuki et al. 2001; Tsuzuki et al. 2011). The *cch* mutant shows ABA insensitivity in all the three major ABA responses with ABA-insensitive phenotypes relatively weak in seed germination and seedling growth but strong in stomatal movement that leads to hypersensitivity to dehydration (Shen et al. 2006; Wu et al. 2009; Du et al. 2012). The *rtl1* mutant is susceptible to drought with significant defect in stomatal response to ABA and displays weak ABA-insensitive phenotype in seed germination but wild-type phenotype in ABA-induced post-germination growth arrest (Tsuzuki et al. 2011; Du et al. 2012). The strong ABA-insensitive phenotype in stomatal movement of the *cch* and *rtl1* mutants suggests that the core function domains of ABAR for regulation of stomatal response to ABA are more lesioned in these mutants. The *abar-2* and *abar-3* mutants both shows ABA insensitivity in ABA-induced post-germination growth arrest, and seed germination of the *abar-2* mutant is insensitive to ABA while that of the *abar-3* mutant is hypersensitive to ABA (Wu et al. 2009). The *abar-2* and *abar-3* mutants show wild-type phenotypes in stomatal movement in response to ABA, likely because these two mutants involve mutations in the N-terminus of ABAR, which may function as a regulatory region but not as a core functional region like the C-terminal half (Wu et al. 2009). Together, the phenotypic analysis of the four *abar* mutants provides genetic evidence that ABAR may regulate seed germination/seedling growth and stomatal movement in response to ABA through different mechanisms by modification of the specific domains in the ABAR molecule (Wu et al. 2009).

As a matter of fact, the *ABAR*-down-expression lines through transgenic RNAi technique show strong ABA-insensitive phenotypes in all the three major ABA responses (Shen et al. 2006), but the *ABAR*-RNAi construct is deleted in the T4 generation RNAi lines in which the *ABAR* gene expression is restored to wild-type level (X.F. Wang and D.P. Zhang, unpublished data). Currently, there is no mutant defective in the C-terminal half of ABAR, which is likely the reason why the above-mentioned mutants show relatively weak ABA-related phenotypes in seed germination and post-germination growth.

Increasing evidence verifies the essential role of ABAR in ABA signaling. Du et al. (2012) showed that ABAR also regulates guard cell signaling in response to ABA in tobacco (*Nicotiana benthamiana*) leaves. Other groups demonstrated that CHLH/ABAR mediates ABA signaling in guard cells in both *Arabidopsis* (Legnaioli et al. 2009; Tsuzuki et al. 2011, 2013) and peach (*Prunus persica*) leaves (Jia et al. 2011b). Recently, Tsuzuki et al. (2013) showed that CHLH/ABAR mediates ABA inhibition of blue light (BL)-induced phosphorylation of H^+-ATPase in *Arabidopsis* guard cells, suggesting that CHLH/ABAR regulates not only ABA-induced stomatal closure but also ABA inhibition of BL-mediated stomatal opening. Interestingly, it has been demonstrated that CHLH/ABAR mediates ABA signaling in fruit ripening of both peach (Jia et al. 2011b) and strawberry (*Fragaria ananassa*) (Jia et al. 2011a). These data demonstrate that CHLH/ABAR is a conserved ABA signaling regulator in plant cells.

6.3.3 ABAR-mediated ABA Signaling Is Distinct from Chlorophyll Biosynthesis

As described earlier, CHLH/ABAR was initially identified as a key enzyme of chlorophyll biosynthesis pathway (Walker and Willows 1997), and a regulator of plastid-to-nucleus signaling (Mochizuki et al. 2001; Surpin et al. 2002; Strand et al. 2003; Nott et al. 2006). Although some *ABAR*-RNAi transgenic lines and *cch* mutant have low chlorophyll contents and show yellow leaves (Shen et al. 2006; Wu et al. 2009), multiple lines of evidence demonstrate that ABAR-mediated ABA signaling is distinct from chlorophyll biosynthesis in *Arabidopsis*. First, the *ch1-1* and *ch1-2* mutants defective in chlorophyll biosynthesis show wild-type response to ABA (Shen et al. 2006; Liu et al. 2013). Second, many *ABAR*-RNAi transgenic lines with a chlorophyll level that is not lower than in wild-type plants show strong ABA-insensitive phenotypes; no correlation of the chlorophyll content with ABA signaling was found in the phenotypic analysis with a large number of *ABAR*-RNAi transgenic lines (Shen et al. 2006). Third, the *abar-2* and *abar-3* mutants defective in ABA signaling show normal chlorophyll content in comparison to wild-type plants (Wu et al. 2009). Fourth, the transgenic lines expressing different truncated ABAR proteins do not alter the chlorophyll content but show significant ABA hypersensitive phenotypes in all the major ABA responses (Wu et al. 2009). Fifth, the transgenic lines expressing the C-terminus linked to or free of the N-terminal chloroplast signal peptide, which leads to localization of the expressed C-terminus of ABAR to the chloroplast or to the cytosol only, show the same ABA hypersensitive phenotypes; similarly, the transgenic expression of the C-terminus free of the N-terminal chloroplast signal peptide in the *cch* mutant, which leads to localization of the expressed C-terminus of ABAR to the cytosol but not to the chloroplast, cannot rescue the slight chlorosis phenotype but show the same ABA hypersensitive phenotypes as the transgenic lines expressing the C-terminus to the chloroplast in the *cch* mutant that is rescued from the chlorosis phenotype (Wu et al. 2009). And last, among the four subunits of Mg-chelatase (CHLH, CHLI, CHLD, and GUN4), CHLH/ABAR and CHLI, but not CHLD and GUN4, are involved in ABA signaling, suggesting that CHLH interacts with CHLI to form a hetero-dimer, which cooperates to regulate ABA signaling, while the function of Mg-chelatase requires all the four components/subunits CHLH, CHLI, CHLD, and GUN4 to form a hetero-tetramer complex, which catalyzes magnesium chelating to protoporphyrin-IX to produce Mg-ProtoIX (Du et al. 2012). This model provides an explanation why ABAR-mediated ABA signaling is distinct from chlorophyll biosynthesis.

6.3.4 An ABAR–WRKY40 Coupled Signaling Pathway

How a chloroplast ABAR protein transduces ABA signal to downstream cytosolic or nuclear components of ABA signaling pathway had been puzzling. Shang et al. (2010) observed that the *Arabidopsis* ABAR is a chloroplast protein, which spans the chloroplast envelope with its N- and C-termini exposed to the cytosol, which provides opportunities for a chloroplast protein to interact with cytosolic proteins that relay ABA signal from ABAR to downstream cascade. Further, Shang et al. (2010) identified a group of WRKY transcription factors (WRKY40, WRKY18, and WRKY60) as ABAR interaction partners, which bind ABAR through the C-terminal half exposed to the cytosolic space. These WRKYs function downstream of ABAR as negative regulators of ABA signaling in seed germination and post-germination growth, of which WRKY40 is a central negative regulator, and inhibit expression of a subset of ABA-responsive genes, such as *ABF4*, *ABI4*, *ABI5*, *DREB1A*, *DREB2A*, *MYB2,* and *RAB18* (Shang et al. 2010; Liu et al. 2012). Besides, WRKY40 inhibits expression of the *LHCB* genes encoding the light-harvesting chlorophyll *a*/*b*-binding proteins (Liu et al. 2013) and the *KAT2* gene encoding 3-ketoacyl-CoA thiolase-2 (an enzyme of fatty acid beta-oxidation; Jiang et al. 2011), both of which are positively involved in ABA signaling (Jiang et al.

2011; Xu et al. 2012; Liu et al. 2013). Further evidence reveals that WRKY18, WRKY40, and WRKY60 transcription factors function upstream of ABI4 and ABI5 and cooperate to repress *ABI4* and *ABI5* genes (Shang et al. 2010; Liu et al. 2012). In addition, these WRKYs may downregulate expression of their own genes in a manner of auto- and cross-repression (Yan et al. 2013).

Recently, a chloroplast co-chaperonin 20 (CPN20) was identified as a novel interaction partner of ABAR in *Arabidopsis*, which negatively regulates ABA signaling at the same node with ABAR but upstream of WRKY40 transcription repressor (Zhang et al. 2013, 2014).

These studies support a model that ABAR antagonizes the WRKY transcription repressors to relieve ABA-responsive genes of inhibition (Shang et al. 2010; Liu et al. 2012; Yan et al. 2013; Zhang et al. 2013, 2014). The cytosolic C-terminus of ABAR interacts with a group of WRKY transcription factors, WRKY40, WRKY18, and WRKY60, which negatively regulate ABA signaling and inhibit expression of ABA-responsive genes, such as *ABI4* and *ABI5*. High level of ABA promotes ABAR–WRKY40 interaction, recruits WRKY40 from the nucleus to the cytosol, and activates an unknown factor or signaling cascade to downregulate *WRKY40* expression. Consistent with our observations (Shang et al. 2010), an independent group observed that the expression level of the *WRKY40* gene is greatly enhanced in the *cch* mutant (Adhikari et al. 2011), supporting that ABAR is required for repression of *WRKY40*. Decrease of WRKY40 relieves *ABI4* and *ABI5* genes of inhibition to induce physiological responses (Shang et al. 2010; Liu et al. 2012; Yan et al. 2013; Fig. 6.1). This ABAR–WRKY40 coupled mechanism may be inhibited by CPN20, a chloroplast interaction partner of ABAR. CPN20 interacts with ABAR tightly at low ABA level, competitively attenuating the interaction between ABAR and WRKY40, which is favorable to keeping a high level of WRKY40 to repress ABA-responsive genes. High level of ABA inhibits the ABAR–CPN20 interaction, which in turn promotes the ABAR–WRKY40 interaction to trigger the downstream signaling to repress WRKY40 expression and finally to relieve ABA-responsive genes of inhibition (Zhang et al. 2013, 2014; Fig. 6.1). These findings describe a unique ABA signaling pathway from the early signaling events to downstream gene expression.

6.3.5 A Pentatricopeptide Repeat Protein SOAR1: A Hub of ABA Signaling Downstream of ABAR

To further explore the mechanism of the ABAR-mediated ABA signaling, a suppressor of the *ABAR*-overexpression 1D (*soar1D*) lines was identified from the population of the *ABAR*-overexpression transgenic lines (Mei et al. 2014). The *SOAR1* gene encodes a pentatricopeptide repeat (PPR) protein, which localizes to both the cytosol and nucleus, but not to chloroplast or mitochondrion. Downregulation of *SOAR1* strongly enhances, but upregulation of *SOAR1* almost completely impairs, ABA responses, revealing that SOAR1 is a critical, negative,

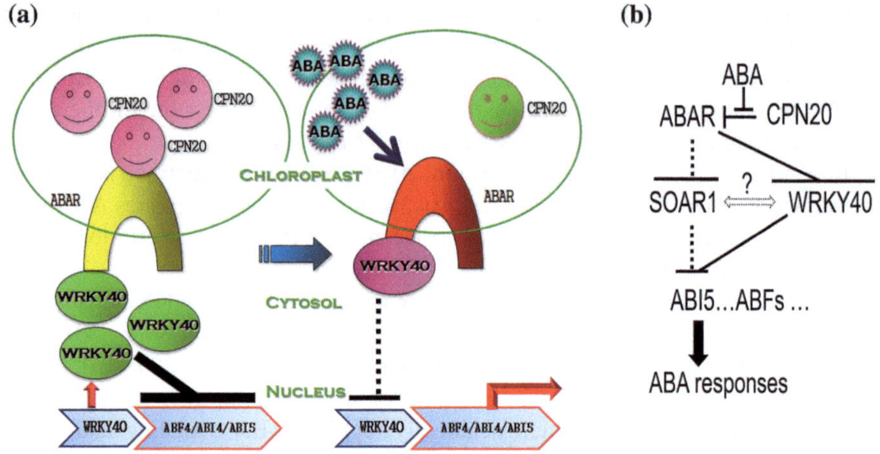

Fig. 6.1 Proposed model for the functional mechanisms of ABAR. **a** A model in which ABAR antagonizes WRKY40 transcription repressor to relieve ABA-responsive genes of inhibition [modified from Shang et al. (2010) and Zhang et al. (2014)]. WRKY40 negatively regulates ABA signaling and inhibits expression of ABA-responsive genes, such as ABA-responsive transcription factors (ABFs) including *ABI4* and *ABI5*. High level of ABA promotes ABAR–WRKY40 interaction and activates an unknown factor or signaling cascade to downregulate *WRKY40* expression. Decrease of WRKY40 relieves *ABI4* and *ABI5* genes of inhibition. This ABAR–WRKY40 coupled mechanism is inhibited by CPN20, which interacts with ABAR tightly at low ABA level, competitively attenuating the interaction between ABAR and WRKY40. High level of ABA inhibits the ABAR–CPN20 interaction, which in turn promotes the ABAR–WRKY40 interaction to trigger the downstream signaling to repress WRKY40 expression and finally to induce physiological responses. **b** A model for the ABAR-mediated signaling network. In addition to the ABAR/CPN20–WRKY40-ABI5-linked signaling as mentioned in the Figure a, SOAR1 is involved in the ABAR-mediated signaling downstream of ABAR and upstream of ABI5. See detailed explanation in the text. *Arrows* denote positive regulation or activation and bars negative regulation or repression. The *solid lines* indicate direct effect, and *dotted lines* indicate indirect effect or yet unknown mechanism. *Question mark* indicates unconfirmed link

regulator of ABA signaling (Mei et al. 2014). The intensity of the ABA overly sensitive phenotypes of the two *SOAR1*-knockdown mutants *soar1-2* and *soar1-3* is similar to, or stronger than, that of the well-characterized *abi1 abi2* double-knockout mutant. The intensity of the ABA-insensitive phenotypes of the *SOAR1*-overexpressing lines is much stronger than that of the *abi1–1* dominant mutant, *abi4*, *abi5* loss-of-function mutants and a strong *ABI2*-overexpressing line (Mei et al. 2014). Surprisingly, the seeds of the *SOAR1*-overexpressors germinate and their post-germination seedlings continue to grow in the medium containing >200 μM (±)ABA (Mei et al. 2014). Previous studies reported that the seeds of the *srk2dei* mutant, triple-knockout mutant of three SnRK2 members SnRK2.2, SnRK2.3, and SnRK2.6, germinate and continue to grow in the presence of 50- or 100-μM exogenous ABA, which is believed to completely impair ABA response (Fujii and Zhu 2009; Nakashima et al. 2009; Umezawa et al. 2009). In this regard, the intensity of ABA insensitivity of the *SOAR1*-overexpression lines is comparable to that of the triple loss-of-function mutant of the SnRK2 members, uncovering that SOAR1 is a hub of ABA signaling pathways.

Further genetic evidence reveals that SOAR1 functions downstream of ABAR and upstream of the ABA-responsive transcription factor ABI5 (Mei et al. 2014), suggesting a possible ABAR-SOAR1-ABI5 linked signaling cascades. Given that PPR proteins are a class of RNA-binding proteins involved in many aspects of RNA processing (Meierhoff et al. 2003; Williams and Barkan 2003; Lurin et al. 2004), the *ABI5* mRNA may be a target of the SOAR1 protein (Mei et al. 2014). The alteration in *SOAR1* expression changes significantly the expression of a subset of genes, of which the encoded proteins have been identified to be directly involved in the PYR/PYL/RCAR-mediated ABA signaling, suggesting the PYR/PYL/RCAR-mediated and ABAR-mediated signaling pathways may crosstalk to regulate ABA signaling through SOAR1. Together, the discovery of SOAR1 as a crucial ABA signaling component suggests a central role of the ABAR-mediated signaling pathway in the highly complicated ABA signaling network, and exploration of the mechanisms of the SOAR1 protein that functions in the nuclear events will be of particular importance to understand the mechanisms of ABA signaling (Fig. 6.1).

6.4 G Protein-coupled Receptors (GPCRs): Receptors for ABA at the Cell Surface

As described earlier, previous experiments suggested that there are plasma membrane-localized ABA receptors to perceive the extracellular ABA signal (Gilroy and Jones 1992; Anderson et al. 1994; Assmann 1994; Jeannette et al. 1999; Finkelstein et al. 2002; Verslues and Zhu 2007). G protein-coupled receptors (GPCRs) are a class of plasma membrane proteins, which constitute, together with their G protein partners, one of the most elaborate receptor–effector-linked signaling cascades and play a vital role in perceiving extracellular signal in eukaryotic cells (Pierce et al. 2002; Offermanns 2003). The core components of the G protein signaling are comprised of Gα, Gβ, and Gγ subunits, and G protein-coupled receptors (GPCRs), among which the three G protein subunit form heterotrimeric G proteins (Pierce et al. 2002; Offermanns 2003). The catalytically active protein of the complex, Gα subunit, which has both GTP-binding and GTPase activity, acts as a bimodal molecular switch, typically with a GDP-bound off mode and a GTP-bound on mode. GDP-bound Gα remains associated with Gβγ (GDP-Gαβγ) and represents the inactive signaling status. Signal perception of GPCRs results in their conformation changes, which leads to exchange of GDP for GTP at the Gα subunit and dissociation of the heterotrimer into free GTP-Gα and Gβγ dimers, both of which can relay signal to trigger downstream signaling events. The G protein signaling is terminated by regeneration of the GDP-bound Gα subunit and re-association of the G protein heterotrimer, thus completing one signaling cycle

(Pierce et al. 2002; Cabrera-Vera et al. 2003; Offermanns 2003; Pandey et al. 2009; Hackenberg et al. 2013; Urano et al. 2013).

G proteins have been shown to regulate plant growth and development and act as nodes for integration and amplification of a host of abiotic, biotic, and hormonal signals (Assmann, 2004). The *Arabidopsis* genome encodes one Gα (GPA1), one Gβ (AGB1), two Gγ (AGG1 and AGG2) subunits and one regulator of G protein signaling (RGS) protein, which may form G protein complexes to function with GPCRs to link ligand perception with downstream effectors (Ma et al. 1990; Ma 1994; Mason and Botella 2000, 2001; Jones and Assmann 2004). Multiple lines of evidence have shown that the G protein complexes regulate ABA signaling (Wang et al. 2001; Coursol et al. 2003; Pandey and Assmann 2004; Pandey et al. 2006), suggesting that GPCR-type plasma membrane receptors for ABA function at the cell surface to perceive ABA signal.

6.4.1 GCR1 and GCR2: Do They Sense ABA at the Cell Surface?

Pandey and Assmann (2004) identified a GPCR, named GCR1, which encodes a protein predicted to possess 7-transmembrane domain structure characteristic of GPCRs. GCR1 negatively regulates ABA signaling in seedling growth and stomatal movement and interacts with GPA1, acting as a negative regulator of GPA1-mediated ABA responses in guard cells (Pandey and Assmann 2004). However, it remains unknown whether GCR1 perceives extracellular ABA signal as a receptor for ABA. GCR2 is the second candidate GPCR reported to be involved in ABA signaling, which was shown to function as a plasma membrane receptor for ABA (Liu et al. 2007a). However, it remains controversial whether GCR2 is a member of GPCRs and regulates ABA signaling as an ABA receptor (Johnston et al. 2007; Liu et al. 2007b; Gao et al. 2007; Chen and Ellis 2008; Guo et al. 2008; Illingworth et al. 2008; Muschietti and McCormick 2010). So, more supporting evidence is needed to clarify whether GCR2 is a bona fide ABA receptor.

6.4.2 GTG1 and GTG2: GPCR-type G Proteins Acting as Cell Surface ABA Receptors

Pandey et al. (2009) identified two homologous GPCR-type G proteins, named GPCR-type G proteins 1 and 2 (GTG1 and GTG2), which show homology to a human orphan vertebrate GPCR (GPR89). GTG1 and GTG2 have classic GTP-binding and GTPase activity of G proteins and interact with GPA1 that in turn affects GTPase activity of the GTG proteins. The double *gtg1 gtg2* mutant exhibits significant ABA hypersensitive phenotypes in ABA-induced inhibition of seed germination, post-germination growth arrest and promotion of stomatal closure, but single *gtg* mutant plants show substantially wild-type phenotypes. The *gtg1 gtg2* double mutant phenotype could be fully complemented with expression of either GTG1 or GTG2 gene, which implies that GTG1 and GTG2 function redundantly in ABA signaling (Pandey et al. 2009). Further quantitative proteomics-based analysis supports a significant role of the GTG proteins in regulation of ABA response in *Arabidopsis* roots and provides clues to their possible links with some of the well-established effectors of the ABA signaling pathways (Alvarez et al. 2013), though a report using independently generated mutant alleles of the GTG genes suggests a normal ABA sensitivity of the *gtg1 gtg2* double mutants (Jaffe et al. 2012).

The ABA-binding assays with ^{3}H-ABA and purified recombinant GTG proteins shows the ABA-binding ability and specificity of the GTG1 and GTG2 proteins, which meets essential criteria of ligand–receptor binding and supports the ABA receptor nature of GTGs (Pandey et al. 2009). The ABA-binding ability of the GTG proteins was confirmed by an independent study (Kharenko et al. 2013). Interestingly, the GTG proteins associated with GDP are more favorable to ABA binding than the GTGs–GTP complexes, and inhibition of GTPase activity of GTGs will abolish ABA binding, revealing that the GTG proteins bind ABA in the GTG–GDP form rather that GTG–GTP form (Pandey et al. 2009).

Based on the analysis, a working model was proposed to explain the GTG-mediated ABA signaling, in which G protein (Gα subunit GPA1) regulates the G protein-coupled receptor: GDP-bound GTGs bind ABA, which initiates the ABA signaling cascade; GPA1 acts as a 'rheostat' on GTG, such that GTP-bound GPA1, inhibiting formation of the active complex GTG–GDP, downregulates ABA binding to the GTGs and represses ABA signaling (Pandey et al. 2009; Fig. 6.2). This model suggests an unusual type of G protein signaling, which is opposite from conventional model for signaling mediated by G proteins, where GDP-bound Gα 'turns off' the signaling system (Pandey et al. 2009; Klingler et al. 2010). Further studies will be needed to identify downstream effectors to elucidate mechanisms of the GTG-mediated ABA signaling.

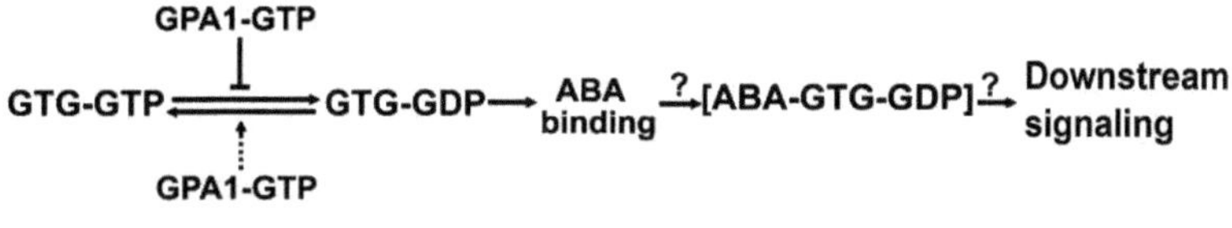

Fig. 6.2 A model for mechanism of the GTG-mediated ABA perception [modified from Pandey et al. (2009)]. GDP-bound GTGs (GTG–GDP) bind ABA. GTP-bound GPA1 (GPA1–GTP) inhibits the GTG–GDP formation, downregulating ABA binding to the GTGs and thus repressing ABA signaling. See detailed explanation in the text. *Arrows* denote activation effects and bars inhibition effects. *Question marks* indicate aspects of the model inferred from, but not directly demonstrated by, experimental results to date

6.5 PYR/PYL/RCAR Proteins: Cytosolic ABA Receptors Mediating a Central Signaling Pathway

6.5.1 START-Domain Proteins PYR/PYL/RCAR: Soluble ABA Receptors Coupled with PP2Cs

Park et al. (2009) developed a chemical genetic strategy to screen candidate receptors for ABA, which may bypass redundancy by inducing phenotypes not revealed by conventional forward genetic approaches with single-locus mutation: A selective agonist can illuminate the function of one member of normally redundant regulators or receptors (Cutler and McCourt 2005). In a forward genetic screen for new signaling components or receptors for ABA, a synthetic ABA agonist specifically inhibiting seed germination, called pyrabactin, was used and allowed to successfully isolate a *PYRABACTIN RESISTANCE 1* mutant allele (*pyr1*), which shows pyrabactin insensitive phenotype in seed germination. *PYR1* encodes a member of the START-domain superfamily soluble ligand-binding proteins (Iyer et al. 2001), and the gene family includes thirteen homologous genes similar to *PYR1*, named *PYR-LIKE 1* to *PYR-LIKE 13* (*PYL1* to *PYL13*) (Park et al. 2009). The *pyr1* mutant responds normally to ABA, suggesting that functional redundancy from other family members could mask PYR1's role in ABA signal transduction. Indeed, triple and quadruple mutants with genotypes *pyr1 pyl1 pyl4* and *pyr1 pyl1 pyl2 pyl4*, respectively, show significant ABA insensitivity in seed germination and seedling growth (Park et al. 2009). The quadruple mutant also shows ABA insensitivity in ABA-induced stomatal closure (Nishimura et al. 2010) and ABA-mediated transcriptional responses (Park et al. 2009). A more recent report showed that a sextuple mutant impaired in six PYR/PYL receptors, namely PYR1, PYL1, PYL2, PYL4, PYL5, and PYL8, displays extremely strong ABA-insensitive phenotypes: It germinates and grows even on 100-μM ABA, and its leaves are overly sensitive to drought (Gonzalez-Guzman et al. 2012). Transgenic overexpression of either PYL1 or PYL4 restores ABA sensitivity in the triple mutant, revealing the functional specificity for ABA signaling (Park et al. 2009). These data illustrate the power of the chemical genetic approach for sidestepping genetic redundancy.

Another group (Ma et al. 2009) used reverse genetic approaches to screen new components of ABA signaling and identified, by a yeast two-hybrid system, an interaction protein of ABI2 type 2C protein phosphatase that has been well characterized as a key, negative, regulator of ABA signaling (Leung et al. 1994, 1997; and see Chap. 8). This interaction partner of ABI2 was named regulatory component of ABA receptor 1 (RCAR1), which is identical to PYL9. Consistent with the observations in the mutants of *PYR/PYL* genes (Park et al. 2009), downregulation of *RCAR1/PYL9* reduces, but upregulation of *RCAR1/PYL9* enhances, ABA sensitivity in the three major ABA responses in seed germination, vegetative growth, and stomatal movement (Ma et al. 2009). Further ABA-binding assays through hetero-nuclear single-quantum coherence nuclear magnetic resonance and isothermal titration calorimetry techniques showed that PYR1 and RCAR1/PYL9 bind ABA (Ma et al. 2009; Park et al. 2009). PYR/PYL/RCAR proteins, upon ABA binding, inhibit activities of the group-A PP2Cs, such as ABI1 and ABI2 and thus antagonize the PP2Cs to positively regulate ABA signaling (Ma et al. 2009; Park et al. 2009). The ABA receptor identity of the PYR/PYL/RCAR proteins and PYR/PYL/RCAR-PP2C coupled ABA signaling was further confirmed by another two groups who identified PYL5 by the yeast two-hybrid assay as an interaction partner of HAB1 (Santiago et al. 2009b), and several PYR/PYL proteins by co-purification technique as interaction partners of ABI1 (Nishimura et al. 2010).

Structural studies provided compelling evidence to support that the PYR/PYL/RCAR proteins interact with ABA and function as ABA receptors (Melcher et al. 2009; Miyazono et al. 2009; Nishimura et al. 2009; Santiago et al. 2009a; Yin et al. 2009; Shibata et al. 2010; see also Chap. 7). Consistent with the biochemical observations that the interaction of PYL9 with ABI2 and PYL5 with HAB1 stimulates ABA binding to PYL9 and PYL5, respectively (Ma et al. 2009; Santiago et al. 2009b), the structural studies provided submolecular evidence that RCAR/PYR1/PYL proteins can form tight complexes with PP2Cs and cooperate with PP2Cs in the ABA binding to PYR/PYL/RCAR, which induces the enzymatic inactivation of the PP2Cs (Melcher et al. 2009; Miyazono et al. 2009; Nishimura et al. 2009; Santiago et al. 2009a; Yin et al. 2009; Shibata et al. 2010; see also Chap. 7). However, the cooperation of PP2Cs with RCAR/PYR1/PYL proteins in the RCAR/PYR1/PYL ABA binding led to a question whether the RCAR/PYR1/PYL-PP2C complex is a co-receptor for ABA. In regard to this question, the structural studies support a model that ABA first binds RCAR/PYR1/PYL proteins, which recruits PP2Cs to bind and stabilize the ABA-RCAR/PYR1/PYL complexes (Melcher et al. 2009; Miyazono et al. 2009; Nishimura et al. 2009; Santiago et al. 2009a; Yin et al. 2009; Shibata et al. 2010; see also Chap. 7), which is consistent with the idea that the PYR/PYL/RCAR proteins, rather than the RCAR/PYR1/PYL-PP2C complexes, are ABA receptors.

6.5.2 A PYR/PYL/RCAR-PP2C–SnRK2-Linked Linear Signaling Pathway

Park and colleagues observed that the ABA-induced activation of SNF1-related protein kinases (SnRK2s) is reduced in the *pyr1 pyl1 pyl2 pyl4* quadruple mutant (Park et al. 2009), which is confirmed in a further study showing that the *pyr1 pyl1 pyl2 pyl4 pyl5 pyl8* sextuple mutant lacks ABA-mediated activation of SnRK2s (Gonzalez-Guzman et al. 2012), revealing that, in addition to PP2Cs, SnRK2s function is dependent on, and downstream of, PYR/PYL/RCAR receptors in ABA signaling pathway. The SnRK2s, mainly including SnRK2.2, SnRK2.3, and SnRK2.6/OST1, are key players in ABA signaling (see Chap. 8) and were shown to be substrates of the clade A PP2Cs in many reports. Several groups found that SnRK2.6/OST1 interacts with ABI1, and the SnRK2 activities induced by ABA are impaired in *abi1–1* mutants (Mustilli et al. 2002; Yoshida et al. 2006; Umezawa et al. 2009). Consistent with this, Umezawa et al. (2009) provided clear evidence that the group-A PP2Cs directly interact with SnRK2s and inactivates and dephosphorylates SnRK2s. In response to ABA, SnRK2s are activated in association with the internal phosphorylation of multiple serine/threonine residues (involving Ser175) in its kinase activation loop. The same sites are dephosphorylated by PP2Cs, resulting in the inactivation of SnRK2s (Umezawa et al. 2009). In addition, Vlad et al. (2009, 2010), using a protein phosphatase profiling strategy to screen for putative substrates of HAB1 PP2C, showed that SnRK2.6/OST1 is one of the substrates of a HAB1 and identified Ser175 in SRK2.6/OST1 as a target site of the PP2C that directly binds in vivo to the regulatory C-terminal domain of SRK2.6/OST1. Further results revealed that ABI1, ABI2, and their mutant forms display very similar substrate preferences as HAB1 and its mutant form hab1^{G246D} (Vlad et al. 2009, 2010). These data provide convincing evidence that SnRK2s are direct targets of the group-A PP2Cs and lead to a model that PYR/PYL/RCAR inhibits the PP2C-dependent negative regulation of SnRK2s in an ABA-modulated fashion (Umezawa et al. 2009). In support of this model, Umezawa et al. (2009) reconstituted these three components in vitro using recombinant ABI1 or *abi1–1*, PYR1, and GFP-tagged SnRK2s from the *Arabidopsis* cells. After incubation with PYR1, PP2Cs, and SnRK2s in the presence or absence of ABA, SnRK2 activity was monitored via an in-gel phosphorylation assay, which showed that PYR1 inhibits ABI1-mediated inactivation of SnRK2 in an ABA-dependent manner; however, a dominant mutation form of ABI1, *abi1–1,* inactivates SnRK2s even in the presence of PYR1 and ABA, consistent with previous data that the dominant mutation forms of ABI1 and ABI2, namely *abi1–1* and *abi2–1,* lack RCAR/PYR-binding ability (Ma et al. 2009; Park et al. 2009). These results also provide a clear explanation for why *abi1–1* and *abi2–1* are dominantly insensitive to ABA (Umezawa et al. 2009; see also Chap. 8). This model was further supported by reconstituting in vitro the PYR/PYL/RCAR-PP2C-SnRK2.6-ABF2-linked signaling cascades in the *Arabidopsis* protoplasts, in which ABA-triggered phosphorylation and activation of the transcription factor depends on the introduction of these four components into plant protoplasts (Fujii et al. 2009). It was testified again that, in the presence of ABA, the PYR/PYL/RCAR receptor proteins (with the exception of PYL13) can disrupt the interaction between the PP2Cs and SnRK2s, thus preventing the PP2C-mediated dephosphorylation of the SnRK2s and resulting in the activation of the SnRK2 kinases (Fujii et al. 2009).

Most recently, the *Arabidopsis* glycogen synthase kinase 3 (GSK3)-like kinases (BIN2 and BIN2-like BIL1/2) that regulate negatively brassinosteroid signaling (He et al. 2002; Li and Nam 2002) were shown to be positive regulators of ABA signaling (Cai et al. 2014). BIN2 kinase interacts with, phosphorylates and activates SnRK2.2 and SnRK2.3 kinases to function in ABA signaling downstream of PYR1/PYL/RCAR receptors and PP2Cs (Cai et al. 2014), suggesting a regulation mechanism of SnRK2s by a reversible phosphorylation process where the BIN2/BIL kinase-catalyzed phosphorylation cooperates with the PP2C-mediated dephosphorylation of SnRK2s in regulating the SnRK activities.

The ABA-responsive bZIP transcription factors (ABFs) were shown to be substrates of SnRK2s, which interact with, are phosphorylate by, and function

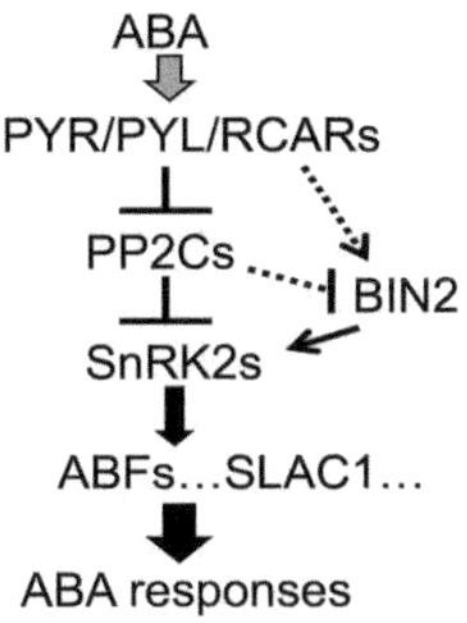

Fig. 6.3 A PYR/PYL/RCAR-mediated linear ABA signaling pathway. Under basal ABA levels, clade A PP2Cs function as negative regulators to repress ABA signaling, through either dephosphorylation of SnRK2s or interaction with other targets. ABA binding to the PYR/PYL/RCAR receptors inactivates PP2Cs, leading to the activation of SnRK2s through phosphorylation by BIN2 kinases, which subsequently phosphorylate downstream targets, such as members of the ABF transcription factors or regulatory components of the stomatal aperture, such as the anion channels SLAC1, which finally induces a diversity of ABA responses. Arrows denote positive regulation or activation, and *bars* denote negative regulation or repression. The solid lines and dotted lines indicate direct effects and indirect effects, respectively. *Question mark* indicates unconfirmed link

downstream of SnRK2s (Furihata et al. 2006; Fujii et al. 2007, 2009; Fujii and Zhu 2009; Fujita et al. 2009, 2013; Sirichandra et al. 2010; see also Chaps. 8 and 11). The SnRK2s directly activate, through phosphorylation, the transcription factors that bind to ABA-responsive promoter elements and regulate gene expression (see Chap. 11). In the guard cells, SnRK2s regulate stomatal aperture by modulating the slow anion channel SLAC1 and SLAH3, K^+ inward channel KAT1 (Lee et al. 2009, 2013; Geiger et al. 2009, 2010, 2011; Brandt et al. 2012; Acharya et al. 2013; see also Chap. 8).

Together, these results establish a core signaling pathway, in which PYR/PYL/RCAR-PP2C-SnRK2 is the core ABA signaling module, which functions to regulate downstream ABFs or ion channels of guard cells to induce a diversity of ABA responses (Cutler et al. 2010; Umezawa et al. 2010; Fig. 6.3).

6.5.3 Functional Diversity of the Members of PYR/PYL/RCAR Family

Among the fourteen members of the PYR/PYL/RCAR family, the receptor nature of PYL13/RCAR7 is questioned. A study showed that PYL13/RCAR7 is lack of the crucial lysine residue for ABA binding and has no physical interaction with PP2Cs (Joshi-Saha et al. 2011a), and another two studies reported that PYL13 did not bind ABA because of changes in conserved amino acid residues involved in ABA binding and selectively inhibits PP2CA independent of ABA (Li et al. 2013; Zhao et al. 2013). However, a recent report showed that PYL13/RCAR7 inhibits the activities of a number of PP2Cs such as PP2CA and ABI2 in an

ABA-dependent manner, although not the structurally related HAB1, thus functions as a bona fide ABA receptor (Fuchs et al. 2014). Further study will be needed to clarify these discrepancies.

The other 13 members of PYR/PYL/RCARs act undoubtedly as ABA receptors, which function redundantly but with diverse functional specialization features to regulate ABA signaling. First, PYR/PYL/RCAR receptors can be characterized on their oligo-merization states; some members of PYR/PYL receptors are homo-dimeric (PYR1, PYL1, and PYL2), while others are monomeric (PYL4, PYL5, PYL6, PYL8, PYL9, and PYL10) (Dupeux et al. 2011; Hao et al. 2011; see also Chap. 7). The homo-dimer assembly is required to regulate strictly the ABA-dependent switching of the signal transduction by PYR/PYL/RCARs. The intrinsic affinities of the receptors for ABA are closely related to their monomeric/dimeric states (Dupeux et al. 2011). The Kd of the monomeric receptors is approximately 1 μM (Santiago et al. 2009a; Szostkiewicz et al. 2010; Dupeux et al. 2011), whereas that of the dimeric receptors (Kd > 50 μM) is almost two orders of magnitude lower (Dupeux et al. 2011). This difference can be explained by the unfavorable contribution of dimer dissociation to the thermodynamics of ABA binding rather than by differences in the ABA-binding cavity (Dupeux et al. 2011). However, both monomer and dimer bind ABA with high affinity when interacting with PP2Cs (Kd ≈ 30–60 nM) (Ma et al. 2009; Santiago et al. 2009a, b; see also Chap. 7). It is noteworthy that the PYL3 receptor is a trans-dimer, which suffers a cis- to trans-dimer transition upon ABA binding to facilitate the subsequent dissociation to monomer (Zhang et al. 2012; see also Chap. 7).

Second, the 14 homologous PYR/PYL/RCARs may selectively interact with six members of clade A PP2Cs that are implicated in ABA signaling (Fujii et al. 2009; Cutler et al. 2010; Raghavendra et al. 2010; see also Chap. 8), which may form numerous combinations of the receptor-PP2C complexes. Importantly, some of the closely related receptor complexes differ in their selectivity and sensitivity to ABA or natural structural derivatives of ABA and are likely to target different downstream components (Ma et al. 2009; Park et al. 2009; Santiago et al. 2009b; Hao et al. 2011; Kepka et al. 2011; Szostkiewicz et al. 2010). For example, in contrast to PYR1 and PYLs 1 to 4, PYL9 and PYL5 show constitutive binding interactions with PP2Cs (Ma et al. 2009; Park et al. 2009; Santiago et al. 2009b). PYR1, PYL1, PYL2, and PYL3 inhibit PP2Cs, such as ABI1, HAB1, HAB2, and PP2CA, in an ABA-dependent manner, while PYL5, PYL6, PYL8, PYL9, and PYL10 inhibit PP2Cs even in the absence of ABA; PYL4 inhibits HAB2 PP2C activity only in the absence of ABA; ABA-independent inhibition of PP2C requires the PYLs to exist in a monomeric state (Hao et al. 2011). HAI1 PP2C (encoded by At5g59220 gene) is selectively inhibited by some PYR/PYL/RCAR receptors, and its close relatives, such as PP2CA/AHG3 and AHG1, show a contrasting sensitivity to PYR/PYL inhibition (Antoni et al. 2012). Different oligo-merization states of PYR/PYL/RCAR receptors and multiple receptor complexes could potentially fine-tune a multitude of ABA responses in different tissues or under various conditions.

Additionally, different expression patterns of the genes encoding different members of PYR/PYL/RCARs suggest substantial functional differences of the different members (Antoni et al. 2012; Gonzalez-Guzman et al. 2012). The expression level of PYL3 and PYL10 to PYL13 is extremely low in the whole plant level, while the expression of PYR1 and the rest of PYL1 to PYL9 could be detected in both vegetative and reproductive tissues at different levels (Antoni et al. 2012; Gonzalez-Guzman et al. 2012).

This functional diversity of different members of PYR/PYL/RCAR receptors has been supported by several recent reports. A study revealed that PYL8 interacts with at least five PP2Cs, namely HAB1, HAB2, ABI1, ABI2, and PP2CA, and plays a non-redundant role for the regulation of root ABA sensitivity; the single *pyl8* mutant shows ABA insensitivity in root growth (Antoni et al. 2013). Further, Zhao et al. (2014) found that PYL8 promotes lateral root growth independently of the core ABA–SnRK2 signaling pathway by enhancing the activities of MYB77 and its paralogs, MYB44 and MYB73, to augment auxin signaling. Yin et al. (2013) provided genetic evidence that four ABA receptors PYR1, PYL1, PYL2, and PYL4 are not sufficient for ABA-induced stomatal opening inhibition in *Arabidopsis*: ABA-induced stomatal closure is impaired in the *pyr1 pyl1 pyl2 pyl4* quadruple ABA receptor mutant, whereas ABA inhibition of the opening of the mutant's stomata remains intact; ABA substantially inhibits blue light-induced phosphorylation of H^{+}-ATPase in guard cells in both the mutant and the wild type (Yin et al. 2013). This suggests that stomatal opening and closure in response to ABA are differentially regulated, and other more members of the PYR/PYL/RCAR receptors, or other different classes of ABA receptors, such as ABAR/CHLH or GTG1/GTG2, are required for ABA-induced inhibition of stomatal opening.

6.6 Summary

Recent identification of the receptors or candidate receptors for ABA, localized to plasma membrane, cytosol, and chloroplast, confirms previous observations that ABA signal is perceived both inter- and intracellularly. ABAR/CHLH is a chloroplast-membrane protein involved in multiple functions including chlorophyll biosynthesis, plastid-to-nucleus retrograde signaling and ABA signaling. Multiple lines of evidence demonstrate ABA-binding ability of ABAR, though structural study will be needed to resolve the interaction of ABAR with ABA at the subcellular level. Increasing evidence well supports that ABAR is a crucial player in ABA signaling, which antagonizes a group of WRKY transcription repressors (WRKY18/40/60) to relieve ABA-responsive genes of inhibition. The ABAR–WRKY40 coupled mechanism may be inhibited by CPN20, a chloroplast interaction partner of ABAR. High level of ABA inhibits the ABAR–CPN20 interaction, which in turn promotes the ABAR–WRKY40 interaction to trigger the downstream signaling to repress WRKY40 expression and finally to relieve ABA-responsive genes of inhibition (Fig. 6.1). Recent discovery of a PPR protein SOAR1 that functions as a key regulator of ABA signaling downstream of ABAR suggests that ABAR regulates a central ABA signaling network. It is intriguing to uncover the mechanism by which SOAR1 functions downstream of ABAR in ABA signaling.

GTG1 and GTG2 are a novel class of G proteins, namely GPCR-type G proteins, which may perceive intercellular ABA signal. In the GTG-mediated ABA signaling, GDP-bound GTGs bind ABA, which initiates the ABA signaling cascade; GTP-bound GPA1 inhibits formation of the active complex GTG–GDP, downregulates ABA binding to the GTGs, and represses ABA signaling (Fig. 6.2). This model suggests an unusual type of G protein signaling, which is opposite to the conventional model for signaling mediated by G proteins, where GDP-bound Gα turns off the signaling system. Further studies will be needed to identify downstream components to elucidate underlying mechanisms of the GTG-mediated ABA signaling.

The PYR/PYL/RCAR-mediated signaling pathway is the currently best-characterized ABA signaling pathway. Under basal ABA levels, clade A PP2Cs function as negative regulators to repress ABA signaling, either through dephosphorylation of SnRK2s or interaction with other targets. ABA binding to the PYR/PYL ABA receptors inactivates PP2Cs, leading to the activation of SnRK2s through phosphorylation by BIN2/BILs kinases, which subsequently phosphorylate downstream targets, such as members of the ABF transcription factors or regulatory components of the stomatal aperture, such as the anion channels SLAC1 and SLAH3. Some PP2C members such as ABI1, working downstream of PYR/PYL/RCAR receptors, also inhibit the calcium-dependent kinases CPK21 and CPK23 (see Chap. 8), which share the anion channels SLAC1 and SLAH3 as substrates with SnRK2s to control stomatal movement. This core signaling pathway connects ABA perception, inactivation of clade A PP2Cs, and activation of SnRK2 protein kinases and their various downstream targets to induce a diversity of ABA responses (Fig. 6.3). However, how to integrate known ABA signal components, especially the ABAR- and GTGs-mediated signaling pathways, into the PYR/PYL/RCAR-mediated signaling pathway, is a challenging question in the future.

Acknowledgements The authors thank all colleagues who provided unpublished results and apologize to those whose research was not discussed due to page limitation. The research on ABAR/CHLH in the authors' laboratory was supported by the grants from the National Key Basic Research Program of China (2012CB114302), National Natural Science Foundation of China, and the Ministry of Agriculture of China (grant 2013ZX08009003).

References

Acharya BR, Jeon BW, Zhang W, Assmann SM. Open Stomata 1 (OST1) is limiting in abscisic acid responses of *Arabidopsis* guard cells. New Phytol. 2013;200:1049–63.

Addicott FT, Lyon JL. Physiology of abscisic acid and related substances. Annu Rev Plant Physiol. 1969;20:139–64.

Adhikari ND, Froehlich JE, Strand DD, Buck SM, Kramer DM, Larkin RM. GUN4-porphyrin complexes bind the ChlH/GUN5 subunit of Mg-chelatase and promote chlorophyll biosynthesis in *Arabidopsis*. Plant Cell. 2011;23:1449–67.

Allan AC, Fricker MD, Ward JL, Beale MH, Trewavas AJ. Two transduction pathways mediate rapid effects of abscisic acid in *Commelina* guard cells. Plant Cell. 1994;6:1319–28.

Alvarez S, Choudhury SR, Hicks LM, Pandey S. Quantitative proteomics-based analysis supports a significant role of GTG proteins in regulation of ABA response in *Arabidopsis* roots. J Proteome Res. 2013;12:1487–501.

Anderson BE, Ward JM, Schroeder JI. Evidence for an extra-cellular reception site for abscisic acid in *Commelina* guard cells. Plant Physiol. 1994;104:1177–83.

Antoni R, Gonzalez-Guzman M, Rodriguez L, Rodrigues A, Pizzio GA, Rodriguez PL. Selective inhibition of clade A phosphatases type 2C by PYR/PYL/RCAR abscisic acid receptors. Plant Physiol. 2012;158:970–80.

Antoni R, Gonzalez-Guzman M, Rodriguez L, Peirats-Llobet M, Pizzio GA, Fernandez MA, Winne ND, De Jaeger G, Dietrich D, Bennett MJ, Rodriguez PL. PYRABACTIN RESISTANCE1-LIKE8 plays an important role for the regulation of abscisic acid signaling in root. Plant Physiol. 2013;161:931–41.

Assmann SM. Ins and outs of guard cell ABA receptors. Plant Cell. 1994;6:1187–90.

Assmann SM. Plant G proteins, phytohormones, and plasticity: three questions and a speculation. Sci STKE. 2004;264:re20.

Brandt B, Brodsky DE, Xue S, Negi J, Iba K, Kangasjärvi J, Ghassemian M, Stephan AB, Hu H, Schroeder JI. Reconstitution of abscisic acid activation of SLAC1 anion channel by CPK6 and OST1 kinases and branched ABI1 PP2C phosphatase action. Proc Natl Acad Sci USA. 2012;109:10593–8.

Cabrera-Vera TM, Vanhauwe J, Thomas TO, Medkova M, Preininger A, Mazzoni MR, Hamm HE. Insights into G protein structure, function, and regulation. Endocr Rev. 2003;24:765–81.

Cai Z, Liu J, Wang H, Yang C, Chen Y, Li Y, Pan S, Dong R, Tang G, de Dios Barajas-Lopez J, Fujii H, Wang X. GSK3-like kinases positively modulate abscisic acid signaling through phosphorylating subgroup III SnRK2s in *Arabidopsis*. Proc Natl Acad Sci USA. 2014;111:9651–6.

Chen JG, Ellis BE. GCR2 is a new member of the eukaryotic lanthionine synthetase component C-like protein family. Plant Signal Behav. 2008;3:307–10.

Chen J, Zhu MJ. Properties of ABA binding proteins in the maize root microsomes and stress-induced effects. Acta Phytophysiologica Sinica. 1996;22:63–8.

Chen R, Wang LR, Lu L, Zhang ZX, Zhang DY. Characterization of the membrane-bound ABA-binding sites in rice seedlings. Acta Botanica Sinica. 1992;34:185–90.

Coursol S, Fan L, Le Stunff H, Spiegel S, Gilroy S, Assmann SM. Sphingolipid signalling in *Arabidopsis* guard cells involves heterotrimeric G proteins. Nature. 2003;423:651–4.

Cutler S, McCourt P. Dude, Where's my phenotype? Dealing with redundancy in signaling networks. Plant Physiol. 2005;138:558–9.

Cutler SR, Rodriguez PL, Finkelstein RR, Abrams SR. Abscisic acid: emergence of a core signaling network. Annu Rev Plant Biol. 2010;61:651–79.

Du SY, Zhang XF, Lu ZK, Xin Q, Wu Z, Jiang T, Yan L, Wang XF, Zhang DP. Roles of the different components of magnesium chelatase in abscisic acid signal transduction. Plant Mol Biol. 2012;80:519–37.

Dupeux F, Santiago J, Betz K, Twycross J, Park SY, Rodriguez L, Gonzalez-Guzman M, Jensen MR, Krasnogor N, Blackledge M, Holdsworth M, Cutler SR, Rodriguez PL, Márquez JA. A thermodynamic switch modulates abscisic acid receptor sensitivity. EMBO J. 2011;30:4171–84.

Finkelstein RR, Gampala SS, Rock CD. Abscisic acid signaling in seeds and seedlings. Plant Cell. 2002;14:S15–45.

Fuchs S, Tischer SV, Wunschel C, Christmann A, Grill E. Abscisic acid sensor RCAR7/PYL13, specific regulator of protein phosphatase coreceptors. Proc Natl Acad Sci USA. 2014;111:5741–6.

Fujii H, Zhu JK. *Arabidopsis* mutant deficient in 3 abscisic acid-activated protein kinases reveals critical roles in growth, reproduction, and stress. Proc Natl Acad Sci USA. 2009;106:8380–5.

Fujii H, Verslues PE, Zhu JK. Identification of two protein kinases required for abscisic acid regulation of seed germination, root growth, and gene expression in *Arabidopsis*. Plant Cell. 2007;19:485–94.

Fujii H, Chinnusamy V, Rodrigues A, Rubio S, Antoni R, Park SY, Cutler SR, Sheen J, Rodriguez PL, Zhu JK. In vitro reconstitution of an abscisic acid signaling pathway. Nature. 2009;462:660–4.

Fujita Y, Nakashima K, Yoshida T, Katagiri T, Kidokoro S, Kanamori N, Umezawa T, Fujita M, Maruyama K, Ishiyama K, Kobayashi M, Nakasone S, Yamada K, Ito T, Shinozaki K, Yamaguchi-Shinozaki K. Three SnRK2 protein kinases are the main positive regulators of abscisic acid signaling in response to water stress in *Arabidopsis*. Plant Cell Physiol. 2009;50:2123–32.

Fujita Y, Yoshida T, Yamaguchi-Shinozaki K. Pivotal role of the AREB/ABF-SnRK2 pathway in ABRE-mediated transcription in response to osmotic stress in plants. Physiol Plant. 2013;147:15–27.

Furihata T, Maruyama K, Fujita Y, Umezawa T, Yoshida R, Shinozaki K, Yamaguchi-Shinozaki K. Abscisic acid-dependent multisite phosphorylation regulates the activity of a transcription activator AREB1. Proc Natl Acad Sci USA. 2006;103:1988–93.

Gao Y, Zeng Q, Guo J, Cheng J, Ellis BE, Chen JG. Genetic characterization reveals no role for the reported ABA receptor, GCR2, in ABA control of seed germination and early seedling development in *Arabidopsis*. Plant J. 2007;52:1001–13.

Geiger D, Maierhofer T, Al-Rasheid KAS, Scherzer S, Mumm P, Liese A, Ache P, Wellmann C, Marten I, Grill E, Romeis T, Hedrich R. Stomatal closure by fast abscisic acid signaling is mediated by the guard cell anion channel SLAH3 and the receptor RCAR1. Sci Signal. 2011;4:ra32. doi:10.1126/scisignal.2001346.

Geiger D, Scherzer S, Mumm P, Stange A, Marten I, Bauer H, Ache P, Matschi S, Liese A, Al-Rasheid KA, Romeis T, Hedrich R. Activity of guard cell anion channel SLAC1 is controlled by drought-stress signaling kinase-phosphatase pair. Proc Natl Acad Sci USA. 2009;106:21425–30.

Geiger D, Scherzer S, Mumm P, Marten I, Ache P, Matschi S, Liese A, Wellmann C, Al-Rasheid KA, Grill E, Romeis T, Hedrich R. Guard cell anion channel SLAC1 is regulated by CDPK protein kinases with distinct Ca^{2+} affinities. Proc Natl Acad Sci USA. 2010;107:8023–8.

Gilroy S, Jones R. Gibberellic acid and abscisic acid coordinately regulate cytoplasmic calcium and secretory activity in barley aleurone protoplasts. Proc Natl Acad Sci USA. 1992;89:3591–5.

Gilroy S, Jones R. Perception of gibberellin and abscisic acid at the external face of the plasma membrane of barley *(Hordeum vulgare L.)* aleurone protoplasts. Plant Physiol. 1994;104:1185–92.

Giraudat J, Hauge BM, Valon C, Smalle J, Parcy F, Goodman HM. Isolation of the *Arabidopsis* ABI3 gene by positional cloning. Plant Cell. 1992;4:1251–61.

Giraudat J, Parcy F, Bertauche N, Gosti F, Leung J, Morris PC, Bouvier-Durand M, Vartanian N. Current advances in abscisic acid action and signaling. Plant Mol Biol. 1994;26:1557–77.

Gonzalez-Guzman M, Pizzio GA, Antoni R, Vera-Sirera F, Merilo E, Bassel GW, Fernández MA, Holdsworth MJ, Perez-Amador MA, Kollist H, Rodriguez PL. *Arabidopsis* PYR/PYL/RCAR receptors play a major role in quantitative regulation of stomatal aperture and transcriptional response to abscisic acid. Plant Cell. 2012;24:2483–96.

Guo J, Zeng Q, Emami M, Ellis BE, Chen JG. The GCR2 gene family is not required for ABA control of seed germination and early seedling development in *Arabidopsis*. PLoS ONE. 2008;3:e2982.

Hackenberg D, Sakayama H, Nishiyama T, Pandey S. Characterization of the heterotrimeric G-protein complex and its regulator from the green alga *Chara braunii* expands the evolutionary breadth of plant G-protein signaling. Plant Physiol. 2013;163:1510–7.

Hallouin M, Ghelis T, Brault M, Bardat F, Cornel D, Miginiac E, Rona JM, Sotta B, Jeannette E. Plasmalemma abscisic acid perception leads to *RAB18* expression via phospholipase D activation in *Arabidopsis* suspension cells. Plant Physiol. 2002;130:265–72.

Hao Q, Yin P, Li W, Wang L, Yan C, Lin Z, Wu JZ, Wang J, Yan SF, Yan N. The molecular basis of ABA-independent inhibition of PP2Cs by a subclass of PYL proteins. Mol Cell. 2011;42:662–72.

Hartung W. The site of action of abscisic acid at the guard cell plasmalemma of *Valerianella locusta*. Plant Cell Environ. 1983;6:427–8.

He JX, Gendron JM, Yang Y, Li J, Wang ZY. The GSK3-like kinase BIN2 phosphorylates and destabilizes BZR1, a positive regulator of the brassinosteroid signaling pathway in *Arabidopsis*. Proc Natl Acad Sci USA. 2002;99:10185–90.

Himmelbach A, Yang Y, Grill E. Relay and control of abscisic acid signaling. Curr Opin Plant Biol. 2003;6:470–9.

Hocking J, Clapham J, Cattell KJ. Abscisic acid binding to subcellular fractions from leaves of *Vicia faba*. Planta. 1978;138:303–4.

Hornberg C, Weiler EW. High affinity binding sites for abscisic acid at the plasmalemma of *Vicia faba* guard cells. Nature. 1984;310:321–4.

Horton RF, Bruce KR. Inhibition by abscisic acid of the light and dark uptake of potassium by slices of *Vicia faba* leaves. Can J Bot. 1972;50:1915–7.

Illingworth CJR, Parkes KE, Snell CR, Mullineaux PM, Reynolds CA. Criteria for confirming sequence periodicity identified by Fourier transform analysis: application to GCR2, a candidate plant GPCR? Biophys Chem. 2008;133:28–35.

Iyer LM, Koonin EV, Aravind L. Adaptations of the helix-grip fold for ligand binding and catalysis in the START domain superfamily. Proteins. 2001;43:134–44.

Jacob T, Ritchie S, Assmann SM, Gilroy S. Abscisic acid signal transduction in guard cells is mediated by phospholipase D activity. Proc Natl Acad Sci USA. 1999;96:12192–7.

Jaffe FW, Freschet GE, Valdes BM, Runions J, Terry MJ, Williams LE. G protein-coupled receptor-type g proteins are required for light-dependent seedling growth and fertility in *Arabidopsis*. Plant Cell. 2012;24:3649–68.

Jeannette E, Rona JP, Bardat F, Cornel D, Sotta B, Miginiac E. Induction of RAB18 gene expression and activation of K + outward rectifying channels depend on an extracellular perception of ABA in *Arabidopsis* thaliana suspension cells. Plant J. 1999;18:13–22.

Jia HF, Chai YM, Li CL, Lu D, Luo JJ, Qin L, Shen YY. Abscisic acid plays an important role in the regulation of strawberry fruit ripening. Plant Physiol. 2011a;157:188–99.

Jia HF, Chai YM, Li CL, Qin L, Shen YY. Cloning and characterization of the H subunit of a magnesium chelatase gene (PpCHLH) in peach. J Plant Growth Regul. 2011b;30:445–55.

Jiang T, Zhang XF, Wang XF, Zhang DP. *Arabidopsis* 3-ketoacyl-CoA thiolase-2 (KAT2), an enzyme of fatty acid beta-oxidation, is involved in ABA signal transduction. Plant Cell Physiol. 2011;52:528–38.

Johnston CA, Temple BR, Chen JG, Gao Y, Moriyama EN, Jones AM, Siderovski DP, Willard FS. Comment on 'A G protein coupled receptor is a plasma membrane receptor for the plant hormone abscisic acid'. Science. 2007; 318:914 (author reply 914).

Jones AM, Assmann SA. Plants: the latest model system for G-protein research. EMBO Rep. 2004;5:572–8.

Joshi-Saha A, Valon C, Leung J. A brand new START: abscisic acid perception and transduction in the guard cell. Sci Signal. 2011; 4:re4.

Kepka M, Benson CL, Gonugunta VK, Nelson KM, Christmann A, Grill E, Abrams SR. Action of natural abscisic acid Precursors and catabolites on abscisic acid receptor complexes. Plant Physiol. 2011;157:2108–19.

Kharenko OA, Choudhary P, Loewen MC. Abscisic acid binds to recombinant *Arabidopsis thaliana* G-protein coupled receptor-type G-protein 1 in *Sacaromycese cerevisiae* and in vitro. Plant Physiol Biochem. 2013;68:32–6.

Klingler JP, Batelli G, Zhu JK. ABA receptors: the START of a new paradigm in phytohormone signaling. J Exp Bot. 2010;61:3199–210.

Lee SC, Lan W, Buchanan BB, Luan S. A protein kinase–phosphatase pair interacts with an ion channel to regulate ABA signaling in plant guard cells. Proc Natl Acad Sci USA. 2009;106:21419–24.

Lee SC, Lim CW, Lan W, He K, Luan S. ABA signaling in guard cells entails a dynamic protein-protein interaction relay from the PYL-RCAR family receptors to ion channels. Mol Plant. 2013;6:528–38.

Legnaioli T, Cuevas J, Mas P. TOC1 functions as a molecular switch connecting the circadian clock with plant responses to drought. EMBO J. 2009;28:3745–57.

Leung J, Giraudat J. Abscisic acid signal transduction. Annu Rev Plant Physiol Plant Mol Biol. 1998;49:199–222.

Leung J, Bouvier-Durand M, Morris PC, Guerrier D, Chefdor F, Giraudat J. *Arabidopsis* ABA-response gene ABI1: features of a calcium-modulated protein phosphatase. Science. 1994;264:1448–52.

Leung J, Merlot S, Giraudat J. The *Arabidopsis* ABSCISIC ACID-INSENSITIVE2 (ABI2) and ABI1 genes encode redundant protein phosphatases 2C involved in abscisic acid signal transduction. Plant Cell. 1997;9:759–71.

Li J, Nam KH. Regulation of brassinosteroid signaling by a GSK3/SHAGGY-like kinase. Science. 2002;295:1299–301.

Li WQ, Li W, Sheng XL, Yan CY, Zhou R, Hang J, Yin P, Yan N. Molecular basis for the selective and ABA-independent inhibition of PP2CA by PYL13. Cell Res. 2013;23:1369–79.

Liu X, Yue Y, Li B, Nie Y, Li W, Wu WH, Ma LG. A G protein-coupled receptor is a plasma membrane receptor for the plant hormone abscisic acid. Science. 2007a;315:1712–6.

Liu X, Yue Y, Li W, Ma LG. Response to comment on "A G protein coupled receptor is a plasma membrane receptor for the plant hormone abscisic acid". Science. 2007b;318:914.

Liu ZQ, Yan L, Wu Z, Mei C, Wang XF, Zhang DP. Cooperation of three WRKY-domain transcription factors WRKY18, WRKY40 and WRKY60 in repressing two ABA-responsive genes ABI4 and ABI5 in *Arabidopsis*. J Exp Bot. 2012;63:6371–92.

Liu R, Xu YH, Jiang SC, Lu K, Lu YF, Feng XJ, Wu Z, Liang S, Yu YT, Wang XF, Zhang DP. Light-harvesting chlorophyll *a/b*-binding proteins, positively involved in abscisic acid signalling, require a transcription repressor, WRKY40, to balance their function. J Exp Bot. 2013;64:5443–56.

Lurin C, Andrés C, Aubourg S, Bellaoui M, Bitton F, Bruyère C, Caboche M, Debast C, Gualberto J, Hoffmann B, Lecharny A, Le Ret M, Martin-Magniette ML, Mireau H, Peeters N, Renou JP, Szurek B, Taconnat L, Small I. Genome-wide analysis of *Arabidopsis* pentatricopeptide repeat proteins reveals their essential role in organelle biogenesis. Plant Cell. 2004;16:2089–103.

Ma H. GTP-binding proteins in plants: new members of an old family. Plant Mol Biol. 1994;26:1611–36.

Ma H, Yanofsky M, Meyerowitz EM. Molecular cloning and characterization of GPAZ, a G protein a subunit gene from *Arabidopsis thaliana*. Proc Natl Acad Sci USA. 1990;87:3821–5.

Ma Y, Szostkiewicz I, Korte A, Moes D, Yang Y, Christman A, Grill E. Regulators of PP2C phosphatase activity function as abscisic acid sensors. Science. 2009;324:1064–8.

MacRobbie EAC. ABA-induced ion efflux in stomatal guard cells: multiple actions of ABA inside and outside the cell. Plant J. 1995;7:565–76.

Mason MG, Botella J. Completing the heterotrimer: isolation and characterization of an *Arabidopsis thaliana* G protein gamma-subunit cDNA. Proc Natl Acad Sci USA. 2000;97:14784–8.

Mason MG, Botella JR. Isolation of a novel G-protein gamma-subunit from *Arabidopsis thaliana* and its interaction with Gbeta. Biochim Biophys Acta. 2001;1520:147–53.

McCourt P, Creelman R. The ABA receptors—we report you decide. Curr Opin Plant Biol. 2008;11:474–8.

Mei C, Jiang SC, Lu YF, Wu FQ, Yu YT, Liang S, Feng XJ, Comeras SP, Lu K, Wu Z, Wang XF, Zhang DP. *Arabidopsis* pentatricopeptide repeat protein SOAR1 plays a critical role in abscisic acid signaling. J Exp Bot. 2014. doi:10.1093/jxb/eru293.

Meidner H, Willmer C. Mechanics and Metabolism of Guard Cells. Curr Adv Plant Sci. 1975;17:1–15.

Meierhoff K, Felder S, Nakamura T, Bechtold N, Schuster G. HCF152, an *Arabidopsis* RNA binding pentatricopeptide repeat protein involved in the processing of chloroplast *psbB-psbT-psbH-petB-petD* RNAs. Plant Cell. 2003;15:1480–95.

Melcher K, Ng L, Zhou XE, Soon F, Xu Y, Suino-Powell KM, Park SY, Weiner JJ, Fujii H, Chinnusamy V, Kovach A, Li J, Wang Y, Li J, Peterson FC, Jensen DR, Yong EL, Volkman BF, Cutler SR, Zhu JK, Xu HE. A gate–latch–lock mechanism for hormone signalling by abscisic acid receptors. Nature. 2009;462:602–8.

Milborrow BV. The Chemistry and Physiology of Abscisic Acid. Ann Rev Plant Physiol. 1974;25:259–307.

Miyazono K, Miyakawa T, Sawano Y, Kubota K, Kang H, Asano A, Miyauchi Y, Takahashi M, Zhi Y, Fujita Y, Yoshida T, Kodaira KS, Yamaguchi-Shinozaki K, Tanokura M. Structural basis of abscisic acid signalling. Nature. 2009;462:609–14.

Mochizuki N, Brusslan JA, Larkin R, Nagatani N, Chory J. *Arabidopsis* genomes uncoupled 5 (GUN5) mutant reveals the involvement of Mg-chelatase H subunit in plastid-to-nucleus signal transduction. Proc Natl Acad Sci USA. 2001;98:2053–8.

Müller AH, Hansson M. The barley magnesium chelatase 150-kD subunit is not an abscisic acid receptor. Plant Physiol. 2009;150:157–66.

Muschietti J, McCormick S. Abscisic acid (ABA) receptors: light at the end of the tunnel. F1000 Biol rep. 2010;2:1–5.

Mustilli A, Merlot S, Vavasseur A, Fenzi F, Giraudat J. *Arabidopsis* OST1 protein kinase mediates the regulation of stomatal aperture by abscisic acid and acts upstream of reactive oxygen species production. Plant Cell. 2002;14:3089–99.

Nakashima K, Fujita Y, Kanamori N, Umezawa T, Kidokoro S, Maruyama K, Yoshida T, Ishiyama K, Kobayashi M, Shinozaki K, Yamaguchi-Shinozaki K. Three *Arabidopsis* SnRK2 protein kinases, SRK2D/SnRK2.2, SRK2E/SnRK2.6/OST1 and SRK2I/SnRK2.3, involved in ABA signaling are essential for the control of seed development and dormancy. Plant Cell Physiol. 2009;50:1345–63.

Nishimura N, Hitomi K, Arvai AS, Rambo RP, Hitomi C, Cutler SR, Schroeder JI, Getzoff ED. Structural mechanism of abscisic acid binding and signaling by dimeric PYR1. Science. 2009;326:1373–9.

Nishimura N, Sarkeshik A, Nito K, Park S, Wang A, Carvalho PC, Lee S, Caddell DF, Cutler SR, Chory J, Yates JR, Schroeder JI. PYR/PYL/RCAR family members are major in-vivo ABI1 protein phosphatase 2C-interacting proteins in *Arabidopsis*. Plant J. 2010;61:290–9.

Nott A, Jung HS, Koussevitzky S, Chory J. Plastid-to-nucleus retrograde signaling. Annu Rev Plant Biol. 2006;57:730–59.

Nyangulu JM, Galka MM, Jadhav A, Gai Y, Graham CM, Nelson KM, Cutler AJ, Taylor DC, Banowetz GM, Abrams SR. An affinity probe for isolation of abscisic acid-binding proteins. J Am Chem Soc. 2005;127:1662–4.

Offermanns S. G-proteins as transducers in transmembrane signalling. Prog Biophys Mol Biol. 2003;83:101–30.

Ohkuma K, Lyon JL, Addicott FT, Smith OE. Abscisin II, an abscission-accelerating substance from young cotton fruit. Science. 1963;142:1592–3.

Pandey S, Assmann SM. The *Arabidopsis* putative G protein coupled receptor GCR1 interacts with the G protein a subunit GPA1 and regulates abscisic acid signaling. Plant Cell. 2004;16:1616–32.

Pandey S, Chen JG, Jones AM, Assmann SM. G-protein complex mutants are hypersensitive to abscisic acid regulation of germination and postgermination development. Plant Physiol. 2006;141:243–56.

Pandey S, Nelson DC, Assmann SM. Two novel GPCR-type G proteins are abscisic acid receptors in *Arabidopsis*. Cell. 2009;136:136–48.

Park SY, Fung P, Nishimura N, Jensen DR, Fujii H, Zhao Y, Lumba S, Santiago J, Rodrigues A, Chow TF, Alfred SE, Bonnetta D, Finkelstein R, Provart NJ, Desveaux D, Rodriguez PL, McCourt P, Zhu JK, Schroeder JI, Volkman BF, Cutler SR. Abscisic acid inhibits type 2C protein phosphatases via the PYR/PYL family of START proteins. Science. 2009;324:1068–71.

Pedron J, Brault M, Nake C, Miginiac E. Detection of abscisic-acid-binding proteins in the microsomal protein fraction of *Arabidopsis thaliana* with abscisic-acid-protein conjugates used as affinity probes. Eur J Biochem. 1998;252:385–90.

Pierce KL, Premont RT, Lefkowitz RJ. Seven-transmembrane receptors. Nat Rev Mol Cell Biol. 2002;3:639–50.

Raghavendra AS, Gonugunta VK, Christmann A, Grill E. ABA perception and signalling. Trends Plant Sci. 2010;15:395–401.

Ritchie S, Gilroy S. Abscisic acid signal transduction in the barley aleurone is mediated by phospholipase D activity. Proc Natl Acad Sci USA. 1998;95:2697–702.

Ritchie S, Gilroy S. Abscisic acid stimulation of phospholipase D in the barley aleurone is G-protein mediated and localized to the plasma membrane. Plant Physiol. 2000;124:639–702.

Santiago J, Dupeux F, Round A, Antoni R, Park SY, Jamin M, Cutler SR, Rodriguez PL, Marquez JA. The abscisic acid receptor PYR1 in complex with abscisic acid. Nature. 2009a;462:665–8.

Santiago J, Rodrigues A, Saez A, Rubio S, Antoni R, Dupeux F, Park SY, Marquez JA, Cutler SR, Rodriguez PL. Modulation of drought resistance by the abscisic acid receptor PYL5 through inhibition of clade A PP2Cs. Plant J. 2009b;60:575–88.

Schwartz A, Wu WH, Tucker EB, Assmann SM. Inhibition of inward K^+ channels and stomatal response by abscisic acid: An intracellular locus of phytohormone action. Proc Natl Acad Sci USA. 1994;91:4019–23.

Shang Y, Yan L, Liu ZQ, Cao Z, Mei C, Xin Q, Wu FQ, Wang XF, Du SY, Jiang T, Zhang XF, Zhao R, Sun HL, Liu R, Yu YT, Zhang DP. The Mg-chelatase H subunit of *Arabidopsis* antagonizes a group of transcription repressors to relieve ABA-responsive genes of inhibition. Plant Cell. 2010;22:1909–35.

Shen YY, Wang XF, Wu FQ, Du SY, Cao Z, Shang Y, Wang SL, Peng CC, Yu XC, Zhu SY, Fan RC, Xu YH, Zhang DP. The Mg-chelatase H subunit is an abscisic acid receptor. Nature. 2006;443:823–6.

Shibata N, Kagiyama M, Nakagawa M, Hirano Y, Hakoshima T. Crystallization of the plant hormone receptors PYL9/RCAR1, PYL5/RCAR8 and PYR1/RCAR11 in the presence of (+)-abscisic acid. Acta Crystallogr, Sect F: Struct Biol Cryst Commun. 2010;66:456–9.

Sirichandra C, Davanture M, Turk BE, Zivy M, Valot B, Leung J, Merlot S. The *Arabidopsis* ABA-activated kinase OST1 phosphorylates the bZIP transcription factor ABF3 and creates a 14-3-3 binding site involved in its turnover. PLoS ONE. 2010;5:e13935.

Strand Å, Asami T, Alonso J, Ecker JR, Chory J. Chloroplast to nucleus communication triggered by accumulation of Mg-protoporphyrinIX. Nature. 2003;421:79–83.

Surpin M, Larkin RM, Chory J. Signal transduction between the chloroplasts and the nucleus. Plant Cell. 2002;14:S327–38.

Szostkiewicz I, Richter K, Kepka M, Demmel S, Ma Y, Korte A, Assaad FF, Christmann A, Grill E. Closely related receptor complexes differ in their ABA selectivity and sensitivity. Plant J. 2010;61:25–35.

Tsuzuki T, Takahashi K, Inoue S, Okigaki Y, Tomiyama M, Hossain MA, Shimazaki K, Murata Y, Kinoshita T. Mg-chelatase H subunit affects ABA signaling in stomatal guard cells, but is not an ABA receptor in *Arabidopsis thaliana*. J Plant Res. 2011;124:527–38.

Tsuzuki T, Takahashi K, Tomiyama M, Inoue S, Kinoshita T. Overexpression of the Mg-chelatase H subunit in guard cells confers drought tolerance via promotion of stomatal closure in *Arabidopsis thaliana*. Frontiers Plant Sci. 2013;4:1–8.

Umezawa T, Sugiyama N, Mizoguchi M, Hayashi S, Myouga F, Yamaguchi-Shinozaki K, Ishihama Y, Hirayama T, Shinozaki K. Type 2C protein phosphatases directly regulate abscisic acid-activated protein kinases in *Arabidopsis*. Proc Natl Acad Sci USA. 2009;106:17588–93.

Umezawa T, Nakashima K, Miyakawa T, Kuromori T, Tanokura M, Shinozaki K, Yamaguchi-Shinozaki K. Molecular basis of the core regulatory network in ABA responses: sensing, signaling and transport. Plant Cell Physiol. 2010;51:1821–39.

Urano D, Chen JG, Botella JR, Jones AM. Heterotrimeric G protein signalling in the plant kingdom. Open Biol. 2013;3:120186.

Veliev SN. Cytosolic high affinity ABA-binding protein from wheat shoot. Biochem (Moscow). 1991;56:420–5.

Venis M. Methods in receptor research. In: Venis M, editor. Hormone binding sites in plants. New York: Longman; 1985. p. 24–40.

Verslues PE, Zhu JK. New developments in abscisic acid perception and metabolism. Curr Opin Plant Biol. 2007;10:447–52.

Vlad F, Rubio S, Rodrigues A, Sirichandra C, Belin C, Robert N, Leung J, Rodriguez PL, Laurière C, Merlot S. Protein phosphatases 2C regulate the activation of the Snf1-related kinase OST1 by abscisic acid in *Arabidopsis*. Plant Cell. 2009;21:3170–84.

Vlad F, Droillard MJ, Valot B, Khafif M, Rodrigues A, Brault M, Zivy M, Rodriguez PL, Merlot S, Laurière C. Phospho-site mapping, genetic and in planta activation studies reveal key aspects of the different phosphorylation mechanisms involved in activation of SnRK2s. Plant J. 2010;63:778–90.

Walker CJ, Willows RD. Mechanism and regulation of Mg-chelatase. Biochem J. 1997;327:321–33.

Wang X. The role of phospholipase D in signal cascades. Plant Physiol. 1999;120:645–51.

Wang XF, Zhang DP. Abscisic acid receptors: Multiple signal-perception sites. Ann Bot. 2008;101:311–7.

Wang XF, Shen YY, Xin Q, Zhang DP. Approaches to the identification of ABAR as an abscisic acid receptor. In: Kermode AR, editor. Methods in molecular biology, vol. 773. New York: Springer Science+Business Media, LLC; 2011. p. 83–97.

Wang XQ, Ullah H, Jones AM, Assmann SM. G protein regulation of ion channels and abscisic acid signaling in *Arabidopsis* guard cells. Science. 2001;292:2070–2.

Williams PM, Barkan A. A chloroplast-localized PPR protein required for plastid ribosome accumulation. Plant J. 2003;36:675–86.

Wu FQ, Xin Q, Cao Z, Liu ZQ, Du SY, Mei C, Zhao CX, Wang XF, Shang Y, Jiang T, Zhang XF, Yan L, Zhao R, Cui ZN, Liu R, Sun HL, Yang XL, Su Z, Zhang DP. The Mg-chelatase H subunit binds abscisic acid and functions in abscisic acid signaling: New evidence in *Arabidopsis*. Plant Physiol. 2009;150:1940–54.

Xu YH, Liu R, Yan L, Liu ZQ, Jiang SC, Shen YY, Wang XF, Zhang DP. Light-harvesting chlorophyll a/b-binding proteins are required for stomatal response to abscisic acid in *Arabidopsis*. J Exp Bot. 2012;63:1095–106.

Yamazaki D, Yoshida S, Asami T, Kuchitsu K. Visualization of abscisic acid-perception sites on the plasma membrane of stomatal guard cells. Plant J. 2003;35:129–39.

Yan L, Liu ZQ, Xu YH, Lu K, Wang XF, Zhang DP. Auto- and cross-repression of three *Arabidopsis* WRKY transcription factors WRKY18, WRKY40 and WRKY60 negatively involved in ABA signaling. J Plant Growth Regul. 2013;32:399–416.

Yin P, Fan H, Hao Q, Yuan X, Wu D, Pang Y, Yan CY, Li WQ, Wang JW, Yan N. Structural insights into the mechanism of abscisic acid signaling by PYL proteins. Nat Struct Mol Biol. 2009;16:1230–6.

Yin Y, Adachi Y, Ye W, Hayashi M, Nakamura Y, Kinoshita T, Mori IC, Murata Y. Difference in abscisic acid perception mechanisms between closure Induction and opening inhibition of stomata. Plant Physiol. 2013;163:600–10.

Yoshida R, Umezawa T, Mizoguchi T, Takahashi S, Takahashi F, Shinozaki K. The regulatory domain of SRK2E/OST1/SnRK2.6 interacts with ABI1 and integrates abscisic acid (ABA) and osmotic stress signals controlling stomatal closure in *Arabidopsis*. J Biol Chem. 2006;281:5310–8.

Zhang DP, Zhang ZL, Chen J, Jia WS. Specific abscisic acid-binding sites in mesocarp of grape berry: properties and sub-cellular localization. J Plant Physiol. 1999;155:324–31.

Zhang DP, Chen SW, Peng YB, Shen YY. Abscisic acid-specific binding sites in the flesh of developing apple fruits. J Exp Bot. 2001;52:2097–103.

Zhang DP, Wu ZY, Li XY, Zhao ZX. Purification and identification of 42-kilodalton abscisic acid-specific-binding protein from epidermis of broadbean leaves. Plant Physiol. 2002;128:714–25.

Zhang X, Zhang Q, Xin Q, Yu L, Wang Z, Wu W, Jiang L, Wang G, Tian W, Deng Z, Wang Y, Liu Z, Long J, Gong Z, Chen Z. Complex structures of the abscisic acid receptor PYL3/RCAR13 reveal a unique regulatory mechanism. Structure. 2012;20:780–90.

Zhang XF, Jiang T, Wu Z, Du SY, Yu YT, Jiang SC, Lu K, Feng XJ, Wang XF, Zhang DP. Cochaperonin CPN20 negatively regulates abscisic acid signaling in *Arabidopsis*. Plant Mol Biol. 2013;83:205–18.

Zhang XF, Jiang T, Yu YT, Wu Z, Jiang SC, Lu K, Feng XJ, Liang S, Lu YF, Wang XF, Zhang DP. *Arabidopsis* co-chaperonin CPN20 antagonizes Mg-chelatase H subunit to derepress ABA-responsive WRKY40 transcription repressor. Sci China Life Sci. 2014;57:11–21.

Zhao Y, Chan Z, Xing L, Liu X, Hou Y, Chinnusamy V, Wang P, Duan C, Zhu JK. The unique mode of action of a divergent member of the ABA-receptor protein family in ABA and stress signaling. Cell Res. 2013;23:1380–95.

Zhao Y, Xing L, Wang, X, Hou YJ, Gao J, Wang P, Duan CG, Zhu X, Zhu JK. The ABA receptor PYL8 promotes lateral root growth by enhancing MYB77-dependent transcription of auxin-responsive genes. Sci Signal. 2014; 7:ra53.

Chapter 7
Structural Basis of ABA Perception by PYR/PYL/RCAR Receptors

Lun Jiang, Xingliang Zhang and Zhongzhou Chen

Abstract Abscisic acid (ABA) has been found to play an important role in the adaptive responses to the abiotic and biotic stresses, such as drought and high salinity, as well as the regulation of plant development, such as seed germination. Recent studies have identified several ABA receptors, including chloroplast envelope-localized receptor (the H subunit of chloroplast Mg^{2+}-chelatase), three plasma membrane-localized receptors (G protein-coupled receptor 2 and two GPCR-type G proteins), and cytosol/nucleus-localized pyrabactin resistant-like (PYLs) (or named regulatory component of ABA receptor (RCAR)), among which PYLs are mostly known and confirmed by many groups. In the ABA signaling pathway, PYLs change their conformations upon binding to the ABA molecule, and then directly bind to and inhibit the downstream A-group protein phosphatases 2C (PP2Cs). PYLs/PP2Cs heterodimer blocks substrates binding to PP2Cs, and thus releases the downstream transcription factors sucrose-non-fermentation kinase subfamily 2 (SnRK2s) whose kinase activity is formerly inhibited by PP2Cs through physical interaction and dephosphorylation, finally allows the plants to adapt to the stress conditions. This chapter will focus on the structural basis of ABA perception by PYR/PYL/RCAR receptors, elucidate its molecular mechanisms and provide some structural details for further design of ABA analogues.

Keywords PYR/PYL/RCAR receptors · Crystal structures · Ternary complexes · Chemical regulation

---

L. Jiang · Z. Chen (✉)
State Key Laboratory of Agrobiotechnology, China Agricultural University,
Beijing 100193, China
e-mail: chenzhongzhou@cau.edu.cn

X. Zhang
Department of Pediatrics, The Affiliated Hospital of Guangdong Medical College,
Zhanjiang 524001, Guangdong, China

D.-P. Zhang (ed.), *Abscisic Acid: Metabolism, Transport and Signaling*,
DOI 10.1007/978-94-017-9424-4_7

7.1 The Detection of ABA Receptors

As we all know, a receptor, which is usually a protein molecule found on the surface of a cell or in the cytoplasm/nucleus, receives chemical signals from outside of the cell. The molecule that binds to a receptor is called a ligand, which can be a peptide or other small molecules like hormone. The receptors have several features such as: high affinity with ligand; saturable and reversible binding; high specificity. Abscisic acid (ABA) is a plant hormone that regulates a lot of processes including seed germination, development, and abiotic stress tolerance, especially drought resistance (Kline et al. 2010). By binding to the receptors of the cell, ABA molecule delivers its signal to the downstream messengers, and finally to the responsive genes, thus regulating the expression of many genes. In the ABA signaling pathway, the perception by the ABA receptors is mostly important.

Usually, through traditional genetic approach, many proteins involved in hormone signaling are identified through screening of *Arabidopis* mutants with reduced or enhanced sensitivity to a hormone. Many hormone receptors were cloned through this method (Santner and Estelle 2009). But for ABA, it did not work, because ABA receptors were endowed with functional redundancy or pleiotropic effects including embryo or gamete lethality (Santiago et al. 2012). By using biochemical techniques, ABA-binding proteins CHLH/ABAR/GUN5 were first found by Zhang group (Shen et al. 2006), then G protein-coupled receptors GCR2 (Liu et al. 2007), GTG1/GTG2 (Pandey et al. 2009), were also found to bind ABA by pharmacological evidence. A chemical genetic approach using a synthetic selective ABA agonist, pyrabactin (4-bromo-*N*-pyridin-2-yl methyl naphthalene-1-sulfonamide), was pursued to identify a family of soluble ABA receptors, which is named *PYRABACTIN RESISTANCE 1* (PYR1)-like (PYL) (Park et al. 2009). Mutants were not strongly abscisic acid resistant until the plants lost at least three members of PYLs family. In an independent study, Ma et al. 2009 also identified the PYLs family and referred it as the *regulatory component of ABA receptors* (RCAR) by a yeast-2-hybrid screen. RCAR interacted with PP2Cs in an abscisic acid-dependent manner and inhibited its phosphatase activity.

PYR/PYL/RCAR receptors belong to the START/Bet v I superfamily, which forms a helix-grip fold structural motif containing a seven-stranded β-sheet and three-to-four α-helices. The helix-grip fold creates a large conserved hydrophobic ligand-binding pocket that can bind diverse lipids, hormones, and antibiotics (Klingler et al. 2010). There are 14 PYR/PYL/RCAR family members, named PYR1, PYL1-PYL13 in *Arabidopis*, showing functional redundancy.

7.2 Structural Insights into PYR/PYL/RCAR Receptors

The crystal structures of ABA receptors, PYR1, PYL1, PYL2, PYL3, PYL5, PYL9, PYL10, and PYL13 (Miyazono et al. 2009; Santiago et al. 2009; Yin et al. 2009; Hao et al. 2011; Zhang et al. 2012, 2013 ; Li et al. 2013), had been reported to date. These data confirmed that the PYLs family belongs to a branch of the Bet v I superfamily, which possesses a large and conserved hydrophobic ligand-binding pocket.

From the published apo-PYLs structures, we know that all the PYLs share a highly similar helix-grip structure, in which the α2 helix and the C-terminal α4 helix enclose a seven-stranded β-sheet to create a ligand-binding pocket, which accommodate the solvent (Fig. 7.1). Two highly conserved loops encompass the ligand-binding pockets: the residues SGLPA 'gate' loop (also refers to CL2 loop) between the β3 and β4 strands; the residues HRL 'latch' loop (also refers to CL3 loop) (Melcher et al. 2009; Yin et al. 2009) between the β5 and β6 strands (Fig. 7.1). These two loops are very important for ABA binding.

Usually, the conformation of the receptor molecule changes when binding to ligand. This alters the interaction of the receptor molecules with associated biochemicals and, in turn, leads to a cellular response mediated by the associated biochemical pathway. From the ABA-bound structures of PYR1, PYL1, PYL2, PYL3, PYL9, PYL10 (Miyazono et al. 2009; Santiago et al. 2009; Yin et al. 2009; Hao et al. 2011; Zhang et al. 2012, 2013; Li et al. 2013), the ABA is encaged in the binding pocket by a combination of hydrophobic interactions and water-mediated hydrogen bonds (see Fig. 7.2b, c). First, the acidic head group of ABA faces innermost, forms a conserved water-mediated hydrogen bond network with polar residues of PYLs, defining the location of the pentadienoic acid moiety and the hydroxyl group of ABA in the binding pocket, and forming charge interactions with a conserved lysine (such as K59 in PYR1, K86 in PYL1, K79 in PYL3) (see Fig. 7.3). These polar interactions help to locate the ABA molecule. However, this lysine residue is replaced by a glutamine in PYL13, consistent with the lack of evidence for binding ABA. In the recently published structure of PYL13 (Li et al. 2013), we know that in addition to the lost lysine residue, there would be a steric clash between the aromatic ring of Phe71 residue in PYL13 and the hydrophobic moiety of ABA. An invariant residue leucine is located in the corresponding site

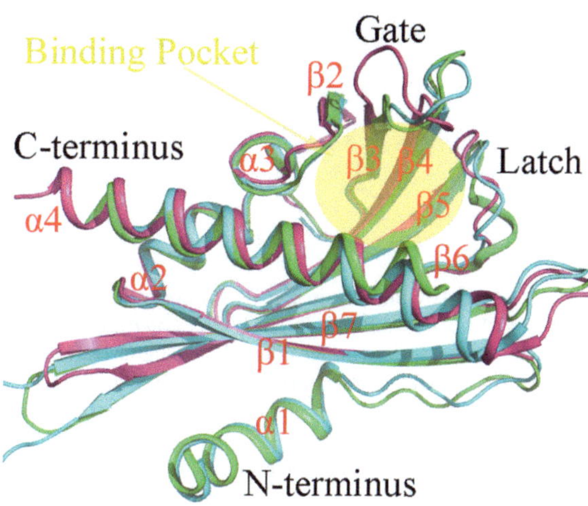

Fig. 7.1 Structure comparison of PYR1, PYL3 and PYL10. PYR1 (pdb ID: 3K90), PYL3 (pdb ID: 3KLX) and PYL10 (pdb ID: 3UQH) are shown in *green*, *cyan* and *magenta*, respectively. Binding pocket is shown in *yellow ellipse*

in other PYLs abrogate completely this steric clash (Li et al. 2013). Second, the isoprene moiety and the cyclohexene ring (see Fig. 7.2d) form a lot of hydrophobic interactions and water-mediated hydrogen bonds with the side chains in PYLs. Among these, the interactions between ABA and residues in the gate or latch loops are most important in stabilizing its closure conformation.

Through the superposition of apo-PYLs and (+)-ABA-bound PYLs structures, it reveals notable conformation changes in the gate-and-latch loops. In apo-form, the proline residue in the gate (such as P88 in PYR1, P115 in PYL1, P92 in PYL2, P112 in PYL3) is flipped out of the binding pocket. Upon binding (+)-ABA, the proline moves toward the ligand to cover the cavity. In addition, the imidazole ring of histidine on the latch (such as H115 in PYR1, H142 in PYL1, H119 in PYL2, H139 in PYL3) also turns into the binding pocket to form van der Waals contacts with the cyclohexene moiety of ABA (see Fig. 7.2a). Besides, the front part of the

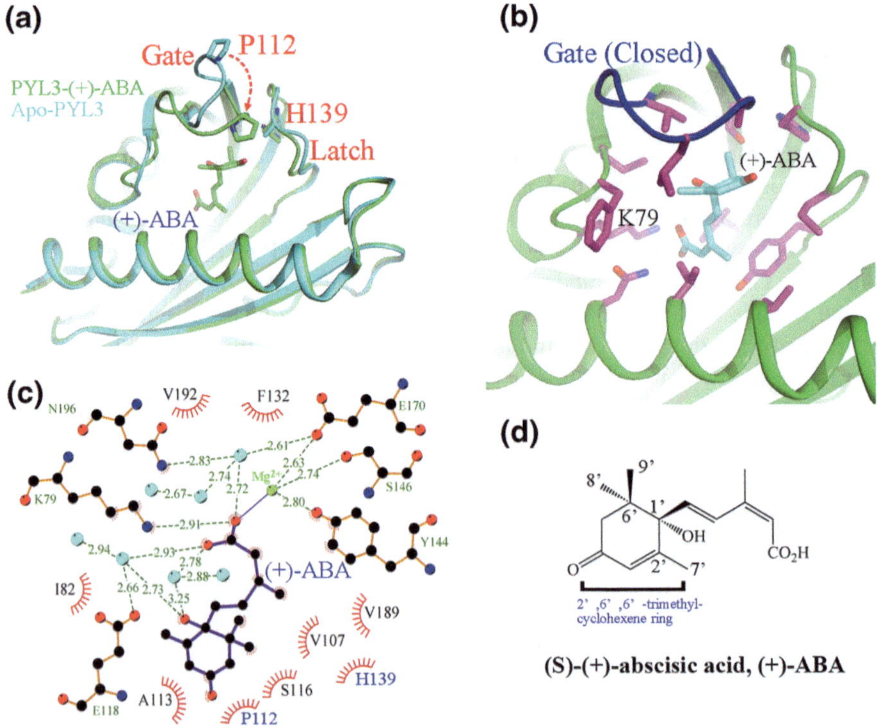

Fig. 7.2 (+)-ABA binding in the PYL3 pocket. **a** Superposition of apo-PYL3 (pdb ID:3KLX, *cyan*) and PYL3-(+)-ABA (pdb ID:4DSC, *green*). Pro112 in gate loop is flipped out of the binding pocket in the apo-form. Residues Pro112 and His139 in latch loop turn into the cavity after binding ABA. **b** ABA (*cyan*) buried in the PYL3 cavity, Lys79 is located at the *bottom* of the cavity, establishes an electrostatic interaction between its amine group and the carboxylate group of the ABA. **c** 2D map of interactions between (+)-ABA and PYL3. Hemispheres represent hydrophobic interactions whereas polar interactions are represented by *lines*. *Cyan spheres* indicate water molecules. Mg^{2+} indicates magnesium ions. **d** The chemical structure of (+)-ABA

C-terminal α-helix is also involved in the gate closing. In PYR1 and PYL10, the front parts of the C-terminal α-helix alter conformation after binding (+)-ABA (see Fig. 7.8a). These conformational alterations facilitate the latch loop lock the closed gate loop by hydrogen bonds and van der Waals contacts, refers to open-to-closed gating mechanism. Altogether, the gate-and-latch loops in apo-PYLs adopt an open conformation that allows the entry of ABA into the binding pocket. Upon binding ABA, the gate-and-latch loops fold over ABA to complete ABA enclosure by conformation changes.

Interestingly, the published PYR1, PYL1, PYL2, and PYL3 structures were homodimers, whereas PYL5, PYL9, and PYL10 were monomers. From gel filtration chromatography, we know that PYR1, PYL1–3 are dimers in solution, PYL4–10 are monomers in solution (Hao et al. 2011). The structural changes by ABA perception mentioned above only consider one protomer of PYLs. Thus we should consider the oligomeric state of PYLs. For homodimeric PYLs, the above mechanisms also can be applied. For PYR1, PYL1, PYL2, except PYL3, the apo-PYLs and ABA-bound PYLs form cis-homodimers by the gate loop and the C-terminal α-helix. Upon binding ABA, the gate loop and the C-terminal α-helix change their conformation just like those in the monomeric PYLs, which alter the relative orientation of the two promoters and weaken the dimeric assembly to form complex with ABA. Similar dimerization of PYL3 (cis-homodimer) has also been found in the apo-PYL3 structure (see Fig. 7.4a), but ABA-bound PYL3 transforms into the trans-homodimer by a protomer rotation of almost 135° compared with the cis-homodimer (see Fig. 7.4b). The gate-and-latch loops in the trans-homodimer are more exposed into the solvent (Zhang et al. 2012) (see Fig. 7.4b). Therefore, based on the solved structures of PYLs-(+)-ABA, the PYLs could be divided into three major subclasses, cis-homodimer for PYR1, PYL1, and PYL2, trans-homodimer for PYL3, monomer for PYL4–PYL13 (Zhang et al. 2012).

The above-mentioned ABA is a naturally S-(+)-ABA form (also referred to as (+)-ABA), with one asymmetric carbon atom at C-1' position. However, the synthetic enantiomer (-)-ABA (also referred to as R-(-)-ABA) has been extensively investigated since the discovery of ABA (Cutler et al. 2010) (see Fig. 7.5a). Previous studies showed that in some assays, the (-)-ABA has same or ever higher activity compared with (+)-ABA (Milborrow 1974). Two hypothesized mechanisms were mentioned for explaining the bioactivity of (-)-ABA; one is the same occupied site between these two ABA molecules by flipping the cyclohexene plane (Milborrow 1974), the other is the dual selectivity of ABA receptors (Nambara et al. 2002). From the reported assay (Zhang et al. 2013), all PYLs inhibited the phosphatase activity of PP2Cs in the presence of (+)-ABA except PYL7, PYL11, and PYL12 that denied soluble expression in *Escherichia coli*. However, the situation for (-)-ABA in all PYLs was different. The inhibition of HAB1 by monomeric PYLs was stronger than that by cis-dimeric PYLs, except monomeric PYL10. Interestingly, trans-dimeric PYL3 showed a powerful inhibition of PP2Cs like monomeric PYLs. PYL5 bound (-)-ABA very strong, while PYL9 did not inhibit PP2Cs activity in the presence of (-)-ABA. Due to the highly sequence conservation, further structural investigations of apo-PYL5, PYL3-(-)-ABA and

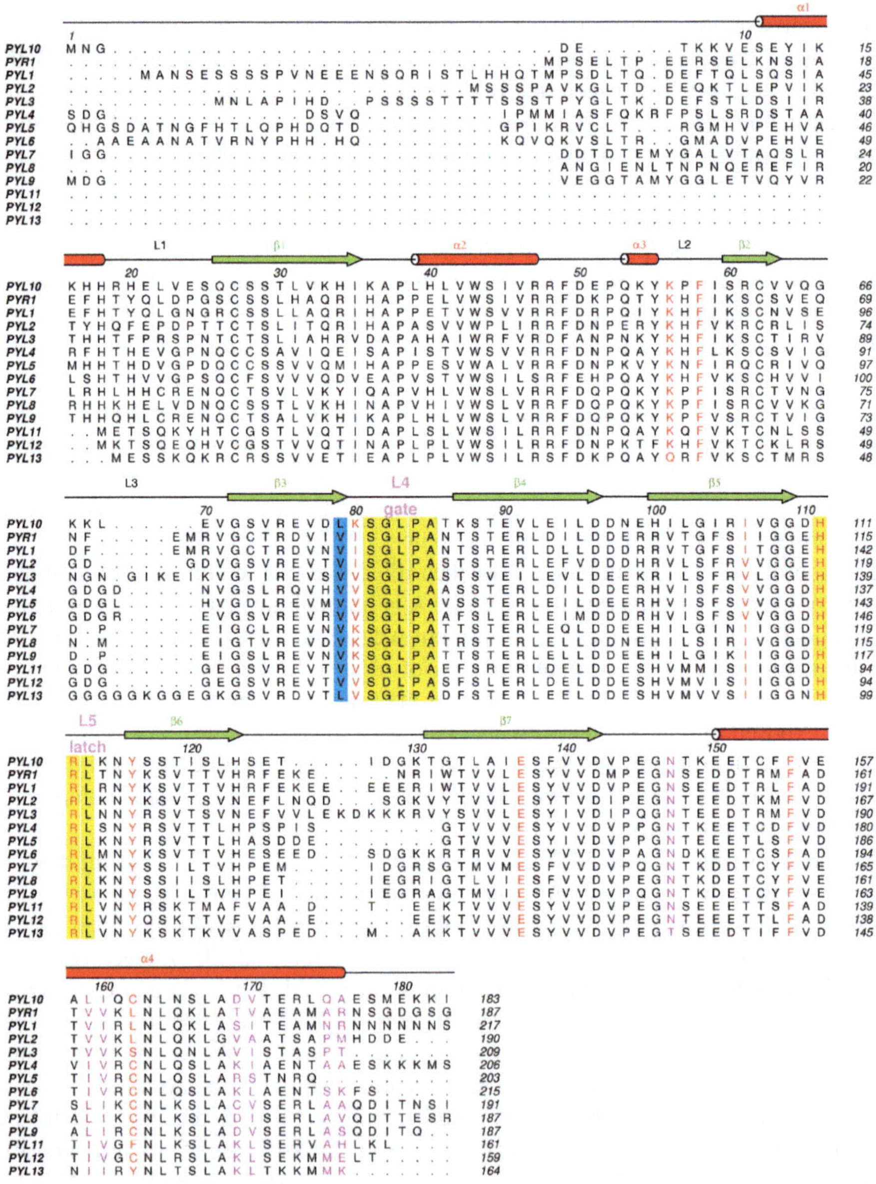

Fig. 7.3 Secondary structure alignment of PYL10 with other PYL/RCAR family members. The secondary structures of PYL10 are shown above the sequence. The structure composed of four helices and seven β strands. The PYLs sequence number is shown in the *right*. This figure was made by the program ALSCRIPT (Barton 1993). The nomenclature of gate (L4 loop) and latch (L5 loop) according to the publication is colored *purple* (Melcher et al. 2009). The gate-and-latch loops residues are colored in *yellow*, the critical residues involved in ABA binding and ABI1 interactions are colored *red* and *purple*. The residues in PYLs affect constitutive binding and inhibition of PP2Cs is colored in *blue*

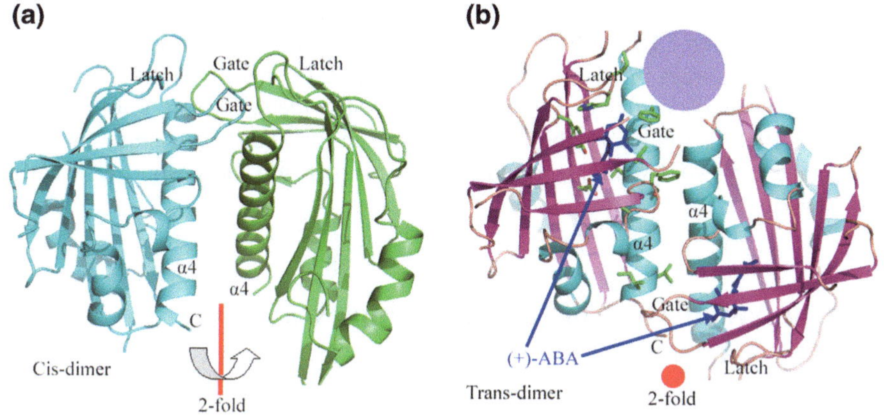

Fig. 7.4 Formation of the PYL3 trans-homodimer with ligands. **a** Apo-PYL3 is a cis-homodimer, two protomers are related by a twofold rotation axis, parallel to the plane of the page. **b** The PYL3-(+)-ABA complex structure is a trans-homodimer. The (+)-ABA (*color blue*) binding pocket in ligand-bound PYL3 is exposed to the solvent and cycled (*purple*). Two protomers are related by a twofold rotation axis, perpendicular to the plane of the page (Zhang et al. 2012)

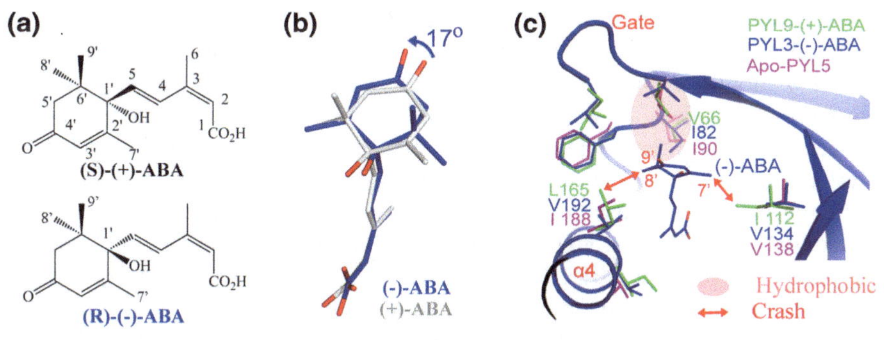

Fig. 7.5 Structure alignment of (-)-ABA with (+)-ABA. **a** Chemical structure of (+)-ABA and (-)-ABA. **b** Superposition of (+)-ABA and (-)-ABA from their complex structures in PYL3. There were partially rotation and shift between the rings in both ABA. **c** Superposition of apo-PYL5, PYL3-(-)-ABA and PYL9-(+)-ABA indicated that the major variant residues underlain the favor of PYL binding with (-)-ABA. Two bulk side chains of I112 and L165 in PYL9 seriously collided to 7′ and 8′ methyl groups of (-)-ABA, respectively. The stereo constraints were vanished in PYL5 because of two corresponding small side chains. On the other hand, V66I in PYL9 would give a strong coordination with 8′ and 9′ methyl groups in (-)-ABA through a strong hydrophobic network (Zhang et al. 2013)

PYL9-(+)-ABA were determined to uncover the reason (Zhang et al. 2013). By superimposing these three structures, we found that the binding orientation and pocket of (-)-ABA in PYLs are obviously different from those of (+)-ABA (see Fig. 7.5b), which might deny the "flip" hypothesis. Structural and biochemical investigations showed the major variable residues surrounding the mono-methyl and di-methyl groups of ABA cyclohexene ring might underlay the preference of

PYL binding to ABA enantiomers, such as steric hindrance by the two bulk side chains of I112 and L165 in PYL9 (see Fig. 7.5c). Moreover, hydrophobic interaction through indirect interaction with 8′, 9′ methyl groups of (-)-ABA also contributes the stereospecificity of PYLs to ABA enantiomers, because V66I mutation increases PYL9's inhibitory ability on PP2Cs (see Fig. 7.5c). This study on PYLs preference to (-)-ABA might provide novel insights into the design of selective ABA agonists.

7.3 The Architecture of Ternary Complexes PYLs–ABA–PP2Cs

As we all know, the important role of PYLs in ABA signaling pathway is to inhibit the phosphatase activity of PP2Cs in an ABA-dependent manner. The crystal structures of four ternary complexes have been reported, PYL1–ABA–ABI1, PYL2–ABA–HAB1, PYR1–ABA–HAB1, and PYL3–ABA–HAB1 (Melcher et al. 2009; Yin et al. 2009; Dupeux et al. 2011; Zhang et al. 2012). Moreover, the structures of PYL10–HAB1 and PYL13–PP2CA in the absence of ABA have also been solved recently (Hao et al. 2011; Li et al. 2013).

In these structures, the PP2Cs catalytic cores (residues: ABI1 125-429, HAB1 172-511) adopt a folding with two central five-stranded β-sheets sandwiched by two pairs of α-helices, and the catalytic site located at the edge of the two central β-sheets (see Fig. 7.6a). The antiparallel β-sheets of PP2Cs provide the binding interface with PYLs, so the activity of PP2Cs is blocked by the closed gate loop of PYLs when binding to each other. Other than the blocked active sites, the PP2Cs contact with the PYLs through a small protruding region containing an important Trp residue (Trp300 in ABI1, Trp385 in HAB1). This Trp residue points into the ABA-binding pocket through the gate-and-latch loops covering ABA molecule, and forms a water-mediated hydrogen bond to the ketone group of ABA. Upon the insertion of the Trp in PP2Cs into the ABA-binding pocket, the gate loop and latch loop undergo conformational changes to close the pocket, generate additional contacts between PYLs and ABA molecule. Thus, the Trp residue acts as a lock to keep the gate-and-latch loops in a more closed conformation, which is referred to gate–latch–lock mechanism. Apart from these interactions, a conserved arginine residue of PP2Cs stacks its guanidinium group with a conserved proline residue on the gate loop (Arg389 in HAB1, Pro112 in PYL3). Moreover, the gate loop of PYLs directly packs against the PP2Cs active site with a conserved serine residue, which exposed upon ABA binding and formed two hydrogen bonds with the conserved residues of PP2Cs (metal-stabilizing residue Glu203 in HAB1), and with the conserved glycine residue on the active site loop of PP2Cs (Gly180 in ABI1, Gly246 in HAB1) (see Fig. 7.6b). Thus, the phosphatase active center of PP2Cs is competitively blocked by PYLs bound. Overall, PP2Cs use their active sites to interact with the gate-and-latch loops of PYLs and induce some conformational changes; these PP2Cs-induced conformational changes make a higher binding

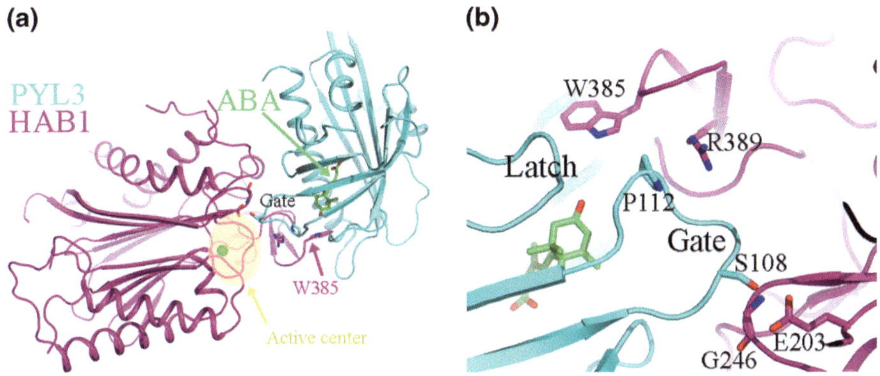

Fig. 7.6 PYL3–ABA–HAB1 complex structure. **a** Overall structure of PYL3–ABA–HAB1 complex (pdb ID: 4DSC). Active center of PP2Cs is shown in *pale yellow*. **b** The interface between PYL3 and HAB1 in the PYL3–ABA–HAB1 complex. S108 of PYL3 contacts with G246 and E203 of HAB1. P112 of PYL3 stacks with R389 of HAB1. W385 of HAB1 inserts the binding pocket. ABA, *green*; PYL3, *cyan*; HAB1, *magenta*

affinity between PYLs and ABA. And finally, the activity of PP2Cs is inhibited by binding to PYLs.

As mentioned above, PYR1, PYL1, PYL2, and PYL3 are homodimers in their apo states, but the binding ratio between PYLs and PP2Cs is 1:1 in the PYLs–PP2Cs complex structure. Even interestingly, the homodimer interface encompasses the region that plays an important role in interacting with PP2Cs. So, how can the dimeric PYLs bind to the PP2Cs? From the analysis of the structures of dimeric PYR1, PYL1, PYL2, PYL3, the dimeric interfaces mainly appeared on the gate loop and the C-terminal α-helix (see Fig. 7.7a, b), showing that fixation of the gate loop in the dimeric interface prevents the binding of PP2Cs to the gate loop. Therefore, before the receptor binds to the downstream proteins (PP2Cs), the dimeric PYLs need to be dissociated. Receptor dimerization may therefore be a mechanism to reduce basal receptors–downstream proteins interaction (PYLs–PP2Cs), by blocking downstream activity. Significantly, the binding of ABA to dimeric PYLs does not disrupt the dimerization. Once ABA binds to dimeric PYLs, the gate loop of PYLs changes its conformation to close the gate, and the relative distance between these two PYLs protomers becomes larger. These two changes rearrange the interface residues, bring about the decrease of van der Waals contact and hydrogen bonds at the dimeric interface (see Figs. 7.3b and 7.7b). Hence, the binding of ABA induces the closure of the gate loop, generates a platform to interact with PP2Cs, further weakens the dimerization interaction, and finally facilitates the dissociation of dimeric PYLs to form a 1:1 PYLs–PP2Cs complex.

In the above-mentioned ABA signaling pathway, ABA first binds to PYLs, and then the PYLs–ABA complex binds to the downstream PP2Cs; this process is called the induced pathway. However, some PYLs do not need ABA to inhibit the phosphatase activity of PP2Cs; this process is called the constitutive pathway.

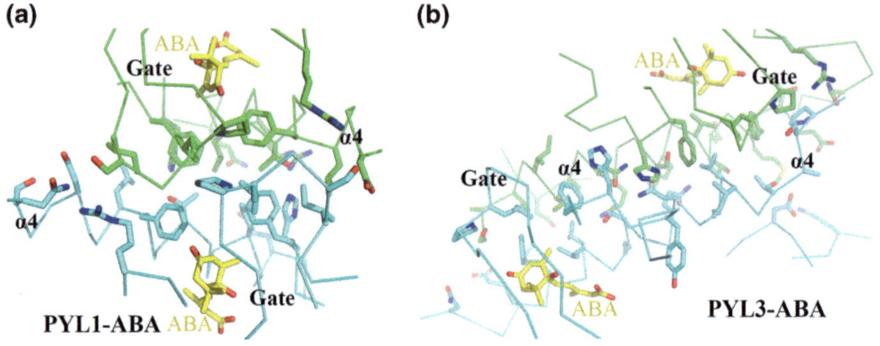

Fig. 7.7 Interaction of two PYLs molecules in PYLs-(+)-ABA structure (*cyan* and *green*). **a** PYL1-(+)-ABA (pdb ID: 3JRS) and **b** PYL3-(+)-ABA (pdb ID: 4DSC). ABA, α4 and gate are shown. Nitrogen: *blue*, oxygen: *red*

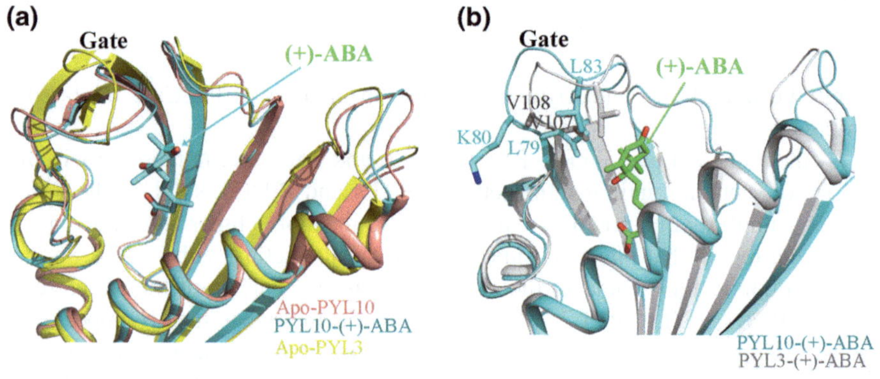

Fig. 7.8 **a** Structural superimposition of the apo-PYL10 with PYL10-(+)-ABA and apo- PYL3 reveals a partially close conformation of the gate loop in the ligand-free PYL10. Apo-PYL10 (PDB: 3RT2) is in *violet*, apo-PYL3 (PDB: 3KLX) in *yellow*, PYL10-(+)-ABA (PDB: 3R6P) and ABA in *cyan*. **b** Structural comparison of ABA-bound PYL10 (*cyan*) and PYL3-(+)-ABA (PDB: 4DSB, *gray*). The ABA molecule in PYL10 is shown as *green sticks*. Leu79 and Lys80 are at the entrance of the ligand-binding pocket, affecting the ABA-independent inhibition of PP2Cs. Both the corresponding residues in PYL3 are valine

The recent research had examined the effect of all PYLs (except for PYL7, 11, 12, 13) on inhibiting the phosphatase activity of four PP2Cs, including ABI1, HAB1, HAB2, and PP2CA, in the presence or absence of ABA (Hao et al. 2011). They found that a subclass of PYLs proteins, such as PYL10, exhibited constitutive binding and inhibition of PP2Cs even in the absence of any ligands. While other PYLs relied on ABA to inhibit PP2Cs. By comparing the structures between apo-PYL10 and apo-PYL3, we know that the gate loop of the apo-PYL10 is partially in the closed conformation (see Fig. 7.8a). Thus, ligand-free PYL10 is able to adopt a compatible conformation for PP2Cs recognition. By comparing the structure of apo-PYL10 with other dimeric PYLs (see Fig. 7.8b), we found that the bulkier hydrophobic residue (Leu79) at the entrance of the ligand-binding pocket in PYL10 provided a platform to dock the pocket-facing, hydrophobic Leu83 of gate loop, thus facilitating a closed conformation of the gate loop. On the contrary, the corresponding residue in other PYLs except PYL13 is valine (see Fig. 7.3), which is smaller than leucine. Moreover, a lysine in PYL10 (Lys80), which prevents PYL10 dimerization in solution (the charged residue is unfavorable in the hydrophobic interface), is replaced by isoleucine or valine residues of dimeric PYLs (see Fig. 7.8b). Therefore, PYL10 exists as a monomer.

Taken together, in order to achieve ABA-independent inhibition of PP2Cs, PYLs should be a monomer first. Then, the residues guarding the ligand-binding pocket of PYLs must be bulky and hydrophobic so that they can make sufficient hydrophobic contacts to stabilize a closed gate loop in the absence of ABA. If the biochemical features of PYLs pockets cannot match these requirements, an amphipathic ligand, such as ABA or pyrabactin, is required to close the gate loop to the binding pocket. Recent studies found that PYL13 selectively inhibited PP2CA independent of ABA, and PYL13 antagonized PYL10 in the ABA-independent inhibition of PP2Cs because PYL13 and PYL10 were the only two PYLs containing a Leu residue instead of Val residue to facilitate the closure of the gate loop (Li et al. 2013).

Although some monomeric PYLs can bind to PP2Cs in the absence of ABA, their inhibitory is greatly reduced compared with the inhibitory in the presence of ABA. This can be structurally explained by the fact that ABA enhances the interaction between PYLs and PP2Cs by a lot of hydrophobic interactions and water-mediated hydrogen bonds.

The structural research have revealed how the binding of ABA to PYLs leads to the inhibition of PP2Cs, but how does this event then convey to other outputs? An important clue came from the finding of Snf1 (Sucrose-non-fermentation 1)-related kinases subfamily 2 (SnRK2s) as a positive regulator of ABA signaling (Park et al. 2009). The SnRK2s subfamily contains 10 members in *Arabidopsis*. In normal growth conditions where cellular ABA levels are low, PP2Cs bind and dephosphorylate SnRK2s to keep them in inactive state. When stresses come, elevated cellular ABA binds to PYLs, which in turn bind and deactivate the phosphatase activity of PP2Cs, and then release the SnRK2s (Boudsocq 2004; Belin et al. 2006; Vlad et al. 2009). The activated SnRK2s could be self-activated via auto-phosphorylation, then phosphorylate different downstream targets, such as the b-ZIP transcription factors and ion channels SLAC1 (slow anion channel 1), KAT (inward rectifying K^+ channel) (Geiger et al. 2009; Lee et al. 2009; Sato et al. 2009), and so on.

SnRK2s contain a Snf1-related kinase domain and a highly acidic C-terminal segment termed ABA box, which is important for their interactions with PP2Cs (Boudsocq et al. 2006). The solved SnRK2s–PP2Cs complex structure reveals marked similarity in PP2Cs recognition by SnRK2s and ABA receptors (Soon et al. 2012). In addition to the highly acidic C-terminal region, SnRK2.6 contributes three separate regions within the kinase domain for binding to HAB1: first, the activation loop inserts deeply into the catalytic cleft of HAB1 and mimics the

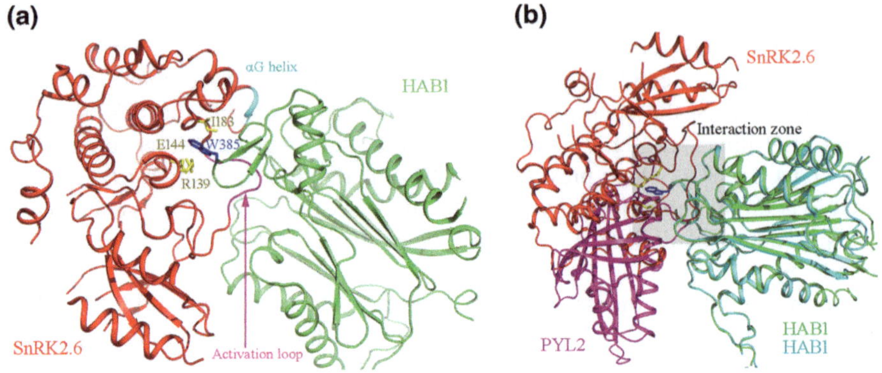

Fig. 7.9 Structures of the SnRK2.6-HAB1 complex (pdb ID: 3UIG). **a** A structure overview. SnRK2.6 is shown in *red*, HAB1 is colored *green*. The three separate regions within the kinase domain of SnRK2.6 for binding to HAB1 are shown, the activation loop is colored *magenta* and the αG helix is colored *cyan*. Besides, residues Arg139, Ile183, and Glu144 (*yellow*) of SnRK2.6 make an interaction with the conserved ABA-sensing tryptophan (Trp385 in HAB1, *blue*). **b** Overlay of the interaction surfaces in the SnRK2.6-HAB1 and PYL2-ABA-HAB1 complexes

gate loop of PYLs; second, the region near residues Arg139, Ile183, and Glu144 of SnRK2.6 mimics the cleft formed by the gate-and-latch loops of PYLs and the conserved tryptophan (such as Trp385 in HAB1) from PP2Cs inserts into this region; third, the SnRK2.6 αG helix binds to the region near the PYLs interaction site in HAB1 (see Fig. 7.9a). The SnRK2s–PP2Cs interface largely overlaps with the PYLs–PP2Cs binding interface; the HAB1 structure in the SnRK2.6–HAB1 complex is nearly identical to that in the PYLs–HAB1 complex. Thus, both SnRK2.6 and PYLs use a similar gate-and-lock mechanism to recognize PP2Cs (see Fig. 7.9b) (Soon et al. 2012).

In *Arabidopsis*, there are 14 members of PYLs, nine members of PP2Cs, and 10 members of SnRK2s, differential binding between PP2Cs and PYLs, PP2Cs and SnRK2s is important for regulating the physiological responses of *Arabidopsis* to the stress conditions. And the common recognition mechanism of gate–latch–lock likely ensures the signaling pathway by preventing the inappropriately regulation.

7.4 Chemical Regulation of PYLs

ABA agonists may be able to improve the yield of crop plants under different stresses as well as to manipulate ABA response in different tissues or in developmental stages. Pyrabactin was first identified as a synthetic and receptor-selective ABA agonist, which led to the identification of PYR1 (Park et al. 2009). Through years' research, the complex structures of PYL1, PYL2, PYL3, and PYR1 with pyrabactin were reported (Hao et al. 2010; Melcher et al. 2010; Yuan et al. 2010;

Zhang et al. 2012). Pyrabactin binds to PYL1 and PYR1 in productive mode as an agonist, whereas to PYL2 or PYL3 in non-productive mode as an antagonist, revealing that pyrabactin can adopt different conformations within the highly conserved ligand-binding pocket of PYLs (Zhang et al. 2012).

Structural analyses have revealed that pyrabactin occupies the similar binding pocket to ABA, although their chemical structures are different. From the superimposition of the structures of PYLs–ABA and PYLs–pyrabactin, several key interactions are conserved between them. The pyridyl nitrogen or the sulfonamide functional group of pyrabactin occupies the position of carboxylate oxygen of ABA, forming hydrogen bonds with Lys86 and Glu121 in PYL1 (Lys64 and Glu98 in PYL2). The amine group of the sulfonamide moiety of pyrabactin occupies the position of the hydroxyl group of ABA, forming a hydrogen bond with Asn197 in PYL1 (Asn173 in PYL2). The naphthalene ring of pyrabactin mimics the 2,6,6-trimethyl-cyclohexene ring of ABA, forming hydrophobic interactions with the gate loop of PYLs and promoting the closure of gate (see Fig. 7.10a).

Moreover, there are also some obvious differences between the solved four PYLs–pyrabactin complexes. Pyrabactin can induce the closure of the gate in PYR1 and PYL1, PYL3 as an agonist or antagonist, but not in PYL2, although it is bound in the same binding pocket. Compared with those in PYL1 or PYR1, the naphthalene and pyridine rings of pyrabactin bound in PYL2 rotate almost

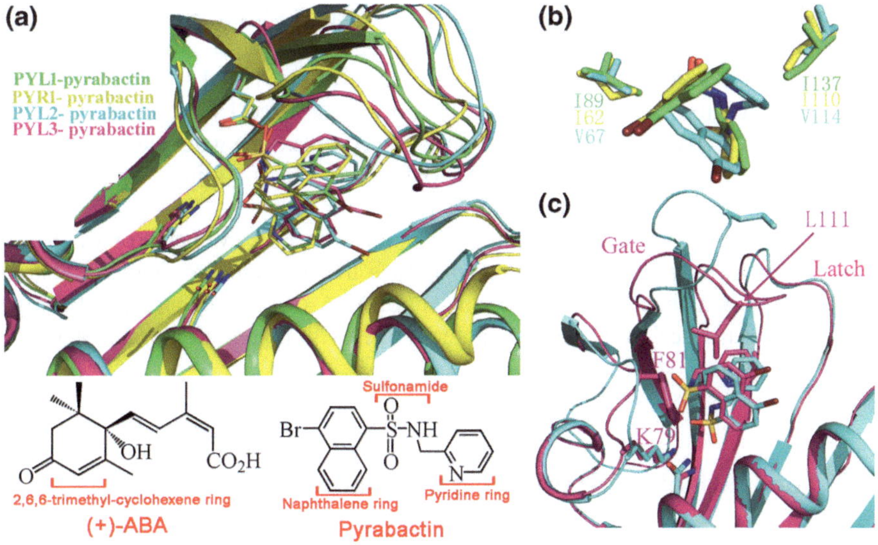

Fig. 7.10 Structural basis for pyrabactin selectivity. **a** Structure superimposition of structures PYL1-pyrabactin (3NEF, *green*), PYR1-pyrabactin (3NJO, *yellow*), PYL2-pyrabactin (3NR4, *cyan*) and PYL3-pyrabactin (3OJI, *magenta*). The lower panel shows the chemical structures of ABA and pyrabactin. **b** The molecular determinant affects the agonism and antagonism of pyrabactin toward PYR1, PYL1 and PYL2. **c** Structure superimposition of PYL2-pyrabactin and PYL3-pyrabactin. The residues affect the gate loop closure are shown in *sticks*

90º, made that the napthalene ring be located too far from the gate loop of PYL2 to close it. Two hydrophobic residues Val114 and Val67 in the binding pocket of PYL2 (corresponding to Ile110 and Ile62 in PYR1, Ile137 and Ile89 in PYL1) that arrange the orientation of the pyrabactin are smaller than those in PYL1 or PYR1. And the position of larger residue Ile137 in PYL1 would collide with the naphthalene ring of pyrabactin in the PYL2 conformation. So, these two smaller valine residues make the naphthalene ring locate deeply in the binding pocket and far from the gate loop. And thus the gate is unable to be closed due to the loss of its interaction with pyrabactin (see Fig. 7.10b).

However, the situation in PYL3 is totally different, the orientations of the naphthalene and pyridine ring of pyrabactin are rotated by 80° compared with those in PYL1–pyrabactin complex. Moreover, the sulfonamide group moves to Phe81 and does not form a hydrogen bond with Lys79, while the conserved Lys residue in other PYLs forms a hydrogen bond with the pyridyl nitrogen to locate pyrabactin inside to the binding pocket. Pyrabactin can induce the closure of the gate in PYL3, but the gate loop moves further toward the binding pocket and tightly closes the latch. On the other hand, Leu111 in PYL3 binds to pyrabactin tighter, which also gives rise to the approach between the gate loop and the binding pocket compared with that in PYL1, PYL2, or PYR1. So, there is no enough space between the gate loop and the latch loop for the insertion of conserved Trp residue from PP2Cs. Taken together, pyrabactin works as an antagonist for PYL3 (see Fig. 7.10c) (Zhang et al. 2012).

From the above analysis, pyrabactin is a receptor-selective ABA agonist and antagonist. By solving the structure of the PYL1–pyrabactin–ABI1 complex, we know that this complex resembles the structure of the PYL1–ABA–ABI1 complex, in which the closure of the gate is further stabilized by the insertion of the locking residue Trp300 of ABI1. In the ternary complex, the gate–latch interface is tightly packed against the active site of ABI1, therefore providing a mechanism of phosphatase inhibition (Melcher et al. 2010).

These different pyrabactin binding orientations in PYLs either provide or lack Van der Waals interactions required to induce the gate closure, and these structural information shed lights on the concept of ABA receptor agonism and antagonism, lead to the design of pyrabactin derivatives with higher affinity and agonistic activity.

7.5 The Outlook of Synthetic ABA Analog

By comparing pyrabactin and (+)-ABA in their structures complexed with PYLs, we know that these two ligands overlap with each other despite the distinct chemical structures. First, both of pyrabactin and ABA ligands are amphipathic, each consisting of polar groups buried deep inside the pocket and hydrophobic groups adjacent to the gate-and-latch loops. Second, the pyridyl nitrogen and the sulfonamide group of pyrabactin mimic the carboxylate and the hydroxyl groups of ABA, respectively, facilitating the ligand locate inside to the binding pocket through water-mediated hydrogen bonds. Third, the bromo-naphthalene ring resembles the 2,6,6-trimethyl-cyclohexene ring of ABA, which makes hydrophobic interaction with the hydrophobic residues of the gate loop to induce a closed conformation (Hao et al. 2010) (see Fig. 7.10a). Therefore, changing the pyridyl ring or the bromo-naphthalene ring is the general principle in designing function ABA analog.

As we all know, the salt bridge between the carboxylate of ABA and the amine group of the conserved Lys in PYLs (except for PYL13) is essential for ABA binding. Due to the lack of this salt bridge, lower efficacy of the PYL1–pyrabactin-mediated PP2Cs inhibition compared with that of ABA was found. Thus, modification of some function groups on pyrabactin may affect the binding efficacies of the ligands (Hao et al. 2010). First, by altering the nitrogen position on the pyridyl ring or adding some groups to the pyridyl ring, it would not form hydrogen bonds with the polar residues or other residues in PYLs, so the synthetic ABA analog could not locate into the binding pocket, thus altering the binding between PYLs and PP2Cs. Second, modifying some chemical groups in the bromo-naphthalene ring, it would affect the binding of the gate loop to the ligand; thus, the gate closes properly or not, so further influencing the binding with downstream PP2Cs.

All these mentioned above are established on the structural analysis. In order to build up the mechanism of synthetic ABA analog, we need more knowledge of synthetic chemistry to guide.

7.6 Other Receptors, Other Hormones

Previous reports have shown that ABA perception happens at different sites in cells, such as plasma membrane and cytoplasm; thus, ABA receptors need to locate in different sites (Hamilton et al. 2000; Levchenko et al. 2005). So, there may exist other ABA receptors except for PYLs.

A report identified a G protein-coupled receptor (GPCR) homologue, GCR2, with a K_d (dissociation constant) value of 2.1 nM to bind ABA. Binding of ABA to GCR2 results in release of the G protein and dissociation of the Gαβγ heterotrimeric complex to activate downstream ABA effectors and to trigger the ABA responses (Liu et al. 2007). However, GCR2 has been controversial with respect to the reproducibility of ABA-related phenotypes (Gao et al. 2007). No published reports from other groups to date have independently reproduced the experimental results. On the other hand, GCR2 lacks the typical seven transmembrane domain of GPCRs by bioinformatics analysis.

Second, GPCR-type G proteins GTG1 and GTG2 have nine predicted transmembrane domains, are localized at the plasma membrane. GTG1 and GTG2 both have specific GTP-binding ability and GTPase activities. They can both bind specifically to the natural S-ABA stereoisomer in a saturable manner with K_d values of approximately 40 nM, similar to the value reported for GCR2 (Pandey et al. 2009). However, they showed an unusual type of G protein signaling, in which the GTP-bound Gα protein 'turns off' signaling while the GDP-bound form allows ABA binding to the GTG receptor and the initiation of a signaling cascade, different from the conventional model for signaling by G proteins. However, the above experiments were unable to show evidence for at least one binding site per GTG protein molecule (Klingler et al. 2010).

Another possible ABA receptor protein in *Arabidopsis* is CHLH (Mg-chelatase H subunit), a chloroplast protein involved in chlorophyll biosynthesis. With the K_d value of 30–40 nM, a lot of overexpression, RNAi, insertion, and point mutations experiments have shown the affinity between CHLH and ABA (Shen et al. 2006).

Compared with those above ABA receptors, PYLs are widely studied with crystallographic studies, site-specific mutations, and so on. Also, ABA-PYLs signaling pathway are investigated in detail. Studies of PYLs give more information for understanding of ABA signaling and give more means to identify other ABA receptors.

Structural studies of other phytohormone receptors have been identified in recent years. Auxin binds to TIR1 and mediates contacts between TIR1 and the transcriptional repressor Aux/IAA, but auxin binding does not induce significant conformational changes in the receptor, so auxin is called a molecule glue (Tan et al. 2007). However, gibberellin induces conformational changes in GID1, providing the binding interface for the transcriptional repressor DELLA, functioned as allosteric effector (Murase et al. 2008; Shimada et al. 2008). Similar to gibberellin, ABA induces conformational changes in PYLs by binding to the pocket, creating a binding interface for PP2Cs. So, ABA regulates PYLs as an allosteric effector. Although the common action of phytohormone toward receptors is to introduce interaction to the downstream proteins, the mechanisms of the action are quite different for each receptors (Umezawa et al. 2010).

These structural investigations have led to major progress in understanding hormone signaling; the mechanistic basis of them provides a rational framework for future design of alternative ligands.

This chapter gives a detailed structural analysis of ABA perception by PYR/PYL/RCAR receptors and the PYLs–ABA–PP2Cs ternary complex structures, explains the gate–latch–lock mode, and elucidates its molecular mechanisms. However, many questions are still unknown, why the ABA signaling pathway in *Arabidopsis* involves so many different PYLs receptors. The structural mechanism of the individual receptors differs in their sensitivity and selectivity in response to ABA, how can we develop new selective ABA agonists and antagonists to use in agriculture? Finally, this is just a beginning to understand the ABA signaling pathway. Much work needs to be done to gain a global mechanistic synthesis of the complete signaling network.

Acknowledgments The work in author's laboratory was supported by National Basic Research Program of China (973 Program, 2011CB965304), and National Natural Science Foundation of China (31370720 and 31222032).

References

Barton GJ. ALSCRIPT: a tool to format multiple sequence alignments. Protein Eng. 1993;6(1):37–40.

Belin C, de Franco PO, Bourbousse C, Chaignepain S, Schmitter JM, Vavasseur A, Giraudat J, Barbier-Brygoo H, Thomine S. Identification of features regulating OST1 kinase activity and OST1 function in guard cells. Plant Physiol. 2006;141(4):1316–27.

Boudsocq M. Identification of nine sucrose nonfermenting 1-related protein kinases 2 activated by hyperosmotic and saline stresses in *Arabidopsis thaliana*. J Biol Chem. 2004;279(40):41758–66.

Boudsocq M, Droillard M-J, Barbier-Brygoo H, Laurière C. Different phosphorylation mechanisms are involved in the activation of sucrose non-fermenting 1 related protein kinases 2 by osmotic stresses and abscisic acid. Plant Mol Biol. 2006;63(4):491–503.

Cutler SR, Rodriguez PL, Finkelstein RR, Abrams SR. Abscisic acid: emergence of a core signaling network. Annu Rev Plant Biol. 2010;61:651–79.

Dupeux F, Antoni R, Betz K, Santiago J, Gonzalez-Guzman M, Rodriguez L, Rubio S, Park SY, Cutler SR, Rodriguez PL, Marquez JA. Modulation of abscisic acid signaling in vivo by an engineered receptor-insensitive protein phosphatase type 2C allele. Plant Physiol. 2011;156(1):106–16.

Gao Y, Zeng Q, Guo J, Cheng J, Ellis BE, Chen JG. Genetic characterization reveals no role for the reported ABA receptor, GCR2, in ABA control of seed germination and early seedling development in Arabidopsis. Plant J. 2007;52(6):1001–13.

Geiger D, Scherzer S, Mumm P, Stange A, Marten I, Bauer H, Ache P, Matschi S, Liese A, Al-Rasheid KAS, Romeis T, Hedrich R. Activity of guard cell anion channel SLAC1 is controlled by drought-stress signaling kinase-phosphatase pair. Proc Natl Acad Sci. 2009;106(50):21425–30.

Hamilton DWA, Hills A, Kohler B, Blatt MR. Ca2 + channels at the plasma membrane of stomatal guard cells are activated by hyperpolarization and abscisic acid. Proc Natl Acad Sci. 2000;97(9):4967–72.

Hao Q, Yin P, Yan C, Yuan X, Li W, Zhang Z, Liu L, Wang J, Yan N. Functional mechanism of the abscisic acid agonist pyrabactin. J Biol Chem. 2010;285(37):28946–52.

Hao Q, Yin P, Li W, Wang L, Yan C, Lin Z, Wu JZ, Wang J, Yan SF, Yan N. The molecular basis of ABA-independent inhibition of PP2Cs by a subclass of PYL proteins. Mol Cell. 2011;42(5):662–72.

Kline KG, Sussman MR, Jones AM. Abscisic acid receptors. Plant Physiol. 2010;154(2):479–82.

Klingler JP, Batelli G, Zhu JK. ABA receptors: the START of a new paradigm in phytohormone signalling. J Exp Bot. 2010;61(12):3199–210.

Lee SC, Lan W, Buchanan BB, Luan S. A protein kinase-phosphatase pair interacts with an ion channel to regulate ABA signaling in plant guard cells. Proc Natl Acad Sci USA. 2009;106(50):21419–24.

Levchenko V, Konrad KR, Dietrich P, Roelfsema MR, Hedrich R. Cytosolic abscisic acid activates guard cell anion channels without preceding Ca2 + signals. Proc Natl Acad Sci USA. 2005;102(11):4203–8.

Li W, Wang L, Sheng X, Yan C, Zhou R, Hang J, Yin P, Yan N. Molecular basis for the selective and ABA-independent inhibition of PP2CA by PYL13. Cell Res. 2013;23(12):1369–79.

Liu X, Yue Y, Li B, Nie Y, Li W, Wu W-H, Ma L. A G Protein-Coupled Receptor Is a Plasma Membrane Receptor for the Plant Hormone Abscisic Acid. Science. 2007;315(5819):1712–6.

Ma Y, Szostkiewicz I, Korte A, Moes D, Yang Y, Christmann A, Grill E. Regulators of PP2C phosphatase activity function as abscisic acid sensors. Science. 2009;324(5930):1064–8.

Melcher K, Ng L-M, Zhou XE, Soon F-F, Xu Y, Suino-Powell KM, Park S-Y, Weiner JJ, Fujii H, Chinnusamy V, Kovach A, Li J, Wang Y, Li J, Peterson FC, Jensen DR, Yong E-L, Volkman BF, Cutler SR, Zhu J-K, Xu HE. A gate-latch-lock mechanism for hormone signalling by abscisic acid receptors. Nature. 2009;462(7273):602–8.

Melcher K, Xu Y, Ng LM, Zhou XE, Soon FF, Chinnusamy V, Suino-Powell KM, Kovach A, Tham FS, Cutler SR, Li J, Yong EL, Zhu JK, Xu HE. Identification and mechanism of ABA receptor antagonism. Nat Struct Mol Biol. 2010;17(9):1102–8.

Milborrow BV. The chemistry and physiology of abscisic acid. Ann Rev Plant Physiol. 1974;25(1):259–307.

Miyazono K, Miyakawa T, Sawano Y, Kubota K, Kang HJ, Asano A, Miyauchi Y, Takahashi M, Zhi Y, Fujita Y, Yoshida T, Kodaira KS, Yamaguchi-Shinozaki K, Tanokura M. Structural basis of abscisic acid signalling. Nature. 2009;462(7273):609–14.

Murase K, Hirano Y, Sun TP, Hakoshima T. Gibberellin-induced DELLA recognition by the gibberellin receptor GID1. Nature. 2008;456(7221):459–63.

Nambara E, Suzuki M, Abrams S, McCarty DR, Kamiya Y, McCourt P. A screen for genes that function in abscisic acid signaling in *Arabidopsis thaliana*. Genetics. 2002;161:1247–55.

Pandey S, Nelson DC, Assmann SM. Two novel GPCR-type G proteins are abscisic acid receptors in Arabidopsis. Cell. 2009;136(1):136–48.

Park S-Y, Fung P, Nishimura N, Jensen DR, Fujii H, Zhao Y, Lumba S, Santiago J, Rodrigues A, Chow T-FF, Alfred SE, Bonetta D, Finkelstein R, Provart NJ, Desveaux D, Rodriguez PL, McCourt P, Zhu J-K, Schroeder JI, Volkman BF, Cutler SR. Abscisic acid inhibits type 2C protein phosphatases via the PYR/PYL family of START proteins. Science. 2009;324(5930):1068–71.

Santiago J, Dupeux F, Round A, Antoni R, Park SY, Jamin M, Cutler SR, Rodriguez PL, Marquez JA. The abscisic acid receptor PYR1 in complex with abscisic acid. Nature. 2009;462(7273):665–8.

Santiago J, Dupeux F, Betz K, Antoni R, Gonzalez-Guzman M, Rodriguez L, Marquez JA, Rodriguez PL. Structural insights into PYR/PYL/RCAR ABA receptors and PP2Cs. Plant Sci. 2012;182:3–11.

Santner A, Estelle M. Recent advances and emerging trends in plant hormone signalling. Nature. 2009;459(7250):1071–8.

Sato A, Sato Y, Fukao Y, Fujiwara M, Umezawa T, Shinozaki K, Hibi T, Taniguchi M, Miyake H, Goto DB, Uozumi N. Threonine at position 306 of the KAT1 potassium channel is essential for channel activity and is a target site for ABA-activated SnRK2/OST1/SnRK2.6 protein kinase. Biochem J. 2009;424(3):439–48.

Shen YY, Wang XF, Wu FQ, Du SY, Cao Z, Shang Y, Wang XL, Peng CC, Yu XC, Zhu SY, Fan RC, Xu YH, Zhang DP. The Mg-chelatase H subunit is an abscisic acid receptor. Nature. 2006;443(7113):823–6.

Shimada A, Ueguchi-Tanaka M, Nakatsu T, Nakajima M, Naoe Y, Ohmiya H, Kato H, Matsuoka M. Structural basis for gibberellin recognition by its receptor GID1. Nature. 2008;456(7221):520–3.

Soon FF, Ng LM, Zhou XE, West GM, Kovach A, Tan MH, Suino-Powell KM, He Y, Xu Y, Chalmers MJ, Brunzelle JS, Zhang H, Yang H, Jiang H, Li J, Yong EL, Cutler S, Zhu JK, Griffin PR, Melcher K, Xu HE. Molecular mimicry regulates ABA signaling by SnRK2 kinases and PP2C phosphatases. Science. 2012;335(6064):85–8.

Tan X, Calderon-Villalobos LI, Sharon M, Zheng C, Robinson CV, Estelle M, Zheng N. Mechanism of auxin perception by the TIR1 ubiquitin ligase. Nature. 2007;446(7136):640–5.

Umezawa T, Nakashima K, Miyakawa T, Kuromori T, Tanokura M, Shinozaki K, Yamaguchi-Shinozaki K. Molecular basis of the core regulatory network in ABA responses: sensing, signaling and transport. Plant Cell Physiol. 2010;51(11):1821–39.

Vlad F, Rubio S, Rodrigues A, Sirichandra C, Belin C, Robert N, Leung J, Rodriguez PL, Lauriere C, Merlot S. Protein phosphatases 2C regulate the activation of the Snf1-related kinase OST1 by abscisic acid in Arabidopsis. Plant Cell. 2009;21(10):3170–84.

Yin P, Fan H, Hao Q, Yuan X, Wu D, Pang Y, Yan C, Li W, Wang J, Yan N. Structural insights into the mechanism of abscisic acid signaling by PYL proteins. Nat Struct Mol Biol. 2009;16(12):1230–6.

Yuan X, Yin P, Hao Q, Yan C, Wang J, Yan N. Single amino acid alteration between valine and isoleucine determines the distinct pyrabactin selectivity by PYL1 and PYL2. J Biol Chem. 2010;285(37):28953–8.

Zhang X, Zhang Q, Xin Q, Yu L, Wang Z, Wu W, Jiang L, Wang G, Tian W, Deng Z, Wang Y, Liu Z, Long J, Gong Z, Chen Z. Complex structures of the abscisic acid receptor PYL3/RCAR13 reveal a unique regulatory mechanism. Structure. 2012;20(5):780–90.

Zhang X, Jiang L, Wang G, Yu L, Zhang Q, Xin Q, Wu W, Gong Z, Chen Z. Structural insights into the abscisic acid stereospecificity by the ABA receptors. PLoS ONE. 2013;8(7):e67477.

Chapter 8
Protein Kinases and Phosphatases Involved in ABA Signaling

Shan Liang and Da-Peng Zhang

Abstract Reversible protein phosphorylation, catalyzed by protein kinases and phosphatases, is involved in almost all key events of cell metabolism and signaling in eukaryotes. A wide array of protein kinases and phosphatases have been identified as crucial players in ABA signaling, among which Ca^{2+}-dependent protein kinases (CDPKs), sucrose non-fermenting-1 (SNF1)-related protein kinases (SnRKs), mitogen-activated protein kinases (MAPKs), receptor-like kinases (RLKs), and type 2C and type 2A protein phosphatases (PP2Cs and PP2As) are relatively best characterized and their functional mechanisms in ABA signaling begin to be understood. In this chapter, we examine these advances in the plant protein kinases and phosphatases as well as the insights of these discoveries into the mechanisms of ABA-signaling network.

Keywords ABA signal transduction · Phosphorylation · Dephosphorylation · Protein kinase · Protein phosphatase

8.1 Introduction

Reversible protein phosphorylation (i.e., protein phosphorylation and de-phosphorylation), catalyzed by protein kinases and phosphatases, represents a major form of reversible posttranslational modification. Molecular genetic and biochemical studies have greatly advanced our knowledge of protein kinases and phosphatases and provide compelling evidence that the reversible protein phosphorylation is involved in almost all key events of cell metabolism and signaling in eukaryotes, serving as an important "on-and-off" switch in the regulation of the cell activities

S. Liang · D.-P. Zhang (✉)
Center for Plant Biology, School of Life Sciences,
Tsinghua University, Beijing 100084, China
e-mail: zhangdp@tsinghua.edu.cn

D.-P. Zhang (ed.), *Abscisic Acid: Metabolism, Transport and Signaling*,
DOI 10.1007/978-94-017-9424-4_8

by modulating the conformation, activity, localization, and stability of substrate proteins (Hunter 1995; Cohen 1997; Smith and Walker 1996; Ichimura et al. 2002; Luan et al. 2002; Hrabak et al. 2003; Luan 2003; Schweighofer et al. 2004; McCormack et al. 2005; Morillo and Tax 2006; Farkas et al. 2007; Colcombet and Hirt 2008; De Smet et al. 2009; Halford and Hey 2009; Luan 2009; Weinl and Kudla 2009; Gish and Clark 2011; Perochon et al. 2011; Boudsocq and Sheen 2013; Schulz et al. 2013; Hamel et al. 2014). Both protein kinases and phosphatases were originally discovered in animal systems in which reversible protein phosphorylation plays important roles in diverse cellular processes such as signal transduction, glycogen metabolism, cell cycle progression, ion transport, and developmental regulation (Hunter 1995). In plants, as in animals, it has been recognized that nearly all aspects of cell function involve reversible phosphorylation, and significant progress has been made in the identification of plant protein kinases and phosphatases and elucidation of their functional mechanisms by which plants respond to developmental and environmental cues (Smith and Walker 1996; Ichimura et al. 2002; Luan et al. 2002; Hrabak et al. 2003; Luan 2003; Schweighofer et al. 2004; McCormack et al. 2005; Morillo and Tax 2006; Farkas et al. 2007; Colcombet and Hirt 2008; De Smet et al. 2009; Halford and Hey 2009; Luan 2009; Weinl and Kudla 2009; Gish and Clark 2011; Perochon et al. 2011; Boudsocq and Sheen 2013; Schulz et al. 2013; Hamel et al. 2014). In the field of ABA signal transduction, the researches of the identification of protein kinases and phosphatases involved in ABA signaling have attracted much attention in the last decades, and an increasing number of protein kinases and phosphatases have been reported to be important components of ABA signaling, covering a wide range of protein kinase and phosphatases, such as Ca^{2+}-dependent protein kinases (CDPK), SNF1-related protein kinases (SnRKs), mitogen-activated protein kinases (MAPKs), receptor-like kinases (RLKs), clade-A protein phosphatase 2C (PP2C), and protein phosphatase 2A (PP2A). Some of these protein kinases and phosphatases play pivotal roles in ABA signaling, forming a reversible protein phosphorylation cascade, which deepens our understanding of highly complicated ABA-signaling pathway. In this chapter, we examine these advances in the plant protein kinases and phosphatases as well as the insights of these discoveries into the mechanisms of ABA-signaling network.

8.2 Ca^{2+}-Dependent Protein Kinases (CDPK) Involved in ABA Signaling

8.2.1 CDPKs: Key Ca^{2+} Sensors in Plant Cells

Ca^{2+} has long been recognized as a major, conserved second messenger in eukaryotic cells in responses to developmental and environmental stimuli. Ca^{2+} signaling is believed to be modulated by specific Ca^{2+} signatures, i.e., the specific patterns of variations in the amplitude, duration, location, and frequency of cytosolic free Ca^{2+}-spikes in response to different stimuli. It has been accepted that the specific

Ca^{2+} signatures are recognized by different Ca^{2+} sensors to transduce specific Ca^{2+}-mediating signals into downstream events (Sanders et al. 1999; Harmon et al. 2000; Rudd and Franklin-Tong 2001), though the molecular, cellular, and genetic links between Ca^{2+} signatures and the multiple downstream signaling events are largely obscure. Plants have several classes of Ca^{2+} sensor proteins, including calmodulin (CaM) and CaM-related proteins (CMLs) (Zielinski 1998; Snedden and Fromm 2001; Luan et al. 2002; McCormack et al. 2005; Perochon et al. 2011), calcineurin B-like (CBL) proteins (Luan et al. 2002; Luan 2009; Weinl and Kudla 2009), and Ca^{2+}-dependent protein kinases (CDPKs) (Harmon et al. 2001; Cheng et al. 2002b; Hrabak et al. 2003; Boudsocq and Sheen 2013; Schulz et al. 2013; Hamel et al. 2014).

CDPKs are Ser/Thr protein kinases that are found in vascular and nonvascular plants, in green algae, and also in certain protozoa, but not found in animals or yeast; CMLs and CBLs are also restricted to plants and some protists, while CaM exists ubiquitously in all eukaryotic cells (Hrabak et al. 2003; Defalco et al. 2010). Unlike CaM and CBL that must relay the Ca^{2+}-induced conformational change to their protein partners, CDPKs are a novel class of Ca^{2+} sensor having the unique feature of both Ca^{2+} sensing and kinase activities within a single protein to directly translate Ca^{2+} signals into phosphorylation events (Cheng et al. 2002b; Ludwig et al. 2004; Harper and Harmon, 2005; Boudsocq and Sheen 2010). CDPKs are the best characterized Ca^{2+} sensor involved in most of Ca^{2+}-stimulated protein kinase activities identified in plants (Harmon et al. 2001; Cheng et al. 2002b; Hrabak et al. 2003; Boudsocq and Sheen 2013; Schulz et al. 2013; Hamel et al. 2014).

CDPKs have a conserved molecular structure: an N-terminal variable domain, followed by a Ser/Thr kinase domain joined to a C-terminal CaM-like domain via a junction region that serves to stabilize and maintain kinase in an auto-inhibited state (Harper et al. 1991, 1994; Harmon et al. 2001; Cheng et al. 2002b. See Fig. 8.1). The junction region is sometimes referred to as "auto-inhibitory junction domain." The CaM-like domain harbors four EF-hand Ca^{2+}-binding motifs, which are organized into two lobes that have distinct Ca^{2+} affinities, resulting in different roles in CDPK regulation (Harper et al. 2004; Harper and Harmon 2005; Boudsocq and Sheen 2013; Liese and Romeis 2013). A model for CDPK activation was proposed in which the auto-inhibitory junction domain interacts with and keeps the kinase domain in an inactive state by acting as a pseudosubstrate that restricts access to the kinase catalytic center in the absence of or at low level of Ca^{2+} (Harper et al. 2004; Harper and Harmon 2005). Ca^{2+} binding to the

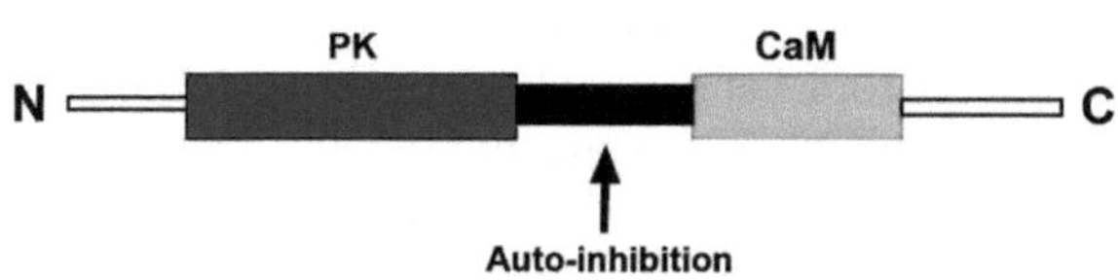

Fig. 8.1 A schema of the CDPK structure. The N-terminal variable domain is followed by a Ser/Thr kinase domain (PK) and the auto-inhibition domain of this kinase; a C-terminal CaM-like domain (CaM) serves as a Ca^{2+} sensor. See detailed explanation in the text

CaM-like domain induces a conformational change that releases the auto-inhibition (Harper et al. 2004; Harper and Harmon 2005). This results in kinase activity, enabling CDPK autophosphorylation as well as the phosphorylation of target substrates (Liese and Romeis 2013). Besides Ca^{2+}, phosphorylation, lipids, and interaction with 14-3-3 proteins have been reported to further modulate CDPKs in vitro (Harper et al. 2004; Harper and Harmon 2005; Klimecka and Muszynska 2007; Cheng et al. 2002b; Ludwig et al. 2004; Witte et al. 2010; Klimecka et al. 2011). This working model was supported and refined by recent structural analysis (Wernimont et al. 2010, 2011; Schulz et al. 2013; Hamel et al. 2014).

The CDPKs are encoded by a large multigene family, which is divided into four subgroups (Harmon et al. 2000; Cheng et al. 2002b). The *Arabidopsis* genome encodes 34 CDPK members, rice (*Oryza sativa*) encodes 29 members, wheat (*Triticum aestivum*) encodes 20 members, maize (*Zea mays*) encodes 35 members, and poplar (*Populus trichocarpa*) encodes 30 members (Harmon et al. 2000; Asano et al. 2005; Li et al. 2008; Ma et al. 2013; Zuo et al. 2013). The large gene-family-encoded CDPKs implicate possible redundancy and/or diversity in their functions (Harmon et al. 2001; Cheng et al. 2002b; Boudsocq and Sheen 2013; Hamel et al. 2014). Growing evidence indicates that CDPKs regulate many aspects in plant growth and development as well as plant adaptation to biotic and abiotic stresses (Bachmann et al. 1995, 1996; McMichael et al. 1995; Pei et al. 1996; Sheen 1996; Li et al. 1998; Sugiyama et al. 2000; Romeis et al. 2001; Hrabak et al. 2003; Shao and Harmon 2003; McCubbin et al. 2004; Choi et al. 2005; Ivashuta et al. 2005; Mori et al. 2006; Yu et al. 2007; Zhu et al. 2007; Zou et al. 2010; Asano et al. 2011, 2012; Brandt et al. 2012; Boudsocq and Sheen 2013; Ding et al. 2013; Liese and Romeis 2013; Liu et al. 2013). In plant hormone signaling, CDPKs are believed to be important regulators to be involved in various signaling pathways (Cheng et al. 2002b; Ludwig et al. 2004; Asano et al. 2012; Boudsocq and Sheen 2013).

The CDPK isoforms are expressed differently in plant organs/tissues (Harper et al. 2004; Ray et al. 2007; Li et al. 2008; Wan et al. 2007) with some members expressed in most organs whereas others specifically in some tissues. The expression profiles or kinase activities of the CDPK members were reported to be modified in response to diverse stimuli, including abscisic acid (ABA), cold, drought, salinity, heat, elicitors, and pathogens (Romeis et al. 2001; Ray et al. 2007; Li et al. 2008; Wan et al. 2007; Abbasi et al. 2004; Yu et al. 2006, Zhu et al. 2007). CDPK members have been showed to be localized to diverse cellular compartments, including the cytosol, nucleus, plasma membrane, endoplasmic reticulum, tonoplast, mitochondria, chloroplasts, oil bodies, and peroxisomes (Boudsocq et al. 2010; Zou et al. 2010; Li et al. 2008; Myers et al. 2009; Yu et al. 2006; Zhu et al. 2007; Lu and Hrabak 2002; Benetka et al. 2008; Mehlmer et al. 2010; Choi et al. 2005; Coca and San Segundo 2010). This diversity of subcellular localization indicates that CDPKs have access to a plethora of potential substrates throughout the plant cell. In recent years, biochemical analyses, together with genetic approaches, have identified CDPK substrates that are involved in diverse cellular processes, such as primary and secondary metabolism, stress responses, ion and water transport, transcription, and cell signaling (Harper and Harmon 2005; Cheng et al. 2002b; Curran et al. 2011; Boudsocq and Sheen 2013; Schulz et al. 2013), which helps to unravel mechanisms of CDPK functions.

In summary, the multigene family-encoded CDPK members and diversity in a broad range of their characteristics, such as expression pattern, subcellular localization, and substrate specificities, are likely to provide functional specificity and redundancy in mediating Ca^{2+} signaling in developmental and environmental responses of plant cells.

8.2.2 CDPKs: Hubs in ABA Signaling

In recent years, major progress has been made to uncover the central roles of CDPKs in triggering downstream cellular events in response to ABA. An earlier report showed that the constitutively ectopic expression of two *Arabidopsis* CDPKs AtCPK10/CDPK1 (*Arabidopsis* gene identifier number At1g18890) and AtCPK30/CDPK1a (At1g74740) from subgroup III (Hrabak et al. 2003; Boudsocq and Sheen 2013) in maize leaf protoplasts activated a stress- and ABA-inducible HVA1 promoter, showing the connection of CDPKs to ABA-signaling pathway (Sheen 1996). Later, the function of the AtCPK10 in ABA signaling was confirmed genetically: AtCPK10 regulates ABA-mediated stomatal movement in response to drought stress possibly by interacting with a heat-shock protein HSP1 (Zou et al. 2010). Constitutively, active AtCPK10 also induces endogenous stress- and ABA-inducible genes in *Arabidopsis* leaf cells (Boudsocq and Sheen 2013), suggesting a conservation of specific CDPK functions in dicots and monocots. This transcriptional induction is likely to be mediated by the ABA-responsive transcription factors, ABFs, which were identified as in vitro substrates for several CDPKs from subgroups I and III (Zhu et al. 2007; Choi et al. 2005). An *Arabidopsis* CDPK, AtCPK32, was shown to interact with and activate an ABA-induced transcription factor ABF4, resulting in the induction of ABF4 target genes, and constitutive overexpression of AtCPK32 resulted in ABA-hypersensitive phenotypes in ABA-induced inhibition of seed germination (Choi et al. 2005). The *Arabidopsis* CPK3 and CPK6 were identified as key, positive regulators in ABA-regulated stomatal signaling (Mori et al. 2006), but it has been unknown whether these two CDPK members regulate the ABFs or other ABA-responsive transcription factors through which they modulate expression of ABA-responsive genes. Yu et al. (2006) isolated an ABA-induced CDPK from grape berry, ACPK1, which phosphorylates a P-type ATPase in vitro, and was further shown to confer ABA-hypersensitive phenotypes and drought tolerance when ectopically expressed in *Arabidopsis* (Yu et al. 2007). The two closet homologs of ACPK1 in *Arabidopsis*, AtCPK4 and AtCPK11 from subgroup I (Hrabak et al. 2003; Boudsocq and Sheen 2013), were shown to function as positive regulators in all the major ABA responses including seed germination, seedling growth, guard cell regulation, and plant tolerance to drought and salt stresses (Zhu et al. 2007). Plants overexpressing AtCPK4 or AtCPK11 showed hypersensitivity to

ABA and increased tolerance to drought, whereas the *cpk4* or *cpk11* mutant exhibits the opposite phenotypes and the *cpk4 cpk11* double mutant shows stronger ABA-related phenotypes than the single mutants. AtCPK4 and AtCPK11 phosphorylate ABF1 and ABF4 (Zhu et al. 2007; Lynch et al. 2012), and AtCPK4 also phosphorylates ABF2 (Lu et al. 2013). The two CDPKs regulate expression of a subset of ABA-responsive genes probably through these ABA-responsive transcription factors, which correlates with ABA-related phenotypes and altered drought and salt tolerance in the overexpressing lines and loss-of-function mutants (Zhu et al. 2007). By contrast, the *Arabidopsis* CPK12, the closest homolog of AtCPK4/AtCPK11 in subgroup I (Hrabak et al. 2003; Boudsocq and Sheen 2013), was reported to function as a negative regulator of ABA signaling in seed germination and postgermination growth, and to phosphorylate ABF1, ABF4 as well as a key, negative ABA-signaling regulator ABA insensitive 2 (ABI2), suggesting that different members of the CDPK family may constitute a regulation loop by functioning positively and negatively in ABA signal transduction (Zhao et al. 2011a, b). Together, all these findings suggest that ABA-induced transcriptional reprogramming via ABFs is likely to be a key feature of the CDPK-mediated ABA signaling (Boudsocq and Sheen 2013).

In crop plants, a homolog of the *Arabidopsis* CPK4/11/12 in maize, ZmCPK11, was reported to regulate ABA-induced antioxidant defense (Ding et al. 2013), suggesting that these CDPK members may regulates multiple ABA responses in both abiotic and biotic stresses. In rice, two members of CDPK, OsCPK12 and OsCPK21, were reported to be positive regulators of ABA signaling in response to salt stresses (Asano et al. 2011, 2012). The rice plants overexpressing OsCPK12 or OsCPK21 showed increased salt tolerance and ABA sensitivity in seedling growth, and the expression levels of several ABA- and high salinity-inducible genes were enhanced in these transgenic plants. However, the loss-of-function *oscpk12* mutant and *OsCPK12* RNAi lines showed wild-type ABA-related phenotypes, suggesting that functional redundancy among OsCPK12 and other CDPKs exists in the ABA-signaling pathway in rice (Asano et al. 2011, 2012). It remains unclear, however, whether these rice CDPKs regulates gene expression through ABA-responsive transcription factors like the *Arabidopsis* ABFs to mediate ABA signaling, which needs studies in the future.

Although the role of CDPK phosphorylation has not been thoroughly investigated, other transcription factors than ABFs and proteins of diverse identity involved in ABA signaling have also been identified as CDPK substrates in vitro. The *Arabidopsis* CPK11 was shown to interact physically with a nuclear zinc finger transcription factor AtDi19-1 (drought-induced 19-1, Rodriguez et al. 2006; Curran et al. 2011; Liu et al. 2013) and a heat-shock protein HSP1 (Uno et al. 2009). The Di19-1 and Di19-2 of cotton (*Gossypium hirsutum*) were reported to regulate plant response to ABA and salt stress (Li et al. 2010). A more recent report showed that AtDi19 functions as a transcriptional activator and was involved in *Arabidopsis* responses to drought stress through upregulation of pathogenesis-related *PR1*, *PR2*, and *PR5* gene expressions, and this transactivation activity of AtDi19 is enhanced by AtCPK11 (Liu et al. 2013). This suggests that AtCPK11 may interact with both ABFs (ABF1/4) and AtDi19 transcription factors to regulate gene expression, supporting the point of view that a single CDPK member may have multiple substrates, possibly integrating different signaling pathways. As mentioned earlier, the HSP1 was identified as a functionally interacting partner of AtCPK10 in ABA signaling of guard cells (Zou et al. 2010), suggesting that different CDPK members may share a common substrate or the same signaling pathway and that AtCPK11 may function similarly to AtCPK10 through the common substrate to regulate guard cell signaling in response to ABA.

Recently, two closet homologous *Arabidopsis* CDPK members, AtCPK21 and AtCPK23 from subgroup II (Hrabak et al. 2003; Boudsocq and Sheen 2013), together with AtCPK3 and AtCPK6, were shown to interact with and phosphorylate the guard cell anion channels SLAC1 and SLAH3 and to be involved in a Ca^{2+}-dependent signaling pathway of guard cells in response to ABA through regulation of these two direct targets (Geiger et al. 2010, 2011; Brandt et al. 2012; Scherzer et al. 2012; Demir et al. 2013). We will describe in the next section this Ca^{2+}-dependent ABA-signaling network in the *Arabidopsis* guard cells.

8.2.3 CDPKs: Key Regulators of Ca^{2+}-Dependent ABA Signaling in Guard Cells

It has been known that, during stomatal opening, inward K^+ channels mediate K^+ uptake for net solute accumulation that drives water influx, guard cell swelling, and pore opening, while during stomatal closure, outward K^+ channels play essential roles in solute removal from guard cells. Efflux of anions through anion channels is also essential for stomatal closure (Pandey et al. 2007; Roelfsema et al. 2012). In response to drought stress, ABA induces stomatal closure. It has been known that ABA regulates ion fluxes involved in stomatal movements in a Ca^{2+}-dependent (Hetherington and Woodward 2003; Hetherington and Brownlee 2004) and Ca^{2+}-independent manner (Levchenko et al. 2005; Marten et al. 2007). Recent studies in *Arabidopsis* showed that the Ca^{2+}-independent activation of ion fluxes across the guard cell membranes is mainly mediated by the SNF1-related protein kinase (SnRK) open stomata 1 (OST1) (described in the next section), while Ca^{2+}-dependent activation of such ion fluxes is directly associated with the CDPK-dependent phosphorylation of ion channels. The ABA activation of slow-type anionic channels and Ca^{2+} permeable channels were reported to be impaired in the *cpk3 cpk6* double mutant that displays ABA-insensitive phenotypes in ABA-induced stomatal closure (Mori et al. 2006). Despite some discrepancies, AtCPK3 and AtCPK6 were shown to phosphorylate the major guard cell anion channel SLAC1 (Negi et al. 2008) and its related homolog SLAH3 anion channel to activate these channels in *Xenopus* (*Xenopus laevis*) oocytes in a Ca^{2+}-dependent fashion and in response to ABA, suggesting that AtCPK3 and AtCPK6 play an important role in the ABA-controlled, Ca^{2+}-dependent activation of SLAC-type anion currents in guard cells (Brandt et al. 2012; Scherzer et al. 2012). The SLAC1 anion channel and its homolog SLAH3 were also shown to be phosphorylated and activated by AtCPK21 and AtCPK23 in response to ABA (Geiger et al. 2010, 2011; Demir et al. 2013). However, AtCPK6 and AtCPK23 preferentially phosphorylate and activate SLAC1 and SLAH3 independently of Ca^{2+}, whereas AtCPK3 and AtCPK21 mainly regulate these channels in a Ca^{2+}-dependent manner (Brandt et al. 2012; Scherzer et al. 2012). This suggests that AtCPK6 and AtCPK23, together with the OST1 kinase (see next section), may be associated with Ca^{2+}-independent signaling steps, whereas AtCPK3 and AtCPK21 may regulate Ca^{2+}-dependent SLAC1 and SLAH3 activation in ABA signaling in guard cells. It remains to explore the underlying mechanism how these CDPKs function distinctly in relation to Ca^{2+} dependence in the ABA-mediated anion channel activation to understand stomatal regulation in response to ABA. Importantly, a signaling pathway has been described from ABA signal perception to the CDPK-mediated activation of the SLAC1 and SLAH3 anion channels, where the PYR/PYL/RCAR family ABA receptor members PYR1 and RCAR1 inhibit a type 2C protein phosphatase member ABI1 to release AtCPK3/6/21/23 from inhibition, leading to Ca^{2+}-dependent activation of the anion channels of guard cells in response to ABA (Geiger et al. 2010, 2011; Brandt et al. 2012; Scherzer et al. 2012; Demir et al. 2013). See Fig. 8.2 and Chap. 6 of this book).

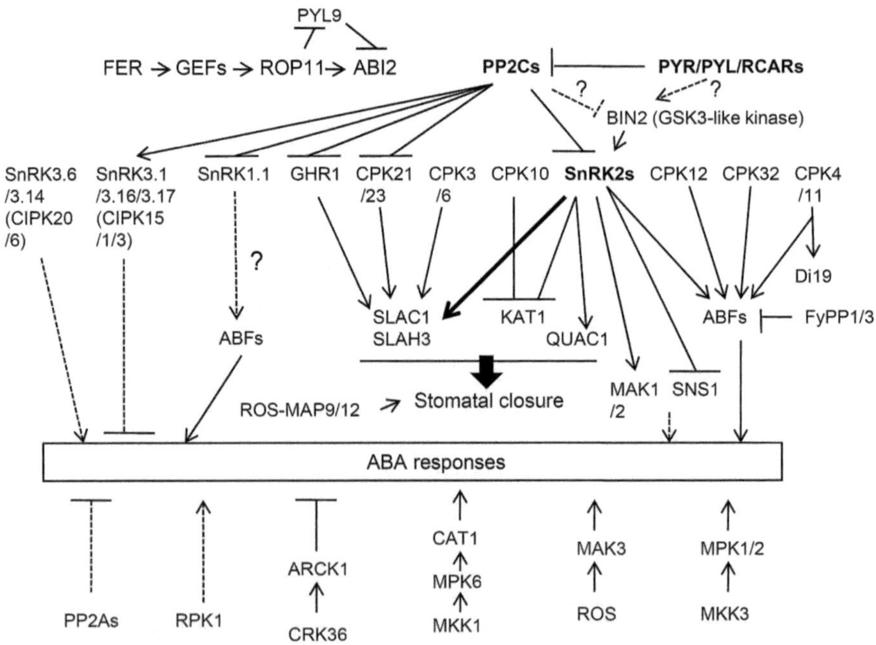

Fig. 8.2 A summary of the protein kinases and phosphatases involved in ABA signaling and their possible functional mechanisms in the model plant *Arabidopsis*. See detailed description in the text. *Arrows* denote positive regulation or activation, and *bars* denote negative regulation or repression. The *solid lines* indicate direct effect and *dotted lines* indirect effect. *Question mark* indicates unconfirmed link

The *Arabidopsis cpk3* and *cpk6* show ABA-insensitive phenotypes in ABA-induced stomatal closure, and the *cpk3 cpk6* double mutant shows stronger ABA-insensitive stomatal phenotypes and Ca^{2+}-insensitive phenotypes in Ca^{2+}-induced stomatal closure (Mori et al. 2006). However, *cpk21* and *cpk23* single mutants do not exhibit any stomatal phenotype in response to ABA (Geiger et al. 2010, 2011). This suggests that AtCPK3 and AtCPK6 may play a more global role in regulation of the ion fluxes involved in ABA-induced stomatal closure. At the whole-plant level, overexpression of *AtCPK6* enhances salt- and drought stress tolerance, whereas no altered stress tolerance is evident in the *cpk6* mutant line (Xu et al. 2010), suggesting a functional redundancy among AtCPK6 and other CDPKs to mediate stress signaling. Surprisingly, *cpk21* and *cpk23* mutants are drought tolerant, whereas overexpression of *AtCPK21* or *AtCPK23* results in plant oversensitivity to drought or osmotic stress (Ma and Wu 2007; Franz et al. 2011), though AtCPK21 and AtCPK23, functioning as ABA-induced, Ca^{2+}-dependent activators of SLAC-type anion currents as described above, should positively regulate ABA-induced stomatal closure, and thus, one may expect that the *cpk21* and *cpk23* mutant decrease, but the *AtCPK21*- or *AtCPK23* overexpressors increase the ABA sensitivity of stomatal movement and tolerance to drought. These opposite effects in different levels (or between short-term response in the guard cells and long-term response at the tissue or whole-plant level) suggest a high complexity of plant stress signaling network where a sophisticated balance between positive and negative roles at the whole-plant level results in a final, physiological output.

CDPKs may also function to inhibit K^+ inward channels, such as KAT1, to regulate ABA-induced stomatal closure. It has been reported that loss of function of the *Arabidopsis CPK10* gene reduced ABA response in stomatal closure by downregulating the ABA-induced inhibition of K^+ inward channels in guard cells, suggesting that AtCPK10 may regulate KAT1 to mediate Ca^{2+}-dependent ABA signaling in guard cells (Zou et al. 2010). However, AtCPK4 and AtCPK11 are important components of ABA signaling in guard cells, and the loss-of-function mutants *cpk4* and *cpk11* display ABA-insensitive phenotypes in ABA-induced stomatal closure (Zhu et al. 2007), but it remains unclear whether these two CDPKs are involved in the Ca^{2+}-dependent activation of the ion channels in guard cells in response to ABA, which merits further studies.

8.2.4 CDPKs and ABA Signaling: A Summary

CDPKs are versatile and evolutionarily conserved Ca^{2+} sensors that function as central regulators of Ca^{2+}-mediated ABA and stress responses that are crucial for plant survival. Recent advances have identified some key players in ABA signaling, including AtCPK3, AtCPK4, AtCPK6, AtCPK10, AtCPK11, AtCPK12, AtCPK21, AtCPK23, AtCPK32, OsCPK12, OsCPK21, and ZmCPK11, which may function as crucial signaling nodes in ABA-signaling network. A broad range of characteristics of CDPKs indicate complex and sophisticated Ca^{2+}-signaling networks via protein phosphorylation to coordinate the dynamic plant cellular processes. Among these characteristics, functional redundancy, mainly due to multiple members of the CDPK family, may provide plants with robust responsiveness to ABA and adaptability to stressful environment, and substrate specificity may be a key feature to confer functional specificity of different CDPK members in ABA and stress signaling (Fig. 8.2).

The ABA-responsive transcription factors ABFs were identified as substrates of several CDPKs in *Arabidopsis*, which may be key downstream players of CDPKs to regulate ABA-mediated gene expression. Other substrates of diverse identity, including transcription factors, heat-shock protein, and ion channels, were also identified as regulators to relay CDPK-mediated ABA signaling. AtCPK4 and AtCPK11 function redundantly to regulate major ABA response, possibly by phosphorylating two ABA-responsive transcription factors ABF1 and ABF4, and AtCPK23, phosphorylating ABF4, may function in concert with AtCPK4 and AtCPK11. AtCPK11 may also interact with a nuclear zinc finger transcription factor AtDi19-1 and a heat-shock protein HSP1 to regulate ABA signaling. AtCPK10 use the same HSP1 as a functionally interacting partner in ABA signaling of guard cells. Four other CDPK members, AtCPK3, AtCPK6, AtCPK21, and AtCPK23, may function redundantly in guard cell signaling in response to ABA, likely by targeting the common substrates anion channels SLAC1 and SLAH3 (Fig. 8.2). Thus, a single CDPK member may have multiple substrates, possibly integrating different signaling pathways, and different CDPK members may also share a common substrate or the same pathway to regulate ABA signaling.

8.3 SNF1-Related Protein Kinases (SnRKs) Involved in ABA Signaling

8.3.1 SnRK Family: SnRK1, SnRK2, and SnRK3

Plants contain a large group of protein kinase related to the classical sucrose non-fermenting-1 (SNF1)-type kinases from yeast where they function as global regulators of carbon metabolism; these kinases are named SNF1-related protein kinases (SnRKs) (Halford and Hardie 1998; Hrabak et al. 2003). SnRKs are a class of Ser/Thr kinases like CDPKs (Hardie 2000; Hrabak et al. 2003). Based on the analysis of their evolutionary origins (sequence similarity and domain structure), SnRKs are recognized as a family member of the CDPK-SnRK superfamily, which can be subdivided into three subfamilies: SnRK1, SnRK2, and SnRK3 (Hrabak et al. 2003). The *Arabidopsis* genome contains 38 SnRKs, of which 3 (SnRK1.1 to SnRK1.3) are SnRK1s; 10 (SnRK2.1 to SnRK2.10) are SnRK2s; and 25 (SnRK3.1 to SnRK3.25) are SnRK3s (Hrabak et al. 2003). The SnRKs of the three subfamilies share strong sequence identity in their kinase domains but have divergent C-terminal domains which are likely to regulate their activities in response to a variety of metabolic and/or stress signaling pathways (Hrabak et al. 2003; Halford and Hey 2009).

The SnRK1 group is the most closely related to SNF1 from yeast and to AMP-activated protein kinases (AMPK) from animals. Plant SnRK1s are likely to form heterotrimeric complexes (consisting of α, β, and γ subunits) and may regulate carbon metabolism and stress signaling like their yeast and animal counterparts (Halford et al. 2000, 2003; Ferrando et al. 2001; Hardie 2000; Hrabak et al. 2003; Polge and Thomas 2006; Halford and Hey 2009; Coello et al. 2011; Robaglia et al. 2012). However, the SnRK2 and SnRK3 groups appear to be plant-specific classes of kinases, which diverge from AMPK and SNF1 than do SnRK1s, and may likely emerge as a result of duplication and then evolve rapidly, taking on new roles to enable plants to link metabolic and stress signaling (Halford et al. 2000; Hrabak et al. 2003; Kolukisaoglu et al. 2004; Halford and Hey 2009; Coello et al. 2011). The SnRK3s are the CBL-interacting protein kinases (CIPKs), which interact with calcium-binding proteins related to animal neuronal calcium sensors and to the regulatory B subunit of the protein phosphatase calcineurin (Kudla et al. 1999; Hrabak et al. 2003; Luan 2009; Stefan and Kudla 2009). Subsequent analyses have identified a complement of 10 CBLs and 25 CIPKs and 10 CBLs and 30 CIPKs in the genomes of *Arabidopsis* and rice (*Oryza sativa*), respectively (Kolukisaoglu et al. 2004). The SnRK members of the three groups have been shown to be involved in plant response to ABA, drought, salt, and nutritional stresses, and disease (Kudla et al. 1999; Hrabak et al. 2003; Kolukisaoglu et al. 2004; Polge and Thomas 2006; Halford and Hey 2009; Luan 2009; Stefan and Kudla 2009; Coello et al. 2011).

8.3.2 SnRK1S: A Convergence Point of ABA and Sugar Signaling Pathways

It has been demonstrated that ABA and sugar signaling pathways may share signaling components (Cheng et al. 2002a; Radchuk et al. 2006). Earlier studies indicated that SnRK1 could be implicated in these interactions (Nemeth et al. 1998; Bradford et al. 2003; Thelander et al. 2004; Radchuk et al. 2006; Lu et al. 2007). Further, a report showed that overexpression of an *Arabidopsis* SnRK1.1, a major form of the AtSnRK1s, confers sugar- and ABA-hypersensitive phenotype in seedling growth (Jossier et al. 2009), and it is suggested that the AtSnRK1s function by phosphorylating ABF (AREB) transcription factors (Zhang et al. 2008). The SnRK1s of wheat (*Triticum aestivum*) were also suggested to be involved in ABA signaling (Coello et al. 2012). Most recently, Rodrigues et al. (2013) revealed that two members of the clade-A PP2Cs acting as major, negative regulators in ABA signaling (reviewed in Cutler et al. 2010, and see the section as follows), ABI1 and PP2CA, inhibit SnRK1.1 activity by dephosphorylating it to repress SnRK1-mediated sugar signaling, and ABA promotes SnRK1-mediated sugar signaling by inhibiting the PP2Cs (Fig. 8.2), which allows the coordination of ABA and energy signaling, strengthening the stress response through the cooperation of two key and complementary pathways.

8.3.3 SnRK2S: A Central Node of ABA-Signaling Pathways

8.3.3.1 Studies in Crop Plants

Compelling evidence has been provided to reveal that members of the SnRK2 subfamily play critical roles in ABA-signaling pathway and regulate plant responses to environmental stresses. Earlier studies in crop plants identified two members of the SnRK2 subfamily, PKABA1 from wheat (*Triticum aestivum*) and AAPK from broad bean (*Vicia faba*), as the ABA-stimulated protein kinases and positive regulators of ABA response (Anderberg and Walker-Simmons 1992; Li and Assmann 1996; Gomez-Cadenas et al. 1999, 2001; Li et al. 2000, 2002; Shen et al. 2001; Johnson et al. 2002). Transcription of *PKABA1* is induced by ABA and occurs in embryos and seedlings, and PKABA1 functions to mediate ABA-suppressed gene expression (Gomez-Cadenas et al. 1999), and phosphorylates and activates the ABA-responsive basic-domain leu-zipper (bZIP) transcription factor TaABF, a seed-specific homolog of ABI5 of *Arabidopsis* (Johnson et al. 2002), suggesting that the TaABF may serve as a physiological substrate for PKABA1 which may regulate ABA signaling through activating this transcription factor in seed development and germination (Shen et al. 2001; Johnson et al. 2002). AAPK mediates ABA-induced stomatal closure and anion channels (Li et al. 2000) and modulates an RNA-binding protein AKIP1 by phosphorylating it and inducing its translocation into subnuclear speckles in guard cells (Li et al. 2002). In rice, expression of three members of the rice SnRK2 subfamily, SAPK8, SAPK9, and SAPK10, homologs of PKABA1/AAPK, is induced by osmotic stress and ABA (Kobayashi et al. 2004), suggesting that these rice SnRK2s may be involved in ABA and stress signaling.

8.3.3.2 SnRK2.2, SnRK2.3, and SnRK2.6 Constitute a Core Signaling Point in *Arabidopsis*

The kinase activities of five *Arabidopsis* SnRK2 members, SnRK2.2, SnRK2.3, SnRK2.6, SnRK2.7, and SnRK2.8, can be activated by ABA, and all of the SnRKs except SnRK2.9 can be activated by osmotic stress (Boudsocq et al. 2004; Furihata et al. 2006; Yoshida et al. 2006b), suggesting that most *Arabidopsis* SnRK2 members may be involved in ABA and stress signaling. The OST1/SnRK2.6 is the first member of the *Arabidopsis* SnRK2 subfamily that was identified by a forward genetic approach as a regulator of ABA signaling; the *snrk2.6* (*ost1*) mutant cannot exhibit ABA-induced stomatal closure, revealing that it positively regulates stomatal response to ABA (Mustilli et al. 2002; Yoshida et al. 2002). However, the *SnRK2.6* gene is mainly expressed in guard cells and in the vascular system, and seed germination and postgermination growth were not affected in *snrk2.6* loss-of-function mutant (Mustilli et al. 2002; Yoshida et al. 2002), suggesting that other SnRK2 protein kinases may function in seed germination and seedling growth. The studies through a reverse genetic approach in

Arabidopsis identified SnRK2.2 and SnRK2.3, two SnRK members most closely related to SnRK2.6 (Hrabak et al. 2003), as key positive regulators of ABA signaling: The *snrk2.2* and *2.3* single mutants show weak, while the *snrk2.2/2.3* double mutant exhibits strong ABA-insensitive phenotypes in terms of seed germination and seedling growth (Fujii et al. 2007). Thus, SnRK2.2 and SnRK2.3 function redundantly in ABA-induced inhibition of seed germination and postgermination growth (Fujii et al. 2007). These results led to a hypothesis of functional segregation between SnRK2.6 and SnRK2.2/2.3, i.e., SnRK2.6 functions in guard cells, whereas SnRK2.2 and SnRK2.3 specialize in seed germination and seedling growth regulation (Fujii et al. 2007). However, further investigations found that, in the *snrk2.2/3/6* triple mutant, nearly all major ABA responses are blocked (Fujii and Zhu 2009; Nakashima et al. 2009; Fujita et al. 2009), demonstrating that these functionally segregated kinases SnRK2.2, SnRK2.3, and SnRK2.6 still work redundantly in ABA regulation of guard cells, seed germination, and seedling growth, acting together as a core signaling point in ABA-signaling pathways. The expression profile of the three *SnRK* genes supports their redundant, cooperative roles in the major ABA responses: *SnRK2.6* is expressed not only in guard cells, but also in several other tissues, and *SnRK2.2* and *SnRK2.3* are expressed in all tissues (Fujii et al. 2007, 2011).

Other members of the *Arabidopsis* SnRK2 subfamily are likely to play roles in ABA and/or stress signaling. Loss-of-function mutations in two other closet homologous *SnRK* genes, *SnRK2.7* and *SnRK2.8*, affect expression of ABA- and drought-responsive genes, suggesting their potential roles in ABA signaling, though the *snrk2.7/2.8* double mutant shows wild-type phenotypes probably because of functional redundancy (Mizoguchi et al. 2010). Fujii et al. (2011) generated an *Arabidopsis* decuple mutant carrying mutations in all 10 members of the SnRK2 subfamily, *snrk2.1/2/3/4/5/6/7/8/9/10* and observed that this mutant grows poorly under hyperosmotic stress conditions but is similar to the wild type in the absence of osmotic stress, revealing critical functions of the SnRK2s in osmotic stress signaling and tolerance.

8.3.3.3 Downstream Targets and Functional Mechanism of *Arabidopsis*SnRK2S

The *Arabidopsis* SnRK2s, like PKABA1 and AAPK (as mentioned earlier) as well as SnRK1s, phosphorylate downstream targets to mediate ABA signaling (Zhang et al.2008, and also see test above). SnRK2s have been shown to phosphorylate and activate ABA-responsive bZIP transcription factor ABFs(ABF1-ABF4 and ABI5)and regulate ABA-responsive genes to mediate ABA signaling(Boudsocq et al. 2004; Kobayashi et al. 2004; Furihata et al. 2006; Yoshida et al. 2006b; Fujii et al. 2007; Fujii and Zhu2009; Fujii et al. 2009). In addition to the ABFs, guard cell anion channel SLAC1 was identified as a substrate of the SnRK2.6/OST1, and the inward K^+ channel is also likely to be a substrate of SnRK2.6 (Lee et al. 2009, 2013; Geiger et al. 2009,

2010; Brandt et al. 2012; Acharya et al. 2013). As described earlier, whereas Ca^{2+}-dependent activation of such ion channels is directly associated with the CDPKs (Mori et al. 2006; Geiger et al. 2010, 2011; Brandt et al. 2012; Scherzer et al. 2012; Demir et al. 2013), the Ca^{2+}-independent activation of the ion channels is mainly mediated by SnRK2.6 (Lee et al. 2009, 2013; Geiger et al. 2010, 2011; Brandt et al. 2012; Acharya et al. 2013).

It has been shown that SnRK2.6/OST1 can interact with and phosphorylate an inward K^+ channel KAT1, indicating that KAT1 is a target of SnRK2.6/OST1 (Sato et al. 2009; Acharya et al. 2013). ABA inhibition of inward K^+ channels and light-induced stomatal opening are reduced in *ost1* mutants, while transgenic plants overexpressing *OST1* show ABA hypersensitivity in these responses, suggesting that SnRK2.6/OST1 negatively regulates KAT1 to induce stomatal closure and inhibit stomatal opening in response to ABA (Acharya et al. 2013). A recent report showed that SnRK2.6/OST1 interacts with and phosphorylates a K^+ uptake transporter KUP6, suggesting that this kinase may also regulate stomatal movement through the KUP6 K^+ transporter (Osakabe et al. 2013). Importantly, like CPK3/6/21/23 (Mori et al. 2006; Geiger et al. 2010, 2011; Brandt et al. 2012; Scherzer et al. 2012), SnRK2.6/OST1 interacts with and phosphorylates guard cell anion channel SLAC1, and OST1 coexpression activates SLAC1 anion channels in a heterologous system (*Xenopus oocytes*) (Geiger et al. 2009, 2010; Lee et al. 2009, 2013; Brandt et al. 2012; Acharya et al. 2013), while *ost1* mutant shows strongly reduced slow anion currents in *Arabidopsis* guard cells and is insensitive to ABA-induced stomatal closure and regulation of slow anion currents, but *OST1* overexpressing lines are hypersensitive for these responses, resulting in accelerated stomatal closure in response to ABA (Geiger et al. 2009; Acharya et al. 2013). Thus, the CDPK (as described earlier) and OST1 branch of ABA signal transduction in guard cells seem to converge on the level of SLAC1. Additionally, the *ost1* mutation reduces ABA activation of rapidly activating (QUAC1) anion currents in guard cells (Imes et al. 2013), consistent with the idea that SnRK2.6 plays a central role in Ca^{2+}-independent ABA regulation of stomatal behavior by targeting key ion channels in guard cells.

Recently, two independent groups, using phosphoproteomic approach and working in *Arabidopsis*, identified a set of new potential substrates of SnRK2 protein kinase, including proteins involved in a variety of cell metabolic and signaling processes such as flowering time regulation, RNA and DNA binding, miRNA and epigenetic regulation, signal transduction, chloroplast function, and many other cellular processes (Wang et al. 2013; Umezawa et al. 2013). Wang and coworkers confirmed that, consistent with the SnRK2 phosphorylation of flowering time regulators, the *snrk2.2/2.3/2.6* triple mutant flowers significantly earlier than the wild-type plants (Wang et al. 2013). Umezaw and coworkers observed that SnRK2 promotes the ABA-induced activation of the mitogen-activated protein kinases AtMPK1 and AtMPK2 and also identified a previously unknown protein, SnRK2-substrate 1 (SNS1), which is phosphorylated by SnRK2s and negatively regulates ABA responses in *Arabidopsis* (Umezawa et al. 2013). These studies suggest a highly complicated SnRK2-mediated phosphorylation network and are helpful to elucidate the functional mechanisms of the SnRK2 protein kinases in ABA-signaling pathways.

A signaling pathway from the *Arabidopsis* PYR/PYL/RCAR family receptors for ABA to ABF transcription factor or SLAC1 anion channel activation through SnRK2.2/2.3/2.6 has been described recently. Similar to the ABA-signaling pathway via CPK3/6/21/23 in guard cells, the PYR/PYL/RCAR receptors bind to ABA and inhibit the activity of phosphatase A-type PP2Cs which dephosphorylate and inactivate SnRK2s in the absence of ABA and thus activate downstream ABF transcription factors or SLAC1 anion channel to induce ABA-responsive gene expression or stomatal closure (Ma et al. 2009; Park et al. 2009; Fujii et al. 2009; Umezawa et al. 2009; Lee et al. 2009, 2013; Cutler et al. 2010; Geiger et al. 2010, 2011; Brandt et al. 2012. See Fig. 8.2 and Chap. 6 of this book). Most recently, SnRK2.2 and SnRK2.3 were shown to be substrats of the *Arabidopsis* glycogen synthase kinase 3 (GSK3)-like kinase, BIN2, which, together with its homologs BIN2-like kinases BILs, regulates negatively brassinosteroid signaling (He et al. 2002; Li and Nam 2002), but is positively involved in ABA signaling (Cai et al. 2014). BIN2 kinase activates SnRK2.2 and SnRK2.3 through phosphorylation, and BIN2/BILs function downstream of PYR1/PYL/RCAR receptors and PP2Cs (Cai et al. 2014). This suggests a reversible phosphorylation process of SnRK2s where phosphorylation and dephosphorylation may be catalyzed by BIN2/BIL kinases and PP2Cs, respectively.

8.3.4 SnRK3S/CIPKs: Mediators of Ca^{2+}-Dependent ABA Signaling

The CBL Ca^{2+} sensor interacts with its target CIPK effector to form a Ca^{2+} sensor–effector complex to sense Ca^{2+} signal and trigger downstream signaling (Halford and Hey 2009; Luan 2009; Stefan and Kudla 2009; Coello et al. 2011). SnRK3s/CIPKs, together with corresponding CBLs, have been suggested to regulate Ca^{2+}-dependent ABA signaling in concert with CDPKs. In *Arabidopsis*, it was reported that CIPK15/SnRK3.1/PKS3 interacts with CBL1/ScaBP5, likely acting as a CBL1-CIPK15 calcium sensor–kinase complex, functions to negatively regulate ABA responses (Guo et al. 2002). The CBL1 closely related calcium sensor CBL9 and CIPK3/SnRK3.17, likely forming a CBL9–CIPK3 complex, were also shown to function as important, negative regulators of ABA signaling: Genetic evidence reveals that the *cipk3* and *cbl9* loss-of-function mutants are hypersensitive to ABA (Kim et al. 2003; Pandey et al. 2004, 2008). Additionally, disruption of the *CBL1* gene renders plants' hypersensitive to drought and salt stress, suggesting that CBL1 functions as an important integrator of abiotic stress responses (Albrecht et al. 2003; Cheong et al. 2003); however, in these studies, the *cbl1* loss-of-function mutant showed no ABA-related phenotypes (Albrecht et al. 2003; Cheong et al. 2003). So it remains controversial whether CBL1 mediates ABA responses. A further study identified CIPK1/SnRK3.16 as a negative regulator of ABA signaling and suggests that alternative complex formation of the kinase CIPK1 with either CBL9 or CBL1 mediates ABA-dependent and ABA-independent signaling responses (D'Angelo et al. 2006). Thus, additional studies will be needed to clarify the role of CBL1 in ABA signaling.

Several efforts have been made to explore functional mechanism of the *Arabidopsis* CIPK15-mediated ABA signaling. It was reported that CIPK15 interacts with two type-A phosphatases 2C ABI1 and ABI2 (Guo et al. 2002), both of which are the most characterized negative regulators of ABA signaling (Leung et al. 1994, 1997; Meyer et al. 1994; Gosti et al. 1999; Merlot et al. 2001; and also see the sections as follows), suggesting that they may possibly act as Ca^{2+}-sensors downstream or upstream of the ABI1 and ABI2 phosphatases when forming a protein CBL–CIPK complex for perceiving calcium signal. Additionally, an AP2 transcription factor ERF7 that negatively regulates ABA response was shown to be a kinase substrate of CIPK15 (Song et al. 2005), suggesting that CIPK15 may regulate ABA signaling more directly by phosphorylating transcription factor and modulating gene expression.

Additionally, SnRK3.11/CIPK24/SOS2, a key regulator of plant response to salt stress (Liu et al. 2000), was shown to interact with the type-A protein phosphatase 2C ABI2, suggesting that the CIPK kinase and ABI2 phosphatase may possibly control phosphorylation status of each other to relay salt signaling (Ohta et al. 2003). The SOS pathway is one of the CBL–CIPK signaling pathways conferring plant salt tolerance. The CBL4/SOS3 calcium sensor (Liu and Zhu 1998) activates the kinase SnRK3.11/SOS2 in response to salt stress, and the SOS2 kinase phosphorylates and activates SOS1 (Shi et al. 2000), a plasma membrane sodium/proton antiporter (Qiu et al. 2002), which keeps potassium homeostasis in plant cells under high salt stress. The interaction between ABI2 and SnRK3.11/SOS2 may constitute a point though which ABA exerts its effect on plant salt tolerance.

In contrast with the CIPK1, CIPK3, and CIPK15 that function as negative ABA-signaling regulators, CIPK6/SnRK3.14 and CIPK20/PKS8/SnRK3.6 were shown to be positive regulators of ABA signaling in *Arabidopsis*, which may antagonizes CIPK1/3/15 in ABA-signaling pathway. CIPK20/SnRK3.6/PKS18 (T169D, constitutively active form) overexpression lines show ABA-hypersensitive phenotypes, whereas RNAi plants show ABA-insensitive phenotypes (Gong et al. 2002). Overexpression of an active form of *Brassica napus* CIPK member BnCIPK6M in *Arabidopsis* confers ABA-hypersensitive phenotypes, while downexpression of its homologous gene the *Arabidopsis CIPK6* results in ABA-insensitive phenotypes (Chen et al. 2012).

As described earlier, *Arabidopsis* genome encodes at least 10 members in CBL and 25 members in CIPK gene family, which provides a high level of diversity, complexity, and flexibility in the function of the CBL–CIPK network. Further researches need to identify new members of CIPK-mediating ABA signaling and particularly upstream regulators/downstream targets to elucidate their functional mechanisms in ABA-signaling pathways.

8.3.5 SnRKs and ABA Signaling: A Summary

SnRKs are a class of Ser/Thr kinases and are recognized as a family member of the CDPK-SnRK superfamily, which can be subdivided into three subfamilies: SnRK1, SnRK2, and SnRK3. The *Arabidopsis* genome contains 38 SnRKs, of which 3 (SnRK1.1 to SnRK1.3) are SnRK1s, 10 (SnRK2.1 to SnRK2.10) are SnRK2s, and 25 are (SnRK3.1 to SnRK3.25) SnRK3s. The SnRK1 group is the most closely related to SNF1 from yeast and to AMP-activated protein kinases (AMPK) from animals, whereas the SnRK2 and SnRK3 groups appear to be plant-specific classes of kinases. The SnRK3s are the CBL-interacting protein kinases CIPKs.

SnRK1s likely function as a convergence point of ABA and sugar signaling pathways. The *Arabidopsis* SnRK1.1, a major form of the AtSnRK1s, positively regulates sugar- and ABA signaling in seedling growth, possibly by phosphorylating ABF transcription factors. The SnRK1s of wheat were also suggested to be involved in ABA signaling. Two members of the clade-A PP2Cs, ABI1 and PP2CA, inhibit SnRK1.1 activity by dephosphorylating it to repress SnRK1-mediated sugar signaling, while ABA promotes SnRK1-mediated sugar signaling by inhibiting the PP2Cs, which allows the coordination of ABA and energy signaling, strengthening the stress response through the cooperation of two key and complementary pathways (Fig. 8.2).

SnRK2s play critical roles in ABA-signaling pathway and regulate plant responses to environmental stresses. A wheat SnRK2, PKABA1, may mediate ABA signaling in seed development and germination, and a broad bean SnRK2, AAPK, regulates guard cell signaling in response to ABA. In *Arabidopsis*, OST1/SnRK2.6 is a key player in ABA regulation of stomatal movement. SnRK2.2 and SnRK2.3 function redundantly in ABA-induced inhibition of seed germination and postgermination growth. However, these functionally segregated kinases SnRK2.2, SnRK2.3, and SnRK2.6 still work redundantly in the three major ABA responses, acting together as a core signaling point in ABA-signaling pathways. Other members of the *Arabidopsis* SnRK2 subfamily, such as SnRK2.7/2.8, are likely to play roles in ABA and/or stress signaling. An *Arabidopsis* decuple mutant carrying mutations in all 10 members of the SnRK2 subfamily grows poorly under hyperosmotic stress conditions, revealing critical functions of the SnRK2s in osmotic stress signaling and tolerance. SnRK2s phosphorylate and activate ABA-responsive bZIP transcription factor ABFs and regulate ABA-responsive genes to mediate ABA signaling. In addition, guard cell anion channel SLAC1 and inward K^+ channel are substrates of the SnRK2.6/OST1. Whereas Ca^{2+}-dependent activation of such ion channels is directly associated with the CDPKs, the Ca^{2+}-independent activation of the ion channels are mainly mediated by SnRK2.6. Phosphoproteomic approaches identified a set of new potential substrates of SnRK2 protein kinase in *Arabidopsis*, suggesting a highly complicated SnRK2-mediated phosphorylation network (Fig. 8.2).

SnRK3s/CIPKs, together with corresponding CBLs, have been suggested to regulate Ca^{2+}-dependent ABA signaling in concert with CDPKs. In *Arabidopsis*, the interactions of CIPK15/SnRK3.1 with CBL1, or CIPK3/SnRK3.17 with CBL9, or CIPK1/SnRK3.16 with CBL9, negatively mediate ABA signaling, thought it still remains controversial whether CBL1 regulates ABA responses. CIPK15 interacts with ABI1 and ABI2 and targets an AP2 transcription factor

ERF7 that negatively regulates ABA response, suggesting that CIPK15 may regulate ABA signaling in cooperation with these two ABA-signaling components. Additionally, SnRK3.11/CIPK24/SOS2, a key regulator of plant response to salt stress, interacts with ABI2, which may constitute a point of ABA action on plant salt tolerance. CIPK6/SnRK3.14 and CIPK20/PKS8/SnRK3.6 are positive regulators of ABA signaling in *Arabidopsis*, which may possibly antagonize CIPK1/3/15 (Fig. 8.2). Multigene family-encoded CBLs and CIPKs provide a high level of diversity, complexity, and flexibility in the function of the CBL–CIPK network in ABA signaling.

8.4 Mitogen-Activated Protein Kinases (MAPKs) Involved in ABA Signaling

Mitogen-activated protein kinase (MAPK) cascades are universal signal transduction modules in eukaryotes, including yeasts, animals, and plants. These protein phosphorylation cascades link extracellular stimuli to a wide range of cell signaling pathways (Widmann et al. 1999; Chen and Thorner 2007; Colcombet and Hirt 2008). MAPKs are serine/threonine kinases that phosphorylate a variety of substrates, such as transcription factors, protein kinases, and cytoskeletal proteins. MAPKKs (or MEKs) are dual-specificity kinases that phosphorylate MAPKs on their threonine and tyrosine residues. MAPKKKs (or MEKKs) are serine/threonine kinases and phosphorylate MAPKKs on two serine/threonine residues. MAPKs form the terminal components of the MAPK sequential phosphorylation cascades and are activated by MAPKKs, which are in turn activated by MAPKKKs (Widmann et al. 1999; Ichimura et al. 2002; Chen and Thorner 2007; Colcombet and Hirt 2008). The *Arabidopsis* genome encodes 20 MAPKs, 10 MAPKs, and 80 MAPKs (Colcombet and Hirt 2008), which provides a potential for functional redundancy and diversity in cell signaling. MAPK cascades have been shown to be involved in responses to various environmental stresses and developmental processes (Ichimura et al. 2002; Colcombet and Hirt 2008).

Earlier studies revealed that some members of the MAPK family could be activated by ABA and potentially involved in ABA signaling. For example, several barley (*Hordeum vulgare*) MAPK isoforms that are activated by ABA (Knetsch et al. 1996), an ABA-inducible MAPK in *Vicia* guard cell protoplasts (Mori and Muto 1997), a *Pisum sativum* epidermal MAPK that may correlate with stomatal responses to ABA (Burnett et al. 2000), a rice (*Oryza sativa*) MAPK, OsMAPK5, which is involved in disease resistance and abiotic stress tolerance (Xiong and Yang 2003), and a maize (*Zea mays*) MAPK that may be involved in ABA-induced antioxidant defense (Zhang et al. 2006).

The studies in *Arabidopsis* help to understand the mechanisms of MAPK signaling in response to ABA. ABA treatment activates AtMPK3, and overexpression of AtMPK3 increases ABA sensitivity in the postgermination arrest of seedling growth (Lu et al. 2002), revealing a role of this MAPK in response of the seedling growth to ABA. In addition, transgenic downregulation of *MPK3* expression impairs ABA inhibition of stomatal opening and H_2O_2-induced stomatal closure, but does not affect ABA promotion of stomatal closure and ABA-induced H_2O_2 production, suggesting that MPK3 may function downstream of ROS in ABA inhibition of stomatal opening but not in ABA-induced stomatal closure (Gudesblat et al. 2007). Overexpression of AtMKK3 or its putative downstream kinases AtMPK1 and AtMPK2 enhance plant sensitivity to ABA, suggesting an AtMKK3-AtMPK1/2-coupled signaling cascade involved in ABA signaling (Hwa and Yang 2008). AtMKK1 and AtMPK6 were reported to link each other and mediate ABA-induced *CAT1* gene expression and H_2O_2 production (Xing et al. 2008). A signaling cascade of ABA-AtMKK1-AtMPK6-H_2O_2 was proposed where the ABA-activated AtMPK6 may be a substrate of AtMKK1 and acts upstream of CAT1 that regulates H_2O_2 production to mediate ABA signaling in seed germination and seedling growth (Xing et al. 2008). Two other MAPK members in *Arabidopsis*, AtMPK9 and AtMPK12, highly and preferentially expressed in guard cells, were shown to regulate guard cell signaling in response to ABA. Mutations in both *MPK9* and *MPK12* reduce ABA sensitivity in ABA-induced stomatal closure and ABA inhibition of stomatal opening, impair ABA and calcium activation of anion channels, and enhance transpiration water loss in leaves. The two MAPKs share functional redundancy and function as positive regulators downstream of ROS in guard cell ABA signaling (Jammes et al. 2009). Based on these findings, a model of ABA-ROS-MPK9/12 signaling cascade was proposed (Jammes et al. 2009), which is essentially consistent with the ABA-ROS-AtMPK3 cascade in ABA inhibition of stomatal opening as described earlier (Gudesblat et al. 2007). Wang et al. (2010) characterized biochemically an apple MAPK signaling cascade MdMKK1-MdMPK1, which is transiently activated by ABA. Expression of MdMKK1 or MdMPK1 in *Arabidopsis* confers ABA hypersensitivity in seed germination and seedling growth, and MdMPK1 phosphorylates the *Arabidopsis* ABI5 protein. They proposed that the apple MdMKK1-MdMPK1-coupled signaling cascade may function positively in ABA signaling by regulating the phosphorylation status of ABI5 or ABI5-like transcription factors (Wang et al. 2010). MdMKK1 and MdMPK1 are homologs of the *Arabidopsis* MKK2 and MPK6, respectively. The MdMKK1-MdMPK1-coupled signaling seems to be consistent with a previously reported AtMKK2-AtMPK4/6-coupled signaling cascade that regulates plant response to cold and salt stresses (Teige et al. 2004). Recently, a phosphoproteomic approach allowed to reveal a possible protein phosphorylation network in the ABA signaling pathway in *Arabidopsis,* where SnRK2 promotes the ABA-induced activation of the mitogen-activated protein kinases AtMPK1 and AtMPK2, suggesting that SnRK2 acts upstream of a MAPK cascade (Umezawa et al. 2013).

Whereas considerable progress has been made in the MAPK-mediated ABA signaling (Fig. 8.2), few upstream components and direct target of MAPK mediating ABA signaling have been identified so far. Therefore, further molecular genetic and cell biological studies are required to identify new members of the MAPK family, their upstream regulators, and direct targets involved in ABA signaling in order to understand the complex network of ABA-signaling pathways.

8.5 Receptor-Like Kinases (RLKs): Can They Sense ABA at the Cell Surface?

Cell surface-localized sensing mechanisms play fundamental roles in both plant and animal cells, which perceive, through plasma membrane receptors, extracellular signals from the environment and from other cells, activating downstream signaling cascades. Receptor-like kinases (RLKs) are a class of plasma membrane associated receptor kinase and found in metazoans and plants. Plant RLKs comprise one of the largest families, with more than 610 and 1131 members in *Arabidopsis* and rice (*Oryza sativa*), respectively (Shiu et al. 2004; Morillo and Tax 2006). Structurally, a typical receptor kinase consists of an extracellular domain, a single transmembrane region, and a cytoplasmic Ser/Thr protein kinase domain. Extracellular domains vary greatly to perceive a wide range of signals or stimuli, while intracellular kinase domains are relatively conserved to transduce signals. The RLKs may be classified into multiple subfamilies on the basis of the structure of their extracellular kinase domains. The biggest subfamily is leucine-rich repeat RLKs (LRR-RLKs), and the other subfamilies include cysteine-rich repeat (CRR) RLKs (CRKs), S-domain RLKs, domain of unknown function26 RLKs, and others (Shiu and Bleecker 2001). RLKs have been shown to regulate a wide range of plant developmental processes and environmental responses, such as the pathogen response, root, shoot and leaf development, meristem maintenance, floral organ abscission, organ shape regulation, cellular differentiation, symbiosis, self-incompatibility, and brassinosteroid signaling (reviewed in Morillo and Tax 2006; De Smet et al. 2009; Gish and Clark 2011).

An earlier study identified an ABA-induced LRR-RLK, RPK1, in *Arabidopsis* (Hong et al. 1997), which contains an extracellular LRR and a Ser/Thr kinase domain and are localized at the plasma membrane. Expression of the *RPK1*gene is also induced by dehydration, high salt, and low temperature (Hong et al. 1997), suggesting that RPK1 may be involved in plant response to environmental stresses. Genetic approaches showed that the loss of function of RPK1 reduces ABA sensitivity in seed germination, plant growth, stomatal closure, and gene expression in *Arabidopsis*, revealing a positive role of RPK1 in ABA signaling (Osakabe et al. 2005). Recently, RPK1 was shown to be involved in reactive oxygen species (ROS) signaling during abiotic stresses, and overproduction of *RPK1* enhances both water and oxidative stress tolerance in *Arabidopsis* (Osakabe et al. 2010). RPK1 was also reported to regulate ABA-induced leaf senescence (Lee et al. 2011), suggesting that RPK1 have multiple functions in ABA and stress signaling.

Hua et al. (2012) identified another member of the *Arabidopsis* LRR-RLK, named GHR1, as a positive regulator of guard cell signaling in response to H_2O_2 and ABA. The *ghr1* mutation impaired H_2O_2 activation of the calcium channel and ABA/H_2O_2 induction of stomatal closure. GHR1 functions downstream of the cytosolic players ABI1 and ABI2 and may control stomatal movement by directly interacting with, phosphorylating, and regulating SLAC1 anion channel, a key anion channel in stomatal signaling (see earlier description in CDPK and SnRK sections). This study shows that GHR1 is an important player in ABA and H_2O_2 signaling for stomatal movement.

Tanaka et al. (2012) identified an *Arabidopsis* receptor-like kinase, ARCK1, as a negative ABA signaling regulator during the postgermination growth. ARCK1 is a receptor-like cytosolic protein kinase, which belongs to the cysteine-rich repeat RLK (CRK) subfamily. The extracellular region of the cytosolic CRK protein contains two copies of the DUF26 domain containing four conserved cysteines, three of which form the motif C-8X-C-2X-C, which is presumed to be involved in formation of the 3D structure of the protein through disulfide bonds and plays roles in protein–protein interactions. Another member of the CRK subfamily, CRK36, interacts with ARCK1 in the plasma membrane, phosphorylates ARCK1 in vitro, and functions as a negative ABA signaling regulator during the postgermination growth (Tanaka et al. 2012). These results suggest that CRK36 and ARCK1 form a functional complex in the plasma membrane to regulate ABA signaling during the postgermination growth (Tanaka et al. 2012).

Sheng Luan's group (Yu et al. 2012) identified a receptor-like kinase, FERONIA (FER), a positive regulator of auxin-promoted growth (Duan et al. 2010), as a negative regulator of ABA signaling. They described a cross talk mechanism between ABA- and auxin-signaling pathways where FER suppress ABA response through activation of ABI2, a negative regulator of ABA signaling. The plasma membrane-localized FER kinase interacts with guanine exchange factors GEFs including GEF1, GEF4, and GEF10, which, in turn, activate GTPase ROP11. The ROP11 protein interacts with the A-type PP2C ABI2 phosphatase and enhances its activity, thereby linking the FER pathway with the inhibition of ABA signaling (for the cytosolic ABA receptors PYR/PYL/RCAR and A-type PP2C-mediated ABA signaling cascades, see Cutler et al. 2010, and Chap. 6 of this book). Interestingly, at the same time, an independent group reports that ROP11 GTPase negatively regulates ABA signaling by protecting ABI1 phosphatase activity from inhibition by the ABA receptor RCAR1/PYL9 in *Arabidopsis* (Li et al. 2012a, b). Thus, this FER–GEFs–ROP11-PP2C linked pathway may function in parallel with the PYR/PYL/RCAR-PP2C-coupled pathway (Fig. 8.2); however, it remains unknown whether FER may function as a cell surface receptor for ABA.

Although candidate plasma membrane receptors for ABA, such as GPCR-type G proteins GTG1/GTG2 (Pandey et al. 2009), and cytosolic ABA receptors PYR/PYL/RCAR-mediated signaling pathway, have been described, an ABA-signaling pathway from cell surface signal perception to downstream signaling processes remains to be elucidated. Can RLKs serve as a plasma membrane ABA receptor? Genetic and cell biological studies are required in the future to further identify members of the RLK family and their targets involved in ABA signaling in order to uncover primary signaling events that take place in the cell surface for ABA perception and thus to understand the complicated ABA-signaling pathways.

8.6 Protein Phosphatases Involved in ABA Signaling: PP2C, PP2A, and Others

8.6.1 Protein Phosphatase Family: PTP, PPP (PP1, PP2A, PP2B), and PPM (PP2C, PDP)

Protein phosphatases can be divided into two major classes: protein tyrosine phosphatases (PTPs) and protein serine/threonine phosphatases. Protein tyrosine phosphatases include PTPs and dual-specificity phosphatases (DSPTPs). The Ser/Thr-specific phosphoprotein phosphatases (PPPs) execute the major phosphatase activities in eukaryotes (Olsen et al. 2006). Based upon differential sensitivity to small molecule inhibitors, the protein serine/threonine phosphatases are classified into the PPP and PPM (for metal ion-dependent protein phosphatase) gene families. The PPP family includes types 1 (PP1), 2A (PP2A), 2B (PP2B), whereas the PPM family includes type 2C (PP2C) and pyruvate dehydrogenase phosphatase (PDP) (Ingebritsen and Cohen 1983; Cohen 1997; Smith and Walker 1996; Luan 2003). Among the PP2 members that vary in their subunit structure, divalent cation requirements, and substrate specificities, PP2A is a heterotrimer consisting of a catalytic C subunit and two distinct regulatory A and B subunits, which does not require divalent cations for its activity; PP2B is activated by Ca^{2+} and is a heterodimer of a catalytic A subunit and a regulatory B subunit; PP2B is an important Ca^{2+}-regulated PPP in other eukaryotes, but is absent in plants; PP2C exists as a monomer and requires Mg^{2+} for its activity (Cohen 1989, 1997; Smith and Walker 1996; Luan 2003). Three major classes of the plant protein phosphatases are homologs of the mammalian type-1, -2A, and -2C protein serine/threonine phosphatases (Smith and Walker 1996; Luan 2003; Schweighofer et al. 2004). The *Arabidopsis* genome encodes more than 100 phosphatase catalytic subunit sequences (Arino et al. 1993; Smith and Walker 1996; Schweighofer et al. 2004), suggesting that the protein phosphatases may function in multiple physiological processes in plants. Indeed, molecular genetic and biochemical studies reveal important roles of these enzymes in plant cell events such as signal transduction, regulation of metabolism, cell cycle progression, hormonal regulation, and stress responses (Smith and Walker 1996; Luan 2003; Schweighofer et al. 2004; Farkas et al. 2007). Especially, some of the PP2C members have been identified as critical components of ABA signaling, working together with a class of the cytosolic PYR/PYL/RCAR receptors for ABA (reviewed in Cutler et al. 2010, and see Chap. 6 of this book).

8.6.2 Group-A PP2Cs: Central, Upstream Repressors of ABA Signaling

The *Arabidopsis* genome encodes 76 PP2C-type phosphatase candidates, which are divided into ten groups (from A to J) (Kerk et al. 2002; Schweighofer et al. 2004). Group A contains nine members, six of which have been identified as important, negative regulators in ABA signaling, including ABI1, ABI2, HAB1, HAB2, AHG1, and AHG3/PP2CA.

Early forward genetic approaches allowed to isolate the *abi1-1* and *abi2-1* mutants that show dominant, strong ABA-insensitive phenotypes in the major ABA responses including reduced seed dormancy, ABA-induced inhibition of seed germination, postgermination growth arrest, promotion of stomatal closure, and inhibition of stomatal opening (Koornneef et al. 1982; Leung et al. 1994, 1997; Meyer et al. 1994). The mutations were shown, by the map-based cloning technique, to take place in the genes encoding two homologous PP2C proteins, named ABI1 and ABI2 (Leung et al. 1994, 1997; Meyer et al. 1994). Further studies revealed that ABI1 and ABI2 function redundantly as negative regulators of ABA signaling essentially because loss of function of these genes results in ABA-hypersensitive phenotypes and double mutation increases the ABA hypersensitivity (Sheen 1998; Gosti et al. 1999; Merlot et al. 2001). ABI1 and ABI2 contribute nearly 50 % of the ABA-induced PP2C activity, suggesting that these two PP2C members play essential roles in the PP2C-mediated ABA signaling, but other PP2Cs are also involved in ABA signaling (Merlot et al. 2001), which is verified by pursuant studies. HAB1, HAB2 (Leonhardt et al. 2004; Saez et al. 2004, 2006; Robert et al. 2006), and AHG1 (Nishimura et al. 2007) and AHG3/PP2CA (Cherel et al. 2002; Kuhn et al. 2006; Yoshida et al. 2006a; Lee et al. 2009) were also identified as negative ABA-signaling regulators, all of which belong to the clade-A PP2Cs (Schweighofer et al. 2004). The *hab1 pp2ca* and *abi1 pp2ca* double mutants show ABA-hypersensitive phenotypes in seed germination, postgermination growth, reduced water loss and enhanced resistance to drought stress, and two triple mutants *hab1 abi1 abi2* and *hab1 abi1 pp2ca* show an extreme ABA hypersensitivity in the major ABA responses, impaired growth and partial constitutive response to endogenous ABA (Rubio et al. 2009). These genetic data reveal that these six members of the clade-A PP2Cs, functioning negatively and redundantly, play central roles in ABA signaling.

The remaining three PP2Cs, HAI1, HAI2/AIP1, and HAI3 (with *Arabidopsis* genome locus At5g59220, At1g07430, and At2g29380, respectively), were recently reported to have both effect on ABA-independent low water potential phenotypes and effect on ABA-sensitivity phenotypes in seed germination and seedling growth; however, double or triple mutants of the *HAI* genes have moderate ABA hypersensitivity in seedling growth, while these mutants have ABA-insensitive seed germination, which contrasts with the phenotypes of other clade-A PP2C mutants (Antoni et al. 2012; Bhaskara et al. 2012). Only was HAI1 identified clearly as a negative regulator of ABA signaling (Antoni et al. 2012). This suggests a level of functional differentiation among the clade-A PP2Cs in ABA signaling (Bhaskara et al. 2012).

Recent advances, especially with discovery of the cytosolic PYR/PYL/RCAR receptors for ABA (reviewed in Cutler et al. 2010, and see Chap. 6 of this book), have uncovered how the clade-A PP2Cs work in ABA signaling. Early studies found that the point mutations of both *abi1-1* and *abi2-1* in ABI1 and ABI2 PP2Cs, respectively, involve substitution of the same amino acid residue (substitution of an aspartic acid residue for a glycine residue) at an equivalent position close to the Mg^{2+}-coordinating center (G180D for *abi1-1* mutation and G168D for *abi2-1* mutation) in the catalytic domain (Leung et al. 1994, 1997; Meyer et al. 1994), but it remained puzzling, for a long time, how this mutation results in strong ABA insensitivity (Schweighofer et al. 2004; Cutler et al. 2010). Currently, it has been known that the ABA-activated PYR/PYL/RCAR receptors interact with and antagonize the clade-A PP2Cs that dephosphorylate and inactivate their substrates SnRK2s and SnRK1s (Ma et al. 2009; Park et al. 2009; Fujii et al. 2009; Cutler et al. 2010; Rodrigues et al. 2013). It has been uncovered that the *abi1-1* and *abi2-1* mutations disrupt the interactions between the PP2Cs and the ABA receptors PYR/PYL/RCAR, but do not affect the interactions between the PP2Cs and their downstream regulatory components SnRK2s (Yoshida et al. 2006b; Ma et al. 2009; Park et al. 2009; Umezawa et al. 2009; Vlad et al. 2009). These mutations reduce PP2C activity of both ABI1 and ABI2 in vitro (Leung et al. 1997; Leube et al. 1998; Gosti et al. 1999), but appear to enhance constitutively the dephosphorylation activity of the PP2Cs for their natural substrates SnRK2s in vivo (Umezawa et al. 2009), which explains the dominant mutation of strong ABA-insensitive phenotypes of the *abi1-1* and *abi2-1* mutants (Leung et al. 1994, 1997; Meyer et al. 1994; Gosti et al. 1999), and is consistent with their central, upstream roles in ABA signaling.

In addition to the SnRK2s, the *Arabidopsis* clade-A PP2Cs ABI1 and/or ABI2 may use CDPK, SnRK1, and SnRK3/CIPK as substrates, such as CPK21, CPK23, SnRK1.1, SnRK3.1/CIPK15, and SnRK3.11/CIPK24/SOS2 (see earlier description in the CDPKs and SnRKs sections). The multiple substrates may be likely to allow the clade-A PP2Cs to execute specific and/or redundant functions in ABA signaling (Fig. 8.2).

It is noteworthy that two homologous members of clade B PP2Cs, PP2C5 and AP2C1, were also reported to regulate ABA signaling (Brock et al. 2010), suggesting that PP2Cs of other groups than A group may also be involved in ABA signaling. Further studies will be needed to identify additional members of PP2Cs involved in ABA signaling and to elucidate their functional mechanisms in the complicated ABA-signaling network.

8.6.3 PP2As and PP2A-Like Proteins: Positive or Negative Regulators in ABA Signaling?

As mentioned above, the PPP family proteins are composed of PP1, PP2A, and PP2B proteins, which are ubiquitous enzymes in all eukaryotes except for PP2B that is absent in plants, while their functions are largely unknown in higher plants. In the *Arabidopsis* genome, twenty-six catalytic subunits C-encoding genes of the PPP family have been found, which are related to PP1, PP2A, and a novel class of phosphatases PP4, PP5, PP6, and PP7, of which PP4 and PP6 are PP2A-like proteins according to their sequence similarities (Farkas et al. 2007). Protein phosphatase 2A is a holoenzyme composed of a scaffolding subunit A, a regulatory subunit B, and a catalytic subunit C. The subunits A and C directly interact to form the AC core enzyme, which binds subunit B to form a variety of heterotrimeric complexes (Cohen 1989, 1997; Mayer-Jaekel and Hemmings 1994; Smith and Walker 1996; Luan 2003). The regulatory subunit B determines the substrate specificity, subcellular localization, and catalytic activity of the PP2As (Virshup 2000; Janssens and Goris 2001; Luan 2003). For PP2A family proteins, the *Arabidopsis* genome encodes five catalytic C subunits (Arino et al. 1993; Smith and Walker 1996; Kerk et al. 2002; Schweighofer et al. 2004), three scaffolding A subunits (Slabas et al. 1994; Garbers et al. 1996), and 17 regulatory B subunits (Haynes et al. 1999; Terol et al. 2002; Farkas et al. 2007). The multiple isoforms may combine to form a wide variety of functional complexes, theoretically accounting for 255 different heterotrimer combinations that could eventually perform distinct functions in the regulation of different processes or redundant functions in specific processes (Janssens et al. 2005; Farkas et al. 2007).

Earlier pharmacological studies suggested that protein phosphatases PP1 and PP2A might function as positive or negative regulators of ABA signaling (Schmidt et al. 1995; Esser et al. 1997; Hey et al. 1997; Wu et al. 1997). Further genetic approaches in *Arabidopsis* allowed to identify a mutant *rcn1* with reduced levels of PP2A activity, harboring a recessive disruption in the *RCN1* gene encoding a guard cell-expressed PP2A scaffolding A subunit (PP2A-A1); the mutant shows ABA insensitivity in ABA inhibition of seed germination, ABA-induced stomatal closing, and ABA activation of slow anion channels, suggesting that the RCN1 is a positive regulator of ABA signal transduction in *Arabidopsis* (Kwak, et al. 2002). However, RCN1 is also involved in auxin transport and ethylene responses whose mutants therefore exhibit pleiotropic phenotypes resulting from a complicated cross talk among the three hormones (Rashotte et al. 2001; Larsen and Cancel 2003). A T-DNA insertional mutant *pp2ac-2* with loss of function of a PP2A catalytic subunit (PP2Ac) was identified more recently, in which the total PP2A activity is reduced (Pernas et al. 2007). The *pp2ac-2* mutant shows ABA hypersensitivity in seed germination, dormancy, seedling growth, and ABA-dependent gene expression, while the *PP2Ac-2* overexpressing lines show ABA-insensitive phenotypes (Pernas et al. 2007). These results reveal that PP2Ac-2 is a negative regulator of ABA signal transduction in *Arabidopsis*. Given that PP2Ac-2 is a catalytic subunit with a specific role in ABA signaling, the identification of PP2Ac-2 as a negative regulator of ABA responses suggests that PP2As are likely to function as a negative, but not positive, component in the ABA-signaling pathway (Pernas et al. 2007).

Recently, a study showed that FyPP1 and FyPP3 (for Phytochrome-associated serine/threonine protein phosphatase1/3), two homologous catalytic subunits of PP6, are negatively involved in ABA signaling in *Arabidopsis* (Dai et al. 2013). It is noteworthy that there are high sequence similarities among the C subunits of PP6 and PP2A phosphatases, while the PP6 activity requires Zn^{2+}, but PP2A does not require divalent cations for its activity as mentioned above (Cohen 1989, 1997; Smith and Walker 1996; Luan 2003; Wang et al. 2007; Dai et al. 2012). Genetic approaches showed that the loss-of-function mutants of *FyPP* display ABA-hypersensitive phenotypes in seed germination and postgermination growth (Dai et al. 2013). The FyPP proteins directly interact with and dephosphorylate ABI5; and further genetic evidence demonstrated an antagonistic interaction between FyPP1/3 with SnRK2 in regulating ABA responses (Dai et al. 2013). These findings describe a model that FyPP1 and FyPP3 function as two negative regulators of ABA signaling through dephosphorylating ABI5 to counteract SnRK2s that phosphorylate ABI5 to promote ABA signaling (Fig. 8.2).

The high degree of genetic redundancy caused by manifold possible combinations of functional complexes may have limited the experimental evidence concerning processes regulated by PP2As. Further identification of more members of PP2As or PP2A heterotrimer combinations involved in ABA signaling will be necessary. Additionally, it merits to be answered whether PP2As may function as the PP2A-like protein PP6 to cooperate with SnRK2s to catalyze reversible ABI5 phosphorylation in ABA signaling, or whether they may function, in a manner similar to PP2Cs, as negative regulators directly downstream of the PYR/PYL/RCAR receptors for ABA and upstream of SnRK2s in ABA signaling. The elucidation of functional mechanisms of PP2As will deepen our understanding of ABA-signaling pathway.

8.7 Concluding Remarks and Perspectives

Extensive research efforts with integrative approaches have led to considerable progresses to understand highly complex ABA-signaling network involving reversible protein phosphorylation catalyzed by protein kinases and phosphatases. A wide array of protein kinases and phosphatases have been identified as crucial players in ABA signaling, among which CDPKs, SnRKs, MAPKs, RLKs, PP2Cs, and PP2As are relatively best characterized and their functional mechanisms in ABA signaling begin to be understood (Fig. 8.2). Particularly, a PYR/PYL/RCAR-PP2C-SnRK2s-ABI5/ABFs-linked protein dephosphorylation–phosphorylation signaling cascade has been described, and most recent studies supplement this pathway with both a reversible phosphorylation of SnRK2s catalyzed by BIN2/BILs and PP2Cs and a reversible ABI5 phosphorylation catalyzed by SnRK2s and FyPP1/3 (Fig. 8.2). Additionally, some intriguing signaling cascades, such as CDPK-SLAC1/SLAH3-coupled signaling and ABA-ROS-MPK9/12 signaling cascade in guard cells, and a plasma membrane RLK-mediated FER–GEFs–ROP11-PP2C linked pathway, have been described (Fig. 8.2). However, many more regulatory interactions of protein kinases and phosphatases in ABA signaling are probably still hidden. Further studies will be needed to provide compelling evidence to link these signaling processes to the bigger picture of ABA-signaling pathway/network from the primary perception events of ABA signal to the most downstream ABA-specific physiological response. Special attention should be attracted to elucidating the possible cooperation of the identified CDPKs, SnRKs, MAPKs, RLKs, PP2Cs, and PP2As, and establishing molecular, cellular, and genetic links among them in ABA-signaling pathways. The multiple cross talks between them should provide additional layers of regulation to fine-tune plant ABA and stress responses. Future molecular, cellular, genetic, genomic, and phosphoproteomic (combined with mass spectrometry) studies should lead to more precise understanding of specific and redundant roles of the identified protein kinases and phosphatases especially through identifying new in vivo substrates of these kinases and phosphatases, and to finding new members of these protein kinase/phosphatase families and additional, new kinase and phosphatase families functioning in the ABA-signaling networks. These future studies should provide new insights into the reversible protein phosphorylation-mediated ABA signaling to understand its integrated roles in diverse biological responses in plants.

Acknowledgments The authors thank all colleagues who provided unpublished results and apologize to those whose research was not discussed due to page limitation. The research on protein kinases and phosphatases in the author's laboratory was supported by the grants from the National Key Basic Research Program of China (2012CB114300-002), National Natural Science Foundation of China, and the Ministry of Agriculture of China (Grant 2013ZX08009003).

References

Abbasi F, Onodera H, Toki S, Tanaka H, Komatsu S. OsCDPK13, a calcium-dependent protein kinase gene from rice, is induced by cold and gibberellin in rice leaf sheath. Plant Mol Biol. 2004;55:541–52.

Acharya BR, Jeon BW, Zhang W, Assmann SM. Open Stomata 1 (OST1) is limiting in abscisic acid responses of *Arabidopsis* guard cells. New Phytol. 2013;200:1049–63.

Albrecht V, Weinl S, Blazevic D, D'Angelo C, Batistic O, Kolukisaoglu U, Bock R, Schulz B, Harter K, Kudla J. The calcium sensor CBL1 integrates plant responses to abiotic stresses. Plant J. 2003;36:457–70.

Anderberg RJ, Walker-Simmons MK. Isolation of a wheat cDNA clone for an ABA-inducible transcript with homology to protein kinases. Proc Natl Acad Sci USA. 1992;89:10183–7.

Arino J, Perez-Callejon E, Cunillera N, Camps M, Posas F, Ferrer A. Protein phosphatases in higher plants: multiplicity of type 2A phosphatases in *Arabidopsis thaliana*. Plant Mol Biol. 1993;21:475–85.

Asano T, Hakata M, Nakamura H, Aoki N, Komatsu S, Ichikawa H, Hirochika H, Ohsugi R. Functional characterisation of OsCPK21, a calcium-dependent protein kinase that confers salt tolerance in rice. Plant Mol Biol. 2011;75:179–91.

Asano T, Hayashi N, Kobayashi M, Aoki N, Miyao A, Mitsuhara I, Ichikawa H, Komatsu S, Hirochika H, Kikuchi S, Ohsugi R. A rice calcium-dependent protein kinase OsCPK12 oppositely modulates salt-stress tolerance and blast disease resistance. Plant J. 2012;69:26–36.

Asano T, Tanaka N, Yang G, Hayashi N, Komatsu S. Genome-wide identification of the rice calcium-dependent protein kinase and its closely related kinase gene families: comprehensive analysis of the CDPKs gene family in rice. Plant Cell Physiol. 2005;46:356–66.

Bachmann M, McMichael RMJ, Huber JL, Kaiser WM, Huber SC. Partial purification and characterization of a calcium-dependent protein kinase and an inhibitor protein required for inactivation of spinach leaf nitrate reductase. Plant Physiol. 1995;108:1083–91.

Bachmann M, Shiraishi N, Campbell WH, Yoo BC, Harmon AC, Huber SC. Identification of a Ser-543 as the major regulatory phosphorylation site in spinach leaf nitrate reductase. Plant Cell. 1996;8:505–17.

Benetka W, Mehlmer N, Maurer-Stroh S, Sammer M, Koranda M, Neumüller R, Betschinger J, Knoblich JA, Teige M, Eisenhabe F. Experimental testing of predicted myristoylation targets involved in asymmetric cell division and calcium-dependent signalling. Cell Cycle. 2008;7:3709–19.

Bhaskara GB, Nguyen TT, Verslues PE. Unique drought resistance functions of the highly ABA-induced clade a protein phosphatase 2Cs. Plant Physiol. 2012;160:379–95.

Boudsocq M, Barbier-Brygoo H, Lauriere C. Identification of nine sucrose nonfermenting 1-related protein kinases 2 activated by hyperosmotic and saline stresses in *Arabidopsis thaliana*. J Biol Chem. 2004;279:41758–66.

Boudsocq M, Sheen J. Stress signaling II: calcium sensing and signaling. In: Pareek A, Sopory SK, Bohnert HJ, Govindjee X, editors. Abiotic stress adaptation in plants: physiological, molecular and genomic foundation. New York: Springer; 2010. p. 75–90.

Boudsocq M, Sheen J. CDPKs in immune and stress signaling. Trends Plant Sci. 2013;18:30–40.

Boudsocq M, Willmann MR, McCormack M, Lee H, Shan L, He P, Bush J, Cheng SH, Sheen J. Differential innate immune signalling via Ca^{2+} sensor protein kinases. Nature. 2010;464:418–22.

Brandt B, Brodsky DE, Xue S, Negi J, Iba K, Kangasjärvi J, Ghassemian M, Stephan AB, Hu H, Schroeder JI. Reconstitution of abscisic acid activation of SLAC1 anion channel by CPK6 and OST1 kinases and branched ABI1 PP2C phosphatase action. Proc Natl Acad Sci USA. 2012;109:10593–8.

Bradford KJ, Downie AB, Gee OH, Alvarado V, Yang H, Dahal P. Abscisic acid and gibberellin differentially regulate expression of genes of the SNF1-related kinase complex in tomato seeds. Plant Physiol. 2003;132:1560–76.

Brock AK, Willmann R, Kolb D, Grefen L, Lajunen HM, Bethke G, Lee J, Nurnberger T, Gust AA. The *Arabidopsis* mitogen-activated protein kinase phosphatase PP2C5 affects seed germination, stomatal aperture, and abscisic acid-inducible gene expression. Plant Physiol. 2010;153:1098–111.

Burnett EC, Desikan R, Moser RC, Neill SJ. ABA activation of an MBP kinase in *Pisum sativum* epidermal peels correlates with stomatal responses to ABA. J Exp Bot. 2000;51:197–205.

Cai Z, Liu J, Wang H, Yang C, Chen Y, Li Y, Pan S, Dong R, Tang G, deDios Barajas-Lopez J, Fujii H, Wang X. GSK3-like kinases positively modulate abscisic acid signaling through phosphorylating subgroup III SnRK2s in *Arabidopsis*. Proc Natl Acad Sci USA. 2014;111:9651–6.

Chen L, Ren F, Zhou L, Wang QQ, Zhong H, Li XB. The *Brassica napus* calcineurin B-like 1/CBL-interacting protein kinase 6 (CBL1/CIPK6) component is involved in the plant response to abiotic stress and ABA signaling. J Exp Bot. 2012;63:6211–22.

Chen RE, Thorner J. Function and regulation in MAPK signaling pathways: lessons learned from the yeast *Saccharomyces cerevisiae*. Biochim Biophys Acta. 2007;1773:1311–40.

Cheng WH, Endo A, Zhou L, Penney J, Chen HC, Arroyo A, Leon P, Nambara E, Asami T, Seo M. A unique short-chain dehydrogenase/reductase in *Arabidopsis* glucose signaling and abscisic acid biosynthesis and functions. Plant Cell. 2002a;14:2723–43.

Cheng SH, Willmann MR, Chen HC, Sheen J. Calcium signaling through protein kinases. The *Arabidopsis* calcium dependent protein kinase gene family. Plant Physiol. 2002b;129:469–85.

Cheong YH, Kim K, Pandey GK, Gupta R, Grat JJ, Luan S. CBL1, a calcium sensor that differentially regulates salt, drought, and cold responses in *Arabidopsis*. Plant Cell. 2003;15:1833–45.

Cherel I, Michard E, Platet N, Mouline K, Alcon C, Sentenac H, Thibaud JB. Physical and functional interaction of the *Arabidopsis* K channel AKT2 and phosphatase AtPP2CA. Plant Cell. 2002;14:1133–46.

Choi H, Park HJ, Park JH, Kim S, Im MY, Seo HH, Kim YW, Hwang I, Kim SY. *Arabidopsis* calcium-dependent protein kinase CPK32 interacts with ABF4, a transcriptional regulator of abscisic acid-responsive gene expression, and modulates its activity. Plant Physiol. 2005;139:1750–61.

Coca M, San Segundo B. AtCPK1 calcium-dependent protein kinase mediates pathogen resistance in *Arabidopsis*. Plant J. 2010;63:526–40.

Coello P, Hey SJ, Halford NG. The sucrose non-fermenting-1-related (SnRK) family of protein kinases: potential for manipulation to improve stress tolerance and increase yield. J Exp Bot. 2011;62:883–93.

Coello P, Hirano E, Hey SJ, Muttucumaru N, Martinez-Barajas E, Parry MA, Halford NG. Evidence that abscisic acid promotes degradation of SNF1-related protein kinase (SnRK) 1 in wheat and activation of a putative calcium-dependent SnRK2. J Exp Bot. 2012;63:913–24.

Cohen P. Novel protein serine/threonine phosphatases: variety is the spice of life. Trends Biochem Sci. 1997;22:245–51.

Cohen P. Structure and regulation of protein phosphatases. Annu Rev Biochem. 1989;58:453–508.

Colcombet J, Hirt H. *Arabidopsis* MAPKs: A complex signalling network involved in multiple biological processes. Biochem J. 2008;413:217–26.

Curran A, Chang IF, Chang CL, Garg S, Miguel RM, Barron YD, Li Y, Romanowsky S, Cushman JC, Gribskov M. Calcium-dependent protein kinases from *Arabidopsis* show substrate specificity differences in an analysis of 103 substrates. Front Plant Sci. 2011;2:36. doi :10.3389/fpls.2011.00036.

Cutler SR, Rodriguez PL, Finkelstein RR, Abrams SR. Abscisic acid: emergence of a core signaling network. Annu Rev Plant Biol. 2010;61:651–79.

Dai M, Zhang C, Kania U, Chen F, Xue Q, Mccray T, Li G, Qin G, Wakeley M, Terzaghi W, Wan J, Zhao Y, Xu J, Friml J, Deng XW, Wang H. A PP6-type phosphatase holoenzyme directly regulates PIN phosphorylation and auxin efflux in *Arabidopsis*. Plant Cell. 2012;24:2497–514.

Dai M, Xue Q, Mccray T, Margavage K, Chen F, Lee JH, Nezames CD, Guo L, Terzaghi W, Wan JM, Deng XW, Wang HY. The PP6 phosphatase regulates ABI5 phosphorylation and abscisic acid signaling in *Arabidopsis*. Plant Cell. 2013;25:517–34.

D'Angelo C, Weinl S, Batistic O, Pandey G, Cheong Y, Schultke S, Albrecht V, Ehlert B, Schulz B, Harter K. Alternative complex formation of the Ca^{2+}-regulated protein kinase CIPK1 controls abscisic acid-dependent and independent stress responses in *Arabidopsis*. Plant J. 2006;48:857–72.

De Smet I, Voss U, Jürgens G, Beeckman T. Receptor-like kinases shape the plant. Nat Cell Biol. 2009;11:1166–73.

DeFalco TA, Bender KW, Snedden WA. Breaking the code: Ca^{2+} sensors in plant signalling. Biochem J. 2010;425:27–40.

Demir F, Horntrich C, Blachutzik JO, Scherzer S, Reinders Y, Kierszniowska S, Schulze WX, Harms GS, Hedrich R, Geiger D. *Arabidopsis* nanodomain-delimited ABA signaling pathway regulates the anion channel SLAH3. Proc Natl Acad Sci USA. 2013;110:8296–301.

Ding Y, Cao J, Ni L, Zhu Y, Zhang A, Tan M, Jiang M. ZmCPK11 is involved in abscisic acid-induced antioxidant defense and functions upstream of ZmMPK5 in abscisic acid signalling in maize. J Exp Bot. 2013;64:871–84.

Duan Q, Kita D, Li C, Cheung AY, Wu HM. FERONIA receptor-like kinase regulates RHO GTPase signaling of root hair development. Proc Natl Acad Sci USA. 2010;107:17821–6.

Esser JE, Liao YJ, Schroeder JI. Characterization of ion channel modulator effects on ABA- and malate-induced stomatal movements: Strong regulation by kinase and phosphatase inhibitors, and relative insensitivity to mastoparans. J Exp Bot. 1997;48:539–50.

Farkas I, Dombradi V, Miskei M, Szabados L, Koncz C. *Arabidopsis* PPP family of serine/threonine phosphatases. Trends Plant Sci. 2007;12:169–76.

Ferrando A, Koncz-Kalman Z, Farras R, Tiburcio A, Schell J, Koncz C. Detection of in vivo protein interactions between Snf1-related kinase subunits with intron-tagged epitope-labelling in plant cells. Nucleic Acids Res. 2001;29:3685–93.

Franz S, Ehlert B, Liese A, Kurth J, Cazale AC, Romeis T. Calcium-dependent protein kinase CPK21 functions in abiotic stress response in *Arabidopsis thaliana*. Mol Plant. 2011;4:83–96.

Fujii H, Chinnusamy V, Rodrigues A, Rubio S, Antoni R, Park SY, Cutler SR, Sheen J, Rodriguez PL, Zhu JK. In vitro reconstitution of an ABA signaling pathway. Nature. 2009;462:660–4.

Fujii H, Verslues PE, Zhu JK. Identification of two protein kinases required for abscisic acid regulation of seed germination, root growth, and gene expression in *Arabidopsis*. Plant Cell. 2007;19:485–94.

Fujii H, Verslues PE, Zhu JK. *Arabidopsis* decuple mutant reveals the importance of SnRK2 kinases in osmotic stress responses in vivo. Proc Natl Acad Sci USA. 2011;108:1717–22.

Fujii H, Zhu JK. *Arabidopsis* mutant deficient in 3 abscisic acid-activated protein kinases reveals critical roles in growth, reproduction, and stress. Proc Natl Acad Sci USA. 2009;106:8380–5.

Furihata T, Maruyama K, Fujita Y, Umezawa T, Yoshida R, Shinozaki K, Yamaguchi-Shinozaki K. Abscisic acid dependent multisite phosphorylation regulates the activity of a transcription activator AREB1. Proc Natl Acad Sci USA. 2006;103:1988–93.

Fujita Y, Nakashima K, Yoshida T, Katagiri T, Kidokoro S, Kanamori N. Three SnRK2 protein kinases are the main positive regulators of abscisic acid signaling in response to water stress in *Arabidopsis*. Plant Cell Physiol. 2009;50:2123–32.

Garbers C, DeLong A, Derue`re J, Bernasconi P, Soll D. A mutation in protein phosphatase 2A regulatory subunit A affects auxin transport in *Arabidopsis*. EMBO J. 1996;15:2115–24.

Geiger D, Maierhofer T, Al-Rasheid KAS, Scherzer S, Mumm P, Liese A, Ache P, Wellmann C, Marten I, Grill E, Romeis T, Hedrich R. Stomatal closure by fast abscisic acid signaling is mediated by the guard cell anion channel SLAH3 and the receptor RCAR1. Sci Signal 4: ra32. 2011.doi:10.1126/scisignal.2001346.

Geiger D, Scherzer S, Mumm P, Marten I, Ache P, Matschi S, Liese A, Wellmann C, Al-Rasheid KAS, Grill E, Romeis T, Hedrich R. Guard cell anion channel SLAC1 is regulated by CDPK protein kinases with distinct Ca^{2+} affinities. Proc Natl Acad Sci USA. 2010;107:8023–8.

Geiger D, Scherzer S, Mumm P, Stange A, Marten I, Bauer H, Ache P, Matschi S, Liese A, Al-Rasheid KA, et al. Activity of guard cell anion channel SLAC1 is controlled by drought-stress signaling kinase-phosphatase pair. Proc Natl Acad Sci USA. 2009;106:21425–30.

Gish LA, Clark SE. The RLK/Pelle family of kinases. Plant J. 2011;66:117–27.

Gomez-Cadenas A, Verhey SD, Holappa LD, Shen Q, Ho THD, Walker-Simmons MK. An abscisic acid-induced protein kinase, PKABA1, mediates abscisic acid-suppressed gene expression in barley aleurone layers. Proc Natl Acad Sci USA. 1999;96:1767–72.

Gomez-Cadenas A, Zentella R, Walker-Simmons MK, Ho THD. Gibberellin/abscisic acid antagonism in barley aleurone cells: site of action of the protein kinase PKABA1 in relation to gibberellin signaling molecules. Plant Cell. 2001;13:667–79.

Gong D, Zhang C, Chen X, Gong Z, Zhu JK. Constitutive activation and transgenic evaluation of the function of an *Arabidopsis* PKS protein kinase. J Biol Chem. 2002;277:42088–96.

Guo Y, Xiong L, Song C, Gong D, Halfter U, Zhu J. A calcium sensor and its interacting protein kinase are global regulators of abscisic acid signaling in *Arabidopsis*. Dev Cell. 2002;3:233–44.

Gosti F, Beaudoin N, Serizet C, Webb AAR, Vartanian N, Giraudat J. ABI1 protein phosphatase 2C is a negative regulator of abscisic acid signaling. Plant Cell. 1999;11:1897–909.

Gudesblat GE, Iusem ND, Morris PC. Guard cell-specific inhibition of *Arabidopsis* MPK3 expression causes abnormal stomatal responses to abscisic acid and hydrogen peroxide. New Phytol. 2007;173:713–21.

Halford NG, Bouly JP, Thomas M. SNF1-related protein kinases (SnRKs): regulators at the heart of the control of carbon metabolism and partitioning. Adv Bot Res. 2000;32:405–34.

Halford NG, Hardie DG. SNF1-related protein kinases: global regulators of carbon metabolism in plants? Plant Mol Biol. 1998;37:735–48.

Halford NG, Hey SJ. SNF1-related protein kinases (SnRKs) act within an intricate network that links metabolic and stress signalling in plants. Biochem J. 2009;419:247–59.

Halford NG, Hey S, Jhurreea D, Laurie S, McKibbin RS, Paul M, Zhang Y. Metabolic signalling and carbon partitioning: role of Snf1-related (SnRK1) protein kinase. J Exp Bot. 2003;54:467–75.

Hamel LP, Sheen J, Seguin A. Ancient signals: comparative genomics of green plant CDPKs. Trends Plant Sci. 2014;19:79–89.

Hardie DG. Plant protein-serine/threonine kinases: classification into subfamilies and overview of function. Adv Bot Res. 2000;32:1–44.

Harmon AC, Gribskov M, Gubrium E, Harper JF. The CDPK superfamily of protein kinases. New Phytol. 2001;151:175–83.

Harmon AC, Gribskov M, Harper JF. CDPKs: a kinase for every Ca^{2+} signal? Trends Plant Sci. 2000;5:154–9.

Harper JF, Huang JF, Lloyd SJ. Genetic identification of an autoinhibitor in CDPK, a protein kinase with a calmodulin-like domain. Biochemistry. 1994;33:7267–77.

Harper JF, Breton G, Harmon A. Decoding Ca^{2+} signals through plant protein kinases. Annu Rev Plant Biol. 2004;55:263–88.

Harper JF, Harmon A. Plants, symbiosis and parasites: a calcium signalling connection. Nat Rev Mol Cell Biol. 2005;6:555–66.

Harper JF, Sussman MR, Schaller GE, Putnam-Evans C, Charbonneau H, Harmon AC. A calcium-dependence protein kinase with a regulatory domain similar to calmodulin. Science. 1991;252:951–4.

He JX, Gendron JM, Yang Y, Li J, Wang ZY. The GSK3-like kinase BIN2 phosphorylates and destabilizes BZR1, a positive regulator of the brassinosteroid signaling pathway in *Arabidopsis*. Proc Natl Acad Sci USA. 2002;99:10185–90.

Hetherington AM, Woodward FI. The role of stomata in sensing and driving environmental change. Nature. 2003;424:901–8.

Hetherington AM, Brownlee C. The generation of Ca^{2+} signals in plants. Annu Rev Plant Biol. 2004;55:401–27.

Hey SJ, Bacon A, Burnett E, Neill SJ. Abscisic acid signal transduction in epidermal cells of *Pisum sativum* L. Argenteum: both dehydrin mRNA accumulation and stomatal responses require protein phosphorylation and dephosphorylation. Planta. 1997;202:85–92.

Haynes JG, Hartung AJ, Hendershot JD, Passingham RS, Rundle SJ. Molecular characterization of the B regulatory subunit gene family of *Arabidopsis* protein phosphatase 2A. Eur J Biochem. 1999;260:127–36.

Hong SW, Jon JH, Kwak JM, Nam HG. Identification of a receptor-like protein kinase gene rapidly induced by abscisic acid, dehydration, high salt, and cold treatments in *Arabidopsis thaliana*. Plant Physiol. 1997;113:1203–12.

Hrabak EM, Chan CWM, Gribskov M, Harper JF, Choi JH, Halford N, Kudla J, Luan S, Nimmo HG, Sussman MR. The *Arabidopsis* CDPK-SnRK superfamily of protein kinases. Plant Physiol. 2003;132:666–80.

Hua D, Wang C, He J, Liao H, Duan Y, Zhu Z, Guo Y, Chen Z, Gong Z. A plasma membrane receptor kinase, GHR1, mediates abscisic acid- and hydrogen peroxide-regulated stomatal movement in *Arabidopsis*. Plant Cell. 2012;24:2546–61.

Hunter T. Protein kinases and phosphatases: the yin and yang of protein phosphorylation and signaling. Cell. 1995;80:225–36.

Hwa CM, Yang XC. The AtMKK3 pathway mediates ABA and salt signaling in *Arabidopsis*. Acta Physiol Plant. 2008;30:277–86.

Ichimura K, Shinozaki K, Tena G, Sheen J, Henry Y, Champion A, Kreis M, Zhang S, Hirt H. Mitogen-activated protein kinase cascades in plants: a new nomenclature. Trends Plant Sci. 2002;7:301–8.

Ingebritsen TS, Cohen P. Protein phosphatases: properties and role in cellular regulation. Science. 1983;221:331–8.

Imes D, Mumm P, Bohm J, Al-Rasheid KA, Marten I, Geiger D, Hedrich R. Open stomata 1 (OST1) kinase controls R-type anion channel QUAC1 in *Arabidopsis* guard cells. Plant J. 2013;74:372–82.

Ivashuta S, Liu J, Liu J, Lohar DP, Haridas S, Bucciarelli B, VandenBosch KA, Vance CP, Harrison MJ, Gantt JS. RNA interference identifies a calcium-dependent protein kinase involved in *Medicago truncatula* root development. Plant Cell. 2005;17:2911–21.

Jammes F, Song C, Shin D, Munemasa S, Takeda K, Gu D, Cho D, Lee S, Giordo R, Sritubtim S, Leonhardt N, Ellis BE, Murata Y, Kwak JM. MAP kinases MPK9 and MPK12 are preferentially expressed in guard cells and positively regulate ROS-mediated ABA signaling. Proc Natl Acad Sci USA. 2009;106:20520–5.

Janssens V, Goris J. Protein phosphatase 2A: a highly regulated family of serine/threonine phosphatases implicated in cell growth and signalling. Biochem J. 2001;353:417–39.

Janssens V, Goris J, Van Hoof C. PP2A: the expected tumor suppressor. Curr Opin Genet Dev. 2005;15:34–41.

Johnson RR, Wagner RL, Verhey SD, Walker-Simmons MK. The abscisic acid-responsive kinase PKABA1 interacts with a seed-specific abscisic acid response element-binding factor, TaABF, and phosphorylates TaABF peptide sequences. Plant Physiol. 2002;130:837–46.

Jossier M, Bouly JP, Meimoun P, Arjmand A, Lessard P, Hawley S, Grahame HD, Thomas M. SnRK1 (SNF1-related kinase 1) has a central role in sugar and ABA signalling in *Arabidopsis thaliana*. Plant J. 2009;59:316–28.

Kerk D, Bulgrien J, Smith DW, Barsam B, Veretnik S, Gribskov M. The complement of protein phosphatase catalytic subunits encoded in the genome of *Arabidopsis*. Plant Physiol. 2002;129:908–25.

Kim K, Cheong YH, Grant JJ, Pandey GK, Luan S. CIPK3, a calcium sensor-associated protein kinase that regulates abscisic acid and cold signal transduction in *Arabidopsis*. Plant Cell. 2003;15:411–23.

Klimecka M, Muszynska G. Structure and functions of plant calcium-dependent protein kinases. Acta Biochim Pol. 2007;54:219–33.

Klimecka M, Szczegielniak J, Godecka L, Lewandowska-Gnatowska E, Dobrowolska G, Muszynska G. Regulation of wound-responsive calcium dependent protein kinase from maize (ZmCPK11) by phosphatidic acid. Acta Biochim Pol. 2011;58:589–95.

Knetsch MLW, Wang M, Snaar-Jagalska BE, Heimovaara-Dijkstra S. Abscisic acid induces mitogen-activated protein kinase activation in barley aleurone protoplasts. Plant Cell. 1996;8:1061–7.

Kobayashi Y, Yamamoto S, Minami H, Kagaya Y, Hattori T. Differential activation of the rice sucrose non-fermenting 1-related protein kinase2 family by hyperosmotic stress and abscisic acid. Plant Cell. 2004;16:1163–77.

Kolukisaoglu U, Weinl S, Blazevic D, Batistic O, Kudla J. Calcium sensors and their interacting protein kinases: genomics of the *Arabidopsis* and rice CBL-CIPK signaling networks. Plant Physiol. 2004;134:43–58.

Koornneef M, Reuling G, Karssen CM. The isolation and characterization of abscisic acid-insensitive mutants of *Arabidopsis thaliana*. Physiol Plant. 1982;61:377–83.

Kudla J, Xu Q, Harter K, Gruissem W, Luan S. Genes for calcineurin B-like proteins in *Arabidopsis* are differentially regulated by stress signals. Proc Natl Acad Sci USA. 1999;96:4718–23.

Kuhn JM, Boisson-Dernier A, Dizon MB, Maktabi MH, Schroeder JI. The protein phosphatase AtPP2CA negatively regulates abscisic acid signal transduction in *Arabidopsis* and effects of *abh1* on At*PP2CA* mRNA. Plant Physiol. 2006;140:127–39.

Kwak JM, Moon JH, Murata Y, Kuchitsu K, Leonhardt N, DeLong A, Schroeder JI. Disruption of a guard cell expressed protein phosphatase 2A regulatory subunit, RCN1, confers abscisic acid insensitivity in *Arabidopsis*. Plant Cell. 2002;14:2849–61.

Larsen PB, Cancel JD. Enhanced ethylene responsiveness in the *Arabidopsis eer1* mutant results from a loss-of-function mutation in the protein phosphatase 2A A regulatory subunit, RCN1. Plant J. 2003;34:709–18.

Lee SC, Lan W, Buchanan BB, Luan S. A protein kinase-phosphatase pair interacts with an ion channel to regulate ABA signaling in plant guard cells. Proc Natl Acad Sci USA. 2009;106:21419–24.

Lee SC, Lim CW, Lan W, He K, Luan S. ABA signaling in guard cells entails a dynamic protein–protein interaction relay from the PYL-RCAR family receptors to ion channels. Mol Plant. 2013;6:528–38.

Lee IC, Hong SW, Whang SS, Lim PO, Nam HG, Koo JC. Age-dependent action of an ABA-inducible receptor kinase, RPK1, as a positive regulator of senescence in *Arabidopsis* leaves. Plant Cell Physiol. 2011;52:651–62.

Leonhardt N, Kwak JM, Robert N, Waner D, Leonhardt G, Schroeder JI. Microarray expression analyses of *Arabidopsis* guard cells and isolation of a recessive abscisic acid hypersensitive protein phosphatase 2C mutant. Plant Cell. 2004;16:596–615.

Leube MP, Grill E, Amrhein N. ABI1 of *Arabidopsis* is a protein serine/threonine phosphatase highly regulated by the proton and magnesium ion concentration. FEBS Lett. 1998;424:100–4.

Leung J, Bouvier-Durand M, Morris PC, Guerrier D, Chefdor F, Giraudat J. *Arabidopsis* ABA response gene ABI1: features of a calcium modulated protein phosphatase. Science. 1994;264:1448–52.

Leung J, Merlot S, Giraudat J. The *Arabidopsis* ABSCISIC ACIDINSENSITIVE2 (ABI2) and ABI1 genes encode homologous protein phosphatases 2C involved in abscisic acid signal transduction. Plant Cell. 1997;9:759–71.

Levchenko V, Konrad KR, Dietrich P, Roelfsema MR, Hedrich R. Cytosolic abscisic acid activates guard cell anion channels without preceding Ca^{2+} signals. Proc Natl Acad Sci USA. 2005;102:4203–8.

Li G, Tai FJ, Zheng Y, Luo J, Gong SY, Zhang ZT, Li XB. Two cotton Cys2/His2-type zinc-finger proteins, GhDi19-1 and GhDi19-2, are involved in plant response to salt/drought stress and abscisic acid signaling. Plant Mol Biol. 2010;74:437–52.

Li J, Assmann SM. An abscisic acid-activated and calcium independent protein kinase from guard cells of Fava bean. Plant Cell. 1996;8:2359–68.

Li J, Kinoshita T, Pandey S, Ng CKY, Gygi SP, Shimazaki K, Assmann SM. Modulation of an RNA-binding protein by abscisic-acid activated protein kinase. Nature. 2002;418:793–7.

Li J, Lee YRJ, Assmann SM. Guard cells possess a calcium-dependent protein kinase that phosphorylates the KAT1 potassium channel. Plant Physiol. 1998;116:785–95.

Li J, Wang XQ, Watson MB, Assmann SM. Regulation of abscisic acid stomatal closure and anion channels by guard cell AAPK kinase. Science. 2000;287:300–3.

Li J, Nam KH. Regulation of brassinosteroid signaling by a GSK3/SHAGGY-like kinase. Science. 2002;295:1299–301.

Li AL, Zhu YF, Tan XM, Wang X, Wei B, Guo HZ, Zhang ZL, Chen XB, Zhao GY, Kong XY, Jia JZ, Mao L. Evolutionary and functional study of the CDPK gene family in wheat (*Triticum aestivum* L.). Plant Mol Biol. 2008;66:429–43.

Liese A, Romeis T. Biochemical regulation of in vivo function of plant calcium-dependent protein kinases (CDPK). Biochim Biophys Acta. 2013;1833:1582–9.

Liu J, Ishitani M, Halfter U, Kim CS, Zhu JK. The *Arabidopsis thaliana* SOS2 gene encodes a protein kinase that is required for salt tolerance. Proc Natl Acad Sci USA. 2000;97:3730–4.

Liu WX, Zhang FC, Zhang WZ, Song LF, Wu WH, Chen YF. *Arabidopsis* Di19 functions as a transcription factor and modulates *PR1*, *PR2*, and *PR5* expression in response to drought stress. Mol Plant. 2013;6:1487–502.

Liu J, Zhu JK. A calcium sensor homolog required for plant salt tolerance. Science. 1998;280:1943–5.

Lu Y, Chen X, Wu Y, Wang Y, He Y, Wu Y. Directly transforming PCR-amplified DNA fragments into plant cells is a versatile system that facilitates the transient expression assay. PLoS ONE. 2013;8:57171.

Lu C, Han MH, Guevara-Garcia A, Fedoroff NV. Mitogen-activated protein kinase signaling in postgermination arrest of development by abscisic acid. Proc Natl Acad Sci USA. 2002;99:15812–7.

Lu SX, Hrabak EM. An Arabidopsis calcium-dependent protein kinase is associated with the endoplasmic reticulum. Plant Physiol. 2002;128:1008–21.

Lu CA, Lin CC, Lee KW, Chen JL, Huang LF, Ho SL, Liu HJ, Hsing YI, Yu SM. The SnRK1A protein kinase plays a key role in sugar signaling during germination and seedling growth of rice. Plant Cell. 2007;19:2484–99.

Luan S. Protein phosphatases in plants. Annu Rev Plant Biol. 2003;54:63–92.

Luan S. The CBL-CIPK network in plant calcium signaling. Trends Plant Sci. 2009;14:37–42.

Luan S, Kudla J, Rodriguez-Concepcion M, Yalovsky S, Gruissem W. Calmodulins and calcineurin B-like proteins: Calcium sensors for specific signal response coupling in plants. Plant Cell. 2002;14 suppl.:S389–S400.

Ludwig AA, Romeis T, Jones JDG. CDPK-mediated signalling pathways: specificity and cross-talk. J Exp Bot. 2004;55:181–8.

Lynch T, Erickson BJ, Finkelstein RR. Direct interactions of ABA insensitive (ABI)-clade protein phosphatase(PP)2Cs with calcium-dependent protein kinases and ABA response element-binding bZIPs may contribute to turning off ABA response. Plant Mol Biol. 2012;80:647–58.

Li Z, Kang J, Sui N, Dong Liu D. a ROP11 GTPase is a negative regulator of multiple ABA responses in *Arabidopsis*. J Integr Plant Biol. 2012a;54:169–79.

Li Z, Li Z, Gao X, Chinnusamy V, Bressan R, Wang ZX, Zhu JK, Wu JW, Liu D. ROP11 GTPase negatively regulates ABA signaling by protecting ABI1 phosphatase activity from inhibition by the ABA receptor RCAR1/PYL9 in *Arabidopsis*. J Integr Plant Biol. 2012b;54:180–8.

Ma P, Liu J, Yang X, Ma R. Genome-wide identification of the maize calcium-dependent protein kinase gene family. Appl Biochem Biotechnol. 2013;169:2111–25.

Ma SY, Wu WH. AtCPK23 functions in Arabidopsis responses to drought and salt stresses. Plant Mol Biol. 2007;65:511–8.

Ma Y, Szostkiewicz I, Korte A, Moes D, Yang Y, Christmann A, Grill E. Regulators of PP2C phosphatase activity function as abscisic acid sensors. Science. 2009;324:1064–8.

Marten H, Konrad KR, Dietrich P, Roelfsema MR, Hedrich R. Ca^{2+}-dependent and -independent abscisic acid activation of plasma membrane anion channels in guard cells of *Nicotiana tabacum*. Plant Physiol. 2007;143:28–37.

Mayer-Jaekel RE, Hemmings BA. Protein phosphatase 2A – a 'menage a trois'. Trends Cell Biol. 1994;4:287–91.

McCormack E, Tsai YC, Braam J. Handling calcium signaling: *Arabidopsis* CaMs and CMLs. Trends Plant Sci. 2005;10:383–9.

McCubbin AG, Ritchie SM, Swanson SJ, Gilroy S. The calcium-dependent protein kinase HvCDPK1 mediates the gibberellic acid response of the barley aleurone through regulation of vacuolar function. Plant J. 2004;39:206–18.

McMichael RMJ, Bachmann M, Huber SC. Spinach leaf sucrose-phosphate synthase and nitrate reductase are phosphorylated/inactivated by multiple protein kinases in vitro. Plant Physiol. 1995;108:1077–82.

Mehlmer N, Wurzinger B, Stael S, Hofmann-Rodrigues D, Csaszar E, Pfister B, Bayer R, Teige M. The Ca^{2+}-dependent protein kinase CPK3 is required for MAPK-independent salt-stress acclimation in *Arabidopsis*. Plant J. 2010;63:484–98.

Meyer K, Leube MP, Grill E. A protein phosphatase 2C involved in ABA signal transduction in *Arabidopsis thaliana*. Science. 1994;264:1452–5.

Merlot S, Gosti F, Guerrier D, Vavasseur A, Giraudat J. The ABI1 and ABI2 protein phosphatases 2C act in a negative feedback regulatory loop of the abscisic acid signaling pathway. Plant J. 2001;25:295–303.

Mizoguchi M, Umezawa T, Nakashima K, Kidokoro S, Takasaki H, Fujita Y, Yamaguchi-Shinozaki K, Shinozaki K. Two closely related subclass II SnRK2 protein kinases cooperatively regulate drought-inducible gene expression. Plant Cell Physiol. 2010;51:842–7.

Mori IC, Murata Y, Yang Y, Munemasa S, Wang YF, Andreoli S, Tiriac H, Alonso JM, Harper JF, Ecker JR, Kwak JM, Schroeder JI. CDPKs CPK6 and CPK3 function in ABA regulation of guard cell S-type anion- and Ca^{2+}-permeable channels and stomatal closure. PLoS Biol. 2006;4:327.

Mori IC, Muto S. Abscisic acid activates a 48-kilodalton protein kinase in guard cell protoplasts. Plant Physiol. 1997;113:833–9.

Morillo SA, Tax FE. Functional analysis of receptor-like kinases in monocots and dicots. Curr Opin Plant Biol. 2006;9:460–9.

Mustilli AC, Merlot S, Vavasseur A, Fenzi F, Giraudat J. Arabidopsis OST1 protein kinase mediates the regulation of stomatal aperture by abscisic acid and acts upstream of reactive oxygen species production. Plant Cell. 2002;14:3089–99.

Myers C, Romanowsky SM, Barron YD, Garg S, Azuse CL, Curran A, Davis RM, Hatton J, Harmon AC, Harper JF. Calcium-dependent protein kinases regulate polarized tip growth in pollen tubes. Plant J. 2009;59:528–39.

Nakashima K, Fujita Y, Kanamori N, Katagiri T, Umezawa T, Kidokoro S, Maruyama K, Yoshida T, Ishiyama K, Kobayashi M, Shinozaki K, Yamaguchi-Shinozaki K. Three *Arabidopsis* SnRK2 protein kinases, SRK2D/SnRK2.2, SRK2E/SnRK2.6/OST1 and SRK2I/SnRK2.3, involved in ABA signaling are essential for the control of seed development and dormancy. Plant Cell Physiol. 2009;50:1345–63.

Negi J, Matsuda O, Nagasawa T, Oba Y, Takahashi H, Kawai-Yamada M, Uchimiya H, Hashimoto M, Iba K. CO_2 regulator SLAC1 and its homologues are essential for anion homeostasis in plant cells. Nature. 2008;452:483–6.

Nemeth K, Salchert K, Putnoky P, Bhalerao R, Koncz-Kálmán Z, Stankovic-Stangeland B, Bakó L, Mathur J, Okrész L, Stabel S, Geigenberger P, Stitt M, Rédei GP, Schell J, Koncz C. Pleiotropic control of glucose and hormone responses by PRL1, a nuclear WD protein, in *Arabidopsis*. Genes Dev. 1998;12:3059–73.

Nishimura N, Yoshida T, Kitahata N, Asami T, Shinozaki K, Hirayama T. ABA-hypersensitive germination1 encodes a protein phosphatase 2C, an essential component of abscisic acid signaling in *Arabidopsis* seed. Plant J. 2007;50:935–49.

Ohta M, Guo Y, Halfter U, Zhu JK. A novel domain in the protein kinase SOS2 mediates interaction with the protein phosphatase 2C ABI2. Proc Natl Acad Sci USA. 2003;100:11771–6.

Olsen JV, Blagoev B, Gnad F, Macek B, Kumar C, Mortensen P, Mann M. Global, in vivo, and sitespecific phosphorylation dynamics in signaling networks. Cell. 2006;127:635–48.

Osakabe Y, Arinaga N, Umezawa T, Katsura S, Nagamachi K, Tanaka H, Ohiraki H, Yamada K, Seo SU, Abo M, Yoshimura E, Shinozaki K, Yamaguchi-Shinozaki K. Osmotic stress responses and plant growth controlled by potassium transporters in *Arabidopsis*. Plant Cell. 2013;25:609–24.

Osakabe Y, Maruyama K, Seki M, Satou M, Shinozaki K, Yamaguchi-Shinozaki K. Leucine-rich repeat receptor-like kinase1 is a key membrane-bound regulator of abscisic acid early signaling in *Arabidopsis*. Plant Cell. 2005;17:1105–19.

Osakabe Y, Mizuno S, Tanaka H, Maruyama K, Osakabe K, Todaka D, Fujita Y, Kobayashi M, Shinozaki K, Yamaguchi-Shinozaki K. Overproduction of the membrane-bound receptor-like protein kinase 1, RPK1, enhances abiotic stress tolerance in *Arabidopsis*. J Biol Chem. 2010;285:9190–201.

Pandey GK, Cheong YH, Kim K, Grant JJ, Li L, Hung W, D'Angelo C, Weinl S, Kudla J, Luan S. The calcium sensor calcineurin B-like 9 modulates abscisic acid sensitivity and biosynthesis in *Arabidopsis*. Plant Cell. 2004;16:1912–24.

Pandey G, Grant J, Cheong YH, Kim BG, Li L, Luan S. Calcineurin B-like protein CBL9 interacts with target kinase CIPK3 in the regulation of ABA response in seed germination. Mol Plant. 2008;2:238–48.

Pandey S, Nelson DC, Assmann SM. Two novel GPCR-type G proteins are abscisic acid receptors in *Arabidopsis*. Cell. 2009;136:136–48.

Pandey S, Zhang W, Assmann SM. Roles of ion channels and transporters in guard cell signal transduction. FEBS Lett. 2007;581:2325–36.

Park SY, Fung P, Nishimura N, Jensen DR, Fujii H, Zhao Y, Lumba S, Santiago J, Rodrigues A, Chow T-F, et al. Abscisic acid inhibits type 2C protein phosphatases via the PYR/PYL family of START proteins. Science. 2009;324:1068–71.

Pernas M, García-Casado G, Rojo E, Solano R, Sánchez-Serrano JJ. A protein phosphatase 2A catalytic subunit is a negative regulator of abscisic acid signalling. Plant J. 2007;51:763–78.

Perochon A, Aldon D, Galaud JP, Ranty B. Calmodulin and calmodulin-like proteins in plant calcium signaling. Biochimie. 2011;93:2048–53.

Pei ZM, Ward JM, Harper JF, Schroeder JI. A novel chloride channel in Vicia faba guard cell vacuoles activated by the serine/threonine kinase, CDPK. EMBO J. 1996;15:6564–74.

Polge C, Thomas M. SNF1/AMPK/SnRK1 kinases, global regulators at the heart of energy control? Trends Plant Sci. 2006;12:20–8.

Qiu QS, Guo Y, Dietrich MA, Schumaker KS, Zhu JK. Regulation of SOS1, a plasma membrane Nat/Ht exchanger in *Arabidopsis thaliana*, by SOS2 and SOS3. Proc Natl Acad Sci USA. 2002;99:8436–41.

Radchuk R, Radchuk V, Weschke W, Borisjuk L, Weber H. Repressing the expression of the SUCROSE NONFERMENTING-1-RELATED PROTEIN KINASE gene in pea embryo causes pleiotropic defects of maturation similar to an abscisic acid-insensitive phenotype. Plant Physiol. 2006;140:263–78.

Rashotte AM, DeLong A, Muday GK. Genetic and chemical reductions in protein phosphatase activity alter auxin transport, gravity response, and lateral root growth. Plant Cell. 2001;13:1683–97.

Ray S, Agarwal P, Arora R, Kapoor S, Tyagi AK. Expression analysis of calcium-dependent protein kinase gene family during reproductive development and abiotic stress conditions in rice (*Oryza sativa* L. ssp. indica). Mol Genet Genomics. 2007;278:493–505.

Robaglia C, Thomas M, Meyer C. Sensing nutrient and energy by SnRK1 and TOR kinases. Curr Opin Plant Biol. 2012;15:301–7.

Robert N, Merlot S, N'Guyen V, Boisson-Dernier A, Schroeder JI. A hypermorphic mutation in the protein phosphatase 2C HAB1 strongly affects ABA signaling in *Arabidopsis*. FEBS Lett. 2006;580:4691–6.

Rodrigues A, Adamo M, Crozet P, Margalha L, Confraria A, Martinho C, Elias A, Rabissi A, Lumbreras V, González-Guzmán M, Antoni R, Rodriguez PL, Baena-Gonzáleza E. ABI1 and PP2CA phosphatases are negative regulators of Snf1-related protein kinase1 signaling in *Arabidopsis*. Plant Cell. 2013;25:3871–84.

Rodriguez MA, Yuichi U, Chang IF, Townsend J, Maher EA, Quilici D, Cushman JC. A novel yeast two-hybrid approach to identify CDPK substrates: characterization of the interaction between AtCPK11 and AtDi19, a nuclear zinc finger protein. FEBS Lett. 2006;580:904–11.

Roelfsema MR, Hedrich R, Geiger D. Anion channels: master switches of stress responses. Trends Plant Sci. 2012;17:221–9.

Romeis T, Ludwig AA, Martin R, Jones JDG. Calcium-dependent protein kinase play an essential role in a plant defense response. EMBO J. 2001;20:5556–67.

Rubio S, Rodrigues A, Saez A, Dizon MB, Galle A, Kim TH, Santiago J, Flexas J, Schroeder JI, Rodriguez PL. Triple loss of function of protein phosphatases type 2C leads to partial constitutive response to endogenous abscisic acid. Plant Physiol. 2009;150:1345–55.

Rudd JJ, Franklin-Tong VE. Unravelling response specificity in Ca2t signaling in plant cells. New Phytol. 2001;151:7–33.

Saez A, Apostolova N, Gonzalez-Guzman M, Gonzalez-Garcia MP, Nicolas C, Lorenzo O, Rodriguez PL. Gain-of-function and loss-of-function phenotypes of the protein phosphatase 2C HAB1 reveal its role as a negative regulator of abscisic acid signalling. Plant J. 2004;37:354–69.

Saez A, Robert N, Maktabi MH, Schroeder JI, Serrano R, Rodriguez PL. Enhancement of abscisic acid sensitivity and reduction of water consumption in *Arabidopsis* by combined inactivation of the protein phosphatases type 2C ABI1 and HAB1. Plant Physiol. 2006;141:1389–99.

Sanders D, Brownlee C, Harper JF. Communicating with calcium. Plant Cell. 1999;11:691–706.

Sato A, Sato Y, Fukao Y, Fujiwara M, Umezawa T, Shinozaki K, Hibi T, Taniguchi M, Miyake H, Goto DB, Uozumi N. Threonine at position 306 of the KAT1 potassium channel is essential for channel activity and is a target site for ABA-activated SnRK2/OST1/SnRK2.6 protein kinase. Biochem J. 2009;424:439–48.

Scherzer S, Maierhofer T, Al-Rasheid KAS, Geiger D, Hedrich R. Multiple calcium-dependent kinases modulate ABA-activated guard cell anion channels. Mol Plant. 2012;5:1409–12.

Schmidt C, Schelle I, Liao YJ, Schroeder JI. Strong regulation of slow anion channels and abscisic acid signaling in guard cells by phosphorylation and dephosphorylation events. Proc Natl Acad Sci USA. 1995;92:9535–9.

Schulz p, Herde M, Romeis T. Calcium-dependent protein kinases: Hubs in plant stress signaling and development. Plant Physiol. 2013;163:523–30.

Schweighofer A, Hirt H, Meskiene I. Plant PP2C phosphatases: emerging functions in stress signaling. Trends Plant Sci. 2004;9:236–43.

Shao JH, Harmon AC. In vivo phosphorylation of a recombinant peptide substrate of CDPK suggests involvement of CDPK in plant stress responses. Plant Mol Biol. 2003;53:731–40.

Sheen J. Ca^{2+}-dependent protein kinases and stress signal transduction in plants. Science. 1996;274:1900–2.

Sheen J. Mutational analysis of protein phosphatase 2C involved in abscisic acid signal transduction in higher plants. Proc Natl Acad Sci USA. 1998;95:975–80.

Shen Q, Gomez-Cadenas A, Zhang P, Walker-Simmons MK, Sheen J, Ho TH. Dissection of abscisic acid signal transduction pathways in barley aleurone layers. Plant Mol Biol. 2001;47:437–48.

Shi H, Ishitani M, Kim CS, Zhu JK. The Arabidopsis thaliana salt tolerance gene SOS1 encodes a putative Na +/H + antiporter. Proc Natl Acad Sci USA. 2000;97:6896–901.

Shiu SH, Bleecker AB. Receptor-like kinases from *Arabidopsis* form a monophyletic gene family related to animal receptor kinases. Proc Natl Acad Sci USA. 2001;98:10763–8.

Shiu SH, Karlowski WM, Pan R, Tzeng YH, Mayer KF, Li WH. Comparative analysis of the receptor-like kinase family in *Arabidopsis* and rice. Plant Cell. 2004;16:1220–34.

Slabas AR, Fordham-Skelton AP, Fletcher D, Martinez-Rivas JM, Swinhoe R, Croy RR, Evans IM. Characterization of cDNA and genomic clones encoding homologues of the 65 kDa regulatory subunit of protein phosphatase 2A in *Arabidopsis thaliana*. Plant Mol Biol. 1994;21:307–16.

Smith RD, Walker JC. Plant protein phosphatases. Annu Rev Plant Physiol Plant Mol Biol. 1996;47:101–25.

Snedden WA, Fromm H. Calmodulin as a versatile calcium signal transducer in plants. New Phytol. 2001;151:35–66.

Song CP, Agarwal M, Ohta M, Guo Y, Halfter U, Wang P, Zhu JK. Role of an *Arabidopsis* AP2/EREBP-type transcriptional repressor in abscisic acid and drought stress responses. Plant Cell. 2005;17:2384–96.

Stefan W, Kudla J. The CBL–CIPK Ca2 + -decoding signaling network: function and perspectives. New Phytol. 2009;184:517–28.

Sugiyama K, Mori IC, Takahashi K, Muto S, Shihira-Ishikawa I. A calcium-dependent protein kinase functions in would healing in *Ventricaria ventricosa* (Chlorophyta). J Phycol. 2000;36:145–1152.

Tanaka H, Osakabe Y, Katsura S, Mizuno S, Maruyama K, Kusakabe K, Mizoi J, Shinozaki K, Yamaguchi-Shinozaki K. Abiotic stress inducible receptor-like kinases negatively control ABA signaling in *Arabidopsis*. Plant J. 2012;70:599–613.

Teige M, Scheikl E, Eulgem T, Doczi R, Ichimura K, Shinozaki K. The MKK2 pathway mediates cold and salt stress signaling in *Arabidopsis*. Mol Cell. 2004;15:141–52.

Terol J, Bargues M, Carrasco P, Perez-Alonso M, Paricio N. Molecular characterization and evolution of the protein phosphatase 2A B regulatory subunit family in plants. Plant Physiol. 2002;129:808–22.

Thelander M, Olsson T, Ronne H. Snf1-related protein kinase 1 is needed for growth in a normal day–night light cycle. EMBO J. 2004;23:1900–10.

Umezawa T, Sugiyama N, Mizoguchi M, Hayashi S, Myouga F, Yamaguchi-Shinozaki K, Ishihama Y, Hirayama T, Shinozaki K. Type 2C protein phosphatases directly regulate abscisic acid-activated protein kinases in *Arabidopsis*. Proc Natl Acad Sci USA. 2009;106:17588–93.

Umezawa T, Sugiyama N, Takahashi F, Anderson JC, Ishihama Y, Peck SC, Shinozaki K. Genetics and phosphoproteomics reveal a protein phosphorylation network in the abscisic acid signaling pathway in *Arabidopsis thaliana*. Sci Signal. 2013;6:270 rs8.

Uno Y, Rodriguez Milla MA, Maher E, Cushman JC. Identification of proteins that interact with catalytically active calcium-dependent protein kinases from *Arabidopsis*. Mol Genet Genomics. 2009;281:375–90.

Virshup DM. Protein phosphatase 2A: a panoply of enzymes. Curr Opin Cell Biol. 2000;12:180–5.

Vlad F, Rubio S, Rodrigues A, Sirichandra C, Belin C, Robert N, Leung J, Rodriguez PL, Lauriere C, Merlot S. Protein phosphatases 2C regulate the activation of the Snf1-related kinase OST1 by abscisic acid in *Arabidopsis*. Plant Cell. 2009;21:3170–84.

Wan B, Lin Y, Mou T. Expression of rice Ca^{2+}-dependent protein kinases (CDPKs) genes under different environmental stresses. FEBS Lett. 2007;581:1179–89.

Wang H, Chevalier D, Larue C, Cho SK, Walker JC. The protein phosphatases and protein kinases of *Arabidopsis thaliana*. The Arabidopsis Book 5: e0106. 2007. doi/10.1199/tab.0106.10.1199/tab.0106.

Wang P, Xue L, Batelli G, Lee S, Hou YJ, Oosten MJV, Zhang H, Tao W, Zhu J. Quantitative phosphoproteomics identifies SnRK2 protein kinase substrates and reveals the effectors of abscisic acid action. Proc Natl Acad Sci USA. 2013;110:11205–10.

Wang XJ, Zhu SY, Lu YF, Zhao R, Xin Q, Wang XF, Zhang DP. Two coupled components of the mitogen-activated protein kinase cascade MdMPK1 and MdMKK1 from apple function in ABA signal transduction. Plant Cell Physiol. 2010;51:754–66.

Weinl S, Kudla J. The CBL-CIPK Ca^{2+}-decoding signaling network: function and perspectives. New Phytol. 2009;184:517–28.

Wernimont AK, Amani M, Qiu W, Pizarro JC, Artz JD, Lin YH, Lew J, Hutchinson A, Hui R. Structures of parasitic CDPK domains point to a common mechanism of activation. Proteins. 2011;79:803–20.

Wernimont AK, Artz JD, Finerty PJ, Lin YH, Amani M, Allali-Hassani A, Senisterra G, Vedadi M, Tempel W, Mackenzie F, Chau I, Lourido S, Sibley LD, Hui R. Structures of apicomplexan calcium-dependent protein kinases reveal mechanism of activation by calcium. Nat Struct Mol Biol. 2010;17:596–601.

Widmann C, Gibson S, Jarpe MB, Johnson GL. Mitogen-activated protein kinase: Conservation of a three-kinase module from yeast to human. Physiol Rev. 1999;79:143–80.

Witte CP, Keinath N, Dubiella U, Demoulière R, Seal A, Romeis T. Tobacco calcium-dependent protein kinases are differentially phosphorylated in vivo as part of a kinase cascade that regulates stress response. J Biol Chem. 2010;285:9740–8.

Wu Y, Kuzma J, Marechal E, Graeff R, Lee HC, Foster R, Chua NH. Abscisic acid signaling through cyclic ADP-ribose in plants. Science. 1997;278:2126–30.

Xing Y, Jia W, Zhang J. AtMKK1 mediates ABA-induced CAT1 expression and H_2O_2 production via AtMPK6-coupled signaling in *Arabidopsis*. Plant J. 2008;54:440–51.

Xiong L, Yang Y. Disease resistance and abiotic stress tolerance in rice are inversely modulated by an abscisic acid inducible mitogen activated protein kinase. Plant Cell. 2003;15:745–59.

Xu J, Tian YS, Peng RH, Xiong AS, Zhu B, Jin XF, Gao F, Fu XY, Hou XL, Yao QH. AtCPK6, a functionally redundant and positive regulator involved in salt/drought stress tolerance in *Arabidopsis*. Planta. 2010;231:1251–60.

Yoshida R, Hobo T, Ichimura K, Mizoguchi T, Takahashi F, Aronso J, Ecker JR, Shinozaki K. ABA-activated SnRK2 protein kinase is required for dehydration stress signaling in *Arabidopsis*. Plant Cell Physiol. 2002;43:1473–83.

Yoshida T, Nishimura N, Kitahata N, Kuromori T, Ito T, Asami T, Shinozaki K, Hirayama T. ABA-hypersensitive germination3 encodes a protein phosphatase 2C (AtPP2CA) that strongly regulates abscisic acid signaling during germination among *Arabidopsis* protein phosphatase 2Cs. Plant Physiol. 2006a;140:115–26.

Yoshida R, Umezawa T, Mizoguchi T, Takahashi S, Takahashi F, Shinozaki K. The regulatory domain of SRK2E/OST1/SnRK2.6 interacts with ABI1 and integrates abscisic acid (ABA) and osmotic stress signals controlling stomatal closure in *Arabidopsis*. J Biol Chem. 2006b;281:5310–8.

Yu XC, Li MJ, Gao GF, Feng HZ, Geng XQ, Peng CC, Zhu SY, Wang XL, Shen YY, Zhang DP. Abscisic acid stimulates a calcium-dependent protein kinase in grape berry. Plant Physiol. 2006;140:558–79.

Yu F, Qian L, Nibau C, Duan Q, Kita D, Levasseur K, Li X, Lua C, Li H, Hou C, Li L, Buchanan BB, Chen L, Cheung AY, Lia D, Luan S. FERONIA receptor kinase pathway suppresses abscisic acid signaling in *Arabidopsis* by activating ABI2 phosphatase. Proc Natl Acad Sci USA. 2012;109:14693–8.

Yu XC, Zhu SY, Gao GF, Wang XJ, Zhao R, Zou KQ, Wang XF, Zhang XY, Wu FQ, Peng CC, Zhang DP. Expression of a grape calcium-dependent protein kinase ACPK1 in *Arabidopsis thaliana* promotes plant growth and confers abscisic acid-hypersensitivity in germination, postgermination growth, and stomatal movement. Plant Mol Biol. 2007;64:531–8.

Zhao R, Wang XF, Zhang DP. CPK12: A Ca^{2+}-dependent protein kinase balancer in abscisic acid signaling. Plant Signal Behav. 2011a;6:1687–90.

Zhao R, Sun HL, Mei C, Wang XJ, Yan L, Liu R, Zhang XF, Wang XF, Zhang DP. The *Arabidopsis* Ca2+-dependent protein kinase CPK12 negatively regulates abscisic acid signaling in seed germination and post-germination growth. New Phytol. 2011b;192:61–73.

Zhang Y, Andralojc PJ, Hey SJ, Primavesi LF, Specht M, Koehler J, Parry MAJ, Halford NG. *Arabidopsis* sucrose non-fermenting-1-related protein kinase-1 and calcium dependent protein kinase phosphorylate conserved target sites in ABA response element binding proteins. Ann Appl Biol. 2008;153:401–9.

Zhang AY, Jiang MY, Zhang JH, Tan MP, Hu ZL. Mitogen-activated protein kinase is involved in abscisic acid induced antioxidant defense and acts downstream of reactive oxygen species production in leaves of maize plants. Plant Physiol. 2006;141:475–87.

Zhu SY, Yu XC, Wang XJ, Zhao R, Li Y, Fan RC, Shang Y, Du SY, Wang XF, Wu FQ, Xu YH, Zhang XY, Zhang DP. Two calcium-dependent protein kinases, CPK4 and CPK11, regulate abscisic acid signal transduction in *Arabidopsis*. Plant Cell. 2007;19:3019–36.

Zielinski RE. Calmodulin and calmodulin-binding proteins in plants. Annu Rev Plant Physiol Plant Mol Biol. 1998;49:697–725.

Zou JJ, Wei FJ, Wang C, Wu JJ, Ratnasekera D, Liu WX, Wu WH. *Arabidopsis* calcium-dependent protein kinase CPK10 functions in abscisic acid- and Ca^{2+}-mediated stomatal regulation in response to drought stress. Plant Physiol. 2010;154:1232–43.

Zuo R, Hu R, Chai G, Xu M, Qi G, Kong Y, Zhou G. Genome-wide identification, classification, and expression analysis of CDPK and its closely related gene families in poplar (*Populus trichocarpa*). Mol Biol Rep. 2013;40:2645–62.

Chapter 9
Protein Ubiquitination and Sumoylation in ABA Signaling

Yaorong Wu, Feifei Yu and Qi Xie

Abstract Abscisic acid (ABA) has been known as a key phytohormone to regulate plant and environmental interaction to modulate plant growth and development under different environmental diversification. ABA signaling could be regulated at different levels, most importantly at transcriptional and post-translational levels. Among them, ubiquitination and sumoylation have been paid attention especially in recent years. Ubiquitination and sumoylation share similar biochemistry pathways, but a three-step enzymatic cascade catalyzed by their specific E1, E2, and E3 enzymes, respectively. Both kinds of post-translational protein modifications can act together or separately to control protein stability, localization, and activity. In this chapter, we will summarize the recent development of protein ubiquitination and sumoylation in control of ABA pathway, including ABA biogenesis and signaling, mainly based on the data from researches in the model plant Arabidopsis.

Keywords Ubiquitination · Sumoylation · ABA signaling

9.1 Introduction

Ubiquitin and small ubiquitin-like modifier (SUMO) are small proteins that can regulate the stability, localization, or function of the target proteins in eukaryotic systems by being covalently attached to substrate proteins. Ubiquitination and sumoylation involve a three-step enzymatic cascade catalyzed by their specific E1, E2, and E3 enzymes, respectively. In many aspects, ubiquitination and

Y. Wu · F. Yu · Q. Xie (✉)
State Key Laboratory of Plant Genomics, National Center for Plant Gene Research, Institute of Genetics and Developmental Biology, Chinese Academy of Sciences, No. 1 West Beichen Road, Chaoyang, Beijing 100101, China
e-mail: qxie@genetics.ac.cn

D.-P. Zhang (ed.), *Abscisic Acid: Metabolism, Transport and Signaling*,
DOI 10.1007/978-94-017-9424-4_9

sumoylation play antagonist roles in protein stability, thus act oppositely in the target protein involved in biological processes. Abscisic acid (ABA) is a key phytohormone to regulate plant growth and development, especially in the aspect of plant and environmental interaction. Recently, a number of discoveries have shown that both ubiquitination and sumoylation participate in many aspects of ABA biology, including regulation of ABA biosynthesis, activation or repression of ABA response, by regulating the stabilization or the degradation of ABA-signaling components.

9.2 Protein Ubiquitination and Sumoylation

Ubiquitin, a highly conserved 76-amino acid protein, was covalently ligated to other proteins to post-translationally regulate the target proteins in eukaryotic systems. Ubiquitin modified the substrate protein by either one molecule (monoubiquitination) or polymers (polyubiquitination) covalently attached to the target. Typically, ubiquitin proteasome system (UPS) firstly requires a three-step sequential enzymatic cascade of E1 (ubiquitin-activating enzyme), E2 (ubiquitin-conjugating enzyme), and E3 (ubiquitin ligase). Among them, E3s are the most abundant and very important to define the substrate specificity. For example, there are 2 E1s-, 37 E2s-, and 8 E2-like proteins, but over 1,500 different E3s in Arabidopsis genome. According to the mechanisms of action and subunit composition, E3s were classified to four types in plants, HECT, really interesting new gene (RING), U-box and Cullin-RING ligases (CRLs). CRLs, the multi-subunit ligases, are further divided into four subtypes based on a different target recognition site, SCF E3s (S-phase kinase-associated protein1-cullin1-F-Box ligases), BTB E3s (bric-a-brac-tramtrak-broad complex), DDB E3s (the DNA damage-binding proteins), and the anaphase-promoting complex (APC) (Vierstra 2009). Most ubiquitinated proteins were broken down by the 26S proteasome system, which degrade the target and release the linked ubiquitin for reuse.

Besides ubiquitin, there are several functional ubiquitin-like proteins (UBLs) in eukaryotes, including SUMO, RUB (related to ubiquitin, or Nedd8 in yeast and animals), ATG8 and ATG12 (autophagy 8 and 12), FUB1 (fau ubiquitin-like protein 1), URM1 (ubiquitin-related modifier 1), UFM1 (ubiquitin-fold modifier1), and HUB1 (homology to ubiquitin1) (Park et al. 2011). Among them, SUMO is a kind of essential post-translation modification, firstly identified in yeast (Meluh and Koshland 1995), and then exclusively reported in animals and plants. Although SUMO shares only about 30 % identity with ubiquitin, three-dimensional structures are highly conserved, constituted by one α helix and four β strands (Fig. 9.1, Downes and Vierstra 2005; Hochstrasser 2009). Only one SUMO coding gene exists in yeast and four in human, while eight forms of SUMOs (encoded by *SUMO1-8*) and one SUMO-like pseudogene (*SUMO9*) are present in Arabidopsis (Miura et al. 2007), implying specific function of various forms of SUMOs in plants. SUMO preproteins (95-111 amino acids) encoded by these

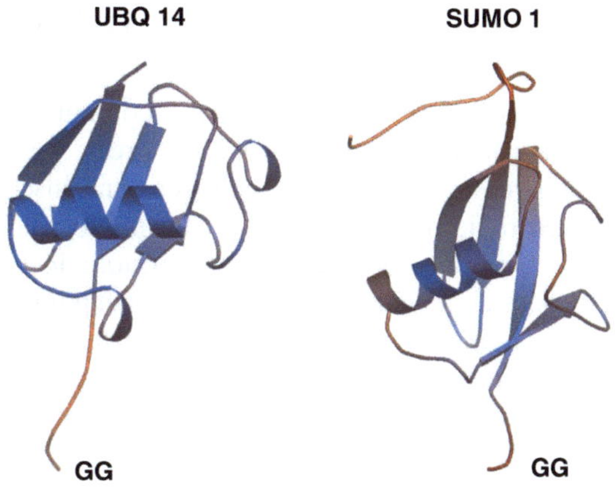

Fig. 9.1 The three-dimensional structure of UBQ 14 (AT4G02890) and SUMO 1 (AT4G26840) is predicted at the Web site http://swissmodel.expasy.org

SUMO genes were processed by ubiquitin-like SUMO-specific proteases (ULP) that delete about 10 amino acids directly after C terminus diglycine (GG) motif, leaving diglycine residues in mature SUMO proteins (Colby et al. 2006). Plant ULP also possesses isopeptidase activity to recycle SUMO from substrates, indicating sumoylation is a kind of reversible post-translational modification.

Similar to ubiquitination process, sumoylation of substrate involves sequential catalytic reaction by a cascade of enzymes: SUMO E1 activating enzyme, SUMO E2 conjugating enzyme, and SUMO E3 ligase. In Arabidopsis, SUMO E1, a heterodimer SAE1/SAE2 (different from ubiquitination with a single E1), activates sumoylation primarily and subsequently transfer SUMO through a thioester linkage from a catalytic cysteine residue in E1 to the single known SUMO-conjugating E2 (SCE1). Whereas more than 1,500 ubiquitin E3s (or complexes) were identified, up to now only two SUMO E3 ligases were identified in Arabidopsis, AtSIZ1 and HPY2/AtMMS21, which facilitate the transfer of SUMO from SUMO E2-conjugating enzyme to substrate protein (Miura and Hasegawa 2010). A covalent bond linkage occurs between diglycine of SUMO and ε amino group of lysine residue located in a conserved sumoylation motif ψKXE/D (ψ, an aliphatic residue, preferably L, I or V; X, any amino acid). Mostly from yeast and mammalian, it has been shown that sumoylation controlled a broad of cellular activities, including roles in gene expression, maintenance of chromatin integrity, signal transduction, protein trafficking, and stabilizing proteins by protecting them from ubiquitination-dependent protein degradation.

Proteomic and genetic analysis mainly in Arabidopsis highlight the importance of ubiquitination and sumoylation action to control the key factors during hormone synthesis, perception, and downstream signaling in the most plant hormone signaling, including auxin, ABA, brassinosteroid (BR), cytokinin, ethylene (ET), gibberellins (GA), jasmonic acid (JA), salicylic acid (SA), strigolactone (SL), and zeatin. Here, we summarize the current knowledge on the protein ubiquitination and sumoylation in ABA signaling in the plants.

9.3 Protein Ubiquitination in ABA Signaling

Being sessile organisms, plants are continuously exposed in complex and volatile environment, including abiotic stress, such as drought, salt, and cold, and biotic stress, for examples, pathogens and insects. At the same time, plants have evolved rapid and efficient mechanisms to adapt changeable conditions. ABA is a key hormone for many plant physiological processes, including senescence, seed dormancy and germination, and adapting to stress conditions. Recently, the understanding of ABA signaling, from the perception to downstream regulation, involves many intermediates and is very complicated (Ma et al. 2009; Park et al. 2009; Cutler et al. 2010). Furthermore, many aspects of ABA biology are regulated by protein ubiquitination.

9.3.1 ABA Biosynthesis Impacted by Protein Ubiquitination

Protein ubiquitination is known to control a broad of protein stability, including signaling molecules and metabolic enzymes. SENESCENCE-ASSOCIATED E3 UBIQUITIN LIGASE1/PLANT U-BOX E3 LIGASE 44 (SAUL1/AtPUB44) has been shown to be a functional E3 ubiquitin ligase and directly target ABSCISIC ALDEHYDE OXIDASES 3 (AAO3), an enzyme to convert abscisic aldehyde to ABA, for ubiquitin-dependent degradation via the 26S proteasome. *saul1* mutants exhibit enhanced ABA level and show premature senescence (Raab et al. 2009). *XERICO*, encoding a small protein with RING-H2 RING finger motif, has been shown to increase ABA levels and drought tolerance of plants through upregulation of *AtNECD3*, which is a key ABA-biosynthesis gene. Furthermore, yeast two-hybrid assay indicates that XERICO interacts with an E2 ubiquitin-conjugating enzyme, AtUBC8 and also AtTLP9, which is an ASK1-interacting F-box protein involved in ABA-signaling pathway (Ko et al. 2006).

9.3.2 ABA Signaling Regulated by Protein Ubiquitination

Recent progress in interpreting the ABA signaling shows a core ABA signal pathway, in which ABA receptor proteins (PYR/PRL/RCAR) bind to the type 2C protein phosphatases (PP2Cs) and inhibit the phosphatase activity in the presence of ABA, subsequently releasing the SnRK2s kinases and the accumulated phosphorylated SnRK2s kinases phosphorylating the ABA-responsive element binding factors (ABFs) (Ma et al. 2009; Park et al. 2009; Cutler et al. 2010). Besides a few of ubiquitin conjugases, the increased numbers of E3 ligases, belonging to different E3 subfamilies, playing important roles in ABA signaling have been identified.

9.3.2.1 RING-type E3 Ligases Involved in ABA Signaling

RING-type E3 ligases were classified due to their conserved cysteine and histidine domain. Various RING fingers exhibit binding activity toward E2 ubiquitin-conjugating enzymes (Ubc's) and also the single RING-type E3 ligase has capacity to bind and ubiquitinate its substrate(s). This family has around 500 members in Arabidopsis, the second large family of E3 in Arabidopsis. A couple of RING finger E3 ligases were discovered in ABA signaling. Abscisic Acid-Insensitive 3 (ABI3), a B3-domain transcription factor, plays a central role in ABA-signaling pathway. Both biochemical and genetic evidences demonstrated that the abundance of ABI3 is post-translationally regulated by AIP2 (ABI3-interacting protein2), a RING-type E3 ligase. AIP2 can directly polyubiquitinate ABI3 in vitro. *AIP2*-overexpression plants show lower levels of ABI3 protein and more resistant to ABA compared with wild type plants, which is similar to the phenotype of *abi3*. Whereas *aip2* mutant contains higher ABI3 protein levels and is hypersensitive to ABA, phenocopying *ABI3*-overexpressing plants. Therefore, AIP2 is a negative regulator through regulating the stability of ABI3 (Zhang et al. 2005).

Plants have evolved protective mechanism to survive from adverse environmental conditions during their transition from heterotrophic to autotrophic growth. This post-germination developmental checkpoint is controlled by ABA, which induces the accumulation of Abscisic Acid-Insensitive 5 (ABI5) in germination stage. ABI5, a basic Leu Zipper transcription factor acting downstream of ABI3, is essential for ABA-dependent post-germinative growth arrest (Lopez-Molina et al. 2001, 2002). ABA-induced ABI5 protein accumulation can be regulated through both activation of transcription and enhanced protein stability. Enhanced ABI5 protein in seedlings treated with 26S proteasome inhibitors indicates that the degradation of ABI5 is dependent on the UPS (Lopez-Molina et al. 2001; Brocard et al. 2002). A RING-type E3 KEEP ON GOING (KEG) is a negative regulator of ABA signaling, required for keeping the low levels of ABI5 in the absence of ABA. KEG can interact with and ubiquitinate ABI5 in vitro. Cytoplasmic degradation of ABI5 is mediated by KEG. *keg* mutant plants accumulated very high levels of ABI5 and resulted in hypersensitivity to ABA, exhibiting plant growth arrest immediately after germination in presence of ABA, similar to the phenotype of *ABI5*-overexpressing plants. Loss of *ABI5* can partially rescue the *keg* phenotype (Stone et al. 2006; Liu and Stone 2013). Furthermore, KEG protein levels are modulated by ABA through self-ubiquitination and subsequent degradation via 26S proteasome (Liu and Stone 2010). Another protein, ABI FIVE BINDING PROTEIN (AFP) can attenuate ABA signals by promoting ABI5 degradation, although the mechanism of this process is not clear (Lopez-Molina et al. 2003).

In addition to ABI5, two other ABI5-related bZIP transcription factors, named ABF1 and ABF3, also regulated seed germination and post-germination growth (Jakoby et al. 2002; Finkelstein et al. 2005). Similar model to ABI5 degradation modulated by KEG, the abundance of ABF1 and ABF3 proteins is also affected by ABA and KEG. KEG can interact with and ubiquitinate ABF1 and ABF3 in vitro, and the loss of *KEG* slows the degradation of ABF1 and ABF3 in vivo and in vitro, suggesting that KEG can also modulate the degradation of ABF1 and ABF3. Similarly, the loss of *ABF1*, *ABF3*, or both of them cannot fully rescue the *keg* phenotype (Chen et al. 2013), which is similar to the *abi5 keg* double mutant. All above evidence indicates that other bZIP transcription factors, such as ABF2 and ABF4, maybe function in the similar way (Jakoby et al. 2002). Calcineurin B-like interacting protein kinase (CIPK26) was identified as another KEG-interacting protein by using a yeast two-hybrid screen. CIPK26 and KEG can interact in vitro and in vivo. The degradation of CIPK26 was increased along with the elevated levels of KEG protein, *CIPK26*-overexpressing plant was more sensitive to ABA compared with wild-type plants. Furthermore, CIPK26 was able to phosphorylate ABI5 in vitro. So the data show that CIPK26 is involved in ABI5- and KEG-mediated ABA-signaling network (Lyzenga et al. 2013). Since no evidence showed the KEG can direct ubiquitinate and promote the degradation of CIPK26, the triadic relation among CIPK26, KEG, and ABI5 still need to be addressed.

Through testing the germination of 100 different Arabidopsis RING finger gene T-DNA insertion mutant lines in the presence of exogenous ABA, Kim's lab found three RING E3 ubiquitin ligases, At Arabidopsis ABA-insensitive RING protein 1 (AIRP1), At AIRP2 and At AIRP3 were involved in ABA response. These three E3 ubiquitin ligases are positive regulators in ABA-dependent response to drought stress, among them AtAIRP1 and AtAIRP2 play combinatory roles in this process. AtAIRP3 can target the Cys proteinase of the papain family, RESPONSIVE TO DEHYDRATION21, for degradation in vitro and in vivo (Ryu et al. 2010; Cho et al. 2011; Kim and Kim 2013).

In addition, a number of other RING-type E3 ligases have been shown to play positive roles in ABA signaling in both seed germination and post-germination growth, although their substrates remain to be identified. These include ATL43 (Serrano et al. 2006), SDIR1 (Zhang et al. 2007), RHA2a (Bu et al. 2009), and RHA2b (Li et al. 2011).

9.3.2.2 U-Box-type E3 Ligases Involved in ABA Signaling

U-box family E3 ligase was classified based on their modified RING finger domain, without the full complement of Zn^{2+}-binding ligands, probable involved in E2-dependent ubiquitination. At least four U-box-type E3 ligases have been characterized in ABA signaling. The expression of *AtPUB9* can be induced by ABA treatment, accompanying the changes in AtPUB9 subcellular localization from the nucleus to the plasma membrane in tobacco BY-2 cells. *atpub9* mutant plants were hypersensitive to ABA-induced seed germination inhibition, which can be rescued by *abi3-6*, indicating that AtPUB9 is involved in ABA response (Bergler and Hoth 2011). AtPUB18 and AtPUB19 are two homologous U-box E3 ubiquitin ligases. Loss-of-function and overexpression assays reveal that AtPUB18 and AtPUB19 are negative regulators of ABA-mediated drought responses. Moreover, the *atpub18atpub19* double mutant displayed more sensitive to ABA and enhanced drought tolerance than did each single mutant plant (Bergler and

Hoth 2011; Liu et al. 2011; Seo et al. 2012). AtCHIP, a U-box-containing E3 ligase, plays important roles in temperature stress tolerance and ABA response. Overexpression of *AtCHIP* enhanced plant more sensitive to ABA and low- and high-temperature treatments compared with wild type. AtCHIP functions as an E3 ligase of the A subunit of protein phosphatase 2A (PP2A), which functions in plant response to ABA and stress. In vitro assay showed it seemed that a single ubiquitin molecule was added to A subunits of PP2A by E3 ligase AtCHIP; author proposed the possibility that the ubiquitylation event might not lead to the degradation of the A subunits of PP2A; instead it might lead to a change in their interaction with the B or C subunit or to the enzymatic activity of PP2A (Yan et al. 2003; Luo et al. 2006). In vivo protein-state analysis should be a key point needed to support their claim.

9.3.2.3 CRLs Involved in ABA Signaling

The multi-subunit ligases CRLs (CULLIN-RING E3 ubiquitin ligases) are belong to the biggest E3 family in Arabidopsis. Upon to now several members of this family were founded to be in ABA pathway. Class I homeobox-Leu Zipper (HD-ZIP) transcription factor, ATHB6, a target of the ABI1, plays a negative role in ABA signal pathway in stomatal closure and germination assays (Himmelbach et al. 2002). MATH/BTB proteins (BPMs), acting as adaptors for Cullin3-based ubiquitin E3 ligase to bind substrate, directly interact with and target ATHB6 for proteasomal degradation. The knockdown of *BPMs* and the overexpression of *ATHB6* result in ABA insensitivity. Reducing the function of $CUL3^{BPM}$ leads to higher accumulation of ATHB6 proteins (Lechner et al. 2011).

Mutants in two DWD genes, *DWA1* and *DWA2* (*DWD hypersensitive to ABA1 and 2*) encoding substrate receptors for CULLIN4-RING E3 ligase, are hypersensitive to ABA and accumulate higher ABI5 proteins. DWA1 and DWA2 can interact with ABI5 in vivo. So, DWA1 and DWA2 may work as the substrate receptors for Cul4 E3 ligase to promote the degradation of ABI5. Furthermore, *cul4* mutant is hypersensitive to ABA and has higher ABI5 protein compared with wild type after ABA treatment (Lee et al. 2010). ABD1 (ABA-hypersensitive DCAF1) is another substrate receptor for CULLIN4-based E3 ubiquitin ligases, which negatively regulates ABA signaling by binding to and affecting the ABI5 stability in the nucleus. *abd1* mutant is hypersensitive to ABA during seed germination and seedling growth and leads to higher accumulation of ABI5 (Seo et al. 2014). In in vivo Co-IP experiments, ABD1 and DWA1/2 can bind different isoforms of ABI5. ABD1 interacts with the two isoforms of ABI5, while DWA1 and DWA2 interact with the slow migrating form of ABI5 (Lee et al. 2010; Seo et al. 2014). Recently, Irigoyen et al. found that ABA receptors, PYL8, as well as PYL4 and PYL9, can interact with DET1-, DDB1-ASSOCIATED1 (DDA1), which is the part of COP10-DET1-DDA1 (CDD) complex and can provide substrate for CRL4 ligase. Furthermore, overexpression of *DDA1* promotes the degradation of PYL8 protein and reduces plant sensitivity of ABA (Irigoyen et al. 2014). So, DDA1 negatively regulates ABA response by targeting the degradation of ABA receptor, which is very important when ABA levels in plant decrease.

Other two F-box proteins are involved in ABA-signaling pathway. DOR, a putative F-box protein, belongs to the S-locus F-box-like family related to AhSLF-S2 and can interact with ASK14 and CUL1. *dor* mutant plants are hypersensitive to ABA-induced stomatal closure and tolerant to drought stress. *DOR*-overexpressing plants were more sensitive to drought stress (Zhang et al. 2008). Based on the physiological analysis about the overexpresser lines and knockdown mutant plants, the EID1-like protein 3 (EDL3) is another F-box protein who acts as a positive regulator in ABA-signaling pathway (Koops et al. 2011). But up to now, no substrate was discovered for both of F-box proteins.

9.3.2.4 E2 Involved in ABA Signaling

Besides ubiquitin E3 ligases control the substrate specificity, ubiquitin conjugases were also discovered to be important in substrate recognition and thus may also works in the specific physiological regulation. The *Mungbean VrUBC1* (*Vignaradiata UBC1*) encodes a functional UBC E2. Overexpression of *VrUBC1* plants enhanced ABA-induced stomatal closing and germination inhibition. A number of key ABA-signaling genes, such as *ABI5*, *ABF2*, *ABF3,* and *ABF4*, were increased in transgenic plants compared to the wild-type plants. VrUBC1 can interact with AtVBP1, a C3HC4-type RING E3 ligase, indicating that VrUBC1 is involved in ABA response possibly through interaction with a RING E3 ligase (Chung et al. 2013). Arabidopsis UBC32 was identified as a stress-induced transmembrane-contained E2, similar to its counterpart in yeast and mammals, it was discovered to be involved in the ER-associated protein degradation of unfolding protein and participate in salt and ABA response in Arabidopsis. The *ubc32* lines were more insensitive than wild-type plants to ABA in both the germination and post-germination stages, while the *35S-UBC32* lines were more sensitive to ABA, indicating that UBC32 is also a positive regulator in ABA-signaling pathway (Cui et al. 2012).

9.4 Protein Sumoylation in ABA Signaling

Many biochemical and genetic evidences have revealed that SUMO post-translation modification participates in numerous crucial biological processes in plants, including plant vegetative growth and development, flowering regulation, embryogenesis, embryogenesis, hormonal response, nitrogen assimilation, pathogen resistance, especially the abiotic stress response, including phosphate-deficiency, cold, drought, salt, heat shock, and copper tolerance (Miura and Hasegawa 2010). Up to now, several sumoylation components in sumoylation pathway were reported to regulate ABA signaling. *AtSUMO1/2* overexpressing plants with increased level of sumoylation are less sensitive to ABA, showing reduced seedling primary root growth inhibition (Lois et al. 2003). AtSCE1a, the unique SUMO-conjugating enzyme in Arabidopsis, is colocalized in nucleus with AtSUMO1/2. Transgenic plants with *AtSCE1a* co-suppression showed ABA-mediated enhanced root growth inhibition compared to wide type (Lois et al. 2003). Both known SUMO E3 ligases were also discovered to be involved in ABA signaling. AtSIZ1, as a multiple biological functional SUMO ligase E3, is confirmed to be related to ABA signaling, with the evidence that the enhanced ABA response was observed in *siz1* mutant (Miura and Hasegawa 2009). The other SUMO E3 ligase, HPY2/MMS21, firstly identified as a modulator of cell cycle progression and meristem maintenance, was also demonstrated to participate in ABA signaling. Similar to *siz1*, enhanced ABA sensitivity was also observed in *mms21* (Zhang et al. 2013). Altogether, it seems that knockout/knockdown of components in sumoylation pathway all increased ABA sensitivity.

How does sumoylation regulate plant ABA response? The effect of sumoylation on key components in ABA-signaling pathway has drawn the attention. Genetic analysis showed mutant *abi5-4* represses ABA sensitivity in *siz1-2* mutant both in seed germination and primary root growth, which indicates the genetic interaction between SIZ1 and ABI5. Biochemical evidence demonstrated that ABI5 was sumoylated by SIZ1 and SIZ1-mediated ABI5 sumoylation protected the protein stability of ABI5 (Miura et al. 2009). Whereas AFP (ABI 5-binding protein) and KEG (KEEP ON GOING, a RING finger ubiquitin E3 ligase) facilitate ABI5 degradation through 26S proteasomes system, it seems that SUMO E3 ligase SIZ1 protects ABI5 protein by sumoylation of ABI5 to against the ubiquitin-mediated ABI5 degradation (Lopez-Molina et al. 2003; Miura et al. 2009; Liu and Stone 2010). The additive ABA-hypersensitive phenotype caused by *siz1-2afp-1* double mutation compared to single mutant suggested that sumoylation and AFP facilitated ABI5 degradation may function in different ways, perhaps other types of protein modifications involved (Miura et al. 2009). Moreover, the recent study found phenotype of double mutants of *abi5-8 myb30-2* under ABA treatment is comparable to wild-type plants during seed germination, so do the expression level of a group of stress-responsive genes, including *COR413*, *COR15B*, *LOX3*, and *BGL2*. It is further found that MYB30 can be sumoylated at residue K283 dependent on SIZ1. Due to the different sets of genes regulated by MYB30 and ABI5 in response to ABA, MYB30 and ABI5 were considered to balance different sets of downstream genes expression upon ABA treatment (Zheng et al. 2012). Additionally, ABI5 is phosphorylated by SnRK2.2 and SnRK2.3 in response to ABA, and then ABI5 was activated as a transcription factor. Up to know, at least three different protein modifications have been found on ABI5 protein, how ABI5 is precisely regulated by these three post-translational modifications, and whether there is some kind of balance needs further investigation.

9.5 Conclusions and Perspective

In fact, the synthesis and destruction of proteins are equally important for plant to grow and survive from adverse conditions. Therefore, by targeting an ABA receptor or ABA-responsive proteins at the right moments, the ubiquitination and sumoylation modification modulate the ABA biosynthesis and the sensitivity of plant responsive to ABA (Fig. 9.2). Based on the observations illustrated above, the following mechanisms are involved, (1) one E3 ligase can target several substrates, (2) one target can be modified by several E3 ligases, (3) the positive regulator and the negative regulator in the same pathway can be regulated by the same E3 ligase.

For instance, the ubiquitin ligase KEG, being the negative regulator of ABA, leads to the degradation of ABF1, ABF3, ABI5, and CIPK6, and four proteins play positive roles in ABA inhibition of seed germination and post-germination growth. ABF1, ABF3, and ABI5 are bZIP-type transcription factors, and CIPK26 is CIPK which can phosphorylate ABI5 in vitro. KEG is stable in the absence of ABA, and ABA can promote KEG auto-ubiquitination and degradation by the 26S proteasome. The reduction of KEG protein leads to the accumulation of ABF1, ABF3, ABI5, and CIPK6, and subsequent growth arrest of plants. Oppositely one substrate also could be directly regulated by different E3 or E3 complexes. Well-studied cell cycle negative regulator, ICK/KRP (p27), both in mammals and plant, is controlled by different types of E3 ligases in different location in cells. In nucleus, it is degraded by SCF complex, while in cytosol, it is targeted by RING-type E3 ligase (Kamura et al. 2004; Ren et al. 2008; Lai et al. 2009). It seems that the key transcriptional factor ABI5 is regulated by similar manners. The degradation of ABI5 is modulated by KEG in cytosol under lower ABA level, and by DWA1, DWA2, and ABD1 complexes in nucleus, evidenced by the exclusive localization of those adaptor proteins in nucleus under higher level of ABA and stress conditions. Another example is the SUMO E3 ligase SIZ1, the negative regulator of ABA by targeting two proteins, ABI5 and MYB30, in plant response to ABA. Sumoylation by SIZ1 can stabilize ABI5 and MYB30. Different from *abi5* mutant, *myb30* mutants are ABA hypersensitive to ABA during ABA-controlled germination. Furthermore, sumoylated ABI5 proteins are inactive, but sumoylated MYB30 proteins are inactive. Therefore, SIZ1 sumoylation of both ABI5 and MYB30 transcription factors regulate plant ABA response in the parallel way. Why the same SUMO E3 ligase has different effects on the activity of different substrates, one possibility is that sumoylation might affect other types of protein modification, for example, phosphorylation of ABI5 is essential for the activity of ABI5. This conception need be addressed experimentally.

Besides the SUMO, different types of E2 and E3 ligases were discovered in ABA signaling, the component of the UPS destruction 26S proteasome complex, RPN10, was also found to be important in ABA signaling. *rpn10-1* was found more sensitive to ABA with the effect on the selective stabilization of the short-lived ABA-signaling protein ABI5 (Smalle et al. 2003). Together with all

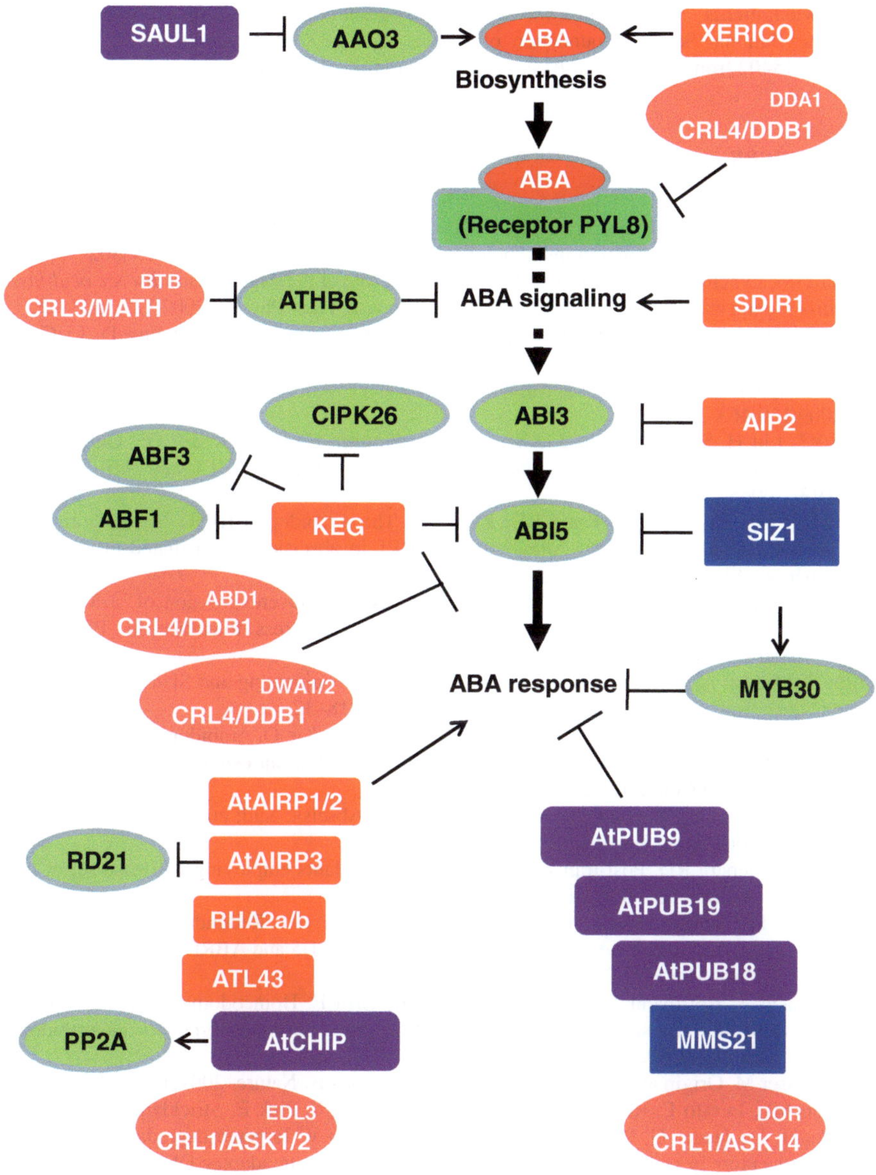

Fig. 9.2 Ubiquitin and SUMO E3 ligases (complexes) and their substrates participate in ABA pathway. *Green color* refers to substrates. *Orange color* highlights RING-type E3 ligase. *Purple color* indicates U-box-type E3 ligase. Cullin-based E3 ligases are in light brown color. SUMO E3 ligases are in *blue color*

evidences analyzed in this chapter, it is believed that protein ubiquitination and sumoylation play vital roles in ABA signaling.

Acknowledgments This research was supported by grant NSFC 91317308/31030047/31270301 from the National Science Foundation of China, and Qi Xie is supported by 973 Program 2011CB915402 from the National Basic Research Program of China.

References

Bergler J, Hoth S. Plant U-box armadillo repeat proteins AtPUB18 and AtPUB19 are involved in salt inhibition of germination in Arabidopsis. Plant Biol (Stuttg). 2011;13:725–30.

Brocard IM, Lynch TJ, Finkelstein RR. Regulation and role of the Arabidopsis *abscisic acid-insensitive 5* gene in abscisic acid, sugar, and stress response. Plant Physiol. 2002;129:1533–43.

Bu Q, Li H, Zhao Q, Jiang H, Zhai Q, Zhang J, Wu X, Sun J, Xie Q, Wang D, Li C. The Arabidopsis RING finger E3 ligase RHA2a is a novel positive regulator of abscisic acid signaling during seed germination and early seedling development. Plant Physiol. 2009;150:463–81.

Chen YT, Liu H, Stone S, Callis J. ABA and the ubiquitin E3 ligase KEEP ON GOING affect proteolysis of the *Arabidopsis thaliana* transcription factors ABF1 and ABF3. Plant J. 2013;75:965–76.

Cho SK, Ryu MY, Seo DH, Kang BG, Kim WT. The Arabidopsis RING E3 ubiquitin ligase AtAIRP2 plays combinatory roles with AtAIRP1 in abscisic acid-mediated drought stress responses. Plant Physiol. 2011;157:2240–57.

Chung E, Cho CW, So HA, Kang JS, Chung YS, Lee JH. Overexpression of, a Mung Bean E2 Ubiquitin-Conjugating enzyme, enhances osmotic stress tolerance in. PLoS ONE. 2013;8:e66056.

Colby T, Matthai A, Boeckelmann A, Stuible HP. SUMO-conjugating and SUMO-deconjugating enzymes from Arabidopsis. Plant Physiol. 2006;142:318–32.

Cui F, Liu L, Zhao Q, Zhang Z, Li Q, Lin B, Wu Y, Tang S, Xie Q. Arabidopsis ubiquitin conjugase UBC32 is an ERAD component that functions in brassin osteroid-mediated salt stress tolerance. Plant Cell. 2012;24:233–44.

Cutler SR, Rodriguez PL, Finkelstein RR, Abrams SR. Abscisic acid: emergence of a core signaling network. Annu Rev Plant Biol. 2010;61:651–79.

Downes B, Vierstra RD. Post-translational regulation in plants employing a diverse set of polypeptide tags. Biochem Soc Trans. 2005;33:393–9.

Finkelstein R, Gampala SS, Lynch TJ, Thomas TL, Rock CD. Redundant and distinct functions of the ABA response loci ABA-INSENSITIVE(ABI)5 and ABRE-BINDING FACTOR (ABF)3. Plant Mol Biol. 2005;59:253–67.

Himmelbach A, Hoffmann T, Leube M, Hohener B, Grill E. Homeodomain protein ATHB6 is a target of the protein phosphatase ABI1 and regulates hormone responses in Arabidopsis. EMBO J. 2002;21:3029–38.

Hochstrasser M. Origin and function of ubiquitin-like proteins. Nature. 2009;458:422–9.

Irigoyen ML, Iniesto E, Rodriguez L, Puga MI, Yanagawa Y, Pick E, Strickland E, Paz-Ares J, Wei N, De Jaeger G, Rodriguez PL, Deng XW, Rubio V. Targeted degradation of Abscisic acid receptors is mediated by the Ubiquitin Ligase substrate adaptor DDA1 in Arabidopsis. Plant Cell. 2014.

Jakoby M, Weisshaar B, Droge-Laser W, Vicente-Carbajosa J, Tiedemann J, Kroj T, Parcy F, ZIPRG B. bZIP transcription factors in Arabidopsis. Trends Plant Sci. 2002;7:106–11.

Kamura T, Hara T, Matsumoto M, Ishida N, Okumura F, Hatakeyama S, Yoshida M, Nakayama K, Nakayama KI. Cytoplasmic ubiquitin ligase KPC regulates proteolysis of p27(Kip1) at G1 phase. Nat Cell Biol. 2004;6:1229–35.

Kim JH, Kim WT. The Arabidopsis RING E3 ubiquitin ligase AtAIRP3/LOG2 participates in positive regulation of high-salt and drought stress responses. Plant Physiol. 2013;162:1733–49.

Ko JH, Yang SH, Han KH. Upregulation of an Arabidopsis RING-H2 gene, XERICO, confers drought tolerance through increased abscisic acid biosynthesis. Plant J. 2006;47:343–55.

Koops P, Pelser S, Ignatz M, Klose C, Marrocco-Selden K, Kretsch T. EDL3 is an F-box protein involved in the regulation of abscisic acid signalling in Arabidopsis thaliana. J Exp Bot. 2011;62:5547–60.

Lai J, Chen H, Teng K, Zhao Q, Zhang Z, Li Y, Liang L, Xia R, Wu Y, Guo H, Xie Q. RKP, a RING finger E3 ligase induced by BSCTV C4 protein, affects gemini virus infection by regulation of the plant cell cycle. Plant J. 2009;57:905–17.

Lechner E, Leonhardt N, Eisler H, Parmentier Y, Alioua M, Jacquet H, Leung J, Genschik P. MATH/BTB CRL3 receptors target the homeodomain-leucine zipper ATHB6 to modulate abscisic acid signaling. Dev Cell. 2011;21:1116–28.

Lee JH, Yoon HJ, Terzaghi W, Martinez C, Dai M, Li J, Byun MO, Deng XW. DWA1 and DWA2, two Arabidopsis DWD protein components of CUL4-based E3 ligases, act together as negative regulators in ABA signal transduction. Plant Cell. 2010;22:1716–32.

Li H, Jiang H, Bu Q, Zhao Q, Sun J, Xie Q, Li C. The Arabidopsis RING finger E3 ligase RHA2b acts additively with RHA2a in regulating abscisic acid signaling and drought response. Plant Physiol. 2011;156:550–63.

Liu H, Stone SL. Abscisic acid increases Arabidopsis ABI5 transcription factor levels by promoting KEG E3 ligase self-ubiquitination and proteasomal degradation. Plant Cell. 2010;22:2630–41.

Liu H, Stone SL. Cytoplasmic degradation of the Arabidopsis transcription factor abscisic acid insensitive 5 is mediated by the RING-type E3 ligase KEEP ON GOING. J Biol Chem. 2013;288:20267–79.

Liu YC, Wu YR, Huang XH, Sun J, Xie Q. AtPUB19, a U-box E3 ubiquitin ligase, negatively regulates abscisic acid and drought responses in Arabidopsis thaliana. Mol Plant. 2011;4:938–46.

Lois LM, Lima CD, Chua NH. Small ubiquitin-like modifier modulates abscisic acid signaling in Arabidopsis. Plant Cell. 2003;15:1347–59.

Lopez-Molina L, Mongrand S, Chua NH. A postgermination developmental arrest checkpoint is mediated by abscisic acid and requires the ABI5 transcription factor in Arabidopsis. Proc Natl Acad Sci U S A. 2001;98:4782–7.

Lopez-Molina L, Mongrand S, McLachlin DT, Chait BT, Chua NH. ABI5 acts downstream of ABI3 to execute an ABA-dependent growth arrest during germination. Plant J. 2002;32:317–28.

Lopez-Molina L, Mongrand S, Kinoshita N, Chua NH. AFP is a novel negative regulator of ABA signaling that promotes ABI5 protein degradation. Genes Dev. 2003;17:410–8.

Luo J, Shen G, Yan J, He C, Zhang H. AtCHIP functions as an E3 ubiquitin ligase of protein phosphatase 2A subunits and alters plant response to abscisic acid treatment. Plant J. 2006;46:649–57.

Lyzenga WJ, Liu H, Schofield A, Muise-Hennessey A, Stone SL. Arabidopsis CIPK26 interacts with KEG, components of the ABA signalling network and is degraded by the ubiquitin-proteasome system. J Exp Bot. 2013;64:2779–91.

Ma Y, Szostkiewicz I, Korte A, Moes D, Yang Y, Christmann A, Grill E. Regulators of PP2C phosphatase activity function as abscisic acid sensors. Science. 2009;324:1064–8.

Meluh PB, Koshland D. Evidence that the MIF2 gene of Saccharomyces cerevisiae encodes a centromere protein with homology to the mammalian centromere protein CENP-C. Mol Biol Cell. 1995;6:793–807.

Miura K, Hasegawa PM. Sumoylation and abscisic acid signaling. Plant Signal Behav. 2009;4:1176–8.

Miura K, Hasegawa PM. Sumoylation and other ubiquitin-like post-translational modifications in plants. Trends Cell Biol. 2010;20:223–32.

Miura K, Jin JB, Hasegawa PM. Sumoylation, a post-translational regulatory process in plants. Curr Opin Plant Biol. 2007;10:495–502.

Miura K, Lee J, Jin JB, Yoo CY, Miura T, Hasegawa PM. Sumoylation of ABI5 by the Arabidopsis SUMO E3 ligase SIZ1 negatively regulates abscisic acid signaling. Proc Natl Acad Sci U S A. 2009;106:5418–23.

Park SY, Fung P, Nishimura N, Jensen DR, Fujii H, Zhao Y, Lumba S, Santiago J, Rodrigues A, Chow TF, Alfred SE, Bonetta D, Finkelstein R, Provart NJ, Desveaux D, Rodriguez PL, McCourt P, Zhu JK, Schroeder JI, Volkman BF, Cutler SR. Abscisic acid inhibits type 2C protein phosphatases via the PYR/PYL family of START proteins. Science. 2009;324:1068–71.

Park HJ, Park HC, Lee SY, Bohnert HJ, Yun DJ. Ubiquitin and ubiquitin-like modifiers in plants. J Plant Biol. 2011;54:275–85.

Raab S, Drechsel G, Zarepour M, Hartung W, Koshiba T, Bittner F, Hoth S. Identification of a novel E3 ubiquitin ligase that is required for suppression of premature senescence in Arabidopsis. Plant J. 2009;59:39–51.

Ren H, Santner A, del Pozo JC, Murray JA, Estelle M. Degradation of the cyclin-dependent kinase inhibitor KRP1 is regulated by two different ubiquitin E3 ligases. Plant J. 2008;53:705–16.

Ryu MY, Cho SK, Kim WT. The Arabidopsis C3H2C3-type RING E3 ubiquitin ligase AtAIRP1 is a positive regulator of an abscisic acid-dependent response to drought stress. Plant Physiol. 2010;154:1983–97.

Seo KI, Lee JH, Nezames CD, Zhong S, Song E, Byun MO, Deng XW. ABD1 Is an Arabidopsis DCAF substrate receptor for CUL4-DDB1-based E3 Ligases that acts as a negative regulator of Abscisic acid signaling. Plant Cell. 2014.

Seo DH, Ryu MY, Jammes F, Hwang JH, Turek M, Kang BG, Kwak JM, Kim WT. Roles of four Arabidopsis U-box E3 ubiquitin ligases in negative regulation of abscisic acid-mediated drought stress responses. Plant Physiol. 2012;160:556–68.

Serrano M, Parra S, Alcaraz LD, Guzman P. The ATL gene family from Arabidopsis thaliana and Oryza sativa comprises a large number of putative ubiquitin ligases of the RING-H2 type. J Mol Evol. 2006;62:434–45.

Smalle J, Kurepa J, Yang P, Emborg TJ, Babiychuk E, Kushnir S, Vierstra RD. The pleiotropic role of the 26S proteasome subunit RPN10 in Arabidopsis growth and development supports a substrate-specific function in abscisic acid signaling. Plant Cell. 2003;15:965–80.

Stone SL, Williams LA, Farmer LM, Vierstra RD, Callis J. KEEP ON GOING, a RING E3 ligase essential for Arabidopsis growth and development, is involved in abscisic acid signaling. Plant Cell. 2006;18:3415–28.

Vierstra RD. The ubiquitin-26S proteasome system at the nexus of plant biology. Nat Rev Mol Cell Biol. 2009;10:385–97.

Yan J, Wang J, Li Q, Hwang JR, Patterson C, Zhang H. AtCHIP, a U-box-containing E3 ubiquitin ligase, plays a critical role in temperature stress tolerance in Arabidopsis. Plant Physiol. 2003;132:861–9.

Zhang X, Garreton V, Chua NH. The AIP2 E3 ligase acts as a novel negative regulator of ABA signaling by promoting ABI3 degradation. Genes Dev. 2005;19:1532–43.

Zhang Y, Yang C, Li Y, Zheng N, Chen H, Zhao Q, Gao T, Guo H, Xie Q. SDIR1 is a RING finger E3 ligase that positively regulates stress-responsive abscisic acid signaling in Arabidopsis. Plant Cell. 2007;19:1912–29.

Zhang Y, Xu W, Li Z, Deng XW, Wu W, Xue Y. F-box protein DOR functions as a novel inhibitory factor for abscisic acid-induced stomatal closure under drought stress in Arabidopsis. Plant Physiol. 2008;148:2121–33.

Zhang S, Qi Y, Liu M, Yang C. SUMO E3 ligase AtMMS21 regulates drought tolerance in Arabidopsis thaliana(F). J Integr Plant Biol. 2013;55:83–95.

Zheng Y, Schumaker KS, Guo Y. Sumoylation of transcription factor MYB30 by the small ubiquitin-like modifier E3 ligase SIZ1 mediates abscisic acid response in Arabidopsis thaliana. Proc Natl Acad Sci U S A. 2012;109:12822–7.

Chapter 10
Reactive Oxygen Species (ROS) and ABA Signalling

Ling Bai, Pengtao Wang and Chun-Peng Song

Abstract Various abiotic and biotic stress conditions result in the accumulation of both abscisic acid (ABA) and reactive oxygen species (ROS) in plants. The adaptation of plants to these stress conditions is a very complex process, which including stress signal perception, transduction and change of genes expression. The signalling pathways between ABA and ROS are independent of each other or crosstalk at various levels. The first part of this chapter focus on ROS turnover in the presence of ABA and under stress conditions, including their generation from photosynthetic system and non-photosynthetic system, the ROS detoxification by enzymatic antioxidant systems and non-enzymatic low-molecular weight metabolites. The second part is attempted to explore the dual role of ROS for toxic to cells and signalling in stress adaptation regulated by ABA. Finally, the third part in detail discusses the crosstalk between ABA and ROS in regulation of stomatal movement, dormancy and germination of seed, development of root system and stress adaptation in the transcription levels.

Keywords Reactive oxygen species · ABA signalling · Biological roles · Crosstalk

Abbreviations

ROS	Reactive oxygen species
ABI1	ABA insensitive 1
ABI2	ABA insensitive 2
AtGPX3	*Arabidopsis* glutathione peroxidase 3
OST1	Open stomata 1
MPK	Mitogen-activated protein kinase
MEKK	Mitogen-activated protein kinase kinase

L. Bai · P. Wang · C.-P. Song (✉)
Institute of Plant Stress Biology, State Key Laboratory of Cotton Biology, Department of Biology, Henan University, Kaifeng 475001, China
e-mail: songcp@henu.edu.cn

D.-P. Zhang (ed.), *Abscisic Acid: Metabolism, Transport and Signaling*,
DOI 10.1007/978-94-017-9424-4_10

GHR1	Guard cell hydrogen peroxide-resistant1
HSFs	Heat shock transcription factor
Zat	Zinc finger protein
WRKY	WRKY transcription factors
SLAC1	S–type anion channel
GPCR	G protein-coupled receptor
PYL/PYL/RCA	ABA receptor
PP2C	Type 2C protein phosphatase
SOD	Superoxide dismutase
APX	Ascorbate peroxidase
GPX	Glutathione peroxidase
GST	Glutathione-S–transferase
CAT	Catalase
PLD	Phospholipase dalpha1
PA	Phosphatidic acid
PI3P	Phosphatidylinositol 3-phosphate
GCA2	Growth controlled by abscisic acid 2
cGMP	Cyclic guanosine monophosphate
ATHK	*Arabidopsis thaliana* histidine kinase
HAB1	Hypersensitive to ABA 1
SWI3B	*Arabidopsis thaliana* homologue of the yeast SWI3 subunit of SWI/SNF chromatin-remodelling complexes
GPA1	Gα subunit
RCN1	Reticulocalbin 1
MEK1/2	Mitogen-activated protein kinase kinase 1/2
OXI1	Oxidative signal-inducible serine/threonine protein kinase 1
ANP1	MAPK kinase kinase
AtNDPK2	Nucleoside diphosphate kinase 2
CBL	Calcineurin B-like
CIPK	CBL-interacting protein kinases
AtRac1	*Arabidopsis* small guanosine triphosphatase (GTPase) protein
SNAC1	Stress-related NAC superfamily of transcription factors
ABP9	ABA-responsive-element (ABRE) binding protein 9

10.1 Introduction

Various abiotic stress conditions, including salinity, extremes of temperature and water shortage, result in the accumulation of phytohormone abscisic acid (ABA) in plants. This signalling molecule plays an important role in the regulation of plant growth and development. These stresses are the principal causes of significant reductions in crop yield. Stress tolerance is a complex phenomenon because plants might undergo multiple stresses at the same time during their development. For example, ABA induces the closure of stomata in the epidermis, which regulates transpiration and photosynthesis in plants. It also modulates the expression of many genes, the products of which might function in stress adaptation and tolerance.

In addition to stimulating the production of ABA, the above stress conditions also lead to the generation of reactive oxygen species (ROS), which are highly active and toxic. ROS that have been investigated in plants include superoxide radical ($O_2^{\cdot-}$), hydrogen peroxide (H_2O_2), singlet oxygen ($^1O_2$), hydroxyl radical ($OH^{\cdot}$) and nitric oxide (NO). H_2O_2, $O_2^{\cdot-}$ and $OH^{\cdot}$ can interconvert into one another (Halliwell 2012; Quan et al. 2008). Such interconversion may occur spontaneously or be catalysed by enzymes (Van Breusegem et al. 2008). Being highly reactive and toxic, ROS have long been considered as compounds that damage cellular components, such as by the destruction of nucleic acids, the oxidation of proteins and the induction of lipid peroxidation (Foyer and Noctor 2005). Unfavourable environmental conditions and ABA induce ROS production from numerous sources; an increase in ROS imposes oxidative stress, which can ultimately result in cell death. Therefore, plants have evolved an elaborate system to control cellular ROS levels (Mittler et al. 2011). ROS act in signal transduction to help the cell to counteract oxidative damage by initiating the expression of certain genes and thus activating related signal transduction pathways; these findings suggest that ROS function as important signals to activate and control various stress responses (Dalton et al. 1999). Not only do ROS function in stress responses, but they are also essential for the maintenance of normal energy and metabolic fluxes, the optimisation of cell function, the acclimation responses activated by retrograde signalling and the regulation of whole-plant systemic signalling pathways (Møller 2001; Jaspers and Kangasjärvi 2010). ROS metabolism and signalling in plants have become a frontier of research in basic and applied plant science, given the differential actions of ROS depending on the cellular compartments in which they function as components of a complex network.

Similarities between ABA and ROS include that both can be induced by various stress conditions, both act as ubiquitous 'signalling molecules' or 'secondary messengers' in normal plant cell function, and both initiate plant stress adaptation and resistance by the modulation of gene transcription. They also exhibit similar spatial and temporal distributions, as well as dynamic changes in plants in response to the alteration of environmental conditions. Several reviews that discuss ABA and ROS signalling have been published (Ahmad et al. 2010; Song et al. 2014). In this chapter, we focus on the turnover of ROS, their biological roles and their crosstalk with ABA signalling, and also attempt to describe the relationship between ROS and ABA signalling.

10.2 ROS Turnover in the Presence of ABA and Stress Conditions

ROS arise in plant cells via a number of routes. It has been estimated that 1 % of O_2 consumed by plants is diverted to produce ROS in various subcellular locations (Del Rio et al. 1992). ROS are produced continuously as by-products of various metabolic pathways that are localised in different cellular compartments, such as chloroplasts, mitochondria and peroxisomes (Del Río et al. 2006; Navrot et al. 2007).

10.2.1 ROS Generation by the Photosynthetic Apparatus

Under steady-state conditions, oxygen generated in the chloroplasts during photosynthesis can accept electrons passing through the photosystems, thus forming $O_2^{\cdot-}$. The accumulation of ROS is a common response to environmental stresses. The chloroplasts and peroxisomes are the major sites of ROS production upon exposure to high light (HL) (Foyer and Noctor 2009). In chloroplasts, $O_2^{\cdot-}$ and $^1O_2$ are mainly produced by photosystems I and II (PSI and PSII). $O_2^{\cdot-}$ arises as a result of single electron transfer to molecular oxygen in electron transfer chains, principally during the Mehler reactions in PSI (Asada 2006). $O_2^{\cdot-}$ can be dismutated to form H_2O_2. Compared with both $O_2^{\cdot-}$ and H_2O_2, $OH^{\cdot}$ is much reactive, which can be formed from $O_2^{\cdot-}$ and H_2O_2 in the presence of Fe, through catalysis by the Haber–Weiss reaction (Bhattacharjee 2010). $^1O_2$, an electronically excited species of O_2 that is also very toxic, is continuously produced during photosynthesis mainly in PSII. The enhanced generation of $^1O_2$ is caused by photoinhibition of PSII under excess photochemical stress or light energy (Hideg et al. 2002). Therefore, excess photochemical energy inevitably increases ROS production. The chloroplast may be function as a sensor of environmental information; its redox signalling allows the plant to acclimatize to the environmental stresses. (Pfannschmidt 2003).

Most (81 %) of 28 HL-responsive genes analysed in one study, including the *ASCORBATE PEROXIDASE 2* (*APX2*), require photosynthetic electron transport for their expression and are responsive to ABA (68 %) (Bechtold et al. 2008). This suggests that both ABA and ROS are crucial in the expression of HL-responsive genes. Furthermore, the analysis of mutants with altered ROS metabolism indicated that the expression of 61 % of these genes, including *APX2*, might be responsive to chloroplast-derived ROS (Bechtold et al. 2008).

10.2.2 ROS Generated from Non-photosynthetic Systems

Besides chloroplasts, mitochondria are the other important organelles that are sites for ROS production; they probably constitute the main source of ROS in plants under dark conditions. The mitochondrial electron transport system can produce $O_2^{\cdot-}$, H_2O_2 and $OH^{\cdot}$. Experiments on isolated mitochondria showed that about 1–5 % of the O_2 used in oxidative respiration leads to H_2O_2 production (Møller 2001). In the mitochondrial electron transport chain, complexes I and III are major sites for ROS production under dark conditions and in tissues that lack mitochondria (Laloi et al. 2004). Impairment of the components in this chain leads to the accumulation of ROS in mitochondria. Given that the mitochondrial electron transport chain also contains an alternative oxidase that can be activated under stress to remove toxic ROS (Maxwell et al. 1999), ROS produced from mitochondria are present at only low concentrations in green tissues (Apel and Hirt 2004). *ABO6* encodes a DEXH box RNA helicase that regulates the splicing of several genes that encode components of complex I in mitochondria. Compared with wild-type (WT) plants, the *abo6* mutant accumulated more ROS in mitochondria, but two dominant-negative mutations in the *ABA-insensitive1* (*abi1-1*) and *abi2-1* mutants greatly reduced ROS production in mitochondria. These findings provide molecular evidence for the interplay between ABA and auxin through the production of ROS from mitochondria (He et al. 2012).

Besides chloroplasts and mitochondria, ROS are also generated extracellularly in apoplasts in plants. pH-dependent cell wall peroxidases, germin-like oxalate oxidases and amine oxidases have been proposed as sources of H_2O_2 in the apoplasts of plant cells (Bolwell and Woftastek 1997). Alkaline pH activates pH-dependent cell wall peroxidases, which produce H_2O_2 in the presence of reductant. Apoplast alkalisation upon elicitor recognition precedes the oxidative burst, and H_2O_2 production by a pH-dependent cell wall peroxidase has been proposed as an alternative pathway of ROS production during biotic stress (Bolwell and Woftastek 1997).

ROS are also generated at the plasma membrane. ROS derived from NADPH-dependent oxidases (NADPH oxidases) in the plasma membrane participate in the oxidative burst associated with localised nonspecific pathogen responses, incompatible plant-pathogen interaction and hormone signalling in plants. ROS generated by NADPH oxidase RbohC, RbohD and RbohF function in plant development, defence and ABA signalling (Foreman et al. 2003; Monshausen et al. 2007). ABA induces the generation of H_2O_2 in *Vicia* guard cells, and the sources of H_2O_2 are chloroplastic and plasma membrane NADPH oxidase (Zhang et al. 2001a, b). Another study suggested that the accumulation of ROS occurred upon ABA-induced stomatal closure in *Arabidopsis* (Pei et al. 2000). Two NADPH oxidase subunit genes, *AtrbohD* and *AtrbohF*, function in ABA-mediated stomatal closure, and the *atrbohD/F* double mutant shows reduced stomatal closure and ROS production compared with wild-type plants upon ABA treatment (Kwak et al. 2003). The ABA-activated SnRK2 protein kinase open stomata 1 (OST1) acts upstream of ROS in guard cell ABA signalling. OST1 interacts with AtrbohF and can phosphorylate its Ser13 and Ser174 to activate NADPH oxidase (Sirichandra et al. 2009), which then causes the production of H_2O_2 (Santiago et al. 2009; Kepka et al. 2011). It was established that ABA-insensitive mutants play different function in ROS production in ABA-regulated stomatal movement. The observation that *abi2*, but not *abi1*, mutants can generate ROS, suggests that the *abi2-1* mutation impairs ABA signalling downstream of ROS production (Murata et al. 2001). *RbohD* is also required for ROS-dependent signal propagation in response to abiotic stresses, such as HL, heat and wounding (Miller et al. 2009).

10.2.3 Regulation of ROS Detoxification by ABA

In all aerobic organisms, the concentration of ROS is tightly controlled by various ROS-scavenging antioxidative defence systems (Foyer and Noctor 2005). However, the equilibrium between the generation and scavenging of ROS is challenged by various stress factors, which disrupt the redox homoeostasis of cells, leading to oxidative damage (Foyer et al. 1994). The antioxidant defence system consists of antioxidant enzymes—such as superoxide dismutase (SOD), catalase (CAT) and peroxidase (POD)—and antioxidant compounds, such as ascorbate and reduced tripeptide glutathione (GSH; γ-glu-cys-gly). The antioxidant system protects plants against oxidative damage by scavenging ROS; the presence of antioxidant enzymes and compounds in almost all cellular compartments suggests the importance of ROS detoxification for protection against various stresses.

Enzymatic antioxidant systems: The enzymatic antioxidant systems include a variety of scavengers, such as SOD, ascorbate peroxidase (APX), glutathione peroxidase (GPX), glutathione-*S*-transferase (GST) and CAT. The discovery of the enzymatic activity of SOD 45 years ago (McCord and Fridovich 1969) opened up the field of ROS biology.

Three classes of SOD are defined by their metal cofactors: copper/zinc (Cu/Zn-SOD), manganese (Mn-SOD) and iron (Fe-SOD) (Mittler 2002). SOD enzymes provide the first line of defence against the toxic effects of elevated levels of ROS. Overexpression of these SOD enhances plant tolerance to various stresses. For example, *Arabidopsis* with Mn-SOD overexpression showed increased salt resistance (Wang et al. 2004). Transgenic tobacco plants that overexpressed Cu/Zn-SOD were tolerant to multiple stresses, e.g. drought, salt and HL (Badawi et al. 2004). In addition, Fe-SOD enzymes have been found to function in early chloroplast development in *Arabidopsis* by protecting the chloroplast nucleoids from ROS (Myouga et al. 2008).

CAT enzymes dismutate H_2O_2 into H_2O and O_2 and are indispensable for ROS detoxification under stress conditions (Garg and Manchanda 2009). APX has a higher affinity for H_2O_2 (μM range) than CAT and POD (mM range), and it may have a more crucial role than CAT and POD in scavenging ROS and protecting cells during stress. The enhanced expression of APX in plants has been demonstrated under different stress conditions. *APX* overexpression in *Nicotiana tabacum* increases plant tolerance of either excess salt or water deficit (Badawi et al. 2004). *Arabidopsis* plants that overexpress *OsAPXa* or *OsAPXb* also exhibited increased salt tolerance (Lu et al. 2007). Transgenic plants that overexpress *SbpAPX*, a peroxisomal ascorbate peroxidase gene cloned from *Salicornia brachiata*, were more tolerant to salt and drought stresses than WT plants (Singh et al. 2013).

Another example of an enzymatic antioxidant, GPX enzymes comprise a large family of diverse isozymes, which can reduce H_2O_2 and organic and lipid hydroperoxides and can protect plant cells from oxidative damage (Noctor et al. 2002). Analysis of the physiological electron donor system for the *Arabidopsis* GPX family demonstrated that this system involves thioredoxin, not GSH (Comtois et al. 2003; Iqbal et al. 2006; Miao et al. 2006). This suggests that GSH does not contribute to the scavenging of ROS via GPX. The seven members of the GPX family in *Arabidopsis* localise to different subcellular sites (Millar et al. 2003). The overexpression of GPX enhances the tolerance of transgenic plants to abiotic stress. Mutation of *AtGPX3* impairs ABA and drought stress responses, and AtGPX3 functions to control water loss from the stomata by its interaction with ABI2 in the ABA signalling pathway (Miao et al. 2006). Recent results indicated that the deficiency of AtGPX8 accelerated the progression of oxidative stress in *AtGPX8* knockout plants (Gaber 2013). Plant GST enzymes (EC 2.5.1.18) catalyse the conjugation of electrophilic xenobiotic substrates with GSH. GST can also reduce peroxides by using GSH. The ability of GST enzymes to remove cytotoxic or genotoxic compounds can protect DNA, RNA and proteins from damage (Noctor et al. 2002). Plant *GST* gene families are large, with 25 members in soyabean, 42 in maize and 54 in *Arabidopsis* (Dixon et al. 2002; Sappl et al. 2004). *GST* overexpression can also enhance plant tolerance to various abiotic stresses. For example, tobacco seedlings that overexpress *GST* and *GPX* showed enhanced growth under thermal- or salt-stress conditions (Roxas et al. 2000).

Numerous reports have demonstrated that ABA affects the abundances of scavenging enzymes at the transcriptional and post-transcriptional levels. Treatment of *Syzygium cumini* with ABA caused a drastic decrease in CAT activity but increase in APX transient and the increase in SOD sustained. H_2O_2 treatment was also very effective in increasing APX, CAT and SOD enzyme activities (Choudhary et al. 2012). On this issue of transcription regulation, maize calcium-dependent protein kinase 11 (ZmCPK11) is involved in ABA-induced up-regulation of the levels of expression and activities of SOD and APX, and in the production of H_2O_2 (Ding et al. 2013). As another example of OsDMI3 encodes a rice Ca^{2+}/CaM-dependent protein kinase, which is required for ABA-induced increases in the expression and activities of SOD and CAT and the production of H_2O_2 (Shi et al. 2012), the ABA-activated mitogen-activated protein kinase (MAPK) OsMPK1 is also involved in this process (Shi et al. 2014).

Increased drought and salt tolerance in transgenic plants is associated with ABA-induced production of H_2O_2 via NADPH oxidase and NO via NOS-like enzymes, which sequentially induce the transcription/translation and activities of SOD, CAT, APX and glutathione reductase (GR) (Zhang et al. 2009a, b). Photosynthetic redox signals and ABA signals integrate antagonistically at a more distally located promoter region, called a redox box. There, the APETALA-2-type transcription factor RAP2.4a, induces *2CPA* transcription upon moderate oxidative stress (Shaikhali et al. 2008). However, conflicting results have indicated that ABA significantly reduced the activities of SOD, APX, POD and CAT, as well as the levels of GSH and ASA, during adventitious rooting (Li et al. 2014).

Non-enzymatic low-molecular-weight metabolites: The non-enzymatic low-molecular-weight metabolites that protect against oxidative stress are ascorbic acid (ASH, vitamin C), GSH, α-tocopherols (vitamin E), carotenoids, flavonoids and proline (Mittler et al. 2004; Chen and Dickman 2005).

ASH is the most abundant and water-soluble antioxidant. Owing to its ability to donate electrons in a number of enzymatic and non-enzymatic reactions, ASH has been considered as the most powerful ROS scavenger for the prevention of oxidative damage in plants (Athar et al. 2008). It can protect membranes by directly scavenging $O_2^{\cdot-}$ and $OH^{\cdot}$ and by regenerating α-tocopherols from tocopheroxyl radicals. ASH mostly remains available in reduced form in leaves and chloroplasts under normal physiological conditions (Smirnoff 2000).

GSH, which is necessary to maintain the normal reduced state of cells, has been considered to be the most important source of intracellular defence against ROS-induced oxidative stress (Meyer 2008). All plant cell compartments contain GSH, and most GSH is in reduced form (Mittler and Zilinskas 1992; Jimenez et al. 1998). GSH and other antioxidants together contribute to the regulation of cellular redox status; these regulatory systems respond rapidly to a complex range of different changes in the environment. The nuclear pool of glutathione changes dynamically in plants (Diaz et al. 2010). GSH plays a pivotal role in the regulation of sulphate transport, signal transduction, conjugation of metabolites, detoxification of xenobiotics and the expression of stress-responsive genes (Xiang et al. 2001). GSH also plays important roles in plant cell differentiation, cell death and senescence, pathogen resistance and enzyme regulation (Rausch and Wachter 2005).

Thiol/disulphide exchange reactions link two glutathione molecules by a disulphide bond to form glutathione disulphide (GSSG), the oxidised form of GSH. The balance between the relative levels of GSH and GSSG is a central component in maintaining the cellular redox state (Foyer and Noctor 2005). Depletion of GSH in *Arabidopsis* cell suspensions renders them susceptible to oxidative damage (May and Leaver 1993). ROS, particularly H_2O_2, can influence the recycling of GSSG and GSH, so they have an impact on the redox status of the cell (by changing the thiol/disulphide ratio), which can ultimately instigate cellular redox signalling. Reversibly oxidised cysteine sulphhydryl groups serve as redox sensors or targets of redox sensing that are important in different physiological processes. Treatment of *Brassica napus* guard cells with ABA identified 65 potential redox-responsive proteins. Of these, a SNRK2 kinase and an isoform of isopropylmalate dehydrogenase were confirmed to be redox-regulated and involved in stomatal movement regulated by ABA (Zhu et al. 2014). However, considerably little is known about the roles of GSH and GSSG in the signal transduction system of plants.

The asserted links between antioxidant metabolism and ABA are derived from the observation that the ASH-deficient mutant *vtc1* (Conklin et al. 2000) has ABA levels 1.6-fold higher than those of the wild type (Pastori et al. 2003). It has been suggested that cellular responses to an alteration in ASH levels might be mediated by ABA (Pastori et al. 2003). Correlation of ABA with regulation of the size of the ASH pool and comparison of the regulation of transcript abundance in the starch-biosynthetic mutant *adg1*, the ascorbate biosynthesis mutant *vtc1* and the ABA biosynthetic mutant *aba2* showed a link between sugar induction of APX and ASH biosynthesis (Heiber et al. 2014). Flavonols are plant metabolites that have been implicated as antioxidants, because it is similar to ASH and flavonols in terms of protection for plants. A recent study also showed that ethylene induced guard cell-specific flavonoid synthesis, suppressed ROS accumulation and decreased the rate of ABA-dependent stomatal closure (Watkins et al. 2014).

10.3 Biological Roles of ROS Related to ABA

ROS play a dual role in plants: at low concentrations, they act as signal molecules involved in acclamatory signalling that triggers tolerance to stresses, and at high concentrations, they lead to damage to biological molecules and cells (Quan et al. 2008). Research has shown that H_2O_2 acts as a key regulator that mediates many physiological processes, such as stomatal movement (Bright et al. 2006), photorespiration and photosynthesis (Noctor and Foyer 1998), senescence (Peng et al. 2005), the cell cycle (Mittler et al. 2004) and growth and development (Foreman et al. 2003). ROS influence the expression of a number of genes and signal transduction pathways that modulate the plant stress-response processes.

Besides reacting with and damaging cellular components, ROS can also participate in signal transduction. Owing to its relatively long life and high permeability across membranes, H_2O_2 has been accepted as a secondary messenger. ROS signalling is a core regulator of plant cell physiology and cellular responses to the environment. For example, ROS act as intermediates in many hormone-regulated plant biology events: auxin, ethylene, MJ and ABA signals all appear to recruit ROS (Acharya and Assmann 2009).

10.3.1 Oxidative Damage

Abiotic stress leads to excessive generation of ROS in plants, which causes reduced crop productivity worldwide (Mittler 2002; Apel and Hirt 2004). Being highly reactive, most ROS can damage cell structures, nucleic acids, lipids and proteins. When the equilibrium between the production of ROS and the antioxidant defence system is perturbed by various biotic and abiotic stresses, increased ROS at the intracellular level can cause significant damage to cell structures.

Damage to DNA molecules induced by ROS includes base deletion, pyrimidine dimers, cross-links, strand breaks and base modification (Tuteja et al. 2001). Being the most reactive, $OH^{\cdot}$ can damage both the purine and pyrimidine bases and also the deoxyribose backbone of the DNA molecule (Halliwell 2012). DNA damage influences many aspects of plant physiology, which include the disturbance of transcription, reduction of protein synthesis, destruction of the cell membrane and genomic instability (Britt 1999); these events affect the growth and development of the whole plant.

ROS or by-products of oxidative stress can also cause protein oxidation by covalent modification. Some of these forms of oxidation are essentially irreversible; the most common is carbonylation, which is a widely used marker of protein oxidation (Møller et al. 2007). Whatever the location of ROS synthesis and action, ROS are likely to target proteins that include sulphur-containing amino acids and thiol groups. A study to investigate protein carbonylation in wheat leaves showed that its level was higher in mitochondria than in chloroplasts and peroxisomes; this indicates that mitochondria are more susceptible to oxidative damage than chloroplasts and peroxisomes (Bartoli et al. 2004). Oxidised peptides are used as secondary ROS signalling molecules by mitochondria, chloroplasts, peroxisomes and possibly other organelles.

It is important to note that the efficiency of antioxidative systems determines the steady-state level of ROS in a cell (Foyer et al. 1994). When the equilibrium is perturbed by various stress factors, an increase in the intracellular level of ROS can cause significant damage to cell structures; oxidative stress is the dominate type of stress in this situation. Therefore, it is the equilibrium between ROS generation and the rate of antioxidant scavenging that determines which response pathway plant cells employ.

10.3.2 Oxidative Signalling

It is well established that ROS signalling is important in the regulation of plant cell physiology and cellular responses to the environment. ROS have been accepted as secondary messengers that participate in many biology processes, and more and more evidence has shown that ROS signalling is intertwined with other types of signalling. ROS have also been accepted as the 'cellular indicators of stress', owing to their increased production upon exposure to many stresses (Mittler 2002). Oxidative stress can activate the expression of defence-related genes. For example, early studies showed that ROS signalling in *Arabidopsis* stimulates antioxidative defence through upregulating the expression of antioxidative genes and activating the genes that encode inducible stress proteins (Santos et al. 1996; Karpinski et al. 1997). In this context, a question arises: What are the advantages of ROS being used as signal molecules? One advantage is that they can rapidly propagate signals along extensive distances throughout the plant. H_2O_2 accumulates in many stress situations, and it has a relatively long life and high permeability across membranes, which facilitate its role as a secondary messenger. The dynamic nature of ROS has been proved in root hair cells, in which the oscillation of ROS was shown to support root hair elongation; in addition, between cells, a ROS burst due to wounding can trigger a rapid ROS signal at a rate of 8.4 centimeters per minute (Miller et al. 2009). This indicates that ROS can act as long-distance cell-to-cell signals. Intracellularly generated H_2O_2 can move into neighbouring cells (Allan and Fluhr 1997). A study by Miller et al. (2009) suggested that all cells along the signal pathway were stimulated to generate ROS and that their capacity to transmit this signal in an autonomous manner enabled transmission of a ROS signal over long distances. It is not known how the signal stimulated ROS production in adhering cells, or what the relationship is between the cell ROS-scavenging system and newly formed ROS in each cell that participates in this long-distance signalling pathway.

The rapid production of ROS in response to different environmental conditions and the ability of cells to scavenge ROS using their antioxidant system provide another advantage for ROS signalling. Given that the production and scavenging of ROS occur at the same time, these two processes produce rapid changes in ROS levels, which respond quickly to environmental signals.

Studies that investigated ROS-mediated signalling and its association with the expression of specific genes have identified a range of elements that respond to oxidative stress, including specific promoters and transcription factors (TFs) (Aslund et al. 1999). However, the redox-sensing mechanisms and the associated signalling pathways induced by ROS are still somewhat unclear, and the sensors for ROS in plant cells have yet to be identified.

Pinheiro and Chaves (2011) undertook a literature survey on reports published between 1995 and February 2010 that revealed strong associations between drought, photosynthesis, ROS, ABA, sucrose and starch at the metabolite level. Analysis of protein and gene networks also revealed important roles for sugars, starch, ROS and ABA pathways in responses to drought and changes in photosynthesis. Some regulatory proteins that are associated with photosynthetic responses and drought are the TFs T6L1.5, HY5, AHBP-1B and GBF3, members of the bZIP (basic leucine zipper domain) family, and a TF that belongs to the abscisic acid-insensitive 3 (ABI3) family. The involvement of the bZIP and ABI3 TFs, both of which are ABA-dependent, in responses to drought and changes in photosynthesis has also been reviewed (Saibo et al. 2009). Binding of ABI4 to the DNA element CCACGT might prevent that promoter sequence from being bound by other TF, which likely prolongs the duration of signalling. The CCAAT binding factor A (CBFA), which is a subunit of the HAP2/HAP3/HAP5 (haeme activator protein) trimeric transcription complex, acts downstream of ABI4. When an emergency occurs (such as herbicide treatment or environmental stress followed by ABA and ROS accumulation), the master transcription factor ABI4 down-regulates certain TFs, such CBFA. Thereafter, some other TF subunits enter the transcription complex, and the transcriptional efficiency of stress-responsive genes (including the transcription co-factor CBP) is improved instantaneously (Zhang et al. 2013).

Whereas H_2O_2 can oxidise cysteine residues in proteins to form cysteine sulphenic acid or disulphide bonds, cellular reductants can reduce them back to cysteine. Depending on these mechanisms, several putative H_2O_2 sensors have been revealed. These include AtGPX3, HSF and two cysteine-rich receptor-like kinase and a leucine-rich repeat (LRR-RLK) proteins in *Arabidopsis* (Miao et al. 2006; Miller and Mittler 2006; Tripathy and Oelmüller 2012), and their redox states are regulated by ROS.

Functional genomics has yet to be used to detect ROS receptors or ROS-sensing TFs in plant extracts. It is important to identify the genes and the mechanisms of oxidative stress that can support the breeding of plants that can resist environmental stress. Studies to reveal ROS signal function in ABA-regulated stomatal closure have revealed some members of the signalling pathway. In the presence of ABA, the ABA receptor PYR/PYL/RCAR (pyrabactin resistance/PYR1-like/regulatory component of ABA receptor) forms a protein complex with PP2C, which activates OST1, making it possible for the ABA signal to be transmitted to the downstream components (Fujii et al. 2009; Park et al. 2009; Santiago et al. 2009; Umezawa et al. 2009). ABA-activated OST1 acts upstream of ROS in guard cell ABA signalling.

To date, no complete description of a global transcriptomic study of any one plant species' response to ROS has been published. The ROS-activated TFs that have been revealed in plants are members of the WRKY, Zat, RAV, GRAS and Myb families (Tripathy and Oelmüller 2012). Zat12 is an important component of the oxidative stress-response signal transduction network of *Arabidopsis*, the expression of which is induced by abiotic stresses and wound-induced systemic signalling (Davletova et al. 2005). Bioinformatics and chromatin immunoprecipitation (ChIP) approaches identified seven potential members of the ROS-responsive elements (ROSEs). ROSE7/GCC was specifically bound by the APETALA2/ethylene-responsive-element binding factor 6 (ERF6), and ERF6 interacts physically with mitogen-activated protein kinase 6 (MPK6) to regulate ROS-responsive gene expression at the transcriptional level (Wang et al. 2013a). Activation of a MAPK cascade mediates responses to exogenous ABA and a variety of abiotic stress conditions in plants (Knetsch et al. 1996; Hirt 2000; Lu et al. 2002). At least three members of the MAPK family in *Arabidopsis*—AtMPK3, AtMPK4 and AtMPK6—can be activated by different stresses, including oxidative stress (Kovtun et al. 2000; Ichimura et al. 2000; Yuasa et al. 2001).

Given that ROS signals are tightly linked with abiotic and biotic stresses, unravelling the mechanism of ROS regulation and the details of the genes induced by ROS is important for deciphering ROS signals, which should ultimately contribute to the development of high-quality crops.

10.4 Crosstalk Between ABA and ROS Signalling

Substantial effort has been expended on attempts to discover possible crosstalk between ABA and ROS in plant stress damage and signalling upon exposure to different stress conditions (Table 10.1). For example, the time course of ABA accumulation in the leaves of maize plants exposed to water stress was monitored at the same time as the increased generation of ROS (such as O_2^- and H_2O_2) and the induction of several antioxidant enzymes (such as SOD, CAT, APX and GR) (Jiang and Zhang 2002).

10.4.1 Stomatal Closure

Guard cells, which form stomata, have been investigated as a single-cell system that is useful for the elucidation of individual signalling mechanisms that function in the cellular signalling network. Given that ABA plays a key role in

Table 10.1 Overview of the roles of both ABA and ROS in stress responses

Stress	Mediated components	Response	Species	References
ABA	AtrbohD, AtrbohF, H_2O_2	Stomatal closure	*Arabidopsis*	Kwak et al. (2003)
ABA	Atrboh C, Atrboh F, H_2O_2	Root growth	*Arabidopsis*	Bai et al. (2007), Foreman et al. (2003)
Ca^{2+}	Atrboh C	Root hair growth	*Arabidopsis*	Takeda et al. (2008)
ABA	H_2O_2, $O_2^{\cdot -}$	Stomatal closure	*Vicia faba*, *Arabidopsis*	Miao et al. (2000), Pei et al. (2000), Zhang et al. (2001a, b), Kohler (2003), An et al. (2008)
ABA	Cytosolic alkalisation, ROS	Stomatal closure	*Arabidopsis*	Gonugunta et al. (2009)
ABA, H_2O_2	Ca^{2+} channels	Stomatal closure	*Arabidopsis*	Hamilton et al. (2000), Pei et al. (2000), Murata et al. (2001)
ABA	H_2O_2, plasma membrane H^+-ATPase	Stomatal opening	*Vicia faba* L	Zhang et al. (2004b)
ABA, H_2O_2	GCA2	Stomatal closure	*Arabidopsis*	Pei et al. (2000)
ABA	H_2O_2, cGMP	Stomatal closure.	*Arabidopsis*	Dubovskaya et al. (2011), Joudoi et al. (2013)
ABA	PI3P, H_2O_2	Stomatal closure	*Vicia faba*	Park et al. (2003)
ABA, H_2O_2	ATHK1, ATHK5	Stomatal closure	*Arabidopsis*	Lü et al. (2013), Desikan et al. (2008)
ABA, H_2O_2	ABI1, ABI2	Stomatal closure	*Arabidopsis*	Murata et al. (2001)
ABA, H_2O_2	SLAC1	Stomatal closure	*Arabidopsis*	Vahisalu et al. (2008), Negi et al. (2008)
ABA	KAT1	Stomatal opening	*Vicia faba*	Sutter et al. (2007)
ABA	SLAH3	Stomatal closure	*Arabidopsis*	Geiger et al. (2011)
Drought	OST1, QUAC1	Stomatal closure	*Arabidopsis*	Imes et al. (2013)
Drought, ABA	GPX3, ABI2	Stomatal closure	*Arabidopsis*	Miao et al. (2006)
ABA	phospholipase D, ABI1, PA	Stomatal closure	*Arabidopsis*	Zhang et al. (2004a, 2009a, b)
ABA	ABI2, GPA1	Stomatal opening	*Arabidopsis*	Zhao and Wang (2004)
ABA	HAB1, SWI3B	Seed germination, root elongation, stomatal closure, gene expression	*Arabidopsis*	Saez et al. (2004)

(continued)

Table 10.1 (continued)

Stress	Mediated components	Response	Species	References
ABA	PP2CA, HAB1	Seed germination, stomatal closure	*Arabidopsis*	Yoshida et al. (2006), Robert et al. (2006)
ABA	RCN1, H_2O_2	Stomatal closure	*Arabidopsis*	Kwak et al. (2002)
ABA	OST1, H_2O_2	Stomatal closure	*Arabidopsis*	Mustilli et al. (2002)
ABA, H_2O_2	MEK1/2, p38-kinase	Stomatal closure	*Vicia faba*	Jiang et al. (2008)
ABA	MKK1, MKK2, MPK3, MPK4, MPK6, MPK9, MPK11, MPK12, MPK15,	Antioxidant defence, stomatal closure, seed germination	*Arabidopsis*	Liu (2012)
Phytopathogens	NaMPK4, ABA, H_2O_2	Stomatal closure	*Nicotiana attenuata*	Hettenhausen et al. (2012)
Heat shock, freezing, salt, ROS	OXI1, ANP1,MPK3, MPK6, AtNDPK2	Plant immunity	*Arabidopsis*	Kovtun et al. (2000), Moon et al. (2003), Rentel et al. (2004)
ABA	AtCPK3, AtCPK4, AtCPK6, AtCPK10, AtCPK11, St CDPK4, St CDPK5, St RBOHB	ROS production, stomatal closure	*Arabidopsis*, Potato	Mori et al. (2006), Kobayashi et al. (2007), Zou et al. (2010), Zhu et al. (2007)
Abiotic, biotic stress	CBL1, CBL9, CIPK26, RBOHF, Ca^{2+}	ROS	*Arabidopsis*	Drerup et al. (2013)
ABA, H_2O_2	ABI1, AtRac1, actin, Ca^{2+} channel	Stomatal movement	*Arabidopsis*	Jiang et al. (2012), Zhang et al. (2007), Eun et al. (2001), Lemichez et al. (2001), Hwang and Lee (2001)
Drought	SNAC1, OsSRO1c, H_2O_2	Stomatal closure	Rice	You et al. (2013)
ABA, H_2O_2, drought, salt	ABP9, ROS	Seed germination, root growth and stomatal closure	Maize	Zhang et al. (2011)
ABA, H_2O_2	GHR1	Stomatal closure	*Arabidopsis*	Hua et al. (2012)
UV-B	ABA, H_2O_2, NO	Attenuate UV-B-derived cell damage	Maize	Tossi et al. (2009)
Drought	ABA, glutathione, ROS	Stomatal closure	*Arabidopsis*	Munemasa et al. (2013)

controlling stomatal closure during drought stress, substantial work has focused on ABA signalling in guard cells (Cutler et al. 2010; Kim et al. 2010). It has also been reported that H_2O_2 acts as a key regulator that mediates stomatal movement (Bright et al. 2006). Considering that ROS signalling has been shown to be intertwined with other hormonal signals, it is likely that ROS are the vital hub that connects the external growth conditions with ABA signals. The relationships between ROS and ABA signals mainly involve regulation of guard cell movement. Many important advances have been made in unravelling the roles of ROS and their relationships with ABA signalling in guard cells (Fig. 10.1), reviewed detially by Song et al. (2014).

H_2O_2 can be induced by ABA and accumulates in guard cells (Zhang et al. 2001a, b; Pei et al. 2000). The function of constitutive ROS accumulation in ABA signalling has been studied by using *CAT1, CAT2* and *CAT3* gene mutants or the application of 3-amino-1,2,4-triazole (3-AT, a CAT inhibitor). These mutants and this treatment cause a constitutively higher ROS level; however, the accumulated ROS do not affect the stomatal aperture in the absence of ABA or methyl jasmonate (MeJA) (Jannat et al. 2011a, b, 2012). However, the enhanced ROS

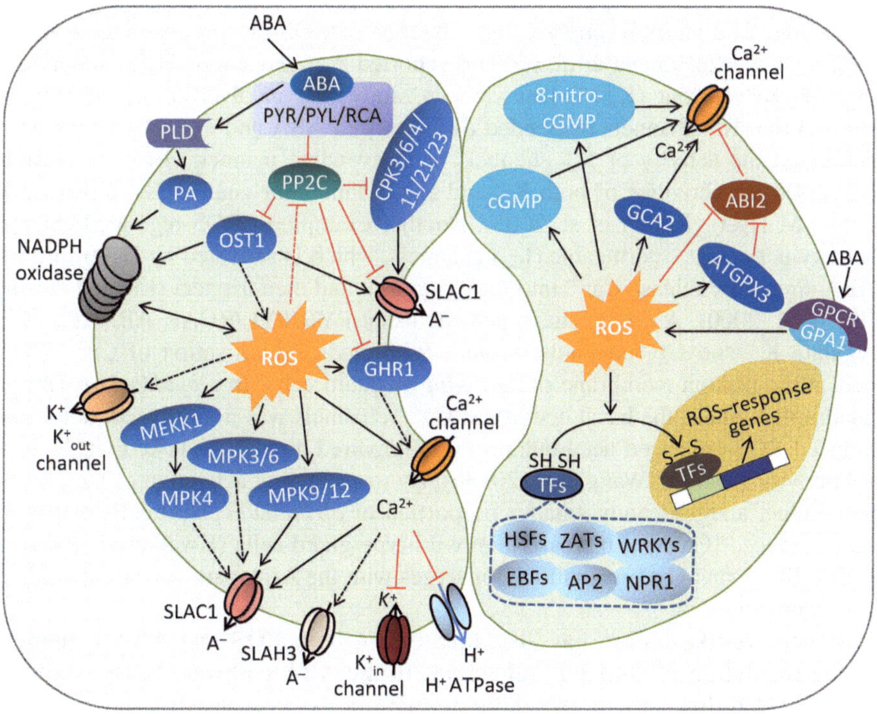

Fig. 10.1 ROS signal in guard cells. ROS sensors and targets not only monitor intra- and extracellular ROS levels, but also respond to ROS signals. Black and red lines indicate activation and suppression, respectively. Dashed line indicates the exact details of the signal pathway are still unknown (this figure has been adapted from a recent review with little modification: Song et al. 2014)

production and oscillations in $[Ca^{2+}]_{cyt}$ during ABA-induced stomatal closure suggest that ABA-induced ROS accumulation—not the constitutive accumulation of ROS—causes ABA-controlled stomatal closure (Jannat et al. 2011a, b). This is not to say that constitutively accumulated ROS have no function; recently, a study on a mutant lacking *APX1* (*KO-APX1*) showed high sensitivity to wounding or MeJA treatment (Maruta et al. 2012). Previously, we discussed that AtGPX3, the redox state of which is regulated by H_2O_2, interacts strongly with ABI2 to control stomatal aperture in an ABA-dependent manner (Miao et al. 2006). Exogenously applied ABA and 12-oxo-phytodienoic acid (12-OPDA, a precursor of jasmonic acid), individually or combined, which promote stomatal closure of ABA and allene oxide synthase biosynthetic mutants, albeit most effectively when combined. The potency of this combination in inducing stomatal closure was verified using tomato (*Solanum lycopersicum*), *Brassica napus* and *Arabidopsis*. These data have identified drought as a stress signal that uncouples the conversion of 12-OPDA to JA and have revealed 12-OPDA as a drought-responsive regulator of stomatal closure, which functions most effectively together with ABA (Savchenko et al. 2014).

Guard cell turgor and the size of the stomatal aperture are determined by the concentration of ions. ROS regulate stomatal channels and transporters in ABA signalling. In a study to analyse the influence of ROS on ion channels by using *Vicia* guard cells, Zhang et al. (2001a) reported that exogenous H_2O_2 suppressed inwards K^+ current (K^+_{in}). Meanwhile, a study that involved *Vicia* guard cells showed that K^+ channels responded differently to ABA and H_2O_2: whereas ABA depressed the activity of K^+_{in} channels in a reversible manner, H_2O_2 irreversibly depressed the activities of both K^+_{in} and K^+_{out}channels for guard cells immersed in 1–50 μM H_2O_2 (Köhler et al. 2003). Another example of ROS-regulated channel activity is the Ca^{2+}-permeable (I_{Ca}) channels, which is activated by H_2O_2 through ABA signalling release Ca^{2+} into the cytoplasm and then triggers stomatal closure (Pei et al. 2000). A recent study indicated that PYR/PYL/RCAR ABA receptors regulate K^+ and Cl^- channels through ROS-mediated activation of Ca^{2+} channels at the plasma membrane of *Arabidopsis* guard cells. In a *pyr1/pyl1/pyl2/pyl4* quadruple mutant, the basal activity of Ca^{2+} channels was not affected, but ABA-induced ROS-enhanced accumulation was impaired and ABA-evoked Ca^{2+} channel activity was lost (Wang et al. 2013b). By investigating activation of the plasma membrane, a significantly smaller proportion of *atrbohD/F* guard cells displayed ABA-induced $[Ca^{2+}]_{cyt}$ increase than wild-type guard cells (Kwak et al. 2003). It is thus likely that H_2O_2 signalling converges with the ABA pathway at the point of Ca^{2+} activation.

Receptors-like kinase has also function in both ROS and ABA signalling in the regulation of stomatal behaviour. In the ABA pathway, H_2O_2 is known to be located downstream of OST1 in guard cells because ABA-induced ROS production does not occur in *ost1* mutants (Mustilli et al. 2002). A plasma membrane receptor-like kinase, guard cell H_2O_2-resistant1 (GHR1), activates ABA- and H_2O_2-regulated S-type anion currents and stomatal closure (Hua et al. 2012). It was also established that GHR1 phosphorylated and activated the anion channel-associated1 (SLAC1). Protein phosphatases that function in ABA-related stomatal movement can be regulated by ROS, and ABI1 and ABI2 activities are significantly inactivated by H_2O_2 (Meinhard and Grill 2001; Meinhard et al. 2002). Recent research showed that the GSH-deficient *Arabidopsis* mutant *cad2-1* showed enhanced H_2O_2-induced stomatal closure and a significant increase of ROS accumulation in whole leaves in response to ABA. The *cad2-1* mutant enhanced the activation of Ca^{2+}-permeable channels by H_2O_2, but did not affect the ABA regulation of S-type anion channels (Munemasa et al. 2013).

ROS also regulate the ABA response of guard cells through MAPK cascades. Time-course analysis showed that the accumulation of H_2O_2 activates MAPK activity, which promotes stomatal closure (Desikan et al. 2004; Jiang et al. 2008). Double-mutant *mpk9/12* showed increased transpirational water loss and decreased sensitivity in terms of the stomatal response to ABA and H_2O_2, when compared with wild-type plants. Both ABA and H_2O_2 can enhance the protein kinase activity of MPK12. These results suggested that these two MPK act upstream of anion channels and downstream of ROS to promote ABA signalling in guard cells (Jammes et al. 2009).

The *siz1* mutant, which impairs the SIZ-type small ubiquitin-related modifier E3 ligase in *Arabidopsis* (Muraoka and Miura 2005), displayed ABA hypersensitivity (Miura and Hasegawa 2009). SIZ1 negatively affects stomatal closure and drought tolerance through the accumulation of SA (Miura et al. 2013). Endogenous ABA is not involved in SA-induced stomatal closure because SA can induce stomatal closure not only in *Arabidopsis* wild-type plants but also in ABA-deficient *aba2* mutants (Issak et al. 2013). Moreover, SA-induced stomatal closure is accompanied by the production of ROS (Issak et al. 2013). Together, all of these results suggest that, although the ABA and SA signalling pathways are independent of each other, the two regulators coordinately regulate ROS production in response to drought and pathogen invasion. Detailed information for crosstalk between SA and ABA signalling in guard cells have yet to be established.

10.4.2 Dormancy and Germination of Seed

ABA is a key regulator of seed dormancy and germination. A recent study found that blue light inhibits the germination of barley by induction of the expression of the ABA biosynthetic gene that encodes 9-cis-epoxycarotenoid dioxygenase (NCED) and dampening of expression of the ABA-degradation gene, 8′-hydroxylase, which increases ABA content in the grain and prevents the completion of germination (Jose et al. 2014). Considerable research has explored the interaction of ABA with ROS in seed dormancy and germination. ROS have recently emerged as key players in seed physiology (Bailly et al. 2008). There are several reasons why the mechanisms by which ROS act as signalling molecules in seeds remain largely unknown. Firstly, ROS are ubiquitous and present in seeds at all stages, from embryogenesis through to germination; they also exist in various forms (e.g. $O_2^{\cdot-}$, H_2O_2, $^1O_2$ and $OH^{\cdot}$) in seeds (Bailly et al. 2008). Secondly, these compounds have a very short life span (Moller et al. 2007). Thirdly, individual ROS can be integrated into several different transduction pathways (Mittler et al. 2011), they show various levels of reactivity towards a wide range of macromolecules that can propagate the oxidative signal (e.g. proteins, Moller et al. 2007) and they have an effect on the cellular redox status (Dietz et al. 2010).

It is well known that ROS are produced at a certain level during seed imbibition. It has been proposed that the germination is completed only when the ROS content is within an oxidative window that allows ROS signalling (Bailly et al. 2008). Levels of ROS above or below this 'oxidative window for germination' would not permit progression towards germination. According to this model, seed dormancy—the inability of seeds to germinate in favourable environmental conditions (Finch-Savage and Leubner-Metzger 2006)—is regulated by ROS signalling. Imbibition of pea seeds in the presence of ABA also reduced the endogenous H_2O_2 contents of pea seedlings in control and thioproline (TP)-treated seeds. The incubation of pea seeds with TP and/or H_2O_2 in the presence or absence of ABA decreased the activity of H_2O_2-scavenging enzymes. The increase of the endogenous H_2O_2 contents observed in TP and/or H_2O_2 treatments in the absence of ABA could be correlated with the decrease in these activities (Barba-Espín et al. 2012).

The crosstalk between ROS and ABA or gibberellin metabolism and signalling might control barley seed dormancy (Bahin et al. 2011). Through its effects on ABA metabolism (Liu et al. 2009), NO promotes the release of seed dormancy in *Arabidopsis* (Bethke et al. 2006, 2007). Exogenous H_2O_2 reduces ABA synthesis and stimulates gibberellin synthesis, thus releasing dormancy (Liu et al. 2010). A membrane-bound enzyme, RBOHB (respiratory burst oxidase homologue B), which produces $O_2^{\cdot-}$, affects seed after-ripening and germination in *Arabidopsis* (Müller et al. 2009). Expression of a set of genes related to dormancy upon imbibition in *cat2-1* and *vet1-1* seeds revealed that their nondormant phenotype was probably not related to ABA or gibberellin metabolism, but suggested that ROS could trigger germination through the activation of gibberellin signalling (Leymarie et al. 2012).

10.4.3 Development of Root System

Plants are anchored in the soil by their roots and depend on their root systems for water and nutrients. The vital functions of roots make it important to understand the development of root systems. Root growth is regulated by hormones and many environmental variables (Teale et al. 2008; Fukaki and Tasaka 2009). Although cell cycle progression is negatively controlled by ROS, H_2O_2 is essential for root gravitropism (Joo et al. 2001). The requirement for ROS in root growth has already been established using a mutation in *RbohC* (*rhd2*), in that the reduced ROS generation in *rhd2* decreased Ca^{2+} channel activities in root hairs, which further led to the inhibition of root hair growth (Foreman et al. 2003). The NADPH oxidases *RbohD* and *RbohF* also regulate root growth, with the double-mutant *rbohD/rbohF* showing marked reductions in root length relative to the wild type (Kwak et al. 2003). ROS control the transition from cell proliferation to differentiation in roots via the auxin signalling pathway, which is mediated by the basic helix-loop-helix TF UPBEAT1 (Tsukagoshi et al. 2010). As highly dynamic signalling molecules, ROS also function as long-distance auto-propagating signals (Miller et al. 2009), with ROS oscillations shown to occur in root hairs (Monshausen et al. 2007; Takeda et al. 2008).

Plant hormones have long been known to play a crucial role in the regulation of root growth (Moubayidin et al. 2009; Ubeda-Tomas et al.2008). Cytokinin, auxin, brassinosteroids (BR), ethylene and ABA all play roles in the control of meristem size and root growth. ABA acts on quiescent centre (QC) and stem cells in order to regulate root meristem size. In addition, QC cells induce division when ABA biosynthesis is blocked in seedlings (Zhang et al. 2010). ABA is required for root growth and ABA-deficient mutants have reduced root systems (Xiong and Zhu 2003); however, high concentrations of exogenous ABA inhibit root growth. It is believed that ABA stimulates the production of ROS through NADPH oxidases (Kwak et al. 2003). In *Arabidopsis*, both the single-mutant *rbohF* and the double-mutant *rbohD/rbohF* are insensitive to ABA-mediated inhibition of root growth (Kwak et al. 2003). However, the mechanisms that underlie this process remain elusive. A recent study showed that the double mutants *atrbohD1/F1* and *atrbohD2/F2* were less sensitive to ABA suppression of root cell elongation than WT plants. Furthermore, the double mutants showed impaired ABA responses in roots, including ROS generation, cytosolic Ca^{2+} increases and activation of plasma membrane Ca^{2+}-permeable channels, compared with WT plants. Exogenous H_2O_2 can activate the Ca^{2+} channel activity in roots of *atrbohD1/F1* plants. In addition, exogenous application of the auxin transport inhibitor naphthylphthalamic acid effectively promoted ABA-mediated inhibition of root growth of the mutants relative to that of WT plants. The ABA-induced decreases in auxin sensitivity of the root tips were more pronounced in the WT than in *atrbohD1/F1*. These findings suggest that both AtrbohD and AtrbohF are essential for ABA-promoted ROS production in roots. ROS activate Ca^{2+} signalling and reduce auxin sensitivity of roots, thus positively regulating ABA-inhibited primary root growth in *Arabidopsis* (Jiao et al. 2013). This suggests that ROS not only play roles in ABA-regulated stomatal movement but also in ABA-regulated root growth.

Besides plasma membrane NADPH oxidases, mitochondrially produced ROS have also been proved to be important in the ABA signalling pathway in root growth. In a study to screen the regulator in ABA-mediated *Arabidopsis* root growth, a mutant overly sensitive to ABA, *abo6*, showed a higher level of ROS in the mitochondria under both ABA treatment and normal conditions. *ABO6* encodes a DEXH box RNA helicase that is localised in the mitochondria and is required for the splicing of several genes that encode components of complex I (He et al. 2012).

Glutathione peroxidases (GPX) fulfil important functions in oxidative signalling and protect against the adverse effects of excessive oxidation. *Arabidopsis gpx1*, *gpx4*, *gpx6*, *gpx7* and *gpx8* mutants had a significantly greater lateral root density (LRD) than the wild type. Conversely, the *gpx2* and *gpx3* mutants had significantly lower LRD values than WT. Auxin increased LRD and synthetic strigolactone GR24 and ABA decreased LRD in all genotypes. These findings demonstrate the importance of redox controls mediated by AtGPX in the control of root architecture (Passaia et al. 2014).

The biological roles of ROS in the ABA-regulated root growth signalling pathway remain largely unknown. However, secondary messenger calcium has been proved to be important in Pro-rich extensin-like receptor kinase 4 (PERK4)-regulated root growth in ABA signalling (Bai et al. 2009). It is possible that changes in Ca^{2+} might be an element common to the signalling pathways downstream of both ROS and ABA.

10.4.4 ROS-dependent Cell Death

Being highly reactive, most ROS can induce oxidative stress, which affects the growth and development of the whole plant. When the equilibrium between antioxidative systems and the production of ROS is disturbed by various stress factors, an intense increase in the levels of ROS at the intracellular level can cause significant damage to cell structures. Increased ROS induces oxidative stress, which ultimately results in cell death.

There are two forms of cell death: programmed cell death (PCD) and necrosis; they are important features of plant development. ROS participate in both of these, with the difference between them being that PCD is genetically regulated, but necrosis is induced by severe and persistent trauma and is considered to not be genetically regulated. Experiments that involved cultured soyabean cells provided the first evidence that ROS act as signals that initiate plant cell death; namely, a short pulse of H_2O_2 was sufficient to induce it (Pennell and Lamb 1997). The precise ROS signal transduction pathway in cell death remains somewhat unclear. However, it is known that H_2O_2-induced cell death needs time for the transcription and translation of certain genes, and exogenous H_2O_2-induced cell death shows concentration- and time-dependent features (Desikan et al. 1998). Exogenous H_2O_2 applied at a concentration over 5 mM initiates cell death in *Arabidopsis* suspension cultures; this concentration is higher than that which stimulates the expression of defence genes (Desikan et al. 1998; Levine et al. 1994). These different responses to different levels of H_2O_2 clearly suggest the complexity of the signal transduction mechanisms for ROS that exist in plant cells. How can cells discriminate and adopt one of these pathways? What are the pivotal factors in the decision of cells to respond to different levels of H_2O_2? Answering these questions requires the analysis of mutants deficient in ROS signalling combined with specific ROS staining to facilitate monitoring of the ROS dynamics in situ.

ROS signalling is highly integrated with the hormonal signalling pathways that regulate many aspects of plant growth and development. The ability of ABA to inhibit the expression of α-amylase gene expression is consistent with a role for the ABA-induced protein kinase ABA-responsive protein kinase (PKABA) as an intermediate in GA signalling. In aleurone cells, the application of ABA results in rapid increase in *PKABA* mRNA levels, and enhancement of *PKABA1* transcription can mimic the effect of exogenous ABA in suppressing the expression of GA-induced genes, such as those that encode GAMyb and α-amylase (Gómez-Cadenas et al. 2001). H_2O_2 released suppression of the accumulation of *GAMyb* mRNA by PKABA and consequently promoted the production of *α-amylase* mRNA; this suggests that the H_2O_2 generated by GA in aleurone cells is a signal molecule that antagonises ABA signalling (Ishibashi et al. 2012).

10.4.5 Stress Adaptation in the Transcription Levels

There has been considerable progress in recent years in understanding how ABA and ROS modulate adaptation to stress at the level of transcription. H_2O_2 has been implicated in the numerous observations of ABA-mediated induction of antioxidant gene expression (Sakamoto et al. 1995; Guan et al. 2000; Fryer et al. 2003; Jiang and Zhang 2003; Yoshida et al. 2003). To reveal the molecular mechanisms that are involved in the crosstalk between the ABA and ROS signalling pathways, several large-scale studies have analysed changes in gene expression in response to ROS and/or ABA (Desikan et al. 2001; Seki et al. 2002; Leonhardt et al. 2004; Takahashi et al. 2004; Vandenabeele et al. 2004). The effects of exogenous ABA and H_2O_2 on gene expression in *Arabidopsis* seedlings were investigated by microarray technology using Affymetrix GeneChips (Wang et al. 2006). Of the 24,000 genes that were analysed following H_2O_2 treatment, we found that the expression of 459 transcripts was significantly increased, whereas the expression of 221 transcripts was decreased when compared with those in the seedlings that were not exposed to H_2O_2. Following ABA treatment, we also found that the levels of transcription of 391 and 322 genes were upregulated and downregulated, respectively. Further analysis indicated that 143 of the upregulated genes and 75 of the downregulated ones responded to both ABA and H_2O_2. These results imply that there is an overlap between ABA- and H_2O_2-induced transcription of genes and that these two signalling molecules regulate many downstream genes in a coordinated manner (Wang et al. 2006). These findings are consistent with previous reports that the responses to ABA and oxidative stress are linked (Guan et al. 2000; Zhang et al. 2001a; Murata et al. 2001; Kwak et al. 2003; Wang and Song 2008).

Comparison of guard cell expression profiles with those of mesophyll cells identified 64 transcripts that were expressed preferentially in guard cells. These genes encode TFs, signal transduction proteins such as protein kinases, receptor protein kinases and metabolic pathway proteins. Expression profiling has revealed that ABA modulates the expression of many ABA signalling components in guard cells at the transcript level (Leonhardt et al. 2004). Furthermore, it is important to identify the function of previously uncharacterised genes that responded to the application of exogenous ABA and H_2O_2. However, the functions of many of these genes and their products remain to be determined; these represent areas of future investigation.

10.5 Concluding Remarks

The present chapter described ROS turnover under ABA treatment and stress conditions, their biological roles in plant growth, development and stress responses. ROS can induce oxidative damage and defence-responsive activities in plants, with the dual roles of ROS depending on the balance between ROS production and the scavenging system. In plant cells and tissues, both ABA and ROS can be induced by drought, cold and salt stresses. Many components in ABA-regulated stress adaptation and resistance in plants have been identified, with a complex regulatory network involving ROS being involved in this biological process.

Some other unanswered questions are: What are the components in both ABA and ROS signalling? Although many genes have been revealed to be involved in such signalling, it is still difficult to draw precise conclusions on the regulatory system. Different genetic responses are induced under different abiotic and biotic stresses, and ROS generation is the central response to both abiotic and biotic stresses. In terms of how ROS signal transduction modulates changes in gene expression upon exposure to different stresses, the underlying molecular mechanisms are largely unknown, especially for the ABA response. For example, are all of the genes regulated by ROS signalling in the ABA pathway sensitive to the redox state of the cell? Answering these questions might require combining the screening of more ROS-response mutants with the use of advanced imaging tools (Song et al. 2014).

Other relevant questions include: How are ABA and ROS signals transmitted a long distance and what happens during cell-to-cell communication? Whether does H_2O_2 replace ABA for long-distance transportation in plant? Although, H_2O_2 has been believed to cross membranes freely studies have indicated that the diffusion is limited and its transport need channel proteins such as aquaporins (Bienert et al. 2006). Therefore, analysis of the mechanism of long-distance transmit will be a new target in ABA and ROS research.

Acknowledgments Research in the author's laboratories was supported by the National Key Basic Special Funds (2012CB114301) and the National Natural Science Foundation of China (90817106).

References

Acharya BR, Assmann SM. Hormone interactions in stomatal function. Plant Mol Biol. 2009;69:451–62.

Ahmad P, Jaleel CA, Salem MA, Nabi G, Sharma S. Roles of enzymatic and nonenzymatic antioxidants in plants during abiotic stress. Crit Rev Biotechnol. 2010;30:161–75.

Allan AC, Fluhr R. Two distinct sources of elicited reactive oxygen species in tobacco epidermal cells. Plant Cell. 1997;9:1559–72.

An Z, Jing W, Liu Y, Zhang W. Hydrogen peroxide generated by copper amine oxidase is involved in abscisic acid-induced stomatal closure in *Vicia faba*. J Exp Bot. 2008;59:815–25.

Apel K, Hirt H. Reactive oxygen species: metabolism, oxidative stress, and signal transduction. Annu Rev Plant Biol. 2004;55:373–99.

Asada K. Production and scavenging of reactive oxygen species in chloroplasts and their functions. Plant Physiol. 2006;141:391–6.

Aslund F, Zheng M, Beckwith J, Storz G. Regulation of the OxyR transcription factor by hydrogen peroxide and the cellular thiol-disulfide status. Proc Natl Acad Sci USA. 1999;96:6161–5.

Athar HUR, Khan A, Ashraf M. Exogenously applied ascorbic acid alleviates salt-induced oxidative stress in wheat. Environ Exp Bot. 2008;63:224–31.

Badawi GH, Kawano N, Yamauchi Y, Shimada E, Sasaki R, Kubo A, Tanaka K. Over-expression of ascorbate peroxidase in tobacco chloroplasts enhances the tolerance to salt stress and water deficit. Physiol Plant. 2004;121:231–8.

Bahin E, Bailly C, Sotta B, Kranner I, Corbineau F, Leymarie J. Crosstalk between reactive oxygen species and hormonal signalling pathways regulates grain dormancy in barley. Plant, Cell Environ. 2011;34:980–93.

Bai L, Zhou Y, Zhang XR, Song CP, Cao MQ. Hydrogen peroxide modulates abscisic acid signaling in root growth and development in *Arabidopsis*. Chin Sci Bull. 2007;52:1142–5.

Bai L, Zhang G, Zhou Y, Zhang Z, Wang W, Du Y, Wu Z, Song CP. Plasma membrane associated proline rich extension like receptor kinase 4, a novel regulator of Ca^{2+} signalling, is required for abscisic acid responses in *Arabidopsis thaliana*. Plant J. 2009;60:314–27.

Bailly C, El Maarouf Bouteau H, Corbineau F. Seed dormancy alleviation and oxidative signaling. J Soc Biol. 2008;202:241–8.

Barba-Espin G, Nicolas E, Almansa MS, Cantero-Navarro E, Albacete A, Hernández JA, Díaz-Vivancos P. Role of thioproline on seed germination: interaction ROS-ABA and effects on antioxidative metabolism. Plant Physiol Biochem. 2012;59:30–6.

Baron KN, Schroeder DF, Stasolla C. Transcriptional response of abscisic acid (ABA) metabolism and transport to cold and heat stress applied at the reproductive stage of development in *Arabidopsis thaliana*. Plant Sci. 2012;188–189:48–59.

Bartoli CG, Gómez F, Martinez DE, Guiamet JJ. Mitochondria are the main target for oxidative damage in leaves of wheat (*Triticum aestivum L.*). J Exp Bot. 2004;55:1663–9.

Bechtold U, Richard O, Zamboni A, Gapper C, Geisler M, Pogson B, Karpinski S, Mullineaux PM. Impact of chloroplastic- and extracellular-sourced ROS on high light-responsive gene expression in *Arabidopsis*. J Exp Bot. 2008;59:121–33.

Bethke PC, Libourel IG, Aoyama N, Chung YY, Still DW, Jones RL. The *Arabidopsis* aleurone layer responds to nitric oxide, gibberellin, and abscisic acid and is sufficient and necessary for seed dormancy. Plant Physiol. 2007;143:1173–88.

Bethke PC, Libourel IG, Reinöhl V, Jones RL. Sodium nitroprusside, cyanide, nitrite, and nitrate break *Arabidopsis* seed dormancy in a nitric oxide-dependent manner. Planta. 2006;223:805–12.

Bhattacharjee S. Sites of generation and physicochemical basis of formation of reactive oxygen species in plant cell. In: Reactive oxygen species and antioxidants in higher plants. Science Publishers; 2010. p. 1–30.

Bienert GP, Schjoerring JK, Jahn TP. Membrane transport of hydrogen peroxide. Biochim Biophys Acta. 2006;1758:994–1003.
Bolwell GP, Woftastek P. Mechanism for the generation of reactive oxygen species in plant defense-Broad perspective. Physiol Mol Plant Pathol. 1997;51:347–9.
Bright J, Desikan R, Hancock JT, Weir IS, Neill SJ. ABA-induced NO generation and stomatal closure in *Arabidopsis* are dependent on H_2O_2 synthesis. Plant J. 2006;45:113–22.
Britt AB. Molecular genetics of DNA repair in higher plants. Trends Plant Sci. 1999;4:20–5.
Chen C, Dickman MB. Proline suppresses apoptosis in the fungal pathogen Colletotrichum trifolii. Proc Natl Acad Sci USA. 2005;102:3459–64.
Choudhary R, Saroha AE, Swarnkar PL. Effect of abscisic acid and hydrogen peroxide on antioxidant enzymes in *Syzygium cumini* plant. J Food Sci Technol. 2012;49:649–52.
Comtois SL, Gidley MD, Kelly DJ. Role of the thioredoxin system and the thiol-peroxidases Tpx and Bcp in mediating resistance to oxidative and nitrosative stress in *Helicobacter pylori.* Microbiology. 2003;149:121–9.
Conklin PL, Saracco SA, Norris SR, Last RL. Identification of ascorbic acid-deficient *Arabidopsis thaliana* mutants. Genetics. 2000;154:847–56.
Cutler SR, Rodriguez PL, Finkelstein RR, Abrams SR. Abscisic acid: emergence of a core signaling network. Annu Rev Plant Biol. 2010;61:651–79.
Dalton TP, Shertzer HG, Puga A. Regulation of gene expression by reactive oxygen. Annu Rev Pharmacol Toxicol. 1999;39:67–101.
Davletova S, Rizhsky L, Liang H, Shengqiang Z, Oliver DJ, Coutu J, Shulaev V, Schlauch K, Mittler R. Cytosolic ascorbate peroxidase 1 is a central component of the reactive oxygen gene network of *Arabidopsis*. Plant Cell. 2005;17:268–81.
Del Río LA, Sandalio LM, Palma JM, Bueno P, Corpas FJ. Metabolism of oxygen radicals in peroxisomes and cellular implications. J Free Radic Biol Med. 1992;13:557–80.
Del Río LA, Sandalio LM, Corpas FJ, Palma JM, Barroso JB. Reactive oxygen species and reactive nitrogen species in peroxisomes. Production, scavenging, and role in cell signaling. Plant Physiol. 2006;141:330–5.
Desikan R, Reynolds A, Hancock J, Neill S. Harpin and hydrogen peroxide both initiate programmed cell death but have differential effects on defence gene expression in *Arabidopsis* suspension cultures. Biochem J. 1998;330:115–20.
Desikan R, A-H-Mackerness S, Hancock JT, Neill SJ. Regulation of the *Arabidopsis* transcriptome by oxidative stress. Plant Physiol. 2001;127:159–72.
Desikan R, Cheung MK, Bright J, Henson D, Hancock JT, Neill SJ. ABA, hydrogen peroxide and nitric oxide signalling in stomatal guard cells. J Exp Bot. 2004;55:205–12.
Desikan R, Horák J, Chaban C, Mira-Rodado V, Witthöft J, Elgass K, Grefen C, Cheung MK, Meixner AJ, Hooley R, Neill SJ, Hancock JT, Harter K. The histidine kinase AHK5 integrates endogenous and environmental signals in *Arabidopsis* guard cells. PLoS ONE. 2008;3:e2491.
Diaz VP, Wolff T, Markovic J, Pallardó F, Foyer C. A nuclear glutathione cycle within the cell cycle. Biochem J. 2010;431:169–78.
Dietz K, Jacquot J, Harris G. Hubs and bottlenecks in plant molecular signalling networks. New Phytol. 2010;188:919–38.
Ding Y, Cao J, Ni L, Zhu Y, Zhang A, Tan M, Jiang M. ZmCPK11 is involved in abscisic acid-induced antioxidant defence and functions upstream of ZmMPK5 in abscisic acid signaling in maize. J Exp Bot. 2013;64:871–84.
Dixon DP, Davis BG, Edwards R. Functional divergence in the glutathione transferase superfamily in plants identification of two classes with putative functions in redox homeostasis in *Arabidopsis thaliana*. J Bio Chem. 2002;277:30859–69.
Drerup MM, Schlücking K, Hashimoto K, Manishankar P, Steinhorst L, Kuchitsu K, Kudla J. The Calcineurin B-like calcium sensors CBL1 and CBL9 together with their interacting protein kinase CIPK26 regulate the *Arabidopsis* NADPH oxidase RBOHF. Mol Plant. 2013;6:559–69.
Dubovskaya LV, Bakakina YS, Kolesneva EV, Sodel DL, McAinsh MR, Hetherington AM, Volotovski ID. cGMP-dependent ABA-induced stomatal closure in the ABA-insensitive *Arabidopsis* mutant *abi1-1*. New Phytol. 2011;191:57–69.

Eun SO, Bae SH, Lee Y. Cortical actin filaments in guard cells respond differently to abscisic acid in wild-type and abi1-1 mutant *Arabidopsis*. Planta. 2001;212:466–9.

Finch-Savage WE, Leubner-Metzger G. Seed dormancy and the control of germination. New Phytol. 2006;171:501–23.

Foreman J, Demidchik V, Bothwell JH, Mylona P, Miedema H, Torres MA, Linstead P, Costa S, Brownlee C, Jones JD, Davies JM, Dolan L. Reactive oxygen species produced by NADPH oxidase regulate plant cell growth. Nature. 2003;422:442–6.

Foyer CH, Noctor G. Redox homeostasis and antioxidant signaling: a metabolic interface between stress perception and physiological responses. Plant Cell. 2005;17:1866–75.

Foyer CH, Noctor G. Redox regulation in photosynthetic organisms: signaling, acclimation, and practical implications. Antioxid Redox Signal. 2009;11:861–905.

Foyer CH, Descourvieres P, Kunert KJ. Protection against oxygen radicals: an important defence mechanism studied in transgenic plants. Plant Cell Environ. 1994;17:507–23.

Fryer MJ, Ball L, Oxborough K, Karpinski S, Mullineaux PM, Baker NR. Control of ascorbate peroxidase 2 expression by hydrogen peroxide and leaf water status during excess light stress reveals a functional organisation of *Arabidopsis* leaves. Plant J. 2003;33:691–705.

Fujii H, Chinnusamy V, Rodrigues A, Rubio S, Antoni R, Park SY, Cutler SR, Sheen J, Rodriguez PL, Zhu JK. In vitro reconstitution of an abscisic acid signalling pathway. Nature. 2009;462:660–4.

Fukaki H, Tasaka M. Hormone interactions during lateral root formation. Plant Mol Biol. 2009;69:437–49.

Gaber A. The importance of *Arabidopsis* glutathione peroxidase 8 for protecting *Arabidopsis* plant and *E. coli* cells against oxidative stress. GM Crops Food: Biotechnol Agric Food Chain. 2013;5:1–7.

Garg N, Manchanda G. ROS generation in plants: boon or bane? Plant Biosyst. 2009;143:81–96.

Geiger D, Maierhofer T, Al-Rasheid KA, Scherzer S, Mumm P, Liese A, Ache P, Wellmann C, Marten I, Grill E, Romeis T, Hedrich R. Stomatal closure by fast abscisic acid signaling is mediated by the guard cell anion channel SLAH3 and the receptor RCAR1. Sci Signal. 2011;4:ra32.

Gómez-Cadenas A, Zentella R, Walker-Simmons MK, Ho TH. Gibberellin/abscisic acid antagonism in barley aleurone cells: site of action of the protein kinase PKABA1 in relation to gibberellin signaling molecules. Plant Cell. 2001;13:667–79.

Gonugunta VK, Srivastava N, Raghavendra AS. Cytosolic alkalinization is a common and early messenger preceding the production of ROS and NO during stomatal closure by variable signals, including abscisic acid, methyl jasmonate and chitosan. Plant Signal Behav. 2009;4:561–4.

Guan LM, Zhao J, Scandalios JG. Cis-elements and trans-factors that regulate expression of the maize Cat1 antioxidant gene in response to ABA and osmotic stress: H_2O_2 is the likely intermediary signaling molecule for the response. Plant J. 2000;22:87–95.

Halliwell B. Free radicals and antioxidants: updating a personal view. Nutr Rev. 2012;70:257–65.

Hamilton DW, Hills A, Kohler B, Blatt MR. Ca^{2+} channels at the plasma membrane of stomatal guard cells are activated by hyperpolarization and abscisic acid. Proc Natl Acad Sci USA. 2000;97:4967–72.

He J, Duan Y, Hua D, Fan G, Wang L, Liu Y, Chen Z, Han L, Qu LJ, Gong Z. DEXH box RNA helicase-mediated mitochondrial reactive oxygen species production in *Arabidopsis* mediates crosstalk between abscisic acid and auxin signaling. Plant Cell. 2012;24:1815–33.

Heiber I, Cai W, Baier M. Linking chloroplast antioxidant defense to carbohydrate availability: the transcript abundance of stromal ascorbate peroxidase is sugar-controlled via ascorbate biosynthesis. Mol Plant. 2014;7:58–70.

Hettenhausen C, Baldwin IT, Wu J. Silencing MPK4 in Nicotiana attenuata enhances photosynthesis and seed production but compromises abscisic acid-induced stomatal closure and guard cell-mediated resistance to *Pseudomonas syringae pv* tomato DC3000. Plant Physiol. 2012;158:759–76.

Hideg É, Barta C, Kálai T, Vass I, Hideg K, Asada K. Detection of singlet oxygen and superoxide with fluorescent sensors in leaves under stress by photoinhibition or UV radiation. Plant Cell Physiol. 2002;43:1154–64.

Hirt H. Connecting oxidative stress, auxin, and cell cycle regulation through a plant mitogen-activated protein kinase pathway. Proc Natl Acad Sci USA. 2000;97:2405–7.

Hua D, Wang C, He J, Liao H, Duan Y, Zhu Z, Guo Y, Chen Z, Gong Z. A plasma membrane receptor kinase, GHR1, mediates abscisic acidand hydrogen peroxide-regulated stomatal movement in *Arabidopsis*. Plant Cell. 2012;24:2546–61.

Hwang JU, Lee Y. Abscisic acid-induced actin reorganization in guard cells of dayflower is mediated by cytosolic calcium levels and by protein kinase and protein phosphatase activities. Plant Physiol. 2001;125:2120–8.

Ichimura K, Mizoguchi T, Yoshida R, Yuasa T, Shinozaki K. Various abiotic stresses rapidly activate *Arabidopsis* MAP kinases ATMPK4 and ATMPK6. Plant J. 2000;24:655–65.

Imes D, Mumm P, Böhm J, Al-Rasheid KA, Marten I, Geiger D, Hedrich R. Open stomata 1 (OST1) kinase controls R-type anion channel QUAC1 in *Arabidopsis* guard cells. Plant J. 2013;74:372–82.

Iqbal A, Yabuta Y, Takeda T, Nakano Y, Shigeoka S. Hydroperoxide reduction by thioredoxin-specific glutathione peroxidase isoenzymes of *Arabidopsis thaliana*. FEBS J. 2006;273:5589–97.

Ishibashi Y, Tawaratsumida T, Kondo K, Kasa S, Sakamoto M, Aoki N, Zheng SH, Yuasa T, Iwaya-Inoue M. Reactive oxygen species are involved in gibberellin/abscisic acid signaling in barley aleurone cells. Plant Physiol. 2012;158:1705–14.

Issak M, Okuma E, Munemasa S, Nakamura Y, Mori IC, Murata Y. Neither endogenous abscisic acid nor endogenous jasmonate is involved in salicylic acid-, yeast elicitor-, or chitosan-induced stomatal closure in *Arabidopsis thaliana*. Biosci Biotechnol Biochem. 2013;77:1111–3.

Jammes F, Song C, Shin D, Munemasa S, Takeda K, Gu D, Cho D, Lee S, Giordo R, Sritubtim S, et al. MAP kinases MPK9 and MPK12 are preferentially expressed in guard cells and positively regulate ROS-mediated ABA signaling. Proc Natl Acad Sci USA. 2009;106:20520–5.

Jannat R, Uraji M, Hossain MA, Islam MM, Nakamura Y, Mori IC, Murata Y. Catalases negatively regulate methyl jasmonate signaling in guard cells. J Plant Physiol. 2012;169:1012–6.

Jannat R, Uraji M, Morofuji M, Hossain MA, Islam MM, Nakamura Y, Mori IC, Murata Y. The roles of CATALASE2 in abscisic acid signaling in *Arabidopsis* guard cells. Biosci Biotechnol Biochem. 2011a;75:2034–6.

Jannat R, Uraji M, Morofuji M, Islam MM, Bloom RE, Nakamura Y, McClung CR, Schroeder JI, Mori IC, Murata Y. Roles of intracellular hydrogen peroxide accumulation in abscisic acid signaling in *Arabidopsis* guard cells. J Plant Physiol. 2011b;168:1919–26.

Jaspers P, Kangasjärvi J. Reactive oxygen species in abiotic stress signaling. Physiol Plant. 2010;138:405–13.

Jiang M, Zhang J. Role of abscissic acid in water stress-induced antioxidant defense in leaves of maize seedlings. Free Radic Res. 2002;36:1001–15.

Jiang M, Zhang J. Cross-talk between calcium and reactive oxygen species originated from NADPH oxidase in abscisic acid-induced antioxidant defense in leaves of maize seedlings. Plant Cell Environ. 2003;26:929–39.

Jiang J, Wang P, An G, Wang P, Song CP. The involvement of a P38-like MAP kinase in ABA-induced and H_2O_2-mediated stomatal closure in *Vicia faba* L. Plant Cell Rep. 2008;27:377–85.

Jiang K, Sorefan K, Deeks MJ, Bevan MW, Hussey PJ, Hetherington AM. The ARP2/3 complex mediates guard cell actin reorganization and stomatal movement in *Arabidopsis*. Plant Cell. 2012;24:2031–40.

Jiao Y, Sun L, Song Y, Wang L, Liu L, Zhang L, Liu B, Li N, Miao C, Hao F. AtrbohD and AtrbohF positively regulate abscisic acid-inhibited primary root growth by affecting Ca^{2+} signalling and auxin response of roots in *Arabidopsis*. J Exp Bot. 2013;64:4183–92.

Jiménez A, Hernández JA, Pastori G, del Río LA, Sevilla F. Role of the ascorbate-glutathione cycle of mitochondria and peroxisomes in the senescence of pea leaves. Plant Physiol. 1998;118:1327–1335.

Joo JH, Bae YS, Lee JS. Role of auxin-induced reactive oxygen species in root gravitropism. Plant Physiol. 2001;126:1055–60.

Jose M, Barrero A, Bruce D, Qian X, Frank G. A role for barley CRYPTOCHROME1 in light regulation of grain dormancy and germination. Plant Cell. 2014;26:1094–104.

Joudoi T, Shichiri Y, Kamizono N, Akaike T, Sawa T, Yoshitake J, Yamada N, Iwai S. Nitrated cyclicGMP modulates guard cell signalling in *Arabidopsis*. Plant Cell. 2013;25:558–71.

Karpinski S, Escobar C, Karpinska B, Creissen G, Mullineaux PM. Photosynthetic electron transport regulates the expression of cytosolic ascorbate peroxidase genes in *Arabidopsis* during excess light stress. Plant Cell. 1997;9:627–40.

Kepka M, Benson CL, Gonugunta VK, Nelson KM, Christmann A, Grill E, Abrams SR. Action of natural abscisic acid precursors and catabolites on abscisic acid receptor complexes. Plant Physiol. 2011;157:2108–19.

Kim TH, Böhmer M, Hu H, Nishimura N, Schroeder JI. Guard cell signal transduction network: advances in understanding abscisic acid, CO_2, and Ca^{2+} signaling. Annu Rev Plant Biol. 2010;61:561–91.

Knetsch ML, Wang M, Snaar-Jagalska BE, Heimovaara-Dijkstra S. Abscisic acid induces mitogen-activated protein kinase activation in barley aleurone protoplasts. Plant Cell. 1996;8:1061–7.

Kobayashi M, Ohura I, Kawakita K, Yokota N, Fujiwara M, Shimamoto K, Doke N, Yoshioka H. Calcium-dependent protein kinases regulate the production of reactive oxygen species by potato NADPH oxidase. Plant Cell. 2007;19:1065–80.

Köhler B, Hills A, Blatt MR. Control of guard cell ion channels by hydrogen peroxide and abscisic acid indicates their action through alternate signaling pathways. Plant Physiol. 2003;131:385–8.

Kovtun Y, Chiu WL, Tena G, Sheen J. Functional analysis of oxidative stress-activated mitogen-activated protein kinase cascade in plants. Proc Natl Acad Sci USA. 2000;97:2940–5.

Kwak JM, Moon JH, Murata Y, Kuchitsu K, Leonhardt N, DeLong A, Schroeder JI. Disruption of a guard cell-expressed protein phosphatase 2A regulatory subunit, RCN1, confers abscisic acid insensitivity in *Arabidopsis*. Plant Cell. 2002;14:2849–61.

Kwak JM, Mori IC, Pei ZM, Leonhardt N, Torres MA, Dangl JL, Bloom RE, Bodde S, Jones JD, Schroeder JI. NADPH oxidase AtrbohD and AtrbohF genes function in ROS-dependent ABA signalling in *Arabidopsis*. EMBO J. 2003;22:2623–33.

LaIoi C, Apel K, Danon A. Reactive oxygen signaling: the latest news. Curr Opin Plant Biol. 2004;7:323–8.

Lemichez E, Wu Y, Sanchez JP, Mettouchi A, Mathur J, Chua NH. Inactivation of AtRac1 by abscisic acid is essential for stomatal closure. Genes Dev. 2001;15:1808–16.

Leonhardt N, Kwak JM, Robert N, Waner D, Leonhardt G, Schroeder JI. Microarray expression analyses of *Arabidopsis* guard cells and isolation of a recessive abscisic acid hypersensitive protein phosphatase 2C mutant. Plant Cell. 2004;16:596–615.

Levine A, Tenhaken R, Dixon R, Lamb C. H_2O_2 from the oxidative burst orchestrates the plant hypersensitive disease resistance response. Cell. 1994;79:583–93.

Leymarie J, Vitkauskaité G, Hoang HH, Gendreau E, Chazoule V, Meimoun P, Corbineau F, El-Maarouf-Bouteau H, Bailly C. Role of reactive oxygen species in the regulation of *Arabidopsis* seed dormancy. Plant Cell Physiol. 2012;53:96–106.

Li SW, Leng Y, Feng L, Zeng XY. Involvement of abscisic acid in regulating antioxidative defense systems and IAA-oxidase activity and improving adventitious rooting in mung bean [*Vigna radiata* (L.) *Wilczek*] seedlings under cadmium stress. Environ Sci Pollut Res Int. 2014;21:525–37.

Liu Y. Roles of mitogen-activated protein kinase cascades in ABA signaling. Plant Cell Rep. 2012;31:1–12.

Liu Y, Shi L, Ye N, Liu R, Jia W, Zhang J. Nitric oxide-induced rapid decrease of abscisic acid concentration is required in breaking seed dormancy in *Arabidopsis*. New Phytol. 2009;183:1030–42.

Liu Y, Ye N, Liu R, Chen M, Zhang J. H_2O_2 mediates the regulation of ABA catabolism and GA biosynthesis in *Arabidopsis* seed dormancy and germination. J Exp Bot. 2010;61:2979–90.
Lü D, Wang W, Miao C. ATHK1 acts downstream of hydrogen peroxide to mediate ABA signaling through regulation of calcium channel activity in *Arabidopsis* guard cells. Chin Sci Bull. 2013;58:336–43.
Lu C, Han MH, Guevara-Garcia A, Fedoroff NV. Mitogen-activated protein kinase signaling in postgermination arrest of development by abscisic acid. Proc Natl Acad Sci India Sect B Biol Sci. 2002;99:15812–7.
Lu Z, Liu D, Liu S. Two rice cytosolic ascorbate peroxidases differentially improve salt tolerance in transgenic *Arabidopsis*. Plant Cell Rep. 2007;26:1909–17.
Maruta T, Inoue T, Noshi M, Tamoi M, Yabuta Y, Yoshimura K, Ishikawa T, Shigeoka S. Cytosolic ascorbate peroxidase 1 protects organelles against oxidative stress by wounding- and jasmonate-induced H_2O_2 in *Arabidopsis* plants. Biochim Biophys Acta. 2012;1820:1901–7.
Maxwell DP, Wang Y, McIntosh L. The alternative oxidase lowers mitochondrial reactive oxygen production in plant cells. Proc Natl Acad Sci USA. 1999;96:8271–6.
May MJ, Leaver CJ. Oxidative stimulation of glutathione synthesis in *Arabidopsis thaliana* suspension cultures. Plant Physiol. 1993;103:621–7.
McCord J, Fridovich I. Superoxide dismutase an enzymic function for erythrocuprein (hemocuprein). J. Bio.chem. 1969;244:6049–6055.
Meinhard M, Grill E. Hydrogen peroxide is a regulator of ABI1, a protein phosphatase 2C from *Arabidopsis*. FEBS Lett. 2001;508:443–6.
Meinhard M, Rodriguez PL, Grill E. The sensitivity of ABI2 to hydrogen peroxide links the abscisic acid-response regulator to redox signalling. Planta. 2002;214:775–82.
Meyer AJ. The integration of glutathione homeostasis and redox signaling. J Plant Physiol. 2008;165:1390–403.
Miao YC, Song CP, Dong FC, Wang XC. ABA-induced hydrogen peroxide generation in guard cells of *Vicia faba*. Acta Phytophysiol Sinica. 2000;26:53–8 (in Chinese).
Miao Y, Lv D, Wang P, Wang XC, Chen J, Miao C, Song CP. An *Arabidopsis* glutathione peroxidase functions as both a redox transducer and a scavenger in abscisic acid and drought stress responses. Plant Cell. 2006;18:2749–66.
Millar AH, Mittova V, Kiddle G, Heazlewood JL, Bartoli CG, Theodoulou FL, Foyer CH. Control of ascorbate synthesis by respiration and its implications for stress responses. Plant Physiol. 2003;133:443–7.
Miller G, Mittler R. Could heat shock transcription factors function as hydrogen peroxide sensors in plants? Ann Bot. 2006;98:279–88.
Miller G, Schlauch K, Tam R, Cortes D, Torres MA, Shulaev V, Dangl JL, Mittler R. The plant NADPH oxidase RBOHD mediates rapid systemic signaling in response to diverse stimuli. Sci Signal. 2009;2:ra45.
Mittler R. Oxidative stress, antioxidants and stress tolerance. Trends Plant Sci. 2002;7:405–10.
Mittler R, Zilinskas BA. Molecular cloning and characterization of a gene encoding pea cytosolic ascorbate peroxidase. J Bio Chem. 1992;267:21802–7.
Mittler R, Vanderauwera S, Gollery M, Van Breusegem F. Reactive oxygen gene network of plants. Trends Plant Sci. 2004;9:490–8.
Mittler R, Vanderauwera S, Suzuki N, Miller G, Tognetti VB, Vandepoele K, Gollery M, Shulaev V, Van Breusegem F. ROS signaling: the new wave? Trends Plant Sci. 2011;16:300–9.
Miura K, Hasegawa PM. Sumoylation and abscisic acid signaling. Plant Signal Behav. 2009;4:1176–8.
Miura K, Okamoto H, Okuma E, Shiba H, Kamada H, Hasegawa PM, Murata Y. SIZ1 deficiency causes reduced stomatal aperture and enhanced drought tolerance via controlling salicylic acid-induced accumulation of reactive oxygen species in *Arabidopsis*. Plant J. 2013;73:91–104.
Møller IM. Plant mitochondria and oxidative stress: electron transport, NADPH turnover, and metabolism of reactive oxygen species. Annu Rev Plant Biol. 2001;52:561–91.

Møller IM, Jensen PE, Hansson A. Oxidative modifications to cellular components in plants. Annu Rev Plant Biol. 2007;58:459–81.

Monshausen GB, Bibikova TN, Messerli MA, Shi C, Gilroy S. Oscillations in extracellular pH and reactive oxygen species modulate tip growth of *Arabidopsis* root hairs. Proc. Natl. Acad. Sci. India Sect. B Biol Sci. 2007;104:20996–1001.

Moon H, Lee B, Choi G, Shin D, Prasad DT, Lee O, Kwak SS, Kim DH, Nam J, Bahk J, et al. NDP kinase 2 interacts with two oxidative stress-activated MAPKs to regulate cellular redox state and enhances multiple stress tolerance in transgenic plants. Proc Natl Acad Sci USA. 2003;100:358–63.

Mori IC, Murata Y, Yang Y, Munemasa S, Wang YF, Andreoli S, Tiriac H, Alonso JM, Harper JF, Ecker JR, et al. CDPKs CPK6 and CPK3 function in ABA regulation of guard cell S-type anion- and Ca^{2+}-permeable channels and stomatal closure. PLoS Biol. 2006;4:e327.

Moubayidin L, Di Mambro R, Sabatini S. Cytokinin–auxin crosstalk. Trends Plant Sci. 2009;14:557–62.

Müller K, Carstens AC, Linkies A, Torres MA, Leubner-Metzger G. The NADPH-oxidase AtrbohB plays a role in *Arabidopsis* seed after-ripening. New Phytol. 2009;184:885–97.

Munemasa S, Muroyama D, Nagahashi H, Nakamura Y, Mori IC, Murata Y. Regulation of reactive oxygen species-mediated abscisic acid signaling in guard cells and drought tolerance by glutathione. Front Plant Sci. 2013;4:472.

Muraoka S, Miura T. Salicylic acid-induced inactivation of creatine kinase in the presence of lactoperoxidase and H_2O_2. Chem Biol Interact. 2005;151:63–70.

Murata Y, Pei ZM, Mori IC, Schroeder J. Abscisic acid activation of plasma membrane Ca^{2+} channels in guard cells requires cytosolic NAD(P)H and is differentially disrupted upstream and downstream of reactive oxygen species production in abi1-1 and abi2-1 protein phosphatase 2C mutants. Plant Cell. 2001;13:2513–23.

Mustilli AC, Merlot S, Vavasseur A, Fenzi F, Giraudat J. *Arabidopsis* OST1 protein kinase mediates the regulation of stomatal aperture by abscisic acid and acts upstream of reactive oxygen species production. Plant Cell. 2002;14:3089–99.

Myouga F, Hosoda C, Umezawa T, Iizumi H, Kuromori T, Motohashi R, Shono Y, Nagata N, Ikeuchi M, Shinozaki K. A heterocomplex of iron superoxide dismutases defends chloroplast nucleoids against oxidative stress and is essential for chloroplast development in *Arabidopsis*. Plant Cell. 2008;20:3148–62.

Navrot N, Rouhier N, Gelhaye E, Jacquot JP. Reactive oxygen species generation and antioxidant systems in plant mitochondria. Physiol Plant. 2007;129:185–95.

Negi J, Matsuda O, Nagasawa T, Oba Y, Takahashi H, Kawai-Yamada M, Uchimiya H, Hashimoto M, Iba K. CO_2 regulator SLAC1 and its homologues are essential for anion homeostasis in plant cells. Nature. 2008;452:483–6.

Noctor G, Foyer CH. A re-evaluation of the ATP: NADPH budget during C3 photosynthesis: a contribution from nitrate assimilation and its associated respiratory activity? J Exp Bot. 1998;49:1895–908.

Noctor G, Gomez L, Vanacker H, Foyer CH. Interactions between biosynthesis, compartmentation and transport in the control of glutathione homeostasis and signalling. J Exp Bot. 2002;53:1283–304.

Park KY, Jung JY, Park J, Hwang JU, Kim YW, Hwang I, Lee Y. A role for phosphatidylinositol 3-phosphate in abscisic acid-induced reactive oxygen species generation in guard cells. Plant Physiol. 2003;132:92–8.

Park SY, Fung P, Nishimura N, Jensen DR, Fujii H, Zhao Y, Lumba S, Santiago J, Rodrigues A, Chow TF, et al. Abscisic acid inhibits type 2C protein phosphatases via the PYR/PYL family of START proteins. Science. 2009;324:1068–71.

Passaia G, Queval G, Bai J, Margis-Pinheiro M, Foyer CH. The effects of redox controls mediated by glutathione peroxidases on root architecture in *Arabidopsis thaliana*. J Exp Bot. 2014;65:1403–13.

Pastori GM, Kiddle G, Antoniw J, Bernard S, Veljovic-Jovanovic S, Verrier PJ, Noctor G, Foyer CH. Leaf vitamin C contents modulate plant defense transcripts and regulate genes that control development through hormone signaling. Plant Cell. 2003;15:939–51.

Pei ZM, Murata Y, Benning G, Thomine S, Klüsener B, Allen GJ, Grill E, Schroeder JI. Calcium channels activated by hydrogen peroxide mediate abscisic acid signalling in guard cells. Nature. 2000;406:731–4.

Peng CL, Ou ZY, Liu N, Lin GZ. Response to high temperature in flag leaves of super high-yielding rice Pei'ai 64S/E32 and Liangyoupeijiu. Rice Sci. 2005;12:179–86.

Pennell RI, Lamb C. Programmed cell death in plants. Plant Cell. 1997;9:1157–68.

Pfannschmidt T. Chloroplast redox signals: how photosynthesis controls its own genes. Trends Plant Sci. 2003;8:33–41.

Pinheiro C, Chaves MM. Photosynthesis and drought: can we make metabolic connections from available data? J Exp Bot. 2011;62:869–82.

Quan LJ, Zhang B, Shi WW, Li HY. Hydrogen peroxide in plants: a versatile molecule of the reactive oxygen species network. J Integr Plant Biol. 2008;50:2–18.

Rausch T, Wachter A. Sulfur metabolism: a versatile platform for launching defence operations. Trends Plant Sci. 2005;10:503–9.

Rentel MC, Lecourieux D, Ouaked F, Usher SL, Petersen L, Okamoto H, Knight H, Peck SC, Grierson CS, Hirt H, Knight MR. OXI1 kinase is necessary for oxidative burst-mediated signalling in *Arabidopsis*. Nature. 2004;427:858–61.

Robert N, Merlot S, N'guyen V, Boisson-Dernier A, Schroeder JI. A hypermorphic mutation in the protein phosphatase 2C HAB1 strongly affects ABA signaling in *Arabidopsis*. FEBS Lett. 2006;580:4691–6.

Roxas VP, Lodhi SA, Garrett DK, Mahan JR, Allen RD. Stress tolerance in transgenic tobacco seedlings that overexpress glutathione S-transferase/glutathione peroxidase. Plant Cell Physiol. 2000;41:1229–34.

Saez A, Apostolova N, Gonzalez-Guzman M, Gonzalez-Garcia MP, Nicolas C, Lorenzo O, Rodriguez PL. Gain-of-function and loss-of-function phenotypes of the protein phosphatase 2C HAB1 reveal its role as a negative regulator of abscisic acid signalling. Plant J. 2004;37:354–69.

Saibo NJ, Lourenco T, Oliveira MM. Transcription factors and regulation of photosynthetic and related metabolism under environmental stresses. Ann Bot. 2009;103:609–23.

Sakamoto A, Okumura T, Kaminaka H, Sumi K, Tanaka K. Structure and differential response to abscisic acid of two promoters for the cytosolic copper/zinc-superoxide dismutase genes, SodCc1 and SodCc2, in rice protoplasts. FEBS Lett. 1995;358:62–6.

Santiago J, Dupeux F, Round A, Antoni R, Park SY, Jamin M, Cutler SR, Rodriguez PL, Márquez JA. The abscisic acid receptor PYR1 in complex with abscisic acid. Nature. 2009;462:665–8.

Santos M, Gousseau H, Lister C, Foyer C, Creissen G, Mullineaux P. Cytosolic ascorbate peroxidase from *Arabidopsis thaliana* L. is encoded by a small multigene family. Planta. 1996;198:64–9.

Sappl PG, Onate-Sanchez L, Singh KB, Millar AH. Proteomic analysis of glutathione S-transferases of *Arabidopsis thaliana* reveals differential salicylic acid-induced expression of the plant-specific phi and tau classes. Plant Mol Biol. 2004;54:205–19.

Savchenko T, Kolla VA, Wang CQ, Nasafi Z, Hicks DR, Phadungchob B, Chehab WE, Brandizzi F, Froehlich J, Dehesh KB. Functional convergence of oxylipin and abscisic acid pathways controls stomatal closure in response to drought. Plant Physiol. 2014;164:1151–60.

Seki M, Ishida J, Narusaka M, Fujita M, Nanjo T, Umezawa T, Kamiya A, Nakajima M, Enju A, Sakurai T, et al. Monitoring the expression pattern of around 7,000 *Arabidopsis* genes under ABA treatments using a full-length cDNA microarray. Funct Integr Genomics. 2002;2:282–91.

Shaikhali J, Heiber I, Seidel T, Ströher E, Hiltscher H, Birkmann S, Dietz KJ, Baier M. The redox-sensitive transcription factor Rap2.4a controls nuclear expression of 2-Cys peroxiredoxin A and other chloroplast antioxidant enzymes. BMC Plant Biol. 2008;8:48.

Shi B, Ni L, Zhang A, Cao J, Zhang H, Qin T, Tan M, Zhang J, Jiang M. OsDMI3 is a novel component of abscisic acid signaling in the induction of antioxidant defense in leaves of rice. Mol Plant. 2012;5:1359–74.

Shi B, Ni L, Liu Y, Zhang A, Tan M, Jiang M. OsDMI3-mediated activation of OsMPK1 regulates the activities of antioxidant enzymes in abscisic acid signalling in rice. Plant Cell Environ. 2014;37:341–52.

Singh, N., Mishra, A., and Jha, B. Over-expression of the peroxisomal ascorbate peroxidase (SbpAPX) gene cloned from halophyte salicornia brachiata confers salt and drought stress tolerance in transgenic tobacco. Mar Biotechnol. 2014;16:321–32.

Sirichandra C, Gu D, Hu HC, Davanture M, Lee S, Djaoui M, Valot B, Zivy M, Leung J, Merlot S, Kwak JM. Phosphorylation of the *Arabidopsis* AtrbohF NADPH oxidase by OST1 protein kinase. FEBS Lett. 2009;583:2982–6.

Smirnoff N. Ascorbic acid: metabolism and functions of a multifaceted molecule. Curr Opin Plant Biol. 2000;3:229–35.

Song Y, Miao Y, Song CP. Behind the scenes: the roles of reactive oxygen species in guard cells. New Phytol. 2014;201:1121–40.

Sutter JU, Sieben C, Hartel A, Eisenach C, Thiel G, Blatt MR. Abscisic acid triggers the endocytosis of the *Arabidopsis* KAT1 K^+ channel and its recycling to the plasma membrane. Curr Biol. 2007;17:1396–402.

Takahashi S, Seki M, Ishida J, Satou M, Sakurai T, Narusaka M, Kamiya A, Nakajima M, Enju A, Akiyama K, et al. Monitoring the expression profiles of genes induced by hyperosmotic, high salinity, and oxidative stress and abscisic acid treatment in *Arabidopsis* cell culture using a full-length cDNA microarray. Plant Mol Biol. 2004;56:29–55.

Takeda S, Gapper C, Kaya H, Bell E, Kuchitsu K, Dolan L. Local positive feedback regulation determines cell shape in root hair cells. Science. 2008;319:1241–4.

Teale WD, Ditengou FA, Dovzhenko AD, Li X, Molendijk AM, Ruperti B, Paponov I, Palme K. Auxin as a model for the integration of hormonal signal processing and transduction. Mol Plant. 2008;1:229–37.

Tossi V, Lamattina L, Cassia R. An increase in the concentration of abscisic acid is critical for nitric oxide-mediated plant adaptive responses to UV-B irradiation. New Phytol. 2009;181:871–9.

Tripathy BC, Oelmüller R. Reactive oxygen species generation and signaling in plants. Plant Signal Behav. 2012;7:1621–33.

Tsukagoshi H, Busch W, Benfey PN. Transcriptional regulation of ROS controls transition from proliferation to differentiation in the root. Cell. 2010;143:606–16.

Tuteja N, Singh MB, Misra MK, Bhalla PL, Tuteja R. Molecular mechanisms of DNA damage and repair: progress in plants. Crit Rev Biochem Mol Biol. 2001;36:337–97.

Ubeda-Tomás S, Swarup R, Coates J, Swarup K, Laplaze L, Beemster GT, Hedden P, Bhalerao R, Bennett MJ. Root growth in *Arabidopsis* requires gibberellin/DELLA signalling in the endodermis. Nature Cell Biol. 2008;10:625–8.

Umezawa T, Sugiyama N, Mizoguchi M, Hayashi S, Myouga F, Yamaguchi-Shinozaki K, Ishihama Y, Hirayama T, Shinozaki K. Type 2C protein phosphatases directly regulate abscisic acid-activated protein kinases in *Arabidopsis*. Proc Natl Acad Sci USA. 2009;106:17588–93.

Vahisalu T, Kollist H, Wang YF, Nishimura N, Chan WY, Valerio G, Lamminmäki A, Brosché M, Moldau H, Desikan R, Schroeder JI, Kangasjärvi J. SLAC1 is required for plant guard cell S-type anion channel function in stomatal signalling. Nature. 2008;452:487–91.

Van Breusegem F, Bailey-Serres J, Mittler R. Unraveling the tapestry of networks involving reactive oxygen species in plants. Plant Physiol. 2008;147:978–84.

Vandenabeele S, Vanderauwera S, Vuylsteke M, Rombauts S, Langebartels C, Seidlitz HK, Zabeau M, Van Montagu M, Inzé D, Van Breusegem F. Catalase deficiency drastically affects gene expression induced by high light in *Arabidopsis thaliana*. Plant J. 2004;39:45–58.

Wang PT, Song CP. Guard-cell signalling for hydrogen peroxide and abscisic acid. New Phytol. 2008;178:703–18.

Wang Y, Ying Y, Chen J, Wang X. Transgenic *Arabidopsis* overexpressing Mn-SOD enhanced salt-tolerance. Plant Sci. 2004;167:671–7.

Wang PC, Du YY, An GY, Zhou Y, Miao C, Song CP. Analysis of global expression profiles of *Arabidopsis* genes under abscisic acid and H_2O_2 applications. J Integr Plant Biol. 2006;48:62–74.

Wang P, Du Y, Zhao X, Miao Y, Song CP. The MPK6-ERF6-ROS-responsive cis-acting element7/GCC box complex modulates oxidative gene transcription and the oxidative response in *Arabidopsis*. Plant Physiol. 2013a;161:1392–408.

Wang Y, Chen ZH, Zhang B, Hills A, Blatt MR. PYR/PYL/RCAR abscisic acid receptors regulate K^+ and Cl^- channels through reactive oxygen species-mediated activation of Ca^{2+} channels at the plasma membrane of intact *Arabidopsis* guard cells. Plant Physiol. 2013b;163:566–77.

Watkins JM, Hechler PJ, Muday GK. Ethylene-induced flavonol accumulation in guard cells suppresses reactive oxygen species and moderates stomatal aperture. Plant Physiol. 2014;164:1707–17.

Xiang C, Werner BL, E'Lise MC, Oliver DJ. The biological functions of glutathione revisited in *Arabidopsis* transgenic plants with altered glutathione levels. Plant Physiol. 2001;126:564–74.

Xiong L, Zhu JK. Regulation of abscisic acid biosynthesis. Plant Physiol. 2003;133:29–36.

Yoshida K, Kaothien P, Matsui T, Kawaoka A, Shinmyo A. Molecular biology and application of plant peroxidase genes. Appl Microbiol Biotechnol. 2003;60:665–70.

Yoshida T, Nishimura N, Kitahata N, Kuromori T, Ito T, Asami T, Shinozaki K, Hirayama T. ABA-hypersensitive germination3 encodes a protein phosphatase 2C (AtPP2CA) that strongly regulates abscisic acid signaling during germination among *Arabidopsis* protein phosphatase 2Cs. Plant Physiol. 2006;140:115–26.

You J, Zong W, Li X, Ning J, Hu H, Li X, Xiao J, Xiong L. The SNAC1-targeted gene OsSRO1c modulates stomatal closure and oxidative stress tolerance by regulating hydrogen peroxide in rice. J Exp Bot. 2013;64:569–83.

Yuasa T, Ichimura K, Mizoguchi T, Shinozaki K. Oxidative stress activates ATMPK6, an *Arabidopsis* homologue of MAP kinase. Plant Cell Physiol. 2001;42:1012–6.

Zhang X, Miao YC, An GY, Zhou Y, Shangguan ZP, Gao JF, Song CP. K^+ channels inhibited by hydrogen peroxide mediate abscisic acid signalling in *Vicia* guard cells. Cell Res. 2001a;11:195–202.

Zhang X, Zhang L, Dong F, Gao J, Galbraith DW, Song CP. Hydrogen peroxide is involved in abscisic acid-induced stomatal closure in *Vicia faba*. Plant Physiol. 2001b;126:1438–48.

Zhang W, Qin C, Zhao J, Wang X. Phospholipase D alpha 1-derived phosphatidic acid interacts with ABI1 phosphatase 2C and regulates abscisic acid signaling. Proc Natl Acad Sci USA. 2004a;101:9508–13.

Zhang X, Wang H, Takemiya A, Song CP, Kinoshita T, Shimazaki K. Inhibition of blue light-dependent H^+ pumping by abscisic acid through hydrogen peroxide-induced dephosphorylation of the plasma membrane H^+-ATPase in guard cell protoplasts. Plant Physiol. 2004b;136:4150–8.

Zhang W, Fan L-M, Wu WH. Osmo-sensitive and stretch activated calcium-permeable channels in *Vicia faba* guard cells are regulated by actin dynamics. Plant Physiol. 2007;143:1140–51.

Zhang Y, Tan J, Guo Z, Lu S, He S, Shu W, Zhou B. Increased abscisic acid levels in transgenic tobacco over-expressing 9 cis-epoxycarotenoid dioxygenase influence H_2O_2 and NO production and antioxidant defences. Plant, Cell Environ. 2009a;32:509–19.

Zhang Y, Zhu H, Zhang Q, Li M, Yan M, Wang R, Wang L, Welti R, Zhang W, Wang X. Phospholipase Dα1 and phosphatidic acid regulate NADPH oxidase activity and production of reactive oxygen species in ABA-mediated stomatal closure in *Arabidopsis*. Plant Cell. 2009b;21:2357–77.

Zhang H, Han W, De Smet I, Talboys P, Loya R, Hassan A, Rong H, Jurgens G, Paul Knox J, Wang MH. ABA promotes quiescence of the quiescent centre and suppresses stem cell differentiation in the *Arabidopsis* primary root meristem. Plant J. 2010;64:764–74.

Zhang X, Wang L, Meng H, Wen H, Fan Y, Zhao J. Maize ABP9 enhances tolerance to multiple stresses in transgenic *Arabidopsis* by modulating ABA signaling and cellular levels of reactive oxygen species. Plant Mol Biol. 2011;75:365–78.

Zhang ZW, Feng LY, Cheng J, Tang H, Xu F, Zhu F, Zhao ZY, Yuan M, Chen YE, Wang JH, et al. The roles of two transcription factors, ABI4 and CBFA, in ABA and plastid signalling and stress responses. Plant Mol Biol. 2013;83:445–58.

Zhao J, Wang X. *Arabidopsis* phospholipase Dalpha1 interacts with the heterotrimeric G-protein alpha-subunit through a motifanalogous to the DRY motif in G-protein-coupled receptors. J Biol Chem. 2004;279:1794–800.

Zhu SY, Yu XC, Wang XJ, Zhao R, Li Y, Shang Y, Du SY, Wang XF, Wu FQ, et al. Two calcium-dependent protein kinases, CPK4 and CPK11, regulate abscisic acid signal transduction in *Arabidopsis*. Plant Cell. 2007;19:3019–36.

Zhu M, Zhu N, Song WY, Harmon AC, Assmann SM, Chen S. Thiol-based redox proteins in Brassica napus guard cell abscisic acid and methyl jasmonate signaling. Plant J. 2014;78:491–515.

Zou JJ, Wei FJ, Wang C, Wu JJ, Ratnasekera D, Liu WX, Wu WH. *Arabidopsis* calcium-dependent protein kinase CPK10 functions in abscisic acid- and Ca^{2+}-mediated stomatal regulation in response to drought stress. Plant Physiol. 2010;154:1232–43.

Chapter 11
Transcription Factors Involved in ABA Signaling

Soo Young Kim

Abstract Abscisic acid plays an important regulatory role in seed development and adaptive responses to abiotic stresses in vegetative tissues. The major part of the ABA function entails gene expression events. Various genome-wide transcriptome analyses indicate that ABA regulates the largest number of genes among major plant hormones. Promoter analyses of the ABA-regulated genes revealed that *cis*-elements sharing the PyACGTGGC core sequence are present in many ABA-responsive genes. The element is generally known as ABA response element (ABRE), and a subfamily of basic leucine zipper (bZIP) proteins has been identified that mediate ABA regulation through the ABRE. Other *cis*-elements and cognate transcription factors (TFs) involved in ABA-dependent transcription have also been identified by traditional approaches. Additionally, numerous TFs have been reported to regulate ABA-responsive gene expression by reverse genetics means. In this article, I will review ABA-dependent transcription, focusing on the Arabidopsis TFs whose in vivo functions have been experimentally demonstrated.

Keywords Transcription factors · Abscisic acid (ABA) · Gene regulation

11.1 Introduction

An analysis of public Affymetrix ATH1 GeneChip microarray data sets revealed that ABA has the most profound effect on gene expression among seven plant hormones (i.e., ABA, GA, IAA, ET, CK, BR, and MJ) (Nemhauser et al. 2006). According to the analysis, ABA affects the expression of 2,936 genes. Matsui

S.Y. Kim (✉)
Department of Molecular Biotechnology and Kumho Life Science Laboratory,
College of Agriculture and Life Sciences, Chonnam National University,
Gwangju 500-757, South Korea
e-mail: sooykim@chonnam.ac.kr

D.-P. Zhang (ed.), *Abscisic Acid: Metabolism, Transport and Signaling*,
DOI 10.1007/978-94-017-9424-4_11

et al. (2008) demonstrated, by whole-genome expression profiling using a tiling array, that more than 6,300 genes are up- or down-regulated by ABA (Matsui et al. 2008). These and other studies (Hoth et al. 2002; Seki et al. 2002; Takahashi et al. 2004; Zeller et al. 2009) show that up to 25 % of the Arabidopsis genes are affected by ABA. ABA plays a key role in seed development and stress response, and the studies indicate that large fraction of the ABA-regulated genes are seed-specific genes, such as storage protein and LEA genes, and that over 60 % of the ABA-regulated genes are also responsive to various abiotic stresses.

Approximately 10 % of the ABA-responsive genes are associated with transcription (Nemhauser et al. 2006; Seki et al. 2002), which include most of the major classes of transcription factors (TFs), such as bZIP, AP2/ERF, MYB, HB, zinc finger, WRKY, etc. Not all ABA-responsive TFs may be involved in ABA regulation of gene expression, and, conversely, it is certain that TFs whose expression is not ABA-dependent are involved in ABA-regulated gene expression. Nonetheless, the genome-wide transcriptome studies suggest that several hundreds of TFs may be responsive to ABA. Furthermore, the list of TFs involved in ABA response and/or stress responses in other plant species is ever-increasing.

Several reviews on ABA-dependent transcription in general and in association with stress response are available (Busk and Pages 1998; Cutler et al. 2010; Fujita et al. 2011; Nakashima et al. 2009b). Also, a large number of general reviews on various classes of TFs are available (see below). In this review, I will focus on the Arabidopsis TFs whose functions in ABA signaling have been experimentally determined and which can be related to the core ABA signaling pathway mentioned in previous chapters.

11.2 *Cis*-Elements

A number of ABA response elements have been determined by promoter analyses of typical ABA-responsive genes, such as *Em*, *RAB*, and *RD29A* (Busk and Pages 1998). Most ubiquitous elements are those possessing the PyACGTGGC core sequence. The core element is similar to the G-box, CACGTG, which is present in many light-regulated genes (Menkens et al. 1995). The conserved core sequence is generally known as ABA response element (ABRE). Originally found in the monocot LEA genes (Guiltinan et al. 1990; Marcotte et al. 1989; Mundy et al. 1990), the element was found to be present in numerous ABA-responsive genes (Gomez-Porras et al. 2007; Yazaki and Kikuchi 2005; Zhang et al. 2005a). Later experiments showed that a single ABRE is not sufficient to induce ABA-dependent gene expression: It usually functions in combination with other *cis*-elements for full ABA induction. The additional elements, named "coupling elements," may be the ABRE itself or one of other *cis*-elements, such as coupling element 1 (CE1) (Shen and Ho 1995), coupling element 3 (CE3) (Shen et al. 1996), Motif III (Ono et al. 1996), and Hex III (Lam and Chua 1991). Except CE1, these coupling elements share the CGCGTG consensus. CE3 has been demonstrated to be functionally equivalent to the ABRE (Hobo et al. 1999), suggesting that the ABRE and coupling elements are likely to be the same class of ABA response element. Interestingly, the drought response element (DRE) and the C-repeat (CRT) sequence (Baker et al. 1994; Yamaguchi-Shinozaki and Shinozaki 1994), which share the CCGAC consensus, function as a coupling element in the ABA-induced expression of RD20A (Narusaka et al. 2003). In the case of maize *RAB17*, DRE itself functions as an ABRE (Kizis and Pages 2002).

Other experimentally determined ABREs are the MYB and the MYC recognition sequences in *RD22* (Abe et al. 2003). Mutations of these elements result in reduced ABA induction of reporter expression in tobacco.

11.3 ABFs/AREBs/ABI5

A small subfamily of bZIP class TFs that bind to the G-box-type ABRE have been isolated by yeast one-hybrid screens (Choi et al. 2000; Uno et al. 2000). Subsequently, the TFs, which are named ABRE-binding factors (ABFs: ABF1, ABF2, ABF3, and ABF4) or ABRE-binding proteins (AREBs: AREB1, AREB2, and AREB3), have been demonstrated to mediate ABA and stress responses in vivo (Fujita et al. 2005b; Kang et al. 2002; Kim et al. 2004; Yoshida et al. 2010). As mentioned in previous chapters, ABFs/AREBs constitute the core ABA signaling pathway (i.e., PYR/PYL/RCAR-ABI1/2-SnRK2s-ABFs/AREBs) (Cutler et al. 2010; Fujii et al. 2009). Numerous other bZIP class TFs had been isolated based on their in vitro interactions with the ABRE prior to the isolation of ABFs/AREBs (Menkens et al. 1995), but their *in planta* functions in ABA response have not been reported.

ABF/AREB family TFs are highly homologous to sunflower *Dc3* promoter-binding factors (DPBFs) (Kim et al. 1997; Kim and Thomas 1998) and their Arabidopsis homologs AtDPBFs (AtDPBF1/ABI5, AtDPBF2, AtDPBF3, and AtDPBF4) (Kim et al. 2002). Homologous bZIP factors have been reported in monocot plants, such as rice, barley, and wheat (Casaretto and Ho 2003; Hobo et al. 1999; Johnson et al. 2002). The Arabidopsis bZIP proteins constitute a clade consisted of 9 members (Bensmihen et al. 2002; Fujita et al. 2005a; Kim et al. 2002; Kim 2006), including ABI5 (see below). The family members share a highly conserved bZIP domain and additional short conserved regions, C1–C4. Among the nine members, *ABF1–ABF4* are expressed in vegetative tissues, and their expression is ABA-inducible. Additionally, their expression is stress inducible; that is, *ABF2/AREB1*, *ABF3,* and *ABF4/AREB2* are highly inducible by high salt and high osmolarity, whereas *ABF1* is cold inducible (Choi et al. 2000; Fujita et al. 2005a). Other five members of the family (i.e., *AtDPBF1/ABI5*, *AtDPBF2*, *AtDPBF3/AREB3*, *AtDPBF4/EEL*, and an unknown gene) are expressed mainly in embryos. A binding site selection assay showed that the preferred binding site of ABF1, one of the ABFs, is the CACGTGGC, the typical ABRE mentioned above (Choi et al. 2000). It also could bind the coupling element CE3. In fact, only

ABFs/AREBs were isolated as CE3-binding proteins in our one-hybrid screens (unpublished observation).

The in vivo functions of ABF2/AREB1, ABF3, and ABF4/AREB2 in ABA-dependent stress response have been demonstrated by transgenic approaches. Constitutive overexpression of the three bZIP proteins resulted in enhanced ABA sensitivity with concomitant increase in stress tolerance, whereas their knockout mutations resulted in decrease in ABA sensitivity and stress tolerance (Finkelstein et al. 2005; Kim 2006; Yoshida et al. 2010). These phenotypic changes were accompanied by changes in the expression levels of a large number of ABA and/or stress-responsive genes (Abdeen et al. 2010; Fujita et al. 2005a; Yoshida et al. 2010).

ABI5, which is the same as AtDPBF1 (Kim et al. 2002), has been isolated based on genetic screens to isolate ABA response mutants (Finkelstein 1994; Finkelstein and Lynch 2000; Lopez-Molina and Chua 2000) and belongs to the ABF/AREB subfamily. ABI5 plays an important role in ABA response in seeds and during early stage of seedling establishment, and its mutations result in ABA-insensitive germination and seedling growth (Finkelstein 1994; Lopez-Molina et al. 2002). Transcriptome analysis of the transgenic plants overexpressing ABI5 demonstrated that 59 gene are up-regulated by ABI5 in ABA-dependent manner (Reeves et al. 2011).

11.4 Modulation of ABFs/AREBs/ABI5 Functions by Phosphorylation

Not only the expression of ABF/AREB/ABI5 family members is induced by ABA, but their activities are also regulated by ABA posttranslationally (Uno et al. 2000; Yoshida et al. 2010). The most important aspect of ABA-dependent posttranslation modification of the bZIP factors is phosphorylation. Various studies indicate that ABFs/AREBs are phosphorylated by SnRK2.2/SnrK2D, SnRK2.3/SnRK2I, and SnRK2.6/SnRK2E/OST1 (Fujii et al. 2007, 2009; Fujii and Zhu 2009; Fujita et al. 2009, 2013; Furihata et al. 2006; Sirichandra et al. 2010). The three SnRK2s, which are key regulators of ABA response, are key players in the phosphorylation of ABFs/AREBs/ABI5. For instance, ABF1, ABF2/AREB1, and ABI5 are phosphorylated by the three SnRKs and by SnRK2.7/SnRK2F and SnRK2.8/SnRK2C as well (Furihata et al. 2006; Fujii et al. 2007). ABF3 is phosphorylated by SnRK2.6/SnRK2E/OST1 in vivo (Sirichandra et al. 2010). ABI5 also is phosphorylated by the three SnRK2s (Fujii et al. 2007; Nakashima et al. 2009a). Recently, FyPP1 and FyPP3, which encode PP6 phosphatase catalytic subunits, have been reported to negatively regulate ABA response (Dai et al. 2013). The phosphatases act antagonistically with SnRKs by dephosphorylating ABI5, and thereby destabilizing it.

Several calcium-dependent protein kinases (CDPKs), which play positive regulatory roles in ABA response, are known to phosphorylate ABFs/AREBs. CPK32, for example, phosphorylates ABF4, and its overexpression enhances

ABA sensitivity (Choi et al. 2005). It also interacts with ABF1, ABF2/AREB1, and ABF3. Two other CDPKs, CPK4 and CPK11, phosphorylates ABF1 and ABF4/AREB2 and act as positive regulators of ABA response (Zhu et al. 2007).

11.5 Modulation of ABFs/AREBs/ABI5 Functions by Protein Degradation

Protein degradation by proteasome is an important mechanism of regulating the abundance of regulatory proteins (Vierstra 2009), and it plays an important role in the regulation of ABI5 activity. ABA promotes the accumulation of ABI5 protein by stabilizing it (Lopez-Molina et al. 2001). This stabilization process involves protein phosphorylation, which is accomplished by above-mentioned kinases. Thus, protein phosphorylation not only activates the transcriptional activity of ABI5 but also increases its stability.

Several components of the ubiquitin-26S proteasome system have been reported to control ABI5 degradation. AFP, which is one of the ABI5-interacting proteins, facilitates ubiquitin-mediated degradation of ABI5 (Lopez-Molina et al. 2003). KEEP ON GOING (KEG), a RING-type E3 ligase, ubiquitinates ABI5 and promotes its proteasomal degradation to negatively regulate ABA signaling (Liu and Stone 2010, 2013; Stone et al. 2006). Recently, ABF1 and ABF3 also have been demonstrated to be the substrates of KEG (Chen et al. 2013b). CUL4-based E3 ligases, DWA1 and DWA2, act as negative regulators of ABA response by targeting ABI5 for proteasomal degradation (Lee et al. 2010a). Although not related to protein degradation directly, ABI5 is a target of the SUMO E3 ligase SIZ1, which inactivates and stabilizes ABI5 (Miura et al. 2009). It has been proposed that reversible sumoylation of ABI5 by SIZ1 protects inactive form of ABI5 from degradation. In this scheme of ABI5 activity regulation, ABA activation of ABI5 is preceded by desumoylation of ABI5.

11.6 ABI3/ABI4

ABI3 and ABI4 have been isolated by map-based cloning (Finkelstein 1994; Giraudat et al. 1992). ABI3, which is an ortholog of the maize VP1 (McCarty et al. 1991), belongs to the B3-domain family of TFs (Swaminathan et al. 2008). *ABI3* expression is seed specific, and it plays a key role in the establishment of seed dormancy and seed maturation, together with two other B3 proteins FUS3 and LEC2 (Holdsworth et al. 2008; Suzuki and McCarty 2008). It also plays an essential role in embryo degreening process (Delmas et al. 2013). ABI3 binds to the Sph/RY element (TCCATGCAT) present in seed-specific gene promoters (Monke et al. 2004; Suzuki et al. 1997), and the genome-wide analysis of ABI3 regulon indicates that

98 genes are the targets of ABI3 (Monke et al. 2012). Most of the ABI3 target genes are ABA-regulated and seed-specific genes, such as storage proteins, oleosin, and LEA protein genes. One of the ABI3 target genes is ABI5, which acts downstream of ABI3 to arrest seedling establishment (Lopez-Molina et al. 2002). Several studies indicate that ABI3 functions synergistically with ABI5 to regulate ABA-responsive genes (Casaretto and Ho 2003; Gampala et al. 2002; Hobo et al. 1999; Lim et al. 2013; Nakashima et al. 2006). ABI3 protein is unstable in the absence of ABA, and it has been shown that AIP2, a RING-type E3 ligase, ubiquitinates ABI3 to promote its proteasomal degradation (Lopez-Molina et al. 2002; Zhang et al. 2005b).

Initially, ABI4 has been isolated as ABA-insensitive mutant with seed-specific defects (Finkelstein 1994). However, later studies showed that it plays diverse roles in sugar signaling, stress response, lipid mobilization, chloroplast retrograde signaling, root development, and biosynthesis of ABA and GA (Leon et al. 2012; Shu et al. 2013; Wind et al. 2013). ABI4 belongs to the DREB/CBF subfamily of the AP2/ERF superfamily (Sakuma et al. 2002), and its binding sites have been determined in vitro employing maize ABI4 (Niu et al. 2002). The study shows that its preferred binding sites contain the CACCG core sequence, which is similar to the coupling element CE1. Reeves et al. (2011) showed that approximately 100 genes are induced by ectopically expressed ABI4. However, CE1 element is not enriched in the promoters of the ABI4 target genes, although ABI4 binds them. The result suggests that ABI4 may bind sequences other than CE1 in vivo, and it is consistent with the observation that other AP2-domain proteins than ABI4 are likely to bind the CE1 element (Shen et al. 2004). Also, strong synergy between ABI4 and ABI5, ABF1, or ABF3 was observed in the activation of the promoters lacking CE1 in yeast.

The functions of ABI3 and ABI4 in ABA signaling have been firmly established. However, their position in the core ABA signaling pathway is not clearly understood. ABI3 is epistatic to ABI5; that is, ABI3 acts upstream of ABI5 (Lopez-Molina et al. 2002), and *ABI5* expression is regulated by ABI3. How ABI3 is related to upstream components of the core ABA signaling pathway and, if it functions in a separate pathway which converges on ABI5 with the core pathway, the identity of the upstream regulatory components remains to be determined. Similarly, ABA signaling components functioning upstream of ABI4 remain to be determined.

11.7 AP2/ERF Domain Proteins

In addition to ABI4, several other AP2/ERF domain proteins have been reported to function in ABA response. Maize DRE-binding proteins, DBF1 and DBF2, are involved in the ABA regulation of *RAB17* (Kizis and Pages 2002). DREB2C interacts with ABFs and positively regulates ABA response (Lee et al. 2010b). Other DREBs, i.e., DREB1A and DREB2A, also interact with ABF2/AREB1 and ABF4/AREB2, suggesting that ABFs/AREBs and DREB family TFs, which function in the ABA-independent stress response pathway (Nakashima et al. 2009b), may cooperate to regulate ABA- and/or stress-responsive genes. Overexpression of *RAP2.6* enhances ABA and stress responses, indicating that it is a positive regulator of ABA response (Zhu et al. 2010). ADAP (ARIA-interacting AP2 domain protein) interacts with ARIA, which, in turn, interacts with ABF2/AREB1 to positively regulate ABA response, and functions as a positive regulator in ABA-dependent growth regulation and stress response (Lee et al. 2009). Another AP2 subfamily protein CHO1 mediates ABA response during germination by repressing GA biosynthesis (Yano et al. 2009).

A group of AP2/ERF family proteins (AtERF1, 2, 5, 13, 15, ERF1, ORA59, RAP2.4, etc.) that bind the coupling element CE1 has been isolated by yeast one-hybrid screen (Lee et al. 2010c). The proteins, named CE1-binding factors (CEBFs), belong to the B-3 or A-6 subfamily of AP2/ERF superfamily. Analysis of *in planta* functions of three CEBFs showed that AtERF13 is a positive regulator of ABA response during seedling growth, whereas RAP2.4 and RAP2.4L are involved in stress response.

AP2/ERF family TFs which negatively regulate ABA response have also been reported. ABR1 represses ABA response during seed germination and down-regulates ABA-responsive genes (Pandey et al. 2005). AtERF7 acts as transcriptional repressor, probably as part of a transcriptional repressor complex containing Sin3-histone deacetylase, and reduces ABA response during seed germination and in guard cells (Song et al. 2005). Similarly, AtERF4 negatively regulates ABA response as well as ethylene response (Yang et al. 2005). Most of the AP2/DREB/ERF family TFs mentioned above bind the DRE, CE1 element, or ethylene response element (i.e., GCC box) and are also involved in abiotic stress or ethylene response.

11.8 WRKY Proteins

Several WRKY family TFs (Eulgem et al. 2000) are known to be involved in ABA response (Rushton et al. 2012). Disruption of *WRKY2* confers ABA hypersensitivity during germination and early seedling growth, and its expression is induced by ABA in an ABI3- and ABI5-dependent manner (Jiang and Yu 2009). WRKY63/ABO3 plays both positive and negative regulatory role (Ren et al. 2010). Its loss of function mutant is hypersensitive to ABA during germination and seedling growth. However, the mutant is impaired in the stomatal regulation and susceptible to water-deficit condition. It has been demonstrated that WRKY63 binds the W-box (PyTGACPy) in the *ABF2* promoter and positively regulates ABF2 expression (Ren et al. 2010). WRKY8, on the other hand, mediates mainly defense response, but it also positively regulates *ABI4* expression (Chen et al. 2013a). Thus, WRKY8 may be involved in the cross talk between ABA and defense signaling.

Chen et al. (2010) showed that WRKY18 and WRKY60 play positive role in ABA inhibition of seed germination and root elongation, whereas WRKY40 has a negative effect on ABA sensitivity (Chen et al. 2010). Previously, these factors have been demonstrated to form hetero- and homocomplexes and to be involved in defense response (Xu et al. 2006). WRKY18 and WRKY40 expression is rapidly induced by ABA, and it has been proposed that the two WRKY proteins regulate the expression of WRKY60, whose ABA induction is delayed compared with WRKY18 and WRKY40. The molecular basis of WRKY40 function in ABA response has been reported (Shang et al. 2010). It binds the W-box in the promoters of major regulatory genes of ABA and stress responses, such as *ABI4*, *ABF4*, *MYB2*, *DREB1A,* and *DREB2A*, and represses their expression. This repression is relieved in the presence of ABA, which suppresses WRKY40 expression and, at the same time, recruits WRKY40 from nucleus to cytosol to promote its binding to the C-terminus of the chloroplast-localized ABA receptor, ABAR (Shen et al. 2006). An elaborate model of cooperative and antagonistic interactions between the three WRKY proteins has been proposed recently (Liu et al. 2012).

11.9 NAC Proteins

NAC family TFs play a variety of roles in plant development and biotic/abiotic stress responses, including hormone signaling (Nakashima et al. 2012; Nuruzzaman et al. 2013). The drought-inducible NAC-domain protein RD26/ANAC72 is inducible by ABA, and its over- or underexpression enhances or represses, respectively, ABA sensitivity, indicating its positive regulatory role in ABA response (Fujita et al. 2004). RD26/NAC72 has been isolated as one of the *ERD1* promoter-binding proteins together with two other NAC TFs, ANAC19 and ANAC55 (Tran et al. 2004). These two TFs interact with RHA2, a RING-type E3 ligase that positively regulates ABA response during germination and early seedling growth (Jiang et al. 2009). Another NAC TF ATAF1 regulates ABA biosynthesis by binding to the promoter region of NCED3 to up-regulate its expression (Jensen et al. 2013). Expression of the NAC TF VNI2 is ABA-inducible, and it regulates ABA- and/or stress-responsive genes. It has been suggested that VNI2 integrates ABA signal into leaf senescence (Yang et al. 2011). Another NAC protein ANAC96 modulates ABA response by cooperatively interacting with ABFs. Specifically, it interacts with ABF2 and ABF4 and synergistically activates ABA-responsive genes (Xu et al. 2013).

11.10 MYB/MYC Proteins

A number of MYB class TFs (Yanhui et al. 2006) have been reported that are involved in ABA response. AtMYB2, which binds the ABA/dehydratin-responsive promoter of *RD22*, is inducible by ABA and various abiotic stresses (Abe et al.

1997). Overexpression of *AtMYB2* up-regulates ABA-responsive genes and confers ABA hypersensitivity during germination and enhanced drought tolerance, indicating that it is a positive regulator of ABA response (Abe et al. 2003). Likewise, *AtMYB15* is ABA/stress-inducible, and its overexpression results in ABA hypersensitivity and enhanced stress tolerance (Ding et al. 2009). Transgenic plants overexpressing another ABA-inducible MYB TF gene *AtMYB44* are also ABA hypersensitive and drought tolerant, whereas its knockout mutants display opposite phenotypes (Jung et al. 2008). Interestingly, salt-induced expression of PP2C genes (i.e., *ABI1*, *ABI2*, *HAB1*, and *HAB2*), which are negative regulators of ABA signaling, is repressed in the *AtMYB44* overexpression lines. Similar observations were also made with *AtMYB20*: Its overexpression enhances salt tolerance and represses the expression of *ABI1* and *ABI2*(Cui et al. 2013). AtMYB20 directly binds the MYB recognition sites in the *ABI1* promoter. Seo et al. demonstrated that *AtMYB96*, which is ABA- and drought-inducible, is involved not only in ABA/drought response but also in the regulation of lateral root development under drought condition (Seo et al. 2009). *AtMYB52* is known to regulate secondary cell wall synthesis, but a recent study showed that it also plays a positive regulatory role in ABA and stress responses (Park et al. 2011). On the other hand, MYC2 binds to the MYC recognition site in the *RD22* promoter mentioned above and positively regulates ABA response and drought tolerance together with AtMYC2 (Abe et al. 2003).

11.11 Homeodomain Proteins

Several homeodomain proteins, especially those belonging to the HD-Zip I subfamily (Mukherjee et al. 2009), are known to mediate ABA response. ATHB6 has been isolated as an ABI1-interacting protein and shown to negatively regulate ABA response in germination and stomatal closure (Himmelbach et al. 2002). It has been further demonstrated to be a target of CUL3-based ubiquitin E3 ligase (Lechner et al. 2011). On the other hand, two highly homologous and ABA/drought-inducible homeodomain proteins, ATHB7 and ATHB12, are positive regulators of ABA response during postgermination growth (Olsson et al. 2004). Recently, ATHB12 has been demonstrated to regulate plant growth by suppressing the *GA20* oxidase gene expression (Son et al. 2010). *ATHB5* is also known as a positive regulator of ABA response (Johannesson et al. 2003). Its expression is ABA-inducible, and the ABA inducibility is compromised in *abi3* and *abi5* mutants. Although its knockout mutant does not exhibit noticeable phenotypes, its overexpression lines are ABA hypersensitive. *ATHB20* has been isolated based on gene expression profiling to identify genes involved in seed dormancy (Barrero et al. 2010). A T-DNA insertion mutant of ATHB2 is more dormant than wild-type plants, whereas the germination of its overexpression lines is partially insensitive to ABA. *ATHB17*, another HD-Zip I subfamily homebox gene, positively regulates ABA response during seedling establishment stage and under water-deficit condition (Park et al. 2013).

11.12 Zinc Finger Proteins

C2H2 Zn finger proteins constitute one of the largest transcription factor families in plants (Ciftci-Yilmaz and Mittler 2008), and several of them negatively regulate ABA response. *AZF1*, *AZF2*, and *STZ* are inducible by abiotic stresses and ABA, and their encoded proteins possess transcriptional repression activity (Sakamoto et al. 2004). Transgenic analyses indicate that they suppress ABA/stress-repressive gene expression and ABA signaling (Drechsel et al. 2010; Kodaira et al. 2011). ABA-repressive genes are down-regulated in their overexpression lines, whereas their knockout lines are ABA hypersensitive. Also, auxin-regulated SAUR family genes are repressed in the *AZF1* and *AZF2* overexpression lines. Another C2H2 protein gene *SAZ* is repressed by abiotic stresses, SA and ABA (Jiang et al. 2008). The analysis of a *SAZ* knockout mutant showed that *RD29B* and *RAB18* are up-regulated in the mutant. Other ABA/stress-responsive genes were not affected, suggesting that only a subset of ABA/stress-responsive genes is affected by *SAZ*.

Members of other subfamily of zinc finger proteins are also involved in ABA response. MARD1, for instance, which possess a unique pattern of zinc finger in its C-terminus, regulates seed dormancy (He and Gan 2004). An enhancer trap mutant, in which *MARD1* expression is abolished, is less dormant than wild-type plants, and its germination is ABA-insensitive. Several CCCH finger proteins have been reported to mediate ABA signaling, although their involvement in transcription has not been demonstrated. AtTZF1, which is a member of TZF (tandem zinc finger) subfamily zinc finger proteins containing two CCCH motifs in tandem, is a positive regulator of ABA response (Lin et al. 2011). Its overexpression results in pleiotropic phenotypes, including ABA and sugar hypersensitivity during germination, enhanced drought cold tolerance, dwarfism, and late flowering. By contrast, its RNAi lines exhibit opposite phenotypes. Similarly, other TZF family members, TZF4, TZF5, and TZF6, are seed-specific and ABA-inducible, and they positively regulate ABA-dependent inhibition of seed germination. The expression of the TZFs is repressed by GA, and, in turn, they negatively regulate GA response (Bogamuwa and Jang 2013). On the other hand, AtTZF2 and AtTZF3 enhance ABA inhibition of seedling growth and drought tolerance when overexpressed (Lee et al. 2012). A number of JA-responsive genes are up-regulated, and leaf senescence is delayed in the transgenic plants, suggesting that it is involved in JA response as well.

11.13 Conclusion and Perspectives

As summarized in this chapter, a large number of TFs that regulate ABA-responsive gene expression have been identified. Numerous TFs with similar functions have been reported in other plant species as well. In the case of ABF/AREB/ABI5 family members, they constitute the core ABA signaling pathway, PYR/PYL/RCAR-PP2C-SnRK2-ABF/AREB/ABI5. The signaling pathway(s) associated with other

TFs is not understood well. However, many of the TFs appear to affect ABA response by modulating the activities or abundance of the core signaling components. Several AP2/ERF family proteins and ABI3 physically interact with ABF/AREB/ABI5 family members and synergistically cooperate to activate ABA-responsive genes. A number of them, on the other hand, activate or repress the expression of the core signaling components, such as ABFs/AREBs/ABI5 and ABI1/2. Some TFs modulate the expression of key ABA biosynthetic genes, thereby affecting the ABA level. In many cases, TFs regulating ABA response also regulates other physiological processes, including other hormone responses (i.e., auxin, ethyelene, gibberellin, jasmonate) and defense responses (Ton et al. 2009). These TFs are likely to mediate cross talk between signaling pathways by integrating different signals.

References

Abdeen A, Schnell J, Miki B. Transcriptome analysis reveals absence of unintended effects in drought-tolerant transgenic plants overexpressing the transcription factor ABF3. BMC Genom. 2010;11:69.

Abe H, Yamaguchi-Shinozaki K, Urao T, Iwasaki T, Hosokawa D, Shinozaki K. Role of arabidopsis MYC and MYB homologs in drought- and abscisic acid-regulated gene expression. Plant Cell. 1997;9(10):1859–68.

Abe H, Urao T, Ito T, Seki M, Shinozaki K, Yamaguchi-Shinozaki K. Arabidopsis AtMYC2 (bHLH) and AtMYB2 (MYB) function as transcriptional activators in abscisic acid signaling. Plant Cell. 2003;15(1):63–78.

Baker SS, Wilhelm KS, Thomashow MF. The 5′-region of Arabidopsis thaliana cor15a has cis-acting elements that confer cold-, drought- and ABA-regulated gene expression. Plant Mol Biol. 1994;24(5):701–13.

Barrero JM, Millar AA, Griffiths J, Czechowski T, Scheible WR, Udvardi M, Reid JB, Ross JJ, Jacobsen JV, Gubler F. Gene expression profiling identifies two regulatory genes controlling dormancy and ABA sensitivity in Arabidopsis seeds. Plant J. 2010;61(4):611–22.

Bensmihen S, Rippa S, Lambert G, Jublot D, Pautot V, Granier F, Giraudat J, Parcy F. The homologous ABI5 and EEL transcription factors function antagonistically to Fine-tune gene expression during late embryogenesis. Plant Cell. 2002;14:1391–403.

Bogamuwa S, Jang JC. The Arabidopsis tandem CCCH zinc finger proteins AtTZF4, 5 and 6 are involved in light-, abscisic acid- and gibberellic acid-mediated regulation of seed germination. Plant Cell Environ. 2013;36(8):1507–19.

Busk PK, Pages M. Regulation of abscisic acid-induced transcription. Plant Mol Biol. 1998;37:425–35.

Casaretto J, Ho TH. The transcription factors HvABI5 and HvVP1 are required for the abscisic acid induction of gene expression in barley aleurone cells. Plant Cell. 2003;15(1):271–84.

Chen H, Lai ZB, Shi JW, Xiao Y, Chen ZX, Xu XP. Roles of arabidopsis WRKY18, WRKY40 and WRKY60 transcription factors in plant responses to abscisic acid and abiotic stress. BMC Plant Biol. 2010;10:281.

Chen L, Zhang L, Li D, Wang F, Yu D. WRKY8 transcription factor functions in the TMV-cg defense response by mediating both abscisic acid and ethylene signaling in Arabidopsis. Proc Natl Acad Sci USA. 2013a;110(21):E1963–71.

Chen YT, Liu HX, Stone S, Callis J. ABA and the ubiquitin E3 ligase KEEP ON GOING affect proteolysis of the *Arabidopsis thaliana* transcription factors ABF1 and ABF3. Plant J. 2013b;75(6):965–76.

Choi H, Hong J, Ha J, Kang J, Kim, SY. ABFs, a family of ABA-responsive element binding factors. J Biol Chem. 2000;21:1723–1730.

Choi HI, Park HJ, Park JH, Kim S, Im MY, Seo HH, Kim YW, Hwang I, Kim SY. Arabidopsis calcium-dependent protein kinase AtCPK32 interacts with ABF4, a transcriptional regulator of abscisic acid-responsive gene expression, and modulates its activity. Plant Physiol. 2005;139(4):1750–61.

Ciftci-Yilmaz S, Mittler R. The zinc finger network of plants. Cell Mol Life Sci. 2008;65(7-8):1150–60.

Cui MH, Yoo KS, Hyoung S, Nguyen HT, Kim YY, Kim HJ, Ok SH, Yoo SD, Shin JS. An Arabidopsis R2R3-MYB transcription factor, AtMYB20, negatively regulates type 2C serine/threonine protein phosphatases to enhance salt tolerance. FEBS Lett. 2013;587(12):1773–8.

Cutler SR, Rodriguez PL, Finkelstein RR, Abrams SR. Abscisic acid: emergence of a core signaling network. Annu Rev Plant Biol. 2010;61:651–79.

Dai M, Xue Q, McCray T, Margavage K, Chen F, Lee JH, Nezames CD, Guo L, Terzaghi W, Wan J, Deng XW, Wang H. The PP6 phosphatase regulates ABI5 phosphorylation and abscisic acid signaling in Arabidopsis. Plant Cell. 2013;25(2):517–34.

Delmas F, Sankaranarayanan S, Deb S, Widdup E, Bournonville C, Bollier N, Northey JG, McCourt P, Samuel MA. ABI3 controls embryo degreening through Mendel's I locus. Proc Natl Acad Sci U S A. 2013;110(40):E3888–94.

Ding Z, Li S, An X, Liu X, Qin H, Wang D. Transgenic expression of MYB15 confers enhanced sensitivity to abscisic acid and improved drought tolerance in *Arabidopsis thaliana*. J Gene Genom. 2009;36(1):17–29.

Drechsel G, Raab S, Hoth S. Arabidopsis zinc-finger protein 2 is a negative regulator of ABA signaling during seed germination. J Plant Physiol. 2010;167(16):1418–21.

Eulgem T, Rushton PJ, Robatzek S, Somssich IE. The WRKY superfamily of plant transcription factors. Trends Plant Sci. 2000;5(5):199–206.

Finkelstein RR. Mutations at two new Arabidopsis ABA response loci are similar to the abi3 mutations. Plant J. 1994;5(6):765–71.

Finkelstein RR, Lynch TJ. The Arabidopsis abscisic acid response gene ABI5 encodes a basic leucine zipper transcription factor. Plant Cell. 2000;12:599–609.

Fujii H, Zhu JK. Arabidopsis mutant deficient in 3 abscisic acid-activated protein kinases reveals critical roles in growth, reproduction, and stress. Proc Natl Acad Sci USA. 2009;106(20):8380–5.

Finkelstein R, Gampala SSL, Lynch TJ, Thomas TL, Rock CD. Redundant and distinct functions of the ABA response loci ABA-INSENSITIVE(ABI)5 and ABRE-BINDING FACTOR (ABF)3. Plant Mol Biol. 2005;59(2):253–67.

Fujii H, Chinnusamy V, Rodrigues A, Rubio S, Antoni R, Park SY, Cutler SR, Sheen J, Rodriguez PL, Zhu JK. In vitro reconstitution of an abscisic acid signalling pathway. Nature. 2009;462(7273):660–4.

Fujii H, Verslues PE, Zhu JK. Identification of two protein kinases required for abscisic acid regulation of seed germination, root growth, and gene expression in Arabidopsis. Plant Cell. 2007;19(2):485–94.

Fujita M, Fujita Y, Maruyama K, Seki M, Hiratsu K, Ohme-Takagi M, Tran LS, Yamaguchi-Shinozaki K, Shinozaki K. A dehydration-induced NAC protein, RD26, is involved in a novel ABA-dependent stress-signaling pathway. Plant J. 2004;39(6):863–76.

Fujita Y, Fujita M, Satoh R, Maruyama K, Parvez M, Seki M, Hiratsu K, Ohme-Takagi M, Shinozaki K, Yamaguchi-Shinozaki K. AREB1, an ABA-induced transcription factor, plays a key role in drought stress response in *Arabidopsis thaliana*. Plant Cell Physiol. 2005a;46:S174–S174.

Fujita Y, Fujita M, Satoh R, Maruyama K, Parvez MM, Seki M, Hiratsu K, Ohme-Takagi M, Shinozaki K, Yamaguchi-Shinozaki K. AREB1 is a transcription activator of novel ABRE-dependent ABA signaling that enhances drought stress tolerance in Arabidopsis. Plant Cell. 2005b;17(12):3470–88.

Fujita Y, Fujita M, Shinozaki K, Yamaguchi-Shinozaki K. ABA-mediated transcriptional regulation in response to osmotic stress in plants. J Plant Res. 2011;124(4):509–25.

Fujita Y, Nakashima K, Yoshida T, Katagiri T, Kidokoro S, Kanamori N, Umezawa T, Fujita M, Maruyama K, Ishiyama K, Kobayashi M, Nakasone S, Yamada K, Ito T, Shinozaki K, Yamaguchi-Shinozaki K. Three SnRK2 protein kinases are the main positive regulators of abscisic acid signaling in response to water stress in Arabidopsis. Plant Cell Physiol. 2009;50(12):2123–32.

Fujita Y, Yoshida T, Yamaguchi-Shinozaki K. Pivotal role of the AREB/ABF-SnRK2 pathway in ABRE-mediated transcription in response to osmotic stress in plants. Physiol Plantarum. 2013;147(1):15–27.

Furihata T, Maruyama K, Fujita Y, Umezawa T, Yoshida R, Shinozaki K, Yamaguchi-Shinozaki K. Abscisic acid-dependent multisite phosphorylation regulates the activity of a transcription activator AREB1. P Natl Acad Sci USA. 2006;103(6):1988–93.

Gampala SS, Finkelstein RR, Sun SS, Rock CD. ABI5 interacts with abscisic acid signaling effectors in rice protoplasts. J Biol Chem. 2002;277(3):1689–94.

Giraudat J, Hauge BM, Valon C, Smalle J, Parcy F, Goodman HM. Isolation of the Arabidopsis ABI3 gene by positional cloning. Plant Cell. 1992;4(10):1251–61.

Gomez-Porras JL, Riano-Pachon DM, Dreyer I, Mayer JE, Mueller-Roeber B. Genome-wide analysis of ABA-responsive elements ABRE and CE3 reveals divergent patterns in Arabidopsis and rice. BMC Genom. 2007;8:260.

Guiltinan MJ, Marcotte WR Jr, Quatrano RS. A plant leucine zipper protein that recognizes an abscisic acid response element. Science. 1990;250(4978):267–71.

He Y, Gan S. A novel zinc-finger protein with a proline-rich domain mediates ABA-regulated seed dormancy in Arabidopsis. Plant Mol Biol. 2004;54(1):1–9.

Himmelbach A, Hoffmann T, Leube M, Hohener B, Grill E. Homeodomain protein ATHB6 is a target of the protein phosphatase ABI1 and regulates hormone responses in Arabidopsis. EMBO J. 2002;21(12):3029–38.

Hobo T, Kowyama Y, Hattori T. A bZIP factor, TRAB1, interacts with VP1 and mediates abscisic acid-induced transcription. Proc Natl Acad Sci USA. 1999;96(26):15348–53.

Holdsworth MJ, Bentsink L, Soppe WJ. Molecular networks regulating Arabidopsis seed maturation, after-ripening, dormancy and germination. New Phytol. 2008;179(1):33–54.

Hoth S, Morgante M, Sanchez JP, Hanafey MK, Tingey SV, Chua N-H. Genome-wide gene expression profiling in *Arabidopsis thaliana* reveals new targets of abscisic acid and largely impaired gene regulation in the abi1-1 mutant. J Cell Sci. 2002;115:4891–900.

Jensen MK, Lindemose S, de Masi F, Reimer JJ, Nielsen M, Perera V, Workman CT, Turck F, Grant MR, Mundy J, Petersen M, Skriver K. ATAF1 transcription factor directly regulates abscisic acid biosynthetic gene NCED3 in *Arabidopsis thaliana*. FEBS open bio. 2013;3:321–7.

Jiang CJ, Aono M, Tamaoki M, Maeda S, Sugano S, Mori M, Takatsuji H. SAZ, a new SUPERMAN-like protein, negatively regulates a subset of ABA-responsive genes in Arabidopsis. Mol Genet Genom. 2008;279(2):183–92.

Jiang H, Li H, Bu Q, Li C. The RHA2a-interacting proteins ANAC019 and ANAC055 may play a dual role in regulating ABA response and jasmonate response. Plant Signal Behav. 2009;4(5):464–6.

Jiang WB, Yu DQ. Arabidopsis WRKY2 transcription factor mediates seed germination and post-germination arrest of development by abscisic acid. BMC Plant Biol. 2009;9:96.

Johannesson H, Wang Y, Hanson J, Engstrom P. The *Arabidopsis thaliana* homeobox gene ATHB5 is a potential regulator of abscisic acid responsiveness in developing seedlings. Plant Mol Biol. 2003;51(5):719–29.

Johnson RR, Wagner RL, Verhey SD, Walker-Simmons MK. The abscisic acid-responsive kinase PKABA1 interacts with a seed-specific abscisic acid response element-binding factor, TaABF, and phosphorylates TaABF peptide sequences. Plant Physiol. 2002;130:837–46.

Jung C, Seo JS, Han SW, Koo YJ, Kim CH, Song SI, Nahm BH, Choi YD, Cheong JJ. Overexpression of AtMYB44 enhances stomatal closure to confer abiotic stress tolerance in transgenic Arabidopsis. Plant Physiol. 2008;146(2):623–35.

Kang J, Choi H, Im M, Kim SY. Arabidopsis basic leucine zipper proteins that mediate stress-responsive abscisic acid signaling. Plant Cell. 2002;14:343–57.

Kim SY. The role of ABF family bZIP class transcription factors in stress response. Physiol Plantarum. 2006;126(4):519–27.

Kim SY, Thomas TL. A family of novel basic leucine zipper proteins bind to seed-specification elements in the carrot Dc3 gene promoter. J Plant Physiol. 1998;152:607–13.

Kim SY, Chung HJ, Thomas TL. Isolation of a novel class of bZIP transcription factors that interact with ABA-responsive and embryo-specification elements in the Dc3 promoter using a modified yeast one-hybrid system. Plant J. 1997;11:1237–51.

Kim SY, Ma J, Perret P, Li Z, Thomas TL. Arabidopsis ABI5 subfamily members have distinct DNA-binding and transcriptional activities. Plant Physiol. 2002;130:688–97.

Kim S, J-y Kang, Cho D-I, Park JH, Kim SY. ABF2, an ABRE-binding bZIP factor, is an essential component of glucose signaling and its overexpression affects multiple stress tolerance. Plant J. 2004;40(1):75–87.

Kizis D, Pages M. Maize DRE-binding proteins DBF1 and DBF2 are involved in rab17 regulation through the drought-responsive element in an ABA-dependent pathway. Plant J. 2002;30(6):679–89.

Kodaira KS, Qin F, Tran LSP, Maruyama K, Kidokoro S, Fujita Y, Shinozaki K, Yamaguchi-Shinozaki K. Arabidopsis Cys2/His2 zinc-finger proteins AZF1 and AZF2 negatively regulate abscisic acid-repressive and auxin-inducible genes under abiotic stress conditions. Plant Physiol. 2011;157(2):742–56.

Lam E, Chua NH. Tetramer of a 21-base pair synthetic element confers seed expression and transcriptional enhancement in response to water stress and abscisic acid. J Biol Chem. 1991;266(26):17131–5.

Lechner E, Leonhardt N, Eisler H, Parmentier Y, Alioua M, Jacquet H, Leung J, Genschik P. MATH/BTB CRL3 receptors target the homeodomain-leucine zipper ATHB6 to modulate abscisic acid signaling. Dev Cell. 2011;21(6):1116–28.

Lee SJ, Cho DI, Kang JY, Kim SY. An ARIA-interacting AP2 domain protein is a novel component of ABA signaling. Mol Cells. 2009;27(4):409–16.

Lee JH, Yoon HJ, Terzaghi W, Martinez C, Dai M, Li J, Byun MO, Deng XW. DWA1 and DWA2, two Arabidopsis DWD protein components of CUL4-based E3 ligases, act together as negative regulators in ABA signal transduction. Plant Cell. 2010a;22(6):1716–32.

Lee SJ, Kang J-y, Park H-J, Kim MD, Bae MS, Choi H-i, Kim SY. DREB2C interacts with ABF2, a bZIP protein regulating abscisic acid-responsive gene expression, and its overexpression affects abscisic acid sensitivity. Plant Physiol. 2010b;153(2):716–27.

Lee SJ, Park JH, Lee MH, Yu JH, Kim SY. Isolation and functional characterization of CE1 binding proteins. BMC Plant Biol. 2010c;10:277.

Lee SJ, Jung HJ, Kang H, Kim SY. Arabidopsis zinc finger proteins AtC3H49/AtTZF3 and AtC3H20/AtTZF2 are involved in ABA and JA responses. Plant Cell Physiol. 2012;53(4):673–86.

Leon P, Gregorio J, Cordoba E. ABI4 and its role in chloroplast retrograde communication. Front Plant Sci. 2012;3:304.

Lim S, Park J, Lee N, Jeong J, Toh S, Watanabe A, Kim J, Kang H, Kim DH, Kawakami N, Choi G. ABA-INSENSITIVE3, ABA-INSENSITIVE5, and DELLAs interact to activate the expression of SOMNUS and other high-temperature-inducible genes in imbibed seeds in Arabidopsis. Plant Cell. 2013;25(12):4863–78.

Lin PC, Pomeranz MC, Jikumaru Y, Kang SG, Hah C, Fujioka S, Kamiya Y, Jang JC. The Arabidopsis tandem zinc finger protein AtTZF1 affects ABA- and GA-mediated growth, stress and gene expression responses. Plant J. 2011;65(2):253–68.

Liu H, Stone SL. Abscisic acid increases Arabidopsis ABI5 transcription factor levels by promoting KEG E3 ligase self-ubiquitination and proteasomal degradation. Plant Cell. 2010;22(8):2630–41.

Liu H, Stone SL. Cytoplasmic degradation of the Arabidopsis transcription factor abscisic acid insensitive 5 is mediated by the RING-type E3 ligase KEEP ON GOING. J Biol Chem. 2013;288(28):20267–79.

Liu ZQ, Yan L, Wu Z, Mei C, Lu K, Yu YT, Liang S, Zhang XF, Wang XF, Zhang DP. Cooperation of three WRKY-domain transcription factors WRKY18, WRKY40, and WRKY60 in repressing two ABA-responsive genes ABI4 and ABI5 in Arabidopsis. J Exp Bot. 2012;63(18):6371–92.

Lopez-Molina L, Chua N-H. A null mutation in a bZIP factor confers ABA-insensitivity in *Arabidopsis thaliana*. Plant Cell Physiol. 2000;41:541–7.

Lopez-Molina L, Mongrand S, Chua NH. A postgermination developmental arrest checkpoint is mediated by abscisic acid and requires the ABI5 transcription factor in Arabidopsis. Proc Natl Acad Sci USA. 2001;98(8):4782–7.

Lopez-Molina L, Mongrand B, McLachlin DT, Chait BT, Chua NH. ABI5 acts downstream of ABI3 to execute an ABA-dependent growth arrest during germination. Plant J. 2002;32(3):317–28.

Lopez-Molina L, Mongrand S, Kinoshita N, Chua NH. AFP is a novel negative regulator of ABA signaling that promotes ABI5 protein degradation. Gene Dev. 2003;17(3):410–8.

Marcotte WR Jr, Russell SH, Quatrano RS. Abscisic acid-responsive sequences from the em gene of wheat. Plant Cell. 1989;1(10):969–76.

Matsui A, Ishida J, Morosawa T, Mochizuki Y, Kaminuma E, Endo TA, Okamoto M, Nambara E, Nakajima M, Kawashima M, Satou M, Kim JM, Kobayashi N, Toyoda T, Shinozaki K, Seki M. Arabidopsis transcriptome analysis under drought, cold, high-salinity and ABA treatment conditions using a tiling array. Plant Cell Physiol. 2008;49(8):1135–49.

McCarty DR, Hattori T, Carson CB, Vasil V, Lazar M, Vasil IK. The viviparous-1 developmental gene of maize encodes a novel transcriptional activator. Cell. 1991;66(5):895–905.

Menkens AE, Schindler U, Cashmore AR. The G-box: a ubiquitous regulatory DNA element in plants bound by the GBF family of bZIP proteins. Trend Biochem Sci. 1995;20:506–12.

Miura K, Lee J, Jin JB, Yoo CY, Miura T, Hasegawa PM. Sumoylation of ABI5 by the Arabidopsis SUMO E3 ligase SIZ1 negatively regulates abscisic acid signaling. Proc Natl Acad Sci USA. 2009;106(13):5418–23.

Monke G, Altschmied L, Tewes A, Reidt W, Mock HP, Baumlein H, Conrad U. Seed-specific transcription factors ABI3 and FUS3: molecular interaction with DNA. Planta. 2004;219(1):158–66.

Monke G, Seifert M, Keilwagen J, Mohr M, Grosse I, Hahnel U, Junker A, Weisshaar B, Conrad U, Baumlein H, Altschmied L. Toward the identification and regulation of the *Arabidopsis thaliana* ABI3 regulon. Nucleic Acids Res. 2012;40(17):8240–54.

Mukherjee K, Brocchieri L, Burglin TR. A comprehensive classification and evolutionary analysis of plant homeobox genes. Mol Biol Evol. 2009;26(12):2775–94.

Mundy J, Yamaguchi-Shinozaki K, Chua NH. Nuclear proteins bind conserved elements in the abscisic acid-responsive promoter of a rice rab gene. Proc Natl Acad Sci USA. 1990;87(4):1406–10.

Nakashima K, Fujita Y, Katsura K, Maruyama K, Narusaka Y, Seki M, Shinozaki K, Yamaguchi-Shinozaki K. Transcriptional regulation of ABI3-and ABA-responsive genes including RD29B and RD29A in seeds, germinating embryos, and seedlings of Arabidopsis. Plant Mol Biol. 2006;60(1):51–68.

Nakashima K, Fujita Y, Kanamori N, Katagiri T, Umezawa T, Kidokoro S, Maruyama K, Yoshida T, Ishiyama K, Kobayashi M, Shinozaki K, Yamaguchi-Shinozaki K. Three Arabidopsis SnRK2 protein kinases, SRK2D/SnRK2.2, SRK2E/SnRK2.6/OST1 and SRK2I/SnRK2.3, involved in ABA signaling are essential for the control of seed development and dormancy. Plant Cell Physiol. 2009a;50(7):1345–63.

Nakashima K, Ito Y, Yamaguchi-Shinozaki K. Transcriptional regulatory networks in response to abiotic stresses in Arabidopsis and grasses. Plant Physiol. 2009b;149(1):88–95.

Nakashima K, Takasaki H, Mizoi J, Shinozaki K, Yamaguchi-Shinozaki K. NAC transcription factors in plant abiotic stress responses. Biochim Biophys Acta. 2012;1819(2):97–103.

Narusaka Y, Nakashima K, Shinwari ZK, Sakuma Y, Furihata T, Abe H, Narusaka M, Shinozaki K, Yamaguchi-Shinozaki K. Interaction between two cis-acting elements, ABRE and DRE, in ABA-dependent expression of Arabidopsis rd29A gene in response to dehydration and high-salinity stresses. Plant J. 2003;34(2):137–48.

Nemhauser JL, Hong F, Chory J. Different plant hormones regulate similar processes through largely nonoverlapping transcriptional responses. Cell. 2006;126(3):467–75.

Niu X, Helentjaris T, Bate NJ. Maize ABI4 binds coupling element1 in abscisic acid and sugar response genes. Plant Cell. 2002;14(10):2565–75.

Nuruzzaman M, Sharoni AM, Kikuchi S. Roles of NAC transcription factors in the regulation of biotic and abiotic stress responses in plants. Front Microbiol. 2013;4:248.

Olsson AS, Engstrom P, Soderman E. The homeobox genes ATHB12 and ATHB7 encode potential regulators of growth in response to water deficit in Arabidopsis. Plant Mol Biol. 2004;55(5):663–77.

Ono A, Izawa T, Chua NH, Shimamoto K. The rab16B promoter of rice contains two distinct abscisic acid-responsive elements. Plant Physiol. 1996;112(2):483–91.

Pandey GK, Grant JJ, Cheong YH, Kim BG, Li L, Luan S. ABR1, an APETALA2-domain transcription factor that functions as a repressor of ABA response in Arabidopsis. Plant Physiol. 2005;139(3):1185–93.

Park MY, Kang JY, Kim SY. Overexpression of AtMYB52 confers ABA hypersensitivity and drought tolerance. Mol Cells. 2011;31(5):447–54.

Park MY, Kim SA, Lee SJ, Kim SY. ATHB17 is a positive regulator of abscisic acid response during early seedling growth. Mol Cells. 2013;35(2):125–33.

Reeves WM, Lynch TJ, Mobin R, Finkelstein RR. Direct targets of the transcription factors ABA-Insensitive (ABI)4 and ABI5 reveal synergistic action by ABI4 and several bZIP ABA response factors. Plant Mol Biol. 2011;75(4–5):347–63.

Ren XZ, Chen ZZ, Liu Y, Zhang HR, Zhang M, Liu QA, Hong XH, Zhu JK, Gong ZZ. ABO3, a WRKY transcription factor, mediates plant responses to abscisic acid and drought tolerance in Arabidopsis. Plant J. 2010;63(3):417–29.

Rushton DL, Tripathi P, Rabara RC, Lin J, Ringler P, Boken AK, Langum TJ, Smidt L, Boomsma DD, Emme NJ, Chen X, Finer JJ, Shen QJ, Rushton PJ. WRKY transcription factors: key components in abscisic acid signalling. Plant Biotechnol J. 2012;10(1):2–11.

Sakamoto H, Maruyama K, Sakuma Y, Meshi T, Iwabuchi M, Shinozaki K, Yamaguchi-Shinozaki K. Arabidopsis Cys2/His2-type zinc-finger proteins function as transcription repressors under drought, cold, and high-salinity stress conditions. Plant Physiol. 2004;136(1):2734–46.

Sakuma Y, Liu Q, Dubouzet JG, Abe H, Shinozaki K, Yamaguchi-Shinozaki K. DNA-binding specificity of the ERF/AP2 domain of Arabidopsis DREBs, transcription factors involved in dehydration- and cold-inducible gene expression. Biochem Biophys Res Commun. 2002;290(3):998–1009.

Seki M, Ishida J, Narusaka M, Fujita M, Nanjo T, Umezawa T, Kamiya A, Nakajima M, Enju A, Sakurai T, Satou M, Akiyama K, Yamaguchi-Shinozaki K, Carninci P, Kawai J, Hayashizaki Y, Shinozaki K. Monitoring the expression pattern of around 7,000 Arabidopsis genes under ABA treatments using a full-length cDNA microarray. Funct Integr Genomics. 2002;2(6):282–91.

Seo PJ, Xiang F, Qiao M, Park JY, Lee YN, Kim SG, Lee YH, Park WJ, Park CM. The MYB96 transcription factor mediates abscisic acid signaling during drought stress response in Arabidopsis. Plant Physiol. 2009;151(1):275–89.

Shang Y, Yan L, Liu ZQ, Cao Z, Mei C, Xin Q, Wu FQ, Wang XF, Du SY, Jiang T, Zhang XF, Zhao R, Sun HL, Liu R, Yu YT, Zhang DP. The Mg-chelatase H subunit of Arabidopsis antagonizes a group of WRKY transcription repressors to relieve ABA-responsive genes of inhibition. Plant Cell. 2010;22(6):1909–35.

Shen Q, Ho TH. Functional dissection of an abscisic acid (ABA)-inducible gene reveals two independent ABA-responsive complexes each containing a G-box and a novel cis-acting element. Plant Cell. 1995;7(3):295–307.

Shen Q, Zhang P, Ho TH. Modular nature of abscisic acid (ABA) response complexes: composite promoter units that are necessary and sufficient for ABA induction of gene expression in barley. Plant Cell. 1996;8(7):1107–19.

Shen QJ, Casaretto JA, Zhang P, Ho TH. Functional definition of ABA-response complexes: the promoter units necessary and sufficient for ABA induction of gene expression in barley (*Hordeum vulgare* L.). Plant Mol Biol. 2004;54(1):111–24.

Shen YY, Wang XF, Wu FQ, Du SY, Cao Z, Shang Y, Wang XL, Peng CC, Yu XC, Zhu SY, Fan RC, Xu YH, Zhang DP. The Mg-chelatase H subunit is an abscisic acid receptor. Nature. 2006;443(7113):823–6.

Shu K, Zhang H, Wang S, Chen M, Wu Y, Tang S, Liu C, Feng Y, Cao X, Xie Q. ABI4 regulates primary seed dormancy by regulating the biogenesis of abscisic acid and gibberellins in Arabidopsis. PLoS Genet. 2013;9(6):e1003577.

Sirichandra C, Davanture M, Turk BE, Zivy M, Valot B, Leung J, Merlot S. The Arabidopsis ABA-activated kinase OST1 Phosphorylates the bZIP transcription factor ABF3 and creates a 14-3-3 binding site involved in its turnover. Plos One. 2010;5(11):e13935.

Son O, Hur YS, Kim YK, Lee HJ, Kim S, Kim MR, Nam KH, Lee MS, Kim BY, Park J, Park J, Lee SC, Hanada A, Yamaguchi S, Lee IJ, Kim SK, Yun DJ, Soderman E, Cheon CI. ATHB12, an ABA-inducible homeodomain-leucine zipper (HD-Zip) protein of Arabidopsis, negatively regulates the growth of the inflorescence stem by decreasing the expression of a gibberellin 20-oxidase gene. Plant Cell Physiol. 2010;51(9):1537–47.

Song CP, Agarwal M, Ohta M, Guo Y, Halfter U, Wang P, Zhu JK. Role of an Arabidopsis AP2/EREBP-type transcriptional repressor in abscisic acid and drought stress responses. Plant Cell. 2005;17(8):2384–96.

Stone SL, Williams LA, Farmer LM, Vierstra RD, Callis J. KEEP ON GOING, a RING E3 ligase essential for Arabidopsis growth and development, is involved in abscisic acid signaling. Plant Cell. 2006;18(12):3415–28.

Suzuki M, Kao CY, McCarty DR. The conserved B3 domain of VIVIPAROUS1 has a cooperative DNA binding activity. Plant Cell. 1997;9(5):799–807.

Suzuki M, McCarty DR. Functional symmetry of the B3 network controlling seed development. Curr Opin Plant Biol. 2008;11(5):548–53.

Swaminathan K, Peterson K, Jack T. The plant B3 superfamily. Trends Plant Sci. 2008;13(12):647–55.

Takahashi S, Seki M, Ishida J, Satou M, Sakurai T, Narusaka M, Kamiya A, Nakajima M, Enju A, Akiyama K, Yamaguchi-Shinozaki K, Shinozaki K. Monitoring the expression profiles of genes induced by hyperosmotic, high salinity, and oxidative stress and abscisic acid treatment in Arabidopsis cell culture using a full-length cDNA microarray. Plant Mol Biol. 2004;56(1):29–55.

Ton J, Flors V, Mauch-Mani B. The multifaceted role of ABA in disease resistance. Trends Plant Sci. 2009;14(6):310–7.

Tran LS, Nakashima K, Sakuma Y, Simpson SD, Fujita Y, Maruyama K, Fujita M, Seki M, Shinozaki K, Yamaguchi-Shinozaki K. Isolation and functional analysis of Arabidopsis stress-inducible NAC transcription factors that bind to a drought-responsive cis-element in the early responsive to dehydration stress 1 promoter. Plant Cell. 2004;16(9):2481–98.

Uno Y, Furihata T, Abe H, Yoshida R, Shinozaki K, Yamaguchi-Shinozaki K. Arabidopsis basic leucine zipper transcription factors involved in an abscisic acid-dependent signal transduction pathway under drought and high-salinity. Proc Natl Acad Sci USA. 2000;97:11632–7.

Vierstra RD. The ubiquitin-26S proteasome system at the nexus of plant biology. Nat Rev Mol Cell Biol. 2009;10(6):385–97.

Wind JJ, Peviani A, Snel B, Hanson J, Smeekens SC. ABI4: versatile activator and repressor. Trends Plant Sci. 2013;18(3):125–32.

Xu X, Chen C, Fan B, Chen Z. Physical and functional interactions between pathogen-induced Arabidopsis WRKY18, WRKY40, and WRKY60 transcription factors. Plant Cell. 2006;18(5):1310–26.

Xu ZY, Kim SY, Hyeon do Y, Kim DH, Dong T, Park Y, Jin JB, Joo SH, Kim SK, Hong JC, Hwang D, Hwang I. The Arabidopsis NAC transcription factor ANAC096 cooperates with bZIP-type transcription factors in dehydration and osmotic stress responses. Plant Cell. 2013;25(11):4708–24.

Yamaguchi-Shinozaki K, Shinozaki K. A novel cis-acting element in an Arabidopsis gene is involved responsiveness to drought, low-temperature, or high-salinity stress. Plant Cell. 1994;6:251–64.

Yang Z, Tian L, Latoszek-Green M, Brown D, Wu K. Arabidopsis ERF4 is a transcriptional repressor capable of modulating ethylene and abscisic acid responses. Plant Mol Biol. 2005;58(4):585–96.

Yang SD, Seo PJ, Yoon HK, Park CM. The Arabidopsis NAC transcription factor VNI2 integrates abscisic acid signals into leaf senescence via the COR/RD genes. Plant Cell. 2011;23(6):2155–68.

Yanhui C, Xiaoyuan Y, Kun H, Meihua L, Jigang L, Zhaofeng G, Zhiqiang L, Yunfei Z, Xiaoxiao W, Xiaoming Q, Yunping S, Li Z, Xiaohui D, Jingchu L, Xing-Wang D, Zhangliang C, Hongya G, Li-Jia Q. The MYB transcription factor superfamily of Arabidopsis: expression analysis and phylogenetic comparison with the rice MYB family. Plant Mol Biol. 2006;60(1):107–24.

Yano R, Kanno Y, Jikumaru Y, Nakabayashi K, Kamiya Y, Nambara E. CHOTTO1, a putative double APETALA2 repeat transcription factor, is involved in abscisic acid-mediated repression of gibberellin biosynthesis during seed germination in Arabidopsis. Plant Physiol. 2009;151(2):641–54.

Yazaki J, Kikuchi S. The genomic view of genes responsive to the antagonistic phytohormones, abscisic acid and gibberellin. Vitam Horm. 2005;72:1–30.

Yoshida T, Fujita Y, Sayama H, Kidokoro S, Maruyama K, Mizoi J, Shinozaki K, Yamaguchi-Shinozaki K. AREB1, AREB2, and ABF3 are master transcription factors that cooperatively regulate ABRE-dependent ABA signaling involved in drought stress tolerance and require ABA for full activation. Plant J. 2010;61(4):672–85.

Zeller G, Henz SR, Widmer CK, Sachsenberg T, Ratsch G, Weigel D, Laubinger S. Stress-induced changes in the Arabidopsis thaliana transcriptome analyzed using whole-genome tiling arrays. Plant J. 2009;58(6):1068–82.

Zhang W, Ruan J, Ho TH, You Y, Yu T, Quatrano RS. Cis-regulatory element based targeted gene finding: genome-wide identification of abscisic acid- and abiotic stress-responsive genes in Arabidopsis thaliana. Bioinformatics. 2005a;21(14):3074–81.

Zhang X, Garreton V, Chua NH. The AIP2 E3 ligase acts as a novel negative regulator of ABA signaling by promoting ABI3 degradation. Genes Dev. 2005b;19(13):1532–43.

Zhu SY, Yu XC, Wang XJ, Zhao R, Li Y, Fan RC, Shang Y, Du SY, Wang XF, Wu FQ, Xu YH, Zhang XY, Zhang DP. Two calcium-dependent protein kinases, CPK4 and CPK11, regulate abscisic acid signal transduction in Arabidopsis. Plant Cell. 2007;19(10):3019–36.

Zhu Q, Zhang J, Gao X, Tong J, Xiao L, Li W, Zhang H. The Arabidopsis AP2/ERF transcription factor RAP2.6 participates in ABA, salt and osmotic stress responses. Gene. 2010;457(1–2):1–12.

Chapter 12
Cross Talk of Signaling Pathways Between ABA and Other Phytohormones

Jiaqiang Sun and Chuanyou Li

Abstract The phytohormone abscisic acid (ABA) regulates many aspects of plant development and adaptation to environmental stresses. Molecular genetics studies have significantly advanced our understanding on the molecular basis of ABA signaling in seeds and seedlings. In this book chapter, we review recent advances on the cross talk of signaling pathways between ABA and other phytohormones.

Keywords ABA · Phytohormones · Cross talk

12.1 Introduction

The phytohormone abscisic acid (ABA) is well known for its regulatory roles in integrating environmental constraints with the developmental programs of plants. ABA promotes seed dormancy, inhibition of germination and lateral root (LR) formation, and reduction of water transpiration through stomatal pores (Hauser et al. 2011). During the last decade, tremendous progresses have been made in identifying mechanisms and genes involved in ABA metabolism, transport, and signal transduction. Here, we aim to review the interaction networks between ABA and other hormone-signaling pathways.

C. Li (✉)
State Key Laboratory of Plant Genomics, National Centre for Plant Gene Research, Institute of Genetics and Developmental Biology, Chinese Academy of Sciences, No. 1 West Beichen Road, Beijing 100101, China
e-mail: cyli@genetics.ac.cn

J. Sun
National Key Facility for Crop Gene Resources and Genetic Improvement, Institute of Crop Science, Chinese Academy of Agricultural Sciences, No. 12 South Zhongguancun Street, Haidian District, Beijing 100081, China

D.-P. Zhang (ed.), *Abscisic Acid: Metabolism, Transport and Signaling*,
DOI 10.1007/978-94-017-9424-4_12

12.2 Cross Talk Between ABA and Auxin

Auxin and ABA modulate many aspects of plant development together, mostly in opposite directions, suggesting that active cross talk occurs between the signaling pathways of the two hormones. For example, several components of auxin or ABA signaling have been identified as the interaction nodes between ABA- and auxin-signaling pathways. Wang et al. (2011a) recently found that the expression of *AUXIN RESPONSE FACTOR 2* (*ARF2*) could be induced by ABA and the *ARF2* mutants showed enhanced ABA sensitivity in seed germination and primary root growth. In contrast, the primary root growth and seed germination of transgenic plants overexpressing *ARF2* were less inhibited by ABA than that of the wild type. Further results showed that ABA treatment altered auxin distribution in the primary root tips and made the relative auxin accumulation or auxin signal around quiescent center (QC) cells and their surrounding columella stem cells to other cells stronger in *ARF2-101* than in the wild type. Together, this study indicates that ARF2 is a novel integrative hub between ABA- and auxin-signaling pathways in regulating plant growth (Wang et al. 2011a).

ABA is known to be the sole plant hormone to maintain seed dormancy, which acts through a gene expression network involving the transcription factor abscisic acid insensitive 3 (ABI3). However, whether the plant growth hormone auxin plays a role in the regulation of seed dormancy in response to environmental and internal signals remains unclear. Liu et al. (2013) recently showed that auxin also plays a critical role in seed dormancy in *Arabidopsis*. Disruptions in auxin signaling in *MIR160*-overexpressing plants, auxin receptor mutants, or auxin biosynthesis mutants dramatically released seed dormancy, whereas increases in auxin signaling or biosynthesis greatly enhanced seed dormancy (Liu et al. 2013). Further results showed that auxin action in seed dormancy requires the ABA-signaling pathway (and vice versa), indicating that the roles of auxin and ABA in seed dormancy are interdependent (Liu et al. 2013). Furthermore, auxin acts upstream of the major regulator of seed dormancy, ABI3, by the auxin response factors ARF10 and ARF16 to control the expression of *ABI3* during seed germination. This study uncovers a novel mechanism underlying the cross talk of auxin- and ABA-signaling pathways that auxin controls seed dormancy through stimulation of ABA signaling by inducing ARF-mediated *ABI3* activation in *Arabidopsis* (Liu et al. 2013).

It is generally believed that auxin plays a dominant role in LR formation. However, ABA negatively affects LR formation. Recently, Shkolnik-Inbar and Bar-Zvi (2010) found that mutation of *ABSCISIC ACID INSENSITIVE 4* (*ABI4*), which encodes an ABA-regulated AP2 domain transcription factor, results in an increased number of LRs. They further showed that *ABI4* overexpression impairs LR development. The expression of *ABI4* in roots is enhanced by ABA and is repressed by auxin. Expression of the auxin efflux carrier protein PIN1 is reduced in *ABI4* overexpressors, whereas enhanced in *abi4* mutants. Transport levels of exogenously applied auxin were elevated in *abi4* mutants and reduced in *ABI4* overexpressors. Accordingly, the *abi4* mutant and *ABI4* overexpressor exhibit enhanced and reduced root acropetal auxin transport, respectively (Shkolnik-Inbar and Bar-Zvi 2010). This study therefore demonstrates that ABI4 integrates ABA with auxin pathways probably through the regulation of polar auxin transport.

Although auxin is a major morphogenesis hormone in plants, little is known about ABA–auxin interactions during early seedling growth. Under unfavorable environmental conditions, ABA inhibits the developmental transition from an embryo into a young seedling. Belin et al. (2009) developed a genetic screen to isolate *Arabidopsis* mutants whose early seedling development was resistant to ABA. Through this approach, they identified a recessive mutation in *AUXIN RESISTANT1* (*AUX1*), encoding an auxin influx carrier. They showed that the *aux1* and *pin2* mutants are insensitive to ABA-dependent repression of embryonic axis elongation. Genetic and physiological experiments showed that this involved auxin transport to the embryonic axis elongation zone, where ABA enhanced the activity of an auxin-responsive promoter. Together, this study suggests that ABA represses embryonic axis elongation by potentiating auxin signaling in its elongation zone (Belin et al. 2009).

On the other hand, several key components of ABA signaling have also been identified in regulating the auxin responses by the transcriptional regulation of downstream genes in *Arabidopsis*. For example, Yang et al. (2011) recently reported the functional identification of rice ABI5-like1 (ABL1), which regulates both ABA and auxin responses. *ABL1* is expressed in various tissues and is induced by both ABA and auxin. The *ABL1* deficiency mutant, *abl1*, showed reduced ABA responses, and *ABL1* expression in the *Arabidopsis abi5* mutant was able to rescue the ABA sensitivity (Yang et al. 2011). The ABL1 protein is localized to the nucleus and can directly bind ABRE (G-box) elements in vitro. The gene expression analysis confirms that the large proportion of downregulated genes of *abl1* is involved in stress responses, consistent with the transcriptional activating effects of ABL1. Most importantly, the *abl1* mutant is hypersensitive to exogenous IAA, suggesting that ABL1 modulates both ABA and auxin responses in rice (Yang et al. 2011).

Rac-like GTPases or Rho-related GTPases from plants (RAC/ROPs) are important components of hormone-signaling pathways in plants. Nibau et al. (2013) recently revealed that AtRAC7/ROP9 functions as a modulator of both auxin and ABA signaling. Plants with reduced levels of *AtRAC7/ROP9* had increased sensitivity to auxin and were less sensitive to ABA (Nibau et al. 2013). On the other hand, overexpressing *AtRAC7/ROP9* activated ABA-induced gene expression but repressed auxin-induced gene expression. In addition, both hormones regulated the activity of the *AtRAC7/ROP9* promoter, suggesting a feedback mechanism to modulate the AtRAC7/ROP9-mediated signaling (Nibau et al. 2013). These results place AtRAC7/ROP9 as an important signal transducer that integrates auxin and ABA signaling in the plant. By contrast, Choi et al. (2013) recently demonstrated that ROP-interactive CRIB motif-containing protein 1 (RIC1) is also involved in the interaction between auxin- and ABA-regulated root development. *RIC1* expression is highly induced by both hormones and expressed in the roots of young seedlings (Choi et al. 2013). Whereas auxin-responsive gene induction and the effect of auxin on root growth and LR formation were suppressed in the *ric1* knockout mutant, ABA-responsive gene induction and the effect of ABA on seed germination, root growth, and LR formation were potentiated (Choi et al. 2013). Thus, this study suggests that RIC1, which positively regulates auxin responses, but negatively regulates ABA responses, is a component of the intricate signaling network that underlies auxin and ABA cross talk (Choi et al. 2013).

12.3 Cross Talk Between ABA and Cytokinin

Cytokinin is a major phytohormone involved in various aspects of plant growth and development. The underlying mechanism remains elusive in which how ABA and cytokinin interplay to control plant development. Cytokinin is known to antagonize the effects of ABA on stomatal movement and to affect ethylene biosynthesis (Tanaka et al. 2006). Considering that ethylene has an antagonistic effect on ABA-induced stomatal closure, a possibility that the antagonistic effects of cytokinin on ABA were mediated through ethylene biosynthesis was proposed by Tanaka et al. (2006). Cytokinin, 6-benzyladenine (BA), antagonized ABA-induced stomatal closure in a manner similar to that following application of the ethylene precursor, 1-aminocyclopropane-1-carboxylic acid (ACC). However, the effects were negated when ethylene signaling, perception, or biosynthesis were blocked (Tanaka et al. 2006). These observations suggest that cytokinin inhibits ABA-induced stomatal closure through the modulation of ethylene biosynthesis (Tanaka et al. 2006).

Wang et al. (2011b) recently undertook a gain-of-function genetic screen to identify ABA-insensitive mutants during seed germination in *Arabidopsis* using an estradiol-inducible approach. In the presence of estradiol, one of these mutants *gim1* (*germination insensitive to ABA mutant 1*) exhibited an elevated level of cytokinin that was attributed to the estradiol-induced expression of *AtIPT8* that encodes an isopentenyltransferase for the biosynthesis of cytokinins (Wang et al. 2011b). Further analyses on alterations of gene transcriptome in the *gim1* mutant demonstrated that the expression of some ABA-inducible genes, including ABI5, was reduced and could not be restored by exogenous ABA treatment (Wang et al. 2011b). In summary, this study suggests that cytokinin antagonizes ABA suppression to seed germination of *Arabidopsis* by downregulating ABI5 expression (Wang et al. 2011b).

Moreover, Ding et al. (2008) showed that ABA also modulates cytokinin signaling during the regulation of nodulation in *Medicago truncatula*. Nodulation is tightly regulated in legumes to ensure appropriate levels of nitrogen fixation without excessive depletion of carbon reserves. Ding et al. showed that application of ABA inhibits nodulation, bacterial infection, and nodulin gene expression in *M. truncatula*. ABA acts in a similar manner as ethylene, regulating Nod factor signaling and affecting the nature of Nod factor-induced calcium spiking. However, this action is independent of the ethylene signal transduction pathway (Ding et al. 2008). These results showed that ABA can suppress Nod factor signal transduction in the epidermis and can regulate cytokinin induction of the nodule primordium in the root cortex. Therefore, this study suggests that ABA is capable of coordinately regulating the nodule formation and cytokinin signaling (Ding et al. 2008).

12.4 Cross Talk Between ABA and Gibberellin

The phytohormones gibberellin (GA) and ABA play essential and often antagonistic roles in regulating plant growth, development, and stress responses. For example, GA is associated with the promotion of germination, growth, and flowering. However, ABA inhibits these processes. Moreover, the antagonistic relationship of the two hormones regulates the transition from embryogenesis to seed germination (Razem et al. 2006). Several different mechanisms have been shown to underlie this antagonistic interaction in different developmental processes. During cereal seed germination, the developing embryo releases GAs to the aleurone cells where they induce the transcription of several genes encoding hydrolytic enzymes, including a-amylase. These enzymes are then secreted to the endosperm and hydrolyze starch and proteins, supplying nutrients to the developing embryo (Weiss and Ori 2007). In contrast, ABA suppresses a-amylase expression. The GA-induced, ABA-suppressed transcription of a-amylase in the aleurone layer of cereal seeds was classically used as an experimental system to study the interaction between GA and ABA. The a-amylase promoter contains a GA response element, required for both its activation by GA and suppression by ABA (Rogers and Rogers 1992). Gubler et al. (1995) identified a GA-induced Myb-like protein (GAMyb) that binds to the GA response element box of the a-amylase promoter (Gubler et al. 1995). Induction of GAMyb and a-amylase transcription was shown to be mediated by the DELLA protein SLR1. How does ABA affect this pathway? The induction of GAMyb and a-amylase by GA is suppressed by an ABA-induced Ser/Thr protein kinase, PKABA1. ABA and PKABA1 inhibited the upregulation of GAMyb and a-amylase in *slr1* mutants as well, suggesting that the inhibition of GAMyb and a-amylase by PKABA1 occurs downstream of DELLA (Gomez-Cadenas et al. 1999, 2001). However, a more recent study has shown that when PKABA1 is suppressed by RNAi, ABA still inhibits the GA-induced a-amylase expression. This finding indicates that ABA affects this process through an additional, PKABA1-independent pathway (Zentella et al. 2002). A candidate alternative ABA-signaling pathway to suppress GA responses in rice may involve two ABA-induced WRKY transcriptional regulators (Xie et al. 2006).

A different mechanism of interaction between GA and ABA in the regulation of root growth was proposed (Achard et al. 2006). In *Arabidopsis*, GA promotes and ABA suppresses root growth, and both effects seem to be mediated by the DELLA proteins (Achard et al. 2006). ABA application increased the stability of RGA and blocked its GA-induced degradation. Moreover, the quadruple-DELLA mutant (loss of GAI, RGA, RGL1, and RGL2) is relatively resistant to the growth-inhibitory effects of ABA (Achard et al. 2006). Therefore, while during seed germination, ABA seems to act downstream of DELLA, it affects *Arabidopsis* growth via DELLA. It is thus possible that distinct mechanisms of interaction between GA and ABA are utilized for different developmental decisions.

Further, the fine-tuning of the cross talk between GA and ABA was recently demonstrated in *Arabidopsis*. Both ABA and GA can induce the expression of microRNA159 (miR159), which targets the *MYB33* mRNA. Interestingly, MYB33 promotes ABA responses in seeds and GA responses in flowers. Thus, the two antagonistic hormones exert their actions through a common mediator, MYB33, and desensitize their signaling through the same homeostatic mechanism, miR159, at different developmental stages (Achard et al. 2004; Reyes and Chua 2007).

Recently, Golldack et al. (2013) uncovered novel antagonizing roles of GA and ABA in integrating growth and development in plants with environmental signaling. GRAS transcription factors of the DELLA and SCARECROW-LIKE (SCL) types play a key role as major growth regulators and have pivotal functions in modulating GA signaling (Golldack et al. 2013). In contrast to a pivotal role of GRAS family transcriptional factors in plant growth regulation, recent work has suggested that the DELLA and SCL proteins integrate generic GA responses into ABA-controlled abiotic stress tolerance (Golldack et al. 2013).

ABA and GA are the primary factors that antagonistically regulate the transition from dormancy to germination. High ABA and low GA content in seeds promote seed dormancy. However, the underlying molecular mechanism involving cross talk between ABA and GA in this process remains to be elucidated. Shu et al. (2013) recently found that ABI4, the key transcription factor in the ABA-signaling pathway, indeed controls primary seed dormancy. This result contradicts the previous conclusion that ABI4 is not involved in the control of seed dormancy. Several lines of evidence support this conclusion. For example, detailed physiological analysis of the germination of *abi4* seeds that were harvested immediately and stored for various periods of time and subjected to various treatments showed that ABI4 negatively regulates primary seed dormancy (Shu et al. 2013). The molecular mechanism responsible for this control is that ABI4 directly or indirectly regulates the key genes of the ABA and GA biogenesis pathways, which then regulates the ABA and GA contents in seeds (Shu et al. 2013). Taken together, this study suggests that ABI4 is a key factor that regulates primary seed dormancy by mediating the regulation of GA biogenesis (Shu et al. 2013).

12.5 Cross Talk Between ABA and Ethylene

The plant hormones ethylene and ABA have antagonistic functions in the control of plant growth and development, including seed germination and early seedling development. Ethylene and ABA may control the biosynthesis, catabolism, or signaling of the other to enhance their antagonistic effects on seed germination and early seedling growth (Cheng et al. 2009). Through screening for mutations that either enhanced or suppressed the ABA-resistant seed germination phenotype of the *Arabidopsis abi1-1* mutant, alleles of the constitutive ethylene response mutant *ctr1* and ethylene-insensitive mutant *ein2* were recovered as enhancer and suppressor mutations, respectively (Beaudoin et al. 2000). Genetic analyses showed that the ethylene-signaling cascade defined by the ETR1, CTR1, and EIN2 genes inhibit ABA signaling in seeds. Epistasis analysis between ethylene- and ABA-insensitive mutations indicated that endogenous ethylene promotes seed germination by decreasing sensitivity to endogenous ABA (Beaudoin et al. 2000). In contrast to the situation in seeds, *ein2* and *etr1-1* roots were resistant to both ABA and ethylene. This study indicates that ABA and ethylene antagonistically or synergistically to regulate seed germination and root growth, respectively (Beaudoin et al. 2000). Similarly, Ghassemian et al. (2000) demonstrated that ethylene appears to be a negative regulator of ABA action during germination and positively regulates some aspects of ABA action in regulation of root growth (Ghassemian et al. 2000). For example, the *era3* mutant that is allelic to *ein2* overaccumulates ABA, suggesting that ethylene signaling negatively regulates ABA biosynthesis (Ghassemian et al. 2000). These findings indicate that interaction between ABA and ethylene is complex and possibly interconnected through multiple feedback mechanisms. In contrast, the identification of ABA-insensitive mutants unveiled the LONG HYPOCOTYL5 (HY5) transcription factor as an important molecular link between ABA and ethylene biosynthesis (Li et al. 2011). A dramatic increase in ethylene levels in the ABA-insensitive *hy5* mutant suggested its role in the repression of ethylene biosynthesis (Li et al. 2011). Taken together, these results suggest ABA and ethylene interact to inhibit each other's biosynthesis.

To examine the cross talk between the ethylene and ABA signal transduction pathways during ABA-induced stomatal closure, Tanaka et al. (2005) examined an ethylene overproducing mutant (*eto1-1*) and the two ethylene-insensitive mutants (*etr1-1* and *ein3-1*). Their results showed that stomata of wild-type plants were closed within a few minutes in response to ABA, whereas stomata of the *eto1-1* mutant showed a similar but less-sensitive ABA response. Moreover, ABA-induced stomatal closure could be inhibited by application of ethylene or the ethylene precursor ACC. In contrast, stomata of the *etr1-1* and *ein3-1* mutants were able to close in response to concomitant ABA and ACC application. This study indicates that the ethylene-signaling pathway delays ABA-induced stomatal closure (Tanaka et al. 2005).

12.6 Cross Talk Between ABA and Jasmonate

The plant hormone jasmonates (JAs) play essential roles in plant defense and development. Anderson et al. (2004) revealed a complex interplay between JA- and ABA-signaling pathways that regulates plant defense gene expression and disease resistance. For example, exogenous ABA treatment could suppress both basal and JA-activated transcription from defense genes (Anderson et al. 2004). By contrast, ABA deficiency as conditioned by the mutations in the *ABA1* and *ABA2* genes, which encode enzymes involved in ABA biosynthesis, resulted in upregulation of JA-responsive defense genes (Anderson et al. 2004). Collectively, these results indicate that the antagonistic interactions between multiple components of JA- and ABA-signaling pathways modulate defense and stress responsive gene expression in response to biotic and abiotic stresses (Anderson et al. 2004).

The basic helix-loop-helix (bHLH) transcription factor MYC2 has been identified as a master regulator of most aspects of the JA-signaling pathway in *Arabidopsis* (Boter et al. 2004; Lorenzo et al. 2004; Dombrecht et al. 2007; Chen et al. 2011; Fernandez-Calvo et al. 2011). However, several studies showed that MYC2 is also involved in ABA signaling in *Arabidopsis.* Abe et al. (2003) showed that transgenic plants overexpressing *MYC2* had higher sensitivity to ABA and several ABA-inducible genes were upregulated in the transgenic plants. By contrast, the mutant of the *MYC2* gene was less sensitive to ABA and showed significantly decreased ABA-induced gene expression. These results indicate that MYC2 also function as a positive regulator of ABA signaling in *Arabidopsis* (Abe et al. 2003). Taken together, we propose that MYC2 acts as an integrative hub for the regulation of both JA and ABA signaling in *Arabidopsis.*

Moreover, Nakata et al. (2013) showed that a bHLH-type transcription factor, ABA-inducible bHLH-type transcription factor/JA-associated MYC2-LIKE1 (JAM1), acts as a transcriptional repressor and negatively regulates JA signaling. Gain-of-function transgenic plants expressing the chimeric repressor for JAM1 exhibited substantial reduction of JA responses, including JA-induced inhibition of root growth, accumulation of anthocyanin, and male fertility. Conversely, *jam1* loss-of-function mutants showed enhanced JA responsiveness, including increased resistance to insect attack. JAM1 and MYC2 competitively bind to the target sequence of MYC2, which likely provides the mechanism for negative regulation of JA signaling and suppression of MYC2 functions by JAM1 (Nakata et al. 2013). Considering the role of JAM1 in ABA signaling (Li et al. 2007), we propose that JAM1 may be a novel integrative hub between ABA- and JA-signaling pathways.

Significantly, Chen et al. (2012) recovered the action mechanisms of the mediator25 (MED25) subunit of the *Arabidopsis* mediator complex in regulating JA- and ABA-triggered gene transcription. Their results showed that during jasmonate signaling, MED25 physically associates with the basic helix-loop-helix transcription factor MYC2 in promoter regions of MYC2 target genes and exerts a positive effect on MYC2-regulated gene transcription (Chen et al. 2012). By contrast, MED25 physically associates with the basic Leu zipper transcription factor *ABA-INSENSITIVE5* (ABI5) in promoter regions of ABI5 target genes and shows a negative effect on ABI5-regulated gene transcription (Chen et al. 2012). This study reveals an important role for MED25 as an integrative hub within the mediator complex during the regulation of JA- and ABA-signaling pathways (Chen et al. 2012).

12.7 Conclusions

ABA is an important plant hormone that regulates plant growth and response to environmental stresses through interacting with different other plant hormones. The molecular mechanisms underlying the complex signaling networks between ABA and other hormones remain to be further elucidated in future studies.

Acknowledgement This work was supported by grants from The Ministry of Science and Technology of China (2011CB915400) and The National Natural Science Foundation of China (31030006, 91117013).

References

Abe H, Urao T, Ito T, Seki M, Shinozaki K, Yamaguchi-Shinozaki K. Arabidopsis AtMYC2 (bHLH) and AtMYB2 (MYB) function as transcriptional activators in abscisic acid signaling. Plant Cell. 2003;15:63–78.

Achard P, Herr A, Baulcombe DC, Harberd NP. Modulation of floral development by a gibberellin-regulated microRNA. Development. 2004;131:3357–65.

Achard P, Cheng H, De Grauwe L, Decat J, Schoutteten H, Moritz T, Van Der Straeten D, Peng J, Harberd NP. Integration of plant responses to environmentally activated phytohormonal signals. Science. 2006;311:91–4.

Anderson JP, Badruzsaufari E, Schenk PM, Manners JM, Desmond OJ, Ehlert C, Maclean DJ, Ebert PR, Kazan K. Antagonistic interaction between abscisic acid and jasmonate-ethylene signaling pathways modulates defense gene expression and disease resistance in Arabidopsis. Plant Cell. 2004;16:3460–79.

Beaudoin N, Serizet C, Gosti F, Giraudat J. Interactions between abscisic acid and ethylene signaling cascades. Plant Cell. 2000;12:1103–15.

Belin C, Megies C, Hauserova E, Lopez-Molina L. Abscisic acid represses growth of the Arabidopsis embryonic axis after germination by enhancing auxin signaling. Plant Cell. 2009;21:2253–68.

Boter M, Ruiz-Rivero O, Abdeen A, Prat S. Conserved MYC transcription factors play a key role in jasmonate signaling both in tomato and Arabidopsis. Genes Dev. 2004;18:1577–91.

Chen Q, Sun J, Zhai Q, Zhou W, Qi L, Xu L, Wang B, Chen R, Jiang H, Qi J, Li X, Palme K, Li C. The basic helix-loop-helix transcription factor MYC2 directly represses PLETHORA expression during jasmonate-mediated modulation of the root stem cell niche in Arabidopsis. Plant Cell. 2011;23:3335–52.

Chen R, Jiang H, Li L, Zhai Q, Qi L, Zhou W, Liu X, Li H, Zheng W, Sun J, Li C. The Arabidopsis mediator subunit MED25 differentially regulates jasmonate and abscisic acid signaling through interacting with the MYC2 and ABI5 transcription factors. Plant Cell. 2012;24:2898–916.

Cheng WH, Chiang MH, Hwang SG, Lin PC. Antagonism between abscisic acid and ethylene in Arabidopsis acts in parallel with the reciprocal regulation of their metabolism and signaling pathways. Plant Mol Biol. 2009;71:61–80.

Choi Y, Lee Y, Kim SY, Hwang JU. Arabidopsis ROP-interactive CRIB motif-containing protein 1 (RIC1) positively regulates auxin signalling and negatively regulates abscisic acid (ABA) signalling during root development. Plant, Cell Environ. 2013;36:945–55.

Ding Y, Kalo P, Yendrek C, Sun J, Liang Y, Marsh JF, Harris JM, Oldroyd GE. Abscisic acid coordinates nod factor and cytokinin signaling during the regulation of nodulation in Medicago truncatula. Plant Cell. 2008;20:2681–95.

Dombrecht B, Xue GP, Sprague SJ, Kirkegaard JA, Ross JJ, Reid JB, Fitt GP, Sewelam N, Schenk PM, Manners JM, Kazan K. MYC2 differentially modulates diverse jasmonate-dependent functions in Arabidopsis. Plant Cell. 2007;19:2225–45.

Fernandez-Calvo P, Chini A, Fernandez-Barbero G, Chico JM, Gimenez-Ibanez S, Geerinck J, Eeckhout D, Schweizer F, Godoy M, Franco-Zorrilla JM, Pauwels L, Witters E, Puga MI, Paz-Ares J, Goossens A, Reymond P, De Jaeger G, Solano R. The Arabidopsis bHLH transcription factors MYC3 and MYC4 are targets of JAZ repressors and act additively with MYC2 in the activation of jasmonate responses. Plant Cell. 2011;23:701–15.

Ghassemian M, Nambara E, Cutler S, Kawaide H, Kamiya Y, McCourt P. Regulation of abscisic acid signaling by the ethylene response pathway in Arabidopsis. Plant Cell. 2000;12:1117–26.

Golldack D, Li C, Mohan H, Probst N. Gibberellins and abscisic acid signal crosstalk: living and developing under unfavorable conditions. Plant Cell Rep. 2013;32:1007–16.

Gomez-Cadenas A, Verhey SD, Holappa LD, Shen Q, Ho TH, Walker-Simmons MK. An abscisic acid-induced protein kinase, PKABA1, mediates abscisic acid-suppressed gene expression in barley aleurone layers. Proc Natl Acad Sci USA. 1999;96:1767–72.

Gomez-Cadenas A, Zentella R, Walker-Simmons MK, Ho TH. Gibberellin/abscisic acid antagonism in barley aleurone cells: site of action of the protein kinase PKABA1 in relation to gibberellin signaling molecules. Plant Cell. 2001;13:667–79.

Gubler F, Kalla R, Roberts JK, Jacobsen JV. Gibberellin-regulated expression of a myb gene in barley aleurone cells: evidence for Myb transactivation of a high-pI alpha-amylase gene promoter. Plant Cell. 1995;7:1879–91.

Hauser F, Waadt R, Schroeder JI. Evolution of abscisic acid synthesis and signaling mechanisms. Curr Biol. 2011;21:R346–55.

Li H, Sun J, Xu Y, Jiang H, Wu X, Li C. The bHLH-type transcription factor AtAIB positively regulates ABA response in Arabidopsis. Plant Mol Biol. 2007;65:655–65.

Li Z, Zhang L, Yu Y, Quan R, Zhang Z, Zhang H, Huang R. The ethylene response factor AtERF11 that is transcriptionally modulated by the bZIP transcription factor HY5 is a crucial repressor for ethylene biosynthesis in Arabidopsis. Plant J. 2011;68:88–99.

Liu X, Zhang H, Zhao Y, Feng Z, Li Q, Yang HQ, Luan S, Li J, He ZH. Auxin controls seed dormancy through stimulation of abscisic acid signaling by inducing ARF-mediated ABI3 activation in Arabidopsis. Proc Natl Acad Sci USA. 2013;110:15485–90.

Lorenzo O, Chico JM, Sanchez-Serrano JJ, Solano R. JASMONATE-INSENSITIVE1 encodes a MYC transcription factor essential to discriminate between different jasmonate-regulated defense responses in Arabidopsis. Plant Cell. 2004;16:1938–50.

Nakata M, Mitsuda N, Herde M, Koo AJ, Moreno JE, Suzuki K, Howe GA, Ohme-Takagi M. A bHLH-type transcription factor, ABA-INDUCIBLE BHLH-TYPE TRANSCRIPTION FACTOR/JA-ASSOCIATED MYC2-LIKE1, acts as a repressor to negatively regulate jasmonate signaling in Arabidopsis. Plant Cell. 2013;25:1641–56.

Nibau C, Tao L, Levasseur K, Wu HM, Cheung AY. The Arabidopsis small GTPase AtRAC7/ROP9 is a modulator of auxin and abscisic acid signalling. J Exp Bot. 2013;64:3425–37.

Razem FA, Baron K, Hill RD. Turning on gibberellin and abscisic acid signaling. Curr Opin Plant Biol. 2006;9:454–9.

Reyes JL, Chua NH. ABA induction of miR159 controls transcript levels of two MYB factors during Arabidopsis seed germination. Plant J. 2007;49:592–606.

Rogers JC, Rogers SW. Definition and functional implications of gibberellin and abscisic acid cis-acting hormone response complexes. Plant Cell. 1992;4:1443–51.

Shkolnik-Inbar D, Bar-Zvi D. ABI4 mediates abscisic acid and cytokinin inhibition of lateral root formation by reducing polar auxin transport in Arabidopsis. Plant Cell. 2010;22:3560–73.

Shu K, Zhang H, Wang S, Chen M, Wu Y, Tang S, Liu C, Feng Y, Cao X, Xie Q. ABI4 regulates primary seed dormancy by regulating the biogenesis of Abscisic acid and gibberellins in Arabidopsis. PLoS Genet. 2013;9:e1003577.

Tanaka Y, Sano T, Tamaoki M, Nakajima N, Kondo N, Hasezawa S. Ethylene inhibits abscisic acid-induced stomatal closure in Arabidopsis. Plant Physiol. 2005;138:2337–43.

Tanaka Y, Sano T, Tamaoki M, Nakajima N, Kondo N, Hasezawa S. Cytokinin and auxin inhibit abscisic acid-induced stomatal closure by enhancing ethylene production in Arabidopsis. J Exp Bot. 2006;57:2259–66.

Wang L, Hua D, He J, Duan Y, Chen Z, Hong X, Gong Z. Auxin response factor2 (ARF2) and its regulated homeodomain gene HB33 mediate abscisic acid response in Arabidopsis. PLoS Genet. 2011a;7:e1002172.

Wang Y, Li L, Ye T, Zhao S, Liu Z, Feng YQ, Wu Y. Cytokinin antagonizes ABA suppression to seed germination of Arabidopsis by downregulating ABI5 expression. Plant J. 2011b;68:249–61.

Weiss D, Ori N. Mechanisms of cross talk between gibberellin and other hormones. Plant Physiol. 2007;144:1240–6.

Xie Z, Zhang ZL, Zou X, Yang G, Komatsu S, Shen QJ. Interactions of two abscisic-acid induced WRKY genes in repressing gibberellin signaling in aleurone cells. Plant J. 2006;46:231–42.

Yang X, Yang YN, Xue LJ, Zou MJ, Liu JY, Chen F, Xue HW. Rice ABI5-Like1 regulates abscisic acid and auxin responses by affecting the expression of ABRE-containing genes. Plant Physiol. 2011;156:1397–409.

Zentella R, Yamauchi D, Ho TH. Molecular dissection of the gibberellin/abscisic acid signaling pathways by transiently expressed RNA interference in barley aleurone cells. Plant Cell. 2002;14:2289–301.

Chapter 13
Cross Talk Between Light and ABA Signaling

Rongcheng Lin and Weijiang Tang

Abstract The endogenous phytohormone abscisic acid (ABA) and the exogenous signal light regulate both distinct and overlapping processes in plant growth and development. This review summarizes recent advances in our understanding of the cross-regulation between light and ABA signaling, and their cooperative interactions in modulating plant responses, including seed germination, seedling growth and development, stomatal movement, and hydrotropic growth.

Keywords ABA · Light · Cross talk · Modulation

13.1 Introduction

Due to their sessile nature, plants are controlled by endogenous hormones and influenced by environmental cues. As one of the most important environmental signals, light plays critical roles in regulating diverse plant growth and developmental processes, ranging from seed germination, seedling de-etiolation, phototropism, shade avoidance, stomatal opening, flowering time, and circadian rhythms. Accumulating evidence indicates that light interacts with many phytohormone signaling, including abscisic acid (ABA), gibberellin (GA), brassinosteroid, and ethylene, in controlling various plant response (for reviews, Seo et al. 2009; Alabadi and Blazquez 2009; Lau and Deng 2010). ABA regulates many plant processes that are also mediated by light, such as seed germination and seedling development. The biosynthesis and function of ABA and its regulatory network were extensively reviewed in the other chapters of this book. The scope of this chapter emphasizes advances in our understanding of interaction between light and ABA

R. Lin (✉) · W. Tang
Key Laboratory of Photobiology, Institute of Botany, The Chinese Academy of Sciences, Beijing 100093, China
e-mail: rclin@ibcas.ac.cn

D.-P. Zhang (ed.), *Abscisic Acid: Metabolism, Transport and Signaling*,
DOI 10.1007/978-94-017-9424-4_13

Signaling, based largely on progress achieved so far using genetic and molecular approaches in *Arabidopsis* as a model system. In this review, we first describe a brief introduction of the light signaling pathway, and then summarize, and discuss the convergence of light and ABA Signaling in regulating plant responses.

13.2 Overview of the Light Signaling Pathway

Plants have evolved an array of photoreceptors to perceive and transduce different spectra of light that ultimately modulate the transcriptomes and trigger plant growth and development. These photoreceptors include the red and far-red light (600–750 nm)-absorbing phytochromes (phys), the blue/ultraviolet-A light (320–500 nm)-absorbing cryptochromes (cry1 and cry2), phototropins (phot1 and phot2), and three newly identified LOV/F-box/Kelch-repeat proteins ZEITLUPE (ZTL), FLAVIN-BINDING KELCH REPEAT F-BOX (FKF), and LOV KELCH REPEAT PROTEIN 2 (LKP2), and UV-B light (282–320 nm)-absorbing UV RESISTANCE LOCUS8 (UVR8) (Chen et al. 2004; Christie 2007; Nagatani 2010; Yu et al. 2010; Heijde and Ulm 2012; Ito et al. 2012). Phytochromes are unique photoreceptors because they exist as two distinct but photoreversible forms in vivo. The biological active Pfr form absorbs far-red light, whereas the inactive Pr form absorbs red light (Li et al. 2011). There are five phytochromes, designated phyA to phyE in *Arabidopsis thaliana*. phyA is light labile, while phyB to phyE are light stable (Li et al. 2011).

Seedling de-etiolation is a light-controlled process that has been extensively studied in the past decades. Accumulating evidence has established that phys and crys control two main branches of light signaling during seedling de-etiolation (Lau and Deng 2010). A group of constitutive photomorphogenic/de-etiolated/fusca (COP/DET/FUS) proteins act as repressors downstream of phys and crys that define the first branch of the light signaling pathway (Lau and Deng 2012). Among these proteins, COP1 is a central repressor that targets a number of positive factors, such as ELONGATED HYPOCOTYL5 (HY5) and LONG HYPOCOTYL IN FAR-RED1 (HFR1), for 26S proteasome-mediated degradation, thus desensitizing light signaling (Henriques et al. 2009; Lau and Deng 2012). *HY5* encodes a basic domain/leucine zipper transcription factor that plays a key role in promoting photomorphogenesis in all light conditions by directly regulating the transcription of a wide range of genes (Oyama et al. 1997; Lee et al. 2007). HY5 is stabilized at the post-translational level by light and inhibits hypocotyl growth (Osterlund et al. 2000). In the second branch, a class of basic helix-loop-helix transcription factors, designated PHYTOCHROME-INTERACTING FACTORs (PIFs), accumulate in darkness and thus regulate gene expression to promote the skotomorphogenic response (Leivar et al. 2008). Under light, PIF proteins interact with photoactivated phys and result in PIFs' phosphorylation and subsequent degradation in an unknown manner (Leivar and Quail 2011). PIF proteins mainly regulate the phytochrome pathway, although they might also effect under blue light. Increasing studies demonstrated the broad function of PIFs as integrators in mediating plant development (Leivar and Monte 2014).

Extensive studies have identified dozens of intermediates in the light signaling pathway and revealed the importance of transcriptional regulatory networks in controlling photomorphogenesis (Jiao et al. 2007; Chory 2010). For example, *FAR-RED ELONGATED HYPOCOTYL3* (*FHY3*) and *FAR-RED IMPAIRED RESPONSE1* (*FAR1*) are two positive transcription factors transducing signals in the far-red light pathway (Hudson et al. 1999; Wang and Deng 2002; Lin et al. 2007). For more information on light signaling regulation, readers may go through some recent reviewer articles (Bou-Torrent et al. 2007; Jiao et al. 2007; Demarsy and Fankhauser 2009; Li et al. 2011).

13.3 Inter-regulation Between Light and ABA

13.3.1 Light Regulates ABA Biosynthesis

The ABA metabolic pathway has been described in detail in Chap. 20. Most of the genes involved in ABA biosynthesis and catabolism have been identified genetically. The oxidative cleavage of cis-epoxycarotenoid to xanthoxin is catalyzed by 9-cis-epoxycarotenoid dioxygenase (NCED) and represents the key regulatory step of ABA biosynthesis in plants. ABA 8′-hydroxylases encoded by cytochrome P450 *CYP707A* genes catalyze the first committed step in the predominant ABA catabolic pathway (Kushiro et al. 2004; Nambara and Marion-Poll 2005). The endogenous ABA level is modulated by the precise balance between its biosynthesis and catabolism. Regulation of *NCED* and *CYP707A* has thus been proposed to significantly determine endogenous ABA level in plants (Nambara and Marion-Poll 2005).

ABA is increasingly accumulated in seeds during their maturation. Studies from seed germination have well demonstrated that light plays a crucial role in regulating ABA metabolic gene expression and subsequent ABA level. Red light decreases, whereas far-red light increases endogenous ABA level in Arabidopsis and lettuce (*Lactuca sativa* L.) seeds (Toyomasu et al. 1994; Seo et al. 2006; Sawada et al. 2008). Among *AtNCED* genes, *AtNCED6* and *AtNCED9* have been shown to play key roles in ABA biosynthesis in developing seeds (Lefebvre et al. 2006). The transcript level of *AtNCED6* remains high after pulse of far-red light irradiation and is reduced by a subsequent red light pulse in Arabidopsis seeds. In agreement with this notion, the *nced6-1* mutant showed enhanced germination ability relative to wild type when treated with FR light (Seo et al. 2006). Similarly, red light down-regulates *LsNCED2* and *LsNCED4* and increases *LsABA8ox4* (encoding ABA 8′-hydroxylase) expression (Sawada et al. 2008). As a catabolic gene, the expression pattern of *CYP707A2* undergoes an opposite manner to that of *AtNCED6* (Seo et al. 2006). However, photoreversible expression of *CYP707A1* and *CYP707A3* appears to be regulated indirectly by light (Seo et al. 2006). Thus, ABA biosynthesis is likely regulated through the photoreversible expression of *AtNCED6* and *CYP707A2* in an opposite manner.

Photoreversible regulation of ABA level during seed germination is mainly mediated by phyB photoreceptor, since reduction of ABA level after red light pulse was not observed in the *phyB* mutant (Seo et al. 2006). When phyB is activated, the ABA anabolic genes, *ABA-DEFICIENT 1* (*ABA1*), *AtNCED6*, and *AtNCED9*, are down-regulated, whereas an ABA catabolic gene, *CYP707A2*, is induced (Kim et al. 2008; Oh et al. 2007). This process is mainly controlled by the transcription factor PIL5/PIF1, which negatively regulates phyB responses (Oh et al. 2007, other ref). PIL5 indirectly regulates the transcript levels of ABA metabolic genes, including *ABA1*, *AtNCED6*, *AtNCED9*, and *CYP707A2*, and some GA metabolic genes, such as *GA3ox1* and *GA2ox2* (Oh et al. 2007). PIL5 can target *SOMNUS* (*SOM*) and directly activates its expression. In the *som* mutant, the expression levels of *ABA1* and *AtNCED6* are reduced, whereas the level of *CYP707A2* is increased compared to the wild type. Consistently, the *som* seeds contain low amount of ABA and high levels of active GA_4 (Kim et al. 2008). Therefore, PIL5 regulates endogenous ABA level largely through SOM. However, the question how SOM regulates the expression of ABA metabolic genes is still unknown.

Besides, at the seed germination stage, the expression of *CYP707A2* is also significantly up-regulated, whereas the transcript levels of *AtNCED* genes (including *AtNCED2*, *3*, *5*, and *9*) are decreased by light during seedling de-etiolation (Charron et al. 2009). Moreover, the expression patterns of tomato *LeZEP1* (encoding zeaxanthinepoxidase) and *LeNCED* are under circadian regulation (Thompson et al. 2000). In addition, the transcript level of *ABA-INSENSITIVE 3* (*ABI3*), encoding a key component in the ABA signaling pathway, is also affected by mutations in *phyB* in Arabidopsis (Mazzella et al. 2005), suggesting that phytochrome regulates both the metabolic and signaling genes of ABA.

13.3.2 ABA Modulates the Expression of Light-Responsive Genes

Light-harvesting chlorophyll a/b-binding proteins (LHCBs) are the apoproteins of the photosystem II complex that absorb and transfer light energy. Expression of these nuclear *LHCB* genes is tightly controlled by light, and therefore, *LHCB*s serve as typical light-responsive genes (Johanningmeier 1988; Johanningmeier and Howell 1984). Being a stress signal, ABA plays an important role in the regulation of *LHCB* expression under environmental stress conditions. For example, exogenously application of high concentrations of ABA inhibits *LHCB* expression in various tissues, including tomato leaves, Arabidopsis seedlings, *Lemma gibba* cells, and developing seeds of soybean (Bartholomew et al. 1991; Staneloni et al. 2008; Weatherwax et al. 1996; Chang and Walling 1991). However, low level of ABA enhances *LHCB1.2* transcript level in Arabidopsis seedlings, and *cab3* (*GmLHCB*) expression in soybean seeds (Voigt et al. 2010; Chang and Walling 1991). This is consistent with a recent study showing that physiological levels of ABA enhance *LHCB* expression in Arabidopsis (Liu et al. 2013). Liu and the coauthors (2013) further found that ABA is required for full expression of different *LHCB* members likely via the WRKY40 transcription factor. ABA may be an inducer to fine-tune *LHCB* expression under stressful conditions in cooperation with light that allows plants to adapt to environmental changes. Moreover, a signal transduction chain consisting of GCR1 (a potential G-protein-coupled receptor), GPA1 (the sole Ga subunit), RPN1 (one of four members of an iron-containing subgroup of the cupin superfamily), and a nuclear factor Y convergences blue light and ABA signals to regulate *LHCB* expression in etiolated Arabidopsis seedlings (Warpeha et al. 2007).

In addition, ABA regulates genes involved in the light signal transduction pathway. For instance, the transcript levels of *FHY3* and *FAR1*, encoding two key positive transcription factors in the phyA pathway, are induced in Arabidopsis seedlings after ABA treatment (Tang et al. 2013).

13.4 Light and ABA Coregulate Plant Responses

Light and phytohormone ABA coordinately regulate many plant developmental processes, including seed germination, seedling growth, stomatal movement, and hydrotropic response, as reviewed below in detail. We focus on the function of signaling factors that were genetically identified in recent studies and their regulatory mechanisms on each distinct response.

13.4.1 Seed Germination

Seed germination is an adaptive trait of higher plants that is controlled by both environmental cues and internal growth regulators, including light, GA, and ABA. GA is known to break seed dormancy and promote germination, whereas ABA is involved in maintaining seed dormancy and inhibiting germination (Koornneef et al. 2002; Finch-Savage and Leubner-Metzger 2006). It is now much clear that GA promotes germination by promoting destruction of DELLA repressors, whereas ABA prevents germination by stimulating the expression of ABI repressors. Endogenous ABA biosynthesis in imbibed seeds is required for the maintenance of seed dormancy in Arabidopsis and tobacco (Ali-Rachedi et al. 2004; Grappin et al. 2000).

Light is a critical determinant environmental factor for seed germination in some small-seeded plants, such as Arabidopsis and lettuce (Shinomura 1997). In the middle of twentieth century, it was discovered that red light promotes, whereas far-red light inhibits lettuce seed germination, and the process is reversible by red and far red (Borthwick et al. 1952). The photoreceptor responsible for the reversible photoreaction was discovered from etiolated *Brassica rapa* and *Zea mays* and was named phytochrome (Butler et al. 1959). It has been well established that phyA and phyB play curical role in the light-mediated seed germination (Shinomura et al. 1994, 1996;

Casal and Sanchez 1998). phyA mediates very low-fluence response (VLFR), while phyB acts via photoreversible low-fluence response (LFR) to promote seed germination. However, continuous far-red light inhibits germination via high-irradiance response in many plant species (Botto et al. 1996).

Light controls seed germination predominantly through regulating the endogenous levels of GA and ABA. ABA inhibits germination of lettuce seeds induced by red light, whereas active GA mimics the effect of red light (Kahn et al. 1957; Sankhla and Sankhla 1968). Extensive studies have identified a number of factors that involve in light-controlled seed germination.

Giltu Choi's laboratory firstly reported that a basic helix-loop-helix transcription factor PIL5 acts as a key negative regulator in phytochrome-mediated seed germination (Oh et al. 2004). PIF5 preferentially interacts with the Pfr forms of phyA and phyB. When activated by light, phytochromes bind to and accelerate the degradation of PIL5 in both seeds and seedlings (Oh et al. 2006; Shen et al. 2005). The destabilization of PIL5 thus releases its repression of seed germination and allows seeds to germinate. As a result, loss-of-function mutant of *pil5* germinates well regardless of far-red light treatment mediated by LFR and VLFR, whereas *PIL5* overexpression transgenic lines fail to germinate under relative low intensity of red light (Oh et al. 2004). It was showed that PIL5 directly binds to the promoters of two GA repressor (DELLA) genes, *REPRESSOR OF GA1-3* (*RGA*) and *GA-INSENSITIVE* (*GAI*), and activates their expression (Oh et al. 2007). Furthermore, chromatin immunoprecipitation (ChIP) chip and microarray analyses helped to identify large amount of PIL5 direct target genes involved in hormone signaling and cell wall modification (Oh et al. 2009). Therefore, PIL5 regulates seed germination not only by mediating GA signaling and coordinating GA and ABA metabolism, but also by modulating cell wall properties in imbibed seeds. Since *pil5* could not fully restore the germination deficiency of *phyB* in the *pil-5phyB* double mutant, other factors must be involved in the phyB-mediated germination process (Oh et al. 2004).

SOM was identified as another negative factor in regulating light-dependent seed germination (Kim et al. 2008). The *SOM* gene encodes a CCCH-type zinc finger protein that probably acts as an RNA-binding factor. The *som* mutants have lower levels of ABA and elevated levels of GA and germinate in darkness independently of various light regimens (Kim et al. 2008). PIL5 directly promotes the expression of *SOM* through binding to its promoter sequence, and the reduced germination rate of a PIL5 overexpression line is rescued by the *som* mutation (Kim et al. 2008). Thus, SOM functions downstream of PIL5 and the PIL5-SOM regulatory pathway likely defines an essential step in integrating ABA and light signaling to control seed germination. In addition to PIL5, ABI3 was also found to be targeted to the RY motifs present in the *SOM* promoter. ABI3 and PIL5 interact and collaboratively activate the expression of *SOM* mRNA in Arabidopsis imbibed seeds, but independently induce *SOM* expression in maturing seeds (Park et al. 2011). However, HFR1 plays a negative role on PIL5 transcriptional activity by interacting with PIL5 and preventing its binding to target DNA. Through the HFR1-PIL5 heterdimer, light regulates expression of numerous genes involved in cell wall loosening, cell division, and hormone pathways to initiate seed germination (Shi et al. 2013). Hence, HFR1 defines a new positive regulator of phyB-dependent seed germination.

Recently, Lim et al. (2013) demonstrated that ABI3, ABI5, and DELLAs form a complex on the *SOM* promoter to activate *SOM* expression in imbibed seeds in response to high temperature. ABI5 is a bZIP transcription factor that plays important role in ABA signaling and ABA responses (Finkelstein and Lynch 2000). Two previous researches identified two types of transcription factors that directly regulate *ABI5* expression (Chen et al. 2008; Tang et al. 2013). A ChIP study indicates that HY5 directly binds to the promoter region of *ABI5*, and the binding ability was significantly enhanced by exogenous ABA treatment. Consistent with this observation, HY5 is required for the expression of ABA-inducible genes, such as *ABI3*, *RAB18*, *AtEM1*, and *AtEM6*, in seeds and during seed germination (Chen et al. 2008). Consequently, *hy5* mutant seeds are less sensitive to the inhibition of ABA and glucose on germination (Chen et al. 2008).

FHY3 is another type of transcription factor that directly binds to the promoter of *ABI5* and activates its expression (Tang et al. 2013). Disruption of *FHY3* and/or its homology gene, *FAR1,* reduces sensitivity to ABA-mediated inhibition of seed germination. Germination of the *fhy3* mutant seeds is also less sensitive to salt and osmotic stress than that of the wild type (Tang et al. 2013). Strikingly, constitutive expression of *ABI5* restores the seed germination response of *fhy3*. Furthermore, the expression of several ABA-responsive genes (e.g., *ABI1*, *ABI2*, *ABF3*, *RAB18*, *KIN2*, *COR47*, *DREB2A*, and *RD22*) is decreased in the *fhy3* and/or *far1* mutants during seed imbibition (Tang et al. 2013).

Although both phyA and phyB photoreceptors are essential for seed germination, the mechanism underlying their distinct roles has long been a mystery. Using a seed coat bedding assay system where dissected embryos are cultured on a layer of dissected seed coats, Lee and coauthors (2012) demonstrated that phyA and phyB spatially control seed germination in embryo and endosperm, respectively, in response to far-red light irradiation. The endosperm mediates far-red repression of phyB-dependent germination, whereas FR stimulation of phyA-dependent germination occurs only in the embryo. These responses specifically involve the light signaling genes PIL5 and RGL2 in the endosperm and PIL5, SOM, GAI, and RGA in the embryo, where they regulate the expression of GA and ABA biosynthetic genes in each tissue (Lee et al. 2012). Therefore, early upon seed imbibition, far-red light inactivation of phyB leads to ABA biosynthesis and releases from the endosperm to prevent phyA-dependent promotion of germination in the embryo. This involves an extended regulatory network where ABA overrides phyA signaling by interfering with the expression of light signaling genes and GA and ABA metabolic genes. Over time, a weakening of ABA-dependent responses takes place, thus allowing phyA-dependent germination after a later light treatment. This results in a phyA-dependent "explosive" germination unlike phyB-dependent germination (Lee et al. 2012). Furthermore, far-red light repression of germination involves stabilized DELLA proteins GAI, RGA, and RGL2 that stimulate endogenous ABA biosynthesis, which in turn blocks germination through ABI3 (Piskurewicz et al. 2009).

IMB1 (for imbibition-inducible 1) defines as a putative bromodomain transcription factor. The imb1 loss-of-function mutant is hypersensitive to ABA-mediated inhibition of cotyledon expansion and greening, and is deficient in the phyA-mediated VLFR of seed germination (Duque and Chua 2003). *IMB1* transcript level is elevated during seed imbibition. This study implicates that IMB1 might link phyA to ABA signaling in seed germination (Duque and Chua 2003). Interestingly, *ABI5* transcript level was up-regulated in *imb1* seed germination in ABA when compared to the wild type, suggesting that IMB1 acts upstream of ABI5 in the ABA pathway (Duque and Chua 2003).

In addition to transcription factors, two JmjC domain-containing proteins, JMJ20 and JMJ22, have been shown as positive regulators of seed germination (Cho et al. 2012). *JMJ20* and *JMJ22* encode histone arginine demethylases, and their expression is directly repressed by SOM. Upon phyB activation by red light, *JMJ20* and *JMJ22* are derepressed, resulting in increased GA levels through the removal of repressive histone arginine methylation at *GA3ox1* and *GA3ox2* loci, which in turn promote germination (Cho et al. 2012). However, the ABA metabolic genes are not regulated by JMJ20/JMJ22. This study adds an additional layer that involves repressive epigenetic mechanism during seed germination.

The F-box protein MORE AXILLARY BRANCHES2 (MAX2) plays an important role in promoting photomorphogenesis through modulating GA and ABA biosynthetic pathways (Shen et al. 2007, 2012). The *max2* mutant seeds are hyposensitive to light-induced seed germination and hypersensitive to ABA. Surprisingly, expression of ABA biosynthetic and catabolic genes and ABA-regulated genes is up-regulated by *max2* mutation (Shen et al. 2012; Bu et al. 2014). Though a genetic study indicated that the seed germination phenotype of *max2* is epistatic to *pil5* (Shen et al. 2012), the molecular mechanism between MAX2 and PIL5 remains to be elucidated.

13.4.2 Seedling Growth and Development

After seed germination, seedling growth and development are also regulated by light and ABA. It has been shown that disruption of *HY5* confers tolerance to the inhibitory effect of ABA on lateral root growth and seedling growth. The *hy5* seedlings were also more susceptible to salt and osmotic stresses than the wild-type plants (Chen et al. 2008). *ABI5::GUS* promoter activity was detected in cotyledons, hypocotyls, roots, flowers, and siliques. However, this activity was greatly reduced in the *hy5* mutant background. Furthermore, light promotes *ABI5* expression in a HY5-dependent manner (Chen et al. 2008). This is because HY5 protein is tightly controlled by the COP1-mediated 26S proteasome degradation pathway in the dark (Osterlund et al. 2000). As a consequence, overexpression of *ABI5* restores ABA sensitivity in *hy5* and enhances light response (hypocotyl elongation) in the wild type (Chen et al. 2008). Since FHY3/FAR1 also bind to *ABI5* promoter sequence, *fhy3* and/or *far1* mutants are hyposensitive to ABA-mediated inhibition of seedling greening. *FHY3* and *FAR1* transcripts are up-regulated by ABA and abiotic stresses (Tang et al. 2013). Thus, HY5 and FHY3/FAR1 transcription activators act upstream of ABI5 to integrate light and ABA signaling during early seedling development. In addition, the *max2* mutant seedlings are hypersensitive to ABA (Shen et al. 2012).

13.4.3 Stomatal Movement

ABA is a stress signal that plays a prominent role in inducing stomatal closure to prevent water loss in response to drought stress and thereby contributes to tolerance for plants (Cutler et al. 2010; Hauser et al. 2011). It has been known that blue light receptor phototropins mediate stomatal opening (Kinoshita et al. 2001). Studies from our laboratory showed that the *fhy3* and *far1* mutants have wider stomata, lose water faster, and are more sensitive to drought than the wild type; therefore, FHY3 and FAR1 confer increased resistance to drought (Tang et al. 2013). The drought-sensitive phenotype of *fhy3* may be partly caused by the reduced sensitivity of guard cell movement under drought stress conditions, which may induce the production of ABA. In agreement with this notion, *FHY3* is highly expressed in guard cells (Tang et al. 2013). MAX2 plays a similar role as FHY3/FAR1 in modulating stomatal movement. The *max2* mutant plants are less sensitive to ABA-induced stomatal closure and display increased water loss and drought-sensitive phenotypes. The expression of ABA biosynthesis, catabolism, transport, and signaling genes was impaired in *max2*, compared to wild type in response to drought stress (Bu et al. 2014).

Down-regulation or disruption of any member of the LHCB family, *LHCB1* to *LHCB6*, reduces responsiveness of stomatal movement to ABA. By contrast, overexpression of *LHCB6* enhances stomatal sensitivity to ABA (Xu et al. 2012). These results demonstrate that LHCBs play a positive role in ABA signaling in stomatal movement and the plant response to drought. Similarly, LHCBs positively regulate seed germination and seedling growth in response to ABA (Liu et al. 2013).

Mg-chelatase catalyzes the formation of Mg-protoporphyrin IX by chelating magnesium to protoporphyrin IX in the chlorophyll biosynthesis pathway (Tanaka and Tanaka 2007). The H subunit of Mg-chelatase was identified as an ABA receptor, and it functions in the ABA signaling pathway (Shen et al. 2006; Wu et al. 2009). ABA specifically binds to CHLH, but not to the other Mg-chelatase subunits, CHLI, CHLD, and GUN4 (Du et al. 2012). *CHLH* and *GUN4* are major targets for light regulation during seedling de-etiolation (Stephenson and Terry 2008). Genetic studies showed that the *rtl1* mutant plants (a *chlh* allele) display ABA-insensitive phenotypes in stomatal movement. Interestingly, down-regulation of *CHLI* also confers ABA insensitivity in stomatal response, while up-regulation of *CHLI1* results in ABA hypersensitivity in seed germination (Du et al. 2012). The involvement of these chlorophyll biosynthesis and binding proteins in stomatal movement might coordinate internal development with external signals for optical air exchange and maximal photosynthesis.

13.4.4 Hydrotropic Response

Plant roots undergo hydrotropic growth in response to moisture gradient that helps plants acquire water and nutrients. ABA is involved in hydrotropism as the hydrotropic response was reduced in the ABA-deficient mutant *aba1* and ABA-responsive genes were induced upon hydro-treatment (Takahashi et al. 2002; Moriwaki et al. 2010). Recent study found that hydrotropism is less pronounced in dark-grown seedling than in light-grown seedling and pointing out that a light signal is required for the hydrotropic response (Moriwaki et al. 2012). A genetic study identified MIZUKUSSEI1 (MIZ1) as an essential factor for hydrotropism (Kobayashi et al. 2007). Blue light, but not red light, induces the localization of MIZ1-GFP fusion protein in the root tip. Light and ABA induce the expression of *MIZ1* (Moriwaki et al. 2012). *MIZ1* transcript level was down-regulated in the *phyAphyB* double and *hy5* mutants. Consistently, the hydrotropic curvature was reduced in these mutants compared to the wild type (Moriwaki et al. 2012). Thus, phyA and phyB photoreceptors and HY5 transcription factor play important roles in blue light-mediated induction of *MIZ1* and hydrotropism. Moreover, application of ABA to *hy5* restored its hydrotropic defect, whereas abamine SG (ABA synthesis inhibitor) treatment further reduced the hydrotropic response of *hy5* (Moriwaki et al. 2012).

13.5 Future Perspectives

The last decade has made promising progress in our understanding of the coregulation of light and ABA in plant developmental programs, especially in regulating seed germination. Although a number of signaling factors in the pathway were identified, the molecular and biochemical functions of some components are less well understood. Some of the known proteins belong to transcription factors and play roles through modulating gene expression. Other regulatory levels, including post-transcription, translation, post-translational modification, and epigenetic regulation, are likely involved as well. Furthermore, are there more components involved in the cross talk between light and ABA? If any, how do they function? Future studies using the combination of genetic and molecular approaches are deserved to answer these questions. Elucidating the model and underlying mechanism between light and ABA will certainly contribute to our better understanding of plants' adaptability and plasticity to changing environments and help to design stress-tolerant crops in agriculture.

Acknowledgments The work in the authors' laboratory was supported by grants from the National Natural Science Foundation of China (31170221, 31325002) and the Chinese Academy of Sciences.

References

Alabdi D, Blazquez MA. Molecular interactions between light and hormone signaling to control plant growth. Plant Mol Biol. 2009;69:409–17.
Ali-Rachedi S, Bouinot D, Wagner MH, Bonnet M, Sotta B, Grappin P, Jullien M. Changes in endogenous abscisic acid levels during dormancy release and maintenance of mature seeds: studies with the Cape VerdeIslands ecotype, the dormant model of Arabidopsis thaliana. Planta. 2004;219:479–88.
Bartholomew DM, Bartley GE, Scolnik PA. Abscisic acid control of rbcS and cab transcription in tomato leaves. Plant Physiol. 1991;96:291–6.
Borthwick HA, Hendricks SB, Parker MW, Toole EH, Toole VK. A reversible photoreaction controlling seed germination. Proc Natl Acad Sci USA. 1952;38:662–666.
Botto JF, Sanchez RA, Whitelam GC, Casal JJ. Phytochrome A mediates the promotion of seed germination by very low fluences of light and canopy shade light in Arabidopsis. Plant Physiol. 1996;110:439–44.
Bou-Torrent J, Roig-Villanova I, Martinez-Garcia JF. Light signaling: back to space. Trends Plant Sci. 2007;13:108–14.
Bu Q, Lv T, Shen H, Luong P, Wang J, Wang Z, Huang Z, Xiao L, Engineer C, Kim TH, Schroeder JI, Huq E. Regulation of drought tolerance by the F-box protein MAX2 in Arabidopsis. Plant Physiol. 2014;164:424–39.
Butler WL, Norris KH, Siegelman HW, Hendricks SB. Detection, assay, and preliminary purification of the pigment controlling photoresponsive development of plants. Proc Natl Acad Sci USA. 1959;45:1703–1708.
Casal JJ, Sanchez RA. Phytochromes and seed germination. Seed Sci Res. 1998;8:317–29.
Chang YC, Walling LL. Abscisic acid negatively regulates expression of chlorophyll a/b binding protein genes during soybean embryogeny. Plant Physiol. 1991;97:1260–4.
Charron JB, He H, Elling AA, Deng XW. Dynamic landscapes of four histone modifications during deetiolation in *Arabidopsis*. Plant Cell. 2009;21:3732–48.
Chen H, Zhang J, Neff MM, Hong SW, Zhang H, Deng XW, Xiong L. Integration of light and abscisic acid signaling during seed germination ans early seedling development. Proc Natl Acad Sci USA. 2008;105:4495–500.
Chen M, Chory J, Fankhauser C. Light signal transduction in higher plants. Annu Rev Genet. 2004;38:87–117.
Cho JN, Ryu JY, Jeong YM, Park J, Song JJ, Amasino RM, Noh B, Noh YS. Control of seed germination by light-induced histone arginine demethylation activity. Dev Cell. 2012;22:736–48.
Chory J. Light signal transduction: an infinite spectrum of possibilities. Plant J. 2010;61:982–91.
Christie JM. Phototropin blue-light receptors. Annu Rev Plant Biol. 2007;58:21–45.
Cutler SR, Rodriguez PL, Finkelstein RR, Abrams SR. Abscisic acid: emergence of a core signaling network. Annu Rev Plant Biol. 2010;61:651–79.
Demarsy E, Fankhauser C. Higher plants use LOV to perceive blue light. Curr Opin Plant Biol. 2009;12:69–74.
Du SY, Zhang XF, Lu Z, Xin Q, Wu Z, Jiang T, Lu Y, Wang XF, Zhang DP. Roles of the different components of magnesium chelatase in abscisic acid signal transduction. Plant Mol Biol. 2012;80:519–37.
Duque P, Chua NH. IMB1, a bromodomain protein induced during seed imbibition, regulates ABA- and phyA-mediated responses of germination in Arabidopsis. Plant J. 2003;35:787–99.
Finch-Savage WE, Leubner-Metzger G. Seed dormancy and the control of germination. New Phytol. 2006;171:501–23.
Finkelstein RR, Lynch TJ. The Arabidopsis abscisic acid response gene ABI5 encodes a basic leucine zipper transcription factor. Plant Cell. 2000;12:599–609.

Grappin P, Bouinot D, Sotta B, Miginiac E, Jullien M. Control of seed dormancy in *Nicotiana plumbaginifolia*: postimbibition abscisic acid synthesis imposes dormancy maintenance. Planta. 2000;210:279–85.

Hauser F, Waadt R, Schroeder JI. Evolution of abscisic acid synthesis and signaling mechanisms. Curr Biol. 2011;21:R346–355.

Heijde M, Ulm R. UV-B photoreceptor-mediated signalling in plants. Trends Plant Sci. 2012;17:230–7.

Henriques R, Jang IC, Chua NH. Regulated proteolysis in light-related signaling pathways. Curr Opin Plant Biol. 2009;12:49–56.

Hudson M, Ringli C, Boylan MT, Quail PH. The FAR1 locus encodes a novel nuclear protein specific to phytochrome A signaling. Genes Dev. 1999;13:2017–27.

Ito S, Song YH, Imaizumi T. LOV domain-containing F-box proteins: light-dependent protein degradation modules in Arabidopsis. Mol Plant. 2012;5:573–82.

Jiao Y, Lau OS, Deng XW. Light-regulated transcriptional networks in higher plants. Nat Rev Genet. 2007;8:217–30.

Johanningmeier U. Possible control of transcript levels by chlorophyll precursors in Chlamydomonas. Eur J Biochem. 1988;177:417–24.

Johanningmeier U, Howell SH. Regulation of light-harvesting chlorophyll-binding protein mRNA accumulation in Chlamydomonasreinhardtii. Possible involvement of chlorophyll synthesis precursors. J Biol Chem. 1984;259:13541–9.

Kahn A, Goss JA, Smith DE. Effect of gibberellin on germination of lettuce seeds. Science. 1957;125:645–6.

Kim DH, Yamaguchi S, Lim S, Oh E, Park J, Hanada A, Kamiya Y, Choi G. SOMNUS, a CCCH-type zinc finger protein in Arabidopsis, negatively regulates light-dependent seed germination downstream of PIL5. Plant Cell. 2008;20:1260–77.

Kinoshita T, Doi M, Suetsugu N, Kagawa T, Wada M, Shimazaki K. Photl and phot2 mediate blue light regulation of stomatal opening. Nature. 2001;414:656–660.

Kobayashi A, Takahashi A, Kakimoto Y, Miyazawa Y, Fujii N, Higashitani A, Takahashi H. A gene essential for hydrotropism. Proc Natl Acad Sci USA104. 2007;4724–4729.

Koornneef M, Bentsink L, Hilhorst H. Seed dormancy and germination. Curr Opin Plant Biol. 2002;5:33–6.

Kushiro T, Okamoto M, Nakabayashi K, Yamagishi K, Kitamura S, Asami T, Hirai N, Koshiba T, Kamiya Y, Nambara E. The Arabidopsis cytochrome P450 CYP707A encodes ABA 8'-hydroxylases: key enzymes in ABA catabolism. EMBO J. 2004;23:1647–56.

Lau OS, Deng XW. Plant Hormone signaling lightens up: integrators of light and hormone. Curr Opin Plant Biol. 2010;13:571–7.

Lau OS, Deng XW. Thephotomorphogenic repressors COP1 and DET1: 20 years later. Trends Plant Sci. 2012;17:584–93.

Lee J, He K, Stolc V, Lee H, Figueroa P, Gao Y, Tongprasit W, Zhao H, Lee I, Deng XW. Analysis of transcription factor HY5 genomic binding sites revealed its hierarchical role in light regulation of development. Plant Cell. 2007;19:731–49.

Lee KP, Piskurewicz U, Tureckova V, Carat S, Chappuis R, Strnad M, Fankhauser C, Lopez-Molina L. Spatially and genetically distinct control of seed germination by phytochromes A and B. Genes Dev. 2012;26:1984–96.

Lefebvre V, Helen N, Frey A, Sotta B, Seo M, Okamoto M, Nambara E, Marion-Poll A. Functional analysis of Arabidopsis NCED6 and NCED9 genes indicates that ABA synthesized in the endosperm is involved in the induction of seed dormancy. Plant J. 2006;45:309–91.

Leivar P, Monte E. PIFs: systems integrators in plant development. Plant Cell in press. 2014;26:56–78.

Leivar P, Monte E, Al-Sady B, Carle C, Storer A, Alonso JM, Ecker JR, Quail PH. The Arabidopsis phytochrome-interacting factor PIF7, together with PIF3 and PIF4, regulates responses to prolonged red light by modulating phyB levels. Plant Cell. 2008;20:337–52.

Leivar P, Quail PH. PIFs: pivotal components in a cellular signaling hub. Trends Plant Sci. 2011;16:19–28.

Lem S, Park J, Lee N, Jeong J, Toh S, Watanabe A, Kim J, Kang H, Kim DH, Kawakami N, Choi G. ABA-INSENSITIVE3, ABA-INSENSITIVE5, and DELLAs interact to activate the expression of SOMNUS and other high-temperature-inducible genes in imbibed seeds in Arabidopsis. Plant Cell. 2013;25:4863–78.

Li J, Li G, Wang H, Deng XW. Phytochrome signaling mechanisms. The Arabidopsis book e0148. 2011. doi:10.1199/tab.0148.

Lin R, Ding L, Casola C, Ripoll DR, Feschotte C, Wang H. Transposase-derived transcription factors regulate light signaling in *Arabidopsis*. Science. 2007;318:1302–5.

Liu R, Xu YH, Jiang SC, Lu K, Lu YF, Feng XJ, Wu Z, Liang S, Yu YT, Wang XF, Zhang DP. Light-harvesting chlorophyll a/b-binding proteins, positively involved in abscisic acid signaling, require a transcription repressor, WRKY40, to balance their fucntion. J Exp Bot. 2013;64:5443–56.

Mazzella MA, Arana MV, Staneloni RJ, Perelman S, Rodriguez Batiller MJ, Muschietti J, Cerdán PD, Chen K, Sánchez RA, Zhu T, Chory J, Casal JJ. Phytochrome control of the Arabidopsis transcriptome anticipates seedling exposure to light. Plant Cell. 2005;17:2507–16.

Moriwaki T, Miyazawa Y, Fujii N, Takahashi H. Light and abacisic acid signaling are integrated by MIZ1 gene expression and regulate hydrotropic response in roots of Arabidopsis thaliana. Plant, Cell Environ. 2012;35:1359–68.

Moriwaki T, Miyazawa Y, Takahashi H. Transcriptome analysis of gene expression during the hydrotropic response in Arabidopsis seedlings. Environ Exp Bot. 2010;69:148–57.

Nagatani A. Phytochrome: structural basis for its functions. Curr Opin Plant Biol. 2010;13:565–70.

Nambara E, Marion-Poll A. Abscisic acid biosynthesis and catabolism. Annu Rev Plant Biol. 2005;56:165–85.

Oh E, Kang H, Yamaguchi S, Park J, Lee D, Kamiya Y, Choi G. Genome-wide analysis of genes targeted by PHYTOCHROME INTERACTING FACTOR 3-LIKE5 during seed germination in Arabidopsis. Plant Cell. 2009;21:403–19.

Oh E, Kim J, Park E, Kim JI, Kang C, Choi G. PIL5, a phytochrome-interacting basic helix-loop-helix protein, is a key negative regulator of seed germination in Arabidopsis thaliana. Plant Cell. 2004;16:3045–58.

Oh E, Yamaguchi S, Hu J, Yusuke J, Jung B, Paik I, Lee HS, Sun TP, Kamiya Y, Choi G. PIL5, a phytochrome-interacting bHLH protein, regulates gibberellin responsiveness by binding directly to the GAI and RGA promoters in Arabidopsis seeds. Plant Cell. 2007;19:1192–208.

Oh E, Yamaguchi S, Kamiya Y, Bae G, Chung WI, Choi G. Light activates the degradation of PIL5 protein to promote seed germination through gibberellin in Arabidopsis. Plant J. 2006;47:124–39.

Osterlund MT, Hardtke CS, Wei N, Deng XW. Targeted destabilization of HY5 during light-regulated development of Arabidopsis. Nature. 2000;405:462–6.

Oyama T, Shimura Y, Okada K. The Arabidopsis HY5 gene encodes a bZIP protein that regulates stimulus-induced development of root and hypocotyl. Genes Dev. 1997;11:2983–95.

Park J, Lee N, Kim W, Lim S, Choi G. ABI3 and PIL5 collaboratively activate the expression of SOMNUS by directly binding to its promoter in imbibed Arabidopsis seeds. Plant Cell. 2011;23:1404–15.

Piskurewicz U, Turečková V, Lacombe E, Lopez-Molina L. Far-red light inhibits germination through DELLA-dependent stimulation of ABA synthesis and ABI3 activity. EMBO J. 2009;28:2259–71.

Sankhla N, Sankhla D. Reversal of (±)-abscisin II-induced inhibition of lettuce seed germination. Physiol Plant. 1968;21:190–5.

Sawada Y, Aoki M, Nakaminami K, Mitsuhashi W, Tatematsu K, Kushiro T, Koshiba T, Kamiya Y, Inoue Y, Nambara E, Toyomasu T. Phytochrome- and gibberellin-mediated regulation of abscisic acid metabolism during germination of photoblastic lettuce seeds. Plant Physiol. 2008;146:1386–96.

Seo M, Hanada A, Kuwahara A, Endo A, Okamoto M, Yamauchi Y, North H, Marion-Poll A, Sun TP, Koshiba T, Kamiya Y, Yamaguchi S, Nambara E. Regulation of hormone metabolism in Arabidopsis seeds: phytochrome regulation of abscisic acid metabolism and abscisic acid regulation of gibberellin metabolism. Plant J. 2006;48:354–66.

Seo M, Nambara E, Choi G, Yamaguchi S. Interaction of light and hormone signals in germination seeds. Plant Mol Biol. 2009;69:463–72.

Shen H, Luong P, Huq E. The F-box protein MAX2 functions as a positive regulator of photomorphogenesis in Arabidopsis. Plant Physiol. 2007;145:1471–83.

Shen H, Moon J, Huq E. PIF1 is regulated by light-mediated degradation through the ubiquitin-26S proteasome pathway to optimize photomorphogenesis of seedlings in Arabidopsis. Plant J. 2005;44:1023–35.

Shen H, Zhu L, Bu QY, Huq E. MAX2 affects multiple hormones to promote photomorphogenesis. Mol Plant. 2012;5:750–62.

Shen YY, Wang XF, Wu FQ, Du SY, Cao Z, Shang Y, Wang SL, Peng CC, Yu XC, Zhu SY, Fan RC, Xu YH, Zhang DP. The Mg-chelatase H subunit is an abscisic acid receptor. Nature. 2006;443:823–6.

Shi H, Zhong S, Mo X, Liu N, Nezames CD, Deng XW. HFR1 sequesters PIF1 to govern the transcriptional network underlying light-initiated seed germination in Arabidopsis. Plant Cell. 2013;25:3770–84.

Shinomura T. Phytochrome regulation of seed germination. J Plant Res. 1997;110:151–61.

Shinomura T, Nagatani A, Chory J, Furuya M. The induction of seed germination in Arabidopsis thaliana isregulated principally by Phytochrome B and secondarily byPhytochrome A. Plant Physiol. 1994;104:363–71.

Shinomura T, Nagatani A, Hanzawa H, Kubota M, Watanabe M, Furuya M. Action spectra for phytochrome A- and B-specific photoinduction of seed germination in Arabidopsis thaliana. Proc Natl Acad Sci USA. 1996;93:8129–33.

Staneloni RT, Rodriguez-Batiller MJ, Casal JJ. Abscisic acid, high-light, and oxidative stress down-regulate a photosynthetic gene via a promoter motif not involved in phytochrome-mediated transcriptional regulation. Mol Plant. 2008;1:75–83.

Stephenson PG, Terry MJ. Light signaling pathways regulating the Mg-chelatasebranchpoint of chlorophyll synthesis during de-etiolation in Arabidopsis thaliana. Photochem Photobiol Sci. 2008;7:1243–52.

Takahashi N, Goto N, Okada K, Takahashi H. Hydrotropism in abscisic acid, wavy, and gravitropic mutants of Arabidopsis thaliana. Planta. 2002;216:203–11.

Tanaka R, Tanaka A. Tetrapyrrole biosynthesis in higher plants. Annu Rev Plant Biol. 2007;58:321–46.

Tang WJ, Ji Q, Huang Y, Jiang Z, Bao M, Wang H, Lin R. FHY3 and FAR1 transcription factors integrate light and abscisic acid signaling in Arabidopsis. Plant Physiol. 2013;163:857–66.

Thompson AJ, Jackson AC, Parker RA, Morpeth DR, Burbidge A, Taylor IB. Abscisic acid biosynthesis in tomato: regulation of zeaxanthin epoxidase and 9-cisepoxycarotenoid dioxygenase mRNAs by light/dark cycles, water stress and abscisic acid. Plant Mol Biol. 2000;42:833–45.

Toyomasu T, Yamane H, Murofushi N, Inoue Y. Effects of exogenously applied gibberellin and red light on the endogenous levels of abscisic acid in photoblastic lettuce seeds. Plant Cell Physiol. 1994;35:127–9.

Voigt C, Oster U, Bornke F, Jahns P, Dietze KJ, Leister D, Kleine T. In-depth analysis of the distinctive effects of norflurazon implies that tetrapyrrole biosynthesis, organellar gene expression and ABA cooperate in the GUN-type of plastid signaling. Physiol Plant. 2010;138:503–19.

Wang H, Deng XW. *Arabidopsis* FHY3 defines a key phytochromeA signaling component directly interacting with its homologous partner FAR1. EMBO J. 2002;21:1339–49.

Warpeha KM, Upadhyay S, Yeh J, Adamiak J, Hawkins SI, Lapik YR, Anderson MB, Kaufman LS. The GCR1, GPA1, PRN1, NF-Y signal chain mediates both blue light ans abscisic acid responses in Arabidopsis. Plant Physiol. 2007;143:1590–600.

Weatherwax SC, Ong MS, Degenhardt J, Bray EA, Tobin EM. The interaction of light and abscisic acid in the regulation of plant gene expression. Plant Physiol. 1996;111:363–70.

Wu FQ, Xin Q, Cao Z, Liu ZQ, Du SY, Mei C, Zhao CX, Wang XF, Shang Y, Jiang T, Zhang XF, Yan L, Zhao R, Cui ZN, Liu R, Sun HL, Yang XL, Su Z, Zhang DP. The Mg-chelatase H subunit binds abscisic acid and functions in abscisic acid signaling: New evidence in Arabidopsis. Plant Physiol. 2009;150:1940–54.

Xu YH, Liu R, Yan L, Liu ZQ, Jiang SC, Shen YY, Wang XF, Zhang DP. Light-harvesting chlorophyll a/b-binding proteins are required for stomatal response to abacisic acid in Arabidopsis. J Exp Bot. 2012;63:1095–106.

Yu X, Liu H, Klejnot J, Lin C. The cryptochrome blue light receptors. The Arabidopsis book. 2010;8:e0135.

Chapter 14
ABA Metabolism and Signaling in Fleshy Fruits

Yuan-Yue Shen and Jocelyn K.C. Rose

Abstract Fleshy fruits represent an important component of the human diet in many parts of the world and often provide valuable sources of vitamins and antioxidant compounds. Much of their nutritional value results from complex biochemical processes that occur during ripening, including changes in color, texture, flavor and aroma. The regulation of fruit development and subsequent ripening is controlled by several phytohormones, of which perhaps the most characterized has been the gaseous hormone ethylene, which triggers the ripening program in climacteric fruits. However, recent studies have reinforced the importance of abscisic acid (ABA) biosynthesis, catabolism, and signaling in both the development and ripening of fleshy fruits. Here, we provide an overview of these advances and summarize the insights that they have provided into the molecular mechanisms for ABA from production to action, with particular emphasis on strawberry fruit.

Keywords ABA metabolism · ABA signaling · Fleshy fruit

14.1 Introduction

Ripened fleshy fruits play important roles in our food supply, nutrition, and health. Fleshy fruits are generally divided into climacteric and non-climacteric types on the basis on variations in respiration intensity and ethylene production during ripening. Many of the complex biochemical, molecular and physiological processes that occur

Y.-Y. Shen (✉)
Beijing Key Laboratory for Agricultural Application and New Technique, Beijing University of Agriculture, 102206 Beijing, China
e-mail: sfmn@tom.com

J.K.C. Rose
Department of Plant Biology, Cornell University, Ithaca, NY 14853, USA
e-mail: jr286@cornell.edu

D.-P. Zhang (ed.), *Abscisic Acid: Metabolism, Transport and Signaling*,
DOI 10.1007/978-94-017-9424-4_14

during the development and ripening of fleshy fruits have been shown to be regulated by plant hormones. Of these, there has been particular focus on the central role played the gaseous hormone ethylene in controlling ripening in climacteric fruits, such as tomato (*Solanum lycopersicum*) (Alexander and Grierson 2002; Adams-Phillips et al. 2004; Prasanna et al. 2007; Gapper et al. 2013; Seymour et al. 2013; Perotti et al. 2014). However, there has been a recent increase in the number of studies that have addressed the roles of other phytohormones in fruit biology, as well as the extent and nature of hormonal cross-talk. This is exemplified by studies of abscisic acid (ABA), which has been demonstrated to play an important role in regulating the ripening of non-climacteric and climacteric fleshy fruits (Zhang et al. 2009; Gambetta et al. 2010; Jia et al. 2011; Chai et al. 2011; Akagi et al. 2012; Sun et al. 2012a, b; Romero et al. 2012; Li et al. 2013; Nicolas et al. 2014). Indeed, it has been proposed that, in addition to ethylene, ABA is a fundamentally important regulator of fruit maturation and senescence (Zhang et al. 2009).

It is well established that ABA is extremely important in the adaptation of plants to adverse environmental conditions, as well as regulating various aspects of plant growth and development, such as seed/bud dormancy and growth in both the seedling, and root (Leung and Giraudat 1998; Finkelstein et al. 2002; Himmelbach et al. 2003; Hirayama and Shinozaki 2007; Gapper et al. 2013; Seymour et al. 2013; Perotti et al. 2014). Moreover, much is now known about the molecular processes associated with ABA metabolism, signaling and downstream responses. ABA receptors, which were sought for by scientists for several decades, have now been identified (Fujii et al. 2009; Ma et al. 2009; Miyazono et al. 2009; Melcher et al. 2009; Nishimura et al. 2009; Santiago et al. 2009; Shang et al. 2010), and studies using the experimental model plant *Arabidopsis thaliana* have led to the discovery of two core ABA signaling pathways: ABA-PYR1-PP2C-SnRK2 (Fujii et al. 2009) and ABA-ABAR-WRKY40-ABI5 (Shang et al. 2010). In addition to these breakthroughs in elucidating ABA perception and signal transduction, much progress has also been made toward understanding ABA signaling specifically in fleshy fruits (Zhang et al. 2009; Gambetta et al. 2010; Jia et al. 2011; Chai et al. 2011; Akagi et al. 2012; Sun et al. 2012a, b; Romero et al. 2012; Li et al. 2013; Nicolas et al. 2014). In this review, we summarize some current ideas and observations regarding the roles of ABA in the regulation of fleshy fruit development and ripening and, where appropriate, recent insights into the underlying molecular mechanisms.

14.2 Effects of ABA on Fleshy Fruit Development and Ripening

Earlier studies of the significance of ABA in fleshy fruit development and ripening primarily targeted the berries of grape (*Vitis vinifera*), a non-climacteric fruit, where evidence was uncovered of a relationship between ABA and the veraison stage of development, which is marked by a berry color change at the onset of ripening (Coombe and Hale 1973; Coombe 1976; Inaba et al. 1976; Scienza et al. 1978; Palejwala et al. 1985; Cawthon and Morris 1982; Davies et al. 1997). Endogenous ABA levels are minimal 1 week prior to veraison and the application of exogenous ABA to grape berries can markedly accelerate the ripening process, including the rapid accumulation of sugars and anthocyanins (Coombe and Hale 1973; Inaba et al. 1976; Palejwala et al. 1985; Kataoka et al. 1982; Matsushima et al. 1989; Jeong et al. 2004). It is now widely accepted that ABA is a key factor in the regulation of diverse events during grape berry ripening including coloration, sugar accumulation, acid decline and flesh softening (Yu et al. 2006; Wheeler et al. 2009; Gambetta et al. 2010; Koyama et al. 2010; Giribaldi et al. 2010; Gagné et al. 2011; Nicolas et al. 2014).

In addition to studies in grape berries, considerable progress has been made recently in understanding the role of ABA in other non-climacteric fruits and environmental factors that alter ABA levels. For example, water deficit treated strawberry (*Fragaria* x *ananassa*) fruits have higher ABA levels, as well as reduced berry size, altered sugar/acid ratios, which affect the taste and levels of health-related compounds (phenolic and antioxidant compounds, Terry et al. 2007). Indeed, there is considerable physiological and molecular evidence to suggest an important role for ABA in the regulation of strawberry fruit ripening: (1) exogenous application of ABA or dimethyl sulfoxide (DMSO, an ABA biosynthesis accelerator) significantly promotes fruit ripening, whereas fluridone (an ABA biosynthesis inhibitor) markedly inhibits fruit ripening; (2) silencing of *FaNCED1*, a key ABA biosynthesis gene in strawberry fruit, results in a reduction in endogenous ABA levels and delayed ripening; (3) exogenous ABA rescues the uncolored phenotype of *FaNCED1*-RNAi fruits, in which *FaNCED1* expression has been suppressed using an RNA interference (RNAi) strategy (Jia et al. 2011). The importance of ABA in strawberry fruit development was also confirmed by silencing of the β-glucosidase gene *FaBG3*, which is involved in ABA synthesis, since the encoded protein hydrolyzes an ABA glucose ester to release active ABA (Li et al. 2013). These findings together with other studies indicate that ABA as a key regulator of strawberry fruit ripening (Kano and Asahira 1981; Manning 1994; Perkins-Veazie 1995; Jiang and Joyce 2003; Terry et al. 2007; Jia et al. 2011; Li et al. 2013). However, it should be noted that in addition to its role in grape berry and strawberry ripening, ABA has been found to influence diverse processes of characteristics in other fleshy fruit species, including the development of astringency in persimmon (*Diospyros kaki*) fruit (Akagi et al. 2012), *citrus* peel development (Romero et al. 2012), and cherry (*Prunus* species) fruit maturation (Ren et al. 2010).

In addition to non-climacteric fruits, it is also now clear that ABA can regulate fruit development and ripening in climacteric fruits. For example, it was reported that application of ABA to detached tomato fruits substantially advances the onset of ripening (Mizrahi et al. 1975), which raises that possibility that ABA may also act to promote tomato fruit ripening in vivo. A role for ABA regulating tomato fruit development was indicated by studies of an ABA deficient tomato mutant (*high-pigment 3*) that accumulates 30 % more carotenoids, has an increased number of fruit plastids and a higher lycopene content (Galpaz et al. 2008) than wild type frutis. In addition, the results of targeted studies with transgenic tomato suggest that ABA regulates the degree of pigmentation, carotenoid composition, and fruit firmness during ripening (Sun et al. 2012a, b). Furthermore, the ABA-deficient *notabilis/flacca* (*not/flc*) tomato double mutant has been used to demonstrate that ABA stimulates cell enlargement and increases fruit size (Nitsch et al. 2012).

In addition, a recent report revealed that ABA may also be involved in suberization-based wound healing processes in tomato fruit stem scar tissue (Leide et al. 2011). Other studies have shown that ABA is involved in the regulation of avocado fruit growth (Cowan et al. 1997) and peach fruit sugar accumulation (Kobashi et al. 2001), suggesting that ABA is likely a common signaling molecule that modulates many processes in both climacteric and non-climacteric fleshy fruits.

14.3 ABA Metabolism in Fleshy Fruits

In land plants, ABA levels in a specific tissue are determined by the balance between ABA biosynthesis, catabolism and conjugation. Biosynthesis results from the oxidative cleavage of carotenoids, where the activity of the enzyme 9-*cis*-epoxycarotenoid dioxygenase (NCED) represents a major rate limiting step. Conversely, ABA catabolism results mainly from 8′-hydroxylation reactions catalyzed by 8′-hydroxylase enzymes, which are encoded by a small family of P450 monooxygenase CYP707A genes (Zeevaart et al. 1989; Nambara and Marion-Poll 2005). In addition, ABA conjugation can result from the action of ABA-induced glucosyltransferases (GTs), which convert free ABA to an ABA-glucosylester (ABA-GE; Xu et al. 2002). Earlier reports suggested that ABA-GE represents an inactive end product of ABA catabolism (Lehmann and Schutte 1984; Zeevaart 1999). However, this idea is called into question as a result of the identification of Arabidopsis beta-glucosidase1 (AtBG1), an enzyme that catalyzes the conversion of ABA-GE back into the pool of biologically active ABA, allowing a rapid change in ABA levels (Lee et al. 2006). A contemporary model might therefore involve the control of plant cellular ABA levels by NCEDs and CYPs in a synthesis/degradation pathway or/and by GTs and BGs in a conjugation/dissociation pathway. Potentially, the one-step pathway catalyzed by GTs or BGs allows rapid dynamic changes in ABA levels to meet developmental and adaptive requirements.

This foundation of knowledge resulting from studies of *A. thaliana* and model crop species has been valuable in elucidating ABA biosynthesis and catabolism in fleshy fruits and in identifying conserved mechanisms of ABA metabolism (Li et al. 2011, 2013; Jia et al. 2011; Sun et al. 2012b). For example, down regulation of *FaNCED1* expression in strawberry using virus induced gene silencing (VIGS) resulted in a significant decrease in ABA levels, as well as uncolored fruits, demonstrating that *FaNCED1* is a key gene for ABA biosynthesis in fruits and plays an important role in fruit ripening (Jia et al. 2011). Moreover, Li et al. (2013) reported the importance of BG enzymes in strawberry, showing that *FaBG3*-RNAi-treated fruit has reduced ABA levels and fruit color development and softening is inhibited, indicating a key role of β-glucosidases in regulation of ripening (Li et al. 2013). Interestingly, similar results were generated through studies of tomato fruit, where ABA levels are regulated by *SlNCED1* and *SlCYP707A1* (Nitsch et al. 2009; Zhang et al. 2009). It was observed that suppressing the expression of the *SlNCED1*using RNAi lead to a 20–50 % decrease in both ABA accumulation and *SlNCED1* transcript compared with control fruit, and prolonged fruit shelf life by up to 3 weeks (Sun et al. 2012b).

In addition to these important breakthroughs made through observations of strawberry and tomato fruits (Li et al. 2011, 2013; Jia et al. 2011; Sun et al. 2012b), similar studies have been reported of ABA biosynthesis and catabolism in other fruits. Water deficit results in a nearly two-fold increase in ABA concentration in berries of a red-wine grape Cabernet Sauvignon, the same stress induces a decrease in ABA levels in the Chardonnay white-wine grape at veraison and shortly thereafter (Deluc et al. 2009). The higher transcript levels of *VvNCED1* and *VvBG1* together with lower transcript abundance of *VvCYP1* and *VvGT* contribute to a continuous accumulation of ABA, especially *VvBG1* transcripts increased more rapidly than that of *VvNCED1* during berry red-coloring. An incubation test in vitro indicated that the *Escherichia coli*-expressed VvBG1 protein had high enzymatic activity, validating that β-glucosidase (VvBG1) has high enzymatic activity and might play a role in berry ripening (Sun et al. 2014). In avocado fruits, *PaNCED1* and *PaNCED3*, but not *PaNCED2*, were strongly induced as the fruit ripened, and a correlation with water stress was also established, since *PaNCED1* was induced under these conditions. Furthermore, recombinant PaNCED1 and PaNCED3, but not PaNCED2 could cleave 9-*cis*-xanthophylls into xanthoxin and C (25)-apocarotenoids in vitro, indicating that ABA biosynthesis in avocado is regulated at the level of carotenoid cleavage (Chernys and Zeevaart 2000). In *citrus* fruit, the accumulation of both ABA and ABA-conjugates occurs during fruit maturation (Harris et al. 1986). In *citrus* peels, high levels of ABA are detected and CsNCED1 is likely to play a primary role in ABA biosynthesis (Rodrigo et al. 2014). Moreover, in the flavedo and juice sacs, expression of *CitNCED2* and/or *CitNCED3* increases coincident with a substantial accumulation of ABA (Kato et al. 2006). In watermelon fruit, the expression of *ClBG1*, *ClNCED4* and *ClCYP707A1* has been shown to increase rapidly along with ripening, and reaches the highest levels at the stage of harvest. This trend was show to be consistent with ABA accumulation, which was also shown to be modulated by the described dynamic balance between biosynthesis and catabolic processes via these genes, indicating the importance of glucosidase genes in the ripening process (Li et al. 2012). Similarly, it was shown that the endogenous ABA content of sweet cherry fruit is regulated by *PacNCED1*, *PacCYP707A1* and *PacCYP707A3* transcripts during maturation (Ren et al. 2010), and in apple fruits, the major portion of the ABA has been shown to pool is conjugated to β-D-glucopyranosyl abscisate (ABA-GE) which leads to markedly lower ABA levels (Rock and Zeevaart 1990). The peach *PpNCED1* gene has been shown to initiate ABA biosynthesis at the onset of fruit ripening (Zhang et al. 2009). These reports again reinforce the idea that ABA accumulation is involved in the regulation of ripeness and senescence in a broad range of fleshy fruits.

In all, the detection of key ABA biosynthesis and catabolism genes in fleshy fruit tissues suggests that although ABA could be transported from the leaves to the fruits via the phloem (Shiozaki et al. 1999), it can also be synthesized in fleshy fruits in situ. ABA levels in fleshy fruits are regulated mainly by NCED1 and CYP707A1, the accumulation of ABA-GE catalyzed by GT enzymes, contributes to a rapid response by one-step dissociation of ABA-EG catalyzed by BG in response to developmental and stress cues. During later stages of fruit development, beta-glucosidases (BGs) appear to play a more important role in ripening.

14.4 ABA Signaling in Fleshy Fruits

The presence of both transmembrane and cytosolic ABA receptors in plants was suggested by early studies using experimental model plants (Hornberg and Weiler 1984; Allan et al. 1994; Schwarz and Schroeder 1998), such as the identification of multiple receptors, secondary messengers, protein kinases and phospholipases, transcription factors, *cis*-elements, and target genes in *A. thaliana*. Known ABA receptors include the plasma membrane localized GTG1/GTG2 (Pandey et al. 2009), a class of cytosolic PYR/PYL/RCAR (Park et al. 2009; Ma et al. 2009) proteins and a plastid/chloroplast magnesium-chelatase H subunit ABAR/CHLH (Shen et al. 2006; Wu et al. 2009; Shang et al. 2010). Signalling components include G proteins, phospholipases, and various protein kinases, such as receptor-like kinases, SNF1-related protein kinases (SnRKs), calcium dependent protein kinases (CDPKs), calcineurin B-like protein kinases (CIPKs), and mitogen-activated protein kinases (MAPKs). Protein phosphatases of type-2C/A protein phosphatase (PP2C/A), various classes of transcription factors, including MYBs/MYCs, B3 domain transcription factors, APETALA2 domain transcription factors, bZIP domain transcription factor, and WRKY transcription factors, are also among important regulators of ABA levels (Nambara and Marion-Poll 2005; Hirayama and Shinozaki 2007; Verslues and Zhu 2007; Wang and Zhang 2008, and Cutler et al. 2010). To date, two core ABA signalling pathways in *A. thaliana* have been proposed: the 'ABA-PYR/PYL/RCAR-PP2C-SnRK2' pathway (Fujii et al. 2009), and the'ABA-ABAR-WRKY40-ABI5'pathway (Shang et al. 2010). At the mechanistic level, a detailed gate latch-lock mechanism has been identified, in which ABA promotes the interaction of PYR1 and PP2C, resulting in PP2C inhibition and SnRK2 activation. This transduces ABA signals through phosphorylation of downstream factors such as AREB/ABF, ion channels, and NADPH oxidases (Park et al. 2009; Ma et al. 2009; Fujii et al. 2009).

14.4.1 ABA Perception in Flesh Fruits

In addition to pioneering studies in model experimental plants, there has also been much interest in understanding the molecular basis of ABA detection by specific receptors and subsequent downstream signalling in fleshy fruits. In studies of grape berry microsomes, a high proportion of specific ABA receptors (which had a measured K_d value of 17.5–50 nM) were shown to be localized mainly in the endomembranes and not in the plasma membrane or in the cytoplasm (Zhang et al. 1999). In contrast, ABA binding activity was scarcely detectable in microsomes derived from the flesh of developing apple fruit, but high ABA binding activity was detected in the cytosolic fraction with both high affinity (K_d 2.3 nM) and low affinity sites (K_d 58.8 nM; Zhang et al. 2001). Progress has also been made in understanding ABA perception in strawberry. Using a newly established tobacco rattle virus (TRV)-induced gene silencing technique in strawberry fruit, down-regulation of the expression levels of *FaPYR1* or *FaCHLH/ABAR*, homologous to the *A. thaliana* ABA receptor genes, was shown to inhibit ripening (Jia et al. 2011; Chai et al. 2011). On the basis of these results, a model was proposed for ABA perception and signalling transduction during non-climacteric fruit ripening (Li et al. 2011). More recently, the Type 2C protein phosphatase FaABI1 was demonstrated to be a negative regulator of strawberry fruit ripening (Jia et al. 2013) and it was suggested that the 'ABA-FaPYR1-FaPP2C-FaSnRK2' signalling pathway, involving the positive regulatory effect of FaPYR1 in combination with the negative regulatory effect of FaABI1, represents a core mechanism by which ABA regulates strawberry fruit ripening (Jia et al. 2013).

Recently, the ABA perception mechanisms and core signalling systems in other fruit tree species have also been studied. The *V. vinifera* proteins VvRCAR6 and VvRCAR5 may be the major receptors involved in ABA perception and signalling in grape, mainly through VvPP2C4 (Boneh et al. 2012), while VvPYL1 may be an ABA receptor that modulates ABA signalling by inhibiting type PP2C activity (Li et al. 2012). In tomato, *SlPYL1* and *SlPYL2* are expressed at high levels throughout fruit development and ripening, and *SlPP2C1* and *SlPP2C5* are both strongly expressed during the breaker stage, while *SlSnRK2.2*, *SlSnRK2.3*, *SlSnRK2.4*, and *SlSnRK2C* are highly expressed at all maturity stages (Sun et al. 2011).

14.4.2 Downstream Components of ABA Signaling in Fleshy Fruits

14.4.2.1 Protein Kinases and Phosphatases

Plant protein kinases and phosphatases include CDPKs, SNF1-related kinases (SnRKs), MAPKs, a receptor-type kinase (RPK1), and protein phosphatase PP2Cs (Hirayama and Shinozaki 2007). As described above, reversible protein phosphorylation has been demonstrated to play key roles in ABA signal transduction through the protein kinase SnRK2 and the phosphataseABI1 (Fujii et al. 2009).

In grape berries, earlier studies revealed high activities of both the calcium-dependent protein kinase (CDPK) and the mitogen-activated protein kinase (MAPK) in the lag phase of fruit growth, prior to the ripening stage (Shen et al. 2004). The subsequent experiments lead to the identification and purification of a 58-kD ABA-stimulated CDPK, ACPK1, which localizes to both the plasma membranes and chloroplasts/plastids. The fact that ACPK1 is expressed in a developmental stage-dependent manner, and that it positively regulates the plasma membrane H^+-ATPase in vitro, together with the observation that ABA stimulates ACPK1 in a dose-dependent manner, suggest that ACPK1 may be involved in the ABA-signalling pathway during berry development (Yu et al. 2006). Similar studies found that an apple MAPK signalling cascade, MdMKK1-MdMPK1, is involved in ABA signalling, and that MdMPK1 phosphorylates the ABI5 protein through a unique residue, Ser314, making ABI5 a potential direct downstream component of MAPK in ABA signalling (Wang et al. 2010). In grape berries, a glycogen synthase kinase 3 protein kinase, VvSK1, is strongly expressed when the berries accumulate glucose, fructose, and ABA, and overexpression of *VvSK1* results in an upregulation of the transcripts of four monosaccharide transporters (*VvHT3*, *VvHT4*, *VvHT5*, and *VvHT6*), an up to 5-fold increase in the rate of glucose uptake, and a doubling in the amount of glucose and sucrose accumulation,. This indicates that VvSK1 controls sugar uptake and accumulation by a sugar/ABA-inducible protein kinase (Lecourieux et al. 2010). More importantly, the core phosphorylation signaling of PP2C-SnRK2 has been established in response to ABA during strawberry fruit ripening (Jia et al. 2013).

14.4.2.2 Transcription Factor, *Cis*-Elements and Target Genes

Many transcription factors, *cis*-elements and target genes related to ABA signaling have been identified in developing fleshy fruit. For example, in persimmon fruit, DkbZIP5 was shown to recognize ABA-responsive elements in the promoter region of DkMyb4 and to act as a direct regulator of DkMyb4 in an ABA-dependent manner. It is reported to suggest that ABA signals may be involved in proanthocyanidin (PA) biosynthesis via DkMyb4 activation by DkbZIP5 (Akagi et al. 2012). In tomato fruit, the higher hexose and acid concentration in *SlAREB1*-overexpressing lines can be correlated with an increased expression of genes encoding a vacuolar invertase and a sucrose synthase, suggesting that an AREB-mediated ABA signal affects the metabolism of acid and sugar in tomato fruit development (Bastas et al. 2011). ABA can also affect cell wall catabolism during tomato fruit ripening via down-regulation of the expression of major catabolic genes (*SlPG*, *SlPME*, *SlTBG*, *SlXET*, *SlCels*, and *SlExp*; Sun et al. 2012b), but an ABA biosynthesis inhibitor can increase transcript levels corresponding to enzyme activities associated with other aspects of fruit ripening, as is the case with an alcohol dehydrogenase from mango (Singh et al. 2010) and an indole-3-acetic acid-amido synthetase, DlGH3.2, in longan (Kuang et al. 2011). It has been proposed that the expression of *FaASR*, a member of the ABA-, stress- and ripening-induced (ASR) set of genes might partially contribute to the acceleration of the fruit ripening (Chen et al. 2011). Alterations in strawberry *FaABI1* expression were reported to regulate the transcripts of a set of both ABA-responsive and ripening-related genes, including *ABI3*, *ABI4*, *ABI5*, *SnRK2*, *ABRE1*, *CHS*, *PG1*, *PL*, *CHI*, *F3H*, *DFR*, *ANS*, and *UFGT* (Jia et al. 2013).

Taken together, much progress has been made toward understanding ABA signaling, including the identification of ABA receptors, protein kinases, protein phosphatases, transcription factors, *cis*-elements, and target genes in fleshy fruit. These include the identification of core signaling components controlling the key links to fleshy fruit development, including strawberry 'ABA-FaPYR1-FaPP2C-FaSnRK2', grape 'ABA-VvACPK1-H^+-ATPase' and 'ABA-VvSK1-VvHTs', apple 'ABA-MdMKK1-MdMPK1-ABI5', and persimmon 'ABA- DkbZIP5-DkMyb4-PA'. To date, the target genes include those involved in fruit softening (*PG*, *PL*, *PME*, *TBG*, *XET*, *Cels*, and *Exp*)-, sugar metabolism (*VvHTs*, *MiADH2*)-, pigmentation (*PA*, *CHS*, *CHI*, *F3H*, *DFR*, *ANS*, and *UFGT*), and ripening (*FaASR*, *DlGH3.2*)-related genes.

14.5 Understanding of the Mechanisms for ABA Regulation of Fleshy Fruit Ripening

There is considerable interest at present in the relationship between ABA, ethylene and sugar content in ripening fleshy fruits. Two previous reports suggested that ABA facilitates the initiation and progression of ethylene-mediated ripening events, possibly by enhancing the sensitivity to ethylene (Jiang et al. 2000) or ethylene levels (Riov et al. 1990). In grape berries, sugar and ABA signaling orthologs are activated at the onset of the ripening, including the putative sucrose sensor SUT2, core G-protein signaling components, GPA1 and RGS1, hexose kinases (Hxk), PP2C protein phosphatases, Snf1-related kinases (SnRK), the sugar-related WRKY, ABA-related homeodomain–leucine zipper, as well as homeobox (HB) transcriptions factors, ABRE-binding factor (ABF), and AP2 transcription factors (Gambetta et al. 2010). Conserved changes in the dynamics of metabolic processes during fruit development and ripening across species were found (Klie et al. 2014). During tomato fruit growth, ABA stimulates cell enlargement by suppressing ethylene synthesis (Nitsch et al. 2012; Sun et al. 2012b) and during the ripening, a significant reduction in ABA levels is thought to promote coloring and firmness, resulting in the carbon that normally channels to ABA biosynthesis and catabolism being targeted to compounds upstream of ABA biosynthesis, including lycopene, carotene, and pectin (Sun et al. 2012a, b). Interestingly, in *citrus* fruit, ABA my induce its own biosynthesis at the transcriptional level and the feedback regulation of ABA has been shown to lead to decreases in carotenoid content in *citrus* juice sacs in vitro (Zhang et al. 2012). In strawberry fruit, sucrose may serve as a signalling molecule to promote mRNA expression levels of *FaNCED1*, as well as playing an important role in ABA accumulation and fruit ripening (Jia et al. 2011, 2013).

Fleshy fruit can synthesize ABA in response to developmental and environmental cues and the accumulation of ABA in fleshy fruit is controlled by four main enzymes classes that are fundamentally important in ABA biosynthesis and catabolism: NCEDs, CYP707As, GTs, and BGs. In the ripening of fleshy fruit, the interaction between ABA, ethylene and sugars occurs at physiological and molecular levels.

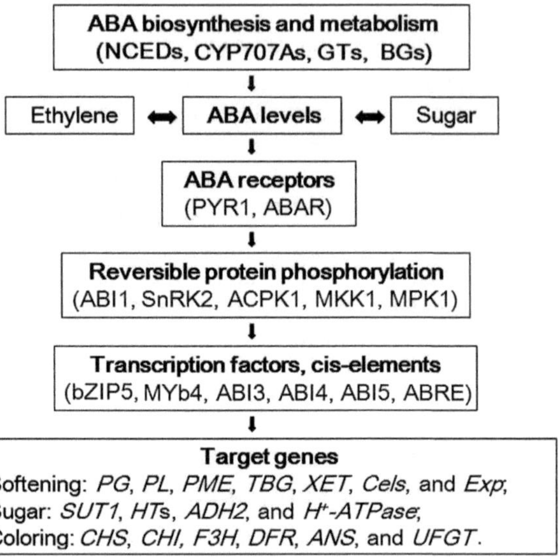

Fig. 14.1 A model for ABA metabolism and signaling in fleshy fruits. In response to developmental and environmental cues, fleshy fruit can synthesize ABA mainly by NCEDs and BGs while catabolize ABA by CYP707As and GTs. ABA levels in fleshy fruits are determined by the balance between ABA biosynthesis and catabolism. One-step dissociation of ABA-EG by BGs may rapidly accumulate ABA. The accumulated ABA in fleshy fruit can be perceived by its receptors, at least including PYR and ABAR. In turn, the initial signals may be relayed by the way of reversible protein phosphorylation, which involves ABI1, SnRK2, ACPK1, MKK1, and MPK1. Subsequently, the cascade signaling of phosphorylation can activate downstream transcript factors and corresponding *cis*-elements, including bZIP5, MYb4, ABI3, ABI4, ABI5, ABRE, then promote the expression of the ripening-related genes: softening-related genes include *PG*, *PL*, *PME*, *TBG*, *XET*, *Cels*, and *Exp*; sugar metabolism-related genes include *SUT1*, *HTs*, *ADH2*, and *H^+-ATPase*, and coloring-related genes include *CHS*, *CHI*, *F3H*, *DFR*, *ANS*, and *UFGT*. Finally these expressed genes accelerate fleshy fruit ripening, to a large extent, the processes of which could be cooperated by the interaction of ABA with sugar and ethylene. Abbreviations: NCDEs, 9-cis-epoxycarotenoid dioxygenases; CYP707As, P450 monooxygenases; GTs, glucosyltransferases; BGs, beta-glucosidases; ABA-GE, ABA-glucosylester; PYR, pyrabactin resistance; ABAR, ABA receptor; SnRK2: sucrose non-fermenting 1-related protein kinase 2; ACPK1, ABA-stimulated calcium-dependent protein kinase1; MKK1, mitogen-activated protein kinase kinase 1; MPK1, mitogen-activated protein kinase 1; bZIP5, basic leucine-zipper 5; MYB4, dehydration responsive element-related transcription factor 4; ABI3-5, abscisic acid insensitive 3-5; ABRE, ABA response element; *PG*, polygalacturonase gene; *PL*, pectate lyase gene; *PME*, pectin methyl esterase gene; *TBG*, β-galactosidase gene; *XET*, xyloglucan endotransglycosylase gene; *Cels*, endo-1,4-β-cellulose gene; *Exp*, expansin gene; *SUT1*, sucrose transporter 1 gene; *HTs*, monosaccharide transporter genes; *ADH2*, alcohol dehydrogenase 2 gene; *H^+-ATPase*, proton-pump ATPase gene; *CHS*, chalcone synthase gene; *CHI*, chalcone isomerase gene; *F3H*, flavanone 3-hydroxylase gene; *DFR*, dihydroflavonol 4-reductase gene; *ANS*, anthocyanidin synthase gene; *UFGT*, UDP-glucose:flavonoid 3-O-glucosyltransferase gene

The initiation of ABA action is triggered by the perception by its receptors, including PYRs and ABAR, causing signaling to be relayed by reversible protein phosphorylation of protein kinases and phosphatases, including ABI1, SnRK2, ACPK1, SK1,

MKK1, and MPK1. The signalling cascade then involves transcription factors and *cis*-elements, including bZIP5, MYb4, ABI3, ABI4, ABI5, ABRE, triggering ripening-related genes, which are involved in modulating softening, and accumulation of sugars and pigments. To a large extent, the processes of which could be cooperated by the interaction of ABA with sugar and ethylene (Fig. 14.1).

14.6 Conclusion and Perspective

Fleshy fruit ripening involves marked physiological and metabolic changes in sugar metabolism, softening, and color development, and it is clear that these processes are substantially regulated by plant hormones. While the role of ethylene in climacteric fruit ripening has been definitively established at the molecular level for many years, it is now emerging that ABA is not only involved in the regulation of climacteric fruit development and ripening, but also in the developmental processes of non-climacteric fruit. Much progress has been made toward understanding the molecular mechanisms involving ABA biosynthesis and signaling in strawberry, a model plant for non-climacteric fruit research. However, given the complexity of non-climacteric fruit ripening, the studies concerning the interaction of ABA with ethylene, of ABA with sugar levels, and of sugars with ethylene, is an important area of future research work.

Acknowledgments The research work in Shen's laboratory was supported by the National Key Basic Research '973' Program of China (grant no. 2012CB126306), the National Science Foundation of China (grant no 31272144), and the Project of Construction of Innovative Teams and Teacher Career Development for Universities and Colleges under Beijing Municipality (grant no. IDHT20140509).

References

Adams-Phillips L, Barry C, Giovannoni J. Signal transduction systems regulating fruit ripening. Trends Plant Sci. 2004;9:331–8.

Akagi T, Katayama-Ikegami A, Kobayashi S, Sato A, Kono A, Yonemori K. Seasonal abscisic acid signal and a basic leucine zipper transcription factor, DkbZIP5, regulate proanthocyanidin biosynthesis in persimmon fruit. Plant Physiol. 2012;158:1089–102.

Alexander L, Grierson D. Ethylene biosynthesis and action in tomato: a model for climacteric fruit ripening. J Exp Bot. 2002;53:2039–55.

Allan AC, Fricker MD, Ward JL, Beale MH, Trewavas AJ. Two transduction pathways mediate rapid effects of abscisic acid in *Commelina* guard cells. Plant Cell. 1994;6:1319–28.

Bastas A, Lpez-Climent M, Valcrcel M, Rosello S, Gmez-Cadenas A, Casaretto JA. Modulation of organic acids and sugar content in tomato fruits by an abscisic acid-regulated transcription factor. Physiol Plant. 2011;141:215–26.

Boneh U, Biton I, Zheng C, Schwartz A, Ben-Ari G. Characterization of potential ABA receptors in *Vitis vinifera*. Plant Cell Rep. 2012;31:311–21.

Cawthon D, Morris J. Relationship of number and maturity to berry development, fruit maturation, hormonal changes and uneven ripening of "Concord" (*Vitis labrusca* L.) grapes. J Am Soc Hort Sci. 1982;107:1097–104.

Chai YM, Jia HF, Li CL, Dong QH, Shen YY. FaPYR1 is involved in strawberry fruit ripening. J Exp Bot. 2011;62:5079–89.

Chen JY, Liu DJ, Jiang YM, Zhao ML, Shan W, Kuang JF, Lu WJ. Molecular characterization of a strawberry *FaASR* gene in relation to fruit ripening. PLoS ONE. 2011;6:e24649.

Chernys JT, Zeevaart JA. Characterization of the 9-*cis*-epoxycarotenoid dioxygenase gene family and the regulation of abscisic acid biosynthesis in avocado. Plant Physiol. 2000;124:343–53.

Coombe B. The development of fleshy fruits. Annu Rev Plant Physiol. 1976;27:507–28.

Coombe BG, Hale CR. The hormone content of ripening grape berries and the effects of growth substance treatments. Plant Physiol. 1973;51:629–34.

Cowan AK, Moore-Gordon CS, Bertling I, Wolstenholme BN. Metabolic control of avocado fruit growth (isoprenoid growth regulators and the reaction catalyzed by 3-hydroxy-3-methylglutaryl coenzyme a reductase). Plant Physiol. 1997;114:511–8.

Cutler SR, Rodriguez PL, Finkelstein RR, Abrams SR. Abscisic acid: emergence of a core signaling network. Annu Rev Plant Biol. 2010;61:651–79.

Davies C, Boss P, Robinson S. Treatment of grape berries, a nonclimateric fruit with a synthetic auxin, retards ripening and alters the expression of developmentally regulated genes. Plant Physiol. 1997;115:1155–61.

Deluc LG, Quilici DR, Decendit A, Grimplet J, Wheatley MD, Schlauch KA, Merillon JM, Cushman JC, Cramer GR. Water deficit alters differentially metabolic pathways affecting important flavor and quality traits in grape berries of cabernet sauvignon and chardonnay. BMC Genom. 2009;10:212.

Finkelstein R, Gampala S, Rock C. Abscisic acid signaling in seeds and seedlings. Plant Cell. 2002;14:S15–45.

Fujii H, Chinnusamy V, Rodrigues A, Rubio S, Aatoni R, Park SY, Cutler SR, Sheen J, Rotrigues PL, Zhu JK. In vitro reconstitution of an abscisic acid signalling pathway. Nature. 2009;462:660–4.

Gagné S, Cluzet S, Merillon JM, Gény L. ABA initiates anthocyanin production in grape cell cultures. J Plant Growth Regul. 2011;30:1–10.

Galpaz N, Wang Q, Menda N, Zamir D, Hirschberg J. Abscisic acid deficiency in the tomato mutant high-pigment 3 leading to increased plastid number and higher fruit lycopene content. Plant J. 2008;53:717–30.

Gambetta GA, Matthews MA, Shaghasi TH, McElrone AJ, Castellarin SD. Sugar and abscisic acid signaling orthologs are activated at the onset of ripening in grape. Planta. 2010;232:219–34.

Gapper NE, McQuinn RP, Giovannoni JJ. Molecular and genetic regulation of fruit ripening. Plant Mol Biol. 2013;82:575–91.

Giribaldi M, Geny L, Delrot S, Schubert A. Proteomic analysis of the effects of ABA treatments on ripening *Vitis vinifera* berries. J Exp Bot. 2010;61:2447–58.

Harris MJ, Dugger WM. The occurrence of abscisic acid and abscisyl-beta-d-glucopyranoside in developing and mature citrus fruit as determined by enzyme immunoassay. Plant Physiol. 1986;82:339–45.

Himmelbach A, Yang Y, Grill E. Relay and control of abscisic acid signaling. Curr Opin Plant Biol. 2003;6:470–9.

Hirayama T, Shinozaki K. Perception and transduction of abscisic acid signals: keys to the function of the versatile plant hormone ABA. Trends Plant Sci. 2007;12:343–51.

HornbergC Weiler EW. High-affinity binding sites for abscisic acid on the plasmalemma of *Vicia faba* guard cells. Nature. 1984;310:321–4.

Inaba A, Ishida M, Sobajima Y. Changes in endogenous hormone concentrations during berry development in relation to the ripening of Delaware grapes. J Jap Soc Hort Sci. 1976;45:245–62.

Jia HF, Chai YM, Li CL, Lu D, Luo JJ, Qin L, Shen YY. Abscisic acid plays an important role in the regulation of strawberry fruit ripening. Plant Physiol. 2011;157:188–99.

Jia HF, Lu D, Sun JH, Li CL, Xing Y, Qin L, Shen YY. Type 2C protein phosphatase ABI1 is a negative regulator of strawberry fruit ripening. J Exp Bot. 2013;64:1677–87.

Jiang Y, Joyce DC. ABA effects on ethylene production, PAL activity, anthocyanin and phenolic contents of strawberry fruit. Plant Growth Regul. 2003;39:171–4.

Jiang Y, Joyce DC, Macnish AJ. Effect of abscisic acid on banana fruit ripening in relation to the role of ethylene. J Plant Growth Regul. 2000;19:106–11.

Jeong S, Goto-Yamamoto N, Kobayashi S, Esaka M. Effects of plant hormones and shading on the accumulation of anthocyanins and the expression of anthocyanin biosynthetic genes in grape berry skins. Plant Sci. 2004;167:247–52.

Kano Y, Asahira T. Roles of cytokinin and abscisic acid in the maturing of strawberry fruits. J Jap Soc Hort Sci. 1981;50:31–6.

Kataoka I, Sugiura A, Utsunomiya N, Tomana T. Effect of abscisic acid and defoliation on anthocyanin accumulation in Kyoho grapes. Vitis. 1982;21:325–32.

Kato M, Matsumoto H, Ikoma Y, Okuda H, Yano M. The role of carotenoid cleavage dioxygenases in the regulation of carotenoid profiles during maturation in *citrus* fruit. J Exp Bot. 2006;57:2153–64.

Klie S, Osorio S, Tohge T, Drincovich MF, Fait A, Giovannoni JJ, Fernie AR, Nikoloski Z. Conserved changes in the dynamics of metabolic processes during fruit development and ripening across species. Plant Physiol. 2014;164:55–68.

Kobashi K, Sugaya S, Gemma H, Iwahori S. Effect of abscisic acid (ABA) on sugar accumulation in the flesh tissue of peach fruit at the start of the maturation stage. Plant Growth Regul. 2001;35:215–23.

Koyama K, Sadamatsu K, Goto-Yamamoto N. Abscisic acid stimulated ripening and gene expression in berry skins of the cabernet sauvignon grape. Funct Integr Genom. 2010;10:367–81.

Kuang JF, Zhang Y, Chen JY, Chen QJ, Jiang YM, Lin HT, Xu SJ, Lu WJ. Two GH3 genes from longan are differentially regulated during fruit growth and development. Gene. 2011;485:1–6.

Lecourieux F, Lecourieux D, Vignault C, Delrot S. A sugar-inducible protein kinase, VvSK1, regulates hexose transport and sugar accumulation in grapevine cells. Plant Physiol. 2010;152:1096–106.

Lee KH, Piao HL, Kim HY, Choi SM, Jiang F, Hartung W, Hwang I, Kwak JM, Lee IJ, Hwang J. Activation of glucosidase via stress-induced polymerization rapidly increases active pools of abscisic acid. Cell. 2006;126:1109–20.

Lehmamm H, Schutte HR. Abscisic-acid metabolism in intact wheat seedlings under normal and stress conditions. J Plant Physiol. 1984;117:201–9.

Leide J, Hildebrandt U, Hartung W, Riederer M, Vogg G. Abscisic acid mediates the formation of a suberized stem scar tissue in tomato fruits. New Phytol. 2011;194:402–15.

Leung J, Giraudat J. Abscisic acid signal transduction. Annu Rev Plant Physiol Plant Mol Biol. 1998;49:199–222.

Li CL, Jia HF, Chai YM, Shen YY. Abscisic acid perception and signaling transduction in strawberry: a model for non-climacteric fruit ripening. Plant Signal & Behav. 2011;6:1950–3.

Li Q, Li P, Sun L, Wang Y, Ji K, Sun Y, Dai S, Chen P, Duan C, Leng P. Expression analysis of glucosidase genes that regulate abscisic acid homeostasis during watermelon (*Citrullus lanatus*) development and under stress conditions. J Plant Physiol. 2012;169:78–85.

Li Q, Ji K, Sun Y, Luo H, Wang H, Leng P. The role of FaBG3 in fruit ripening and *B. cinerea* fungal infection of strawberry. Plant J. 2013;76:24–35.

Ma Y, Szostkiewicz I, Korte A, Moes D, Yang Y, Christmann A, Grill E. Regulators of PP2C phosphatase activity function as abscisic acid sensors. Science. 2009;324:1064–8.

Manning K. Changes in gene expression during strawberry fruit ripening and their regulation by auxin. Planta. 1994;94:62–8.

Matsushima J, Hiratsuka S, Taniguchi N, WSada R, Suzaki N. Anthocyanin accumulation and sugar content in the skin of grape cultivar "Olympia" treated with ABA. J Jap Soc Hort Sci. 1989;58:551–5.

Melcher K, Ng LM, Zhou XE, Soon FF, Xu Y, Suino-Powell KM, Park SY, Weiner JJ, Fujii H, Chinnusamy V, Kovach A, Li J, Wang Y, Li J, Peterson FC, Jensen DR, Yong EL, Volkman BF, Cutler SR, Zhu JK, Xu HE. A gate-latch-lock mechanism for hormone signalling by abscisic acid receptors. Nature. 2009;462:602–8.

Miyazono K, Miyakawa T, Sawano Y, Kubota K, Kang HJ, Asano A, Miyauchi Y, Takahashi M, Zhi Y, Fujita Y, Yoshida T, Kodaira KS, Yamaguchi-Shinozaki K, Tanokura M. Structural basis of abscisic acid signalling. Nature. 2009;462:609–14.

Mizrahi Y, Dostal HC, McGlasson WB, Cherry JH. Effects of abscisic acid and benzyladenine on fruits of normal and rin mutant tomatoes. Plant Physiol. 1975;56:544–6.

Nambara E, Marion-Poll A. Abscisic acid biosynthesis and catabolism. Annu Rev Plant Biol. 2005;56:165–85.

Nicolas P, Lecourieux D, Kappel C, Cluzet S, Cramer GR, Delrot S, Lecourieux F. The bZIP transcription factor VvABF2 is an important transcriptional regulator of ABA-dependent grape berry ripening processes. Plant Physiol. 2014;164:365–83.

Nishimura N, Hitomi K, Arvai AS, Rambo RP, Hitomi C, Cutler SR, Schroeder JI, Getzoff ED. Structural mechanism of abscisic acid binding and signaling by dimeric PYR1. Science. 2009;326:1373–9.

Nitsch LM, Oplaat C, Feron R, Ma Q, Wolters-Arts M, Hedden P, Mariani C, Vriezen WH. Abscisic acid levels in tomato ovaries are regulated by LeNCED1 and SlCYP707A1. Planta. 2009;229:1335–46.

Nitsch L, Kohlen W, Oplaat C, Charnikhova T, Cristescu S, Michieli P, Wolters-Arts M, Bouwmeester H, Mariani C, Vriezen WH, Rieu I. ABA-deficiency results in reduced plant and fruit size in tomato. J Plant Physiol. 2012;169:878–83.

Palejwala V, Himanshu P, Modi P, Modi V. The role of abscisic acid in the ripening of grapes. Physiol Plant. 1985;65:498–502.

Pandey S, Nelson DC, Assmann SM. Two novel GPCR-type G proteins are abscisic acid receptors in *Arabidopsis*. Cell. 2009;136:136–48.

Park SY, Fung P, Nishimura N, Jensen DR, Fujii H, Zhao Y, Lumba S, Santiago J, Rodrigues A, Chow TF, Alfred SE, Bonetta D, Finkelstein R, Provart NJ, Desveaux D, Rodriguez PL, McCourt P, Zhu JK, Schroeder JI, Volkman BF, Cutler SR. Abscisic acid inhibits type 2C protein phosphatases via the PYR/PYL family of START proteins. Science. 2009;324:1068–71.

Perkins-Veazie P. Growth and ripening of strawberry fruit. Hort Rev. 1995;17:267–97.

Perotti VE, Moreno AS, Podestá FE. Physiological aspects of fruit ripening: the mitochondrial connection. Mitochondrion. 2014;17C:1–6.

Prasanna V, Prabha TN, Tharanathan RN. Fruit ripening phenomena–an overview. Crit Rev Food Sci Nutr. 2007;47:1–19.

Ren J, Sun L, Wu J, Zhao S, Wang C, Wang Y, Ji K, Leng P. Cloning and expression analysis of cDNAs for ABA 8′-hydroxylase during sweet cherry fruit maturation and under stress conditions. J Plant Physiol. 2010;67:1486–93.

Riov J, Dagan E, Goren R, Yang SF. Characterization of abscisic acid-induced ethylene production in citrus leaf and tomato fruit tissues. Plant Physiol. 1990;92:48–53.

Rock CD, Zeevaart JA. Abscisic (ABA)-Aldehyde is a precursor to, and 1′, 4′-trans-ABA-Diol a catabolite of, ABA in apple. Plant Physiol. 1990;93:915–23.

Rodrigo MJ, Alquezar B, Zacarias L. Cloning and characterization of two 9-*cis*-epoxycarotenoid dioxygenase genes, differentially regulated during fruit maturation and under stress conditions, from orange (*Citrus sinensis* L. Osbeck). J Exp Bot. 2014;57:633–43.

Romero P, Rodrigo MJ, Alfrez F, Ballester AR, Gonzlez-Candelas L, Zacaras L, Lafuente MT. Unravelling molecular responses to moderate dehydration in harvested fruit of sweet orange (*Citrus sinensis* L. Osbeck) using a fruit-specific ABA-deficient mutant. J Exp Bot. 2012;63:2753–67.

Santiago J, Dupeux F, Round A, Antoni R, Park SY, Jamin M, Cutler SR, Rodriguez PL, Márquez JA. The abscisic acid receptor PYR1 in complex with abscisic acid. Nature. 2009;462:665–8.

SchwarzM SchroederJI. Abscisic acid maintains S-type anion channel activity in ATP-depleted *Vicia faba* guard cells, indicating phosphatase inhibition. FEBS Lett. 1998;428:177–82.

Scienza A, Mieavalle R, Visai C, Fregoni M. Relationships between seed number, gibberellin and abscisic acid levels and ripening in cabernet sauvignon grape berries. Vitis. 1978;17:361–8.

Seymour GB, Østergaard L, Chapman NH, Knapp S, Martin C. Fruit development and ripening. Annu Rev Plant Biol. 2013;64:219–41.

Shang Y, Yan L, Liu ZQ, Cao Z, Mei C, Xin Q, Wu FQ, Wang XF, Jiang T DuSY, Zhang XF, Zhao R, Sun HL, Liu R, Yu YT, Zhang DP. The Mg-chelatase H subunit of Arabidopsis antagonizes a group of transcription repressors to relieve ABA-responsive genes of inhibition. Plant Cell. 2010;22:1909–35.

Shen YY, Duan CQ, Liang XE, Zhang DP. Membrane-associated protein kinase activities in the developing mesocarp of grape berry. J Plant Physiol. 2004;161:15–23.

Shen YY, Wang XF, Wu FQ, Du SY, Cao Z, Shang Y, Zhang DP. The Mg-chelatase H subunit is an abscisic acid receptor. Nature. 2006;443:823–6.

Sun L, Wang YP, Chen P, Ren J, Ji K, Li Q, Li P, Dai SJ, Leng P. Transcriptional regulation of SlPYL, SlPP2C, and SlSnRK2 gene families encoding ABA signal core components during tomato fruit development and drought stress. J Exp Bot. 2011;62:5659–69.

Sun L, Yuan B, Zhang M, Wang L, Cui M, Wang Q, Leng P. Fruit-specific RNAi-mediated suppression of SlNCED1 increases both lycopene and carotene contents in tomato fruit. J Exp Bot. 2012a;63:3097–108.

Sun L, Sun Y, Zhang M, Wang L, Ren J, Cui M, Wang Y, Ji K, Li P, Li Q, Chen P, Dai S, Duan C, Wu Y, Leng P. Suppression of 9-*cis*-epoxycarotenoid dioxygenase, which encodes a key enzyme in abscisic acid biosynthesis, alters fruit texture in transgenic tomato. Plant Physiol. 2012b;158:283–98.

Sun JH, Dong YH, Li CL, Shen YY. Transcription and enzymatic analysis of beta-glucosidase VvBG1 in grape berry ripening. Plant Growth Regul. 2014;. doi:10.1007/s10725-014-9932-x.

Shiozaki S, Kamata Y, Ogata T, Horiuchi S, kawase K. Localization of abscisic acid in grape berry by immunohistochemical techniques. J Jap Soc Hortic Sci. 1999;68:1–9.

Singh RK, Sane VA, Misra A, Ali SA, Nath P. Differential expression of the mango alcohol dehydrogenase gene family during ripening. Phytochemistry. 2010;71:1485–94.

Terry LA, Chope GA, Bordonaba JG. Effect of water deficit irrigation and inoculation with *Botrytis cinerea* on strawberry (*Fragaria* x *ananassa*) fruit quality. J Agric Food Chem. 2007;55:10812–9.

Verslues PE, Zhu JK. New developments in abscisic acid perception and metabolism. Curr Opin Plant Biol. 2007;10:447–52.

Wheeler S, Loveys B, Ford C, Davies C. The relationship between the expression of abscisic acid biosynthesis genes, accumulation of abscisic acid and the promotion of *Vitis vinifera* L. berry ripening by abscisic acid. Aust J Grape Wine Res. 2009;15:195–204.

Wang XF, Zhang DP. Abscisic acid receptors: multiple signal-perception sites. Ann Bot. 2008;101:311–7.

Wang XJ, Zhu SY, Lu YF, Zhao R, Xin Q, Wang XF, Zhang DP. Two coupled components of the mitogen-activated protein kinase cascade MdMPK1 and MdMKK1 from apple function in ABA signal transduction. Plant Cell Physiol. 2010;51:754–66.

Wu FQ, Xin Q, Cao Z, Liu ZQ, Du SY, Mei C, Zhao CX, Wang XF, Shang Y, Jiang T, Zhang XF, Yan L, Zhao R, Cui ZN, Liu R, Sun HL, Yang XL, Su Z, Zhang DP. The Mg-chelatase H subunit binds abscisic acid and functions in abscisic acid signaling: new evidence in *Arabidopsis*. Plant Physiol. 2009;150:1940–54.

Xu ZJ, Nakajima M, Suzuk Y, Yamaguchi I. Cloning and characterization of the abscisic acid-specific glucosyltransferase gene from adzuki bean seedlings. Plant Physiol. 2002;129:1285–95.

Yu XC, Li MJ, Gao GF, Feng HZ, Geng XQ, Peng CC, Zhu SY, Wang XJ, Shen YY, Zhang DP. Abscisic acid stimulates a calcium-dependent protein kinase in grape berry. Plant Physiol. 2006;140:558–79.

Zhang DP, Zhang ZL, Chen J, Jia WS. Specific abscisic acid-binding sites in mesocarp of grape berry: properties and subcellular localization. J Plant Physiol. 1999;155:324–31.

Zhang DP, Chen SW, Peng YB, Shen YY. Abscisic acid-specific binding sites in the flesh of developing apple fruit. J Exp Bot. 2001;52:2097–103.

Zhang M, Yuan B, Leng P. The role of ABA in triggering ethylene biosynthesis and ripening of tomato fruit. J Exp Bot. 2009;60:1579–88.

Zhang L, Ma G, Kato M, Yamawaki K, Takagi T, Kiriiwa Y, Ikoma Y, Matsumoto H, Yoshioka T, Nesumi H. Regulation of carotenoid accumulation and the expression of carotenoid metabolic genes in citrus juice sacs in vitro. J Exp Bot. 2012;63:871–86.

Zeevaart JAD. Abscisic acid metabolism and its regulation. In: Hooykaas PPJ, Hall MA, Libbenga KR, editors. Biochemistry and molecular biology of plant hormones. Amsterdam: Elsevier Science; 1999. p. 189–207.

Zeevaart JA, Heath TG, Gage DA. Evidence for a universal pathway of abscisic acid biosynthesis in higher plants from o incorporation patterns. Plant Physiol. 1989;91:1594–601.

Chapter 15
ABA Regulation of Stomatal Movement

Yong-Fei Wang

Abstract Plants loss water and intake CO_2 mainly through stomata pores, which are surrounded by pairs of guard cells. Drought stress induces the production of ABA, and ABA functions as a main endogenous hormone to close stomata to prevent water loss. In the past decades, researchers have made exciting research progresses in ABA signaling transduction in guard cells, including the identification and isolation of a number of important components of ABA signaling network from upstream ABA receptors to downstream ion channels, and an uninterrupted and relatively complete ABA signaling pathway for ABA-induced stomatal closure has emerged. In this ABA signaling cascade, guard cells first sense ABA by the binding of ABA to ABA receptors to trigger the protein interaction between ABA receptors with PP2Cs, then protein kinases inhibited by PP2Cs can be released and further activate downstream S-type anion channel SLAC1, and the efflux of cations and anions through ion channels drives the stomatal closure. At the same time, a number of known components, small molecular messengers and newly identified components involving in ABA-induced stomatal closure have not been integrated into the ABA signaling network yet, and many scientific questions remain to be answered. This chapter will review the main progresses in the ABA regulation of stomatal movement, and some remaining questions will be discussed.

Keywords Stomatal movement · ABA signaling · Second messengers · Ion channels · Regulation

Y.-F. Wang (✉)
National Key Laboratory of Plant Molecular Genetics,
Institute of Plant Physiology and Ecology,
Shanghai Institutes for Biological Sciences, Chinese Academy of Sciences,
300 Fenglin Road, Shanghai 200032, China
e-mail: wangyongfei@sibs.ac.cn

D.-P. Zhang (ed.), *Abscisic Acid: Metabolism, Transport and Signaling*,
DOI 10.1007/978-94-017-9424-4_15

15.1 Introduction

Stomata locate mainly in the surface of plant leaves, and a stomata pore is surrounded by two guard cells. Stomatal movement, including stomatal opening and closure, is driven by the changes of volume and osmotic turgor pressure in the two guard cells. Stomatal opening results from the accumulation of inorganic ions, including K^+, Cl^-, NO_3^-, and osmotic organic molecules such as malate, which is mainly from the hydrolyzing of sugar and starch. Stomatal closure is driven by the release of the inorganic ions from guard cells plus the gluconeogenic conversion of malate into osmotic inactive starch. Multiple abiotic and biotic stimuli can regulate the stomatal movement. When sun rises at morning, light induces stomatal opening, and plants intake CO_2 for photosynthesis, loss water by transpiration, and release oxygen into atmosphere through stomatal pores. During night time, darkness induces stomatal closure. Further research found that both blue light and red light can induce stomatal opening. Blue light induces stomatal opening by activating plasma membrane H^+-ATPase, which pumps H^+ out of guard cells to hyperpolarize the plasma membrane (Kasahara et al. 2002; Briggs and Christie 2002), and consequently activate plasma membrane inward-rectifying K^+ channels (K_{in}). The identification of OST2 (open stomata 2), a H^+-ATPase in *Arabidopsis*, provides genetic evidences showing the functions of H^+-ATPase in stomatal opening. *ost2* mutants show a constitutive activity of H^+-ATPase, which consequently leads to open stomata and an insensitivity to ABA-induced stomatal closure (Merlot et al. 2007). Therefore, OST2 is also named as AHA1 (*Arabidopsis* H^+ ATPase). Red light is not only a signal, but also an energy source for mesophyll and guard cells. Red light can induce stomatal opening by reducing CO_2 concentration in guard cell (Shimazaki et al. 2007). Stomatal movement can be regulated by changes of atmosphere CO_2 level. Elevated CO_2 induces stomatal closure, and low CO_2 induces stomatal opening to ensure an efficient CO_2 uptake for photosynthesis in a variable environmental CO_2 level. Ozone can enter plants through stomata pores, and ozone and reactive oxygen species (ROS) derived from ozone can damage plant cells. Stomata are capable to be closed in response to ozone to protect plants from the damage. The changes of relative air humidity can regulate stomatal movement as well, the decreases of relative air humidity close stomata, and the increases of relative air humidity can open stomata. Besides environmental stimuli, phytohormone, including ABA, cytokinin, auxin, ethylene, and jasmonates, can all have regulatory effects on stomatal movement, especially ABA, which is the most specific natural hormone than any others to induce stomatal closure to prevent water loss and protect plants from environmental harmful invasion. ABA-induced stomatal closure is the most well-studied signaling mechanism, but it is also well known that some key components of ABA signaling pathway in stomata play essential roles in other stimuli-induced stomatal movement, such as SLAC1 and GCA2. The mutations of SLAC1 and GCA2 impair ABA-, light-, and high CO_2-induced stomatal closure (Negi et al. 2008; Vahisalu et al. 2008). On the contrary, the disruption of HT1 (high leaf temperature 1), β-carbonic anhydrases β-CA1 and β-CA4, and the major regulators of CO_2 responses, impairs CO_2-induced stomatal closure, but retains ABA-induced stomatal closure and blue light-induced stomatal opening in *Arabidopsis* (Hashimoto et al. 2006; Hu et al. 2010). Therefore, the regulatory networks of stomatal movement for different phytohormones and environmental stimuli are not completely independent from each other. They share some key components and crosstalk through these shared components. Guard cells can perceive and integrate different stimulating signals together (Merilo et al. 2014), and regulate the stomata pores precisely to ensure an optimal gas exchange. The elucidation of ABA signaling network in guard cells not only facilitates our understanding about ABA-induced stomatal movement, but also will potentially help us to understand how other signals involves in stomatal movement.

Drought stress induces ABA production, and ABA closes stomata as a signal of drought stress. Almost 30 years ago, the fantastic experiment of dividing the roots of maize between two containers, in which soil was dry in one container, but well-watered in another, resulted in partial closure of stomata and reduced water loss while the plant could grow normally (Blackman and Davies 1985). This technique was called partial root drying (PRD), which shed a light into the mind of plant scientists, and made them to believe that it is possible to reduce water loss significantly without disrupting the normal growth of plants, especially the yield of crops by engineering. Considering the short of fresh water globally, saving fresh water in plant growth, including crops, is meaningful for the survival and development of human society. Numerous ABA signaling researchers from different countries dedicate their energy to unravel the ABA signaling network in stomata for decades, their goals are to understand the signaling network and use the knowledge and techniques they develop to reduce the water loss and improve the resistant ability of plants against environmental stresses. In the last two to three decades, huge advances were made in this research field. This chapter will mainly focus on ABA signaling in stomata.

15.2 Stomata Development

Guard cells are derived from epidermal stem cells. Protodermal cells differentiate into meristemoid mother cells (MMCs), from which two daughter cells in different sizes are derived by asymmetrical cell division. The smaller one (meristemoids) has stem cell-like capabilities and is generally engaged for further asymmetric cell division. The larger one is called guard mother cell (GMC) and divides symmetrically to yield two guard cells, which surround a pore to form a stomata. Stomata stem cells persist mounting in epidermis until all cells complete differentiation. Stomata are separated by at least one epidermis pavement cell, and this pattern is being kept through the whole period of plant development. Stomata development and the formation of correct pattern require an appropriate timing and orientation of asymmetric and symmetric cell divisions, cell fate selection, and cell–cell communication, which is possibly mainly through small peptide molecules to ensure a correct cell positioning. This process is regulated and influenced by a number of genetic components identified in the last decades, mainly including cell surface receptors, secreted small signaling peptides, a series of transcription factors, MAK kinase modules, and microRNA (Nadeau 2009; Dong and Bergmann 2010). The up- and down-regulation of the associated genes of the regulatory components result in the changes of stomata index compared to wild-type plants. In addition, emerging evidences show that stomatal development is also regulated by environmental factors, including CO_2, light density, and water stress. High carbon dioxide (HIC) was identified as a positive regulator of CO_2 signaling for stomatal development; the *hic* mutant growing in elevated CO_2 showed an increased stomata density and index (Gray et al. 2000; Rae et al. 2006). On the other side, the stomata density of some plants decreased significantly compared to their ancients living in hundreds of years ago, and the elevated atmosphere CO_2 resulted from the human activity and industrial development was believed to be the major cause. This manner of stomata regulation by atmosphere CO_2 takes hundreds of years and therefore is "supper slow" compared to CO_2-induced stomatal movement, but accumulating and significant. Stomata development is also regulated by light. An increase in light intensity results in an increase in stomatal index in some plants. The light signal is perceived by mature leaves, and stomata patterning changes in young developing leaves (Casson and Gray 2008). How the signal is transited from mature leaves to young leaves is still unknown. Water condition is a main factor regulating stomata movement, and it is well known that ABA functions as the main signal to trigger stomatal closure in response to drought stress. It has been reported a long time ago that a reduced soil-watering condition resulted in a reduction in stomata index in *Caltha palustris* and wheat (Quarrie and Jones 1977; Salisbury 1928; Casson and Gray 2008). But whether drought stress/ABA functions as a general regulator for stomatal development and patterning in plant kingdom is still unknown. Nevertheless, stomata development is an important aspect of gas exchange for plants, in addition to the stomatal movement.

15.3 ABA Triggering Stomatal Closure Comes Mainly from Roots

Root is the organ of plants sensing water loss of soil firstly. Then, ABA is synthesized in roots, and the ABA stored in roots as inactive ABA-glucose ester (ABA-GE) can also be released at the same time (Lee et al. 2006; Schroeder and Nambara 2006; Xie et al. 2006). Free ABA synthesized in roots is released into xylem from parenchyma cells and transported through xylem into leaves to close stomata. Glucose-conjugated ABA released by roots can be transported through xylem into leaves as well, and it is believed that conjugated ABA needs to be released as a free ABA by β-glucosidases in leaves to be active. Leaves have the ability to synthesize ABA when leaf turgor pressure is very low because of water loss. However, ABA synthesis begins in root when soil is just beginning to loss water, and leaf turgor pressure is still unaffected obviously at that time (Hartung et al. 2002). Therefore, it has been proposed that ABA in leaves triggering stomata closure in response to dehydration of soil is mainly imported from roots, not synthesized locally in leaves. Indeed both free ABA and conjugated ABA were detected in xylem flow (Hartung et al. 2002). On the contrary, ABA level can increase significantly in response to a reduced atmospheric relative humidity while roots are not stressed by drought (Xie et al. 2006), suggesting that ABA can be produced or released in leaves in response to dry atmosphere. Further study found that β-glucosidase plays essential roles in leaves in catalyzing ABA release from ABA-GE, which is originally imported from roots, and this can explain at least partially the increase of ABA level in leaves in response to dry atmosphere. The free/active ABA can be degraded quickly by P450 protein in leaves after acting as a signal to close stomata, and does not deposit and accumulate in guard cells after plants are rescued from drought stress (Jiang and Hartung 2008; Schroeder and Nambara 2006; Peuke et al. 2002). Therefore, ABA signaling intensity for stomatal movement regulation is regulated by ABA synthesis, long-distance transportation of ABA and conjugated ABA, release of active ABA from ABA-GE, and ABA breakdown.

15.4 ABA Receptors, the Key Proteins from Where ABA Signaling Initiates for Stomatal Closure

Numerous studies have been done regarding ABA signaling transduction for stomatal movement. The simple and easiness of stomata movement assay and the identification of many genetic mutants (mainly in *Arabidopsis*) are the main reasons. For stomata, ABA perceiving is the first step for ABA signaling transduction, and ABA receptors are the key proteins sensing ABA stimulus for guard cells. Despite the identification and isolation of a large number of downstream components of ABA signaling pathway in guard cells in the last few decades, ABA receptors were completely unknown until recent few years. Using biochemical approach, several proteins binding to ABA in vitro were identified and suggested as ABA receptors, including CHLH (H subunit of Mg-chelatase) and three G protein-coupled receptors GCR2, GTG1, and GTG2. CHLH (H subunit of Mg-chelatase) plays important roles in chlorophyll biosynthesis and plastid-to-nucleus signaling (Shen et al. 2006). GCR2, GTG1, and GTG2 were suggested as ABA receptors by interacting with GPA1, the α subunit of heterotrimeric G protein, in guard cells (Pandey et al. 2009; Liu et al. 2007). The functions of heterotrimeric G protein in ABA-induced stomatal closure have been reported repeatedly (Fan et al. 2008; Wang et al. 2001), supporting the conclusion of that the three G protein-coupled receptors are ABA receptors. As ABA receptors should be where ABA signaling start in guard cells, a strong ABA-insensitive phenotype of stomatal movement was expected in the ABA receptor mutants. However, the phenotype of ABA-induced stomatal closure is controversial for *gcr2* (Gao et al. 2007; Liu et al. 2007) and partial in *gtg1/gtg2* (Pandey et al. 2009). *gun5* mutant with a point mutation in the *CHLH* gene seems to show wild-type phenotypes in the major ABA responses (Zhang DP, personnel communication), but two other mutant alleles in the *CHLH* gene, *cch* and *rtl1* (for *rapid transcription* from *leaves*), show strong ABA-insensitive phenotypes in ABA-induced stomatal closure and inhibition of stomatal opening (Shen et al. 2006; Wu et al. 2009; Du et al. 2012; Tsuzuki et al. 2011, 2013; Zhang et al. 2013). These data suggest that multiple ABA receptors are functioning in guard cells. Using three different methods, yeast two hybrid, chemical genetics, and coimmunoprecipitation analysis, a cluster of homologous genes called RCAR/PYR/PYL was identified as ABA receptors in *Arabidopsis*, and this family is composed of fourteen members (Park et al. 2009; Nishimura et al. 2010; Ma et al. 2009). Using biochemical technique, ABA binding to RCAR/PYR/PRL was detected (Ma et al. 2009; Park et al. 2009). Crystallographic structural study found that ABA enters a large internal cavity of PYR1 for ABA binding (Nishimura et al. 2010). The overexpression of *RCAR/PYR/PYL* leads to a slightly reduced stomata aperture in the absent of ABA, and a strong hypersensitive phenotype in ABA-induced stomatal closure in an ABA concentration-dependent manner compared to wild type in *Arabidopsis* (Ma et al. 2009). The quadruple mutant *pyr1/pyl1/pyl2/pyl4* shows a strong ABA-insensitive phenotype in ABA-induced stomatal closure, but the single mutant *pyr1* has no ABA-induced stomatal closure phenotype, indicating the presence of functional redundancy within PYR/PYL/RCAR family. Further study is needed to explore whether different types of ABA receptors function in guard cells in a corporative manner to provide a mechanism to adjust the sensitivity of ABA perception for precise responses to drought and other stresses for an optimal stomatal aperture and gas exchange (Ma et al. 2009).

To investigate how ABA receptors deliver ABA signaling to downstream components after binding to ABA, protein interaction analysis was performed, and the direct protein interactions between ABA receptors RCAR/PYR/PYL and phosphatases, PP2Cs, including HAB1, ABI1, ABI2, and PP2CA, were detected by different groups (Ma et al. 2009; Nishimura et al. 2010; Park et al. 2009), and the phosphatase activity of PP2Cs was dramatically inhibited by the protein interactions in an ABA concentration-dependent manner (Ma et al. 2009; Park et al. 2009). Direct protein interactions between PP2C phosphatases and SnRK family were also observed, the protein kinase activity of SnRKs was constitutively released by the protein interactions, and the mutated PP2C phosphatases failed to interact with and inhibit SnRKs (Umezawa et al. 2009; Vlad et al. 2009; Weiner et al. 2010; Dupeux et al. 2011). Structural analysis confirmed the protein interaction of PP2C phosphatases with upstream ABA receptors and downstream SnRK kinases in protein complexes (Soon et al. 2012; Santiago et al. 2012). These series fantastic research formed a complete ABA signaling cascade from ABA receptors to downstream ion channels for ABA-induced stomatal closure. In this ABA signaling pathway, guard cells perceive ABA stimulus by the binding of ABA to ABA receptors to further trigger the protein interaction between ABA receptors and PP2Cs; this protein interaction inhibits the phosphatase activity of PP2Cs and consequently releases the kinase activity of SnRKs and CKPKs, which further activate anion channel SLAC1 by protein phosphorylation; the anion efflux further activates outward-rectifying plasma membrane K^+ channel GORK; finally, the efflux of anions and cations leads to the stomatal closure. In this signaling pathway, ABA receptors are the "starting line" for ABA signaling transduction in guard cells.

15.5 Components of ABA Signaling Network Between ABA Receptors and Downstream Plasma Membrane Ion Channels

15.5.1 Components of ABA Signaling in Guard Cells

Protein phosphorylation and dephosphorylation were found being important for ABA signaling a few decades ago, and several protein phosphatases from protein phosphatase 2C (PP2C) family, including ABI1, ABI2, HAB1, and PP2CA, were then identified as essential regulators of ABA signaling in *Arabidopsis* guard cells, in which ABI1 and ABI2 are the most well-studied phosphatases. *Arabidopsis* mutants *abi1* (ABA insensitive 1) to a*bi5* (ABA insensitive 5) were isolated over 30 years ago by screening seed germination and seedling growth in the presence of 10 μM ABA (Koornneef et al. 1984). *ABI3* to *ABI5* encode transcriptional factors, and mainly function in other cell types, not guard cells (Finkelstein and Lynch 2000; Giraudat et al. 1992; Shkolnik-Inbar et al. 2013). *ABI1* and *ABI2* are known as two essential negative regulators of ABA signaling transduction in guard cells. The *abi1* and *abi2* mutants show excessive water loss phenotype (Finkelstein and Somerville 1990), indicating their functions in ABA-induced stomatal closure. Interestingly, it turned out that *abi1* and *abi2* mutants are resulted from the same point mutation of a substitution of a Gly by an Asp in the conserved catalytic domain in two different PP2C genes (Leung et al. 1994, 1997). This point mutation in the two homologs reduces the phosphatase activity (Leung et al. 1997). ABI1 and ABI2 have a calcium-binding domain (EF hand) and were suggested to be a Ca^{2+}-regulated component of ABA signaling in guard cells (Leung et al. 1994). However, ABA-induced cytosolic Ca^{2+} increases were impaired in *abi1* and *abi2* guard cells, but Ca^{2+} activation of anion channel currents and external Ca^{2+}-induced stomatal closure were not affected in the two mutants (Allen et al. 1999), suggesting that ABI1 and ABI2 function upstream of Ca^{2+} signal in ABA signaling pathway. HAB1 and PP2CA are another two phosphatases from PP2C family. HAB1 was initially named as AtP2C-HA, the disruption of AtP2C-HA leads to an ABA hypersensitive phenotype in seed germination, but stomatal closure is not obviously affected (Leonhardt et al. 2004; Saez et al. 2004). However, the *abi1hab1* double mutants showed a reduced transpirational water loss compared to *abi1* and *hab1* single mutants under a drought condition, suggesting a cooperative negative regulation of the two phosphatases for ABA signaling in guard cells (Saez et al. 2006). PP2CA is more active than other three phosphatases in seed germination, and the mutation of *PP2CA* leads to a strongest

ABA hypersensitive phenotype in seed germination compared to *abi1*, *abi2,* and *hab1* (Yoshida et al. 2006; Kuhn et al. 2006). The four PP2Cs involve in ABA signaling transduction in guard cells by interacting with upstream ABA receptors and downstream protein kinases as described above.

As the counterparts of PP2Cs, some protein kinases are the positive regulators of ABA signaling network in guard cells. Currently, the known protein kinases for ABA-induced stomatal movement are mainly from calcium-dependent protein kinase (CPK) family and SNF1-related kinases (SnRK) family. SnRK2.2, 2.3 and 2.6 are the three SnRK members functioning in guard cells in *Arabidopsis*. The *Arabidopsis* OST1/SnRK2.6 is the ortholog of AAPK from *Vicia faba*, and the mutant *ost1/snrk2.6* shows a strong ABA-insensitive phenotype in stomatal closure (Mustilli et al. 2002; Yoshida et al. 2002). The double mutant of *snrk2.2/2.3* shows a weaker ABA-insensitive phenotype compared to *ost1/snrk2.6* in stomatal closure (Fujii et al. 2007), and ABA insensitivity of the triple mutant *snrk2.2/2.3/2.6* is stronger than *ost1* and *snrk2.2/2.3* (Fujii et al. 2011), indicating the presence of functional redundancy within the three SnRKs for stomatal closure regulation in *Arabidopsis*. In the absence of ABA, SnRKs can be dephosphorylated/inactivated by PP2Cs and cannot phosphorylate downstream S-type anion channel SLAC1 after being dephosphorylated (Belin et al. 2006; Umezawa et al. 2009; Vlad et al. 2009; Weiner et al. 2010; Dupeux et al. 2011). The increase of ABA level in leaves leads to the binding of ABA to ABA receptors PYR/PYL/RCARA, which in turn inhibit the activity of PP2Cs by direct protein interaction. SnRKs are then released from PP2C, activated by autophosphorylation, activate downstream S-type anion channel SLAC1 by phosphorylation, and consequently, stomata are closed. The activation and inhibition of downstream S-type anion channel SLAC1 by SnRKs and PP2Cs in an ABA-dependent manner were verified by direct anion current recordings in *Xenopus* oocyte (Geiger et al. 2009) and structural analysis of protein complexes (Soon et al. 2012; Nishimura et al. 2009). CPKs are a protein kinase family, which contain a Ca^{2+}-binding domain, can be activated by the binding of Ca^{2+}, and further phosphorylate downstream targets. The genome of model plant *Arabidopsis* encodes 34 CDPK members, four of which were found functioning in stomatal closure regulation, including CPK3, 4, 6, and 11, and the double mutants *cpk3/cpk6* and *cpk4/cpk11* show ABA insensitivity in stomatal closure (Mori et al. 2006; Zhu et al. 2007). Further research confirmed the functions of CPK6 as an activator of anion channel SLAC1 by protein phosphorylation in *Xenopus* oocytes, and the phosphatases ABI1, ABI2, and PP2CA can inhibit CPK6-mediated activation of SLAC1 (Brandt et al. 2012), suggesting a different signaling branch from SnRKs. Interestingly, SLAC1 can also be activated by CPK21 and CPK23 in *Xenopus* oocytes, but the *Arabidopsis* mutant *cpk21* and *cpk23* showed a strong drought tolerance phenotype rather than drought hypersensitive phenotype compared to wild type (Franz et al. 2011; Ma and Wu 2007), indicating that these two CPKs involve in stomatal movement regulation in a different manner compared to CPK3, 4, 6, and 11. Guard cell hydrogen peroxide-resistant 1 (GHR1) is a receptor-like kinase localized in the plasma membrane of *Arabidopsis* guard cells (Hua et al. 2012).

The mutant *ghr1* was isolated in a genetic mutant screening in a low-pH growth medium. This mutant losses more water and wilts faster than wild-type plants under drought condition, and ABA- and H_2O_2-induced stomatal closure were impaired in this mutant (Hua et al. 2012). Patch clamping experimental results showed that the mutation of *GHR1* impaired H_2O_2 activation of Ca^{2+} channels, and ABA and H_2O_2 activation of S-type anion channel currents (Hua et al. 2012). Further analysis showed that GHR1 functions downstream of ABI1, ABI2, OST1, and H_2O_2, but upstream of SLAC1 (Hua et al. 2012).

Calcineurin-B-like proteins (CBLs) are calcium sensors, interact with, and modulate the activity of CBL-interacting protein kinases (CIPKs), which further regulate download targets by phosphorylation. *Arabidopsis* genome has 10 CBLs and 25 CIPKs, and CBL1 and CBL9 were found functioning in guard cells by interacting with and targeting CIPK23 to the plasma membrane of guard cells (Cheong et al. 2007). The loss-of-function mutant *cpk23* showed a reduced transpirational water loss and ABA hypersensitive phenotypes in stomatal opening and closure (Cheong et al. 2007).

AtMRP5, a members of ATP-binding cassette (ABC) family, has been suggested functioning in guard cells as a channel (Gaedeke et al. 2001), and the stomatal closure in *atmrp5* mutant is insensitive to ABA and external Ca^{2+} (Klein et al. 2003). But patch clamping analysis showed that both depolarization-activated S-type anion currents and hyperpolarization-activated Ca^{2+} currents are impaired in *atmrp5* mutant guard cells simultaneously (Suh et al. 2007), indicating that AtMRP5 is a general regulator of ABA signaling rather than an ion channel.

To identify loci involved in ABA signaling, Erwin Grill's group screened a large number of chemically mutagenized *Arabidopsis* seeds in the presence of ABA, and 8 loci GCA1 to GCA8 (growth control exerted by ABA) were identified (Himmelbach et al. 1998). Further research found that two of them GCA1 and GCA2 involve in ABA signaling in guard cells (Himmelbach et al. 1998). GCA2 was later found functioning upstream of ABA-induced ROS production, the activation of hyperpolarization-activated inward Ca^{2+} channels, and cytosolic Ca^{2+} elevation in guard cells (Pei et al. 2000). Ca^{2+}-imaging analysis showed that GCA2 involves in CO_2- and ABA-induced stomatal closure by regulating the cytosolic Ca^{2+} oscillation pattern (Young et al. 2006; Allen et al. 2001). Recently, Ca^{2+} imaging and patch clamping analysis showed that ABA induces stomatal closure in both Ca^{2+}-dependent and Ca^{2+}-independent manners, and GCA2 is important for Ca^{2+}-dependent ABA response for stomatal closure, not for Ca^{2+}-independent response (Siegel et al. 2009). Clearly, GCA2 is correlated to Ca^{2+} and functions as an important component for both ABA and CO_2 signaling in guard cells. However, GCA2 has not been cloned so far. The genetic identification of GCA2 will facilitate our understanding about ABA signaling in guard cells.

ABA hypersensitive 1 (*abh1*) is an ABA hypersensitive mutant which was isolated in an *Arabidopsis* seed germination screening in the presence of 0.3 μM ABA, a concentration of ABA allowing wild-type seeds to germinate (Hugouvieux et al. 2001). *abh1* mutant shows an ABA hypersensitive phenotype in drought-induced stomatal closure, but ABA content in leaves shows no significant difference between wild-type and *abh1* mutant, suggesting that ABH1 is related to ABA signaling transduction in guard cells, but not in ABA level regulation (Hugouvieux et al. 2001). Patch clamping analysis showed that K_{in} currents were down-regulated, but anion currents were up-regulated in *abh1* mutant guard cells (Hugouvieux et al. 2002), supporting an important roles of ABH1 in stomatal movement. ABH1 is a nuclear mRNA cap-binding protein, which may function in guard cells by regulating the transcript level of some components of ABA signaling pathway. Further research found that the expression of *PP2CA* in *abh1* can partially suppress the ABA hypersensitive phenotype of *abh1* (Kuhn et al. 2006), suggesting a connection between ABH1 and PP2CA in stomatal movement regulation. Using *abh1* mutant as a background line, a new screening of EMS mutagenized *Arabidopsis* seeds was recently conducted and leads to the isolation of two new mutants called *soa2* (*suppressor of abh1 hypersensitivity to ABA 2*) and *soa3* (Daszkowska-Golec et al. 2013). Both *soa2* and *soa3* showed drought-tolerant phenotypes, suggesting a function of these two proteins in stomatal movement (Daszkowska-Golec et al. 2013). But the genetic identification of SOA2 and SOA3 is still unknown currently.

15.5.2 Second Messengers in Guard Cells

Several types of small molecules function as second messengers and regulators for ABA-induced stomatal movement, including free Ca^{2+}, ROS, NO, CO, cyclic nucleotides (cAMP and cGMP), and cADP ribose.

Ca^{2+} is an important second messenger for signaling transduction in both mammalian and plant cells, and the changes/oscillation of cytosolic free Ca^{2+} concentration encode signals to mediate the signaling from diverse upstream stimuli to downstream targets. In guard cells, ABA can trigger Ca^{2+} increases and oscillation, for which Ca^{2+} are from both external Ca^{2+} influx mediated by hyperpolarization-activated plasma membrane Ca^{2+} channels and Ca^{2+} release from intracellular Ca^{2+} stores in an IP_3-dependent manner. The imposed external Ca^{2+}-induced cytosolic Ca^{2+} changes can close stomata in *Arabidopsis* (Allen et al. 2001), and ABA can induce the production of IP_3 (Lee et al. 1996), which involves in Ca^{2+} oscillation by inducing intracellular Ca^{2+} release in guard cells. Therefore, the Ca^{2+} oscillation patterns encode the ABA signaling in guard cells. Cytosolic Ca^{2+} has diverse functions as a second messenger and can mediate the signaling from many different stimuli, including ABA, CO_2, and ozone for stomatal movement (Young et al. 2006; Vahisalu et al. 2008). Several core components of ABA signaling network, including PP2Cs and CPKs as well as SnRKs, contain Ca^{2+}-binding domain and may be regulated by Ca^{2+}. Patch clamping analysis showed that the activation of both anion channel SLAC1 and Ca^{2+} channels in *cpk3cpk6* double mutant is impaired, suggesting a Ca^{2+}-dependent activation of SLAC1 by CPK3 and CPK6 (Mori et al. 2006). The direct protein interaction between CPK6 and SLAC1 as well as the activation of SLAC1 by CPK6 was further confirmed (Brandt et al. 2012). On the contrary, Ca^{2+} increases and the activation of hyperpolarization-activated Ca^{2+} channels were also impaired or abnormal in the guard cells of *gca2*, *ost1*, *abi1,* and *abi2* mutants (Allen et al. 1999; Mustilli et al. 2002), suggesting that these components may function upstream of Ca^{2+}. Recent research found that ABA induces stomatal closure in Ca^{2+}-dependent and Ca^{2+}-independent manners (Siegel et al. 2009). ABA can only induce 30 % stomatal closure response in the absent of cytosolic Ca^{2+} elevation, the exposure of guard cells to ABA can enhance the ability of Ca^{2+} to activate S-type anion channel currents and down-regulate inward K^+ channels, and the ABA-insensitive mutant *ost1*, *abi2,* and *gca2* showed only partial ABA response correlated to cytosolic Ca^{2+} (Siegel et al. 2009). These data suggest an ABA-priming hypothesis for ABA-induced stomata closure response.

ROS were thought to be toxic for plant cells. But research in the last two decades found that ROS can function in guard cells as important messengers mediating ABA signaling, and hydrogen peroxide (H_2O_2) is the most effective one because of its long life time and stability. ABA-induced ROS production was first observed in the guard cells of *Vicia faba* (Song et al. 2014) and *Arabidopsis* (Pei et al. 2000). Further research identified two NADPH oxidases AtrbohD and AtrbohF as the key enzymes for ROS production in *Arabidopsis* guard cells, and the disruption of these two genes impairs ABA-induced ROS production, cytosolic Ca^{2+} increases, activation of hyperpolarization-activated inward Ca^{2+} channels, and stomatal closure (Kwak et al. 2003). AtrbohD and AtrbohF localize in the plasma membrane of guard cells, but ROS production in chloroplast of guard cells was observed (Zhang et al. 2001), suggesting that ROS can be produced in multiple area of guard cells. In ABA signaling pathway, ROS production locates downstream of ABI1 and OST1, but upstream of ABI2 and GCA2 (Murata et al. 2001; Mustilli et al. 2002; Pei et al. 2000). Because ROS is such small molecules with simple structures, how essential core proteins of ABA signaling network in guard cells, such as PP2Cs, GCA2 and kinases (SnRKs, CDPKs, and GHR1), perceive the small molecule without specific structural character is still unknown, and further work will be needed to answer the question (Song et al. 2014).

Nitric oxide (NO) plays roles in ABA-induced stomatal closure as a second messenger. NO can trigger the increase of cytosolic Ca^{2+}, the efflux of both anion and cation, and stomatal closure (Gayatri et al. 2013). The application of either ABA or ROS can induce the production of NO, and the disruption of AtrBOHD/F, the enzyme catalyzing the production of ROS, impairs the production of both ROS and NO (Gayatri et al. 2013). NO scavenger can impair ABA- and ROS-induced stomatal closure, and NO donor can mimick ABA- and ROS-induced stomatal closure responses. Therefore, NO functions downstream of ROS in guard cells. Nitric oxide synthase (NOS)-like enzyme is supposed to be the essential protein for NO production, but the existence of NOS in plants is still under debate. Together, the current results show that NO is an important second messenger-mediating ABA signaling in guard cells and localizes downstream of ROS but upstream of cytosolic Ca^{2+} in ABA signaling cascade. Further work will be needed to unravel every detail of the mechanisms of NO in ABA-induced stomatal closure.

cGMP and cAMP are well known as important second messengers in mammalian cells for several decades. Cyclic nucleotide-gated channels (CNGCs) were proposed as the downstream targets of cGMP and cAMP because CNGCs contain a nucleotide-binding domain and a calmodulin-binding domain, and can be activated by cyclic nucleotides by direct binding in mammalian cells. cGMP was reported functioning in guard cells downstream of ROS for the activation of Ca^{2+} influx in *Arabidopsis* (Dubovskaya et al. 2011). But whether cyclic nucleotides involve in ABA signaling transduction in guard cells is still under debate because guanylate cyclase and adenylate cyclase could not be identified in plants so far. Recently, 8-nitro-cGMP was reported as a signaling molecule and involves in ABA signaling in *Arabidopsis* guard cells (Joudoi et al. 2013). The ABA- and NO-induced production of 8-nitro-cGMP was detected in guard cells in the presence of ROS (Joudoi et al. 2013). 8-nitro-cGMP can induce stomatal closure alone, but 3′,5′-cyclic monophosphate (8-bromo-cGMP) did not (Joudoi et al. 2013). CNGCs are the most possible downstream targets of cyclic nucleotides in plants. CNGC homologs in plants were identified many years ago, and the model plant *Arabidopsis* has a large CNGC family which is composed of 20 members. The functions of cyclic nucleotides and CNGCs will be reviewed below in Ca^{2+} channel section because CNGCs are proposed as Ca^{2+} channel candidates in plant cells.

15.6 Channels in the Membrane Systems of Guard Cells Function as a Motor as Well as Regulators for ABA-induced Stomatal Closure

Stomatal movement is driven by the changes of osmotic turgor pressure in guard cells. The changes of turgor pressure result from ion flux through plasma membrane and vacuole membrane of guard cells and consequent water flow. Different types of ion channels are the main functional proteins mediating ion flux across the biological membranes in guard cells. ABA is a main stimulating factor inducing stomatal closure and preventing stomatal opening in response to drought stress through a signaling network, in which ion channels are the key components. ABA first induces cytosolic free Ca^{2+} increases and oscillation mainly by activating hyperpolarization-activated Ca^{2+} channels in the plasma membrane of guard cells (Pei et al. 2000; Allen et al. 2000), and partially by triggering Ca^{2+} release from intracellular Ca^{2+} stores (Tang et al. 2007). Cytosolic Ca^{2+} functions as an essential second messenger to mediate ABA signaling to further stimulate downstream components, including outward anion channels and potassium channels. Anion channels and outward-rectifying K^{+} channels mediate osmotic ion efflux which consequently leads to the water loss and stomatal closure. Therefore, ion channels play essential roles for ABA-induced stomatal closure and ABA-related inhibition of stomatal opening. In the past decades, different channels and associated genes were identified, and the channel properties were thoroughly analyzed using electrophysiological and other techniques.

15.6.1 Plasma Membrane Ca^{2+} Channels

Ca^{2+} imaging and electrophysiological techniques provide powerful tools for the analysis of Ca^{2+} ion channels in guard cells. The application of external ABA and Ca^{2+} can trigger cytosolic Ca^{2+} increases and oscillation, which can be monitored using calcium sensor yellow cameleon expressing in vivo (Allen et al. 2000, 2001). Hyperpolarization voltage-dependent inward Ca^{2+} currents have been recorded in guard cell protoplasts, and ABA can activate the Ca^{2+} channel currents by shifting activating voltage to more positive potential (Hamilton et al. 2000; Pei et al. 2000), suggesting the presence of ABA-activated inward Ca^{2+} channels in the plasma membrane of *Arabidopsis* guard cells. Further research found that ABA can induce the production of ROS, which consequently activate the hyperpolarization-activated Ca^{2+} channels in Arabdiopsis guard cells (Pei et al. 2000). Two NADPH oxidases AtrBOHD and AtrBOHF were identified as the essential players for ROS production in *Arabidopsis* guard cells by transferring electrons from NADPH to electron receptors (Kwak et al. 2003). The disruption of the two protein encoding genes impaired ABA-induced ROS production, cytosolic Ca^{2+} increases and hyperpolarization-activated inward Ca^{2+} currents (Kwak et al. 2003), suggesting that ROS are important mediators for ABA signaling as Ca^{2+} channel activators in guard cells. In addition, glutathione peroxidase 3 can regulate ROS homeostasis in guard cells and consequently affect the activity of Ca^{2+} channels, and the mutation of this gene in *Arabidopsis* impaired the ABA activation of hyperpolarization-activated Ca^{2+} channel currents (Miao et al. 2006). Some calcium/calmodulin-dependent protein kinases (CDPKs) (Mori et al. 2006; Zhu et al. 2007), and phosphatases (Miao et al. 2006; Kohler and Blatt 2002) are important ABA signaling regulators in guard cells as described above. The mutations of *CPK3* and *CPK6* impaired the ABA- and hyperpolarization-activated inward Ca^{2+} channel currents in *Arabidopsis* guard cells (Mori et al. 2006), and the application of phosphatase antagonist okadaic acid (OA) and calyculin A (CA) increased the open probability of the Ca^{2+} channels in the excised plasma membrane patches of *Vicia faba* guard cell protoplast (Kohler and Blatt 2002). Despite the numerous analysis on the Ca^{2+} channel activity and the importance of cytosolic Ca^{2+} as a second messenger for ABA signaling, the identities of plasma membrane Ca^{2+} channels activated by ABA in guard cells are still unknown. How the regulating factors involve in the activation of ABA/ROS-activated inward Ca^{2+} channels in plant guard cells remains to be addressed.

The most interesting candidates for the plasma membrane Ca^{2+} channels in plant guard cells are mainly from three channel families, including CNGCs, glutamate receptors (GLR), and annexins (Ma 2011; Ward et al. 2009). It has been reported that cGMP involves in ABA-induced stomatal closure and functions downstream of ROS to activate cytosolic Ca^{2+} increases in *Arabidopsis* guard cells (Dubovskaya et al. 2011), suggesting a role of CNGCs as Ca^{2+} channels. Few CNGCs were identified as cyclic nucleotide-activated cation channels in *Arabidopsis* in the past few years, including CNGC2, CNGC5, CNGC6, and

CNGC18 (Ali et al. 2007; Wang et al. 2013; Gao et al. 2014). CNGC2 is mainly permeable to monovalent cation except Na^+, and functions mainly in innate immunity responses (Ali et al. 2007; Leng et al. 2002). CNGC5 and CNGC6 are mainly permeable to divalent cation, including Ca^{2+} (Wang et al. 2013). However, the disruption of *CNGC5* and *CNGC6*, two genes highly expressed in *Arabidopsis* guard cells, does not obviously inhibit the activity of ABA-activated Ca^{2+} channels (Wang et al. 2013). For GLRs, it has been reported that the overexpression of *Arabidopsis GLR3.1* led to the impairment of external Ca^{2+}-induced stomatal closure and imposed-Ca^{2+} oscillation-programmed long-term stomatal closure without affecting cytosolic Ca^{2+} kinetics (Cho et al. 2009). However, external Ca^{2+}-induced stomatal closure differs from ABA-induced stomatal closure, because the disruption of Ca^{2+} sensor (CAS) blocks external Ca^{2+}-induced stomatal closure without affecting ABA responses in *Arabidopsis* guard cells (Han et al. 2003). Therefore, GLR3.1 and the three CNGCs described above seem not to be essential for ABA signaling and may not be ABA-activated hyperpolarization-activated inward Ca^{2+} channels. Annexins are ubiquitous soluble proteins, some of them can behave as Ca^{2+} channels, and potentially involve in cytosolic Ca^{2+} and ROS signaling in plant kingdom (Mortimer et al. 2008). But annexins are less analyzed compared to CNGC and GLR in plants. Annexin 1 involves in drought-stress responses in *Arabidopsis* (Konopka-Postupolska et al. 2009), and the expression of annexins in *Mimosa pudica* is sensitive to ABA (Hoshino et al. 2004). The identity of annexins as plasma membrane Ca^{2+} channels in guard cells has not been reported yet, and more work is needed to characterize the functions of annexins. Together, the journey for seeking ABA-activated Ca^{2+} channels located in the plasma membrane of guard cells as well as the associated encoding genes just began, and we may have a long way to go.

There are two sources for cytosolic Ca^{2+} increases and oscillation: the influx of external Ca^{2+} through plasma membrane Ca^{2+} channels and Ca^{2+} release from intracelullar Ca^{2+} stores, such as endocytoplasmic reticulum and vacuole in plant cells. ABA can trigger cytosolic Ca^{2+} increase in an InsP6-dependent manner in guard cells (Lemtiri-Chlieh et al. 2003; Lemtiri-Chlieh and Berkowitz 2004), indicating the involvement of intracellular Ca^{2+} release in ABA-induced stomata closure. Several types of ion channels, including TPC1 (two pore channel 1) and chloride channel (CLC) family, were identified as vacuole membrane channels, and some of them involve in intracellular Ca^{2+} release in *Arabidopsis* guard cells (De Angeli et al. 2006; Peiter et al. 2005). The vacuole membrane-located ion channels will be discussed in "Channels in the intracellular membrane" section below.

15.6.2 Plasma Membrane K^+ Channels

Several channel families being permeable to K^+ were found in plants, especially in *Arabidopsis*, including HKT (high-affinity K^+ transporter), KUP/HAK/KTs, CNGCs (cyclic nucleotide-gated channels), and Kir1- and shaker-type K^+ channel family. The main plasma membrane K^+ channels in *Arabidopsis* guard cells are mainly from shaker-type K^+ channel family, and the functions of other channel families as K^+ channels in guard cells are less studied. Shaker-type K^+ channel family is composed of nine members in *Arabidopsis*. They share protein structural similarities, including six transmembrane (TM) domains, a voltage sensor located in the fourth TM domain, and a P loop between TM five and six. Shaker channel members can form homo- or heterotetramers. In *Arabidopsis* guard cells, the expression of six shaker-type K^+ channels, including *KAT1*, *KAT2*, *AKT1*, *AKT2*, *GORK,* and *KC1*, can be detected. GORK is the main outward-rectifying K^+ channels, which is activated by the depolarization of the plasma membrane of guard cells. The T-DNA insertional knockout mutation of *GORK* abolishes outward K^+ channel currents of guard cells completely and impairs dark- and ABA-induced stomatal closure (Hosy et al. 2003). In ABA signaling network, outward anion channels are first activated by upstream kinases as described above, the efflux of anion through anion channels results in a depolarization of guard cell plasma membrane, and consequently activate GORK. For stomata opening, K^+ influx through inward K^+ channels is required, and the inward K^+ channels in *Arabidopsis* guard cells can be activated by hyperpolarization voltage. KAT1, KAT2, AKT1, AKT2, and KC1 are the five proteins forming inward-rectifying K^+ channels in *Arabidopsis* guard cells (Szyroki et al. 2001). The five inward-rectifying K^+ channel proteins can form homotetramers alone or heterotetramers. KC1 does not form functional K^+ channels alone, but can integrate into inward K^+ channels as one of the four subunits and function as an inhibitory regulating subunit. KAT1 is the first inward-rectifying K^+ channels identified in plant, and also the main inward-rectifying K^+ channels in guard cells because the expression level of KAT1 is much higher than other four members, and the disruption of KAT1 leads to more than 50 % reduction of inward K^+ channel activity in *Arabidopsis* guard cells (Szyroki et al. 2001). Dominant negative repressive mutants of *KAT1* and *KAT2* suppress light- and low-CO_2-induced stomatal opening (Kwak et al. 2001; Lebaudy et al. 2008). These data provided genetic evidences showing the importance of inward K^+ channels for stomatal opening, and the presence of five different subunits in K^+ channel tetramers provide a regulation mechanism for inward-rectifying K^+ channel activity. Another regulation mechanism for K^+ channels is protein trafficking. ABA-induced internalization of KAT1 was observed in *Arabidopsis* guard cells during stomatal closure, and KAT1 can go back to the plasma membrane of guard cells for stomatal opening (Sutter et al. 2007; Latorre et al. 2003). ABA can induce stomatal closure by activating outward K^+ channels and inhibiting inward K^+ channels simultaneously. Using patch clamping technique, the inhibition of inward K^+ channel currents by a pre-incubation of guard cell protoplast with ABA was observed in *Arabidopsis* (Fan et al. 2008), supporting the inhibitory effects of ABA on inward K^+ channels. Interestingly, the inhibition of inward K^+ channels by the disruption of outward anion channel SLAC1 was reported recently in *Arabidopsis* (Laanemets et al. 2013; Wang et al. 2013), and the decrease of cytosolic Ca^{2+} concentration releases partially the inhibitory effects of the mutation of SLAC1 on inward K^+

channels (Laanemets et al. 2013). Together, these results show that ABA may regulate inward K^+ channel currents in different levels, including transcriptional level of gene expression, protein trafficking, K^+ channel activity, and coordination of different types of channels in guard cells. These regulating mechanisms work together to facilitate a fast and efficient stomatal movement regulation.

15.6.3 Plasma Membrane Anion Channels

Anion efflux causes the depolarization of the plasma membrane of guard cells, further opens the outward-rectifying K^+ channel, and so stomata can be closed. Therefore, anion channels in the plasma membrane of guard cells play core roles for stomatal closure. Over 20 years ago, the activity of anion channel currents was recorded using patch clamping technique in *Vicia faba* guard cell protoplast (Schroeder and Hagiwara 1989). Based on the electrophysiological analysis, two different types of depolarization-activated anion channel currents were dissected electrophysiologically in *Vicia faba* guard cells (Schroeder and Keller 1992). One was activated by depolarization voltage and rapidly deactivated by hyperpolarization voltage, and so was named as R-type (rapid) anion current (Schroeder and Keller 1992). The second depolarization-activated current is activated and deactivated very slowly, and so was named as S-type (slow) anion currents (Schroeder and Keller 1992). Further study revealed that S-type anion channels in *Arabidopsis* guard cells can be activated by cytosolic Ca^{2+} and external ABA (Pei et al. 1997). However, the genetic identities of the two types of anion channels remained unknown for many years. Recent work from different groups identified slow anion channel-associated 1 (SLAC1) as an important component for stomatal closure in *Arabidopsis* (Negi et al. 2008; Vahisalu et al. 2008). *SLAC1* mainly expresses in guard cells, occasionally in vascular tissue (Negi et al. 2008; Vahisalu et al. 2008). SLAC1 belongs to a small family with five members in *Arabidopsis*, and the other four members were named as SLAC1 homologs (SLAHs), which mainly express in transmitting tissue in root and stem in *Arabidopsis* (Negi et al. 2008). SLAC1 has ten TM domains, a hydrophilic N-terminal tail and a hydrophilic C-terminal tail, and these structural characters suggest that SLAC1 is an ion channel. The loss-of-function mutants of *SLAC1* show strong insensitive phenotypes in stomatal closure in response to ABA, ozone, ROS, darkness, CO_2 elevation, and decrease of relative air humidity (Negi et al. 2008; Vahisalu et al. 2008), indicating that SLAC1 is a key downstream component for a variety of signals (including ABA) in guard cells. The S-type anion channel currents activated by ABA and cytosolic Ca^{2+} in guard cells were completely abolished in *slac1* mutants (Vahisalu et al. 2008), and the fused SLAC1 with GFP can light up the plasma membrane of *Arabidopsis* guard cells and onion epidermal cells (Negi et al. 2008; Vahisalu et al. 2008), suggesting SLAC1 is the long-sought S-type anion channel in guard cells. Coexpression of the genes of SLAC1 and protein kinases OST1 or CPKs heterologously in *Xenopus* oocytes results in S-type anion currents, and coexpression of *SLAC1* and *OST1* with the negative regulators of ABA signaling, phosphatases *ABI1* or *ABI2*, inhibits OST1-activated SLAC1 currents in *Xenopus* oocytes (Geiger et al. 2009; Lee et al. 2009). Ion selectivity analysis showed that SLAC1 has a large permeability to NO_3^- and Cl^-, but its permeability to malate is quite weak (Geiger et al. 2009). These results finally ensured the identity of SLAC1 as a plasma membrane S-type anion channel in guard cells, and also uncovered a mechanism of how SLAC1 is regulated in ABA-induced stomatal closure. Further research found that *Arabidopsis ost1* mutant guard cells showed partially reduced S-type anion channel currents (Geiger et al. 2009), supporting that the regulating mechanisms revealed in *Xenopus* oocytes function in guard cells. *slac1* mutants are insensitive to ABA strongly, but ABA still induces a slight stomatal closure slowly (Negi et al. 2008; Vahisalu et al. 2008), suggesting that other plasma membrane anion channels are functioning in guard cells. Further study found that a homolog of SLAC1, SLAH3, involves in stomatal closure by mediating NO_3^- efflux (Geiger et al. 2011).

As described above, R-type anion channels are another type of anion channels mediating anion efflux in guard cells, but the genetic identities are completely unknown for decades. Recently, ALMT12, a member of aluminum-activated malate transporter (ALMT) family, was found to be required for stomatal closure (Sasaki et al. 2010), and further analysis revealed that ALMT12 represents a R-type anion channel in *Arabidopsis* guard cells (Meyer et al. 2010). ALMT12 locates in the plasma membrane of guard cells, and the stomatal closure of *atalmt12* mutant is partially insensitive to darkness, CO_2, and ABA (Meyer et al. 2010). Patch clamping analysis showed a reduced R-type anion channel currents in *atalmt12* guard cell protoplast in the presence of external malate, and R-type anion currents were also recorded in *Xenopus* oocytes expressing *ALMT12* gene (Meyer et al. 2010). These studies ensured the identity of ALMT12 as a R-type anion channel in *Arabidopsis* guard cells. ALMT9, another member of ALMT family, involves in stomatal closure and was identified as a malate channel (Kovermann et al. 2007). But ALMT9 locates in the vacuole membrane, not plasma membrane of guard cells and other types of plant cells (Kovermann et al. 2007). Whether more members of ALMT family function in guard cells as plasma membrane anion channels needs further analysis.

SLAC1 and SLAH3 are mainly permeable to Cl^- and nitrate, not malate; ALMT9 is mainly permeable to malate, not Cl^-; and ALMT12 is permeable to Cl^- and can be activated by external malate. These data suggest that different anion channels, including SLAC1, SLAH3, ALMT12, and ALMT9, function cooperatively to drive anion efflux and stomatal closure in response to ABA.

15.6.4 Channels in the Intracellular Membranes

Ion channels in the plasma membrane of guard cells have been extensively analyzed. However, ion channels are also present in intracellular biological membrane, especially in vacuole membrane. Vacuole is the main intracellular organ involving in stomatal movement, and the ion uptake from cytoplasm to vacuole and release from vacuole to cytoplasm drives the changes of guard cell turgor pressure, which is essential for stomatal closure and opening. Over 90 % osmotic active solutes released from vacuole to cytoplasm are further released to apoplast through plasma mambrane channels (MacRobbie 1998). Therefore, ion channels and ion pumps are the main functional proteins mediating ion flux through vacuole membrane. Using patch clamping technique, three types of Ca^{2+}-regulated channels were identified electrophysiologically in *Vicia faba* guard cell vacuole, including vacuolar K^+ (VK) channel, slow vacuolar (SV) channel, and fast vacuole (FV) channel (Allen and Sanders 1996). But these three types of vacuole channels are regulated by cytosolic Ca^{2+} differentially (Allen and Sanders 1996) and were proposed functioning in stomatal movement in response to different stimulus. TPK1 (twin-pore K^+ channel 1) is a vacuolar K^+ (VK) channel identified in *Arabidopsis* guard cells. TPK1 is from a family called two-pore-domain K^+ channel family, which has five members in *Arabidopsis*, and shares the structure topology of four TM domains: two pore domains and two EF hand Ca^{2+}-binding domains (Ward et al. 2009). TPK1 can be activated by elevated cytosolic Ca^{2+} and low pH during stomatal closure (Gobert et al. 2007). The disruption of TPK1 in *Arabidopsis* leads to the lack of VK channel activity and slower stomatal closure in response to ABA (Gobert et al. 2007), suggesting a role of TPK1 in ABA-induced stomatal movement.

SV channel activity is present in multiple tissues in plants, and TPC1 (two pore channel 1) was identified to be a SV channel (Peiter et al. 2005). TPC1 has two pore domains and can be activated by cytosolic Ca^{2+} in a voltage-dependent manner. TPC1 is a Ca^{2+}-permeable channel with a large conductance and thought to involve in Ca^{2+} release from vacuole in guard cells. The knockout mutation of *TPC1* leads to an insensitive phenotype to the inhibition of stomatal opening induced by external Ca^{2+} without affecting ABA-induced stomatal closure (Peiter et al. 2005). High extracellular Ca^{2+} can cause the elevation of cytosolic Ca^{2+} in guard cells and further close stomata. Calcium sensor 1 (CAS1) is the sensor to sense the changes of extracellular Ca^{2+} concentration for *Arabidopsis* guard cells (Han et al. 2003). The down-regulation of *CAS1* expression level impairs extracellular Ca^{2+}-induced stomatal closure, but leaves ABA response intact in guard cells (Han et al. 2003). ABA induces cytosolic Ca^{2+} increases and oscillation, and Ca^{2+} release from intracellular Ca^{2+} stores involves in cytosolic Ca^{2+} oscillation. Vacuole is believed to be important for the intracellular Ca^{2+} release and cytosolic Ca^{2+} oscillation. But the intact ABA response in the guard cells of *tpc1* mutant suggests that the Ca^{2+} channels mediating intracellular Ca^{2+} release for ABA-induced Ca^{2+} oscillation differ from TPC1.

FV channel is another type of ion channel located in the membrane of vacuole. FV channel is a nonselective cation channel, which is permeable to cation K^+, Ca^{2+}, and Mg^{2+} with a strong ion selectivity over anion (Allen and Sanders 1996). Mg^{2+} can inhibit FV channel activity, but the presence of Mg^{2+} is required for the activation of FV channels by cytosolic Ca^{2+} (Pei et al. 1999). FV channel is insensitive to cytosolic K^+, but can be regulated by luminal K^+. It has been proposed that FV channels function as K^+ channel mediating K^+ release from vacuole to cytoplasm, and the luminal K^+ in vacuole regulates the activity of FV channels in a feedback manner. The more K^+ is released from vacuole, the lower the FV channel activity is (Pottosin and Martínez-Estévez 2003). So far, the genetic identities of FV channels are still unknown, and further study is needed to find out the genetic identity and whether FV involves in ABA signaling in guard cells.

Malate and nitrate are the most abundant anions mainly stored in vacuole in plant cells, and both anions involve in stomatal movement as osmotic solutes. Therefore, anion channels or transporters must exist in the vacuole membrane of guard cells. In tonoplast, anion channels are mainly from two channel families: CLC and ALMT. For ALMT family, ALMT9 was identified as a malate-activated chloride-permeable channel located in the tonoplast of guard cells in *Arabidopsis* (Kovermann et al. 2007; De Angeli et al. 2013), and the disruption of ALMT9 impairs stomata opening without affecting ABA- and dark-induced stomatal closure (De Angeli et al. 2013). CLC family has seven members in *Arabidopsis*. Four CLC members localize in vacuole, two CLC members localize in Golgi, and one localizes in chloroplast. The *Arabidopsis* vacuole CLC channels include CLCa, CLCb, CLCc, and CLCg. AtCLCc is a real channel being permeable to nitrate, chloride, and citrate and involves in stomatal movement (Harada et al. 2004). The disruption of *AtCLCc* gene inhibits vacuole chloride currents and impairs light-induced stomatal opening and ABA-induced stomatal closure (Jossier et al. 2010). AtCLCa is the most well-analyzed CLC member. The disruption of AtCLCa leads to 50 % reduction of nitrate in root and shoot (Geelen et al. 2000). Further research confirmed that AtCLCa functions as a tonoplast NO_3^-/H^+ exchanger, not a channel (De Angeli et al. 2006). Whether CLCa, CLCb, and CLCg involve in stomatal movement needs further investigation.

Vacuole is important for ABA-induced stomatal closure, but not all vacuole channels are related to ABA signaling cascade because the disruption of some vacuole channel/transporter genes does not impair ABA-induced stomatal closure. Nevertheless, it is still reasonable to conclude that vacuole channels play critical roles in guard cells.

15.7 Conclusion

In the past few decades, fantastic research identified many important components for ABA signaling in guard cells and finally revealed the simple but relatively complete ABA signaling chain in guard cells. Briefly, the ABA signaling chain for ABA-induced stomatal closure can be summarized as the following: First, ABA binds to ABA receptors to trigger the protein interaction between ABA receptors and PP2Cs to inhibit the phosphatase activity of PP2Cs; second, the inhibition of PP2Cs by ABA receptors releases SnRKs and CPKs from the inhibition by PP2Cs, and the kinases are then autophosphorylated and auto-activated; third, the activated kinases activate SLAC1 by phosphorylation and trigger the efflux of anion through SLAC1 and other anion channels; forth, the efflux of anion causes a depolarization of the plasma membrane of guard cells, and the depolarization of plasma membrane activates outward K^+ channel GORK; finally, guard cells loss salt and water, and stomata can close. This relative complete ABA signaling pathway is so exciting, because it is the first time that we can see an uninterrupted ABA signaling pathway and understand much better than ever how ABA closes stomata. However, we also should be aware that many other components involving in the stomatal closure induced by ABA and stomatal opening inhibited by ABA have not been integrated into the ABA signaling network yet, including the known components as we describe above, some others components not mentioned in this chapter, as well as the unknown components such as plasma membrane Ca^{2+} channels that will probably be identified in the future. Moreover, it was known for many years that ABA also regulates stomatal movement in transcription level in guard cells. The application of ABA can up- and down-regulate the expression levels of some genes in *Arabidopsis* guard cells and may consequently regulate stomatal movement. Much more work will be needed to draw the complete picture of ABA signaling network for stomatal movement.

On the other hand, stomatal closure and opening are regulated by diverse signals, including ABA, ozone, CO_2, ROS, and changes of air humidity and pathogen. ABA is only one of the stimuli. Emerging results already showed that the cross talks between ABA and other signaling are complicated; they share some key components and also have their own specific components. For example, the mutations of *GCA2* and *SLAC1* impair the stomatal closure induced by ABA, ozone, ROS, CO_2, and changes of air humidity, but the mutation of *HT1* only impairs CO_2-induced stomata movement without affecting ABA signaling as we mentioned above. To understand how guard cells integrate different regulating signals together will need us to unravel drought/ABA signaling network as well as the signaling networks of other stimuli, and the research progress in one signaling network may unexpectively help us to understand other signaling network for stomatal movement regulation.

Acknowledgments This work was supported by Hundred Talents Program of Chinese Academy of Sciences (2010OHTP06) and the National Basic Research Program of China (973 program; grant No. 2012CB114300). The author apologizes to colleagues whose publications were not cited and discussed in this chapter because of space limitation.

References

Ali R, Ma W, Lemtiri-Chlieh F, Tsaltas D, Leng Q, von Bodman S, Berkowitz GA. Death don't have no mercy and neither does calcium: *Arabidopsis* cyclic nucleotide gated channel 2 and innate immunity. Plant Cell. 2007;19:1081–95.

Allen GJ, Sanders D. Control of ionic currents in guard cell vacuoles by cytosolic and luminal calcium. Plant J. 1996;10:1055–69.

Allen GJ, Kuchitsu K, Chu SP, Murata Y, Schroeder JI. *Arabidopsis abi1-1* and *abi2-1* phosphatase mutations reduce abscisic acid-induced cytoplasmic calcium rises in guard cells. Plant Cell. 1999;11:1785–98.

Allen TW, Kuyucak S, Chung SH. Molecular dynamics estimates of ion diffusion in model hydrophobic and KcsA potassium channels. Biophys Chem. 2000;86:1–14.

Allen GJ, Chu SP, Harrington CL, Schumacher K, Hoffmann T, Tang YY, Grill E, Schroeder JI. A defined range of guard cell calcium oscillation parameters encodes stomatal movement. Nature. 2001;411:1053–7.

Belin C, de Franco PO, Bourbousse C, Chaignepain S, Schmitter JM, Vavasseur A, Giraudat J, Barbier-Brygoo H, Thomine S. Identification of features regulating OST1 kinase activity and OST1 function in guard cells. Plant Physiol. 2006;141:1316–27.

Blackman PG, Davies WJ. Root to shoot communication in maize plants of the effects of soil drying. J Exp Bot. 1985;36:39–48.

Brandt B, Brodsky DE, Xue S, Negi J, Iba K, Kangasjarvi J, Ghassemian M, Stephan AB, Hu H, Schroeder JI. Reconstitution of abscisic acid activation of SLAC1 anion channel by CPK6 and OST1 kinases and branched ABI1 PP2C phosphatase action. Proc Natl Acad Sci USA. 2012;109:10593–8.

Briggs WR, Christie JM. Phototropins 1 and 2: versatile plant blue-light receptors. Trends Plant Sci. 2002;7:204–10.

Casson S, Gray JE. Influence of environmental factors on stomatal development. New Phytol. 2008;178:9–23.

Cheong YH, Pandey GK, Grant JJ, Batistic O, Li L, Kim BG, Lee SC, Kudla J, Luan S. Two calcineurin B-like calcium sensors, interacting with protein kinase CIPK23, regulate leaf transpiration and root potassium uptake in *Arabidopsis*. Plant J. 2007;52:223–39.

Cho D, Kim SA, Murata Y, Lee S, Jae S-K, Nam HG, Kwak JM. De-regulated expression of the plant glutamate receptor homolog AtGLR3.1 impairs long-term Ca^{2+}-programmed stomatal closure. Plant J. 2009;58:437–49.

Daszkowska-Golec A, Chorazy E, Maluszynski M, Szarejko I. Towards the identification of new genes involved in ABA-dependent abiotic stresses using *Arabidopsis* suppressor mutants of *abh1* hypersensitivity to ABA during seed germination. Int J Mol Sci. 2013;14:13403–32.

De Angeli A, Monachello D, Ephritikhine G, Frachisse JM, Thomine S, Gambale F, Barbier-Brygoo H. The nitrate/proton antiporter AtCLCa mediates nitrate accumulation in plant vacuoles. Nature. 2006;442:939–42.

De Angeli A, Zhang J, Meyer S, Martinoia E. AtALMT9 is a malate-activated vacuolar chloride channel required for stomatal opening in *Arabidopsis*. Nature Commun. 2013;4:1804.

Dong J, Bergmann DC. Stomatal patterning and development. Curr Top Dev Biol. 2010;91:267–97.

Du SY, Zhang XF, Lu Z, Xin Q, Wu Z, Jiang T, Lu Y, Wang XF, Zhang DP. Roles of the different components of magnesium chelatase in abscisic acid signal transduction. Plant Mol Biol. 2012;80:519–37.

Dubovskaya LV, Bakakina YS, Kolesneva EV, Sodel DL, McAinsh MR, Hetherington AM, Volotovski ID. cGMP-dependent ABA-induced stomatal closure in the ABA-insensitive *Arabidopsis* mutant *abi1-1*. New Phytol. 2011;191:57–69.

Dupeux F, Antoni R, Betz K, Santiago J, Gonzalez-Guzman M, Rodriguez L, Rubio S, Park SY, Cutler SR, Rodriguez PL, Marquez JA. Modulation of abscisic acid signaling *in vivo* by an engineered receptor-insensitive protein phosphatase type 2C allele. Plant Physiol. 2011;156:106–16.

Fan LM, Zhang W, Chen JG, Taylor JP, Jones AM, Assmann SM. Abscisic acid regulation of guard-cell K^+ and anion channels in Gbeta- and RGS-deficient *Arabidopsis* lines. Proc Natl Acad Sci USA. 2008;105:8476–81.

Finkelstein RR, Somerville CR. Three classes of abscisic acid (ABA)-insensitive mutations of *Arabidopsis* define genes that control overlapping subsets of ABA responses. Plant Physiol. 1990;94:1172–9.

Finkelstein RR, Lynch TJ. The *Arabidopsis* abscisic acid response gene *ABI5* encodes a basic leucine zipper transcription factor. Plant Cell. 2000;12:599–609.

Franz S, Ehlert B, Liese A, Kurth J, Cazale AC, Romeis T. Calcium-dependent protein kinase CPK21 functions in abiotic stress response in *Arabidopsis thaliana*. Mol Plant. 2011;4:83–96.

Fujii H, Verslues PE, Zhu JK. Identification of two protein kinases required for abscisic acid regulation of seed germination, root growth, and gene expression in *Arabidopsis*. Plant Cell. 2007;19:485–94.

Fujii H, Verslues PE, Zhu J-K. *Arabidopsis* decuple mutant reveals the importance of SnRK2 kinases in osmotic stress responses in vivo. Proc Natl Acad Sci USA. 2011;108:1717–22.

Gaedeke N, Klein M, Kolukisaoglu U, Forestier C, Muller A, Ansorge M, Becker D, Mamnun Y, Kuchler K, Schulz B, Mueller-Roeber B, Martinoia E. The *Arabidopsis thaliana* ABC transporter AtMRP5 controls root development and stomata movement. EMBO J. 2001;20:1875–87.

Gao Y, Zeng Q, Guo J, Cheng J, Ellis BE, Chen JG. Genetic characterization reveals no role for the reported ABA receptor, GCR2, in ABA control of seed germination and early seedling development in *Arabidopsis*. Plant J. 2007;52:1001–13.

Gao QF, Fei CF, Dong JY, Gu LL, Wang YF. *Arabidopsis* CNGC18 is a Ca^{2+}-permeable channel. Mol Plant. 2014;7:739–43.

Gayatri G, Agurla S, Raghavendra AS. Nitric oxide in guard cells as an important secondary messenger during stomatal closure. Front Plant Sci. 2013;4:425.

Geelen D, Lurin C, Bouchez D, Frachisse JM, Lelievre F, Courtial B, Barbier-Brygoo H, Maurel C. Disruption of putative anion channel gene AtCLC-a in *Arabidopsis* suggests a role in the regulation of nitrate content. Plant J. 2000;21:259–67.

Geiger D, Scherzer S, Mumm P, Stange A, Marten I, Bauer H, Ache P, Matschi S, Liese A, Al-Rasheid KA, Romeis T, Hedrich R. Activity of guard cell anion channel SLAC1 is controlled by drought-stress signaling kinase-phosphatase pair. Proc Natl Acad Sci USA. 2009;106:21425–30.

Geiger D, Maierhofer T, Al-Rasheid KA, Scherzer S, Mumm P, Liese A, Ache P, Wellmann C, Marten I, Grill E, Romeis T, Hedrich R. Stomatal closure by fast abscisic acid signaling is mediated by the guard cell anion channel SLAH3 and the receptor RCAR1. Sci Signal. 2011;4(173):ra32.

Giraudat J, Hauge BM, Valon C, Smalle J, Parcy F, Goodman HM. Isolation of the *Arabidopsis ABI3* gene by positional cloning. Plant Cell. 1992;4:1251–61.

Gobert A, Isayenkov S, Voelker C, Czempinski K, Maathuis FJ. The two-pore channel *TPK1* gene encodes the vacuolar K^+ conductance and plays a role in K^+ homeostasis. Proc Natl Acad Sci USA. 2007;104:10726–31.

Gray JE, Holroyd GH, van der Lee FM, Bahrami AR, Sijmons PC, Woodward FI, Schuch W, Hetherington AM. The HIC signalling pathway links CO_2 perception to stomatal development. Nature. 2000;408:713–6.

Hamilton DW, Hills A, Kohler B, Blatt MR. Ca^{2+} channels at the plasma membrane of stomatal guard cells are activated by hyperpolarization and abscisic acid. Proc Natl Acad Sci USA. 2000;97:4967–72.

Han S, Tang R, Anderson LK, Woerner TE, Pei ZM. A cell surface receptor mediates extracellular Ca^{2+} sensing in guard cells. Nature. 2003;425:196–200.

Harada H, Kuromori T, Hirayama T, Shinozaki K, Leigh RA. Quantitative trait loci analysis of nitrate storage in Arabidopsis leading to an investigation of the contribution of the anion channel gene, AtCLC-c, to variation in nitrate levels. J Exp Bot. 2004;55:2005–14.

Hartung W, Sauter A, Hose E. Abscisic acid in the xylem: where does it come from, where does it go to? J Exp Bot. 2002;53:27–32.

Hashimoto M, Negi J, Young J, Israelsson M, Schroeder JI, Iba K. *Arabidopsis* HT1 kinase controls stomatal movement in response to CO_2. Nature Cell Biol. 2006;8:391–7.

Himmelbach A, Iten M, Grill E. Signalling of abscisic acid to regulate plant growth. Philos Trans R Soc Lond B Biol Sci. 1998;353:1439–44.

Hoshino D, Hayashi A, Temmei Y, Kanzawa N, Tsuchiya T. Biochemical and immunohistochemical characterization of *Mimosa* annexin. Planta. 2004;219:867–75.

Hosy E, Vavasseur A, Mouline K, Dreyer I, Gaymard F, Poree F, Boucherez J, Lebaudy A, Bouchez D, Very AA, Simonneau T, Thibaud JB, Sentenac H. The *Arabidopsis* outward K^+

channel GORK is involved in regulation of stomatal movement and plant transpiration. Proc Natl Acad Sci USA. 2003;100:5549–54.

Hu H, Boisson-Dernier A, Israelsson-Nordstrom M, Bohmer M, Xue S, Ries A, Godoski J, Kuhn JM, Schroeder JI (2010) Carbonic anhydrases are upstream regulators of CO_2-controlled stomatal movement in guard cells. Nature Cell Biol 12(81–88) 87-93; sup pp 81-18.

Hua D, Wang C, He J, Liao H, Duan Y, Zhu Z, Guo Y, Chen Z, Gong Z. A plasma membrane receptor kinase, GHR1, mediates abscisic acid- and hydrogen peroxide-regulated stomatal movement in *Arabidopsis*. Plant Cell. 2012;24:2546–61.

Hugouvieux V, Kwak JM, Schroeder JI. An mRNA cap binding protein, ABH1, modulates early abscisic acid signal transduction in *Arabidopsis*. Cell. 2001;106:477–87.

Hugouvieux V, Murata Y, Young JJ, Kwak JM, Mackesy DZ, Schroeder JI. Localization, ion channel regulation, and genetic interactions during abscisic acid signaling of the nuclear mRNA cap-binding protein, ABH1. Plant Physiol. 2002;130:1276–87.

Jiang F, Hartung W. Long-distance signalling of abscisic acid (ABA): the factors regulating the intensity of the ABA signal. J Exp Bot. 2008;59:37–43.

Jossier M, Kroniewicz L, Dalmas F, Le Thiec D, Ephritikhine G, Thomine S, Barbier-Brygoo H, Vavasseur A, Filleur S, Leonhardt N. The *Arabidopsis* vacuolar anion transporter, AtCLCc, is involved in the regulation of stomatal movement and contributes to salt tolerance. Plant J. 2010;64:563–76.

Joudoi T, Shichiri Y, Kamizono N, Akaike T, Sawa T, Yoshitake J, Yamada N, Iwai S. Nitrated cyclic GMP modulates guard cell signaling in *Arabidopsis*. Plant Cell. 2013;25:558–71.

Kasahara M, Swartz TE, Olney MA, Onodera A, Mochizuki N, Fukuzawa H, Asamizu E, Tabata S, Kanegae H, Takano M, Christie JM, Nagatani A, Briggs WR. Photochemical properties of the flavin mononucleotide-binding domains of the phototropins from *Arabidopsis*, rice, and *Chlamydomonas reinhardtii*. Plant Physiol. 2002;129:762–73.

Klein M, Perfus-Barbeoch L, Frelet A, Gaedeke N, Reinhardt D, Mueller-Roeber B, Martinoia E, Forestier C. The plant multidrug resistance ABC transporter AtMRP5 is involved in guard cell hormonal signalling and water use. Plant J. 2003;33:119–29.

Kohler B, Blatt MR. Protein phosphorylation activates the guard cell Ca^{2+} channel and is a prerequisite for gating by abscisic acid. Plant J. 2002;32:185–94.

Konopka-Postupolska D, Clark G, Goch G, Debski J, Floras K, Cantero A, Fijolek B, Roux S, Hennig J. The role of annexin 1 in drought stress in *Arabidopsis*. Plant Physiol. 2009;150:1394–410.

Koornneef M, Reuling G, Karssen CM. The isolation and characterization of abscisic acid-insensitive mutants of *Arabidopsis thaliana*. Physiol Plant. 1984;61:377–83.

Kovermann P, Meyer S, Hortensteiner S, Picco C, Scholz-Starke J, Ravera S, Lee Y, Martinoia E. The *Arabidopsis* vacuolar malate channel is a member of the ALMT family. Plant J. 2007;52:1169–80.

Kuhn JM, Boisson-Dernier A, Dizon MB, Maktabi MH, Schroeder JI. The protein phosphatase AtPP2CA negatively regulates abscisic acid signal transduction in *Arabidopsis*, and effects of *abh1* on AtPP2CA mRNA. Plant Physiol. 2006;140:127–39.

Kwak JM, Murata Y, Baizabal-Aguirre VM, Merrill J, Wang M, Kemper A, Hawke SD, Tallman G, Schroeder JI. Dominant negative guard cell K^+ channel mutants reduce inward-rectifying K^+ currents and light-induced stomatal opening in *Arabidopsis*. Plant Physiol. 2001;127:473–85.

Kwak JM, Mori IC, Pei ZM, Leonhardt N, Torres MA, Dangl JL, Bloom RE, Bodde S, Jones JD, Schroeder JI. NADPH oxidase AtrbohD and AtrbohF genes function in ROS-dependent ABA signaling in *Arabidopsis*. EMBO J. 2003;22:2623–33.

Laanemets K, Wang YF, Lindgren O, Wu J, Nishimura N, Lee S, Caddell D, Merilo E, Brosche M, Kilk K, Soomets U, Kangasjarvi J, Schroeder JI, Kollist H. Mutations in the SLAC1 anion channel slow stomatal opening and severely reduce K^+ uptake channel activity via enhanced cytosolic $[Ca^{2+}]$ and increased Ca^{2+} sensitivity of K^+ uptake channels. New Phytol. 2013;197:88–98.

Latorre R, Olcese R, Basso C, Gonzalez C, Munoz F, Cosmelli D, Alvarez O. Molecular coupling between voltage sensor and pore opening in the *Arabidopsis* inward rectifier K^+ channel KAT1. J Gen Physiol. 2003;122:459–69.

Lebaudy A, Hosy E, Simonneau T, Sentenac H, Thibaud J-B, Dreyer I. Heteromeric K^+ channels in plants. Plant J. 2008;54:1076–82.

Lee Y, Choi YB, Suh S, Lee J, Assmann SM, Joe CO, Kelleher JF, Crain RC. Abscisic acid-induced phosphoinositide turnover in guard cell protoplasts of *Vicia faba*. Plant Physiol. 1996;110:987–96.

Lee KH, Piao HL, Kim HY, Choi SM, Jiang F, Hartung W, Hwang I, Kwak JM, Lee IJ, Hwang I. Activation of glucosidase via stress-induced polymerization rapidly increases active pools of abscisic acid. Cell. 2006;126:1109–20.

Lee SC, Lan W, Buchanan BB, Luan S. A protein kinase-phosphatase pair interacts with an ion channel to regulate ABA signaling in plant guard cells. Proc Natl Acad Sci USA. 2009;106:21419–24.

Lemtiri-Chlieh F, Berkowitz GA. Cyclic adenosine monophosphate regulates calcium channels in the plasma membrane of *Arabidopsis* leaf guard and mesophyll cells. J Biol Chem. 2004;279:35306–12.

Lemtiri-Chlieh F, MacRobbie EA, Webb AA, Manison NF, Brownlee C, Skepper JN, Chen J, Prestwich GD, Brearley CA. Inositol hexakisphosphate mobilizes an endomembrane store of calcium in guard cells. Proc Natl Acad Sci USA. 2003;100:10091–5.

Leng Q, Mercier RW, Hua BG, Fromm H, Berkowitz GA. Electrophysiological analysis of cloned cyclic nucleotide-gated ion channels. Plant Physiol. 2002;128:400–10.

Leonhardt N, Kwak JM, Robert N, Waner D, Leonhardt G, Schroeder JI. Microarray expression analyses of *Arabidopsis* guard cells and isolation of a recessive abscisic acid hypersensitive protein phosphatase 2C mutant. Plant Cell. 2004;16:596–615.

Leung J, Merlot S, Giraudat J. The *Arabidopsis ABSCISIC ACID-INSENSITIVE2* (*ABI2*) and *ABI1* genes encode homologous protein phosphatases 2C involved in abscisic acid signal transduction. Plant Cell. 1997;9:759–71.

Leung J, Bouvier-Durand M, Morris PC, Guerrier D, Chefdor F, Giraudat J. *Arabidopsis* ABA response gene ABI1: features of a calcium-modulated protein phosphatase. Science. 1994;264:1448–52.

Liu X, Yue Y, Li B, Nie Y, Li W, Wu WH, Ma L. A G protein-coupled receptor is a plasma membrane receptor for the plant hormone abscisic acid. Science. 2007;315:1712–6.

Ma W. Roles of Ca^{2+} and cyclic nucleotide gated channel in plant innate immunity. Plant Sci. 2011;181:342–6.

Ma SY, Wu WH. AtCPK23 functions in *Arabidopsis* responses to drought and salt stresses. Plant Mol Biol. 2007;65:511–8.

Ma Y, Szostkiewicz I, Korte A, Moes D, Yang Y, Christmann A, Grill E. Regulators of PP2C phosphatase activity function as abscisic acid sensors. Science. 2009;324:1064–8.

MacRobbie EA. Signal transduction and ion channels in guard cells. Philos Trans R Soc London. 1998;353:1475–88.

Merilo E, Jõesaar I, Brosché M, Kollist H. To open or to close: species-specific stomatal responses to simultaneously applied opposing environmental factors. New Phytol. 2014;202:499–508.

Merlot S, Leonhardt N, Fenzi F, Valon C, Costa M, Piette L, Vavasseur A, Genty B, Boivin K, Muller A, Giraudat J, Leung J. Constitutive activation of a plasma membrane H^+-ATPase prevents abscisic acid-mediated stomatal closure. EMBO J. 2007;26:3216–26.

Meyer S, Mumm P, Imes D, Endler A, Weder B, Al-Rasheid KA, Geiger D, Marten I, Martinoia E, Hedrich R. AtALMT12 represents an R-type anion channel required for stomatal movement in *Arabidopsis* guard cells. Plant J. 2010;63:1054–62.

Miao Y, Lv D, Wang P, Wang XC, Chen J, Miao C, Song CP. An *Arabidopsis* glutathione peroxidase functions as both a redox transducer and a scavenger in abscisic acid and drought stress responses. Plant Cell. 2006;18:2749–66.

Mori IC, Murata Y, Yang Y, Munemasa S, Wang YF, Andreoli S, Tiriac H, Alonso JM, Harper JF, Ecker JR, Kwak JM, Schroeder JI. CDPKs CPK6 and CPK3 function in ABA regulation of guard cell S-type anion- and Ca^{2+}-permeable channels and stomatal closure. PLoS Biol. 2006;4:e327.

Mortimer JC, Laohavisit A, Macpherson N, Webb A, Brownlee C, Battey NH, Davies JM. Annexins: multifunctional components of growth and adaptation. J Exp Bot. 2008;59:533–44.

Murata Y, Pei ZM, Mori IC, Schroeder J. Abscisic acid activation of plasma membrane Ca^{2+} channels in guard cells requires cytosolic NAD(P)H and is differentially disrupted upstream and downstream of reactive oxygen species production in *abi1-1* and *abi2-1* protein phosphatase 2C mutants. Plant Cell. 2001;13:2513–23.

Mustilli AC, Merlot S, Vavasseur A, Fenzi F, Giraudat J. *Arabidopsis* OST1 protein kinase mediates the regulation of stomatal aperture by abscisic acid and acts upstream of reactive oxygen species production. Plant Cell. 2002;14:3089–99.

Nadeau JA. Stomatal development: new signals and fate determinants. Curr Opin Plant Biol. 2009;12: 29–35.

Negi J, Matsuda O, Nagasawa T, Oba Y, Takahashi H, Kawai-Yamada M, Uchimiya H, Hashimoto M, Iba K. CO_2 regulator SLAC1 and its homologues are essential for anion homeostasis in plant cells. Nature. 2008;452:483–6.

Nishimura N, Hitomi K, Arvai AS, Rambo RP, Hitomi C, Cutler SR, Schroeder JI, Getzoff ED. Structural mechanism of abscisic acid binding and signaling by dimeric PYR1. Science. 2009;326:1373–9.

Nishimura N, Sarkeshik A, Nito K, Park SY, Wang A, Carvalho PC, Lee S, Caddell DF, Cutler SR, Chory J, Yates JR, Schroeder JI. PYR/PYL/RCAR family members are major *in-vivo* ABI1 protein phosphatase 2C-interacting proteins in *Arabidopsis*. Plant J. 2010;61:290–9.

Pandey S, Nelson DC, Assmann SM. Two novel GPCR-type G proteins are abscisic acid receptors in *Arabidopsis*. Cell. 2009;136:136–48.

Park SY, Fung P, Nishimura N, Jensen DR, Fujii H, Zhao Y, Lumba S, Santiago J, Rodrigues A, Chow TF, Alfred SE, Bonetta D, Finkelstein R, Provart NJ, Desveaux D, Rodriguez PL, McCourt P, Zhu JK, Schroeder JI, Volkman BF, Cutler SR. Abscisic acid inhibits type 2C protein phosphatases via the PYR/PYL family of START proteins. Science. 2009;324:1068–71.

Pei ZM, Ward JM, Schroeder JI. Magnesium sensitizes slow vacuolar channels to physiological cytosolic calcium and inhibits fast vacuolar channels in fava bean guard cell vacuoles. Plant Physiol. 1999;121:977–86.

Pei ZM, Kuchitsu K, Ward JM, Schwarz M, Schroeder JI. Differential abscisic acid regulation of guard cell slow anion channels in *Arabidopsis* wild-type and *abi1* and *abi2* mutants. Plant Cell. 1997;9:409–23.

Pei ZM, Murata Y, Benning G, Thomine S, Klusener B, Allen GJ, Grill E, Schroeder JI. Calcium channels activated by hydrogen peroxide mediate abscisic acid signalling in guard cells. Nature. 2000;406:731–4.

Peiter E, Maathuis FJ, Mills LN, Knight H, Pelloux J, Hetherington AM, Sanders D. The vacuolar Ca^{2+}-activated channel TPC1 regulates germination and stomatal movement. Nature. 2005;434:404–8.

Peuke AD, Jeschke WD, Hartung W. Flows of elements, ions and abscisic acid in *Ricinus communis* and site of nitrate reduction under potassium limitation. J Exp Bot. 2002;53:241–50.

Pottosin II, Martínez-Estévez M. Regulation of the fast vacuolar channel by cytosolic and vacuolar potassium. Biophy J. 2003;84:977–86.

Quarrie SA, Jones HG. Effects of abscisic acid and water stress on development and morphology of wheat. J Exp Bot. 1977;28:192–203.

Rae AM, Ferris R, Tallis MJ, Taylor G. Elucidating genomic regions determining enhanced leaf growth and delayed senescence in elevated CO_2. Plant Cell Environ. 2006;29:1730–41.

Saez A, Robert N, Maktabi MH, Schroeder JI, Serrano R, Rodriguez PL. Enhancement of abscisic acid sensitivity and reduction of water consumption in *Arabidopsis* by combined inactivation of the protein phosphatases type 2C ABI1 and HAB1. Plant Physiol. 2006;141:1389–99.

Saez A, Apostolova N, Gonzalez-Guzman M, Gonzalez-Garcia MP, Nicolas C, Lorenzo O, Rodriguez PL. Gain-of-function and loss-of-function phenotypes of the protein phosphatase 2C HAB1 reveal its role as a negative regulator of abscisic acid signalling. Plant J. 2004;37:354–69.

Salisbury EJ. On the causes and ecological significance of stomatal frequency, with special reference to the woodland flora. Phil Trans R Soc Lond B. 1928;216:1–65.

Santiago J, Dupeux F, Betz K, Antoni R, Gonzalez-Guzman M, Rodriguez L, Marquez JA, Rodriguez PL. Structural insights into PYR/PYL/RCAR ABA receptors and PP2Cs. Plant Sci. 2012;182:3–11.

Sasaki T, Mori IC, Furuichi T, Munemasa S, Toyooka K, Matsuoka K, Murata Y, Yamamoto Y. Closing plant stomata requires a homolog of an aluminum-activated malate transporter. Plant Cell Physiol. 2010;51:354–65.

Schroeder JI, Hagiwara S. Cytosolic calcium regulates ion channels in the plasma membrane of *Vicia faba* guard cells. Nature. 1989;338:427–30.

Schroeder JI, Keller BU. Two types of anion channel currents in guard cells with distinct voltage regulation. Proc Natl Acad Sci USA. 1992;89:5025–9.

Schroeder JI, Nambara E. A quick release mechanism for abscisic acid. Cell. 2006;126:1023–5.

Shen YY, Wang XF, Wu FQ, Du SY, Cao Z, Shang Y, Wang XL, Peng CC, Yu XC, Zhu SY, Fan RC, Xu YH, Zhang DP. The Mg-chelatase H subunit is an abscisic acid receptor. Nature. 2006;443:823–6.

Shimazaki K-I, Doi M, Assmann SM, Kinoshita T. Light regulation of stomatal movement. Ann Rev Plant Biol. 2007;58:219–47.

Shkolnik-Inbar D, Adler G, Bar-Zvi D. ABI4 downregulates expression of the sodium transporter HKT1; 1 in *Arabidopsis* roots and affects salt tolerance. Plant J. 2013;73:993–1005.

Siegel RS, Xue S, Murata Y, Yang Y, Nishimura N, Wang A, Schroeder JI. Calcium elevation-dependent and attenuated resting calcium-dependent abscisic acid induction of stomatal closure and abscisic acid-induced enhancement of calcium sensitivities of S-type anion and inward-rectifying K^+ channels in *Arabidopsis* guard cells. Plant J. 2009;59:207–20.

Song Y, Miao Y, Song CP. Behind the scenes: the roles of reactive oxygen species in guard cells. New Phytol. 2014;201:1121–40.

Soon FF, Ng LM, Zhou XE, West GM, Kovach A, Tan MH, Suino-Powell KM, He Y, Xu Y, Chalmers MJ, Brunzelle JS, Zhang H, Yang H, Jiang H, Li J, Yong EL, Cutler S, Zhu JK, Griffin PR, Melcher K, Xu HE. Molecular mimicry regulates ABA signaling by SnRK2 kinases and PP2C phosphatases. Science. 2012;335:85–8.

Suh SJ, Wang YF, Frelet A, Leonhardt N, Klein M, Forestier C, Mueller-Roeber B, Cho MH, Martinoia E, Schroeder JI. The ATP binding cassette transporter AtMRP5 modulates anion and calcium channel activities in *Arabidopsis* guard cells. J Biol Chem. 2007;282:1916–24.

Sutter JU, Sieben C, Hartel A, Eisenach C, Thiel G, Blatt MR. Abscisic acid triggers the endocytosis of the *Arabidopsis* KAT1 K^+ channel and its recycling to the plasma membrane. Curr Biol. 2007;17:1396–402.

Szyroki A, Ivashikina N, Dietrich P, Roelfsema MR, Ache P, Reintanz B, Deeken R, Godde M, Felle H, Steinmeyer R, Palme K, Hedrich R. KAT1 is not essential for stomatal opening. Proc Natl Acad Sci USA. 2001;98:2917–21.

Tang RH, Han S, Zheng H, Cook CW, Choi CS, Woerner TE, Jackson RB, Pei ZM. Coupling diurnal cytosolic Ca^{2+} oscillations to the CAS-IP3 pathway in *Arabidopsis*. Science. 2007;315:1423–6.

Tsuzuki T, Takahashi K, Inoue S-I, Okigaki Y, Tomiyama M, Hossain M, Shimazaki K-I, Murata Y, Kinoshita T. Mg-chelatase H subunit affects ABA signaling in stomatal guard cells, but is not an ABA receptor in *Arabidopsis thaliana*. J Plant Res. 2011;124:527–38.

Tsuzuki T, Takahashi K, Tomiyama M, Inoue S, Kinoshita T. Overexpression of the Mg-chelatase H subunit in guard cells confers drought tolerance via promotion of stomatal closure in *Arabidopsis thaliana*. Front Plant Sci. 2013;4:440.

Umezawa T, Sugiyama N, Mizoguchi M, Hayashi S, Myouga F, Yamaguchi-Shinozaki K, Ishihama Y, Hirayama T, Shinozaki K. Type 2C protein phosphatases directly regulate abscisic acid-activated protein kinases in *Arabidopsis*. Proc Natl Acad Sci U S A. 2009;106:17588–93.

Vahisalu T, Kollist H, Wang YF, Nishimura N, Chan WY, Valerio G, Lamminmaki A, Brosche M, Moldau H, Desikan R, Schroeder JI, Kangasjarvi J. SLAC1 is required for plant guard cell S-type anion channel function in stomatal signalling. Nature. 2008;452:487–91.

Vlad F, Rubio S, Rodrigues A, Sirichandra C, Belin C, Robert N, Leung J, Rodriguez PL, Lauriere C, Merlot S. Protein phosphatases 2C regulate the activation of the Snf1-related kinase OST1 by abscisic acid in *Arabidopsis*. Plant Cell. 2009;21:3170–84.

Wang XQ, Ullah H, Jones AM, Assmann SM. G protein regulation of ion channels and abscisic acid signaling in *Arabidopsis* guard cells. Science. 2001;292:2070–2.

Wang YF, Munemasa S, Nishimura N, Ren HM, Robert N, Han M, Puzorjova I, Kollist H, Lee S, Mori I, Schroeder JI. Identification of cyclic GMP-activated nonselective Ca^{2+}-permeable cation channels and associated *CNGC5* and *CNGC6* genes in *Arabidopsis* guard cells. Plant Physiol. 2013;163:578–90.

Ward JM, Maser P, Schroeder JI. Plant ion channels: gene families, physiology, and functional genomics analyses. Ann Rev Physiol. 2009;71:59–82.

Weiner JJ, Peterson FC, Volkman BF, Cutler SR. Structural and functional insights into core ABA signaling. Curr Opin Plant Biol. 2010;13:495–502.

Wu FQ, Xin Q, Cao Z, Liu ZQ, Du SY, Mei C, Zhao CX, Wang XF, Shang Y, Jiang T, Zhang XF, Yan L, Zhao R, Cui ZN, Liu R, Sun HL, Yang XL, Su Z, Zhang DP. The magnesium-chelatase H subunit binds abscisic acid and functions in abscisic acid signaling: new evidence in *Arabidopsis*. Plant Physiol. 2009;150:1940–54.

Xie X, Wang Y, Williamson L, Holroyd GH, Tagliavia C, Murchie E, Theobald J, Knight MR, Davies WJ, Leyser HM, Hetherington AM. The identification of genes involved in the stomatal response to reduced atmospheric relative humidity. Curr Biol. 2006;16:882–7.

Yoshida R, Umezawa T, Mizoguchi T, Takahashi S, Takahashi F, Shinozaki K. The regulatory domain of SRK2E/OST1/SnRK2.6 interacts with ABI1 and integrates abscisic acid (ABA) and osmotic stress signals controlling stomatal closure in *Arabidopsis*. J Biol Chem. 2006;281:5310–8.

Yoshida R, Hobo T, Ichimura K, Mizoguchi T, Takahashi F, Aronso J, Ecker JR, Shinozaki K. ABA-activated SnRK2 protein kinase is required for dehydration stress signaling in *Arabidopsis*. Plant Cell Physiol. 2002;43:1473–83.

Young JJ, Mehta S, Israelsson M, Godoski J, Grill E, Schroeder JI. CO_2 signaling in guard cells: calcium sensitivity response modulation, a Ca^{2+}-independent phase, and CO_2 insensitivity of the *gca2* mutant. Proc Natl Acad Sci USA. 2006;103:7506–11.

Zhang X, Zhang L, Dong F, Gao J, Galbraith DW, Song CP. Hydrogen peroxide is involved in abscisic acid-induced stomatal closure in *Vicia faba*. Plant Physiol. 2001;126:1438–48.

Zhang XF, Jiang T, Wu Z, Du SY, Yu YT, Jiang SC, Lu K, Feng XJ, Wang XF, Zhang DP. Cochaperonin CPN20 negatively regulates abscisic acid signaling in *Arabidopsis*. Plant Mol Biol. 2013;83(3):205–18.

Zhu SY, Yu XC, Wang XJ, Zhao R, Li Y, Fan RC, Shang Y, Du SY, Wang XF, Wu FQ, Xu YH, Zhang XY, Zhang DP. Two calcium-dependent protein kinases, CPK4 and CPK11, regulate abscisic acid signal transduction in *Arabidopsis*. Plant Cell. 2007;19:3019–36.

Chapter 16
ABA Regulation of Plant Responses to Drought and Salt Stresses

Yun Ma and Feng Qin

Abstract The phytohormone abscisic acid (ABA) plays an essential role in the abiotic stress response and tolerance of plants, especially during water-related stresses. Thus, ABA is also known as a plant stress hormone. In response to drought and/or salinity stresses, the expression of many ABA-responsive genes is16.1 induced. Several transcription factors and their corresponding DNA target sequences have been identified to control this biological response on the molecular level. Drought stress can directly induce ABA biosynthesis, transportation, and release from its storage form. The activity of ion channels, located on the guard cell membrane, is modulated by drought and/or salinity stresses to regulate stomata movement, which can be simulated by treatment with ABA. In this chapter, we focus on the ABA regulation of drought and salt responses in plants, examining gene expression, stress signaling, and ion homeostasis under these stresses.

Keywords Drought stress · Salt stress · ABA · Plant responses

16.1 Introduction

Drought and salt stresses are two major environmental constrains threatening plant growth and distribution. More than 10 % of the arable land has suffered from drought and salinity, resulting in more than 50 % crop losses worldwide (Bray et al. 2000; Roychoudhury et al. 2013). Dry soil during a drought causes photosynthetic decline, which affects nutrient availability. Salinity can result in physiological drought and ion toxicity (Zhu 2002). Most of the high salinity-inducible

Y. Ma · F. Qin (✉)
Key Laboratory of Plant Molecular Physiology, Institute of Botany,
Chinese Academy of Sciences, Beijing 100093, China
e-mail: qinfeng@ibcas.ac.cn

D.-P. Zhang (ed.), *Abscisic Acid: Metabolism, Transport and Signaling*,
DOI 10.1007/978-94-017-9424-4_16

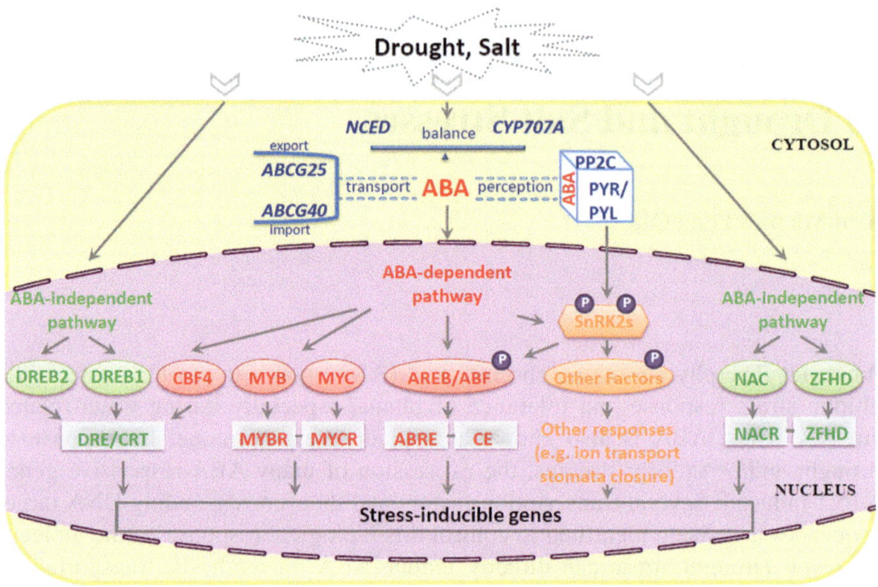

Fig. 16.1 A model for abscisic acid (ABA)-involved responses to drought and salt stresses in *Arabidopsis*. Drought and high salinity induce increased levels of endogenous ABA, which is regulated by the balance of ABA biosynthesis and catabolism mediated by *NCED* and *CYP707A* genes, respectively. The increased endogenous ABA levels result in the formation of the ABA-PYR/PYL-PP2C ternary complex. This complex activates SNF1-related protein kinase 2 (SnRK2s) to phosphorylate downstream components in response to cell dehydration, leading to ion transport and stomata closure, expression of stress-responsive genes, etc. *ABCG25* and *ABCG40*, as major ABA transporters, play central roles in response to drought and salt stresses. Transcription regulatory circuit of major cis-acting elements, transcription factors responding to drought and high salinity in cell nuclei are shown in the light purple region. Transcription factors involved in regulation of stress-inducible gene expression are shown as colored ellipses. Cis-acting elements are shown in gray boxes. Small deep purple circles reveal phosphorylation modifications of transcription factors in response to stress signals. SnRK2s are shown in the hexagon. Red and orange characters represent the transcription regulatory circuit that responds to the drought and salt stresses in an ABA-dependent manner. Green characters represent the transcription regulatory circuit in response to stress conditions through an ABA-independent pathway. (Figure is modified from the following sources: Miyakawa et al. 2013; Nakashima et al. 2009; Roychoudhury et al. 2013; Umezawa et al. 2010).

genes are also induced by drought, indicating extensive overlapping mechanisms between drought and salt stresses (Roychoudhury et al. 2013). Both stresses can induce cellular dehydration, causing osmotic stress that triggers the biosynthesis and increases the endogenous levels of the phytohormone abscisic acid (ABA). Conversely, exogenous application of ABA induces many genes that normally respond to drought and high salinity (Shinozaki et al. 2003; Zhu 2002). ABA controls stress adaptation responses and regulates water balance and is the most studied stress-responsive hormone (Peleg and Blumwald 2011). In this chapter, we will examine the roles of ABA in tolerance resulting from drought and salt stresses using the probable working model shown in Fig. 16.1.

16.2 Drought and Salt Stresses Induce ABA-Responsive Gene Expression

As sessile organisms, plants cannot adjust their location during unfavorable circumstances such as drought and high salinity. Facing these stressful triggers, a long-distance signaling transduction event in plants will start with signal perception, move to signal relay, and end with the stress response, that is, gene expression in the cell nuclei. The expression of stress-responsive genes will finally determine whether plants adapt and survive or succumb to the stressor.

Both drought and high salinity induce the expression of osmotic stress-responsive (OR) genes, which are activated under stressful conditions but silent under normal conditions (Kawasaki et al. 2001; Roychoudhury et al. 2013; Seki et al. 2001, 2002). The ABA-responsive genes increased by drought and high salinity can be classified into two general groups. The first group includes genes that produce important metabolic proteins to protect cells and tissues from stressful conditions. These genes encode enzymes for the biosynthesis of osmolyte, acid metabolism, and cell detoxification, as well as those for water channel proteins, membrane transporters, proteins for protection of macromolecules such as late embryogenesis abundant (LEA) proteins and osmotins, etc. The second group includes genes primarily encoding regulatory proteins for modulating downstream genes expression in response to stressful conditions, such as protein kinases, protein phosphatases, transcription factors, and so on. The transcription factors interact with cis-acting elements to regulate the expression of stress-inducible genes and finally confer tolerance to stress on the plants (Agarwal et al. 2006; Shinozaki and Yamaguchi-Shinozaki 2007). Epigenetic mechanisms have also been reported to participate in controlling the transcription of stress-inducible gene expression, including DNA methylation, histone modifications, histone variants, and non-coding RNA (Chinnusamy and Zhu 2009).

In the first group of genes discussed above, *Rab16, Salt,* and *Osem* from rice, *Rab18* from *Arabidopsis,* and *Rab17* and dehydrins from maize all belong to *LEA* or *LEA*-like genes. Overexpressing *Rab16A*, which originated from a salt-tolerant rice variety called pokkali, in tobacco resulted in enhanced tolerance to salt stress as well as increased gene expression under ABA, drought, and high salinity (Roychoudhury et al. 2007). Similarly, overexpressing the *Rab16A* gene in indica rice also conferred enhanced salt tolerance (Ganguly et al. 2012). ABA-activated protein kinase (AAPK) binds to dehydrin mRNA in response to both drought and ABA. The OPEN STOMATA1 (*OST1*) gene is induced by exogenous ABA in protoplasts of guard cells. Both AAPK and OST1 belong to ABA-activated SNF1-related protein kinases (SnRKs), which function in stomatal closure under drought stress (Himmelbach et al. 2003). *Arabidopsis* overexpressing the *AtHD2C* gene, a member of the histone deacetylases (HDACs) family, show enhanced drought and salt tolerance in addition to an up-regulation of the *LEA* class of genes. This indicates that histone deacetylation plays a role in ABA and osmotic stress responses (Sridha and Wu 2006).

The expression of ABA-responsive genes is important for plant adaptation and survival under drought and salt stresses. Fine regulatory mechanisms are necessary for a correct expression of these stress-inducible genes, and the transcriptional regulation of gene expression responsive to drought and high salinity depends on the interaction of transcription factors with regulatory elements, which usually results in the expression of multiple stress-inducible genes, thus conferring improved plant stress tolerance. The transcription regulatory circuits, which comprise the main cis-acting elements, transcription factors, and other necessary components, in response to drought and salt stresses in the dicotyledonous model plant *Arabidopsis,* are shown in Fig. 16.1. The primary regulatory elements and transcription factors will be described next.

16.2.1 ABRE Element and bZIP-Type Transcription Factors

Many ABA-inducible genes contain a conserved ABA-responsive element, ABRE, an 8-bp cis-acting sequence (PyACGTGGC), in the promoter regions. ABRE, often together with a coupling element (CE), forms an ABA-responsive complex (ABRC) (Hobo et al. 1999a). This complex is the most important cis-acting DNA element involved in ABA-responsive gene expression.

The ABRE element interacting with transcription factors, also known as ABRE-binding factors (ABFs) or ABRE-binding proteins (AREBs), belong to the basic leucine zipper (bZIP) family (Choi et al. 2000; Uno et al. 2000). In *Arabidopsis*, approximately 75 distinct bZIP-type transcription factors have been identified (Jakoby et al. 2002). Among them, AREB1/ABF2, AREB2/ABF4, and ABF3 were induced by drought, high salinity, and ABA in vegetative tissues (Koornneef et al. 1982; Uno et al. 2000). Overexpression of ABF3 and ABF4 in *Arabidopsis* led to ABA-hypersensitive and enhanced drought-tolerant phenotypes, as well as an up-regulated expression of some ABA-responsive genes (Kang et al. 2002). Overexpression of the active form of ABF2/AREB1 also resulted in significantly improved drought tolerance in transgenic *Arabidopsis* plants (Fujita et al. 2005; Kim et al. 2004). By contrast, the *areb1 areb2 abf3* triple mutant displays an ABA-insensitive phenotype and reduced drought tolerance. Accordingly, the expression of stress-responsive genes is largely impaired in the triple mutant (Yoshida et al. 2010). Another bZIP-type transcription factor ABA-insensitive 5 (ABI5) induced by drought and ABA in seeds and seedlings was cloned (Finkelstein and Lynch 2000; Lopez-Molina and Chua 2000). The transcription factor responsible for ABA regulation1 (*TRAB1*), a homolog of ABI5 in rice, was induced by ABA, drought, and salt stress (Hobo et al. 1999b). In soybeans, "Group A"-type bZIP proteins play important roles in ABA and stress signaling. For instance, the *GmbZIP* gene endowed soybeans with salt tolerance (Liao et al. 2008). In short, the bZIP-ABRE system is the key ABA-dependent signal transduction pathway under drought and high salinity stresses.

16.2.2 Additional Cis-Acting Elements and Transcription Factors

In addition to ABRE, MYC and MYB elements (Abe et al. 2003) and the NAC recognition sites (NACRS) (Fujita et al. 2004; Tran et al. 2004) have also been identified as functioning in ABA-inducible gene expression under drought and salt stress conditions.

MYC and MYB recognition sites, present in the promoter region of the *RD22* gene, play key roles in ABA-mediated dehydration-inducible expression of *RD22*. Transgenic *Arabidopsis* overexpressing *AtMYC2, AtMYB2,* or the two genes simultaneously were hypersensitive to ABA and showed enhanced tolerance to osmotic stress (Abe et al. 2003). Overexpressing *OsMYB3R-2,* an R1R2R3-type MYB transcription factor from rice, in transgenic *Arabidopsis* enhanced resistance to drought and salt stresses as well as decreased sensitivity to ABA (Dai et al. 2007).

The cis-acting elements, which usually contain CATGT and harbor CACG core sequences, that are bound by NAC transcription factors are called NACRS. The *Arabidopsis RD26/ANAC072* gene encodes a NAC protein, which can recognize NACRSs in the promoters of some stress-inducible genes and activate their expression. Moreover, the promoter itself also contains a NACRS that is induced by drought, high salinity, and ABA. Transgenic plants overexpressing *RD26* were hypersensitive to ABA and showed up-regulated expression of ABA and stress-inducible genes, whereas plants with the *RD26* gene repressed exhibited the converse phenotype (Fujita et al. 2004).

Additionally, an ABA-independent pathway exists in plants, which is mediated by the interaction of other transcription factors and cis-acting elements in response to drought and salt stresses. It has been widely reported that the dehydration-responsive element/C-repeat (DRE/CRT), with a 9-bp core sequence (TACCGACAT), acts in response to drought and high salinity through an ABA-independent signal transduction pathway (Yamaguchi-Shinozaki and Shinozaki 1994). *DREB2*, an APETALA2 (AP2)-type transcription factor induced by drought and high salinity, transactivates the promoter containing the DRE element of stress-responsive genes. Transgenic *Arabidopsis* plants overexpressing the constitutively active DREB2A (DREB2A-CA) showed enhanced drought tolerance (Sakuma et al. 2006). Maize *ZmDREB2A* overexpression in *Arabidopsis* conferred drought stress tolerance as did DREB2A-CA overexpression (Qin et al. 2007). Another AP2-type transcription factor, *DREB1/CBF*, has downstream genes that always contain DRE/CRT elements (Maruyama et al. 2004). Overexpressing *DREB1A* in transgenic rice increased stress tolerance to drought and high salinity (Oh et al. 2005). The homolog of *DREB1* in rice, *OsDREB1A,* was induced by high salinity but not by ABA, and its overexpression in transgenic *Arabidopsis* improved salt tolerance (Dubouzet et al. 2003). The homolog of *DREB1* in maize, *ZmDREB1A,* was also identified (Qin et al. 2004), and *ZmDREB1A* overexpression in *Arabidopsis* led to enhanced drought tolerance. CBF4 is the one *DREB1* homolog in *Arabidopsis* to function in an ABA-dependent manner; it is increased by drought and ABA. Overexpression of *CBF4* in transgenic *Arabidopsis* enhanced drought tolerance and up-regulated DRE containing stress-responsive genes (Chinnusamy et al. 2004; Haake et al. 2002). The early response to dehydration1 gene (*ERD1*), a NAC family member, responded to drought and high salinity but not to ABA (Nakashima et al. 1997). There are two types of cis-acting elements in the promoter region of *ERD1*. One, a MYC-like sequence (CATGTG) named NACRS, is recognized and bound by NAC-type transcription factors (Tran et al. 2004). The other, a 14-bp rps1 site 1-like sequence, binds to the zinc finger homeodomain (ZFHD) transcription factor. Both cis-acting elements are necessary for the expression of *ERD1* under dehydration stress. The NAC protein functions as a transcription activator alone or in cooperation with the ZFHD protein to regulate the expression of the *ERD1* gene.

16.3 Comparison of *Arabidopsis* with Crops

In order to understand the regulatory mechanisms underlying ABA-inducible gene expression under drought and salt stresses, transcription factors and international cis-acting elements involved in ABA signaling were initially investigated and reported in the dicotyledonous model plant *Arabidopsis*. Later, the survey was extended to important agricultural crops such as rice, maize, wheat, and barley because of their significant commercial value. The functions and mechanisms of transcription factors regulating gene expression in *Arabidopsis* and crops have some similarities and some differences.

The most important transcription factors, AREB/ABFs, regulate ABA-dependent gene expression by binding to the cis-acting elements like ABREs and CEs in the promoter regions of target genes. In *Arabidopsis*, activation of AREB1 requires multisite phosphorylation by ABA-responsive 42-KDa kinases. The overexpression of the phosphorylated active form of AREB1 in transgenic plants up-regulated the expression of ABA-inducible genes in the absence of ABA or under non-stressed conditions (Furihata et al. 2006; Uno et al. 2000). In addition, AAPKs, such as SnRK2.2, SnRK2.3, and SnRK2.6, that belong to the subclass β SnRK2 family, a central hub of ABA signaling, phosphorylate AREB1 to activate ABRE-regulated gene expression (Fujii et al. 2007; Furihata et al. 2006; Kobayashi et al. 2004; Umezawa et al. 2010). In rice, TRAB1, an ABF transcription factor, was rapidly phosphorylated in response to ABA (Kagaya et al. 2002). Kobayashi et al. (2005) reported that the ABA-activated SnRK2 in rice phosphorylated TRAB1. OSRK1, a protein kinase belonging to the SnRK2 rice family, phosphorylated OREB1, an ABRE-binding factor (Chae et al. 2007). In wheat, PKAB1, the ortholog of wheat SnRK2, phosphorylated TaABF, a wheat AREB protein (Johnson et al. 2002). Hence, phosphorylation modifications by protein kinases to activate ABF-type transcription factors play similar roles in the expression of AREB/ABF-regulated ABA-responsive genes under drought and salt stresses in both *Arabidopsis* and crops.

In *Arabidopsis*, a negative regulatory domain occurs within the DREB2A protein. Deletion of this domain transforms DREB2A into the constitutively active

DREB2A-CA, which is more stable in the cell nucleus than the full DREB2A protein. Transgenic *Arabidopsis* overexpressing DREB2A-CA rather than full length of DREB2A show strong drought tolerance as well as up-regulated expression of stress-inducible genes. This indicates that the negative regulatory domain is responsible for regulating the stability of DREB2A, and it is important for DREB2A to function under drought stress conditions (Qin et al. 2008). The *DREB2* gene homologs have also been identified in crops such as rice, wheat, barley, maize, and pearl millet (Agarwal et al. 2007; Dubouzet et al. 2003; Egawa et al. 2006; Qin et al. 2007; Shen et al. 2003; Xue and Loveridge 2004), and most of these genes respond to drought and high salinity. In contrast to the *DREB2A* in *Arabidopsis*, the functional forms of the alternatively spliced *DREB2A* transcripts in wheat, barley, and maize are inducible by stress conditions, although the non-functional transcripts are abundant (Egawa et al. 2006; Qin et al. 2007; Xue and Loveridge 2004). Hence, different regulatory mechanisms underlie *DREB2A* function in *Arabidopsis* and crops.

Although *DREB1A/CBF3* and *DREB1B/CBF1* are not rapidly induced by drought and high salinity, overexpression of *DREB1A* or *DREB1B* in transgenic *Arabidopsis* improves stress tolerance to drought and salt stresses (Jaglo-Ottosen et al. 1998; Kasuga et al. 1999; Liu et al. 1998). This indicates that *DREB1* targets multiple genes. Most of the downstream target genes contain DRE/CRT elements in their promoter regions and respond to stress tolerance, such as genes encoding phospholipase C, sugar transport protein, LEA protein, and osmoprotectant biosynthesis proteins (Maruyama et al. 2004). *DREB1* homolog genes have been isolated from rice, wheat, barley, maize, and sorghum (Brautigam et al. 2005; Dubouzet et al. 2003; James et al. 2008; Qin et al. 2004; Skinner et al. 2005; Vagujfalvi et al. 2005; Xiong and Fei 2006; Xue 2003; Zhao and Bughrara 2008). Transgenic *Arabidopsis* or tobacco plants overexpressing *DREB1* homolog genes derived from other species exhibit enhanced drought stress tolerance as well as a significant increase in the expression of stress-inducible genes under control conditions (Dubouzet et al. 2003; Qin et al. 2004; Skinner et al. 2005; Xiong and Fei 2006; Zhao and Bughrara 2008). Overexpressing transgenic rice *DREB1A* in *Arabidopsis* accumulates osmoprotectants under non-stress conditions. The transgenic tall fescue, harboring *DREB1A* driven by the stress-inducible *RD29A* promoter, showed a marked accumulation of proline under drought stress (Zhao et al. 2007). Transgenic rice overexpressing rice *OsDREB1A* or *Arabidopsis DREB1B* caused expression of downstream stress-inducible genes (Ito et al. 2006; Lee et al. 2004). In summary, DREB1 proteins derived from crops function similar to those in *Arabidopsis*.

In *Arabidopsis*, three NAC-type proteins, ANAC19 (At1g52890), ANAC055 (At3g15500), and ANAC072 (RD26, At4g27410), were isolated using yeast one-hybrid screening. These three NAC proteins, together with the ZHFD1 protein, as transcription activators, cooperatively regulate the expression of the *ERD1* gene, which causes acquired stress tolerance to drought and high salinity in transgenic plants (Fujita et al. 2004; Tran et al. 2004). In rice, six NAC transcription factors have been identified (Ooka et al. 2003). Among them, overexpression of *SNAC1* and *SNAC2* in transgenic rice resulted in enhanced drought and salt tolerances

(Hu et al. 2006, 2008). In addition, the yield of transgenic rice plants expressing *SNAC1* was not penalized under favorable growing conditions (Hu et al. 2006), and under severe drought stress conditions, these plants still showed stronger drought tolerance compared with control plants. *OsNAC6*, another NAC transcription factor, is induced not only by drought, high salinity, and ABA but also by blast disease, jasmonic acid, and wounding. Therefore, *OsNAC6* overexpression in transgenic rice confers enhanced stress tolerance to drought and high salinity as well as a strong tolerance to blast disease. The NAC transcription factor gene, *ONAC010,* regulated senescence and improved grain protein and ion content in wheat (Uauy et al. 2006). These results indicate that compared with *Arabidopsis*, NAC transcription factors in crops not only function in response to drought and salt stresses but are also involved in functions such as seed quality and disease resistance.

In conclusion, in the regulatory circuits responding to drought and salt stresses, many common mechanisms are shared by *Arabidopsis* and crops. However, it is essential to attend to their differences when effectively improving a corresponding trait in agriculturally important crops using gene transfer technology from *Arabidopsis*. The transcription factors, as central regulators of gene expression, will likely become preferential targets for genetic engineering.

16.4 Enhanced ABA Perception, Biosynthesis, and Transportation Improve Plant Dehydration Stress Tolerance

The discovery of the novel ABA-soluble receptor pyrabactin resistance1 (PYR)/PYR1-like (PYL)/regulatory components of ABA receptors (RCAR) by two independent research groups is known as the crucial breakthrough in recent ABA signaling studies. ABA binds these receptors and inhibits the activity of protein phosphatases 2C (PP2Cs), including ABI1, ABI2, and HAB1, abrogating the inhibition of subclass III sucrose non-fermenting-1 (SNF1)-related protein kinase 2 (SnRK2s), switching on the expression of downstream stress-responsive genes. PYR/PYL/RCAR together with ABA, PP2Cs, and SnRK2s constitute the core ABA signaling components that play essential roles in plants for responding to cellular dehydration. The overexpression of PYR/PYL/PCAR proteins, such as PYL5/RCAR3, PYL8/RCAR3, and RCAR1/PYL9, in transgenic *Arabidopsis* enhanced the response to ABA and afforded higher resistance to drought by inhibiting the activity of downstream PP2Cs (Ma et al. 2009; Park et al. 2009; Saavedra et al. 2010; Santiago et al. 2009). By contrast, the *pyr1 pyl1 pyl2 pyl4* quadruple mutant exhibited less sensitivity to ABA-induced stomatal closure and ABA-inhibited stomatal opening (Nishimura et al. 2010). Similarly, the *pyr1 pyl1 pyl2 pyl4 pyl5 pyl8* sextuple mutant was insensitive to ABA and showed increased stomatal aperture and leaf transpiration, as well as severely impaired expression of ABA-responsive genes (Gonzalez-Guzman et al. 2012). With the discovery of the ABA receptor, quinabactin, a sulfonamide ABA agonist, was found to form a hydrogen bond with the PYL2/PP2C complex, which elicited ABA responses in vegetative tissues, such as promoting stomata closure, reducing water loss, and enhancing drought tolerance in *Arabidopsis* and soybean (Okamoto et al. 2013). Cao et al. (2013) recently identified a small-molecule ABA mimic (AM1) that activates multiple members of ABA receptors in *Arabidopsis,* according to crystal structural evidence and the layer of gene expression, to protect plants from drought stress and decreased water loss. These reports provide new technologies for improving stress tolerance of plants mediated through enhanced ABA signaling by receptors or ABA agonists/mimics.

Many of the genes encoding enzymes involved in ABA biosynthesis have been identified, such as 9-cis-epoxycarotenoid dioxygenase (NCED), abscisic aldehyde oxidase (AAO3), MoCo sulfurase (AtABA3), and zeaxanthin epoxidase (AtZEP), which all show increased expression during plant dehydration (Iuchi et al. 2001; Nambara and Marion-Poll 2005; Seo et al. 2000; Vogel et al. 2005; Xiong et al. 2001). Among them, NCED is the key enzyme responsible for ABA biosynthesis; the expression of the *NCED* gene in either *Arabidopsis* or tomato increased the level of ABA (Iuchi et al. 2001; Thompson et al. 2000b). Furthermore, *NCED* mRNA accumulates mainly in vascular tissues, consistent with ABA biosynthetic sites. *NCED* is reportedly induced by drought stress in several plant species, including *Arabidopsis*, tomato, maize, bean, and cow pea (Iuchi et al. 2000; Qin and Zeevaart 1999; Tan et al. 1997; Thompson et al. 2000a, b). In *Arabidopsis*, *AtNCED3* is induced by drought and high salinity. The overexpression of *AtNCED3* in transgenic plants enhanced dehydration stress tolerance, whereas knockouts of *AtNCED3* resulted in a dehydration-sensitive phenotype (Iuchi et al. 2001). Transgenic tobacco plants overexpressing *SgNCED1* originating from *Stylosanthes guianensis* exhibit increased accumulation of ABA in leaves as well as enhanced stress resistance to drought and high salinity. The overexpression of *LeNCED1* in tomato plants also resulted in ABA accumulation (Thompson et al. 2007). Under drought stress, the induction of *VuNCED1* in cowpeas precedes ABA accumulation. This suggests that the transcriptional regulation of the *VuNCED1* gene is responsible for the drought-induced ABA accumulation (Iuchi et al. 2000). The regulation of endogenous ABA levels depends not only on the rate of ABA synthesis but also on the rate of its catabolism. The *CYP707A3* gene encoding an ABA 8'-hydroxylase in *Arabidopsis* is highly induced by ABA and by rehydration. It is the major enzyme responsible for catalyzing ABA degradation (Saito et al. 2004). The *cyp707a3* mutant plants show reduced transpiration rates, enhanced drought tolerance, higher ABA accumulation, and hypersensitivity to exogenous ABA compared with the wild type. By contrast, plant constitutive expression of the *CYP707A3* gene exhibited an opposite phenotype (Umezawa et al. 2006). In light of the finding that *NCED* overexpression in plants also hyperaccumulates phaseic acid, it is likely that ABA regulates the accumulation of phaseic acid by activating its catabolism (Qin and Zeevaart 2002). Therefore, the levels of ABA are tightly modulated by the balance between ABA biosynthesis and catabolism under stressful conditions, and it is feasible to regulate the level of endogenous ABA, which can change the drought tolerance of plants.

ABA-induced stomata closure to reduce transpiration and prevent water loss is crucial for plants under osmotic stresses, such as those induced by drought and high salinity. Many signaling components participate in this reaction, including Ca^{2+}, nitric oxide, reactive oxygen species, protein kinases, protein phosphatases, G protein, G protein-coupled receptors, and phosphatidic acid. The effect of ABA on stomatal closure is to reduce turgor and volume in guard cells, which is mediated by cation and anion effluxes and leads to the depolarization of guard cells (Roychoudhury and Paul 2012). ABA is known to function in stomata consisting of two guard cells, but is mainly biosynthesized in vascular tissues (Cheng et al. 2002; Koiwai et al. 2004; Okamoto et al. 2009). This indicates that ABA transport is crucial for proper ABA function and thus for correct physiological responses. *AtABCG40*, an ATP-binding cassette (ABC) transporter gene, was identified in *Arabidopsis* as an ABA importer in plant cells (Kang et al. 2010). Yeast or BY2 cells expressing *AtABCG40* show an increase in ABA uptake. By contrast, protoplasts of *atabcg40* mutants show a decrease in ABA uptake. In the *atabcg40* mutant, ABA-induced stomata closure is slower than that in the wild type and this directly results in reduced drought tolerance. The expression of the ABA-responsive genes is accordingly delayed in response to exogenous ABA. *AtABCG22* has been identified as a member of *AtABCG* family gene encoding a half-size ABC transporter in *Arabidopsis* and is responsible mainly for stomatal regulation. The *atabcg22* mutants were sensitive to drought stress, with lower leaf temperatures and increased water loss compared with the wild type (Kuromori et al. 2011). Another ABA importer, the ABA-importing transporter1 (*AIT1)* initially named *NRT1.2/NPF4.6*, has also been characterized (Huang et al. 1999; Kanno et al. 2012). *AIT1* is mainly expressed in vascular tissues, where ABA is biosynthesized, and acts as an ABA importer responsible for ABA transport from its site of synthesis to its site of function, where it facilitates and regulates stomata closure responding to drought stress. The *ait1* mutants showed decreased sensitivity to exogenously applied ABA, whereas plants overexpressing *AIT1* were more sensitive to ABA (Kanno et al. 2012). In addition, AtABCG25, another ABC transporter, is responsible for exporting ABA from inside to outside of cells in vascular tissues, playing an important role in stomatal closure. Thus, the overexpression of *AtABCG25* causing an ABA-insensitive phenotype as well as an increase in leaf surface temperature may be a result of stomatal closure (Kuromori et al. 2010).

In conclusion, newly discovered ABA receptors and genes involved in the regulation of the levels of ABA, as well as ABA transporters responsible for translocating ABA from sites of biosynthesis to functional sites have greatly enriched our knowledge regarding plant responses to drought and salt stresses and have broadened our gene resources and generated new technologies in genetic engineering to improve stress tolerance in plants.

16.5 ABA and Maintenance of Ion Homeostasis Under Drought and Salt Stresses

High salinity stress usually causes osmotic stress when ions accumulate in the space outside plant cells, whereas it may cause ionic stress when ions concentrate inside plant cells (Galvan-Ampudia and Testerink 2011). Plants suffer ionic stress not only from the toxic effects of the high cytosolic Na^+ concentration but also from the disrupted K^+ homeostasis resulting from Na^+ competing for K^+ binding sites, which leads to a high Na^+/K^+ ratio. Maintenance of ionic homeostasis by plants is particularly important for adaptation and survival under salt stress conditions. Hence, plant cells must initiate ion transport systems to maintain ion homeostasis and avoid ion toxicity. The maintenance of ionic homeostasis during drought stress is equally as important as that in salt stress, and ABA participates in regulating ion transporters/channels responding to both high salinity and drought.

The most well-known family involved in ion transport is the NHX family, which belongs to the monovalent cation–proton antiporter superfamily (Chanroj et al. 2012; Manohar et al. 2011; Sze et al. 2004). Among NHX family members, the salt overly sensitive (SOS) pathway has been intensively studied. The salt overly sensitive1 (SOS1)/NHX7, a Na^+/H^+ antiporter, was identified by screening *Arabidopsis* salt-sensitive mutants (Wu et al. 1996). SOS1 is localized at the plasma membrane, and its transcript level is up-regulated by salt stress but not by ABA (Oh et al. 2010). SOS1 is responsible for Na^+ extrusion from the cytosol to the extracellular space to maintain the intracellular Na^+ concentration at an appropriate level under high salinity stress (Shi et al. 2002; Wu et al. 1996). SOS1 requires SOS2 and SOS3 to facilitate proper function. SOS2 encodes a SnRK3 protein kinase with an N-terminal catalytic domain and a C-terminal regulatory domain, which phosphorylates the serine residue (S1138) in the C-terminal domain of SOS1 localized in the cytosol (Liu et al. 2000). SOS3, a calcium sensor, encodes a myristoylated calcium-binding protein that senses cytosolic Ca^{2+}. Under high salinity stress, high Na^+ induces SOS3 binding to cytoplasmic Ca^{2+}, facilitates the formation of the SOS3-SOS2 complex, and recruits SOS2 to the plasma membrane to phosphorylate and activate SOS1. The activated SOS1 promotes Na^+ efflux to reduce the cytosolic Na^+ concentration and mitigate the toxic effects of high salinity (Qiu et al. 2002; Quintero et al. 2011; Zhu 2003). In addition, the MAP kinase MPK6, CBL10, SOS3-like SCaBP8, and GIGANTEA (GI) play roles in the SOS pathway. NaCl-induced phosphatidic acid accumulation activates MPK6, which phosphorylates the C-terminal of SOS1, and the activated SOS1 responds to the salt stress (Yu et al. 2010). CBL10 interacts with SOS2 and further regulates the activation of SOS1. GI is reported to be involved in the SOS pathway, as its degradation releases SOS2 to activate SOS1 under salt stress (Kim et al. 2013). SOS1 functions especially at epidermal tissues, and *SOS1* mRNA accumulation in salt-tolerant *T. salsuginea* is higher than that in *Arabidopsis* (Dinneny et al. 2008; Oh et al. 2009). In addition, other NHX family members, for instance, NHX5 and

NHX6, are also involved in cellular Na^+/K^+ homeostasis as well as pH balance. The *nhx5nhx6* double mutant exhibited hypersensitivity to high salinity stress (Bassil et al. 2011). The homolog of *SOS1* in the salinity-tolerant species *Chrysanthemum crassum* (CcSOS1) was identified and found to share approximately 62 % amino acid sequence identity with the SOS1 in *Arabidopsis*. The *CcSOS1* transcript abundance is increased by drought stress. However, its transcript level is down-regulated in the leaves of *C. crassum* in the presence of ABA, which indicates that ABA may participate in the expression of *CcSOS1* (Song et al. 2012). In addition to SOS3, the protein phosphatase 2C, ABI2, has been recognized as another SOS2-interacting protein based on a yeast two-hybrid screening. Their interaction depended on the conserved protein phosphatase interaction (PPI) motif that presented in SOS2 family proteins with 37 amino acid residues. The *abi2-1* mutant showed increased salt tolerance, decreased ABA sensitivity, and a disrupted interaction between SOS2 and ABI2 (Ohta et al. 2003). At least, 23 SOS2-like protein kinase (PKSs) and 8 SOS3-like binding proteins (SCaBPs) exist in the *Arabidopsis* genome (Guo et al. 2001). Guo et al. reported that *pks3* and *scabp5* mutants displayed an ABA hyperinsensitive phenotype as well as an up-regulated expression of drought stress-responsive genes, such as *RD29A*. Furthermore, the ScaBP5-PKS3 complex specifically responds to ABA-induced Ca^{2+} signaling. This complex together with ABI2 constitutes part of a negative regulatory loop that controls ABA sensitivity (Guo et al. 2002). Another SOS2-like protein kinase gene *PKS18* was identified in *Arabidopsis*. Transgenic plants expressing constitutively active forms of PKS18 (PKS18T/D) were hypersensitive to ABA, and the *PKS18T/D* knockdown plant showed an ABA-insensitive phenotype (Gong et al. 2002).

AtHKT1, a Na^+ transporter localized at the plasma membrane and expressed especially in vascular tissues, is putatively responsible for the Na^+ efflux from xylem vessels under high salinity (Kato et al. 2001; Rus et al. 2001). The *hkt1-1* mutant was sensitive to salt stress because of the excessive accumulation of Na^+ accumulated in the shoots (Berthomieu et al. 2003). Hence, AtHKT1 is involved in Na^+ translocation in roots and shoots and in Na^+ accumulation in leaves (Horie et al. 2007; Moller et al. 2009; Sunarpi et al. 2005). HKT transporter genes have also been identified in crop plants, such as wheat and the wheat-relative *Triticum monococcum* and have similar functions as those in *Arabidopsis* (Munns et al. 2012). Other types of HKT family members, such as TaHKT2;1 from *Triticum aestivum*, TsHKT1;2 from *Thellungiella salsuginea*, and EcHKT1;2 from *Eucalyptus camaldulensis*, also participate in K^+ transport to maintain K^+/Na^+ balance under salt stress (Ali et al. 2012; Ardie et al. 2009; Laurie et al. 2002; Liu et al. 2001; Schachtman and Schroeder 1994). Shkolnik-Inbar et al. reported that ABI4 reduced the expression of *AtHKT1* to further affect salt tolerance. This indicates that ABA-signaling components are involved in the regulation and expression of ion transporters (Shkolnik-Inbar et al. 2013).

Additionally, K^+ channels such as AKT2 and KAT2, both encoding K^+ inward rectifying channels, and SKOR, a K^+ outward rectifying channel, are all expressed in vascular tissue and play important roles in K^+ homeostasis (Gaymard et al.

1998; Lebaudy et al. 2010; Marten et al. 1999; Pilot et al. 2001; Xicluna et al. 2007). ABA up-regulated the expression of *AKT2* and decreased the expression of *SKOR* under drought and high salinity stresses (Deeken et al. 2002; Gaymard et al. 1998; Lacombe et al. 2000; Marten et al. 1999). Cherel et al. (2002) provided evidence that AtPP2CA, a protein phosphatase involved in ABA signaling, interacts with AKT2 and is considered a partner of AKT2. It was reported that SRK2E/OST1/SnRK2.6, one of the subclass β SnRK2 family members, global regulators of ABA signaling, inhibits the activity of KAT1, an inward K^+ channel, through phosphorylation of the threonine site at the C-terminal domain of KAT1. This in turn enhances drought tolerance (Ichida et al. 1997; Sato et al. 2009; Szyroki et al. 2001). GORK encodes a K^+-release channel that is expressed mainly in guard cells, roots, and vascular tissue and is up-regulated by ABA, drought, and high salinity. It is responsible for the K^+ efflux from guard cells, a main mechanism to reduce cell turgor and induce stomata closure. PP2Cs, core ABA-signaling components, regulate the expression of GORK (Becker et al. 2003). The multiple gene mutants of the K^+ uptake transporter gene, *kup268* and *kup68 gork,* exhibit decreased drought stress tolerance. Furthermore, SRK2E phosphorylates KUP6, indicating that ion transport system directly regulated by ABA signaling plays important roles in response to drought stress (Osakabe et al. 2013a).

In conclusion, drought and high salinity stresses may lead to an imbalance in ion homeostasis. Maintenance of ion homeostasis is vital for the adaptation and survival of plants under stressful conditions. Ion transporters or channels are mainly responsible for ion transport and flow to maintain ion homeostasis and to avoid ion toxicity. ABA may participate by the interaction of ABA-signaling components with ion transporters/channels. According to relatively recent reports, ABA is primarily involved in the modulation of ion transporters/channels by phosphorylation/dephosphorylation modifications in the cytosolic domains of the ion transporters/channels through protein kinase SnRK2s, the ABA "core signaling components" (Hubbard et al. 2010; Osakabe et al. 2013b; Umezawa et al. 2010).

References

Abe H, Urao T, Ito T, Seki M, Shinozaki K, Yamaguchi-Shinozaki K. Arabidopsis AtMYC2 (bHLH) and AtMYB2 (MYB) function as transcriptional activators in abscisic acid signaling. Plant Cell. 2003;15:63–78.

Agarwal PK, Agarwal P, Reddy MK, Sopory SK. Role of DREB transcription factors in abiotic and biotic stress tolerance in plants. Plant Cell Rep. 2006;25:1263–74.

Agarwal P, Agarwal PK, Nair S, Sopory SK, Reddy MK. Stress-inducible DREB2A transcription factor from *Pennisetum glaucum* is a phosphoprotein and its phosphorylation negatively regulates its DNA-binding activity. Mol Genet Genomics. 2007;277:189–98.

Ali Z, Park HC, Ali A, Oh DH, Aman R, Kropornicka A, Hong H, Choi W, Chung WS, Kim WY, Bressan RA, Bohnert HJ, Lee SY, Yun DJ. TsHKT1;2, a HKT1 homolog from the extremophile Arabidopsis relative *Thellungiella salsuginea*, shows K(+) specificity in the presence of NaCl. Plant Physiol. 2012;158:1463–74.

Ardie SW, Xie L, Takahashi R, Liu S, Takano T. Cloning of a high-affinity K+ transporter gene PutHKT2;1 from *Puccinellia tenuiflora* and its functional comparison with OsHKT2;1 from rice in yeast and Arabidopsis. J Exp Bot. 2009;60:3491–502.

Bassil E, Ohto MA, Esumi T, Tajima H, Zhu Z, Cagnac O, Belmonte M, Peleg Z, Yamaguchi T, Blumwald E. The Arabidopsis intracellular Na+/H+ antiporters NHX5 and NHX6 are endosome associated and necessary for plant growth and development. Plant Cell. 2011;23:224–39.

Becker D, Hoth S, Ache P, Wenkel S, Roelfsema MR, Meyerhoff O, Hartung W, Hedrich R. Regulation of the ABA-sensitive Arabidopsis potassium channel gene GORK in response to water stress. FEBS Lett. 2003;554:119–26.

Berthomieu P, Conejero G, Nublat A, Brackenbury WJ, Lambert C, Savio C, Uozumi N, Oiki S, Yamada K, Cellier F, Gosti F, Simonneau T, Essah PA, Tester M, Very AA, Sentenac H, Casse F. Functional analysis of AtHKT1 in Arabidopsis shows that Na(+) recirculation by the phloem is crucial for salt tolerance. EMBO J. 2003;22:2004–14.

Brautigam M, Lindlof A, Zakhrabekova S, Gharti-Chhetri G, Olsson B, Olsson O. Generation and analysis of 9792 EST sequences from cold acclimated oat, *Avena sativa*. BMC Plant Biol. 2005;5:18.

Bray EA, B-S J, Weretilnyk E. Responses to abiotic stresses. In: Gruissem W, Buchnnan B, Jones R, editors. Biochemistry and molecular biology of plants. Rockville, MD: American Society of Plant Physiologists; 2000. p. 1158–249.

Cao M, Liu X, Zhang Y, Xue X, Zhou XE, Melcher K, Gao P, Wang F, Zeng L, Zhao Y, Deng P, Zhong D, Zhu JK, Xu HE, Xu Y. An ABA-mimicking ligand that reduces water loss and promotes drought resistance in plants. Cell Res. 2013;23:1043–54.

Chae MJ, Lee JS, Nam MH, Cho K, Hong JY, Yi SA, Suh SC, Yoon IS. A rice dehydration-inducible SNF1-related protein kinase 2 phosphorylates an abscisic acid responsive element-binding factor and associates with ABA signaling. Plant Mol Biol. 2007;63:151–69.

Chanroj S, Wang G, Venema K, Zhang MW, Delwiche CF, Sze H. Conserved and diversified gene families of monovalent cation/h(+) antiporters from algae to flowering plants. Frontiers Plant Sci. 2012;3:25.

Cheng WH, Endo A, Zhou L, Penney J, Chen HC, Arroyo A, Leon P, Nambara E, Asami T, Seo M, Koshiba T, Sheen J. A unique short-chain dehydrogenase/reductase in Arabidopsis glucose signaling and abscisic acid biosynthesis and functions. Plant Cell. 2002;14:2723–43.

Cherel I, Michard E, Platet N, Mouline K, Alcon C, Sentenac H, Thibaud JB. Physical and functional interaction of the Arabidopsis K(+) channel AKT2 and phosphatase AtPP2CA. Plant Cell. 2002;14:1133–46.

Chinnusamy V, Zhu JK. Epigenetic regulation of stress responses in plants. Curr Opin Plant Biol. 2009;12:133–9.

Chinnusamy V, Schumaker K, Zhu JK. Molecular genetic perspectives on cross-talk and specificity in abiotic stress signalling in plants. J Exp Bot. 2004;55:225–36.

Choi H, Hong J, Ha J, Kang J, Kim SY. ABFs, a family of ABA-responsive element binding factors. J Biol Chem. 2000;275:1723–30.

Dai X, Xu Y, Ma Q, Xu W, Wang T, Xue Y, Chong K. Overexpression of an R1R2R3 MYB gene, OsMYB3R-2, increases tolerance to freezing, drought, and salt stress in transgenic Arabidopsis. Plant Physiol. 2007;143:1739–51.

Deeken R, Geiger D, Fromm J, Koroleva O, Ache P, Langenfeld-Heyser R, Sauer N, May ST, Hedrich R. Loss of the AKT2/3 potassium channel affects sugar loading into the phloem of Arabidopsis. Planta. 2002;216:334–44.

Dinneny JR, Long TA, Wang JY, Jung JW, Mace D, Pointer S, Barron C, Brady SM, Schiefelbein J, Benfey PN. Cell identity mediates the response of Arabidopsis roots to abiotic stress. Science. 2008;320:942–5.

Dubouzet JG, Sakuma Y, Ito Y, Kasuga M, Dubouzet EG, Miura S, Seki M, Shinozaki K, Yamaguchi-Shinozaki K. OsDREB genes in rice, *Oryza sativa* L., encode transcription activators that function in drought-, high-salt- and cold-responsive gene expression. Plant J Cell Mol Biol. 2003;33:751–63.

Egawa C, Kobayashi F, Ishibashi M, Nakamura T, Nakamura C, Takumi S. Differential regulation of transcript accumulation and alternative splicing of a DREB2 homolog under abiotic stress conditions in common wheat. Genes Genet Syst. 2006;81:77–91.

Finkelstein RR, Lynch TJ. The Arabidopsis abscisic acid response gene ABI5 encodes a basic leucine zipper transcription factor. Plant Cell. 2000;12:599–609.

Fujii H, Verslues PE, Zhu JK. Identification of two protein kinases required for abscisic acid regulation of seed germination, root growth, and gene expression in Arabidopsis. Plant Cell. 2007;19:485–94.

Fujita M, Fujita Y, Maruyama K, Seki M, Hiratsu K, Ohme-Takagi M, Tran LS, Yamaguchi-Shinozaki K, Shinozaki K. A dehydration-induced NAC protein, RD26, is involved in a novel ABA-dependent stress-signaling pathway. Plant J Cell Mol Biol. 2004;39:863–76.

Fujita Y, Fujita M, Satoh R, Maruyama K, Parvez MM, Seki M, Hiratsu K, Ohme-Takagi M, Shinozaki K, Yamaguchi-Shinozaki K. AREB1 is a transcription activator of novel ABRE-dependent ABA signaling that enhances drought stress tolerance in Arabidopsis. Plant Cell. 2005;17:3470–88.

Furihata T, Maruyama K, Fujita Y, Umezawa T, Yoshida R, Shinozaki K, Yamaguchi-Shinozaki K. Abscisic acid-dependent multisite phosphorylation regulates the activity of a transcription activator AREB1. Proc Natl Acad Sci USA. 2006;103:1988–93.

Galvan-Ampudia CS, Testerink C. Salt stress signals shape the plant root. Curr Opin Plant Biol. 2011;14:296–302.

Ganguly M, Datta K, Roychoudhury A, Gayen D, Sengupta DN, Datta SK. Overexpression of Rab16A gene in indica rice variety for generating enhanced salt tolerance. Plant Signal Behav. 2012;7:502–9.

Gaymard F, Pilot G, Lacombe B, Bouchez D, Bruneau D, Boucherez J, Michaux-Ferriere N, Thibaud JB, Sentenac H. Identification and disruption of a plant shaker-like outward channel involved in K+ release into the xylem sap. Cell. 1998;94:647–55.

Gong D, Zhang C, Chen X, Gong Z, Zhu JK. Constitutive activation and transgenic evaluation of the function of an arabidopsis PKS protein kinase. J Biol Chem. 2002;277:42088–96.

Gonzalez-Guzman M, Pizzio GA, Antoni R, Vera-Sirera F, Merilo E, Bassel GW, Fernandez MA, Holdsworth MJ, Perez-Amador MA, Kollist H, Rodriguez PL. Arabidopsis PYR/PYL/RCAR receptors play a major role in quantitative regulation of stomatal aperture and transcriptional response to abscisic acid. Plant Cell. 2012;24:2483–96.

Guo Y, Halfter U, Ishitani M, Zhu JK. Molecular characterization of functional domains in the protein kinase SOS2 that is required for plant salt tolerance. Plant Cell. 2001;13:1383–400.

Guo Y, Xiong L, Song CP, Gong D, Halfter U, Zhu JK. A calcium sensor and its interacting protein kinase are global regulators of abscisic acid signaling in Arabidopsis. Dev Cell. 2002;3:233–44.

Haake V, Cook D, Riechmann JL, Pineda O, Thomashow MF, Zhang JZ. Transcription factor CBF4 is a regulator of drought adaptation in Arabidopsis. Plant Physiol. 2002;130:639–48.

Himmelbach A, Yang Y, Grill E. Relay and control of abscisic acid signaling. Curr Opin Plant Biol. 2003;6:470–9.

Hobo T, Asada M, Kowyama Y, Hattori T. ACGT-containing abscisic acid response element (ABRE) and coupling element 3 (CE3) are functionally equivalent. Plant J. 1999a;19:679–89.

Hobo T, Kowyama Y, Hattori T (1999b) A bZIP factor, TRAB1, interacts with VP1 and mediates abscisic acid-induced transcription. Proc Nat Acad Sci USA. 1999b; 96:15348–53.

Horie T, Costa A, Kim TH, Han MJ, Horie R, Leung HY, Miyao A, Hirochika H, An G, Schroeder JI. Rice OsHKT2;1 transporter mediates large Na+ influx component into K+-starved roots for growth. EMBO J. 2007;26:3003–14.

Hu H, Dai M, Yao J, Xiao B, Li X, Zhang Q, Xiong L. Overexpressing a NAM, ATAF, and CUC (NAC) transcription factor enhances drought resistance and salt tolerance in rice. Proc Natl Acad Sci USA. 2006;103:12987–92.

Hu H, You J, Fang Y, Zhu X, Qi Z, Xiong L. Characterization of transcription factor gene SNAC2 conferring cold and salt tolerance in rice. Plant Mol Biol. 2008;67:169–81.

Huang NC, Liu KH, Lo HJ, Tsay YF. Cloning and functional characterization of an Arabidopsis nitrate transporter gene that encodes a constitutive component of low-affinity uptake. Plant Cell. 1999;11:1381–92.

Hubbard KE, Nishimura N, Hitomi K, Getzoff ED, Schroeder JI. Early abscisic acid signal transduction mechanisms: newly discovered components and newly emerging questions. Genes Dev. 2010;24:1695–708.

Ichida AM, Pei ZM, Baizabal-Aguirre VM, Turner KJ, Schroeder JI. Expression of a Cs(+)-resistant guard cell K+ channel confers Cs(+)-resistant, light-induced stomatal opening in transgenic arabidopsis. Plant Cell. 1997;9:1843–57.

Ito Y, Katsura K, Maruyama K, Taji T, Kobayashi M, Seki M, Shinozaki K, Yamaguchi-Shinozaki K. Functional analysis of rice DREB1/CBF-type transcription factors involved in cold-responsive gene expression in transgenic rice. Plant Cell Physiol. 2006;47:141–53.

Iuchi S, Kobayashi M, Yamaguchi-Shinozaki K, Shinozaki K. A stress-inducible gene for 9-cis-epoxycarotenoid dioxygenase involved in abscisic acid biosynthesis under water stress in drought-tolerant cowpea. Plant Physiol. 2000;123:553–62.

Iuchi S, Kobayashi M, Taji T, Naramoto M, Seki M, Kato T, Tabata S, Kakubari Y, Yamaguchi-Shinozaki K, Shinozaki K. Regulation of drought tolerance by gene manipulation of 9-cis-epoxycarotenoid dioxygenase, a key enzyme in abscisic acid biosynthesis in Arabidopsis. Plant J Cell Mol Biol. 2001;27:325–33.

Jaglo-Ottosen KR, Gilmour SJ, Zarka DG, Schabenberger O, Thomashow MF. Arabidopsis CBF1 overexpression induces COR genes and enhances freezing tolerance. Science. 1998;280:104–6.

Jakoby M, Weisshaar B, Droge-Laser W, Vicente-Carbajosa J, Tiedemann J, Kroj T, Parcy F. bZIP transcription factors in Arabidopsis. Trends Plant Sci. 2002;7:106–11.

James VA, Neibaur I, Altpeter F. Stress inducible expression of the DREB1A transcription factor from xeric, *Hordeum spontaneum* L. in turf and forage grass (*Paspalum notatum* Flugge) enhances abiotic stress tolerance. Transgenic Res. 2008;17:93–104.

Johnson RR, Wagner RL, Verhey SD, Walker-Simmons MK. The abscisic acid-responsive kinase PKABA1 interacts with a seed-specific abscisic acid response element-binding factor, TaABF, and phosphorylates TaABF peptide sequences. Plant Physiol. 2002;130:837–46.

Kagaya Y, Hobo T, Murata M, Ban A, Hattori T. Abscisic acid-induced transcription is mediated by phosphorylation of an abscisic acid response element binding factor, TRAB1. Plant Cell. 2002;14:3177–89.

Kang JY, Choi HI, Im MY, Kim SY. Arabidopsis basic leucine zipper proteins that mediate stress-responsive abscisic acid signaling. Plant Cell. 2002;14:343–57.

Kang J, Hwang JU, Lee M, Kim YY, Assmann SM, Martinoia E, Lee Y. PDR-type ABC transporter mediates cellular uptake of the phytohormone abscisic acid. Proc Natl Acad Sci USA. 2010;107:2355–60.

Kanno Y, Hanada A, Chiba Y, Ichikawa T, Nakazawa M, Matsui M, Koshiba T, Kamiya Y, Seo M. Identification of an abscisic acid transporter by functional screening using the receptor complex as a sensor. Proc Nat Acad Sci USA. 2012;109:9653–58.

Kasuga M, Liu Q, Miura S, Yamaguchi-Shinozaki K, Shinozaki K. Improving plant drought, salt, and freezing tolerance by gene transfer of a single stress-inducible transcription factor. Nat Biotechnol. 1999;17:287–91.

Kato Y, Sakaguchi M, Mori Y, Saito K, Nakamura T, Bakker EP, Sato Y, Goshima S, Uozumi N. Evidence in support of a four transmembrane-pore-transmembrane topology model for the Arabidopsis thaliana Na+/K+ translocating AtHKT1 protein, a member of the superfamily of K+ transporters. Proc Nat Acad Sci USA. 2001;98:6488–93.

Kawasaki S, Borchert C, Deyholos M, Wang H, Brazille S, Kawai K, Galbraith D, Bohnert HJ. Gene expression profiles during the initial phase of salt stress in rice. Plant Cell. 2001;13:889–905.

Kim S, Kang JY, Cho DI, Park JH, Kim SY. ABF2, an ABRE-binding bZIP factor, is an essential component of glucose signaling and its overexpression affects multiple stress tolerance. Plant J. 2004;40:75–87.

Kim WY, Ali Z, Park HJ, Park SJ, Cha JY, Perez-Hormaeche J, Quintero FJ, Shin G, Kim MR, Qiang Z, Ning L, Park HC, Lee SY, Bressan RA, Pardo JM, Bohnert HJ, Yun DJ. Release of SOS2 kinase from sequestration with GIGANTEA determines salt tolerance in Arabidopsis. Nat Commun. 2013;4:1352.

Kobayashi Y, Yamamoto S, Minami H, Kagaya Y, Hattori T. Differential activation of the rice sucrose nonfermenting1-related protein kinase2 family by hyperosmotic stress and abscisic acid. Plant Cell. 2004;16:1163–77.

Kobayashi Y, Murata M, Minami H, Yamamoto S, Kagaya Y, Hobo T, Yamamoto A, Hattori T. Abscisic acid-activated SNRK2 protein kinases function in the gene-regulation pathway of ABA signal transduction by phosphorylating ABA response element-binding factors. Plant J Cell Mol Biol. 2005;44:939–49.

Koiwai H, Nakaminami K, Seo M, Mitsuhashi W, Toyomasu T, Koshiba T. Tissue-specific localization of an abscisic acid biosynthetic enzyme, AAO3, in Arabidopsis. Plant Physiol. 2004;134:1697–707.

Koornneef M, Jorna ML, van der Swan Brinkhorst DL, Karssen CM. The isolation of abscisic acid (ABA) deficient mutants by selection of induced revertants in non-germinating gibberellin sensitive lines of *Arabidopsis thaliana* (L.) heynh. Theor Appl Genet. 1982;61:385–93.

Kuromori T, Miyaji T, Yabuuchi H, Shimizu H, Sugimoto E, Kamiya A, Moriyama Y, Shinozaki K (2010) ABC transporter AtABCG25 is involved in abscisic acid transport and responses. Proc Natl Acad Sci USA. 2010;107:2361–66.

Kuromori T, Sugimoto E, Shinozaki K. Arabidopsis mutants of AtABCG22, an ABC transporter gene, increase water transpiration and drought susceptibility. Plant J. 2011;67:885–94.

Lacombe B, Pilot G, Michard E, Gaymard F, Sentenac H, Thibaud JB. A shaker-like K(+) channel with weak rectification is expressed in both source and sink phloem tissues of Arabidopsis. Plant Cell. 2000;12:837–51.

Laurie S, Feeney KA, Maathuis FJ, Heard PJ, Brown SJ, Leigh RA. A role for HKT1 in sodium uptake by wheat roots. Plant J. 2002;32:139–49.

Lebaudy A, Pascaud F, Very AA, Alcon C, Dreyer I, Thibaud JB, Lacombe B. Preferential KAT1-KAT2 heteromerization determines inward K+ current properties in Arabidopsis guard cells. J Biol Chem. 2010;285:6265–74.

Lee SC, Huh KW, An K, An G, Kim SR. Ectopic expression of a cold-inducible transcription factor, CBF1/DREB1b, in transgenic rice (*Oryza sativa* L.). Mol Cells. 2004;18:107–14.

Liao Y, Zhang JS, Chen SY, Zhang WK. Role of soybean GmbZIP132 under abscisic acid and salt stresses. J Integr Plant Biol. 2008;50:221–30.

Liu J, Ishitani M, Halfter U, Kim CS, Zhu JK. The Arabidopsis thaliana SOS2 gene encodes a protein kinase that is required for salt tolerance. Proc Natl Acad Sci USA. 2000;97:3730–34.

Liu Q, Kasuga M, Sakuma Y, Abe H, Miura S, Yamaguchi-Shinozaki K, Shinozaki K. Two transcription factors, DREB1 and DREB2, with an EREBP/AP2 DNA binding domain separate two cellular signal transduction pathways in drought- and low-temperature-responsive gene expression, respectively, in Arabidopsis. Plant Cell. 1998;10:1391–406.

Liu W, Fairbairn DJ, Reid RJ, Schachtman DP. Characterization of two HKT1 homologues from *Eucalyptus camaldulensis* that display intrinsic osmosensing capability. Plant Physiol. 2001;127:283–94.

Lopez-Molina L, Chua NH. A null mutation in a bZIP factor confers ABA-insensitivity in Arabidopsis thaliana. Plant Cell Physiol. 2000;41:541–7.

Ma Y, Szostkiewicz I, Korte A, Moes D, Yang Y, Christmann A, Grill E. Regulators of PP2C phosphatase activity function as abscisic acid sensors. Science. 2009;324:1064–8.

Manohar M, Shigaki T, Hirschi KD. Plant cation/H+ exchangers (CAXs): biological functions and genetic manipulations. Plant Biol (Stuttg). 2011;13:561–9.

Marten I, Hoth S, Deeken R, Ache P, Ketchum KA, Hoshi T, Hedrich R. AKT3, a phloem-localized K+ channel, is blocked by protons. Proc Natl Acad Sci USA. 1999;96:7581–86.

Maruyama K, Sakuma Y, Kasuga M, Ito Y, Seki M, Goda H, Shimada Y, Yoshida S, Shinozaki K, Yamaguchi-Shinozaki K. Identification of cold-inducible downstream genes of the Arabidopsis DREB1A/CBF3 transcriptional factor using two microarray systems. Plant J Cell Mol Biol. 2004;38:982–93.

Miyakawa T, Fujita Y, Yamaguchi-Shinozaki K, Tanokura M. Structure and function of abscisic acid receptors. Trends Plant Sci. 2013;18:259–66.

Moller IS, Gilliham M, Jha D, Mayo GM, Roy SJ, Coates JC, Haseloff J, Tester M. Shoot Na + exclusion and increased salinity tolerance engineered by cell type-specific alteration of Na+ transport in Arabidopsis. Plant Cell. 2009;21:2163–78.

Munns R, James RA, Xu B, Athman A, Conn SJ, Jordans C, Byrt CS, Hare RA, Tyerman SD, Tester M, Plett D, Gilliham M. Wheat grain yield on saline soils is improved by an ancestral Na(+) transporter gene. Nat Biotechnol. 2012;30:360–4.

Nakashima K, Kiyosue T, Yamaguchi-Shinozaki K, Shinozaki K. A nuclear gene, erd1, encoding a chloroplast-targeted Clp protease regulatory subunit homolog is not only induced by water stress but also developmentally up-regulated during senescence in *Arabidopsis thaliana*. Plant J. 1997;12:851–61.

Nakashima K, Ito Y, Yamaguchi-Shinozaki K. Transcriptional regulatory networks in response to abiotic stresses in Arabidopsis and grasses. Plant Physiol. 2009;149:88–95.

Nambara E, Marion-Poll A. Abscisic acid biosynthesis and catabolism. Annu Rev Plant Biol. 2005;56:165–85.

Nishimura N, Sarkeshik A, Nito K, Park SY, Wang A, Carvalho PC, Lee S, Caddell DF, Cutler SR, Chory J, Yates JR, Schroeder JI. PYR/PYL/RCAR family members are major in-vivo ABI1 protein phosphatase 2C-interacting proteins in Arabidopsis. Plant J. 2010;61:290–9.

Oh SJ, Song SI, Kim YS, Jang HJ, Kim SY, Kim M, Kim YK, Nahm BH, Kim JK. Arabidopsis CBF3/DREB1A and ABF3 in transgenic rice increased tolerance to abiotic stress without stunting growth. Plant Physiol. 2005;138:341–51.

Oh DH, Leidi E, Zhang Q, Hwang SM, Li Y, Quintero FJ, Jiang X, D'Urzo MP, Lee SY, Zhao Y, Bahk JD, Bressan RA, Yun DJ, Pardo JM, Bohnert HJ. Loss of halophytism by interference with SOS1 expression. Plant Physiol. 2009;151:210–22.

Oh DH, Lee SY, Bressan RA, Yun DJ, Bohnert HJ. Intracellular consequences of SOS1 deficiency during salt stress. J Exp Bot. 2010;61:1205–13.

Ohta M, Guo Y, Halfter U, Zhu JK. A novel domain in the protein kinase SOS2 mediates interaction with the protein phosphatase 2C ABI2. Proc Natl Acad Sci USA. 2003;100:11771–76.

Okamoto M, Peterson FC, Defries A, Park SY, Endo A, Nambara E, Volkman BF, Cutler SR. Activation of dimeric ABA receptors elicits guard cell closure, ABA-regulated gene expression, and drought tolerance. Proc Nat Acad Sci USA. 2013;110:12132–37.

Okamoto M, Tanaka Y, Abrams SR, Kamiya Y, Seki M, Nambara E. High humidity induces abscisic acid 8'-hydroxylase in stomata and vasculature to regulate local and systemic abscisic acid responses in Arabidopsis. Plant Physiol. 2009;149:825–34.

Ooka H, Satoh K, Doi K, Nagata T, Otomo Y, Murakami K, Matsubara K, Osato N, Kawai J, Carninci P, Hayashizaki Y, Suzuki K, Kojima K, Takahara Y, Yamamoto K, Kikuchi S. Comprehensive analysis of NAC family genes in Oryza sativa and *Arabidopsis thaliana*. DNA Res Int J Rapid Publ Rep Genes Genomes. 2003;10:239–47.

Osakabe Y, Arinaga N, Umezawa T, Katsura S, Nagamachi K, Tanaka H, Ohiraki H, Yamada K, Seo SU, Abo M, Yoshimura E, Shinozaki K, Yamaguchi-Shinozaki K. Osmotic stress responses and plant growth controlled by potassium transporters in Arabidopsis. Plant Cell. 2013a;25:609–24.

Osakabe Y, Yamaguchi-Shinozaki K, Shinozaki K, Tran LS. ABA control of plant macroelement membrane transport systems in response to water deficit and high salinity. The New phytologist. 2013b.

Park SY, Fung P, Nishimura N, Jensen DR, Fujii H, Zhao Y, Lumba S, Santiago J, Rodrigues A, Chow TF, Alfred SE, Bonetta D, Finkelstein R, Provart NJ, Desveaux D, Rodriguez PL, McCourt P, Zhu JK, Schroeder JI, Volkman BF, Cutler SR. Abscisic acid inhibits type 2C protein phosphatases via the PYR/PYL family of START proteins. Science. 2009;324:1068–71.

Peleg Z, Blumwald E. Hormone balance and abiotic stress tolerance in crop plants. Curr Opin Plant Biol. 2011;14:290–5.

Pilot G, Lacombe B, Gaymard F, Cherel I, Boucherez J, Thibaud JB, Sentenac H. Guard cell inward K+ channel activity in arabidopsis involves expression of the twin channel subunits KAT1 and KAT2. J Biol Chem. 2001;276:3215–21.

Qin X, Zeevaart JA. The 9-cis-epoxycarotenoid cleavage reaction is the key regulatory step of abscisic acid biosynthesis in water-stressed bean. Proc Natl Acad Sci USA. 1999;96:15354–61.

Qin X, Zeevaart JA. Overexpression of a 9-cis-epoxycarotenoid dioxygenase gene in *Nicotiana plumbaginifolia* increases abscisic acid and phaseic acid levels and enhances drought tolerance. Plant Physiol. 2002;128:544–51.

Qin F, Sakuma Y, Li J, Liu Q, Li YQ, Shinozaki K, Yamaguchi-Shinozaki K. Cloning and functional analysis of a novel DREB1/CBF transcription factor involved in cold-responsive gene expression in *Zea mays* L. Plant Cell Physiol. 2004;45:1042–52.

Qin F, Kakimoto M, Sakuma Y, Maruyama K, Osakabe Y, Tran LS, Shinozaki K, Yamaguchi-Shinozaki K. Regulation and functional analysis of ZmDREB2A in response to drought and heat stresses in Zea mays L. Plant J Cell Mol Biol. 2007;50:54–69.

Qin F, Sakuma Y, Tran LS, Maruyama K, Kidokoro S, Fujita Y, Fujita M, Umezawa T, Sawano Y, Miyazono K, Tanokura M, Shinozaki K, Yamaguchi-Shinozaki K. Arabidopsis DREB2A-interacting proteins function as RING E3 ligases and negatively regulate plant drought stress-responsive gene expression. Plant Cell. 2008;20:1693–707.

Qiu QS, Guo Y, Dietrich MA, Schumaker KS, Zhu JK. Regulation of SOS1, a plasma membrane Na+/H+ exchanger in *Arabidopsis thaliana*, by SOS2 and SOS3. Proc Natl Acad Sci USA. 2002;99:8436–41.

Quintero FJ, Martinez-Atienza J, Villalta I, Jiang X, Kim WY, Ali Z, Fujii H, Mendoza I, Yun DJ, Zhu JK, Pardo JM. Activation of the plasma membrane Na/H antiporter Salt-Overly-Sensitive 1 (SOS1) by phosphorylation of an auto-inhibitory C-terminal domain. Proc Natl Acad Sci USA. 2011;108:2611–16.

Roychoudhury A, Paul A. Abscisic acid-inducible genes during salinity and drought stress. In: Berhardt LV, editor. Advances in medicine and biology. New York: Nova Publishers; 2012. p. 1–78.

RoyChoudhury A, Roy C, Sengupta DN. Transgenic tobacco plants overexpressing the heterologous lea gene Rab16A from rice during high salt and water deficit display enhanced tolerance to salinity stress. Plant Cell Rep. 2007;26:1839–59.

Roychoudhury A, Paul S, Basu S. Cross-talk between abscisic acid-dependent and abscisic acid-independent pathways during abiotic stress. Plant Cell Rep. 2013;32:985–1006.

Rus A, Yokoi S, Sharkhuu A, Reddy M, Lee BH, Matsumoto TK, Koiwa H, Zhu JK, Bressan RA, Hasegawa PM. AtHKT1 is a salt tolerance determinant that controls Na(+) entry into plant roots. Proc Natl Acad Sci USA. 2001;98:14150–55.

Saavedra X, Modrego A, Rodriguez D, Gonzalez-Garcia MP, Sanz L, Nicolas G, Lorenzo O. The nuclear interactor PYL8/RCAR3 of *Fagus sylvatica* FsPP2C1 is a positive regulator of abscisic acid signaling in seeds and stress. Plant Physiol. 2010;152:133–50.

Saito S, Hirai N, Matsumoto C, Ohigashi H, Ohta D, Sakata K, Mizutani M. Arabidopsis CYP707As encode (+)-abscisic acid 8′-hydroxylase, a key enzyme in the oxidative catabolism of abscisic acid. Plant Physiol. 2004;134:1439–49.

Sakuma Y, Maruyama K, Osakabe Y, Qin F, Seki M, Shinozaki K, Yamaguchi-Shinozaki K. Functional analysis of an Arabidopsis transcription factor, DREB2A, involved in drought-responsive gene expression. Plant Cell. 2006;18:1292–309.

Santiago J, Rodrigues A, Saez A, Rubio S, Antoni R, Dupeux F, Park SY, Marquez JA, Cutler SR, Rodriguez PL. Modulation of drought resistance by the abscisic acid receptor PYL5 through inhibition of clade A PP2Cs. Plant J. 2009;60:575–88.

Sato A, Sato Y, Fukao Y, Fujiwara M, Umezawa T, Shinozaki K, Hibi T, Taniguchi M, Miyake H, Goto DB, Uozumi N. Threonine at position 306 of the KAT1 potassium channel is essential for channel activity and is a target site for ABA-activated SnRK2/OST1/SnRK2.6 protein kinase. Biochem J. 2009;424:439–48.

Schachtman DP, Schroeder JI. Structure and transport mechanism of a high-affinity potassium uptake transporter from higher plants. Nature. 1994;370:655–8.

Seki M, Narusaka M, Abe H, Kasuga M, Yamaguchi-Shinozaki K, Carninci P, Hayashizaki Y, Shinozaki K. Monitoring the expression pattern of 1300 Arabidopsis genes under drought and cold stresses by using a full-length cDNA microarray. Plant Cell. 2001;13:61–72.

Seki M, Ishida J, Narusaka M, Fujita M, Nanjo T, Umezawa T, Kamiya A, Nakajima M, Enju A, Sakurai T, Satou M, Akiyama K, Yamaguchi-Shinozaki K, Carninci P, Kawai J, Hayashizaki Y, Shinozaki K. Monitoring the expression pattern of around 7,000 Arabidopsis genes under ABA treatments using a full-length cDNA microarray. Funct Integr Genomics. 2002;2:282–91.

Seo M, Peeters AJ, Koiwai H, Oritani T, Marion-Poll A, Zeevaart JA, Koornneef M, Kamiya Y, Koshiba T. The Arabidopsis aldehyde oxidase 3 (AAO3) gene product catalyzes the final step in abscisic acid biosynthesis in leaves. Proc Natl Acad Sci USA. 2000;97:12908–13.

Shen YG, Zhang WK, He SJ, Zhang JS, Liu Q, Chen SY. An EREBP/AP2-type protein in *Triticum aestivum* was a DRE-binding transcription factor induced by cold, dehydration and ABA stress. TAG Theoret Appl Genet (Theoretische und angewandte Genetik). 2003;106:923–30.

Shi H, Quintero FJ, Pardo JM, Zhu JK. The putative plasma membrane Na(+)/H(+) antiporter SOS1 controls long-distance Na(+) transport in plants. Plant Cell. 2002;14:465–77.

Shinozaki K, Yamaguchi-Shinozaki K. Gene networks involved in drought stress response and tolerance. J Exp Bot. 2007;58:221–7.

Shinozaki K, Yamaguchi-Shinozaki K, Seki M. Regulatory network of gene expression in the drought and cold stress responses. Curr Opin Plant Biol. 2003;6:410–7.

Shkolnik-Inbar D, Adler G, Bar-Zvi D. ABI4 downregulates expression of the sodium transporter HKT1;1 in Arabidopsis roots and affects salt tolerance. Plant J. 2013;73:993–1005.

Skinner JS, von Zitzewitz J, Szucs P, Marquez-Cedillo L, Filichkin T, Amundsen K, Stockinger EJ, Thomashow MF, Chen TH, Hayes PM. Structural, functional, and phylogenetic characterization of a large CBF gene family in barley. Plant Mol Biol. 2005;59:533–51.

Song A, Lu J, Jiang J, Chen S, Guan Z, Fang W, Chen F. Isolation and characterisation of Chrysanthemum crassum SOS1, encoding a putative plasma membrane Na(+) /H(+) antiporter. Plant Biol (Stuttg). 2012.

Sridha S, Wu K. Identification of AtHD2C as a novel regulator of abscisic acid responses in Arabidopsis. Plant J. 2006;46:124–33.

Sunarpi Horie T, Motoda J, Kubo M, Yang H, Yoda K, Horie R, Chan WY, Leung HY, Hattori K, Konomi M, Osumi M, Yamagami M, Schroeder JI, Uozumi N. Enhanced salt tolerance mediated by AtHKT1 transporter-induced Na unloading from xylem vessels to xylem parenchyma cells. Plant J. 2005;44:928–38.

Sze H, Padmanaban S, Cellier F, Honys D, Cheng NH, Bock KW, Conejero G, Li X, Twell D, Ward JM, Hirschi KD. Expression patterns of a novel AtCHX gene family highlight potential roles in osmotic adjustment and K+ homeostasis in pollen development. Plant Physiol. 2004;136:2532–47.

Szyroki A, Ivashikina N, Dietrich P, Roelfsema MR, Ache P, Reintanz B, Deeken R, Godde M, Felle H, Steinmeyer R, Palme K, Hedrich R. KAT1 is not essential for stomatal opening. Proc Natl Acad Sci USA. 2001;98:2917–21.

Tan BC, Schwartz SH, Zeevaart JA, McCarty DR. Genetic control of abscisic acid biosynthesis in maize. Proc Nat Acad Sci USA. 1997;94:12235–40.

Thompson AJ, Jackson AC, Parker RA, Morpeth DR, Burbidge A, Taylor IB. Abscisic acid biosynthesis in tomato: regulation of zeaxanthin epoxidase and 9-cis-epoxycarotenoid dioxygenase mRNAs by light/dark cycles, water stress and abscisic acid. Plant Mol Biol. 2000a;42:833–45.

Thompson AJ, Jackson AC, Symonds RC, Mulholland BJ, Dadswell AR, Blake PS, Burbidge A, Taylor IB. Ectopic expression of a tomato 9-cis-epoxycarotenoid dioxygenase gene causes over-production of abscisic acid. Plant J Cell Mol Biol. 2000b;23:363–74.

Thompson AJ, Andrews J, Mulholland BJ, McKee JM, Hilton HW, Horridge JS, Farquhar GD, Smeeton RC, Smillie IR, Black CR, Taylor IB. Overproduction of abscisic acid in tomato increases transpiration efficiency and root hydraulic conductivity and influences leaf expansion. Plant Physiol. 2007;143:1905–17.

Tran LS, Nakashima K, Sakuma Y, Simpson SD, Fujita Y, Maruyama K, Fujita M, Seki M, Shinozaki K, Yamaguchi-Shinozaki K. Isolation and functional analysis of Arabidopsis stress-inducible NAC transcription factors that bind to a drought-responsive cis-element in the early responsive to dehydration stress 1 promoter. Plant Cell. 2004;16:2481–98.

Uauy C, Distelfeld A, Fahima T, Blechl A, Dubcovsky J. A NAC Gene regulating senescence improves grain protein, zinc, and iron content in wheat. Science. 2006;314:1298–301.

Umezawa T, Okamoto M, Kushiro T, Nambara E, Oono Y, Seki M, Kobayashi M, Koshiba T, Kamiya Y, Shinozaki K. CYP707A3, a major ABA 8′-hydroxylase involved in dehydration and rehydration response in *Arabidopsis thaliana*. Plant J. 2006;46:171–82.

Umezawa T, Nakashima K, Miyakawa T, Kuromori T, Tanokura M, Shinozaki K, Yamaguchi-Shinozaki K. Molecular basis of the core regulatory network in ABA responses: sensing, signaling and transport. Plant Cell Physiol. 2010;51:1821–39.

Uno Y, Furihata T, Abe H, Yoshida R, Shinozaki K, Yamaguchi-Shinozaki K. Arabidopsis basic leucine zipper transcription factors involved in an abscisic acid-dependent signal transduction pathway under drought and high-salinity conditions. Proc Nat Acad Sci USA. 2000;97:11632–7.

Vagujfalvi A, Aprile A, Miller A, Dubcovsky J, Delugu G, Galiba G, Cattivelli L. The expression of several Cbf genes at the Fr-A2 locus is linked to frost resistance in wheat. Mol Genet Genomics. 2005;274:506–14.

Vogel JT, Zarka DG, Van Buskirk HA, Fowler SG, Thomashow MF. Roles of the CBF2 and ZAT12 transcription factors in configuring the low temperature transcriptome of Arabidopsis. Plant J Cell Mol Biol. 2005;41:195–211.

Wu SJ, Ding L, Zhu JK. SOS1, a genetic locus essential for salt tolerance and potassium acquisition. Plant Cell. 1996;8:617–27.

Xicluna J, Lacombe B, Dreyer I, Alcon C, Jeanguenin L, Sentenac H, Thibaud JB, Cherel I. Increased functional diversity of plant K+ channels by preferential heteromerization of the shaker-like subunits AKT2 and KAT2. J Biol Chem. 2007;282:486–94.

Xiong Y, Fei SZ. Functional and phylogenetic analysis of a DREB/CBF-like gene in perennial ryegrass (*Lolium perenne* L.). Planta. 2006;224:878–88.

Xiong L, Ishitani M, Lee H, Zhu JK. The Arabidopsis LOS5/ABA3 locus encodes a molybdenum cofactor sulfurase and modulates cold stress- and osmotic stress-responsive gene expression. Plant Cell. 2001;13:2063–83.

Xue GP. The DNA-binding activity of an AP2 transcriptional activator HvCBF2 involved in regulation of low-temperature responsive genes in barley is modulated by temperature. Plant J. 2003;33:373–83.

Xue GP, Loveridge CW. HvDRF1 is involved in abscisic acid-mediated gene regulation in barley and produces two forms of AP2 transcriptional activators, interacting preferably with a CT-rich element. Plant J Cell Mol Biol. 2004;37:326–39.

Yamaguchi-Shinozaki K, Shinozaki K. A novel cis-acting element in an Arabidopsis gene is involved in responsiveness to drought, low-temperature, or high-salt stress. Plant Cell. 1994;6:251–64.

Yoshida T, Fujita Y, Sayama H, Kidokoro S, Maruyama K, Mizoi J, Shinozaki K, Yamaguchi-Shinozaki K. AREB1, AREB2, and ABF3 are master transcription factors that cooperatively regulate ABRE-dependent ABA signaling involved in drought stress tolerance and require ABA for full activation. Plant J. 2010;61:672–85.

Yu L, Nie J, Cao C, Jin Y, Yan M, Wang F, Liu J, Xiao Y, Liang Y, Zhang W. Phosphatidic acid mediates salt stress response by regulation of MPK6 in *Arabidopsis thaliana*. New Phytol. 2010;188:762–73.

Zhao H, Bughrara SS. Isolation and characterization of cold-regulated transcriptional activator LpCBF3 gene from perennial ryegrass (*Lolium perenne* L.). Mol Genet Genomics. 2008;279:585–94.

Zhao J, Ren W, Zhi D, Wang L, Xia G. Arabidopsis DREB1A/CBF3 bestowed transgenic tall fescue increased tolerance to drought stress. Plant Cell Rep. 2007;26:1521–8.

Zhu JK. Salt and drought stress signal transduction in plants. Annu Rev Plant Biol. 2002;53:247–73.

Zhu JK. Regulation of ion homeostasis under salt stress. Curr Opin Plant Biol. 2003;6:441–5.

Chapter 17
ABA Regulation of the Cold Stress Response in Plants

Yiting Shi and Shuhua Yang

Abstract Low temperature is a major environmental factor that limits plant growth, productivity, and distribution. To ensure optimal growth and survival, plants must respond and adapt to cold stress using a variety of biochemical and physiological processes. Currently, the most thoroughly understood cold-signalling pathway is the C-repeat binding factor/DRE-binding factor (CBF/DREB) transcriptional regulatory cascade. Abscisic acid (ABA) is an important stress hormone in plants that has been demonstrated to be involved in the cold stress response through regulation of a set of specific stress-responsive genes. The current consensus is that both ABA-dependent and ABA-independent pathways are involved in plant responses to cold stress. This chapter summarises recent progress made in our understanding of cold signalling and the role of ABA in cold stress, and we also address cross talk between ABA and several classical phytohormones that integrate with cold signalling.

Keywords Cold signalling · ABA regulation · Plant responses · Cross talk

17.1 Introduction

As sessile organisms, plants suffer from a variety of abiotic environmental stresses, including low temperature, which can limit plant growth and the geographic distribution of a plant species, potentially impacting the reproduction of economically important crops. To adapt to adverse environments, plants have

Y. Shi · S. Yang (✉)
State Key Laboratory of Plant Physiology and Biochemistry, College of Biological Sciences, China Agricultural University, Beijing 100193, China
e-mail: yangshuhua@cau.edu.cn

D.-P. Zhang (ed.), *Abscisic Acid: Metabolism, Transport and Signaling*,
DOI 10.1007/978-94-017-9424-4_17

evolved a number of complex mechanisms to avoid or tolerate cold stress. Cold acclimation is one of the most thoroughly understood mechanisms contributing to increased cold tolerance in plants. During cold acclimation, plants modulate multiple morphological and physiological processes in order to establish a new state of cellular and metabolic homeostasis, which also involves in changes in phytohormone homeostasis, allowing for adaptation to stressful conditions (Hirayama and Shinozaki 2010; Theocharis et al. 2012; Zhu et al. 2007).

Abscisic acid (ABA) is a well-known stress hormone that plays a crucial role in dehydration stress. However, it has also been suggested that ABA plays a role in cold acclimation by triggering specific cellular and molecular osmotic responses. This chapter highlights the latest advances in our understanding of cold-response mechanisms that are either directly or indirectly influenced by ABA signalling. In particular, we focus on the regulatory networks used for signal perception, signal transduction, and the regulation of stress-responsive genes. Furthermore, both antagonistic and synergistic cross talk between ABA and other plant hormones involved in cold signalling will be discussed.

17.2 Cold-Signal Sensing and Transduction Pathways

Low temperatures can severely disrupt the metabolism and physiological homeostasis of plant cells and can even lead to plant death. Cold stress can be divided into non-freezing chilling stress (below 15 °C and above freezing point) and freezing stress (below freezing point). Chilling stress inhibits the activities of enzymes involved in photosynthesis, respiration, and biochemical processes such as reactive oxygen species (ROS) scavenging, leading to oxidative damage that can result in toxic compounds accumulation and the inhibition of metabolic reactions (O'Kane et al. 1996; Yang et al. 2005). Freezing stress results in the formation of intracellular ice crystals that induce cellular dehydration and osmotic stress, leading to membrane damage, and ultimately, to death of tissues (Uemura et al. 1995; Webb and Steponkus 1993). Most temperate plants are able to tolerate freezing stress following prior exposure to chilling, non-freezing temperatures, which is referred to as cold acclimation (Thomashow 1999). During cold acclimation, plants initiate global transcriptome changing and become tolerant to freezing temperatures by increasing the accumulation of osmolytes (such as soluble sugars) and antifreezing proteins, as well as by altering membrane composition, which together protect plant cells from dehydration and metabolic disruption (Yamada et al. 2002).

It has been noted that a reduction in the fluidity of the plasma membrane appears to be a primary event for the cold signalling sensing (Yamada et al. 2002). For example, in *Synechocystis*, histidine kinases (Hiks) are used to percept decreased levels of unsaturated fatty acids in the plasma membrane and activate the cold-induced *des* genes, which are responsible for the feedback maintenance of membrane lipid composition to modulate cold tolerance (Suzuki et al. 2000).

Lipid composition of membranes, especially the portion of galactolipids containing unsaturated fatty acids, is crucial for high plant species during cold acclimation. In *Arabidopsis*, the expression of *FAD2*, which encodes an enzyme that is essential for polyunsaturated lipid synthesis. *fad2* mutants show irregular membrane composition and cannot survive at low temperature (Miquel et al. 1993). Similar chilling-sensitive phenotypes are also observed in loss-of-function *fad5* (Hugly and Somerville 1992). A recent study showed that a lipid desaturase, acyl-lipid desaturase2 (ADS2), is required for chilling and freezing tolerance in *Arabidopsis* and functions by altering membrane lipid composition (Chen and Thelen 2013). Furthermore, the *Arabidopsis* sensitive to freezing2 (SFR2) protein was identified as a galactolipid-remodelling enzyme that is localised to the outer chloroplast membrane, and it is essential for membrane lipid remodelling of chloroplast envelope under freezing stress (Moellering et al. 2010; Fourrier et al. 2008).

Under cold stress, membrane rigidity triggers second-messenger molecules such as Ca^{2+} and to activate complex signaling pathways involved in protein kinases or transcription-factor cascades (Viswanathan and Zhu 2002). However, the underlying mechanisms that plants use to perceive and transduce cold signals remain elusive. Currently, the most thoroughly understood cold-signalling pathway is the ICE–CBF–COR transcriptional cascade, which plays a crucial role in the activation of multiple downstream *cold-regulated* (*COR*) genes (Thomashow 2010) (Fig. 17.1). Acting as central nodes in the cold acclimation pathway, the C-repeat binding factors (CBFs) gene family is conserved in many plant species. In *Arabidopsis*, three CBF transcription factors belonging to the AP2/ERF (apetala 2/ethylene-responsive factor) superfamily have been identified. CBFs are also known as dehydration-responsive element binding factors (DREBs), which recognise the conserved cold- and dehydration-responsive C-repeat/DRE DNA motifs found within the promoter regions of *COR* genes (Liu et al. 1998). *CBF1* overexpression triggers constitutive expression of the *COR* genes and induces freezing tolerance in *Arabidopsis* (Jaglo-Ottosen et al. 1998). Expression of the *CBFs* is rapidly induced by cold temperatures, and a growing number of studies have demonstrated that transcriptional regulation of the *CBF* genes is controlled by multiple mechanisms. In *Arabidopsis*, CBF2 is a negative regulator of the expression of *CBF1* and *CBF3*, and it plays an important role in freezing tolerance (Novillo et al. 2004; Novillo et al. 2007). To date, several key upstream regulators of the *CBF* genes have been identified and characterised. CAMTA3 encodes a calmodulin-binding protein and acts upstream to activate *CBF2* expression by binding to the CG-element in its promoter region in response to cold signals (Doherty et al. 2009). The MYC-type, basic helix-loop-helix transcriptional activator inducer of CBF expression 1 (ICE1) has been identified as a transcriptional activator that binds to MYC *cis*-elements within the *CBF3* promoter, and the expression of *CBF3* and its target *COR* genes is impaired in *ice1*-mutant plants during cold acclimation. In addition, overexpression of ICE1 in *Arabidopsis* results in increased freezing tolerance, supporting a pivotal role for ICE1 in the cold stress response (Chinnusamy et al. 2003). The transcription

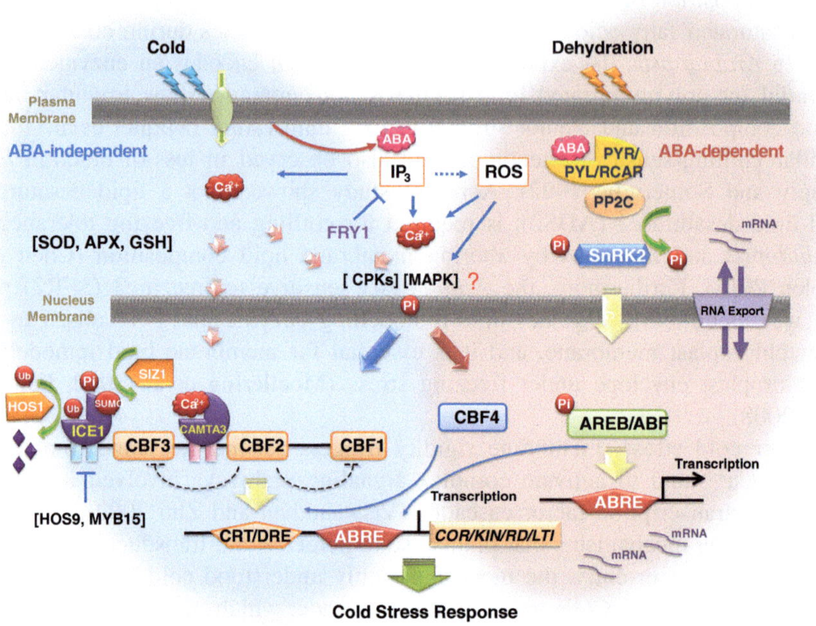

Fig. 17.1 Schematic illustration of the cold and dehydration response regulatory networks in *Arabidopsis*. IP_3, Ca^{2+}, and ROS act as second messengers in signalling networks to transduce signals through protein kinases or transcription-factor cascades. CBFs and AREB/ABFs transcription factors are responsible for the regulation of *COR* genes containing CRT/DRE (CCGAC) and ABRE (ACGT) motifs in their promoters, respectively. In *Arabidopsis*, approximately 10 % of the cold-induced transcriptome contains both CRT and ABRE motifs in their promoter regions, which are upregulated by both cold and dehydration stresses. CBFs are activated by ICE1 and CAMTA transcription factors, whereas repressed by MYB15. HOS1 and SIZ1 encode RING E3 ligase and SUMO E3 ligase, respectively, which antagonistically regulate the abundance of ICE1 protein. Cold activates ICE1 protein and AREB/ABFs are induced by ABA-mediated dehydration signalling pathway. *IP_3* inositol 1,4,5-triphosphate, *ROS* reactive oxygen species, *CPK* calcium-dependent protein kinase, *MAPK* Ras-mitogen-activated protein kinase, *Ub* ubiquitin moiety, *SUMO* Small ubiquitin-related modifier, *Pi* phosphoryl group

factor EIN3 was shown to negatively regulate expression of the *CBFs* by directly binding to *CBF* promoters to modulate the cold stress response in plants. Furthermore, a loss-of-function *ein3* mutant shows enhanced freezing tolerance, whereas overexpression of *EIN3* leads to decreased freezing tolerance (Shi et al. 2012). Using a genetic screening strategy involving changes in the expression of a luciferase construct under the control of a stress-inducible *RD29A* promoter, high expression of osmotically responsive genes1 (HOS1), which encodes a RING-type ubiquitin E3 ligase, was isolated as an upstream negative regulator of the *CBFs*. HOS1 interacts with and ubiquitinates ICE1 to negatively regulate the stability of the ICE1 protein (Dong et al. 2006). The R2R3-type MYB transcription factor AtMYB15 (MYB domain protein 15) was found to physically interact with ICE1. *MYB15* overexpression leads to reduced expression of the *CBF* genes, whereas a*myb15* mutant shows increased *CBF* expression. MYB15 binds to the *CBF3* promoter to repress its *CBF* expression and negatively regulates freezing tolerance (Agarwal et al. 2006). Subsequently, a SUMO E3 ligase, SAP and Miz (SIZ1), was identified as a positive regulator of the ICE1 protein. SIZ1 can sumoylate ICE1 and repress the polyubiquitination of ICE1, which, in turn, enhances ICE1 stability (Miura et al. 2007). Therefore, regulation of the ICE1-CBF cascade at both transcriptional and post-translational levels demonstrates the existence of a complex network of CBF-dependent cold-signalling pathways.

17.3 The Role of ABA Biosynthesis and Signalling in Cold Stress

Investigations into the role of ABA in cold stress were originally based on observations made in the 1960s in woody species, showing that application of the gibberellic acid (GA) inhibitor dormin—which was later identified as ABA-resulted in increased freezing tolerance in trees, equivalent to that observed in plants that had undergone cold acclimation (Chrispeels and Varner 1967; Thomas et al. 1965). It has been shown that cold tolerance is usually accompanied by increased endogenous ABA levels in various plant species (Daie and Campbell 1981; Lang et al. 1994; Mantyla et al. 1995). Continuous application of ABA induces chilling tolerance in chilling-sensitive plant species, such as maize, rice, cucumber, and pepper. Furthermore, exogenous ABA application in temperate plants such as poplar, barely, wheat, and *Arabidopsis* can partially mimic cold acclimation and enhance freezing tolerance (Kadlecová et al. 2000; Smoleńska-Sym et al. 1995; Zhu et al. 2000; Thomashow 1999). Transcriptome analyses in *Arabidopsis* showed that a number of ABA-responsive genes can be induced by cold treatment (Zeevaart and Creelman 1988). However, it has also been shown that ABA application in several plant species has little effect on freezing tolerance, or if there are any changes, they are much less significant than those induced by cold acclimation (Gusta et al. 1982; Fayyaz et al. 1978; Holubowicz et al. 1982), suggesting the natural diversity of ABA responses in different plant species.

The primary source of increased ABA levels under stress conditions is a de novo biosynthesis pathway that converts carotenoids into bioactive ABA. ABA biosynthesis pathway may be required for full development of the cold response, as defects in both basal and acquired freezing tolerance have been observed in ABA-deficient mutants. For instance, *ABA1* and *ABA3* are identified as genes encoding enzymes involved in ABA biosynthetic pathway. Cold induction of the *COR* genes is reduced in the *aba3* mutants *aba3/los5/frs1* (Llorente et al. 2000;

Xiong et al. 2002b). Cold acclimation was shown to be impaired in an *aba1* mutant, which has also reduced expression of specific *COR* genes (Mantyla et al. 1995). The expression of none of the ABA biosynthesis genes is affected by cold treatment in *Arabidopsis* (Lee et al. 2005). Therefore, ABA biosynthesis is not an early event in response to cold stress. Consistent with this, foliar application of ABA to a wide range of plant species does not induce cold hardness or freezing tolerance (Gusta et al. 2005). A recent study showed that cold stress specifically activate the expression of ABA biosynthesis genes in reproductive organs, such as the inflorescence meristem, with only slightly increased expression in the leaves and vegetable organs (Baron et al. 2012). This finding supports the notion of ABA function in stimulating and hastening plant harvests under adverse environmental conditions. Taken together, ABA may function late in the development of cold-induced metabolic changes and is required for determining the maximum levels of cold tolerance in plants.

ABA activates a wide array of genes associated with stress response, seed dormancy, and stomatal movement through the ABA signalling pathway (Yamaguchi-Shinozaki and Shinozaki 2006). The PYR/PYL/RCAR family of proteins are ABA receptors that can inhibit the activity of the type-2C protein phosphatase PP2C in the presence of ABA, relieving the repression of the SnRK2 kinases activity, which in turn phosphorylates other downstream regulators (Furihata et al. 2006; Ma et al. 2009; Park et al. 2009). Overexpression of *TaSnRK2.3* in *Arabidopsis* results in enhanced root-system architecture and significantly enhances the tolerance of this species to drought, salt, and freezing stresses (Tian et al. 2013), whereas downregulation of PP2CA proteins increases ABA sensitivity and accelerates cold acclimation in transgenic plants (Tahtiharju and Palva 2001). Promoter analysis found that some DRE/CRT motif-containing *COR* genes are induced by cold and dehydration and impaired in ABA-deficient mutants (Thomashow 1999; Shinozaki and Yamaguchi-Shinozaki 2000). As mentioned earlier, CBFs/DREB1s and their homolog, CBF4, were shown to bind to CRT/DRE motifs in vivo. *CBF4* expression is induced by drought and ABA, but not by cold, which is quite distinctive from *CBF1–CBF3*. Overexpression of *CBF4* results in increased freezing and drought tolerance (Haake et al. 2002). It appears that at least three separate regulatory signalling pathways exist that are mediated by the binding of transcription factors to CRT/DRE motifs. Among these, cold-induced CBF1–CBF3 and drought-induced DREB2 are the primary transcription factors that control gene expression through the ABA-independent pathway, whereas CBF4 acts as a regulator of the ABA response (Thomashow 1999; Shinozaki and Yamaguchi-Shinozaki 2000). In addition to the CRT/DRE motif, many cold-response gene promoters harbour ABRE *cis*-elements that can be activated in vivo by ABA through direct binding of bZIP transcription-factor proteins (AREBs/ABFs, ABRE-binding proteins/factors) (Uno et al. 2000). This evidence explains that expression of the *COR* genes is regulated by both ABA-dependent and ABA-independent pathways (Fig. 17.1).

17.4 ABA-Dependent Signalling Pathway for Cold Acclimation

Genetic and physiological analyses indicate that ABA-dependent signalling pathway plays an essential role in stress-responsive gene expression during cold-induced osmotic stress. ABA biosynthesis genes contribute to the enrichment of *COR* genes, such as *RD29A*, *RD22*, *COR15A*, *COR47*, and *P5CS* via activity of their ABRE *cis*-element (Xiong et al. 2001a, 2002b). In the study on *Arabidopsis*, four AREB/ABF transcription factors are found to be required for the activation of ABA-mediated signalling (Uno et al. 2000). Among them, *ABF1* expression is significantly induced by cold, but not by osmotic stress (Choi et al. 2000), whereas *AREB1/ABF2*, *AREB2/ABF4*, and *ABF3* are specifically induced by ABA and drought stress (Abdeen et al. 2010; Kang et al. 2002; Fujita et al. 2005). AREB proteins do not appear to play a role in CBF-dependent cold signalling; however, one study showed that ABF2, ABF3, and ABF4 interact with DREB2C in vitro (Lee et al. 2010). In addition, approximately 10 % of ABA-responsive genes are also responsive to cold stress (Kreps et al. 2002). Thus, although ABA- and cold stress-induced transcription factors show distinct regulation, some of the *COR* genes appear to be induced by both of these stimuli. Therefore, one could speculate that limited, cold-induced ABA production could enhance *COR* gene activation through the ABA-dependent pathway during cold acclimation.

In addition to AREBs, other bZIP transcription factors belonging to the ABI5 subfamily have been characterised as ABRE-binding proteins and were shown to affect seed germination as well as the cold stress response. For example, ectopic expression of the seed-specific transcriptional activator *ABI3* confers the ability to express *COR* genes in vegetative tissues and enhances freezing tolerance in *Arabidopsis* (Tamminen et al. 2001). ABI5 has been shown to act downstream of and interact with ABI3 (Suzuki and McCarty 2008; Finkelstein and Lynch 2000, and it can be sumoylated and degraded by the SUMO E3 ligase SIZ1, which attenuates ABA signalling (Miura et al. 2009). Indeed, SIZ1-mediated SUMO conjugation is also required for the expression of *CBF3* by modulating the activity of ICE1 (Miura et al. 2007). *siz1* mutants are impaired in cold acclimation and are sensitive to ABA during seed germination (Miura et al. 2007, 2009). A recent study revealed that SIZ1 is able to sumoylate and stabilise the negative regulator of ABA signalling MYB30 (Zheng et al. 2012). Therefore, SIZ1 appears to be an integration node that mediates ABA and cold responses through post-transcriptional modifications.

Several MYC and MYB family genes have been reported to be important regulators of ABA-responsive gene expression under cold stress. For instance, MYB96 is induced by ABA and drought, and it enhances ABA-mediated drought and freezing tolerance (Guo et al. 2013; Seo et al. 2009). Expression of *OsMYB3R-2* in rice is induced by cold, drought, and salt stresses, and overexpression of *OsMYB3R-2* increases tolerance to freezing, drought, and salt stresses in transgenic *Arabidopsis* (Dai et al. 2007). Recent studies suggest that some NAC-domain family members are involved in stress responses (Olsen et al. 2005). For example, expression of a rice NAC gene, *OsNAC5*, is induced by osmotic stress and ABA, and overexpression of *OsNAC5* increases stress-induced proline and soluble-sugar levels and enhances tolerance to cold, salt, and drought stresses (Takasaki et al. 2010; Song et al. 2011). Thus, it can be assumed that cold induction of some ABA-responsive genes is regulated by different kinds of transcription factors.

17.5 Second Messengers Integrate Cold and ABA Signalling

Both cold stress and ABA treatment induce the production and accumulation of IP3, Ca^{2+}, and ROS inside the cell. These molecules act as second messengers in signalling networks to amplify and transduce signals through activating protein kinases or transcription-factor cascades. Increased inositol 1,4,5-triphosphate (IP_3) levels, which are catalysed by phospholipase C (PLC) under cold stress, are known to release Ca^{2+} from vacuoles into the cytosol (Allen and Sanders 1995; Munnik et al. 1998). In *Arabidopsis*, *FRY1* encodes an inositol polyphosphate 1—phosphatase that is involved in IP_3 metabolism. *fry1* mutant plants show significantly higher IP_3 levels when treated with ABA, and they display defects in cold acclimation and hypersensitivity to osmotic stress, with lowered expression of certain stress-responsive genes (Xiong et al. 2001b).

In plants, Ca^{2+} is a widely used and important signalling molecule during early responses to abiotic stresses, as it functions by activating protein kinase cascades, which in turn can activate transcription factors and stress-responsive genes. More specifically, a transient influx of cytoplasmic Ca^{2+} occurs in response to cold shock, and it has been suggested that low-temperature-induced changes in membrane fluidity are the primary thermosensor signal that activates the Ca^{2+} influx in plants (Knight et al. 2004). Furthermore, calcium is required for the full expression of the *COR* genes in *Arabidopsis*. *CAX1*, which encodes a vacuolar membrane-located Ca^{2+}/H^+ antiporter, participates in the regulation of cold acclimation. Loss-of-function mutations in *CAX1* increase *CBF* expression and enhance freezing tolerance following acclimation in *Arabidopsis* (Catala et al. 2003). One calmodulin proteins in *Arabidopsis*, AtCML9, has recently been demonstrated to be responsive to ABA, cold, and salt stress through an ABA-dependent pathway. The *cml9* mutant exhibits hypersensitivity to ABA and enhanced tolerance to stress and drought (Magnan et al. 2008). CaM methylation is also involved in the early growth of *Arabidopsis* and responds to ABA, cold, and heat stress during germination (Banerjee et al. 2013). As mentioned earlier, the calmodulin-binding transcriptional activator *CAMTA3* is involved in the regulation of freezing tolerance (Doherty et al. 2009). More importantly, a recent study suggested that calcium channels may involve in temperature perception in land plants, based on the observation that plants lacking the membrane protein cyclic nucleotide gated calcium channel (CNGC) showed constitutively high levels of Ca^{2+} influx and impaired thermoperception in a moss and in *Arabidopsis* (Finka et al. 2012). Moreover, a homologue of the synaptic Ca^{2+}-binding protein synaptotagmin (SYT1), which is a calcium sensor that can prevent Ca^{2+}-dependent membrane, damages in *Arabidopsis* protoplasts under freezing conditions (Yamazaki et al. 2008). These results suggest the central role of cytosolic calcium to trigger cold signalling. Ca^{2+} may function through calmodulin or Ca^{2+} sensors to modulate the activation of downstream calcium-dependent protein kinases (CPKs). Calcineurin B-like proteins (CBLs), one kind of calcium sensors, were also shown to regulate expression of the *CBF* genes, based on the observations that overexpression of *CBL1* resulted in reduced freezing tolerance, whereas *cbl1* null mutants showed increased cold-induced expression of stress genes and enhanced freezing tolerance (Albrecht et al. 2003; Cheong et al. 2003).

Several CBL-interacting protein kinases (CIPKs) and CPKs have been implicated in the responses to ABA, cold, and high-salt stress. In *Arabidopsis*, the expression of *CIPK3* is strongly induced by cold stress and ABA treatment, but not by drought stress. In addition, the expression of various cold-induced marker genes is all significantly delayed in a *cipk3* mutant, and the induction of ABA-responsive genes is also significantly inhibited (Kim et al. 2003). Therefore, CIPK3 may function as a cross talk node between signalling pathways for ABA, cold, and other abiotic stresses. It has also been suggested that CIPK1 is a convergence point for the ABA-dependent and ABA-independent stress responses. For example, CIPK1 can interact with CBL1 and CBL9, resulting in the formation of CBL1/CIPK1 and CBL9/CIPK1 complexes that mediate the ABA-independent and ABA-dependent responses, respectively, during osmotic stress (D'Angelo et al. 2006). Rice *OsCDPK7* is a cold- and salt-inducible gene, and overexpression of *OsCDPK7* confers cold and salt/drought tolerance on rice plants (Abbasi et al. 2004). Collectively, these results indicate a central role for calcium in sensing and transducing the cold and ABA signals.

17.6 ABA and ROS Under Cold Stress

The induction of ROS-scavenging systems is common to many stress pathways including ABA and cold stress (Fujita et al. 2006). The roles of ROS in cold stress are considered in two ways. On the one hand, excessive ROS levels are toxic and can lead to lipid peroxidation, and ultimately, to oxidative stress. On the other hand, increased ROS levels can also play a benefit signalling role and activate scavenging enzyme systems to eliminate excess ROS in plants subjected to low temperatures. Numerous studies have shown that ROS-scavenging enzymes such as superoxide dismutase (SOD), ascorbate peroxidise (APX), catalase (CAT), and glutathione peroxidise (GPX) are required for protection against cold-induced oxidative damage (Burdon et al. 1996; O'Kane et al. 1996). For example, overexpression of antioxidant-defensive genes enhances chilling tolerance in maize and soybean (Van Breusegem et al. 1999; Kocsy et al. 2001), whereas repression of catalase gene expression reduces chilling tolerance (Kerdnaimongkol and Woodson 1999). *Arabidopsis chs* (*chilling-sensitive*) mutants display increased sensitivity to chilling temperature, which is at least partially due to the excessive accumulation of H_2O_2 in *chs* mutant seedlings (Huang et al. 2010a, b; Wang et al. 2013; Yang et al. 2010). Recent studies also show that ROS act as signal molecules mediating cold stress-signal transduction. For instance, the *Arabidopsis* mutant *frostbite1* (*fro1*) has a defective form of mitochondrial complex I NADH dehydrogenase, leading to constitutively lower ROS levels compared with the wild type, and this mutant shows reduced expression of cold stress-responsive genes and decreased cold acclimation (Lee et al.2002). *ZAT12* encodes a ROS-responsive zinc-finger transcription factor that functions as a positive regulator of the cold-response pathway. Overexpression of *ZAT12* results in enhanced freezing and oxidative tolerance (Davletova et al. 2005), which suggests that ROS signalling plays a beneficial role in mediating cold acclimation in plants.

It has been shown that the application of exogenous ABA effectively alleviates the symptoms of cold injury in many species. One explanation for this phenomenon is that ABA signalling is able to induce the transcription of ROS-scavenging enzymes. Indeed, one study showed that stress-induced ABA accumulation stimulates ROS scavenging to help maintain the cellular redox state (Guan et al. 2000). Furthermore, ABA-dependent proline accumulation under abiotic stress is regulated by the ROS-scavenging-mediated cellular redox state, which is also important for the cold response (Kishor et al. 2005). In particular, ABA treatment enhances ROS accumulation by activating plasma membrane-bound NADPH oxidases (Kwak et al. 2003). *ABO5* and *ABO6*, which encode a PPR (pentatricopeptide repeat) protein and DEXH box RNA helicase, respectively, are required for ABA-induced ROS production in mitochondria (He et al. 2012; Liu et al. 2010). The activities of ABI1 and ABI2 are inhibited by H_2O_2 in vitro (Meinhard and Grill 2001; Meinhard et al. 2002), and further studies demonstrated that ABI2 interacts with glutathione peroxidase 3 (GPX3), which is an ROS-scavenging enzyme that functions as a redox transducer and scavenger to modulate ABA signalling (Miao et al. 2006). Glucosamine (GlcN) is a naturally occurring amino sugar that inhibits plant growth by significantly increasing the production of ROS. Ectopic overexpression of GlcN induces cell death, whereas scavenging of endogenous GlcN can enhance tolerance to oxidative, drought, and cold stresses in *Arabidopsis* (Chu et al. 2010).

As well-known signalling components downstream of ROS, several mitogen-activated protein kinases (MAPKs) have been shown to be activated by cold and ABA. ABA application induces MAPK activation within only a few minutes (Knetsch et al. 1996). The MAPK cascade is an important signalling pathway that enables the transmission of environmental and hormone signals to activate regulatory components within the cytoplasm and initiate cellular-responsive processes.

In vivo assays data revealed that the MKK2–MPK4/MPK6 complex functions downstream of MEKK1 during salt and cold stress. MPK4 and MPK6 are strongly phosphorylated by MKK2 in response to cold and salt treatment, and null allele of *mkk2* is impaired in ability of cold acclimation (Teige et al. 2004). Direct evidence of MAPK activation by ABA was observed in guard cells. MPK9 and MPK12 function downstream of ROS as positive regulators of ABA signalling, and *mpk9 mpk12* double mutants are partially but specifically impaired in cold-induced stomatal closure, suggesting that they play a role in the cold-mediated ABA signalling cascade (Jammes et al. 2009). *Arabidopsis* NDP kinase 2 (NDPK2) was shown to interact with two oxidative stress-related MAPKs, MPK3, and MPK6, and overexpression of *NDPK2* enhances cold tolerance (Moon et al. 2003). Taken together, these findings raise the possibility that the MAPK cascade acts downstream of ROS in cold and ABA signalling.

17.7 The Effects of Circadian Signals on Cold and ABA Signalling

The circadian clock is regulated by a central oscillator consisting of transcription/translation feedback loops that are entrained to light and temperature cues. The MYB transcription factors circadian clock associated1 (CCA1) and late elongated hypocotyl (LHY) can directly bind to the promoters of the evening-loop components *timing of CAB expression1* (*TOC1*), *pseudo response regulator* (*PRR7*), and *PRR9* to repress their expression after dawn; conversely, PRR7 and PRR9 act as negative regulators of *CCA1* and *LHY* expression in the morning (Harmer 2009). Interestingly, *CBF* gene expression is under circadian regulation, and the cold induction of *CBFs* is gated by the circadian clock (Fowler et al. 2005). It has been shown that CCA1 and LHY are positive regulators of *CBF* expression that directly bind to *CBF* promoters (Lee and Thomashow 2012; Dong et al. 2011). In the morning, *CBF* expression increases following induction of *CCA1* and *LHY*, and *CBF* transcript levels peak shortly after those of *CCA1* and *LHY*. It has also been suggested that CAMTA3 and ICE1 act synergistically with CCA1 and LHY to induce *CBF* expression during the day (Dong et al. 2011). During the evening, the evening element TOC1 interacts with phytochrome interacting factor 7 (PIF7) to repress expression of *CBF2* by directly binding to its promoter. Therefore, the circadian-gated regulation of the *CBF* genes appears to involve the action of CCA/LHY complex and PIF7/TOC1 evening complex (Dong et al. 2011). A recent study showed that TOC1 functions as a molecular switch that connects the circadian clock to the ABA-mediated drought stress response (Legnaioli et al. 2009). ABA signalling is closely related with circadian oscillator period, for example, ABA-mediated stomatal movement varies during day and night (Tallman 2004). It was proposed that the circadian clock may modulate stomatal closure under high-temperature conditions when water supplies are limited in order to optimise water-usage efficiency (Robertson et al. 2009). TOC1 has been found to negatively regulate the expression of ABA-related genes by directly interacting with ABI3 (Kurup et al. 2000). Moreover, TOC1 can also bind to the promoter of *ABAR*, a chloroplastic ABA binding protein, to control its circadian expression, and the gated-induction of *TOC1* by ABA is abolished in *ABAR* RNAi plants. Furthermore, overexpression of *TOC1* causes significant changes in plant responses to ABA and drought stress, which are indicative of a feedback mechanism linking the circadian clock to the ABA response (Legnaioli et al. 2009). Collectively, these findings illustrate the mechanisms by which the circadian clock regulates the cold and ABA responses.

17.8 Post-Transcriptional Modification Mediates Cold and ABA Signalling

RNA splicing and transport can lead to tissue-specific differences in gene expression as well as affect mRNA stability and turnover via nonsense-mediated decay (McGlincy and Smith 2008). By performing a genetic screen for deregulated expression of *RD29A* promoter-driven luciferase reporter gene, several genes encoding RNA processing proteins were identified. *Loss of osmotic responsiveness 4* (*LOS4*) encodes a putative DEAD-box RNA helicase that has been implicated in nucleo-cytoplasmic mRNA transport. The recessive mutations *los4-1* and *los4-2* affect cold-induced *CBF* transcription and lead to changes in chilling and freezing tolerance (Gong et al. 2002, 2005), suggesting that mRNA translocation may also be a control point in cold acclimation. It was also found that a *los4-2* mutant was hypersensitive to ABA (Gong et al. 2005). A recent study identified a *regulator of CBF2 gene expression1* (*rcf1-1*) mutant that was hypersensitive to cold stress. *RCF1* also encodes a cold-inducible DEAD-box RNA helicase. However, unlike LOS4, RCF1 maintains the proper splicing of *COR* pre-mRNAs under cold stress conditions (Guan et al. 2013). Another pre-mRNA splicing factor *Stabilized1* (*STA1*) which is required for pre-mRNA splicing and mRNA turnover is upregulated by cold stress, but not by ABA or NaCl (Lee et al. 2006). A *sta1* mutant displays a chilling-sensitive phenotype and hypersensitivity to ABA-induced root growth, as well as defects in splicing of the cold-induced *COR15A* gene (Lee et al. 2006). *Arabidopsis FIERY2* (*FRY2*), which encodes an RNA polymerase II C-terminal domain (CTD) phosphatase, may function in mRNA processing. FRY2 plays a negative role in regulating salt stress and the ABA response during seed germination, and it has been shown to be a repressor of the CBF transcription factors. However, a *fry2* mutant was shown to be hypersensitive to freezing stress (Xiong et al. 2002a). Sensitive to ABA and drought2 (SAD2) was found to be involved in nucleo-cytoplasmic trafficking during ABA treatment, and *sad2* mutant plants show a hypersensitive response to ABA and enhanced expression of stress genes in response to ABA, salt, and low temperatures (Verslues et al. 2006). Therefore, post-transcriptional regulation also serves as a major mechanism in the global control of the ABA and cold responses.

17.9 Epigenetic Modifications in Cold and ABA Signalling

DNA and histones are epigenetically modified through acetylation and methylation following exposure to stress (Kim et al. 2010; Feng and Jacobsen 2011). *HOS15* encodes a protein similar to the human WD-40 repeat protein TBL1 (transducin-like protein-1), which is involved in histone deacetylation. Both ABA and cold induce the expression of *HOS15*, and *hos15-1* mutants exhibit hypersensitivity to ABA-induced inhibition of germination and freezing stress. Furthermore, HOS15 interacts with histone H4, and levels of acetylated histone H4 are higher in *hos15* mutants than in the wild type (Zhu et al. 2008). Therefore, HOS15 is involved in the regulation of ABA and the cold stress response through H4 deacetylation-dependent chromatin remodelling in *Arabidopsis*. Consistently, increased levels of acetylated histone H4 are consistently associated with increased expression of *RD29A* in a *hos15* mutant compared with the wild type under cold stress. On the other hand, histone deacetylation results in a non-permissive chromatin conformation that represses transcription. In *Arabidopsis*, histone deacetylase 6 (HDA6) has been reported to be involved in the ABA response and low-temperature-mediated flowering (Luo et al. 2012; Chen and Wu 2010). In *Arabidopsis*, *HDA6* encodes an RPD3-type histone deacetylase, and a loss-of-function mutation in*HDA6*, *axe1–5*, displays hypersensitivity to ABA and salt stress (Luo et al. 2012). Interestingly, HOS1 was discovered as a chromatin remodelling factor for cold-induced transcriptional silencing of the *FLC* gene through an interaction with HDA6, thereby antagonising the actions of FVE and resulting in the inhibition of flowering in these plants (He and Amasino 2005; Jung et al. 2013). These findings suggest that plant flowering or stress response processes might be epigenetically controlled through similar mechanisms.

17.10 Cross Talk Between Phytohormones and the Cold Response

Emerging evidence indicates that cold acclimation is influenced by changes in the homeostasis of various hormones and that hormone signalling is crucial for regulation of the CBF pathway. Genetic analyses of *Arabidopsis* mutants with compromised ABA biosynthesis or signalling have identified a complex interplay between ABA and other phytohormone-signalling pathways. Generally, when plants are challenged by stress conditions, the stress hormones ABA and jasmonate acid (JA) inhibit growth by modulating the actions of auxin, GA, and cytokinin (Achard et al. 2006; Wolters and Jurgens 2009; Peleg and Blumwald 2011). The growth inhibition decreases the capacity for energy utilisation, which in turn, results in cold acclimation processes.

17.10.1 GA and Cold Stress

As an important plant phytohormone, GA affects plant abiotic stress responses, including those to salt, oxidative, and cold stresses during germination and seedling development. Acting as GA-signalling repressors, DELLA proteins have been identified as the central modulators of growth under a variety of stress conditions (Achard et al. 2006). ABA promotes the accumulation of DELLA proteins, which induces growth repression in plant lateral roots under salt stress through an ABA-dependent pathway (Achard et al. 2006). Intriguingly, GA treatment rescues normal growth and abolishes the late-flowering phenotype of *CBF1-ox* plants (Achard et al. 2008). Further investigation revealed that cold stress causes increased expression of the biosynthetic enzyme genes *AtGA2ox3* and *AtGA2ox6*, which reduce endogenous levels of bioactive GA. Consistently, overexpression of *CBFs* represses plant growth via the accumulation of DELLA proteins (Achard et al. 2008). ABA is known to act as an antagonist of GA by inhibiting the expression of *GA20ox* and *GA3ox* to reduce GA levels during seed germination (Razem et al. 2006) (Fig. 17.2a). It has been demonstrated that the existence of a novel mechanism through which CBFs may regulate dormancy in parallel to their functions in CBF-mediated cold signalling, by regulation of *DOG1* expression, which is a positive regulator of GA catabolism and ABA biosynthesis that promotes seed maturation at cool temperatures (Chiang et al. 2011; Kendall et al. 2011). A recent study showed that ABI3 and ABI5 interact with DELLA proteins to inhibit seed germination by repressing the expression of high-temperature-induced genes (Lim et al. 2013). Thus, DELLA-dependent growth restraint is necessary for cold acclimation and temperature-mediated seed germination.

17.10.2 Cytokinin and Cold Stress

Cytokinin signalling belongs to the two-component signalling system, which involves plant adaptation to environmental stress (Argueso et al. 2009; Nishimura et al. 2004). The accumulated body of evidence indicates that cytokinin acts as an antagonist of ABA during environmental stress responses (Peleg and Blumwald 2011; Hwang et al. 2012). Based on studies involving mutants or transgenic plants with altered cytokinin biosynthesis, it has been postulated that cytokinin acts as a negative regulator to modulate abiotic stress signalling (Nishiyama et al. 2011). In *Arabidopsis*, the cytokinin receptor histidine kinases AHK2, AHK3, and CRE1 have been shown to play important roles in the regulation of plant abiotic

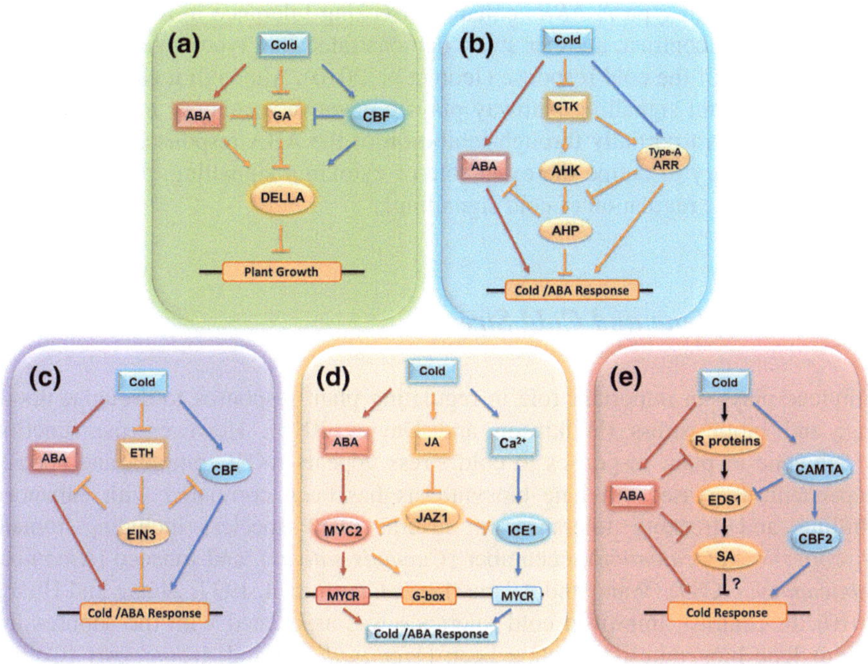

Fig. 17.2 Models of multiple interactions between ABA and other plant hormones in cold stress response. There are multiple points of interaction between ABA and other plant hormones including GA (**a**), cytokinin (**b**), ethylene (**c**), JA (**d**), and SA (**e**) in regulation of plant response to cold stress. *CTK* cytokinin, *ETH* ethylene, *DELLA* DELLA protein, *AHK Arabidopsis* histidine kinase, *AHP Arabidopsis* histidine phosphor-transferase, *ARR Arabidopsis* response regulator, *EIN3* Ethylene insensitive3, *JAZ* jasmonate-zim-domain protein, *MYC2* MYC-related transcriptional activator2

responses in both ABA-dependent and ABA-independent signalling pathways (Tran et al. 2007); on the other hand, the homologous gene *AHK1* was identified as an osmotic stress sensor that positively regulates abiotic stress (Wohlbach et al. 2008; Kumar et al. 2013; Tran et al. 2007). Increasing evidence suggests a negative role for cytokinin signalling in cold- and ABA-mediated abiotic stress. For instance, *ahk2 ahk3* double mutants are highly tolerant to cold, drought, and salt stress and show strong expression of ABA-responsive genes (Jeon et al. 2010; Tran et al. 2007). However, whether cytokinin receptors can perceive stress signals is still unclear. AHP2, AHP3, and AHP5 function as redundant negative regulators of the drought stress response in both ABA-dependent and ABA-independent manners (Nishiyama et al. 2013). These studies support the notion that AHK2–AHK4 and AHPs play negative roles under unfavourable environmental conditions and may integrate with ABA signalling to modulate stress responses (Fig. 17.2b). Type-A *ARR* genes are negative response regulators in cytokinin signalling pathway. They are induced by cytokinin as well as by cold stress (To et al. 2004).

Overexpression of type-A *ARRs* enhances freezing tolerance in *Arabidopsis* (Shi et al. 2012). In contrast, another study demonstrated that type-A ARRs are negative regulators of the cold response (Jeon et al. 2010). Although it can be assumed that the cytokinin signalling pathway plays a negative regulatory role in freezing tolerance, at least partially through inhibition of the ABA response, further study will be necessary to define the exact role of cytokinin signalling, particularly for the ARRs, in the regulation of cold signalling.

17.10.3 Ethylene and Cold Stress

Ethylene plays an important role in regulating plant responses to drought, flooding, and biotic stress (Wilkinson and Davies 2010). However, the function of ethylene in plant responses to cold stress appears to be complex and species dependent. Increased ethylene biosynthesis has been correlated with enhanced chilling and freezing tolerance in several plant species, including tomato (*Lycopersicones culentum*), cucumber (*Cucumis sativus*), and tobacco (*Nicotiana tabacum* cv. NC89) (Wang and Adams 1982; Ciardi et al. 1997; Zhang and Huang 2010). In contrast, improved cold tolerance was associated with the suppression of ethylene biosynthesis in mung bean (*Vignaradiata*) and *Arabidopsis* (Collins et al. 1995; Shi et al. 2012). In *Arabidopsis*, application of the ethylene precursor ACC decreases freezing tolerance, whereas application of the ethylene biosynthesis inhibitor AVG and the ethylene receptor antagonist $AgNO_3$ promote freezing tolerance. Consistent with these findings, several ethylene-insensitive mutants, including *etr1-1*, *ein4-1*, *ein2-5*, *ein3-1*, and *ein3 eil1*, exhibit enhanced freezing tolerance. Genetic and biochemical analyses revealed that ethylene negatively regulates cold signalling, at least partially, through direct transcriptional control of the COR *CBFs* and type-A *ARR* genes via EIN3 (Shi et al. 2012) (Fig. 17.2c).

It is known that ethylene can promote seed germination and repress seed dormancy by antagonising ABA (Wilkinson and Davies 2010). Previous studies indicated that *etr1*-1 and *ein2-5* mutants show enhanced seed dormancy, and their germination is hypersensitive to ABA (Chiwocha et al. 2005; Wang et al. 2007). As ABA accumulates in *etr1* and *ein2* mutants (Ghassemian et al. 2000; Chiwocha et al. 2005; Wang et al. 2007), ethylene may negatively regulate ABA biosynthesis during germination. Consistent with this notion, microarray data showed that a group of genes involved in ABA signalling are upregulated in an *ein3 eil1* double mutant (Shi et al. 2012). Interestingly, a recent study revealed the mechanism of how ABA inhibits root growth by increasing ethylene biosynthesis in Arabidopsis. ABA-activated calcium-dependent protein kinases, CPK4 and CPK11 can phosphorylate ethylene biosynthesis synthase ACS6 and enhance its stability, thus promoting ethylene production (Luo et al. 2014). Although the mechanisms underlying the complex cross talk between ethylene, ABA, and cold signalling remain unclear, it appears likely that the upregulation of ABA-responsive genes also contributes to the freezing tolerance of ethylene-insensitive mutants.

17.10.4 JA and Cold Stress

JA has recently been found to positively modulate cold tolerance. In *Arabidopsis*, exogenous application of JA significantly improves freezing tolerance, whereas blockage of JA biosynthesis and signalling decreases freezing tolerance. Further study revealed that JAZ1 and JAZ4 proteins physically interact with the ICE1-proteins to repress their transcriptional activity, thereby repressing expression of *CBF*s and their regulons (Hu et al. 2013) (Fig. 17.2d). Therefore, it appears that JA modulates cold stress primarily through an ABA-independent pathway. However, an antagonistic interaction between ABA and JA signalling has also been reported. In particular, the transcription factor MYC2 was shown to act as a node for the integration of ABA and JA signalling. MYC2 can directly interfere with defence signalling through JAZ-mediated repression, whereas MYC2 acts as an important positive regulator of the ABA-dependent signalling pathway (Abe et al. 2003; Fernandez-Calvo et al. 2011) (Fig. 17.2d). Furthermore, recent studies showed that an ABA-inducible bHLH transcription factor, JAM1, acts as a repressor that negatively regulates JA signalling in *Arabidopsis* (Nakata et al. 2013).

17.10.5 SA and Cold Stress

Salicylic acid (SA) is important for the activation of plant defence responses (Raskin 1992). Multiple studies have demonstrated a link between cold signalling and the SA-mediated defence response. Cold stress induces the accumulation of SA in *Arabidopsis*, whereas SA-deficient mutants show increased growth in response to cold (Scott et al. 2004). Gain of function of R/R-like protein mutants that overproduce SA, such as*chs1*, *chs2*, and *chs3*, display chilling-sensitive phenotypes that are partially dependent upon SA (Huang et al. 2010a; Yang et al. 2010; Wang et al. 2013). Indeed, there appears to be an optimal threshold of endogenous SA that is crucial for the chilling-dependent inhibition of plant growth. Nevertheless, the exact role of SA in freezing tolerance remains unclear. For example, several SA-overproducing mutants, including *cpr1*, *cpr5*, and *slh1*, show enhanced freezing tolerance (Yang et al. 2010), whereas overexpression of DEAR1 (DREB and EAR motif protein) induces SA accumulation and freezing sensitivity in *Arabidopsis* (Tsutsui et al. 2009). The master regulator of cold signalling, ICE1, was shown to directly regulate the expression of *BON1-associated protein* (*BAP1*), which encodes a C2-domain protein that negatively mediates the SA-dependent defence responses (Zhu et al. 2011; Yang et al. 2006). CAMTA3/AtSR1 recognises the *CBF2* promoter to positively regulate *CBF2* expression during cold stress, but it also interacts with the *EDS1* promoter to repress SA-dependent disease resistance (Kim et al. 2013; Doherty et al. 2009; Du et al. 2009).

ABA has been shown to play a broad role in the regulation of plant defence responses. For example, ABA acts a positive regulator of the defence response during pathogen invasion, whereas it acts as a negative regulator of the defence response following pathogen invasion (Yasuda et al. 2008; Ton et al. 2009; Robert-Seilaniantz et al. 2011). A recent study revealed that ABA deficiencies antagonise the high-temperature inhibition of disease resistance by enhancing the nuclear accumulation of the resistance proteins SNC1 and RPS4 in *Arabidopsis* (Mang et al. 2012). Therefore, ABA and SA may act antagonistically to affect plant development and the defence response, as well as the cold stress response (Fig. 17.2e).

17.11 Conclusions

Cold acclimation is a complex physiological process that is affected by various developmental and environmental factors, including growth stage, tissue type, photoperiod, hydration status, and low temperatures. ABA has long been thought to play an important role in mediating the cold response in plants, and accumulation of ABA may indeed be required for full cold acclimation. However, the extent of its role in vivo has been controversial. It appears that ABA biosynthesis is not a major aspect of the early cold stress response but rather that it contributes to the maximum induction of cold-responsive genes during later stages. It is also possible that ABA plays diverse roles in the regulation of multiple physiological processes that are affected by chilling or freezing stress. However, the majority of the molecular mechanisms underlying how ABA affects freezing tolerance remain to be investigated. In particular, elucidating whether the key components of ABA signalling are involved in CBF-regulated cold acclimation should be a priority for those studying the role of ABA in cold signalling. In addition, understanding the signalling networks between ABA and the other relevant plant hormones during the cold stress response will be crucial for determining any potential applications for ABA in enhancing plant tolerance to cold stress.

References

Abbasi F, Onodera H, Toki S, Tanaka H, Komatsu S. OsCDPK13, a calcium-dependent protein kinase gene from rice, is induced by cold and gibberellin in rice leaf sheath. Plant Mol Biol. 2004;55(4):541–52.

Abdeen A, Schnell J, Miki B. Transcriptome analysis reveals absence of unintended effects in drought-tolerant transgenic plants overexpressing the transcription factor ABF3. BMC Genom. 2010;11:69.

Abe H, Urao T, Ito T, Seki M, Shinozaki K, Yamaguchi-Shinozaki K. *Arabidopsis* AtMYC2 (bHLH) and AtMYB2 (MYB) function as transcriptional activators in abscisic acid signaling. Plant Cell. 2003;15(1):63–78.

Achard P, Cheng H, De Grauwe L, Decat J, Schoutteten H, Moritz T, Van Der Straeten D, Peng J, Harberd NP. Integration of plant responses to environmentally activated phytohormonal signals. Science. 2006;311(5757):91–4.

Achard P, Gong F, Cheminant S, Alioua M, Hedden P, Genschik P. The cold-inducible CBF1 factor-dependent signaling pathway modulates the accumulation of the growth-repressing DELLA proteins via its effect on gibberellin metabolism. Plant Cell. 2008;20(8):2117–29.

Agarwal M, Hao Y, Kapoor A, Dong CH, Fujii H, Zheng X, Zhu JK. A R2R3 type MYB transcription factor is involved in the cold regulation of CBF genes and in acquired freezing tolerance. J Biol Chem. 2006;281(49):37636–45.

Albrecht V, Weinl S, Blazevic D, D'Angelo C, Batistic O, Kolukisaoglu U, Bock R, Schulz B, Harter K, Kudla J. The calcium sensor CBL1 integrates plant responses to abiotic stresses. Plant J. 2003;36(4):457–70.

Allen GJ, Sanders D. Calcineurin, a type 2B protein phosphatase, modulates the Ca^{2+}-permeable slow vacuolar ion channel of stomatal guard cells. Plant Cell. 1995;7(9):1473–83.

Argueso CT, Ferreira FJ, Kieber JJ. Environmental perception avenues: the interaction of cytokinin and environmental response pathways. Plant Cell Environ. 2009;32(9):1147–60.

Banerjee J, Magnani R, Nair M, Dirk LM, DeBolt S, Maiti IB, Houtz RL. Calmodulin-mediated signal transduction pathways in *Arabidopsis* are fine-tuned by methylation. Plant Cell. 2013;25(11):4493–511.

Baron KN, Schroeder DF, Stasolla C. Transcriptional response of abscisic acid (ABA) metabolism and transport to cold and heat stress applied at the reproductive stage of development in *Arabidopsis thaliana*. Plant Sci. 2012;188:48–59.

Van Breusegem F, Slooten L, Stassart JM, Moens T, Botterman J, Van Montagu M, Inze D. Overproduction of *Arabidopsis thaliana* FeSOD confers oxidative stress tolerance to transgenic maize. Plant Cell Physiol. 1999;40(5):515–23.

Burdon RH, O'Kane D, Fadzillah N, Gill V, Boyd PA, Finch RR. Oxidative stress and responses in *Arabidopsis thaliana* and *Oryza sativa* subjected to chilling and salinity stress. Biochem Soc Trans. 1996;24(2):469–72.

Catala R, Santos E, Alonso JM, Ecker JR, Martinez-Zapater JM, Salinas J. Mutations in the Ca^{2+}/H^{+} transporter CAX1 increase CBF/DREB1 expression and the cold-acclimation response in *Arabidopsis*. Plant Cell. 2003;15(12):2940–51.

Chen M, Thelen JJ. ACYL-LIPID DESATURASE2 is required for chilling and freezing tolerance in *Arabidopsis*. Plant Cell. 2013;25(4):1430–44.

Chen LT, Wu K. Role of histone deacetylases HDA6 and HDA19 in ABA and abiotic stress response. Plant Signal Behav. 2010;5(10):1318–20.

Cheong YH, Kim KN, Pandey GK, Gupta R, Grant JJ, Luan S. CBL1, a calcium sensor that differentially regulates salt, drought, and cold responses in *Arabidopsis*. Plant Cell. 2003;15(8):1833–45.

Chiang GC, Bartsch M, Barua D, Nakabayashi K, Debieu M, Kronholm I, Koornneef M, Soppe WJ, Donohue K, De Meaux J. *DOG1* expression is predicted by the seed-maturation environment and contributes to geographical variation in germination in *Arabidopsis thaliana*. Mol Ecol. 2011;20(16):3336–49.

Chinnusamy V, Ohta M, Kanrar S, Lee BH, Hong X, Agarwal M, Zhu JK. ICE1: a regulator of cold-induced transcriptome and freezing tolerance in *Arabidopsis*. Genes Dev. 2003;17(8):1043–54.

Chiwocha SD, Cutler AJ, Abrams SR, Ambrose SJ, Yang J, Ross AR, Kermode AR. The *etr1-2* mutation in *Arabidopsis thaliana* affects the abscisic acid, auxin, cytokinin and gibberellin metabolic pathways during maintenance of seed dormancy, moist-chilling and germination. Plant J. 2005;42(1):35–48.

Choi H, Hong J, Ha J, Kang J, Kim SY. ABFs, a family of ABA-responsive element binding factors. J. Biol. Chem. 2000;275:1723–30.

Chrispeels MJ, Varner JE. Hormonal control of enzyme synthesis: on the mode of action of gibberellic acid and abscisin in aleurone layers of barley. Plant Physiol. 1967;42(7):1008–16.

Chu SH, Noh HN, Kim S, Kim KH, Hong SW, Lee H. Enhanced drought tolerance in *Arabidopsis* via genetic manipulation aimed at the reduction of glucosamine-induced ROS generation. Plant Mol Biol. 2010;74(4–5):493–502.

Ciardi JA, Deikman J, Orzolek MD. Increased ethylene synthesis enhances chilling tolerance in tomato. Physiol Plant. 1997;101(2):333–40.

Collins GG, Nie X, Saltveit ME. Heat shock increases chilling tolerance of mung bean hypocotyls tissues. Physiol Plant. 1995;89(1):117–24.

Dai X, Xu Y, Ma Q, Xu W, Wang T, Xue Y, Chong K. Overexpression of an R1R2R3 MYB gene, OsMYB3R-2, increases tolerance to freezing, drought, and salt stress in transgenic *Arabidopsis*. Plant Physiol. 2007;143(4):1739–51.

Daie J, Campbell WF. Response of tomato plants to stressful temperatures: increase IN abscisic acid concentrations. Plant Physiol. 1981;67(1):26–9.

Davletova S, Schlauch K, Coutu J, Mittler R. The zinc-finger protein Zat12 plays a central role in reactive oxygen and abiotic stress signaling in Arabidopsis. Plant Physiol. 2005;139:847–56.

Doherty CJ, Van Buskirk HA, Myers SJ, Thomashow MF. Roles for *Arabidopsis* CAMTA transcription factors in cold-regulated gene expression and freezing tolerance. Plant Cell. 2009;21(3):972–84.

Dong CH, Agarwal M, Zhang Y, Xie Q, Zhu JK. The negative regulator of plant cold responses, HOS1, is a RING E3 ligase that mediates the ubiquitination and degradation of ICE1. Proc Natl Acad Sci USA. 2006;103(21):8281–6.

Dong MA, Farre EM, Thomashow MF. Circadian clock-associated 1 and late elongated hypocotyl regulate expression of the C-repeat binding factor (CBF) pathway in *Arabidopsis*. Proc Natl Acad Sci USA. 2011;108(17):7241–6.

Du L, Ali GS, Simons KA, Hou J, Yang T, Reddy AS, Poovaiah BW. Ca(2+)/calmodulin regulates salicylic-acid-mediated plant immunity. Nature. 2009;457(7233):1154–8.

D'Angelo C, Weinl S, Batistic O, Pandey GK, Cheong YH, Schultke S, Albrecht V, Ehlert B, Schulz B, Harter K, Luan S, Bock R, Kudla J. Alternative complex formation of the Ca-regulated protein kinase CIPK1 controls abscisic acid-dependent and independent stress responses in *Arabidopsis*. Plant J. 2006;48(6):857–72.

Fayyaz MM, McCown BH, Beck GE. Effect of temperature, photoperiod and several growth substances on the cold hardiness of *Chrysanthemum morifolium* rhizome. Physiol Plant. 1978;44:73–6.

Feng S, Jacobsen SE. Epigenetic modifications in plants: an evolutionary perspective. Curr Opin Plant Biol. 2011;14(2):179–86.

Fernandez-Calvo P, Chini A, Fernandez-Barbero G, Chico JM, Gimenez-Ibanez S, Geerinck J, Eeckhout D, Schweizer F, Godoy M, Franco-Zorrilla JM, Pauwels L, Witters E, Puga MI, Paz-Ares J, Goossens A, Reymond P, De Jaeger G, Solano R. The *Arabidopsis* bHLH transcription factors MYC3 and MYC4 are targets of JAZ repressors and act additively with MYC2 in the activation of jasmonate responses. Plant Cell. 2011;23(2):701–15.

Finka A, Cuendet AF, Maathuis FJ, Saidi Y, Goloubinoff P. Plasma membrane cyclic nucleotide gated calcium channels control land plant thermal sensing and acquired thermotolerance. Plant Cell. 2012;24(8):3333–48.

Finkelstein RR, Lynch TJ. The *Arabidopsis* abscisic acid response gene *ABI5* encodes a basic leucine zipper transcription factor. Plant Cell. 2000;12(4):599–609.

Fourrier N, Bedard J, Lopez-Juez E, Barbrook A, Bowyer J, Jarvis P, Warren G, Thorlby G. A role for sensitive to freezing2 in protecting chloroplasts against freeze-induced damage in *Arabidopsis*. Plant J. 2008;55(5):734–45.

Fowler SG, Cook D, Thomashow MF. Low temperature induction of *Arabidopsis* CBF1, 2, and 3 is gated by the circadian clock. Plant Physiol. 2005;137(3):961–8.

Fujita M, Fujita Y, Noutoshi Y, Takahashi F, Narusaka Y, Yamaguchi-Shinozaki K, Shinozaki K. Crosstalk between abiotic and biotic stress responses: a current view from the points of convergence in the stress signaling networks. Curr Opin Plant Biol. 2006;9(4):436–42.

Fujita Y, Fujita M, Satoh R, Maruyama K, Parvez MM, Seki M, Hiratsu K, Ohme-Takagi M, Shinozaki K, Yamaguchi-Shinozaki K. AREB1 is a transcription activator of novel ABRE-dependent ABA signaling that enhances drought stress tolerance in *Arabidopsis*. Plant Cell. 2005;17(12):3470–88.

Furihata T, Maruyama K, Fujita Y, Umezawa T, Yoshida R, Shinozaki K, et al. Abscisic acid-dependent multisite phosphorylation regulates the activity of a transcription activator AREB1. Proc. Natl. Acad. Sci. U. S. A. 2006;103:1988–93.

Ghassemian M, Nambara E, Cutler S, Kawaide H, Kamiya Y, McCourt P. Regulation of abscisic acid signaling by the ethylene response pathway in *Arabidopsis*. Plant Cell. 2000;12(7):1117–26.

Gong Z, Dong CH, Lee H, Zhu J, Xiong L, Gong D, Stevenson B, Zhu JK. A DEAD Box RNA helicase is essential for mRNA export and important for development and stress responses in *Arabidopsis*. Plant Cell. 2005;17(1):256–67.

Gong Z, Lee H, Xiong L, Jagendorf A, Stevenson B, Zhu JK. RNA helicase-like protein as an early regulator of transcription factors for plant chilling and freezing tolerance. Proc Natl Acad Sci USA. 2002;99(17):11507–12.

Guan Q, Wu J, Zhang Y, Jiang C, Liu R, Chai C, Zhu J. A DEAD Box RNA helicase is critical for Pre-mRNA splicing, cold-responsive gene regulation, and cold tolerance in *Arabidopsis*. Plant Cell. 2013;25(1):342–56.

Guan LM, Zhao J, Scandalios JG. *Cis*-elements and trans-factors that regulate expression of the maize Cat1 antioxidant gene in response to ABA and osmotic stress: H_2O_2 is the likely intermediary signaling molecule for the response. Plant J. 2000;22(2):87–95.

Guo L, Yang H, Zhang X, Yang S. Lipid transfer protein 3 as a target of MYB96 mediates freezing and drought stress in *Arabidopsis*. J Exp Bot. 2013;64(6):1755–67.

Gusta LV, Fowler DB, Tyler NJ. The effect of abscisic acid and cytokinins on the cold hardiness of winter wheat. Can J Bot. 1982;60:301–5.

Haake V, Cook D, Riechmann J, Pineda O, Thomashow MF, Zhang JZ. Transcription factor CBF4 is a regulator of drought adaptation in *Arabidopsis*. Plant Physiol. 2002;130(2):639–48.

Harmer SL. The circadian system in higher plants. Annu Rev Plant Biol. 2009;60:357–77.

He Y, Amasino RM. Role of chromatin modification in flowering-time control. Trends Plant Sci. 2005;10(1):30–5.

He J, Duan Y, Hua D, Fan G, Wang L, Liu Y, Chen Z, Han L, Qu LJ, Gong Z. DEXH box RNA helicase-mediated mitochondrial reactive oxygen species production in *Arabidopsis* mediates crosstalk between abscisic acid and auxin signaling. Plant Cell. 2012;24(5):1815–33.

Hirayama T, Shinozaki K. Research on plant abiotic stress responses in the post-genome era: past, present and future. Plant J. 2010;61(6):1041–52.

Holubowicz T, Cummins JN, Forsline PL. Responses of Malus clones to programmed low-temperature stresses in late winter. J Am Soc Hortic Sci. 1982;107:492–6.

Hu Y, Jiang L, Wang F, Yu D. Jasmonate regulates the inducer of *CBF* expression-C-repeat binding factor/DRE binding factor1 cascade and freezing tolerance in *Arabidopsis*. Plant Cell. 2013;25(8):2907–24.

Huang X, Li J, Bao F, Zhang X, Yang S. A gain-of-function mutation in the *Arabidopsis* disease resistance gene *RPP4* confers sensitivity to low temperature. Plant Physiol. 2010a;154:796–809.

Huang X, Li Y, Zhang X, Zuo J, Yang S. The *Arabidopsis LSD1* gene plays an important role in the regulation of low temperature-dependent cell death. New Phytol. 2010b;187(2):301–12.

Hugly S, Somerville C. A role for membrane lipid polyunsaturation in chloroplast biogenesis at low temperature. Plant Physiol. 1992;99(1):197–202.

Hwang I, Sheen J, Muller B. Cytokinin signaling networks. Annu Rev Plant Biol. 2012;63:353–80.

Jaglo-Ottosen KR, Gilmour SJ, Zarka DG, Schabenberger O, Thomashow MF. *Arabidopsis* CBF1 overexpression induces COR genes and enhances freezing tolerance. Science. 1998;280(5360):104–6.

Jammes F, Song C, Shin D, Munemasa S, Takeda K, Gu D, Cho D, Lee S, Giordo R, Sritubtim S. MAP kinases MPK9 and MPK12 are preferentially expressed in guard cells and positively regulate ROS-mediated ABA signaling. Proc Natl Acad Sci USA. 2009;106(48):20520–5.

Jeon J, Kim NY, Kim S, Kang NY, Novak O, Ku SJ, Cho C, Lee DJ, Lee EJ, Strnad M, Kim J. A subset of cytokinin two-component signaling system plays a role in cold temperature stress response in *Arabidopsis*. J Biol Chem. 2010;285(30):23371–86.

Jung JH, Park JH, Lee S, To TK, Kim JM, Seki M, Park CM. The cold signaling attenuator HIGH EXPRESSION OF OSMOTICALLY RESPONSIVE GENE1 activates FLOWERING LOCUS C transcription via chromatin remodeling under short-term cold stress in *Arabidopsis*. Plant Cell. 2013.

Kadlecová Z, Faltus M, Prášil I. Relationship between abscisic acid content, dry weight and freezing tolerance in barley cv. Lunet. J Plant Physiol. 2000;157(3):291–7.

Kang JY, Choi HI, Im MY, Kim SY. *Arabidopsis* basic leucine zipper proteins that mediate stress-responsive abscisic acid signaling. Plant Cell. 2002;14(2):343–57.

Kendall SL, Hellwege A, Marriot P, Whalley C, Graham IA, Penfield S. Induction of dormancy in *Arabidopsis* summer annuals requires parallel regulation of DOG1 and hormone metabolism by low temperature and CBF transcription factors. Plant Cell. 2011;23(7):2568–80.

Kerdnaimongkol K, Woodson WR. Inhibition of catalase by antisense RNA increases susceptibility to oxidative stress and chilling Injury in transgenic tomato plants. J Am Soc Hortic Sci. 1999;124(4):330–6.

Kim KN, Cheong YH, Grant JJ, Pandey GK, Luan S. CIPK3, a calcium sensor-associated protein kinase that regulates abscisic acid and cold signal transduction in *Arabidopsis*. Plant Cell. 2003;15(2):411–23.

Kim Y, Park S, Gilmour SJ, Thomashow MF. Roles of CAMTA transcription factors and salicylic acid in configuring the low-temperature transcriptome and freezing tolerance of *Arabidopsis*. Plant J. 2013;75(3):364–76.

Kim SY, Zhu T, Sung ZR. Epigenetic regulation of gene programs by EMF1 and EMF2 in *Arabidopsis*. Plant Physiol. 2010;152(2):516–28.

Kishor PK, Sangam S, Amrutha R, Laxmi PS, Naidu K, Rao K, Rao S, Reddy K, Theriappan P, Sreenivasulu N. Regulation of proline biosynthesis, degradation, uptake and transport in higher plants: its implications in plant growth and abiotic stress tolerance. Curr Sci. 2005;88(3):424–38.

Knetsch M, Wang M, Snaar-Jagalska BE, Heimovaara-Dijkstra S. Abscisic acid induces mitogen-activated protein kinase activation in Barley aleurone protoplasts. Plant Cell. 1996;8(6):1061–7.

Knight H, Zarka DG, Okamoto H, Thomashow MF, Knight MR. Abscisic acid induces CBF gene transcription and subsequent induction of cold-regulated genes via the CRT promoter element. Plant Physiol. 2004;135(3):1710–7.

Kocsy G, Toth B, Berzy T, Szalai G, Jednakovits A, Galiba G. Glutathione reductase activity and chilling tolerance are induced by a hydroxylamine derivative BRX-156 in maize and soybean. Plant Sci. 2001;160(5):943–50.

Kreps JA, Wu Y, Chang H-S, Zhu T, Wang X, Harper JF. Transcriptome changes for *Arabidopsis* in response to salt, osmotic, and cold stress. Plant Physiol. 2002;130(4):2129–41.

Kumar MN, Jane WN, Verslues PE. Role of the putative osmosensor *Arabidopsis* histidine kinase1 in dehydration avoidance and low-water-potential response. Plant Physiol. 2013;161(2):942–53.

Kurup S, Jones HD, Holdsworth MJ. Interactions of the developmental regulator ABI3 with proteins identified from developing *Arabidopsis* seeds. Plant J. 2000;21(2):143–55.

Kwak JM, Mori IC, Pei ZM, Leonhardt N, Torres MA, Dangl JL, Bloom RE, Bodde S, Jones JD, Schroeder JI. NADPH oxidase *AtrbohD* and *AtrbohF* genes function in ROS-dependent ABA signaling in *Arabidopsis*. EMBO J. 2003;22(11):2623–33.

Lang V, Mantyla E, Welin B, Sundberg B, Palva ET. Alterations in water status, endogenous abscisic acid content, and expression of rab18 gene during the development of freezing tolerance in *Arabidopsis thaliana*. Plant Physiol. 1994;104(4):1341–9.

Lee BH, Henderson DA, Zhu J-K. The *Arabidopsis* cold-responsive transcriptome and its regulation by ICE1. Plant Cell. 2005;17(11):3155–75.

Lee SJ, Kang JY, Park HJ, Kim MD, Bae MS, Choi HI, Kim SY. DREB2C interacts with ABF2, a bZIP protein regulating abscisic acid-responsive gene expression, and its overexpression affects abscisic acid sensitivity. Plant Physiol. 2010;153(2):716–27.

Lee BH, Kapoor A, Zhu J, Zhu JK. STABILIZED1, a stress-upregulated nuclear protein, is required for pre-mRNA splicing, mRNA turnover, and stress tolerance in *Arabidopsis*. Plant Cell. 2006;18(7):1736–49.

Lee BH, Lee H, Xiong L, Zhu JK. A mitochondrial complex I defect impairs cold-regulated nuclear gene expression. Plant Cell. 2002;14(6):1235–51.

Lee CM, Thomashow MF. Photoperiodic regulation of the C-repeat binding factor (CBF) cold acclimation pathway and freezing tolerance in *Arabidopsis thaliana*. Proc Natl Acad Sci USA. 2012;109(37):15054–9.

Legnaioli T, Cuevas J, Mas P. TOC1 functions as a molecular switch connecting the circadian clock with plant responses to drought. EMBO J. 2009;28(23):3745–57.

Lim S, Park J, Lee N, Jeong J, Toh S, Watanabe A, Kim J, Kang H, Kim DH, Kawakami N, Choi G. ABA-INSENSITIVE3, ABA-INSENSITIVE5, and DELLAs interact to activate the expression of SOMNUS and other high-temperature-inducible genes in imbibed seeds in *Arabidopsis*. Plant Cell. 2013;25(12):4863–78.

Liu Y, He J, Chen Z, Ren X, Hong X, Gong Z. *ABA overly-sensitive 5* (*ABO5*), encoding a pentatricopeptide repeat protein required for cis-splicing of mitochondrial nad2 intron 3, is involved in the abscisic acid response in *Arabidopsis*. Plant J. 2010;63(5):749–65.

Liu Q, Kasuga M, Sakuma Y, Abe H, Miura S, Yamaguchi-Shinozaki K, Shinozaki K. Two transcription factors, DREB1 and DREB2, with an EREBP/AP2 DNA binding domain separate two cellular signal transduction pathways in drought- and low-temperature-responsive gene expression, respectively, in *Arabidopsis*. Plant Cell. 1998;10(8):1391–406.

Llorente F, Oliveros JC, Martinez-Zapater JM, Salinas J. A freezing-sensitive mutant of *Arabidopsis*, *frs1*, is a new *aba3* allele. Planta. 2000;211(5):648–55.

Luo X, Chen Z, Gao J, Gong Z. Abscisic acid inhibits root growth in Arabidopsis through ethylene biosynthesis. Plant J. 2014;79:44–55.

Luo M, Wang YY, Liu X, Yang S, Lu Q, Cui Y, Wu K. HD2C interacts with HDA6 and is involved in ABA and salt stress response in *Arabidopsis*. J Exp Bot. 2012;63(8):3297–306.

Ma Y, Szostkiewicz I, Korte A, Moes D, Yang Y, Christmann A, et al. Regulators of PP2C phosphatase activity function as abscisic acid sensors. Science. 2009;324:1064–8.

Magnan F, Ranty B, Charpenteau M, Sotta B, Galaud JP, Aldon D. Mutations in AtCML9, a calmodulin-like protein from *Arabidopsis thaliana*, alter plant responses to abiotic stress and abscisic acid. Plant J. 2008;56(4):575–89.

Mang HG, Qian W, Zhu Y, Qian J, Kang HG, Klessig DF, Hua J. Abscisic acid deficiency antagonizes high-temperature inhibition of disease resistance through enhancing nuclear accumulation of resistance proteins SNC1 and RPS4 in *Arabidopsis*. Plant Cell. 2012;24(3):1271–84.

Mantyla E, Lang V, Palva ET. Role of abscisic acid in drought-induced freezing tolerance, cold acclimation, and accumulation of LT178 and RAB18 proteins in *Arabidopsis thaliana*. Plant Physiol. 1995;107(1):141–8.

McGlincy NJ, Smith CWJ. Alternative splicing resulting in nonsense-mediated mRNA decay: what is the meaning of nonsense? Trends Biochem Sci. 2008;33(8):385–93.

Meinhard M, Grill E. Hydrogen peroxide is a regulator of ABI1, a protein phosphatase 2C from *Arabidopsis*. FEBS Lett. 2001;508(3):443–6.

Meinhard M, Rodriguez P, Grill E. The sensitivity of ABI2 to hydrogen peroxide links the abscisic acid-response regulator to redox signalling. Planta. 2002;214(5):775–82.

Miao Y, Lv D, Wang P, Wang XC, Chen J, Miao C, Song CP. An *Arabidopsis* glutathione peroxidase functions as both a redox transducer and a scavenger in abscisic acid and drought stress responses. Plant Cell. 2006;18(10):2749–66.

Miquel M, James D Jr, Dooner H, Browse J. *Arabidopsis* requires polyunsaturated lipids for low-temperature survival. Proc Natl Acad Sci USA. 1993;90(13):6208–12.

Miura K, Jin JB, Lee J, Yoo CY, Stirm V, Miura T, Ashworth EN, Bressan RA, Yun DJ, Hasegawa PM. SIZ1-mediated sumoylation of ICE1 controls CBF3/DREB1A expression and freezing tolerance in *Arabidopsis*. Plant Cell. 2007;19(4):1403–14.

Miura K, Lee J, Jin JB, Yoo CY, Miura T, Hasegawa PM. Sumoylation of ABI5 by the *Arabidopsis* SUMO E3 ligase SIZ1 negatively regulates abscisic acid signaling. Proc Natl Acad Sci USA. 2009;106(13):5418–23.

Miyazono KI, Miyakawa T, Sawano Y, Kubota K, Kang H-J, Asano A, Miyauchi Y, Takahashi M, Zhi Y, Fujita Y. Structural basis of abscisic acid signalling. Nature. 2009;462(7273):609–14.

Moellering ER, Muthan B, Benning C. Freezing tolerance in plants requires lipid remodeling at the outer chloroplast membrane. Science. 2010;330(6001):226–8.

Moon H, Lee B, Choi G, Shin D, Prasad DT, Lee O, Kwak SS, Kim DH, Nam J, Bahk J, Hong JC, Lee SY, Cho MJ, Lim CO, Yun DJ. NDP kinase 2 interacts with two oxidative stress-activated MAPKs to regulate cellular redox state and enhances multiple stress tolerance in transgenic plants. Proc Natl Acad Sci USA. 2003;100(1):358–63.

Munnik T, Irvine RF, Musgrave A. Phospholipid signalling in plants. Biochim Biophys Acta. 1998;1389(3):222–72.

Nakata M, Mitsuda N, Herde M, Koo AJ, Moreno JE, Suzuki K, Howe GA, Ohme-Takagi M. A bHLH-type transcription factor, ABA-INDUCIBLE BHLH-TYPE TRANSCRIPTION FACTOR/JA-ASSOCIATED MYC2-LIKE1, acts as a repressor to negatively regulate jasmonate signaling in *Arabidopsis*. Plant Cell. 2013;25(5):1641–56.

Nishimura C, Ohashi Y, Sato S, Kato T, Tabata S, Ueguchi C. Histidine kinase homologs that act as cytokinin receptors possess overlapping functions in the regulation of shoot and root growth in *Arabidopsis*. Plant Cell. 2004;16(6):1365–77.

Nishiyama R, Watanabe Y, Fujita Y, Le DT, Kojima M, Werner T, Vankova R, Yamaguchi-Shinozaki K, Shinozaki K, Kakimoto T, Sakakibara H, Schmulling T, Tran LS. Analysis of cytokinin mutants and regulation of cytokinin metabolic genes reveals important regulatory roles of cytokinins in drought, salt and abscisic acid responses, and abscisic acid biosynthesis. Plant Cell. 2011;23(6):2169–83.

Nishiyama R, Watanabe Y, Leyva-Gonzalez MA, Ha CV, Fujita Y, Tanaka M, Seki M, Yamaguchi-Shinozaki K, Shinozaki K, Herrera-Estrella L, Tran LS. *Arabidopsis* AHP2, AHP3, and AHP5 histidine phosphotransfer proteins function as redundant negative regulators of drought stress response. Proc Natl Acad Sci USA. 2013;110(12):4840–5.

Novillo F, Alonso JM, Ecker JR, Salinas J. CBF2/DREB1C is a negative regulator of CBF1/DREB1B and CBF3/DREB1A expression and plays a central role in stress tolerance in *Arabidopsis*. Proc Natl Acad Sci USA. 2004;101(11):3985–90.

Novillo F, Medina J, Salinas J. *Arabidopsis* CBF1 and CBF3 have a different function than CBF2 in cold acclimation and define different gene classes in the CBF regulon. Proc Natl Acad Sci USA. 2007;104(52):21002–7.

Olsen AN, Ernst HA, Leggio LL, Skriver K. NAC transcription factors: structurally distinct, functionally diverse. Trends Plant Sci. 2005;10(2):79–87.

O'Kane D, Gill V, Boyd P, Burdon R. Chilling, oxidative stress and antioxidant responses in *Arabidopsis thaliana* callus. Planta. 1996;198(3):371–7.

Park SY, Fung P, Nishimura N, Jensen DR, Fujii H, Zhao Y, et al. Abscisic acid inhibits type 2C protein phosphatases via the PYR/PYL family of START proteins. Science. 2009;324:1068–71.

Peleg Z, Blumwald E. Hormone balance and abiotic stress tolerance in crop plants. Curr Opin Plant Biol. 2011;14(3):290–5.

Raskin I. Role of salicylic acid in plants. Annu Rev Plant Biol. 1992;43(1):439–63.

Razem FA, Baron K, Hill RD. Turning on gibberellin and abscisic acid signaling. Curr Opin Plant Biol. 2006;9(5):454–9.

Robert-Seilaniantz A, Grant M, Jones JD. Hormone crosstalk in plant disease and defense: more than just jasmonate-salicylate antagonism. Annu Rev Phytopathol. 2011;49:317–43.

Robertson FC, Skeffington AW, Gardner MJ, Webb AA. Interactions between circadian and hormonal signalling in plants. Plant Mol Biol. 2009;69(4):419–27.

Scott IM, Clarke SM, Wood JE, Mur LA. Salicylate accumulation inhibits growth at chilling temperature in *Arabidopsis*. Plant Physiol. 2004;135(2):1040–9.

Seo PJ, Xiang F, Qiao M, Park JY, Lee YN, Kim SG, Lee YH, Park WJ, Park CM. The MYB96 transcription factor mediates abscisic acid signaling during drought stress response in *Arabidopsis*. Plant Physiol. 2009;151(1):275–89.

Shi Y, Tian S, Hou L, Huang X, Zhang X, Guo H, Yang S. Ethylene signaling negatively regulates freezing tolerance by repressing expression of *CBF* and type-A *ARR* genes in *Arabidopsis*. Plant Cell. 2012;24(6):2578–95.

Shinozaki K, Yamaguchi-Shinozaki K. Molecular responses to dehydration and low temperature: differences and cross-talk between two stress signaling pathways. Curr Opin Plant Biol. 2000;3(3):217–23.

Smoleńska-Sym G, Gawrońska H, Kacperska A. Modifications of abscisic acid level in winter oilseed rape leaves during acclimation of plants to freezing temperatures. Plant Growth Regul. 1995;17(1):61–5.

Song SY, Chen Y, Chen J, Dai XY, Zhang WH. Physiological mechanisms underlying OsNAC5-dependent tolerance of rice plants to abiotic stress. Planta. 2011;234(2):331–45.

Suzuki I, Los DA, Murata N. Perception and transduction of low-temperature signals to induce desaturation of fatty acids. Biochem Soc Trans. 2000;28(6):628–30.

Suzuki M, McCarty DR. Functional symmetry of the B3 network controlling seed development. Curr Opin Plant Biol. 2008;11(5):548–53.

Tahtiharju S, Palva T. Antisense inhibition of protein phosphatase 2C accelerates cold acclimation in *Arabidopsis thaliana*. Plant J. 2001;26(4):461–70.

Takasaki H, Maruyama K, Kidokoro S, Ito Y, Fujita Y, Shinozaki K, Yamaguchi-Shinozaki K, Nakashima K. The abiotic stress-responsive NAC-type transcription factor OsNAC5 regulates stress-inducible genes and stress tolerance in rice. Mol Genet Genomics. 2010;284(3):173–83.

Tallman G. Are diurnal patterns of stomatal movement the result of alternating metabolism of endogenous guard cell ABA and accumulation of ABA delivered to the apoplast around guard cells by transpiration? J. Exp. Bot. 2004;55:1963–76.

Tamminen I, Makela P, Heino P, Palva ET. Ectopic expression of *ABI3* gene enhances freezing tolerance in response to abscisic acid and low temperature in *Arabidopsis thaliana*. Plant J. 2001;25(1):1–8.

Teige M, Scheikl E, Eulgem T, Doczi R, Ichimura K, Shinozaki K, Dangl JL, Hirt H. The MKK2 pathway mediates cold and salt stress signaling in *Arabidopsis*. Mol Cell. 2004;15(1):141–52.

Theocharis A, Clement C, Barka EA. Physiological and molecular changes in plants grown at low temperatures. Planta. 2012;235:1091–105.

Thomas T, Wareing P, Robinson P. Chemistry and physiology of 'dormins' in sycamore: action of the sycamore 'dormin' as a gibberellin antagonist. Nature. 1965;205:1270–2.

Thomashow MF. Plant cold acclimation: freezing tolerance genes and regulatory mechanisms. Annu Rev Plant Biol. 1999;50(1):571–99.

Thomashow MF. Molecular basis of plant cold acclimation: insights gained from studying the CBF cold response pathway. Plant Physiol. 2010;154(2):571–7.

Tian S, Mao X, Zhang H, Chen S, Zhai C, Yang S, Jing R. Cloning and characterization of *TaSnRK2.3*, a novel SnRK2 gene in common wheat. J Exp Bot. 2013;64(7):2063–80.

To JP, Haberer G, Ferreira FJ, Deruere J, Mason MG, Schaller GE, Alonso JM, Ecker JR, Kieber JJ. Type-A *Arabidopsis* response regulators are partially redundant negative regulators of cytokinin signaling. Plant Cell. 2004;16(3):658–71.

Ton J, Flors V, Mauch-Mani B. The multifaceted role of ABA in disease resistance. Trends Plant Sci. 2009;14(6):310–7.

Tran LS, Urao T, Qin F, Maruyama K, Kakimoto T, Shinozaki K, Yamaguchi-Shinozaki K. Functional analysis of AHK1/ATHK1 and cytokinin receptor histidine kinases in response to abscisic acid, drought, and salt stress in *Arabidopsis*. Proc Natl Acad Sci USA. 2007;104(51):20623–8.

Tsutsui T, Kato W, Asada Y, Sako K, Sato T, Sonoda Y, Kidokoro S, Yamaguchi-Shinozaki K, Tamaoki M, Arakawa K, Ichikawa T, Nakazawa M, Seki M, Shinozaki K, Matsui M, Ikeda A, Yamaguchi J. DEAR1, a transcriptional repressor of DREB protein that mediates plant defense and freezing stress responses in *Arabidopsis*. J Plant Res. 2009;122(6):633–43.

Uemura M, Joseph RA, Steponkus PL. Cold acclimation of *Arabidopsis thaliana* (effect on plasma membrane lipid composition and freeze-induced lesions). Plant Physiol. 1995;109(1):15–30.

Uno Y, Furihata T, Abe H, Yoshida R, Shinozaki K, Yamaguchi-Shinozaki K. *Arabidopsis* basic leucine zipper transcription factors involved in an abscisic acid-dependent signal transduction pathway under drought and high-salinity conditions. Proc Natl Acad Sci USA. 2000;97(21):11632–7.

Verslues PE, Guo Y, Dong CH, Ma W, Zhu JK. Mutation of SAD2, an importin beta-domain protein in *Arabidopsis*, alters abscisic acid sensitivity. Plant J. 2006;47(5):776–87.

Viswanathan C, Zhu JK. Molecular genetic analysis of cold-regulated gene transcription. Philos Trans R Soc Lond B Biol Sci. 2002;357(1423):877–86.

Wang CY, Adams DO. Chilling-induced ethylene production in cucumbers (*Cucumis sativus* L.). Plant Physiol. 1982;69:424–7.

Wang Y, Liu C, Li K, Sun F, Hu H, Li X, Zhao Y, Han C, Zhang W, Duan Y, Liu M, Li X. *Arabidopsis EIN2* modulates stress response through abscisic acid response pathway. Plant Mol Biol. 2007;64(6):633–44.

Wang Y, Zhang Y, Wang Z, Zhang X, Yang S. A missense mutation in CHS1, a TIR-NB protein, induces chilling sensitivity in *Arabidopsis*. Plant J. 2013;75(4):553–65.

Webb MS, Steponkus PL. Freeze-induced membrane ultrastructural alterations in Rye (*Secale cereale*) leaves. Plant Physiol. 1993;101(3):955–63.

Wilkinson S, Davies WJ. Drought, ozone, ABA and ethylene: new insights from cell to plant to community. Plant, Cell Environ. 2010;33(4):510–25.

Wohlbach DJ, Quirino BF, Sussman MR. Analysis of the *Arabidopsis* histidine kinase ATHK1 reveals a connection between vegetative osmotic stress sensing and seed maturation. Plant Cell. 2008;20(4):1101–17.

Wolters H, Jurgens G. Survival of the flexible: hormonal growth control and adaptation in plant development. Nat Rev Genet. 2009;10(5):305–17.

Xiong L, Ishitani M, Lee H, Zhu JK. The *Arabidopsis LOS5/ABA3* locus encodes a molybdenum cofactor sulfurase and modulates cold stress- and osmotic stress-responsive gene expression. Plant Cell. 2001a;13(9):2063–83.

Xiong L, Lee B, Ishitani M, Lee H, Zhang C, Zhu JK. *FIERY1* encoding an inositol polyphosphate 1-phosphatase is a negative regulator of abscisic acid and stress signaling in *Arabidopsis*. Genes Dev. 2001b;15(15):1971–84.

Xiong L, Lee H, Ishitani M, Zhu JK. Regulation of osmotic stress-responsive gene expression by the *LOS6/ABA1* locus in *Arabidopsis*. J Biol Chem. 2002;277(10):8588–96.

Xiong L, Lee H, Ishitani M, Tanaka Y, Stevenson B, Koiwa H, Bressan RA, Hasegawa PM, Zhu JK. Repression of stress-responsive genes by FIERY2, a novel transcriptional regulator in *Arabidopsis*. Proc Natl Acad Sci USA. 2002a;99(16):10899–10904.

Yamada T, Kuroda K, Jitsuyama Y, Takezawa D, Arakawa K, Fujikawa S. Roles of the plasma membrane and the cell wall in the responses of plant cells to freezing. Planta. 2002;215(5):770–8.

Yamaguchi-Shinozaki K, Shinozaki K. Transcriptional regulatory networks in cellular responses and tolerance to dehydration and cold stresses. Annu. Rev. Plant Biol. 2006;57:781–803.

Yamazaki T, Kawamura Y, Minami A, Uemura M. Calcium-dependent freezing tolerance in *Arabidopsis* involves membrane resealing via synaptotagmin SYT1. Plant Cell. 2008;20(12):3389–404.

Yang MT, Chen SL, Lin CY, Chen YM. Chilling stress suppresses chloroplast development and nuclear gene expression in leaves of mung bean seedlings. Planta. 2005;221(3):374–85.

Yang H, Li Y, Hua J. The C2 domain protein BAP1 negatively regulates defense responses in *Arabidopsis*. Plant J. 2006;48(2):238–48.

Yang H, Shi Y, Liu J, Guo L, Zhang X, Yang S. A mutant CHS3 protein with TIR-NB-LRR-LIM domains modulates growth, cell death and freezing tolerance in a temperature-dependent manner in *Arabidopsis*. Plant J. 2010;63(2):283–96.

Yasuda M, Ishikawa A, Jikumaru Y, Seki M, Umezawa T, Asami T, Maruyama-Nakashita A, Kudo T, Shinozaki K, Yoshida S, Nakashita H. Antagonistic interaction between systemic acquired resistance and the abscisic acid-mediated abiotic stress response in *Arabidopsis*. Plant Cell. 2008;20(6):1678–92.

Zeevaart J, Creelman R. Metabolism and physiology of abscisic acid. Annu Rev Plant Physiol Plant Mol Biol. 1988;39(1):439–73.

Zhang Z, Huang R. Enhanced tolerance to freezing in tobacco and tomato overexpressing transcription factor *TERF2/LeERF2* is modulated by ethylene biosynthesis. Plant Mol Biol. 2010;73(3):241–9.

Zheng Y, Schumaker KS, Guo Y. Sumoylation of transcription factor MYB30 by the small ubiquitin-like modifier E3 ligase SIZ1 mediates abscisic acid response in *Arabidopsis thaliana*. Proc Natl Acad Sci USA. 2012;109(31):12822–7.

Zhu B, Choi D-W, Fenton R, Close T. Expression of the barley dehydrin multigene family and the development of freezing tolerance. Mol Gen Genet. 2000;264(1–2):145–53.

Zhu J, Dong CH, Zhu JK. Interplay between cold-responsive gene regulation, metabolism and RNA processing during plant cold acclimation. Curr. Opin. Plant Biol. 2007;10:290–5.

Zhu J, Jeong JC, Zhu Y, Sokolchik I, Miyazaki S, Zhu JK, et al. Involvement of Arabidopsis HOS15 in histone deacetylation and cold tolerance. Proc. Natl. Acad. Sci. U. S. A. 2008;105:4945–50.

Zhu Y, Yang H, Mang HG, Hua J. Induction of *BAP1* by a moderate decrease in temperature is mediated by ICE1 in *Arabidopsis*. Plant Physiol. 2011;155(1):580–8.

Chapter 18
ABA and the Floral Transition

Lucio Conti, Massimo Galbiati and Chiara Tonelli

Abstract Plants use endogenous and environmental cues to trigger flowering. While variations in day length and temperature play a major role in controlling the transition to flowering, little is known about water stress-derived signals. Drought conditions cause early flowering in various plant species. Since it is well recognized that drought conditions also stimulate abscisic acid (ABA) accumulation, ABA signalling might underpin the observed early flowering response. Experiments have shown that exogenous applications of ABA cause flowering time alterations, suggesting that ABA might be an endogenous component affecting the floral transition. Confirming a role for endogenous ABA in flowering, mutants impaired in ABA production display flowering alterations, although a consensus as to the precise mode of action of ABA in plants is lacking. ABA activates flowering in several plant species and in Arabidopsis it promotes activation of the key floral gene *FLOWERING LOCUS T*. However, how ABA signalling is integrated in the floral network remains poorly understood. The evidence reviewed here suggests that ABA activates a complex network of signalling components including transcription factors with contrasting effects in flowering. Lesions in these ABA-Photoperiod interaction signalling genes produce alterations in flowering but their regulation and site of action have been so far elusive. ABA is mainly known as a stress hormone and its role in flowering is only beginning to emerge. In this chapter we will review the evidence for and against ABA controlling the floral transition.

Keywords ABA · Floral transition · ABA-Photoperiod interaction

L. Conti (✉) · M. Galbiati · C. Tonelli
Department of Biosciences, University of Milan, Milan, Italy
e-mail: lucio.conti@unimi.it

M. Galbiati
Fondazione Filarete, Milan, Italy

D.-P. Zhang (ed.), *Abscisic Acid: Metabolism, Transport and Signaling*,
DOI 10.1007/978-94-017-9424-4_18

18.1 Introduction

Plasticity in the switch to flowering shows the ability of plants to modify an inbuilt developmental programme according to environmental cues. To maximise reproductive success, the environmentally-sensitive processes of fertilization, embryogenesis and seed development must occur in coincidence with the most favourable environmental conditions.

Variations in day length (photoperiod) or temperature play an important role in the switch to flowering. In species like Arabidopsis (i.e. long day plants) flowering is strongly accelerated under long day photoperiods (LDs). While photoperiod represents a major environmental cue affecting the floral transitions, less is known regarding the role of water-derived signals. This is interesting as water availability plays a major role in plant adaptation and distribution. Water scarcity triggers ABA accumulation in plants, which in turn causes stomatal closure and induces expression of stress-related genes (Shinozaki and Yamaguchi-Shinozaki 2007). ABA thus coordinates different networks of environmental and developmental pathways leading to stress avoidance. Water deprivation is also known to trigger flowering in different plant species, also referred to as drought escape-DE- (Xu et al. 2005; Lafitte et al. 2006; Sherrard and Maherali 2006; Ivey and Carr 2012; Franks et al. 2007; Franks 2011; Sharp et al. 2009; Srivastava et al. 2000). The onset of DE forces plants to flower early and produce seeds before the environment becomes too dry to survive (Verslues and Juenger 2011). Since it is well recognized that drought conditions trigger ABA accumulation in plants, ABA signalling might underpin the observed variations in DE responses.

The transition to flowering occurs at the shoot apical meristem (SAM), a dome-like structure comprising a group of undifferentiated cells responsible for the generation of all lateral primordial (leaves and buds) as well as the internodes (Sussex 1989; Huala and Sussex 1993). Plants are able to detect internal and external cues and integrate these signals in the SAM. At the floral transition, the Arabidopsis vegetative SAM shifts from producing leaves/shoots to flowers (Schultz and Haughn 1993). Four main pathways have been shown to contribute to the floral transition including the photoperiodic, the vernalization, the autonomous and the gibberellin pathways (Simpson et al. 1999; Bernier 1993; Martínez-Zapater et al. 1994). Furthermore, temperature and plant age play an important role in fine-tuning flowering according to environmental fluctuations or endogenous cues (Blázquez et al. 2003; Balasubramanian et al. 2006; Wang et al. 2009; Kumar and Wigge 2010; Seo et al. 2009). These signals are integrated and share common components with the four major pathways described above.

In this chapter, we will critically analyse the evidence for and against ABA controlling the floral transition. We will highlight flowering genes and pathways emphasising their relevance to ABA signalling. Most data on ABA action derive from experiments in the model organism *Arabidopsis thaliana* (Thale cress) although data from other plant species will also be included in our discussion.

18.2 Is ABA a Flowering Hormone?

Traditionally ABA has been mainly associated with the control of germination and different stress responses, principally those connected to drought stress (Fujita et al. 2011; Finkelstein et al. 2002). Experiments have shown that exogenous applications of ABA cause flowering time alterations, suggesting that ABA might be an endogenous component affecting the floral transition (Domagalska et al. 2010; Blazquez et al. 1998; Wang et al. 2013b). However, from a genetic standpoint, flowering alterations caused by ABA signalling or biosynthetic mutants are not dramatic, although this does not rule out an important role for ABA in flowering. Thus, how ABA participates in the floral transition has been elusive thus far.

Hormone signalling ensures developmental flexibility to plants (Santner et al. 2009). It is therefore not surprising that several hormonal pathways have been shown to modulate the floral transition. In the context of the floral transitions ABA is generally considered as an inhibitor of flowering, although in some experiments it appeared to be a promoter. For example, endogenous ABA is a positive regulator of the DE response via the upregulation of the key floral gene *FLOWERING LOCUS T(FT)* (Riboni et al. 2013). However, exogenous ABA applications inhibit flowering in Arabidopsis and other plants including darnel (*Lolium temulentum*), (Blazquez et al. 1998; Domagalska et al. 2010; King and Evans 1977). A genome-wide transcript analysis showed that *FT* is downregulated following ABA applications. However a similar downregulation occur also in *ABA insensitive 1-1* (*abi1-1*) mutants (Hoth et al. 2002). Whole plant ABA applications thus negatively regulate *FT* expression, independent of the well-established endogenous ABA signalling. Alternatively, the negative effect of ABA on *FT* is non specific and/or indirect. One hypothesis is that ABA applications in Arabidopsis might be hampered by the intrinsic spatial and temporal-dependent regulation of ABA signalling, thus making it difficult to distinguish which experimental design fully mimics endogenous ABA action. In this respect, root applications of ABA promote flowering, which is in contrast with other experiments using exogenous ABA (Koops et al. 2011).

Circadian fluctuation in ABA levels, gating of ABA signalling as well as tissue specific ABA signalling and transport are well documented in plants (Lee et al. 2006; Legnaioli et al. 2009; Covington et al. 2008; Endo et al. 2008). Perhaps whole plant applications of ABA subvert such fine tuned mechanism of endogenous ABA metabolism. This could be indeed be the case as in the short day plant *Pharabitis nil* exogenous ABA applications during the first part of the inductive short day are inhibitory whilst ABA promotes flowering when applications occurs at the end of the night (Wilmowicz et al. 2011). Quantifications of endogenous ABA levels indicate that the hormone reach a maximum at the end of the inductive night, coincidently with the temporal window when ABA applications are most effective. The pharmacological manipulation of endogenous ABA by treatments with ABA biosynthesis inhibitors at the end of the night results in a delay in flowering (Wilmowicz et al. 2008). These data suggest a gating mechanism whereby

ABA promotes flowering in *Pharabitis*, but only in coincidence with a competence window. Therefore, the location and timing of ABA application could play a crucial role in determining how ABA affects flowering.

18.2.1 Clues from ABA Deficient Mutants in Arabidopsis

Extensive genetic screens coupled with physiological and biochemical analyses have defined the biosynthetic steps required for ABA production (Nambara and Marion-Poll 2005; Seo and Koshiba 2002). Unfortunately the phenotypes of ABA biosynthetic mutants do not provide clues as to unambiguously assign a role for endogenous ABA in flowering. Originally the ABA insensitive mutants *abi1* were shown to flower earlier under SDs, although this was not apparent under LDs (Martínez-Zapater et al. 1994; Chandler et al. 2000). Similar conclusions were drawn for *aba1* alleles impaired in ABA production grown under LDs. However these experiments were carried out under in vitro conditions, not in soil (Barrero et al. 2005). *aba2/glucose insensitive 1* mutants (Wassilewskija background) exhibit a small early flowering phenotype under long and short days conditions suggesting an inhibitory role for ABA during the floral transition (Domagalska et al. 2010; Cheng et al. 2002). However, transgenic lines that overexpress *9-CIS-EPOXYCAROTENOID DIOXYGENASE 3* (NCED3, a key enzyme in the ABA biosynthesis) do not produce any consistent alterations in flowering and, if anything, are slightly earlier flowering than wild type (Domagalska et al. 2010). Thus, increased levels of endogenous ABA do not recapitulate the effect of exogenous ABA applications in producing the expected delay in flowering (Wang et al. 2013b). Further complicating the topic, flowering time alterations in ABA deficient mutants appear to be ecotype-dependent in Arabidopsis. Under LDs *low expression of osmotic stress-responsive genes 6* (*los6*) mutants (allelic to *aba1*, C24 background) are early flowering and a similar observation was made for *los5* (allelic to *aba3*, C24 background) (Xiong et al. 2001, 2002a). In contrast, in the Columbia background, *aba1/npq1*, *aba2-4* or *aba1* T-DNA insertion mutants are late flowering indicating that ABA is a positive regulator of flowering (Riboni et al. 2013). In summary, the action of endogenous ABA appears to be highly background specific although the cause for such specificity is currently unknown.

18.2.2 ABA Deficient Mutants in Other Plants

Studies from ABA-deficient mutants from tomato provide additional hints as to the role of ABA in regulating the floral transition. The *notabilis* (*not*) mutant harbours a null allele of the tomato 9-CIS EPOXYCAROTENOID DIOXYGENASE1 (*Le*NCED1), the enzyme that catalyses the first dedicated step of ABA biosynthesis (Burbidge et al. 1999). The *flacca* (*flc*) mutant contains a 6 base pairs deletion in a molybdenum-cofactor sulfurase, which reduces the oxidizing capacity of aldehyde oxidases involved in the conversion of abscisic aldehyde to ABA, the final step in ABA biosynthesis (Sagi et al. 1999). Despite their reduced ability to synthesize ABA, *not* and *flc* mutants still contain considerable levels of ABA, amounting to 47 and 21 % of wild type, respectively (Herde et al. 1999). Nevertheless, these two mutants show an extended vegetative phase compared to the wild type, indicating a positive role for ABA in the transition between vegetative and reproductive growth in tomato (Carvalho et al. 2011).

More conclusive evidence in support of a promoting role for ABA in the transition to flowering comes from the analysis of SD plants (El-Antably and Waering 1966). In soybean, a preferential short day crop, ABA-related transcripts accumulate in the SAM upon transferring to floral inductive conditions. Up-regulation of biosynthetic genes, including *NCED1*, translate into a significant increase in ABA levels in SAMs from SD-treated plants, prior to the induction of floral genes (Wong et al. 2009). This correlative evidence not only implies a floral-promoting role for ABA in soybean, but also suggests the striking overlap of the floral signalling pathway with that of abiotic stress responses.

18.3 ABA Interaction with the Photoperiod

Photoperiodic flowering is the result of complex interactions between the circadian clock (an endogenous timekeeping mechanism) and external cues, which ultimately results in the activation of a set of floral genes (Imaizumi 2010). Central to photoperiod-dependent flowering is the pattern of accumulation of the flowering protein CONSTANS (CO) (Putterill et al. 1995). *CO* expression is regulated transcriptionally by the circadian clock through the GIGANTEA (GI)-FLAVIN-BINDING, KELCH REPEAT, F-BOX (FKF1) complex (Imaizumi et al. 2005; Sawa et al. 2007; Fornara et al. 2009; Song et al. 2012). However, CO protein accumulates only under LDs when the CO mRNA coincides with the light phase at the end of the day (Valverde et al. 2004; Suarez-Lopez et al. 2001). Stabilization of CO depends on photoreceptors PHYTOCROME A, CRYPTOCHROME 1 and 2 (CRY1 and 2) which promote CO stability at the end of a long day, while PHYTOCHROME B (PhyB) destabilizes CO protein in the morning (Valverde et al. 2004). CO is the transcriptional regulator that promotes flowering by inducing expression of the florigens genes *FT* and *TWIN SISTER OF FT* (*TSF*) in the phloem companion cells of the leaves (An et al. 2004; Yamaguchi et al. 2005; Kardailsky et al. 1999; Kobayashi et al. 1999). While favourable photoperiod triggers *FT* transcription, it is the FT protein that moves through the phloem from the leaves to the SAM to initiate the floral transition (Corbesier et al. 2007; Mathieu et al. 2007; Jaeger and Wigge 2007; Notaguchi et al. 2008). In the shoot apex, FT interacts with a set of b-ZIP transcription factors (FLOWERING D, FD and FD PARALOG, FDP), whose expression is largely SAM-specific (Wigge et al. 2005; Abe et al. 2005; Jaeger et al. 2013). Here, the FT/FD heterodimer activates several MADS box-type transcription factors, namely *SUPPRESSOR OF OVEREXPRESSION OF CONSTANS 1* (*SOC1*), *APETALA1*, and *FRUITFUL*, responsible for triggering the floral transition in the SAM (Abe et al. 2005; Wigge et al. 2005).

Genetic and expression data suggest a role for ABA in the activation of the florigen genes (Riboni et al. 2013). First, drought conditions promote *FT* upregulation, resulting in a DE response (Riboni et al. 2013; Su et al. 2013). Second, *aba1* mutants display a reduced *FT* expression, especially under drought conditions. Consistent with the idea of ABA stimulating flowering in a photoperiodic-dependent manner, *aba1* mutants do not show any obvious flowering phenotype under SDs, when no photoperiodic-dependent activation of *FT* occurs. Also, no DE response occurs under SDs, despite a substantial increase in endogenous ABA, nor in *gi* mutants under LDs. Interestingly, ABA accumulates primarily in the vascular tissue in Arabidopsis (Endo et al. 2008; Koiwai et al. 2004; Cheng et al. 2002), overlapping with the site of *FT* expression. In addition to *FT*, *FT*-like genes are present in Arabidopsis including *TSF* and *MOTHER OF FT AND TFL1* (*MFT*) and they all appear to be positively regulated by ABA (Xi et al. 2010; Riboni et al. 2013). Taken together these data argue in favour for a positive role for endogenous ABA in flowering via potentiation of florigen-like genes in a photoperiodic manner.

It is unclear how ABA might affect photoperiodic signalling. Drought stress results in an increase in *FT* expression without affecting the physiological circadian oscillation of *FT* (Riboni et al. 2013; Su et al. 2013). Because the pattern of *FT* transcript accumulation is mainly dictated by variations in CO protein levels, ABA might directly affect CO protein levels and/or activity. Supporting an increase in *CO* transcription under drought conditions, *FLOWERING BHLH 1* (*FBH1*), a basic helix-loop-helix-type transcription factor and *CO* positive activator is phosphorylated in vivo following ABA signalling activation (Ito et al. 2012; Wang et al. 2013a). However the role of phosphorylation on FBH1 activity is unclear. An alternative ABA target is *EID1-like protein 3* (*EDL3*), a positive regulator of ABA signalling and an activator of *CO*. *EDL3* transcript is upregulated following ABA applications. EDL3 encodes an F-box type protein that acts upstream of *CO* and positively increase its mRNA levels, thus providing a link between ABA and photoperiodic flowering (Koops et al. 2011). These observations suggest ABA acting upstream of CO. However, *CO* transcripts upon drought conditions are slightly decreased despite *GI* and *FKF1* being upregulated (Han et al. 2013). A further possibility could be that ABA promotes GI activity by facilitating its direct action on *FT* promoter, independent of CO. Other light sensing proteins could be involved in this mechanism. Cryptochromes are emerging as important general components of ABA-dependent signalling since the overexpression of *CRY1* from wheat (*TaCRY1a*) in Arabidopsis results in an ABA hypersensitive phenotype (Xu et al. 2009). Similarly to *gi*, *cry2* mutants have defective DE response, despite accumulating increased ABA levels (Riboni et al. 2013; Boccalandro et al. 2012). Most importantly, CRY2 positively regulates *FT* expression through different mechanisms, including stabilization of GI (Saijo et al. 2008; Zuo et al. 2011; Liu et al. 2008). Therefore, it is intriguing to speculate that CRY2 may participate in the GI- and ABA-dependent activation of *FT*.

18.4 Possible Drought Independent Roles of ABA in Flowering

Besides drought, ABA could mediate several endogenous and environmental stimuli that affect flowering via regulation of *FT* levels. One example is warm ambient temperature, an important cue for flowering in several plant species. *FT* expression is upregulated following an increase in ambient temperature and this is partially independent of CO (Balasubramanian et al. 2006; Kumar et al. 2012). Warm temperature causes the upregulation of several ABA-related transcripts and ABA levels are elevated under warm temperature in germinating seeds (Balasubramanian et al. 2006; Toh et al. 2008). These correlative observations may point to a role for ABA in warm ambient temperature-mediated flowering response. ABA could also play an important role in carbon signalling. Expression analysis have shown a significant overlap between glucose- and ABA-regulated gene expression (Li et al. 2006). Also, genes involved in trehalose metabolism are upregulated upon ABA applications. This is interesting as trehalose metabolism plays a key role in flowering (Wahl et al. 2013). Trehalose-6-phosphate acts at two sites: it is required for the optimal induction of *FT* in the leaves and it affects the expression of flowering genes in the SAM independent of *FT*.

18.5 Flowering Time Phenotypes Associated with Lesions in ABA Signalling Components

Genetic screens identified several mutants altered in both ABA and flowering responses. Supporting a negative role for ABA in flowering, some ABA hypersensitive mutants display a late flowering phenotype. For instance mutants for the β subunit of farnesyl transferase *era1* (*enhanced response to ABA 1*) are late flowering under both LDs and SDs (Yalovsky et al. 2000). Similarly, *hyponastic leaves 1* mutants combine hypersensitivity to ABA and late flowering (Lu and Fedoroff 2000). Genetic screens for *hos* (*high expression of osmotically regulated genes*) mutants identified two CTD phosphatase-like (*CPL*-like) genes (dubbed *cpl1* and *cpl3*) (Koiwa et al. 2002). Both *cpl1* (also known as *fiery2*) and *cpl3* mutants are characterized by increased sensitivity to ABA (Koiwa et al. 2002; Xiong et al. 2002b). However, whilst *cpl1* mutants are late flowering, *cpl3* are early flowering. *At*CPL1 and *At*CPL3 phosphatases are believed to be general regulators of gene expression that modulate the phosphorylation of RNA Polymerase II.

Mutations in the gene encoding the large subunit of the nuclear mRNA cap-binding protein complex (CBC), *ABA hypersensitive 1* (*ABH1*), result in an early flowering phenotype, irrespective of the day length (Kuhn et al. 2007; Bezerra et al. 2004). The early flowering of *abh1* mutants can be largely attributed to a dramatic reduction of the strong flowering repressor *FLOWERING LOCUS C* (*FLC*) (Bezerra et al. 2004). The *FLC* messenger RNA is heavily processed and interacts with three proteins with RNA recognition motifs including FCA, FLK and FPA (Lim et al. 2004; Macknight et al. 1997; Schomburg et al. 2001). Thus CBC may modulate the sensitive step of interaction between FLC mRNA and the regulators of FLC. Despite highlighting the importance of RNA processing in both ABA signalling and *FLC* regulation, ABH1 action does not provide clues as to whether and how ABA might play a specific regulatory role in *FLC* transcript processing.

The above-mentioned mutants show clear defects related to ABA responses, yet their pleiotropic nature does not provide clues as to the precise role of ABA in flowering. The recent identification of PYRABACTIN RESISTANCE (PYR)/ REGULATORY COMPONENT OF ABA RECEPTOR (RCAR) proteins as ABA receptors (Ma et al. 2009; Park et al. 2009) define the early events in ABA signal transduction.

Three classes of protein represent key ABA signalling nodes: PYR/RCARs, PROTEIN PHOSPHATASE 2Cs (PP2Cs) and SUCROSE NON-FERMENTING1-SNF1-RELATED PROTEIN KINASE 2s (SnRK2s). In ABA signalling, PYR/RCARs act as ABA receptors, PP2Cs act as negative regulators of the pathway and SnRK2s act as positive regulators of downstream signalling (Cutler et al. 2010; Umezawa et al. 2010). ABA binds to the PYR/RCAR receptors and this complex inhibits the action of group A PP2C proteins allowing SnRK2s activation and phosphorylation of their targets. In the absence of ABA, PP2Cs dephosphorylate SnRK2s thus preventing their activity (Umezawa et al. 2009; Vlad et al. 2009; Soon et al. 2012). Interestingly, biochemical differences in the PYR/RCAR-PP2C interaction exist, depending on the PYR/RCAR isoform and this suggests that the mechanism of ABA perception and signalling is more complex than anticipated (Hubbard et al. 2010).

In support for a positive role of ABA in flowering, triple *pp2c* mutants are significantly earlier flowering compared to wild type under LDs (Riboni et al. 2013). There are increased levels of *FT* transcripts accumulation in *pp2c* triple mutants, suggesting that ABA acts upstream of *FT* transcription. SnRK2s activity is up-regulated following phosphatase down regulation, implying that SnRK2-type kinases could mediate ABA signalling upstream of *FT* upregulation (Fujii et al. 2009; Umezawa et al. 2009). In agreement with this model *SnRK2.6/OST1* overexpressing plants produce a small flowering acceleration under LDs (Zheng et al. 2010). However, the triple *snrk2* mutants are early flowering under LDs (Wang et al. 2013a, b), arguing in favour of a negative role of ABA in flowering. The similar early flowering phenotype of triple *snrk2* and triple *pp2c* mutants argues against a simple signalling cascade converging onto the activation of *FT*. Since SnRK2s targets are diverse and involved in several independent processes one could speculate that their deregulation might produce pleiotropic effects (Wang et al. 2013a; Umezawa et al. 2013). A more detailed description of the SnRK2s and PP2Cs regulative network would therefore help in discriminating ABA-direct from ABA-indirect effects. Biochemical and genetic studies indicate that several kinds of reversible protein phosphorylation events play an important role in the regulation of ABA signalling. Phosphoproteomic screens indicate that several flowering proteins are targeted by SnRK2s, although less is known about the regulatory role of phosphorylation on the activity of these proteins (Wang et al. 2013a). ABA induces a rapid activation of mitogen-activated protein kinases (MAPKs), and of calcium-dependent protein kinases (CDPKs), some of which are also targeted by SnRK2s for phosphorylation (Jammes et al. 2009; Lu et al. 2002; Mori et al. 2006; Zhu et al. 2007; Wang et al. 2013a). AGAMUS-like MADS-box transcription factors were shown to be MPK substrates (Popescu et al. 2009). Additionally, the flower repressor FLC (flowering locus C), the FLC paralog MAF5.2 (MADS AFFECTING FLOWERING 5variant 2), and REF6 (RELATIVE OF EARLY FLOWERING) are phosphorylated by MAPKs. Thus MPK signalling may play an important role in the control of flowering time as well as adding further complexity to the ABA-regulated network of phosphorylation.

The early flowering phenotype of triple *snrk2* mutants is much more pronounced under SDs conditions compared to LDs (Wang et al. 2013a). Also, despite exhibiting an early flowering phenotype under LDs, triple *pp2c* mutants are considerably late under SDs (Riboni et al. 2013). These independent findings are consistent with a negative role of ABA signalling in flowering under SDs. Because the photoperiodic pathway plays a marginal role in flowering under SDs, it is highly unlikely that the negative effect of ABA signalling in flowering under SDs is exerted through the downregulation of *FT*. Since several pathways promote flowering under SDs directly at the shoot apex, ABA might counteract other floral promoting signals directly in the SAM (Porri et al. 2012; Galvão et al. 2012; Wang et al. 2009). Nevertheless ABA biosynthetic mutants do not present altered flowering under SDs (Riboni et al. 2013). Thus, ABA signalling, rather than ABA per se, is likely to be responsible for the floral inhibition observed under SDs. Shedding some light on this paradox is the recent discovery that a subclass of PYR proteins can bind and inhibit PP2Cs in the absence of any ligand. This finding supports the notion that a basal level of ABA signalling is constitutively present in Arabidopsis, even when ABA is absent (Hao et al. 2011). Inhibition of flowering might be an important role fulfilled by this particular clade of PYR proteins.

18.6 Role of ABA Downstream Effectors in Flowering

ABA plays an important role in gene expression by affecting the activity of several transcription factors, which, in turn, mediate ABA responses. *ABA INSENSITIVE 3* and *5* genes encode B3-type and basic Leucine zipper (bZIP)-type transcription factors, respectively (Finkelstein et al. 2002). ABI3, and ABI5 play a prominent role during seed germination and early seedling development (when these genes are highly expressed). However, low levels of expression of these genes can be found in vegetative tissues (Rohde et al. 1999; Brocard et al. 2002). Knockout and gain-of-function alleles in these ABA transcription factors result in alterations in flowering time suggesting a role for ABA during vegetative development through modulation of these transcription factors. Besides affecting individual transcription factor activity, ABA is likely to play a more general role in control of gene expression by affecting chromatin remodelling complexes (Saez et al. 2008; Han et al. 2012).

The over expression of *ABI5* results in a late flowering phenotype, while *abi5* loss-of-function mutants are early flowering under LDs (Wang et al. 2013b). ABI5 acts as a transcriptional activator and binds to an ABA-responsive element (ABRE), which are common elements in the promoters found in many ABA-induced genes (Busk and Pagès 1998; Zhang et al. 2005). A number of ABRE binding factors (ABFs; also referred to as AREBs) were identified, including ABF1, ABF2/AREB1, ABF3, ABF4/AREB2 (Uno et al. 2000; Choi et al. 2000). The ABFs genes are upregulated following drought stress and their overexpression results in plants with enhanced drought tolerance and altered expression of ABA/stress-regulated genes (Fujita et al. 2005; Kang et al. 2002). Similarly to ABI5, the overexpression of ABFs also results in late flowering, pointing to a negative role for this clade of bZIPs in flowering (Fujita et al. 2005). However multiple ABFs knock-out mutants do not display evident flowering alterations (Yoshida et al. 2010). Posttranslational modifications play a key role in ABI5 and AREB/ABF activity since SnRK2s phosphorylate ABI5 and ABI5-like bZIP transcription factors in an ABA-dependent manner to promote gene expression (Furihata et al. 2006; Fujii et al. 2007, 2009; Lopez-Molina et al. 2001; Wang et al. 2013b).

ABI5 acts as a positive regulator of *FLC*, a MADS-box protein that suppresses flowering (Michaels 1999). *FLC* levels are negatively regulated by the autonomous and vernalization pathways that promote flowering indirectly by repressing *FLC* levels (Simpson 2004). ABA-dependent stimulation of SnRK2s promotes ABI5 activity and *FLC* upregulation. Deletions of different phosphorylation sites through targeted mutagenesis results in lack of ABI5 activity in vivo and no *FLC* activation even in the presence of exogenous ABA applications. Transcript analysis indicates that two important floral integrators (*FT* and *SOC1*) are upregulated in *abi5* mutants, which is in agreement with a reduced *FLC* activity. Likewise, *FT* and *SOC1* are decreased in *35S::ABI5* plants. However, neither *35S::ABI5* nor *abi5* mutant plants display changes in flowering under SDs.

Other important layers of regulation are likely to play a role in controlling bZIPs activity and, as a result, regulation of flowering by ABA. *NUCLEAR FACTOR-Y*-type transcription factors interact with ABI5 and ABFs and participate in *FT* expression by interacting with CO (Kumimoto et al. 2010, 2013; Wenkel et al. 2006). Furthermore, mutations in the *FyPP1* and *FyPP3* (*PHYTOCHROME-ASSOCIATED SERINE/THREONINE PROTEIN PHOSPHATASE 1* and *3*) cause an ABA hypersensitive phenotype and an accumulation of hyper phosphorylated ABI5 forms (Dai et al. 2013). Despite over accumulating ABI5 protein, *fypp1* mutants are early flowering although it is not clear whether this can be directly related to ABI5 or defective signalling in other pathways (Kim et al. 2002).

Other transcription factors mediate ABA response, but independent of SnRK2-mediated phosphorylation. One such transcription factor is ABSCISIC ACID INSENSITIVE 3 (ABI3), a B1/B3 transcription factor responsible for the positive and negative regulation of a plethora of ABA responses (Parcy et al. 1994; Giraudat et al. 1992; Rohde et al. 2000a, b, 2002; Mönke et al. 2012; Suzuki et al. 2003). Besides the well established role in seed germination and early seedling development, ABI3 has broader function in vegetative development, including control of flowering time (Kurup et al. 2001; Rohde et al. 2000a). *abi3* mutants are early flowering under both SDs and LDs while *ABI3* overexpression results in an increased vegetative phase under LDs (Kurup et al. 2001; Rohde et al. 2000a; Zhang et al. 2009). The question arises as to how ABI3 affects the floral transition. The role of ABI3 in transcriptional reprogramming is complex (Mönke et al. 2012; Suzuki et al. 2009). Following spray with ABA, *abi3* mutants display high levels of *TSF*, suggesting a repressive role for ABI3 on *TSF* expression (Suzuki et al. 2003). In germinating seeds, expression of another florigen gene (*MFT*) is downregulated by ABI3. *MFT* is then involved in a negative feedback regulation of ABA signalling by directly repressing *ABI5* (Xi et al. 2010). It is unclear whether this model can be extended to the regulation of the vegetative phase. However, these data could provide a molecular framework for how florigen-like proteins, ABI3 and ABA related bZIPs interact. Second they point to a model where ABI3 acts as a negative regulator of flowering through downregulation of florigen-like genes. Intriguingly, the physical interaction between ABI3 and CO might explain the ABI3-mediated regulation of the florigen genes. ABI3 binds to the CO CCT (CO, CO-like, TOC1) domain (Kurup et al. 2001). The CCT domain appears to be involved in the recruitment of the CO protein to the promoter of *FT* (Tiwari et al. 2010). Thus, binding of ABI3 may interfere with CO (and perhaps other CCT-domain containing proteins) binding to chromatin. A more precise understanding of ABI3-action warrants further investigation and could provide valuable clues as to how ABA and photoperiod interact to affect flowering.

Combined genetic and expression data suggest a model where *ABI3* acts upstream of *ABI5*. First the *abi3* germination defects can be rescued by the ectopic expression of *ABI5*. Conversely the ectopic expression of *ABI3* cannot rescue the germination defect of *abi5* mutants. Second elevated *ABI3* levels trigger ABI5 protein expression in the presence of ABA (Lopez-Molina et al. 2002). ABI3 also interacts with ABI5, suggesting a mode of cross talk between ABI3 and ABI5 (Nakamura et al. 2001). Thus, besides negatively regulating florigen-like genes, ABI3 may also delay flowering indirectly, through upregulation of *ABI5*.

ABI3 activity is regulated through ubiquitination and subsequent proteasome-dependent degradation (Zhang et al. 2009). An ABI3-interacting protein (AIP2) can polyubiquitinate ABI3 and affect ABI3 levels in vivo. A model has emerged where increasing ABA levels trigger *AIP2* transcriptional activation. AIP2 in turn negatively regulates ABA signalling by targeting ABI3 for destruction. These data argue against the idea that the flowering repressive role of ABA is dependent on ABI3, because ABA would promote its degradation.

Other transcription factors play a role in ABA signalling and have been reported to alter the floral transition. ABA modulates the nuclear import and activity of a set of transcription factor to regulate gene expression. ABA can positively regulate the transcription of *SOC1* through the action of the *OXIDATIVE STRESS 2* (*OXS2*) zinc-finger transcription factor family. These proteins are able bind in an ABA-dependent manner the promoter of *SOC1* and multiple mutants in the *OXS2* family are late flowering, consistent with the proposed *OXS2* role as activators of flowering (Blanvillain et al. 2011).

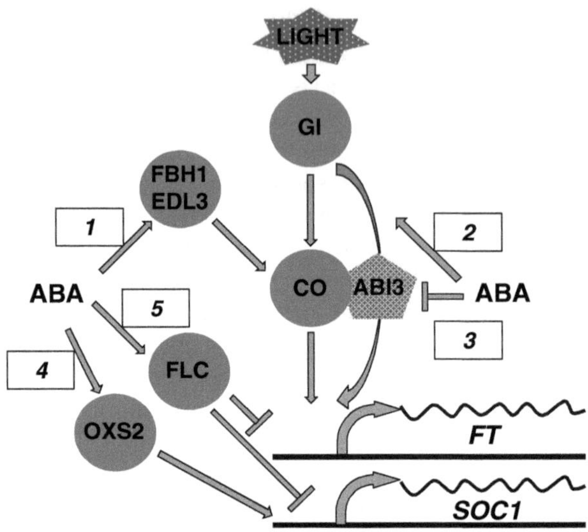

Fig. 18.1 A schematic model summarizing different modes of action of ABA during the floral transition. *1* ABA upregulates the transcription of *EDL3* and affects FBH1 activity in an unknown manner. Both these proteins are positive regulators of CONSTANS (CO). *2* ABA potentiates photoperiod-stimulated GI activity, independent of CO. *3* ABA promotes degradation of ABI3, an interactor of CO, which might lead to an increase in CO activity. *1–3* lead to activation of *FT*. *4* ABA promotes the activity of OXS2, a direct activator of *SOC1*. *5* ABA induces *FLC* expression via upregulation of ABI5 activity. FLC causes downregulation of *FT* and *SOC1* expression

In conclusion, rather than operating in a hierarchal cascade of signalling events, these data suggest ABA acting at multiple (transcriptional and post-transcriptional) levels in the regulation of several transcription factors. Each class of transcription factor positively or negatively regulates the floral transition by acting on a specific floral integrator; OXS2-factors converge on *SOC1,* ABI3 on *TSF*/*MFT* and ABI5 on *FLC*. A future goal will be to understand how and where ABA co-ordinately activates such a complex network of transcription factors, the underlying regulatory logic and ultimately its physiological significance (Fig. 18.1).

18.7 ABA Interaction with Other Hormonal Pathways During the Floral Transition

Several phytohormones participate in the floral transition including Gibberellic acid (GAs), Cytokinin, Salicylic Acid, Ethylene, Nitric Oxide and Brassinosterodis (D'Aloia et al. 2011; Wilson et al. 1992; Achard et al. 2007; Martínez et al. 2004; Domagalska et al. 2007; He et al. 2004). Due to the well-established hormone cross-talk, ABA could affect flowering indirectly, e.g. by modifying other hormonal pathways (Gazzarrini and McCourt 2003). The levels of the floral repressing hormone Ethylene are increased in *aba* mutants plants, and may contribute to their late flowering (LeNoble et al. 2004; Riboni et al. 2013). Ethylene in turn suppresses flowering through the downregulation of GAs biosynthesis (Achard et al. 2007). ABA-GAs cross-talk plays a fundamental role during the germination process where ABA counteracts GA biosynthesis and signalling, thus triggering seed dormancy as GAs beak dormancy to allow germination (Weiss and Ori 2007). Interestingly in the context of the transition to flowering, ABA may act in parallel to GAs with respect to *FT* upregulation (Riboni et al. 2013). GAs are absolutely required for flowering under SDs, when the inductive photoperiodic pathway is not active (Wilson et al. 1992; Reeves and Coupland 2001). Under LDs the role of GAs is less pronounced but still present. Different lines of evidence suggest that GAs promote the floral transition at different sites, by positively regulating *FT* and *TSF* transcription (in parallel with CO) in leaves and by stimulating the *SQUAMOSA PROMOTER BINDING PROTEIN-like* (*SPL*) genes, upstream of *SOC1* in the shoot apical meristem (Porri et al. 2012; Galvão et al. 2012; Hisamatsu and King 2008). Both ABA and GAs appear to share a strong photoperiodic dependency in the way they can activate *FT* expression. ABA does not activate *FT* or *TSF* under SDs. Similarly, the upregulation of *FT* following applied GA is far more effective under LDs compared to SDs (Hisamatsu and King 2008). Taken together these results suggest that ABA and GAs require a photoperiodic component for their ability to upregulate *FT*, although ABA appears more stringent than GAs in this requirement. Whilst ABA and GAs are proposed to both converge to florigen transcription under LDs, the role of ABA signalling under SDs is opposite to that of GAs. ABA could play an important role in the SAM by counteracting positive flowering signals from GAs. The molecular basis for these mechanisms is unknown and further experiments are needed to reveal the underlying photoperiodic component(s) and the spatial context involved.

18.8 Concluding Remarks

ABA is mainly known as a stress hormone and its role in flowering is only beginning to emerge. Independent reports suggest ABA acting as a promoter of flowering in SD plants. In contrast a consensus as to the precise mode of action of ABA in LD plants is lacking. This begs to the question as to whether the ABA-mediated effects on flowering are conserved in plants and if so, their relationship with photoperiod. Ecotype/species specific effects, but also cultivation practices and conditions might affect the observed variability in LD plants/backgrounds and need to be investigated further.

Although progresses have been made in linking ABA signalling to *FT* expression the molecular basis fora photoperiod-ABA cross talk is unknown. Similarly, the mechanism and physiological significance of ABA in the transcriptional/post transcriptional control of *FLC* remain poorly understood. Key questions remain as to how early phosphorylation signals are integrated in the photoperiodic or autonomous floral pathways, whether they involve other components alongside the PYR/RCAR module and their spatial/temporal modes of regulation. Downstream of these early phosphorylation events, the evidence reviewed here suggests that ABA activates a complex network of transcription factors, some of which having opposing function in flowering. A more precise understanding of the early downstream targets of ABA signalling as well as the regulatory logic of the ABA-related network of transcription factors warrant further investigations.

Recent advancements in hormone signalling reveal how their site of production, mode of transport and action is strictly controlled; ABA may thus be no exception. A key goal is therefore to have a better understanding of the spatial context where ABA acts (perhaps in combination with other hormones) during vegetative development. Spatial and temporal constraints might affect the final output of ABA signalling. Tissue specific and temporal manipulation of ABA signalling could inform the nature of these constraints. Knowledge obtained through these experiments will provide concepts that help to understand how enormously variable water-dependent signals are translated into developmental adaptations in plants.

References

Abe M, et al. FD, a bZIP protein mediating signals from the floral pathway integrator FT at the shoot apex. Science. 2005;309(5737):1052–6.

Achard P, et al. The plant stress hormone ethylene controls floral transition via DELLA-dependent regulation of floral meristem-identity genes. Proc Nat Acad Sci USA. 2007;104(15):6484–9.

An H, et al. CONSTANS acts in the phloem to regulate a systemic signal that induces photoperiodic flowering of Arabidopsis. Development. 2004;131(15):3615–26.

Balasubramanian S, et al. Potent induction of Arabidopsis thaliana flowering by elevated growth temperature. PLoS Genet. 2006;2(7):e106.

Barrero JM, et al. A mutational analysis of the ABA1 gene of Arabidopsis thaliana highlights the involvement of ABA in vegetative development. Arabidopsis thaliana. 2005;56(418):2071–83.

Bernier G. Physiological signals that induce flowering. Plant Cell. 1993;5(10):1147–55.

Bezerra IC, et al. Lesions in the mRNA cap-binding gene ABA HYPERSENSITIVE 1 suppress FRIGIDA-mediated delayed flowering in Arabidopsis. Plant J Cell Mol Biol. 2004;40(1):112–9.

Blanvillain R, et al. Stress tolerance to stress escape in plants: role of the OXS2 zinc-finger transcription factor family. EMBO J. 2011;30(18):3812–22.

Blazquez M, et al. Gibberellins promote flowering of arabidopsis by activating the LEAFY promoter. Plant Cell. 1998;10(5):791–800.

Blázquez MA, et al. A thermosensory pathway controlling flowering time in Arabidopsis thaliana. Nat Genet. 2003;33(2):168–71.

Boccalandro HE, et al. Phototropins but not cryptochromes mediate the blue light-specific promotion of stomatal conductance, while both enhance photosynthesis and transpiration under full sunlight. Plant Physiol. 2012;158(3):1475–84.

Brocard IM, et al. Regulation and role of the Arabidopsis abscisic acid-insensitive 5 gene in abscisic acid, sugar, and stress response. Plant Physiol. 2002;129(4):1533–43.

Burbidge A, et al. Characterization of the ABA-deficient tomato mutant *notabilis* and its relationship with maize *Vp14*. Plant J. 1999;17(4):427–31.

Busk PK, Pagès M. Regulation of abscisic acid-induced transcription-Springer. Plant Mol Biol. 1998;37(3):425–35.

Carvalho RF, et al. Convergence of developmental mutants into a single tomato model system: "Micro-Tom" as an effective toolkit for plant development research. Plant Methods. 2011;7(1):18.

Chandler J, et al. Mutations causing defects in the biosynthesis and response to gibberellins, abscisic acid and phytochrome B do not inhibit vernalization in Arabidopsis fca-1. Planta. 2000;210(4):677–82.

Cheng W-H, et al. A unique short-chain dehydrogenase/reductase in Arabidopsis glucose signaling and abscisic acid biosynthesis and functions. Plant Cell. 2002;14(11):2723–43.

Choi H, et al. ABFs, a family of ABA-responsive element binding factors. J Biol Chem. 2000;275(3):1723–30.

Corbesier L, et al. FT protein movement contributes to long-distance signaling in floral induction of Arabidopsis. Science. 2007;316(5827):1030–3.

Covington MF, et al. Global transcriptome analysis reveals circadian regulation of key pathways in plant growth and development. Genome Biol. 2008;9(8):R130.

Cutler SR, et al. Abscisic acid: emergence of a core signaling network. Annu Rev Plant Biol. 2010;61:651–79.

D'Aloia M, et al. Cytokinin promotes flowering of Arabidopsis via transcriptional activation of the FT paralogue TSF. Plant J. 2011;65(6):972–9.

Dai M, et al. The PP6 phosphatase regulates ABI5 phosphorylation and abscisic acid signaling in arabidopsis. Plant Cell. 2013;25(2):517–34.

Domagalska MA, et al. Attenuation of brassinosteroid signaling enhances FLC expression and delays flowering. Development. 2007;134(15):2841–50.

Domagalska MA et al. (2010) Genetic analyses of interactions among gibberellin, abscisic acid, and brassinosteroids in the control of flowering time in Arabidopsis thaliana. Plos One 5(11): p.e14012.

El-Antably HMM, Waering PF. Stimulation of flowering in certain short-day Plants by Abscisin. Nature. 1966;210(5033):328–9.

Endo A, et al. Drought induction of arabidopsis 9-cis-Epoxycarotenoid Dioxygenase occurs in vascular parenchyma cells. Plant Physiol. 2008;147(4):1984–93.

Finkelstein RR, Gampala SSL, Rock CD. Abscisic acid signaling in seeds and seedlings. Plant Cell. 2002; 14:S15–S45.

Fornara F, et al. Arabidopsis DOF transcription factors act redundantly to reduce CONSTANS expression and are essential for a photoperiodic flowering response. Dev Cell. 2009;17(1):75–86.

Franks SJ. Plasticity and evolution in drought avoidance and escape in the annual plant Brassica rapa. New Phytol. 2011;190(1):249–57.

Franks SJ, et al. Rapid evolution of flowering time by an annual plant in response to a climate fluctuation. Proc Nat Acad Sci USA. 2007;104(4):1278–82.

Fujii H, et al. In vitro reconstitution of an abscisic acid signalling pathway. Nature. 2009;462(7273):660–4.

Fujii H, et al. Identification of two protein kinases required for abscisic acid regulation of seed germination, root growth, and gene expression in Arabidopsis. Plant Cell. 2007;19(2):485–94.

Fujita Y, et al. ABA-mediated transcriptional regulation in response to osmotic stress in plants. J Plant Res. 2011;124(4):509–25.

Fujita Y, et al. AREB1 is a transcription activator of novel ABRE-dependent ABA signaling that enhances drought stress tolerance in Arabidopsis. Plant Cell. 2005;17(12):3470–88.

Furihata T, et al. Abscisic acid-dependent multisite phosphorylation regulates the activity of a transcription activator AREB1. Proc Nat Acad Sci USA. 2006;103(6):1988–93.

Galvão VC, et al. Spatial control of flowering by DELLA proteins in Arabidopsis thaliana. Development. 2012;139(21):4072–82.

Gazzarrini S, McCourt P. Cross-talk in plant hormone signalling: what Arabidopsis mutants are telling us. Ann Bot. 2003;91(6):605–12.

Giraudat J, et al. Isolation of the Arabidopsis ABI3 gene by positional cloning. Plant Cell. 1992;4(10):1251–61.

Han S-K, et al. The SWI2/SNF2 chromatin remodeling ATPase BRAHMA represses abscisic acid responses in the absence of the stress stimulus in Arabidopsis. Plant Cell. 2012;24(12):4892–906.

Han Y, et al. The suppression of WRKY44 by GIGANTEA-miR172 pathway is involved in drought response of arabidopsis thaliana. PLoS One. 2013;8(11):e73541.

Hao Q, et al. The molecular basis of ABA-independent inhibition of PP2Cs by a subclass of PYL proteins. Mol Cell. 2011;42(5):662–72.

He Y, et al. Nitric oxide represses the Arabidopsis floral transition. Science. 2004;305(5692):1968–71.

Herde O, et al. Electric signaling and pin2 gene expression on different abiotic stimuli depend on a distinct threshold level of endogenous abscisic acid in several abscisic acid-deficient tomato mutants. Plant Physiol. 1999;119(1):213–8.

Hisamatsu T, King RW. The nature of floral signals in Arabidopsis. II. Roles for FLOWERING LOCUS T (FT) and gibberellin. J Exp Bot. 2008;59(14):3821–9.

Hoth S, et al. Genome-wide gene expression profiling in Arabidopsis thaliana reveals new targets of abscisic acid and largely impaired gene regulation in the abi1-1 mutant. J Cell Sci. 2002;115(24):4891–900.

Huala E, Sussex IM. Determination and cell interactions in reproductive meristems. Plant Cell. 1993;5(10):1157–65.

Hubbard KE, et al. Early abscisic acid signal transduction mechanisms: newly discovered components and newly emerging questions. Genes Dev. 2010;24(16):1695–708.

Imaizumi T. Arabidopsis circadian clock and photoperiodism: time to think about location. Curr Opin Plant Biol. 2010;13(1):83–9.

Imaizumi T, et al. FKF1 F-box protein mediates cyclic degradation of a repressor of CONSTANS in Arabidopsis. Science. 2005;309(5732):293–7.

Ito S, et al. FLOWERING BHLH transcriptional activators control expression of the photoperiodic flowering regulator CONSTANS in Arabidopsis. Proc Nat Acad Sci USA. 2012;109(9):3582–7.

Ivey CT, Carr DE. Tests for the joint evolution of mating system and drought escape in Mimulus. Ann Bot. 2012;109(7):1381.

Jaeger KE, Wigge PA. FT protein acts as a long-range signal in Arabidopsis. Current Biol CB. 2007;17(12):1050–4.

Jaeger KE, et al. Interlocking feedback loops govern the dynamic behavior of the floral transition in Arabidopsis. Plant Cell. 2013;25:820–33.

Jammes F, et al. MAP kinases MPK9 and MPK12 are preferentially expressed in guard cells and positively regulate ROS-mediated ABA signaling. Proc Nat Acad Sci USA. 2009;106(48):20520–5.

Kang J-Y, et al. Arabidopsis basic leucine zipper proteins that mediate stress-responsive abscisic acid signaling. Plant Cell. 2002;14(2):343–57.

Kardailsky I, et al. Activation tagging of the floral inducer FT. Science. 1999;286(5446):1962–5.

Kim DH, et al. A phytochrome-associated protein phosphatase 2A modulates light signals in flowering time control in arabidopsis. Plant Cell. 2002;14(12):3043–56.

King RW, Evans LT. Inhibition of flowering in Lolium temulentum L. By water stress: a role for Abscisic acid. Aust J Plant Physiol. 1977;4(2):225–33.

Kobayashi Y, et al. A pair of related genes with antagonistic roles in mediating flowering signals. Science. 1999;286(5446):1960–2.

Koiwa H, et al. C-terminal domain phosphatase-like family members (AtCPLs) differentially regulate Arabidopsis thaliana abiotic stress signaling, growth, and development. Proc Nat Acad Sci USA. 2002;99(16):10893–8.

Koiwai H, et al. Tissue-specific localization of an abscisic acid biosynthetic enzyme, AAO3, in Arabidopsis. Plant Physiol. 2004;134(4):1697–707.

Koops P, et al. EDL3 is an F-box protein involved in the regulation of abscisic acid signalling in Arabidopsis thaliana. J Exp Bot. 2011;62(15):5547–60.

Kuhn JM, et al. mRNA metabolism of flowering-time regulators in wild-type Arabidopsis revealed by a nuclear cap binding protein mutant, abh1. Plant J. 2007;50(6):1049–62.

Kumar SV, Wigge PA. H2A.Z-containing nucleosomes mediate the thermosensory response in Arabidopsis. Cell. 2010;140(1):136–47.

Kumar SV, et al. Transcription factor PIF4 controls the thermosensory activation of flowering. Nature. 2012;484(7393):242–5.

Kumimoto RW, et al. NF-YC3, NF-YC4 and NF-YC9 are required for CONSTANS-mediated, photoperiod-dependent flowering in Arabidopsis thaliana. Plant J. 2010;63:379–91.

Kumimoto RW, et al. NUCLEAR FACTOR Y transcription factors have both opposing and additive roles in ABA-mediated seed germination. PLoS One. 2013;8(3):e59481.

Kurup S, et al. Interactions of the developmental regulator ABI3 with proteins identified from developing Arabidopsis seeds. Plant J. 2001;21(2):143–55.

Lafitte HR, et al. Improvement of rice drought tolerance through backcross breeding: Evaluation of donors and selection in drought nurseries. Field Crops Res. 2006;97(1):77–86.

Lee KH, et al. Activation of glucosidase via stress-induced polymerization rapidly increases active pools of abscisic acid. Cell. 2006;126(6):1109–20.

Legnaioli T, et al. TOC1 functions as a molecular switch connecting the circadian clock with plant responses to drought. EMBO J. 2009;28(23):3745–57.

LeNoble ME, et al. Maintenance of shoot growth by endogenous ABA: genetic assessment of the involvement of ethylene suppression. J Exp Bot. 2004;55(395):237–45.

Li Y, et al. Establishing glucose- and ABA-regulated transcription networks in Arabidopsis by microarray analysis and promoter classification using a relevance vector machine. Genome Res. 2006;16(3):414–27.

Lim M-H, et al. A new Arabidopsis gene, FLK, encodes an RNA binding protein with K homology motifs and regulates flowering time via FLOWERING LOCUS C. Plant Cell. 2004;16(3):731–40.

Liu H, et al. Photoexcited CRY2 interacts with CIB1 to regulate transcription and floral initiation in Arabidopsis. Science. 2008;322(5907):1535–9.

Lopez-Molina L, et al. ABI5 acts downstream of ABI3 to execute an ABA-dependent growth arrest during germination. Plant J Cell Mol Biol. 2002;32(3):317–28.

Lopez-Molina L, et al. A postgermination developmental arrest checkpoint is mediated by abscisic acid and requires the ABI5 transcription factor in Arabidopsis. Proc Nat Acad Sci USA. 2001;98(8):4782–7.

Lu C, Fedoroff N. A mutation in the Arabidopsis HYL1 gene encoding a dsRNA binding protein affects responses to abscisic acid, auxin, and cytokinin. Plant Cell. 2000;12(12):2351–66.

Lu C, et al. Mitogen-activated protein kinase signaling in post germination arrest of development by abscisic acid. Proc Nat Acad Sci USA. 2002;99(24):15812–7.

Ma Y, et al. Regulators of PP2C phosphatase activity function as abscisic acid sensors. Science. 2009;324(5930):1064–8.

Macknight R, et al. FCA, a gene controlling flowering time in arabidopsis, encodes a protein containing RNA-binding domains. Cell. 1997;89(5):737–45.

Martínez C, et al. Salicylic acid regulates flowering time and links defence responses and reproductive development. Plant J. 2004;37(2):209–17.

Martínez-Zapater JM, et al. The transition to flowering in arabidopsis. Cold Spring Harbor Monogr Arch. 1994;27:403–33.

Mathieu J, et al. Export of FT protein from phloem companion cells is sufficient for floral induction in Arabidopsis. Current Biol CB. 2007;17(12):1055–60.

Michaels SD, Amasino RM. FLOWERING LOCUS C encodes a novel MADS domain protein that acts as a repressor of flowering. Plant Cell. 1999;11(5):949–56.

Mori IC, et al. CDPKs CPK6 and CPK3 function in ABA regulation of guard cell S-type anion- and $Ca(2^{+})$-permeable channels and stomatal closure. PLoS Biol. 2006;4(10):e327.

Mönke G, et al. Toward the identification and regulation of the Arabidopsis thaliana ABI3 regulon. Nucleic Acids Res. 2012;40(17):8240–54.

Nakamura S, et al. Physical interactions between ABA response loci of Arabidopsis. Plant J Cell Mol Biol. 2001;26(6):627–35.

Nambara E, Marion-Poll A. Abscisic acid biosynthesis and catabolism. Annu Rev Plant Biol. 2005;56(1):165–85.

Notaguchi M et al. Long-distance, graft-transmissible action of arabidopsis FLOWERING LOCUS T protein to promote flowering. Plant Cell Physiol. 2008; 49(11):1645–1658.
Parcy F, et al. Regulation of gene expression programs during Arabidopsis seed development: roles of the ABI3 locus and of endogenous abscisic acid. Plant Cell. 1994;6(11):1567–82.
Park S-Y, et al. Abscisic acid inhibits type 2C protein phosphatases via the PYR/PYL family of START proteins. Science. 2009;324(5930):1068–71.
Popescu SC, et al. MAPK target networks in Arabidopsis thaliana revealed using functional protein microarrays. Genes Dev. 2009;23(1):80–92.
Porri A et al. Spatially distinct regulatory roles for gibberellins in the promotion of flowering of Arabidopsis under long photoperiods. Development. 2012;139(12):2198–9.
Putterill J, et al. The constans gene of arabidopsis promotes flowering and encodes a protein showing similarities to Zinc-Finger transcription factors. Cell. 1995;80(6):847–57.
Reeves P, Coupland G. Analysis of flowering time control in arabidopsis by comparison of double and triple mutants. Plant Physiol. 2001;126(3):1085–91.
Riboni M, et al. GIGANTEA enables drought escape response via abscisic acid-dependent activation of the florigens and SUPPRESSOR OF OVEREXPRESSION OF CONSTANS1. Plant Physiol. 2013;162(3):1706–19.
Rohde A, et al. PtABI3 impinges on the growth and differentiation of embryonic leaves during bud set in poplar. Plant Cell. 2002;14(8):1885–901.
Rohde A, et al. ABI3 affects plastid differentiation in dark-grown Arabidopsis seedlings. Plant Cell. 2000a;12(1):35–52.
Rohde A, et al. ABI3 emerges from the seed. Trends Plant Sci. 2000b;5(10):418–9.
Rohde A, Van Montagu M, Boerjan W. The ABSCISIC ACID-INSENSITIVE 3 (ABI3) gene is expressed during vegetative quiescence processes in Arabidopsis. Plant Cell Environ. 1999;22(3):261–70.
Saez A, et al. HAB1-SWI3B interaction reveals a link between abscisic acid signaling and putative SWI/SNF chromatin-remodeling complexes in Arabidopsis. Plant Cell. 2008;20(11):2972–88.
Sagi M, et al. Aldehyde oxidase and xanthine dehydrogenase in a flacca tomato mutant with deficient abscisic acid and wilty phenotype. Plant Physiol. 1999;120(2):571–8.
Saijo Y, et al. Arabidopsis COP1/SPA1 complex and FHY1/FHY3 associate with distinct phosphorylated forms of phytochrome A in balancing light signaling. Mol Cell. 2008;31(4):607–13.
Santner A, et al. Recent advances and emerging trends in plant hormone signalling. Nature. 2009;459(7250):1071.
Sawa M, et al. FKF1 and GIGANTEA complex formation is required for day-length measurement in arabidopsis. Science. 2007;318(5848):261–5.
Schomburg F, et al. FPA, a gene involved in floral induction in Arabidopsis, encodes a protein containing RNA-recognition motifs. Plant Cell. 2001;13(6):1427–36.
Schultz E, Haughn G. Genetic-analysis of the floral initiation process (Flip) in arabidopsis. Development. (1993);119(3):745–5.
Seo E, et al. Crosstalk between cold response and flowering in Arabidopsis is mediated through the flowering-time gene SOC1 and its upstream negative regulator FLC. Plant Cell. 2009;21(10):3185–97.
Seo M, Koshiba T. Complex regulation of ABA biosynthesis in plants. Trends Plant Sci. 2002;7(1):41–8.
Sharp RG, et al. Water deficits promote flowering in Rhododendron via regulation of pre and post initiation development. Sci Hortic. 2009;120(4):511–7.
Sherrard ME, Maherali H. The adaptive significance of drought escape in Avena barbata, an annual grass. Evolution. 2006;60(12):2478–89.
Shinozaki K, Yamaguchi-Shinozaki K. Gene networks involved in drought stress response and tolerance. J Exp Bot. 2007;58(2):221–7.

Simpson G. The autonomous pathway: epigenetic and post-transcriptional gene regulation in the control of Arabidopsis flowering time. Curr Opin Plant Biol. 2004;7(5):570–4.
Simpson GG, Gendall AR, Dean C. When to switch to flowering. Annu Rev Cell Dev Biol. 1999;15:519–50.
Song YH, et al. FKF1 conveys timing information for CONSTANS stabilization in photoperiodic flowering. Science. 2012;336(6084):1045–9.
Soon FF, et al. Molecular mimicry regulates ABA signaling by SnRK2 kinases and PP2C phosphatases. Science. 2012;335(6064):85–8.
Srivastava AK, et al. An analysis on citrus flowering-A review. Agric Rev Agri Res Commun Centre India. 2000;21(1):1–15.
Su Z, et al. Flower development under drought stress: morphological and transcriptomic analyses reveal acute responses and long-term acclimation in arabidopsis. Plant Cell. 2013;25(10):3785–807.
Suarez-Lopez P, et al. CONSTANS mediates between the circadian clock and the control of flowering in Arabidopsis. Nature. 2001;410(6832):1116–20.
Sussex I. Developmental programming of the shoot meristem. Cell. 1989;56(2):225–9.
Suzuki H, et al. Differential expression and affinities of Arabidopsis gibberellin receptors can explain variation in phenotypes of multiple knock-out mutants. Plant J. 2009;60(1):48–55.
Suzuki M, et al. Viviparous1 alters global gene expression patterns through regulation of abscisic acid signaling. Plant Physiol. 2003;132(3):1664–77.
Tiwari SB et al. The flowering time regulator CONSTANS is recruited to the FLOWERING LOCUS T promoter via a uniquecis-element. New Phytol 2010;1–10.
Toh S, et al. High temperature-induced abscisic acid biosynthesis and its role in the inhibition of gibberellin action in Arabidopsis seeds. Plant Physiol. 2008;146(3):1368–85.
Umezawa T et al. Genetics and phosphoproteomics reveal a protein phosphorylation network in the abscisic acid signaling pathway in arabidopsis thaliana. Science Signal 2013;6(270), p.rs8.
Umezawa T, et al. Molecular basis of the core regulatory network in ABA responses: sensing, signaling and transport. Plant Cell Physiol. 2010;51(11):1821–39.
Umezawa T et al. Type 2C protein phosphatases directly regulate abscisic acid-activated protein kinases in Arabidopsis. Proc Natl Acad Sci USA 2009;106(41):17588–593.
Uno Y, et al. Arabidopsis basic leucine zipper transcription factors involved in an abscisic acid-dependent signal transduction pathway under drought and high-salinity conditions. Proc Nat Acad Sci USA. 2000;97(21):11632–7.
Valverde F, et al. Photoreceptor regulation of CONSTANS protein in photoperiodic flowering. Science. 2004;303(5660):1003–6.
Verslues PE, Juenger TE. Drought, metabolites, and Arabidopsis natural variation: a promising combination for understanding adaptation to water-limited environments. Curr Opin Plant Biol. 2011;14(3):240–5.
Vlad F, et al. Protein phosphatases 2C regulate the activation of the Snf1-related kinase OST1 by abscisic acid in arabidopsis. Plant Cell. 2009;21(10):3170–84.
Wahl V, et al. Regulation of flowering by trehalose-6-phosphate signaling in arabidopsis thaliana. Science. 2013;339(6120):704–7.
Wang J-W, et al. miR156-Regulated SPL transcription factors define an endogenous flowering pathway in arabidopsis thaliana. Cell. 2009;138(4):738–49.
Wang P et al. Quantitative phosphoproteomics identifies SnRK2 protein kinase substrates and reveals the effectors of abscisic acid action. Proc Natl Acad Sci USA 2013a;110(27):11205–10.
Wang Y, et al. The inhibitory effect of ABA on floral transition is mediated by ABI5 in Arabidopsis. J Exp Bot. 2013b;64(2):675–84.
Weiss D, Ori N. Mechanisms of cross talk between gibberellin and other hormones. Plant Physiol. 2007;144(3):1240–6.

Wenkel S, et al. CONSTANS and the CCAAT box binding complex share a functionally important domain and interact to regulate flowering of Arabidopsis. Plant Cell. 2006;18(11):2971–84.

Wigge PA, et al. Integration of spatial and temporal information during floral induction in Arabidopsis. Science. 2005;309(5737):1056–9.

Wilmowicz E, et al. Involvement of aba in flower induction of Pharbitis nil. Acta Societatis Botanicorum Poloniae. 2011;80(1):21.

Wilmowicz E, et al. Ethylene and ABA interactions in the regulation of flower induction in Pharbitis nil. J Plant Physiol. 2008;165(18):1917–28.

Wilson R, et al. Gibberellin is required for flowering in arabidopsis-thaliana under short days. Plant Physiol. 1992;100(1):403–8.

Wong CE, et al. Molecular processes underlying the floral transition in the soybean shoot apical meristem. Plant J. 2009;57(5):832–45.

Xi W, et al. MOTHER OF FT AND TFL1 regulates seed germination through a negative feedback loop modulating ABA signaling in Arabidopsis. Plant Cell. 2010;22(6):1733–48.

Xiong L, et al. The arabidopsis LOS5/ABA3 locus encodes a molybdenum cofactor sulfurase and modulates cold stress–and osmotic stress–responsive gene expression. Plant Cell. 2001;13(9):2063–83.

Xiong L, et al. Regulation of osmotic stress-responsive gene expression by the LOS6/ABA1 locus in arabidopsis. J Biol Chem. 2002a;277(10):8588–96.

Xiong L, Lee H, Ishitani M, Tanaka Y, et al. Repression of stress-responsive genes by FIERY2, a novel transcriptional regulator in Arabidopsis. Proc Natl Acad Sci USA. 2002b;99(16):10899–904.

Xu JL, et al. QTLs for drought escape and tolerance identified in a set of random introgression lines of rice. Theor Appl Genet. 2005;111(8):1642–50.

Xu P, et al. Wheat cryptochromes: subcellular localization and involvement in photomorphogenesis and osmotic stress responses. Plant Physiol. 2009;149(2):760–74.

Yalovsky S, et al. Functional requirement of plant farnesyltransferase during development in Arabidopsis. Plant Cell. 2000;12(8):1267–78.

Yamaguchi A, et al. TWIN SISTER OF FT (TSF) acts as a floral pathway integrator redundantly with FT. Plant Cell Physiol. 2005;46(8):1175–89.

Yoshida T, et al. AREB1, AREB2, and ABF3 are master transcription factors that cooperatively regulate ABRE-dependent ABA signaling involved in drought stress tolerance and require ABA for full activation. Plant J. 2010;61(4):672–85.

Zhang W, et al. Cis-regulatory element based targeted gene finding: genome-wide identification of abscisic acid- and abiotic stress-responsive genes in Arabidopsis thaliana. Bioinformatics (Oxford, England). 2005;21(14):3074–81.

Zhang X, et al. The AIP2 E3 ligase acts as a novel negative regulator of ABA signaling by promoting ABI3 degradation. Genes Dev. 2009;276(1673):1532–43.

Zheng Z, et al. The protein kinase SnRK2.6 mediates the regulation of sucrose metabolism and plant growth in arabidopsis. Plant Physiol. 2010;153(1):99–113.

Zhu S-Y, et al. Two calcium-dependent protein kinases, CPK4 and CPK11, regulate abscisic acid signal transduction in Arabidopsis. Plant Cell. 2007;19(10):3019–36.

Zuo Z, et al. Blue light-dependent interaction of CRY2 with SPA1 regulates COP1 activity and floral initiation in Arabidopsis. Curr Biol. 2011;21(10):841–7.

Chapter 19
ABA Signaling and Circadian Clock

Sergi Portolés and Da-Peng Zhang

Abstract The circadian clock is an endogenous timing mechanism that controls a variety of physiological, metabolic, and developmental process. It serves as a time-keeper allowing the organisms to predict recurrent environmental variations, ensuring that the onset of key processes coincides with appropriate conditions. In plants, the use of genome-wide approaches revealed that clock-regulated genes are overrepresented in several plant hormones and stress-responsive pathways, especially those related with abscisic acid (ABA). Concretely, the circadian clock controls the daily fluctuations in ABA concentration and modulates plant sensitivity to this hormone in a process called gating. In turn, changes in ABA levels due to stressful conditions feed back the clock by modifying its pace. Emerging findings have proven that ABA and circadian clock reciprocal regulation is essential to integrate internal and external cues to properly adjust key physiology and development processes such as flowering transition, seed dormancy and germination, and abiotic and biotic stress responses. This mutual regulation is crucial for plant fitness and survival as allows to anticipate and precisely respond to predictable stressful conditions.

Keywords ABA signaling · Circadian clock · Cross talk

19.1 Circadian Clock

Day and night cycles, as a result of the Earth's rotation over its axis, lead to drastic changes in the environmental conditions, challenging the life since the beginning of the evolution. For this reason, the living organisms have adapted their physiological,

S. Portolés (✉) · D.-P. Zhang
School of Life Sciences, Center for Plant Biology,
Tsinghua University, Beijing 100084, China
e-mail: spcgmc@cid.csic.es

D.-P. Zhang
e-mail: zhangdp@tsinghua.edu.cn

D.-P. Zhang (ed.), *Abscisic Acid: Metabolism, Transport and Signaling*,
DOI 10.1007/978-94-017-9424-4_19

developmental, and metabolic process following a rhythmic pattern within a period of 24 h, in clear coordination with the diurnal cycle (Hubbard et al. 2009; de Montaigu et al. 2010). These diurnal oscillations are known as circadian rhythms and the endogenous mechanism responsible for generating and maintaining this rhythmicity is called circadian clock (Wijnen and Young 2006). Until the present day the circadian clock has been found to be present in almost all organisms studied from photosynthetic bacteria to higher plants and mammals (Harmer 2009; Más 2008; McClung 2008). Some examples of circadian rhythms include the sleep–wake cycles in animals, locomotor activity in insects, or photosynthesis and leaf movement in plants. The presence of circadian rhythms has been suggested to be advantageous to adaptation (Dodd et al. 2005; Ouyang et al. 1998; Woelfle et al. 2004), allowing the organisms to coordinate their physiology and metabolism anticipating the environmental changes result of the day/night transitions. Furthermore, the circadian clock allows to temporally separate biological, physiological, or metabolic incompatible process such as feeding and sleeping, nitrogen fixation and photosynthesis or anabolism and catabolism (Golden et al. 1997; Vollmers et al. 2009; Yang 2010).

Circadian rhythms exhibit a 24-h cycle that can be represented as a sinusoidal function, defined by certain parameters such as amplitude, phase, and period (Fig. 19.1). This oscillation is sustained even in the absence of environmental

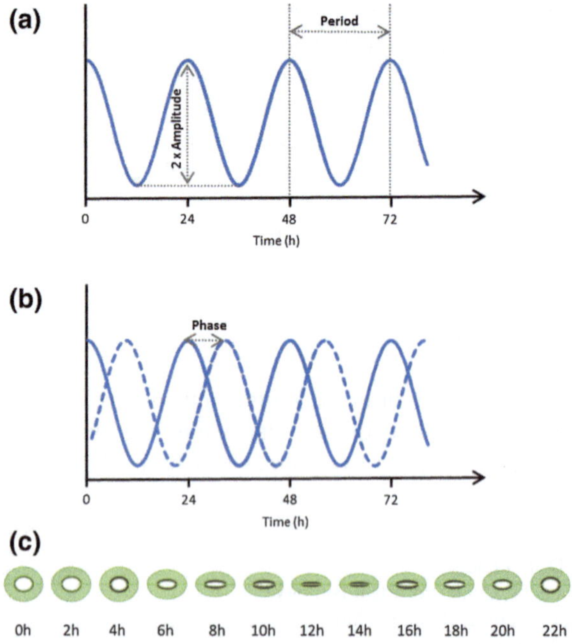

Fig. 19.1 Schematic representation of the mathematical parameters used to define the circadian oscillation. **a** The period is defined as the time between a specific phase of the cycle and when this is repeated again. The amplitude is defined as the difference between the maximum (or minimum) and the average value of the rhythmic parameter studied. **b** The phase is defined as the state of an oscillation relative to another reference rhythmic oscillation. **c** Schematic representation of the circadian oscillation of stomatal aperture

cues (for example, continuous light and constant temperature) with periods close to 24 h. Daily changes in light and temperature are able to synchronize these rhythms every day, so the rhythmicity exactly matches the 24-h cycle (Liu and Bell-Pedersen 2006; Más and Yanovsky 2009; McClung 2006). Furthermore, the circadian rhythms remain almost unchanged over a wide range of physiological temperatures, thus they are temperature compensated (Bruce and Pittendrigh 1956). Hence, the clock has an interesting duality. On the one hand, it is a flexible mechanism capable of being synchronized by environmental changes, and on the other hand, it is a robust and precise timekeeper maintained at different temperatures (Pittendrigh 1954).

Classically, the clock signaling pathway has been represented as linear and unidirectional with three functional modules: input, central oscillator, and output. The input is the module capable of sensing and transmitting the environmental information to the central oscillator, synchronizing it; thus, the central oscillation is in resonance with the external environment. The central oscillator is the mechanism responsible for generating and maintaining the rhythmicity that is going to be transmitted to the output. The output is defined by those rhythmic processes controlled by the clock. However, the present knowledge indicates that this linear and unidirectional idea is an oversimplification and the clock signaling pathway is more complex and branched.

19.2 Evolution of the Circadian Clock Model

The generation of circadian rhythms is based on negative feedback loops, involving repressors and activators elements that regulate each other's expression, localization, and/or activity (Bell-Pedersen et al. 2005; Wijnen and Young 2006). In *Arabidopsis thaliana*, the first model described involves the MYB transcription factors *CIRCADIAN CLOCK ASSOCIATED 1* (*CCA1*) (Wang and Tobin 1998) and *LATE ELONGATED HYPOCOTYL* (*LHY*) (Schaffer et al. 1998) that downregulate the expression of the pseudo response regulator *TIMING OF CAB EXPRESSION 1* (*TOC1*) (Makino et al. 2002; Strayer et al. 2000). In turn, TOC1 activates *CCA1* and *LHY*. This mutual regulation allows to generate an antiphasic and rhythmic expression on *TOC1* and *CCA1*/*LHY* (Alabadi et al. 2001). While *CCA1* and *LHY* expression peaks early in the day, *TOC1* has its maximum expression during the evening. *CCA1* and *LHY* therefore are classified as morning genes and *TOC1* as evening gene (Fig. 19.2).

The functional involvement of these components in the circadian clock has been proven in numerous studies. In plants constitutively expressing *CCA1* (Wang and Tobin 1998) or *LHY* (Schaffer et al. 1998), the clock is no longer functional, whereas loss-of-function mutations of these components results in a clock period shortening (Green and Tobin 1999; Mizoguchi et al. 2002). Although *LHY* and *CCA1* display high sequence homology, similar phenotypes, and the same phase of expression, their function is only partially redundant (Mizoguchi et al. 2002). Similar studies showed that the constitutive overexpression of *TOC1* leads to

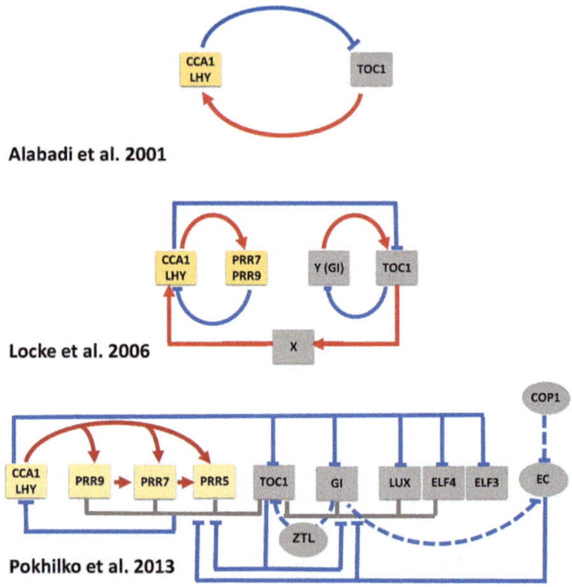

Fig. 19.2 Historical evolution of the proposed model of the *Arabidopsis* circadian clock. Schematic representation of the three most relevant models (Alabadi et al. 2001; Locke et al. 2006; Pokhilko et al. 2013) proposed for the *Arabidopsis* circadian clock core regulation. *Red arrows* and *blue broken lines* represent activation and repression, respectively. Elements belonging to the morning or evening loop are shown in *yellow* and *gray*, respectively. For simplicity, only elements in circles represent proteins (EC, ZTL and COP1). Post-translational regulation is indicated by *dotted lines*

arrhythmia (Más et al. 2003a). In contrast, *TOC1* loss-of-function mutant showed a clear period shortening under white light conditions (Más et al. 2003a) and complete arrhythmia under constant darkness or red light. Therefore, TOC1 shows a dual function dependent and independent of light, suggesting a possible role as molecular connector between the clock input and the processes rhythmically controlled by the oscillator (Más et al. 2003a).

After the initially described feedback loop between *TOC1* and *CCA1/LHY*, early computational models postulated the presence of at least two additional feedback loops interconnected between them (Locke et al. 2006; Rand et al. 2004; Zeilinger et al. 2006). In the so-called morning loop, CCA1/LHY activates the expression of *PSEUDO RESPONSE REGULATOR 7* (*PRR7*) and *PSEUDO RESPONSE REGULATOR 9* (*PRR9*), both members of TOC1 family. In turn, PRR7 and PRR9 act to repress the expression of *CCA1/LHY* (Farre et al. 2005; Nakamichi et al. 2005; Salome and McClung 2005). The evening loop involved an unknown factor called "Y" that activates *TOC1* expression, whereas TOC1 feeds back by repressing Y expression. The authors proposed *GIGANTEA* (*GI*), a multiple-domain protein with several independent functions, as a possible factor "Y", although genetic data were not entirely consistent with it (Ito et al. 2009; Martin-Tryon et al. 2007). Additionally, they suggested the presence of a component called "X" modulating TOC1 activation of *CCA1* and *LHY*, as *TOC1* and *CCA1*/*LHY* expressions are too far temporally (Fig. 19.2).

This model represented a breakthrough on the understanding of how the rhythmicity is generated and also inspired most of the works until now. However, the model was incomplete as many other essential elements for a proper clock function had been described, and also did not include the important post-transcriptional and post-translational regulation described by some of these components.

Among the several components of the clock that are not included in this model should be highlighted *CCA1 HIKING EXPEDITION CCA1*/*LHY* (*CHE*) (Pruneda-Paz et al. 2009), a *CCA1* repressor by direct binding to its promoter region. In turn, CCA1 and probably LHY repress *CHE* expression, forming a new feedback loop. Also, *LUX ARRHYTHMO* (*LUX*), a MYB factor that is expressed at dusk, is essential for maintaining the rhythmicity (Hazen et al. 2005; Onai and Ishiura 2005). Mutations in this gene lead to the absence of rhythmicity in plants under constant conditions. In terms of its function, LUX interacts with two essential clock proteins, EARLY FLOWERING 3 (ELF3) and EARLY FLOWERING 4 (ELF4), forming a DNA-binding protein complex-denominated Evening Complex (Nusinow et al. 2011).

The actual model of the circadian clock has undergone a significant revision recently. TOC1 and the other PRRs proteins were shown to be transcription factors binding DNA through their conserved CCT domain (Gendron et al. 2012). Unexpectedly, the PRRs proteins act as transcriptional repressors. Indeed, experiments where TOC1 was acutely induced by either ABA or ethanol-responsive transgenes led to immediate downregulation of CCA1 and LHY transcription (Huang et al. 2012). These results challenged the well-establish model as TOC1 does not act to promote *LHY* and *CCA1* transcription but to repress it. Furthermore, the identification of the Evening Complex allowed to postulate the mechanism responsible for LHY and CCA1 activation. Mutations in either *lux*, *elf3* or *elf4* lead to higher TOC1 levels (Dixon et al. 2011; Helfer et al. 2011; Kikis et al. 2005; Kolmos et al. 2009), suggesting that the Evening Complex might allow the CCA1 and LHY expression at dawn by downregulating TOC1 and therefore lift the repression (Pokhilko et al. 2012). Incorporating all these new evidence, a new model arose where CCA1/LHY lead the sequential expression and repression of the PRRs genes *PRR9*, *PRR7*, *PRR5*, and *TOC1* that keep *CCA1*/*LHY* expression levels down and finally the Evening Complex release the repression before dawn allowing the cycle to restart (Fig. 19.2).

19.3 Interaction Between the Circadian Clock and ABA Signaling Networks

Many physiological processes show circadian rhythms, including essential processes such as stomatal and leaf movements (Seung et al. 2012; Harmer et al. 2000; Dodd et al. 2005), osmotic and cold responses (Bieniawska et al. 2008;

Matsui et al. 2008), hormone signaling and metabolism (Covington et al. 2008), calcium ion fluxes (Johnson et al. 1995), flowering (Song et al. 2013), sugars signaling (Blasing et al. 2005), nitrogen assimilation (Gutierrez et al. 2008) and cyclic adenosine diphosphate ribose (cADPR) signaling (Dodd et al. 2007a). Therefore, it is not surprising that a large proportion of the transcriptome is subjected to circadian regulation. Indeed, microarray expression analysis estimated that at least one-third of the transcriptome is under clock control (Covington et al. 2008). The use of tilling arrays also allowed to identify intragenic regions, introns, and natural antisense transcripts with circadian regulation, extending the clock control far beyond the protein coding transcripts (Hazen et al. 2009).

The functional clustering of clock microarray datasets revealed that circadian-regulated genes were overrepresented in several hormones and stress-responsive pathways (Covington et al. 2008). Concretely, clock datasets were found to extensively overlap with ABA datasets (Mizuno and Yamashino 2008). The relationship between the circadian clock and ABA is interesting because, while the circadian clock anticipate environmental changes, ABA controls many environmental stress responses, such as water use and drought tolerance, as well as cold responses (Finkelstein et al. 2002). Indeed, genes under clock regulation were also found to be overrepresented in microarray datasets from osmotic, salt, and drought stress (Covington et al. 2008). Considering the significant overlap between clock, ABA, and ABA-related stress-responsive pathways, it is not surprising that ABA levels show circadian rhythms and the clock is able to modulate plant sensitivity to ABA (Robertson et al. 2009). Furthermore, the regulation between ABA and the circadian clock is bidirectional as ABA feeds back by modifying the circadian pace (Hanano et al. 2006).

19.4 Circadian Regulation of ABA Metabolism

ABA-induced responses are regulated by the amount of biologically active hormone present in the cells or tissues. This amount is a result of rates of biosynthesis, catabolism, transport, and activation/deactivation. Transcriptomical analysis showed that several key enzymes involved in ABA biosynthesis and its precursors and are clock-regulated. For instance, *ABA DEFICIENT 1* (*ABA1*) and *ABA DEFICIENT 2* (*ABA2*), *CLOROPLASTOS ALTERADOS 1* (*CLA1*), *PHYTOENE SYNTHASE* (*PSY*), *9-CIS-EPOXYCAROTENOID DIOXYGENASE* (*NCED*), which are involved in ABA biosynthesis, synthesis of isoprenoid precursors and carotenoid synthesis, respectively (Covington et al. 2008; Seung et al. 2012). Similarly, some of the components in ABA degradation and transport also exert circadian regulation (Seung et al. 2012). Additionally, day/night cycles and probably the circadian clock regulate the activity of an ABA-specific β-glucosidase, AtBG1, implicated in the production of bioactive ABA through hydrolysis of glucose-conjugated ABA (ABA-GE) in response to stress (Lee et al. 2006).

Overall, the circadian clock seems to modulate ABA metabolism at all levels. Indeed, ABA levels in leaves show diurnal fluctuations, with a major abundance peak at the end of the day (Fukushima et al. 2009). Interestingly, NCED, one of the limiting steps in ABA synthesis *de novo* in response to drought and salt stress, showed diurnal oscillation in its abundance similar to ABA (Thompson et al. 2000). Although the circadian regulation of ABA metabolism is clear, little is known about the molecular mechanisms implicated.

The first evidences of a molecular link between ABA metabolism and the circadian clock were provided by an extensive metabolomic analysis of the *prr9/prr7/prr5* triple mutant (Fukushima et al. 2009). This mutant displays defects in biosynthetic pathways involved in chlorophyll, carotenoid, tocopherol, and ABA. Particularly, it shows a significant induction of the carotenoid and ABA biosynthetic pathways, resulting in constitutive high ABA content, thus suppressing diurnal oscillation (Fukushima et al. 2009). Not surprisingly, *prr9/prr7/prr5* mutant displayed drought resistance, freezing tolerance, and upregulation of cold-responsive genes (Nakamichi et al. 2009).

19.5 Circadian Modulation of the ABA Core Signaling Network

Although the transcriptomical analysis indicated an important role of the clock modulating ABA-mediated responses, the molecular mechanisms involved in that regulation are largely unknown. Some key elements in ABA signaling exert robust diurnal oscillation, most likely due to clock regulation. Among them should be highlighted elements implicated in the ABA perception such as *PYL3*, *RCAR1*, and *ABAR/CHLH*, and essential ABA signal transduction elements such as *SnRK2.6*, *ABI1*, *ABF3*, and *HAB1* (Seung et al. 2012).

Molecularly, *TOC1* has emerged as a central component acting as a molecular switch between clock and ABA pathways, especially in response to drought stress. TOC1 represses the candidate ABA receptor ABAR/CHLH (Legnaioli et al. 2009; Shen et al. 2006; Wu et al. 2009) by direct binding to its promoter region, thus regulating its diurnal expression. Furthermore, TOC1 directly interacts and modulates the activity of key proteins downstream ABA perception such as the phytochrome-interacting factor (PIF) family proteins and the CBF/DREB1 transcription factors, both implicated in drought and cold responses (Kidokoro et al. 2009), as well as ABA INSENSITIVE3 (ABI3), an essential factor in seed germination and drought tolerance (Kurup et al. 2000).

Although TOC1 has been proven to have a pervasive role connecting circadian clock with ABA signaling, the presence of CCA1 binding elements in the promoters of genes encoding several ABA signaling components, including *ABAR/CHLH*, suggests that other circadian clock components might be implicated in this regulation (Pokhilko et al. 2013). Indeed, mutations in *CCA1* and *LHY* genes result in plant hypersensitivity to salt, osmotic, and heat stress (Kant et al. 2008). Hence, more exhaustive studies are still needed to undercover the intricate mechanisms that allow the circadian clock to precisely modulate the ABA responses.

19.6 ABA Regulation of the Circadian Clock Function

The interaction between ABA and the circadian clock is not a unidirectional process. Conversely, a mutual regulation exists. While the circadian clock modulates ABA levels and sensitivity according to time, ABA in turn set the clock pace in response to environmental stresses. Indeed, external addition of ABA lengthens the period of bioluminescence in seedlings expressing the promoters of chlorophyll a/b-binding protein (CAB2/LHCB1*1, a marker gene for clock output), COLD AND CIRCADIAN REGULATED 2 (CCR2/AtGRP7; a possible slave oscillator component), and CCA1 (Hanano et al. 2006) fused to the luciferase. Consistent with these results, the period of CCR2 promoter activity is shortened in the ABA-deficient mutant *aba2* (Hanano et al. 2006).

Interestingly, *TOC1* expression is acutely induced by ABA, and this induction is gated by the clock (Legnaioli et al. 2009). ABA treatment in the most sensible window advances the phase of TOC1 binding to *ABAR/CHLH* promoter and acutely repress its expression. In addition, ABA upregulation of *TOC1* expression needs the presence of a functional ABAR/CHLH, establishing a feedback loop between ABA and the circadian clock, where *TOC1* and *ABAR/CHLH* play a central role (Legnaioli et al. 2009).

Precise regulation of TOC1 expression had been proven to be essential for proper circadian clock function. Slightly increases in TOC1 expression, either by chemical inhibition of deacetylation or by the presence of additional copies of TOC1 gene, resulted in a period lengthening of the clock (Más et al. 2003b; Perales and Mas 2007). Similarly, ABA induces *TOC1* expression at certain circadian phases and might lead to an overall increase of TOC1 levels. Therefore, it is suggested that TOC1 may be responsable for the period lengthening caused by external addition of ABA. However, other mechanisms may also be implicated. For instance, cADRP levels play an important role in molecular and physiological ABA responses and are modulated both by ABA and the circadian clock (Dodd et al. 2007b; Sanchez et al. 2004). Additionally, prolonged high levels of cADRP shorten the clock period, indicating a mutual regulation between cADRP and the circadian clock (Dodd et al. 2007b).

Another possible molecular connection that might be important in the ABA and the circadian clock cross talk is represented by the SNW/Ski-interacting protein (SKIP). SKIP interacts with *PRR7* and *PRR9* pre-mRNA and was suggested to be important in their splicing processes. Mutations in this gene lead to an increase of *PRR7* and *PRR9* non-functional splicing variants and a temperature-dependent lengthening of the circadian clock period (Wang et al. 2012). Additionally, SKIP expression was shown to be induced by salt, mannitol, and ABA treatments and reduced SKIP protein levels lead to altered tolerance to abiotic stress (Lim et al. 2010). Considering that altered ratios between the *PRR9* splicing variants generate impaired circadian function (Sanchez et al. 2010). It is possible that ABA-mediated induction of *SKIP* expression might produce an increase in *PRR7* and *PRR9* functional splicing variants and therefore a setting on the clock pace.

19.7 Circadian Gating on ABA Responses

An essential aspect in circadian clock function lies in its ability to respond differentially to external stimuli depending on the circadian time through a process called gating (Millar and Kay 1996). A clear example of this capability is the light input. While during the dusk and dawn, the circadian clock is extremely sensible to light changes, allowing a fine-tuning of the period, in the middle of the night, light signals (such as a thunder) are gated, thus not significantly affecting the circadian clock pace. This phenomenon can be explained by the diurnal oscillation of photoreceptors along with other factors. However, the gating phenomenon is not only important in clock synchronization, but also is essential mediating the response to stimuli in some downstream processes such as ABA signaling.

Although ABA has been typically associated with responses to rather random environmental changes, most of these stressful conditions are more likely to occur in a certain moment during the diurnal cycle (Fig. 19.3). Multiple lines of emerging evidence indicate that the capacity to respond to those environmental challenges is gated by the clock. This idea is in agreement with a more efficient use of the resources and consequently and increased plant fitness. Two of the most well-studied physiological processes where the ABA signaling is clearly gated by the clock are the stomatal movements and the cold response.

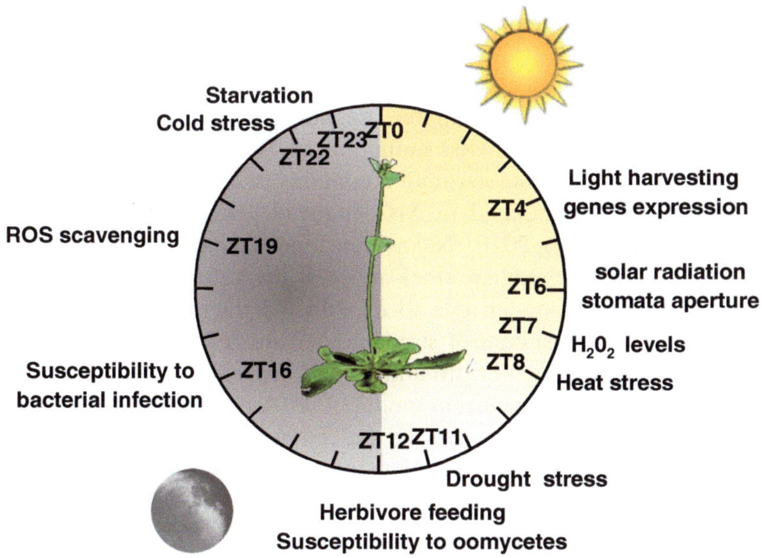

Fig. 19.3 Circadian oscillation of biotic and abiotic stress (adapted from Spoel and Van Ooijen 2014). Many biotic and abiotic stresses fluctuate over the day cycle showing a circadian oscillation. This figure represents the approximate Zeitgeber time (*ZT*) where those stresses reach their maximum. ZT0 and ZT12 represent dawn and dusk, respectively

19.7.1 Stomatal Movements

Circadian regulation in stomatal aperture had been largely reported in well-watered plants. The circadian clock allows to anticipate dawn and dusk transitions, promoting stomata opening before dawn, to maximize the CO_2 uptake necessary for photosynthesis, and stomata closure at dusk, avoiding unnecessary water loss without any increase in carbon fixation. For instance, disruption of the circadian function by CCA1 overexpression results in impaired stomatal movements failing to anticipate dawn and dusk. In these plants, stomata do not start to open before dawn and remain open at dusk, until the dark induces closure (Dodd et al. 2005). Additionally, CCA1-overexpressing plants showed increased water consumption and reduced carbon fixation levels, indicating the importance of this regulation for plant fitness (Dodd et al. 2005).

During water-deficit stress, ABA reduces the size of the stomatal pore. However, ABA is less effective at closing stomata in the morning than in the afternoon (Correia et al. 1995). Thus, the ABA-mediated stomata closure might be gated by the circadian clock. This may ensure that stomata are open to facilitate CO_2 uptake in the cool of the morning, when transpiration is lower, but are closed in the heat of the afternoon if water supply is limiting.

Although the circadian gating of the ABA-mediated stomatal closure is essential to enhance fitness in plants, the molecular mechanisms implicated are still largely unknown. The first molecular evidences were provided by studies using plants miss-expressing *TOC1*. These plants show impaired ABA-mediated stomata closure and altered drought tolerance. While *TOC1* overexpression leads to drought hypersensitivity and reduced sensitivity to ABA, plants with reduced *TOC1* expression show increased drought tolerance and ABA hypersensitivity. TOC1 effect on stomata aperture requires a functional ABAR/CHLH as TOC1 directly binds to its promoter (Legnaioli et al. 2009). ABAR/CHLH is also necessary for ABA-mediated induction of *TOC1* expression (see above). Interestingly, the *TOC1* promoter region contains two ABA-responsive elements (ABRE), commonly associated to ABA-induced genes through PP2C/SnRK2 pathway (Yoshida et al. 2010; Nakashima and Yamaguchi-Shinozaki 2013). Together with current circadian clock models, these findings allowed to build a simple mathematical model able to explain the interaction between the circadian clock and ABA to control stomata aperture (Fig. 19.4) (Pokhilko et al. 2013). This model, although is still simple and might suffer important revisions in the future, challenges the current model of ABA perception (see Chap. 7) and represents an important step toward the understanding of this crucial process for plant adaptation to the changing environmental conditions in water supply.

19.7.2 Cold Response

The exposure to non-damaging low temperature triggers the cold response and the induction of freezing tolerance. Clock gating of the cold responses was

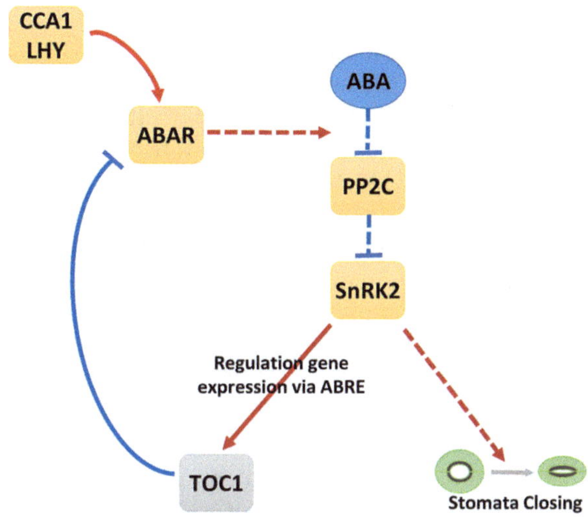

Fig. 19.4 Simplified model defining the regulation of the stomatal aperture by ABA and circadian clock (adapted from Pokhilko et al. 2013). *Red arrows* and *blue broken lines* represent activation and repression, respectively. Transcriptional and protein regulation are shown by *solid* and *dotted lines*, respectively

first postulated after a series of wide cold-responsive transcriptomal analysis. These studies showed significant differences depending on the sampling time (Bieniawska et al. 2008). More detailed analysis including the promoter region of cold-induced genes showed a significantly overrepresentation of evening elements (EE), CCA1/LHY binding elements commonly associated with genes expressed during the evening.

Molecularly, ABA and cold induce the transcription factors *DEHYDRATION-RESPONSIVE ELEMENT-BINDING PROTEIN 1b* (*DREB1b*), *DREB1c* and *DREB1a* (also known as *CBF*s genes), which are partly responsible for cold response and the induction of freezing tolerance (Vogel et al. 2005). Analysis of their promoter regions revealed the presence of EEs often coupled to ABA-responsive elements (ABRE), and both elements might be necessary for cold induction (Mikkelsen and Thomashow 2009). Indeed, DREB1/CBF expression shows diurnal oscillation and its cold induction exert circadian gating (Fowler et al. 2005). In this sense, cold induction studies at different circadian times showed a time-dependent induction with its maximum 4 h after dawn. Similar studies using *CCA1*-overexpressing plants showed no diurnal variation in the cold response, clearly in agreement with the circadian gating of *DREB1/CBF* cold induction (Fowler et al. 2005). The importance of CCA1 and LHY was further studied by using double *cca1/lhy* mutants that exhibit significantly low levels of *DREB1/CBF* expression and reduced freezing tolerance (Dodd et al. 2007b). Furthermore, *CCA1* and *LHY* exhibit temperature-dependent splicing; while the functional splicing variants of *CCA1* are produced preferentially at low temperatures and are important for cold induction of *DREB1*/CBF, LHY active variant is preferentially transcribed at high temperatures (James et al. 2012; Seo et al. 2012).

Other clock-related genes also have been described to participate in the cold-induced freezing tolerance. The *gi-3* mutant is susceptible to freezing, probably due to a decreased endogenous sugar concentration (Cao et al. 2007). Additionally, PRR5 and TOC1 bind the *DREB1/CBF* promoter to repress its expression (Nakamichi et al. 2012), most likely in collaboration with PHYTOCHROME-INTERACTING PROTEIN 7 (PIF7) (Maruyama et al. 2004). Furthermore, the arrhythmic triple mutant *prr5/prr7/prr9* displays constant high DREB1/CBF transcription levels and shows increased freezing tolerance (Nakamichi et al. 2009). These results are in agreement with the diurnal oscillation of the DREB1 genes and may be indicative of an ABA-independent transcriptional regulation.

ABA also mediates increases in cADPR and [Ca^{2+}]cyt in response to cold, and these stimuli are necessary for the induction of a numerous cold-responsive genes such as *RD29A* (*RESPONSIVE TO DESICCATION29A*) (Viswanathan and Zhu 2002; Wu et al. 2003). Abundance of the *RD29A* transcripts, [Ca^{2+}]cyt and [cADPR] has been shown to be circadian-regulated (Dodd et al. 2006, 2007; Johnson et al. 1995). Furthermore, cold response experiments at different times have shown that *RD29A* and [Ca^{2+}]cyt are also gated by the circadian clock, with maximum response during the day (Dodd et al. 2006). Reciprocal regulation between the clock and cADRP suggests that the gating of ABA and the cold signaling might be due to a control of the abundance of the transducing second messengers by the circadian clock.

Gating experiments showed that cold induction of *DREB1/CBF*, [Ca^{2+}]cyt and *RD29A* is maximized during the day period. This is indicative that this pathway may be different to the cold response resulting from the temperature drop during the night. Together with the important role of the circadian clock in photoperiod recognition in seasonal processes such as flowering (Andres and Coupland 2012), these findings strongly suggest that this cold induction may account for an onset of long-term responses such as induction of freezing tolerance in preparation for winter.

19.8 Circadian Regulation of Other ABA-related Physiological Processes

Many aspects of plant growth, development, and stress responses in plants are regulated by the circadian clock, including diurnal processes such as photosynthesis, stomatal opening, hypocotyl growth and leaf movement, or seasonal processes, such as floral transition, seed dormancy, and freezing tolerance (McClung 2006; Yakir et al. 2007). Some of these processes are also known to be modulated by ABA (Finkelstein and Rock 2002). However, the interconnections between clock and ABA regulation are in most of the cases still unknown. The role of ABA in these processes have been extensively reviewed in previous chapters, the following section briefly describes some of the most relevant aspects of the circadian regulation and the possible connections with the ABA signaling.

19.8.1 Seed Dormancy and Germination

Seed studies in order to determine the moment during the development when the circadian clock starts revealed that rhythmic gene expression begins on the second day after inducing germination by hydration, even in the absence of light (Penfield and Hall 2009; Salome et al. 2008; Zhong et al. 1998). These results confirm that seed hydration is sufficient to initiate the circadian clock. However, studies combining hydration with light cycles showed a considerable increase in the robustness and amplitude of the circadian rhythms, suggesting that a light signal is critical to the clock initiation and progression in seeds (Kikis et al. 2005).

Seed dormancy allows plants to germinate only when suitable environmental conditions are present. Environmental cues have been shown to regulate seed dormancy and germination by modulating levels of two phytohormones with opposite action. ABA establishes and maintains dormancy inhibiting germination, while gibberellins promote the breaking of dormancy and induce germination (Finch-Savage and Leubner-Metzger 2006). Taking in to account that several enzymes involved in gibberellin and ABA metabolism are controlled by the circadian clock, it is not surprising that mutations that alter circadian rhythms also affect germination (Penfield and Hall 2009). In this sense, *ztl* (ZEITLUPE, an F-box and blue receptor protein that mediates proteasomal degradation of TOC1 protein) and *lux* mutants exhibit reduced germination (Penfield and Hall 2009). Furthermore, stratification induces germination less efficiently in these mutants than wild-type seeds, while *lhy/cca1* double mutants germinate more efficiently. Germination defects in these mutants may be due to alterations in hormone signaling as happens in other stages of plant development (Covington and Harmer 2007; Dodd et al. 2007b; Hanano et al. 2006; Legnaioli et al. 2009; Mizuno and Yamashino 2008).

A possible mechanistic link connecting the circadian clock with seed dormancy and germination was postulated by the observation that TOC1 interact in yeast two hybrid with the central regulator of dormancy ABI3 (Kurup et al. 2000). This notion was recently been reinforced in studies showing that proper TOC1 levels are necessary for the ABA-mediated inhibition of germination (Castells et al. 2010). Thus, a proper function of the circadian clock in combination with the external signals is an essential mechanism that regulates the germination process.

19.8.2 Biotic Stress

Raising evidences have linked the circadian clock with the biotic responses, specifically pathogen resistance (Roden and Ingle 2009; Walley et al. 2007).

Concretely, *A. thaliana* sensitivity to *oomycete* and *Pseudomonas syringae* infection shows diurnal variations with an enhanced resistance at dawn, when the infection is more likely to occur, and the circadian clock seems to play an important role (Wang et al. 2011; Bhardwaj et al. 2011). Additionally, recent studies showed that not only the clock may modulate biotic stress responses, but the immune response can also input the clock and modify its pace (Zhang et al. 2013). Although the connections between biotic stress and the circadian clock are unknown, recent findings indicate an intricate and complex cross talk between these two pathways.

Many genes implicated in plant defense against pathogens show circadian regulation and a large number of them also contain an EE in their promoter regions (Wang et al. 2011). In agreement with this observation, resistance to infection is compromised at dawn in *cca1* mutants, whereas CCA1 overexpression results in an increased resistance (Wang et al. 2011). Similarly, diurnal variation in the resistance to *Pseudomonas* infection was lost after suppression of circadian rhythms, by either CCA1 overexpression or *elf3* mutation (Bhardwaj et al. 2011). This induced resistance at dawn is believed to be due to the higher expression levels of pathogen-associated molecular patterns (PAMPs)-triggered immunity genes.

To infect a plant, *Pseudomonas syringae* and other pathogens should first invade the leaf lamina through the stomata. The first layer of defense is constituted by the recognition of PAMPs by specific receptors and is known as pattern-triggered immunity (PTI). Activation of PTI induces rapid stomatal closure to prevent further access of the pathogen into the leaf intercellular space. PTI-induced stomatal closure requires components from the ABA signaling pathways such as reactive oxygen species (ROS) (Cho et al. 2009). Indeed, cellular ROS levels are enhanced by ABA in guard cells (Pei et al. 2000) and required for stomata closure. Both stomata closure and ROS levels have been shown to be clock-regulated (Lai et al. 2012). Circadian modulation of ROS levels is likely to be controlled through CCA1 direct regulation of genes implicated in ROS homeostasis. Interestingly, exogenous applications of ROS are able to modulate the circadian clock progression (Lai et al. 2012), adding another pathway whether ABA signaling can turn to regulate the clock pace.

19.8.3 Floral Transition

Flowering is a critical developmental switch in the plant life cycle. Establishing the correct timing for this transition is essential to determine reproductive success. The molecular mechanisms implicated in the control of flowering time have been most extensively studied in *Arabidopsis* and classically divided in four main pathways: vernalization, photoperiodic, autonomous, and gibberellin (Ream et al. 2012; Milec et al. 2014; Andres and Coupland 2012; Mouradov et al. 2002). Particularly, the circadian clock plays a major role in the regulation of photoperiodic flowering pathway, acting as an internal mechanism able to discriminate the duration of the day.

Concretely, *Arabidopsis* is a facultative long-day plant capable of flowering in both long-day and short-day conditions, but the floral transition occurs much earlier in long-day conditions (Hayama and Coupland 2004). The interactions between the light signals and the circadian clock converge in *CONSTANS* (*CO*), a key gene in the photoperiodic regulation of the flowering transition. *CO* gene expression, protein stability, and activity are regulated by light/dark cycles and the circadian clock (Suarez-Lopez et al. 2001; Yanovsky and Kay 2002). In long-day conditions, the expression of *CO* is restricted to the afternoon, whereas under short-day conditions its expression is mainly shifted to the dark period (Suarez-Lopez et al. 2001). This *CO* oscillatory expression has been proven to be essential for the photoperiodic discrimination (Yanovsky and Kay 2002) and therefore extensively studied. To precisely control the diurnal *CO* expression, in the morning, *CO* is repressed by CYCLING DOF FACTORs (CDFs) through direct binding to its promoter region (Fornara et al. 2009; Imaizumi et al. 2005). At the same time, CCA1/LHY represses the *CO* activators *FLAVIN-BINDING, KELCH REPEAT, F-BOX 1* (*FKF1*) and *GI*. In the afternoon, repression *of FKF1* and *GI* is released, which form a blue light-dependent complex. The FKF1-GI complex removes CDF repressors via proteasomal degradation (Sawa et al. 2007). At the same time, the *CDF* expression is downregulated sequentially by PRR9, PRR7, PRR5, and TOC1 (Nakamichi et al. 2009, 2012; Ito et al. 2008), thus ensuring *CO* expression.

Functionally, CO directly promotes *FLOWERING LOCUS T* (*FT*) expression, one of the major florigens. FT is synthesized in the leaf vascular tissue and transported through the phloem to the meristem where, together with other factors, it promotes the transcription reprogramming that is going to allow the floral transition (Kobayashi and Weigel 2007; Turck et al. 2008). CO activation of *FT* depends at least in part on the action of the photoreceptors, PHYA, CRY1, and CRY2 (Kardailsky et al. 1999; Kobayashi et al. 1999; Samach et al. 2000). These photoreceptors promote the stabilization of CO protein and stability, whereas PHYB and SUPRESSOR OF PHYA-105 (SPA1) stimulate CO degradation (Ishikawa et al. 2006; Laubinger et al. 2006; Valverde et al. 2004).

Preventing photoperiodic flowering induction under non-favorable environmental conditions is essential for the reproductive success. Therefore, ABA, being the major stress-response hormone, might play an important role. Indeed, ABA is considered a flowering repressor. External addition of ABA delays flowering, whereas ABA-deficient mutants show early flowering phenotype (Achard et al. 2006; Barrero et al. 2005). Considering the pervasiveness of ABA and circadian clock interaction, it is most likely that ABA may play also a role in the photoperiodic flowering pathway. In this sense, exogenous ABA lengthens the clock period (Hanano et al. 2006) and therefore is possible that, under photoperiodic inductive conditions, ABA might delay *CO* expression by shifting its peak to the dark period, thus inhibiting flowering. This is also consistent with the ABA acute induction of *TOC1* because its precise expression has been shown to be essential to sense variations on the photoperiod (Perales and Mas 2007).

Although ABA may delay flowering partially by its interaction with the photoperiodic pathway, its inhibitory role has been majorly associated with

FLOWERING LOCUS T (FLC) (Wang et al. 2013), a potent flowering repressor. ABA promotes SnRK2-mediated phosphorylation of the transcription factor ABI5, and phosphorylated ABI5 is able to bind FLC promoter and induce its expression, thus extending vegetative growth (Wang et al. 2013). Conversely, *FLC* expression is repressed by vernalization and autonomous pathways through modulation of the chromatin structure (Michaels 2009). Furthermore, some studies indicated that FLC might have circadian regulation (Fujiwara et al. 2010), indicating that FLC might act by integrating ABA, vernalization, autonomous pathway, and circadian signals to regulate flowering. Interestingly, vernalization also inputs the circadian clock by modifying its pace provably through *FLC* expression (Salathia et al. 2006). Consistently, FLC was shown to lengthen the circadian period at higher temperatures, contributing to temperature compensation, most likely through regulation of *LUX* expression (Edwards et al. 2005). Altogether, these results implicate a complex cross talk among ABA, circadian clock, and flowering pathways in which FLC plays a central role. Also, FLC emerges as another possible link through which ABA signaling might turn to regulate clock progression.

19.9 Concluding Remarks

The circadian clock involves a complex signaling network that is able to coordinate external stimuli and an internal time mechanism to optimize multiple processes all along the developmental process of plants. Given the important role of ABA in regulating many physiological and developmental processes in response to the changing environment, the integration of ABA signaling with the circadian clock network increases the robustness and the plasticity of the response, thus enhancing plant fitness and survival.

During the last few years, a series of exiting discoveries have been contributed substantially to a better understanding of the molecular mechanisms implicated in ABA signaling and the circadian clock organization. However, most of these studies have considered these pathways independently. Multiple lines of emerging evidence indicate an intricate network connecting ABA and the circadian clock involving multiple interactions at all levels. Thus, extending our knowledge on the links between these two essential signaling pathways will be crucial to understand at deeper level how plants cope with their changing environment.

References

Achard P, Cheng H, De Grauwe L, Decat J, Schoutteten H, Moritz T, Van Der Straeten D, Peng J, Harberd NP. Integration of plant responses to environmentally activated phytohormonal signals. Science. 2006;311(5757):91–4.

Alabadi D, Oyama T, Yanovsky MJ, Harmon FG, Mas P, Kay SA. Reciprocal regulation between TOC1 and LHY/CCA1 within the *Arabidopsis* circadian clock. Science. 2001;293(5531):880–3.

Andres F, Coupland G. The genetic basis of flowering responses to seasonal cues. Nat Rev Genet. 2012;13(9):627–39.

Barrero JM, Piqueras P, Gonzalez-Guzman M, Serrano R, Rodriguez PL, Ponce MR, Micol JL. A mutational analysis of the ABA1 gene of *Arabidopsis thaliana* highlights the involvement of ABA in vegetative development. J Exp Bot. 2005;56(418):2071–83.

Bell-Pedersen D, Cassone VM, Earnest DJ, Golden SS, Hardin PE, Thomas TL, Zoran MJ. Circadian rhythms from multiple oscillators: lessons from diverse organisms. Nat Rev Genet. 2005;6(7):544–56.

Bhardwaj V, Meier S, Petersen LN, Ingle RA, Roden LC. Defence responses of *Arabidopsis thaliana* to infection by *Pseudomonas syringae* are regulated by the circadian clock. PLoS ONE. 2011;6(10):e26968.

Bieniawska Z, Espinoza C, Schlereth A, Sulpice R, Hincha DK, Hannah MA. Disruption of the *Arabidopsis* circadian clock is responsible for extensive variation in the cold-responsive transcriptome. Plant Physiol. 2008;147(1):263–79.

Blasing OE, Gibon Y, Gunther M, Hohne M, Morcuende R, Osuna D, Thimm O, Usadel B, Scheible WR, Stitt M. Sugars and circadian regulation make major contributions to the global regulation of diurnal gene expression in *Arabidopsis*. Plant Cell. 2005;17(12):3257–81.

Bruce VG, Pittendrigh CS. Temperature Independence in a Unicellular "Clock". Proc Natl Acad Sci USA. 1956; 42(9):676–82.

Cao SQ, Song YQ, Su L. Freezing sensitivity in the *gigantea* mutant of *Arabidopsis* is associated with sugar deficiency. Biol Plant. 2007;51(2):359–62.

Castells E, Portoles S, Huang W, Mas P. A functional connection between the clock component TOC1 and abscisic acid signaling pathways. Plant Signal Behav. 2010;5(4):409–11.

Correia MJ, Pereira JS, Chaves MM, Rodrigues ML, Pacheco CA. ABA xylem concentrations determine maximum daily leaf conductance of field-grown *Vitis vinifera* L. plants. Plant Cell Environ. 1995;18(5):511–21.

Covington MF, Harmer SL. The circadian clock regulates auxin signaling and responses in *Arabidopsis*. PLoS Biol. 2007;5:e222.

Covington MF, Maloof JN, Straume M, Kay SA, Harmer SL. Global transcriptome analysis reveals circadian regulation of key pathways in plant growth and development. Genome Biol. 2008;9(8):R130.

Cho D, Shin D, Jeon B, Kwak J. ROS-mediated ABA signaling. J Plant Biol. 2009;52(2):102–13.

De Montaigu A, Toth R, Coupland G. Plant development goes like clockwork. Trends Genet. 2010;26(7):296–306.

Dixon LE, Knox K, Kozma-Bognar L, Southern MM, Pokhilko A, Millar AJ. Temporal repression of core circadian genes is mediated through EARLY FLOWERING 3 in *Arabidopsis*. Curr Biol. 2011;21(2):120–5.

Dodd AN, Gardner MJ, Hotta CT, Hubbard KE, Dalchau N, Love J, Assie JM, Robertson FC, Jakobsen MK, Goncalves J, Sanders D, Webb AA. The *Arabidopsis* circadian clock incorporates a cADPR-based feedback loop. Science. 2007a;318(5857):1789–92.

Dodd AN, Gardner MJ, Hotta CT, Hubbard KE, Dalchau N, Love J, Assie JM, Robertson FC, Jakobsen MK, Goncalves J, Sanders D, Webb AAR. The *Arabidopsis* circadian clock incorporates a cADPR-based feedback loop. Science. 2007b;318:1789–92.

Dodd AN, Jakobsen MK, Baker AJ, Telzerow A, Hou SW, Laplaze L, Barrot L, Poethig RS, Haseloff J, Webb AA. Time of day modulates low-temperature Ca signals in *Arabidopsis*. Plant J. 2006;48(6):962–73.

Dodd AN, Salathia N, Hall A, Kevei E, Toth R, Nagy F, Hibberd JM, Millar AJ, Webb AA. Plant circadian clocks increase photosynthesis, growth, survival, and competitive advantage. Science. 2005;309(5734):630–3.

Edwards KD, Lynn JR, Gyula P, Nagy F, Millar AJ. Natural allelic variation in the temperature-compensation mechanisms of the *Arabidopsis thaliana* circadian clock. Genetics. 2005;170(1):387–400.

Farre EM, Harmer SL, Harmon FG, Yanovsky MJ, Kay SA. Overlapping and distinct roles of PRR7 and PRR9 in the *Arabidopsis* circadian clock. Curr Biol. 2005;15:47–54.

Finch-Savage WE, Leubner-Metzger G. Seed dormancy and the control of germination. New Phytol. 2006;171(3):501–23.

Finkelstein RR, Gampala SS, Rock CD. Abscisic acid signaling in seeds and seedlings. Plant Cell. 2002;14 Suppl:S15–45.

Finkelstein RR, Rock CD. Abscisic Acid biosynthesis and response. Arabidopsis Book/Am Soc Plant Biol. 2002;1:e0058.

Fornara F, Panigrahi KCS, Gissot L, Sauerbrunn N, Rühl M, Jarillo JA, Coupland G. *Arabidopsis* DOF transcription factors act redundantly to reduce CONSTANS expression and are essential for a photoperiodic flowering response. Dev Cell. 2009;17(1):75–86.

Fowler SG, Cook D, Thomashow MF. Low temperature induction of *Arabidopsis* CBF1, 2, and 3 is gated by the circadian clock. Plant Physiol. 2005;137(3):961–8.

Fujiwara S, Nakagawa M, Oda A, Kato K, Mizoguchi T. Photoperiod pathway regulates expression of MAF5 and FLC that encode MADS-box transcription factors of the FLC family in *Arabidopsis*. Plant Biotechnol. 2010;27(5):447–54.

Fukushima A, Kusano M, Nakamichi N, Kobayashi M, Hayashi N, Sakakibara H, Mizuno T, Saito K. Impact of clock-associated *Arabidopsis* pseudo-response regulators in metabolic coordination. Proc Natl Acad Sci USA. 2009; 106(17):7251–56.

Gendron JM, Pruneda-Paz JL, Doherty CJ, Gross AM, Kang SE, Kay SA. *Arabidopsis* circadian clock protein, TOC1, is a DNA-binding transcription factor. Proc Natl Acad Sci USA. 2012; 109(8):3167–72.

Golden SS, Ishiura M, Johnson CH, Kondo T. Cyanobacterial circadian rhythms. Annu Rev Plant Physiol Plant Mol Biol. 1997;48:327–54.

Green RM, Tobin EM. Loss of the circadian clock-associated protein 1 in *Arabidopsis* results in altered clock-regulated gene expression. Proc Natl Acad Sci USA. 1999; 96(7):4176–79.

Gutierrez RA, Stokes TL, Thum K, Xu X, Obertello M, Katari MS, Tanurdzic M, Dean A, Nero DC, McClung CR, Coruzzi GM. Systems approach identifies an organic nitrogen-responsive gene network that is regulated by the master clock control gene CCA1. Proc Natl Acad Sci USA. 2008; 105(12):4939–44.

Hanano S, Domagalska MA, Nagy F, Davis SJ. Multiple phytohormones influence distinct parameters of the plant circadian clock. Genes Cells. 2006;11(12):1381–92.

Harmer SL. The circadian system in higher plants. Annu Rev Plant Biol. 2009;60:357–77.

Harmer SL, Hogenesch JB, Straume M, Chang HS, Han B. Orchestrated transcription of key pathways in *Arabidopsis* by the circadian clock. Science. 2000;290:2110–3.

Hayama R, Coupland G. The molecular basis of diversity in the photoperiodic flowering responses of *Arabidopsis* and rice. Plant Physiol. 2004;135(2):677–84.

Hazen SP, Naef F, Quisel T, Gendron JM, Chen H, Ecker JR, Borevitz JO, Kay SA. Exploring the transcriptional landscape of plant circadian rhythms using genome tiling arrays. Genome Biol. 2009;10(2):R17.

Hazen SP, Schultz TF, Pruneda-Paz JL, Borevitz JO, Ecker JR, Kay SA. LUX ARRHYTHMO encodes a Myb domain protein essential for circadian rhythms. Proc Natl Acad Sci USA. 2005; 102:10387–92.

Helfer A, Nusinow DA, Chow BY, Gehrke AR, Bulyk ML, Kay SA. LUX ARRHYTHMO encodes a nighttime repressor of circadian gene expression in the *Arabidopsis* core clock. Curr Biol. 2011;21(2):126–33.

Huang W, Perez-Garcia P, Pokhilko A, Millar AJ, Antoshechkin I, Riechmann JL, Mas P. Mapping the core of the *Arabidopsis* circadian clock defines the network structure of the oscillator. Science. 2012;336(6077):75–9.

Hubbard KE, Robertson FC, Dalchau N, Webb AA. Systems analyses of circadian networks. Mol BioSyst. 2009;5(12):1502–11.

Imaizumi T, Schultz TF, Harmon FG, Ho LA, Kay SA. FKF1 F-box protein mediates cyclic degradation of a repressor of CONSTANS in *Arabidopsis*. Science. 2005;309(5732):293–7.

Ishikawa M, Kiba T, Chua NH. The *Arabidopsis* SPA1 gene is required for circadian clock function and photoperiodic flowering. Plant J. 2006;46(5):736–46.

Ito S, Kawamura H, Niwa Y, Nakamichi N, Yamashino T, Mizuno T. A genetic study of the *Arabidopsis* circadian clock with reference to the TIMING OF CAB EXPRESSION 1 (TOC1) gene. Plant Cell Physiol. 2009;50(2):290–303.

Ito S, Niwa Y, Nakamichi N, Kawamura H, Yamashino T, Mizuno T. Insight into missing genetic links between two evening-expressed pseudo-response regulator genes TOC1 and PRR5 in the circadian clock-controlled circuitry in *Arabidopsis thaliana*. Plant Cell Physiol. 2008;49(2):201–13.

James AB, Syed NH, Bordage S, Marshall J, Nimmo GA, Jenkins GI, Herzyk P, Brown JW, Nimmo HG. Alternative splicing mediates responses of the *Arabidopsis* circadian clock to temperature changes. Plant Cell. 2012;24(3):961–81.

Johnson CH, Knight MR, Kondo T, Masson P, Sedbrook J, Haley A, Trewavas A. Circadian oscillations of cytosolic and chloroplastic free calcium in plants. Science. 1995;269(5232):1863–5.

Kant P, Gordon M, Kant S, Zolla G, Davydov O, Heimer YM, Chalifa-Caspi V, Shaked R, Barak S. Functional-genomics-based identification of genes that regulate *Arabidopsis* responses to multiple abiotic stresses. Plant Cell Environ. 2008;31(6):697–714.

Kardailsky I, Shukla VK, Ahn JH, Dagenais N, Christensen SK, Nguyen JT, Chory J, Harrison MJ, Weigel D. Activation tagging of the floral inducer FT. Science. 1999;286(5446):1962–5.

Kidokoro S, Maruyama K, Nakashima K, Imura Y, Narusaka Y, Shinwari ZK, Osakabe Y, Fujita Y, Mizoi J, Shinozaki K, Yamaguchi-Shinozaki K. The phytochrome-interacting factor PIF7 negatively regulates DREB1 expression under circadian control in *Arabidopsis*. Plant Physiol. 2009;151(4):2046–57.

Kikis EA, Khanna R, Quail PH. ELF4 is a phytochrome-regulated component of a negative-feedback loop involving the central oscillator components CCA1 and LHY. Plant J. 2005;44(2):300–13.

Kobayashi Y, Kaya H, Goto K, Iwabuchi M, Araki T. A pair of related genes with antagonistic roles in mediating flowering signals. Science. 1999;286(5446):1960–2.

Kobayashi Y, Weigel D. Move on up, it's time for change–mobile signals controlling photoperiod-dependent flowering. Genes Dev. 2007;21(19):2371–84.

Kolmos E, Nowak M, Werner M, Fischer K, Schwarz G, Mathews S, Schoof H, Nagy F, Bujnicki JM, Davis SJ. Integrating ELF4 into the circadian system through combined structural and functional studies. HFSP J. 2009;3(5):350–66.

Kurup S, Jones HD, Holdsworth MJ. Interactions of the developmental regulator ABI3 with proteins identified from developing *Arabidopsis* seeds. Plant J. 2000;21(2):143–55.

Lai AG, Doherty CJ, Mueller-Roeber B, Kay SA, Schippers JH, Dijkwel PP. CIRCADIAN CLOCK-ASSOCIATED 1 regulates ROS homeostasis and oxidative stress responses. Proc Natl Acad Sci USA. 2012; 109(42):17129–34.

Laubinger S, Marchal V, Le Gourrierec J, Wenkel S, Adrian J, Jang S, Kulajta C, Braun H, Coupland G, Hoecker U. *Arabidopsis* SPA proteins regulate photoperiodic flowering and interact with the floral inducer CONSTANS to regulate its stability. Development. 2006;133(16):3213–22.

Lee KH, Piao HL, Kim HY, Choi SM, Jiang F, Hartung W, Hwang I, Kwak JM, Lee IJ, Hwang I. Activation of glucosidase via stress-induced polymerization rapidly increases active pools of abscisic acid. Cell. 2006;126(6):1109–20.

Legnaioli T, Cuevas J, Mas P. TOC1 functions as a molecular switch connecting the circadian clock with plant responses to drought. EMBO J. 2009;28:3745–57.

Lim GH, Zhang X, Chung MS, Lee DJ, Woo YM, Cheong HS, Kim CS. A putative novel transcription factor, AtSKIP, is involved in abscisic acid signalling and confers salt and osmotic tolerance in *Arabidopsis*. New Phytol. 2010;185(1):103–13.

Liu Y, Bell-Pedersen D. Circadian rhythms in *Neurosporacrassa* and other filamentous fungi. Eukaryot Cell. 2006;5(8):1184–93.

Locke JC, Kozma-Bognar L, Gould PD, Feher B, Kevei E, Nagy F, Turner MS, Hall A, Millar AJ. Experimental validation of a predicted feedback loop in the multi-oscillator clock of *Arabidopsis thaliana*. Mol Syst Biol. 2006;2:59.

Makino S, Matsushika A, Kojima M, Yamashino T, Mizuno T. The APRR1/TOC1 quintet implicated in circadian rhythms of *Arabidopsis thaliana*: I. Characterization with APRR1-overexpressing plants. Plant Cell Physiol. 2002;43(1):58–69.

Martin-Tryon EL, Kreps JA, Harmer SL. GIGANTEA acts in blue light signaling and has biochemically separable roles in circadian clock and flowering time regulation. Plant Physiol. 2007;143:473–86.

Maruyama K, Sakuma Y, Kasuga M, Ito Y, Seki M, Goda H, Shimada Y, Yoshida S, Shinozaki K, Yamaguchi-Shinozaki K. Identification of cold-inducible downstream genes of the *Arabidopsis* DREB1A/CBF3 transcriptional factor using two microarray systems. Plant J. 2004;38(6):982–93.

Más P. Circadian clock function in *Arabidopsis thaliana*: time beyond transcription. Trends Cell Biol. 2008;18(6):273–81.

Más P, Alabadi D, Yanovsky MJ, Oyama T, Kay SA. Dual role of TOC1 in the control of circadian and photomorphogenic responses in *Arabidopsis*. Plant Cell. 2003a;15(1):223–36.

Más P, Kim WY, Somers DE, Kay SA. Targeted degradation of TOC1 by ZTL modulates circadian function in *Arabidopsis thaliana*. Nature. 2003b;426(6966):567–70.

Más P, Yanovsky MJ. Time for circadian rhythms: plants get synchronized. Curr Opin Plant Biol. 2009;12(5):574–9.

Matsui A, Ishida J, Morosawa T, Mochizuki Y, Kaminuma E, Endo TA, Okamoto M, Nambara E, Nakajima M, Kawashima M, Satou M, Kim JM, Kobayashi N, Toyoda T, Shinozaki K, Seki M. *Arabidopsis* transcriptome analysis under drought, cold, high-salinity and ABA treatment conditions using a tiling array. Plant Cell Physiol. 2008;49(8):1135–49.

McClung CR. Plant circadian rhythms. Plant Cell. 2006;18(4):792–803.

McClung CR. Comes a time. Curr Opin Plant Biol. 2008;11(5):514–20.

Michaels SD. Flowering time regulation produces much fruit. Curr Opin Plant Biol. 2009;12(1):75–80.

Mikkelsen MD, Thomashow MF. A role for circadian evening elements in cold-regulated gene expression in *Arabidopsis*. Plant J. 2009;60(2):328–39.

Milec Z, Valarik M, Bartos J, Safar J. Can a late bloomer become an early bird? Tools for flowering time adjustment. Biotechnol Adv. 2014;32(1):200–14.

Millar AJ, Kay SA. Integration of circadian and phototransduction pathways in the network controlling CAB gene transcription in *Arabidopsis*. Proc Natl Acad Sci USA. 1996; 93:15491–96.

Mizoguchi T, Wheatley K, Hanzawa Y, Wright L, Mizoguchi M, Song HR, Carre IA, Coupland G. LHY and CCA1 are partially redundant genes required to maintain circadian rhythms in *Arabidopsis*. Dev Cell. 2002;2(5):629–41.

Mizuno T, Yamashino T. Comparative transcriptome of diurnally oscillating genes and hormone-responsive genes in *Arabidopsis thaliana*: insight into circadian clock-controlled daily responses to common ambient stresses in plants. Plant Cell Physiol. 2008;49:481–7.

Mouradov A, Cremer F, Coupland G. Control of flowering time: interacting pathways as a basis for diversity. Plant Cell. 2002;14 Suppl:S111–30.

Nakamichi N, Kiba T, Kamioka M, Suzuki T, Yamashino T, Higashiyama T, Sakakibara H, Mizuno T. Transcriptional repressor PRR5 directly regulates clock-output pathways. Proc Natl Acad Sci USA. 2012; 109(42):17123–28.

Nakamichi N, Kita M, Ito S, Yamashino T, Mizuno T. PSEUDO-RESPONSE REGULATORS, PRR9, PRR7 and PRR5, together play essential roles close to the circadian clock of *Arabidopsis thaliana*. Plant Cell Physiol. 2005;46(5):686–98.

Nakamichi N, Kusano M, Fukushima A, Kita M, Ito S, Yamashino T, Saito K, Sakakibara H, Mizuno T. Transcript profiling of an *Arabidopsis* PSEUDO RESPONSE REGULATOR arrhythmic triple mutant reveals a role for the circadian clock in cold stress response. Plant Cell Physiol. 2009;50(3):447–62.

Nakashima K, Yamaguchi-Shinozaki K. ABA signaling in stress-response and seed development. Plant Cell Rep. 2013;32(7):959–70.

Nusinow DA, Helfer A, Hamilton EE, King JJ, Imaizumi T, Schultz TF, Farre EM, Kay SA. The ELF4-ELF3-LUX complex links the circadian clock to diurnal control of hypocotyl growth. Nature. 2011;475(7356):398–402.

Onai K, Ishiura M. PHYTOCLOCK 1 encoding a novel GARP protein essential for the *Arabidopsis* circadian clock. Genes Cells. 2005;10(10):963–72.

Ouyang Y, Andersson CR, Kondo T, Golden SS, Johnson CH. Resonating circadian clocks enhance fitness in cyanobacteria. Proc Natl Acad Sci USA. 1998; 95(15):8660–4.

Pei ZM, Murata Y, Benning G, Thomine S, Klusener B, Allen GJ, Grill E, Schroeder JI. Calcium channels activated by hydrogen peroxide mediate abscisic acid signalling in guard cells. Nature. 2000;406(6797):731–4.

Penfield S, Hall A. A role for multiple circadian clock genes in the response to signals that break seed dormancy in *Arabidopsis*. Plant Cell. 2009;21(6):1722–32.

Perales M, Mas P. A functional link between rhythmic changes in chromatin structure and the *Arabidopsis* biological clock. Plant Cell. 2007;19(7):2111–23.

Pittendrigh CS. On temperature independence in the clock system controlling emergence time in *Drosophila*. Proc Natl Acad Sci USA. 1954; 40(10):1018–29.

Pokhilko A, Fernandez AP, Edwards KD, Southern MM, Halliday KJ, Millar AJ. The clock gene circuit in *Arabidopsis* includes a repressilator with additional feedback loops. Mol Syst Biol. 2012;8:574.

Pokhilko A, Mas P, Millar AJ. Modelling the widespread effects of TOC1 signalling on the plant circadian clock and its outputs. BMC Syst Biol. 2013;7:23.

Pruneda-Paz JL, Breton G, Para A, Kay SA. A functional genomics approach reveals CHE as a component of the *Arabidopsis* circadian clock. Science. 2009;323(5920):1481–5.

Rand DA, Shulgin BV, Salazar D, Millar AJ. Design principles underlying circadian clocks. J R Soc Interface. 2004;1(1):119–30.

Ream TS, Woods DP, Amasino RM. The molecular basis of vernalization in different plant groups. Cold Spring Harb Symp Quant Biol. 2012;77:105–15.

Robertson FC, Skeffington AW, Gardner MJ, Webb AA. Interactions between circadian and hormonal signalling in plants. Plant Mol Biol. 2009;69(4):419–27.

Roden LC, Ingle RA. Lights, rhythms, infection: the role of light and the circadian clock in determining the outcome of plant-pathogen interactions. Plant Cell. 2009;21(9):2546–52.

Salathia N, Davis SJ, Lynn JR, Michaels SD, Amasino RM, Millar AJ. FLOWERING LOCUS C-dependent and -independent regulation of the circadian clock by the autonomous and vernalization pathways. BMC Plant Biol. 2006;6:10.

Salome PA, McClung CR. PSEUDO-RESPONSE REGULATOR 7 and 9 are partially redundant genes essential for the temperature responsiveness of the *Arabidopsis* circadian clock. Plant Cell. 2005;17:791.

Salome PA, Xie Q, McClung CR. Circadian timekeeping during early *Arabidopsis* development. Plant Physiol. 2008;147(3):1110–25.

Samach A, Onouchi H, Gold SE, Ditta GS, Schwarz-Sommer Z, Yanofsky MF, Coupland G. Distinct roles of CONSTANS target genes in reproductive development of *Arabidopsis*. Science. 2000;288(5471):1613–6.

Sanchez JP, Duque P, Chua NH. ABA activates ADPR cyclase and cADPR induces a subset of ABA-responsive genes in *Arabidopsis*. Plant J. 2004;38(3):381–95.

Sanchez SE, Petrillo E, Beckwith EJ, Zhang X, Rugnone ML, Hernando CE, Cuevas JC, Godoy Herz MA, Depetris-Chauvin A, Simpson CG, Brown JW, Cerdan PD, Borevitz JO, Mas P, Ceriani MF, Kornblihtt AR, Yanovsky MJ. A methyl transferase links the circadian clock to the regulation of alternative splicing. Nature. 2010;468(7320):112–6.

Sawa M, Nusinow DA, Kay SA, Imaizumi T. FKF1 and GIGANTEA complex formation is required for day-length measurement in *Arabidopsis*. Science. 2007;318(5848):261–5.

Schaffer R, Ramsay N, Samach A, Corden S, Putterill J, Carre IA, Coupland G. The late elongated hypocotyl mutation of *Arabidopsis* disrupts circadian rhythms and the photoperiodic control of flowering. Cell. 1998;93(7):1219–29.

Seo PJ, Park MJ, Lim MH, Kim SG, Lee M, Baldwin IT, Park CM. A self-regulatory circuit of CIRCADIAN CLOCK-ASSOCIATED1 underlies the circadian clock regulation of temperature responses in *Arabidopsis*. Plant Cell. 2012;24(6):2427–42.

Seung D, Risopatron JP, Jones BJ, Marc J. Circadian clock-dependent gating in ABA signalling networks. Protoplasma. 2012;249(3):445–57.

Shen YY, Wang XF, Wu FQ, Du SY, Cao Z, Shang Y, Wang SL, Peng CC, Yu XC, Zhu SY, Fan RC, Xu YH, Zhang DP. The Mg-chelatase H subunit is an abscisic acid receptor. Nature. 2006;443:823–6.

Song YH, Ito S, Imaizumi T. Flowering time regulation: photoperiod- and temperature-sensing in leaves. Trends Plant Sci. 2013;18(10):575–83.

Spoel SH, van Ooijen G. Circadian redox signaling in plant immunity and abiotic stress. Antioxid Redox Signal 2014; 20(18):3024−3039.

Strayer C, Oyama T, Schultz TF, Raman R, Somers DE, Mas P, Panda S, Kreps JA, Kay SA. Cloning of the *Arabidopsis* clock gene TOC1, an autoregulatory response regulator homolog. Science. 2000;289(5480):768–71.

Suarez-Lopez P, Wheatley K, Robson F, Onouchi H, Valverde F, Coupland G. CONSTANS mediates between the circadian clock and the control of flowering in *Arabidopsis*. Nature. 2001;410(6832):1116–20.

Thompson AJ, Jackson AC, Parker RA, Morpeth DR, Burbidge A, Taylor IB. Abscisic acid biosynthesis in tomato: regulation of zeaxanthine oxidase and 9-cis-epoxycarotenoid dioxygenase mRNAs by light/dark cycles, water stress and abscisic acid. Plant Mol Biol. 2000;42(6):833–45.

Turck F, Fornara F, Coupland G. Regulation and identity of florigen: FLOWERING LOCUS T moves center stage. Annu Rev Plant Biol. 2008;59:573–94.

Valverde F, Mouradov A, Soppe W, Ravenscroft D, Samach A, Coupland G. Photoreceptor regulation of CONSTANS protein in photoperiodic flowering. Science. 2004;303(5660):1003–6.

Viswanathan C, Zhu JK. Molecular genetic analysis of cold-regulated gene transcription. Philos Trans R Soc Lond B Biol Sci. 2002;357(1423):877–86.

Vogel JT, Zarka DG, Van Buskirk HA, Fowler SG, Thomashow MF. Roles of the CBF2 and ZAT12 transcription factors in configuring the low temperature transcriptome of *Arabidopsis*. Plant J. 2005;41(2):195–211.

Vollmers C, Gill S, DiTacchio L, Pulivarthy SR, Le HD, Panda S. Time of feeding and the intrinsic circadian clock drive rhythms in hepatic gene expression. Proc Natl Acad Sci USA. 2009; 106(50):21453–8.

Walley JW, Coughlan S, Hudson ME, Covington MF, Kaspi R, Banu G, Harmer SL, Dehesh K. Mechanical stress induces biotic and abiotic stress responses via a novel cis-element. PLoS Genet. 2007;3(10):1800–12.

Wang W, Barnaby JY, Tada Y, Li H, Tor M, Caldelari D, Lee DU, Fu XD, Dong X. Timing of plant immune responses by a central circadian regulator. Nature. 2011;470(7332):110–4.

Wang X, Wu F, Xie Q, Wang H, Wang Y, Yue Y, Gahura O, Ma S, Liu L, Cao Y, Jiao Y, Puta F, McClung CR, Xu X, Ma L. SKIP is a component of the spliceosome linking alternative splicing and the circadian clock in *Arabidopsis*. Plant Cell. 2012;24(8):3278–95.

Wang Y, Li L, Ye T, Lu Y, Chen X, Wu Y. The inhibitory effect of ABA on floral transition is mediated by ABI5 in *Arabidopsis*. J Exp Bot. 2013;64(2):675–84.

Wang ZY, Tobin EM. Constitutive expression of the CIRCADIAN CLOCK ASSOCIATED 1 (CCA1) gene disrupts circadian rhythms and suppresses its own expression. Cell. 1998;93(7):1207–17.

Wijnen H, Young MW. Interplay of circadian clocks and metabolic rhythms. Annu Rev Genet. 2006;40:409–48.

Woelfle MA, Ouyang Y, Phanvijhitsiri K, Johnson CH. The adaptive value of circadian clocks: an experimental assessment in cyanobacteria. Curr Biol. 2004;14(16):1481–6.

Wu FQ, Xin Q, Cao Z, Liu ZQ, Du SY, Mei C, Zhao CX, Wang XF, Shang Y, Jiang T, Zhang XF, Yan L, Zhao R, Cui ZN, Liu R, Sun HL, Yang XL, Su Z, Zhang DP. The magnesium-chelatase H subunit binds abscisic acid and functions in abscisic acid signaling: new evidence in *Arabidopsis*. Plant Physiol. 2009;150(4):1940–54.

Wu Y, Sanchez JP, Lopez-Molina L, Himmelbach A, Grill E, Chua NH. The *abi1-1* mutation blocks ABA signaling downstream of cADPR action. Plant J. 2003;34(3):307–15.

Yakir E, Hilman D, Harir Y, Green RM. Regulation of output from the plant circadian clock. FEBS J. 2007;274(2):335–45.

Yang X. A wheel of time: the circadian clock, nuclear receptors, and physiology. Genes Dev. 2010;24(8):741–7.

Yanovsky MJ, Kay SA. Molecular basis of seasonal time measurement in *Arabidopsis*. Nature. 2002;419(6904):308–12.

Yoshida T, Fujita Y, Sayama H, Kidokoro S, Maruyama K, Mizoi J, Shinozaki K, Yamaguchi-Shinozaki K. AREB1, AREB2, and ABF3 are master transcription factors that cooperatively regulate ABRE-dependent ABA signaling involved in drought stress tolerance and require ABA for full activation. Plant J. 2010;61(4):672–85.

Zeilinger MN, Farre EM, Taylor SR, Kay SA, Doyle FJ 3rd. A novel computational model of the circadian clock in *Arabidopsis* that incorporates PRR7 and PRR9. Mol Syst Biol. 2006;2:58.

Zhang C, Xie Q, Anderson RG, Ng G, Seitz NC, Peterson T, McClung CR, McDowell JM, Kong D, Kwak JM, Lu H. Crosstalk between the circadian clock and innate immunity in *Arabidopsis*. PLoS Pathog. 2013;9(6):e1003370.

Zhong HH, Painter JE, Salome PA, Straume M, McClung CR. Imbibition, but not release from stratification, sets the circadian clock in *Arabidopsis* seedlings. Plant Cell. 1998;10(12):2005–17.

Chapter 20
ABA Regulation of Plant Response to Biotic Stresses

Ligang Chen and Diqiu Yu

Abstract Plants live in complicated environments in which they are obliged to defend against a broad range of attackers. In order to protect themselves, plants have evolved complex regulatory signaling networks where multiple hormonal pathways antagonistically or synergistically interact and influence plant defense responses. Beside of its prominent roles in abiotic stress tolerance, the plant hormone abscisic acid (ABA) has also been emerged as crucial regulator in biotic stresses. Accumulated studies have shown that ABA can exert both positive and negative influence on host defense, and its efficacy is dependent on the specific plant–attacker combination. In this chapter, we mainly focused on recent literature dealing with the roles of ABA in modulating plant defense responses against various attackers.

Keywords ABA · Plant–pathogen interaction · Biotic and abiotic stress · Cross talk

20.1 Introduction

Being unable to escape from the surrounding complex circumstances, plants are obligated to respond more effectively than animals to various living-threaten stresses in order to successfully survive and reproduce. These stresses can negatively or even harmfully affect the growth and productivity of plants and can be broadly divided into two categories: abiotic and biotic. Abiotic stresses include all the non-living environmental factors, such as water deficiency (drought), excessive salt (salinity), threshold temperatures (from freezing to scorching), decreased availability of essential nutrients (nutrient starvation), and variable

L. Chen · D. Yu (✉)
Key Laboratory of Tropical Forest Ecology, Xishuangbanna Tropical Botanical Garden,
Chinese Academy of Sciences, Kunming 650223, Yunnan, China
e-mail: ydq@xtbg.ac.cn

D.-P. Zhang (ed.), *Abscisic Acid: Metabolism, Transport and Signaling*,
DOI 10.1007/978-94-017-9424-4_20

light conditions. Biotic stresses are those that occurs as a result of damage done to plants by other living organisms, such as bacteria, viruses, fungi, oomycetes, parasites, and herbivores These stresses can occur at multiple stages of plant development and often more than one stress simultaneously affects the plant, potentially restricting plant growth, plant development, or even determining plant species distribution across different types of environments.

The adaption to various stresses of all living organisms is the outcome of interaction between their genome and the ever-changing environment. In order to better adapt to adverse conditions and finally complete their life cycles, plants have evolved plenty of complex signaling systems to respond to external and internal cues to regulate growth and development under stressful environments. These responses are mediated by plant growth regulators (phytohormones), compounds derived from plant biosynthetic pathways that can act both locally and systemically. Numerous studies have demonstrated that plant hormones play critical regulatory roles in both plant growth and development and the responses to adverse stresses. In general, it is accepted that the plant hormones consist of five major classes of plant hormones, namely abscisic acid (ABA), ethylene (ET), cytokinin (CK), auxin (IAA), gibberellin (GA), as well as jasmonate (JA), brassinosteroids (BR), salicylic acid (SA), nitric oxide (NO), strigolactone (SL), karrikins, polyamines, and plant peptide hormones, and it is imaginable that additional growth regulators would be discovered in the future. So far, thanks to so many excellent researchers' efforts in these fields, we have been made significant research progresses associated with the biosynthesis of plant hormones, their metabolism, as well as their roles in signaling. In this chapter, we will mainly discuss the role of ABA in different plant–attacker interactions.

20.2 Hormonal Modulation of Plant Immunity

During the long-term and constant interactions with various microbial pathogens or herbivores, resistant plants have successfully evolved sophisticated defense mechanisms to protect themselves. Normally, plants use both constitutive and inducible defense responses to fend off various attackers. Constitutive defenses were based on many preformed barriers, such as cell walls, waxy epidermal cuticles, and bark; while inducible defenses include the production of toxic chemicals, pathogen-degrading enzymes, and deliberate cell suicide, which were activated upon the recognition of pathogen attack. The constitutive and inducible defenses constitute a multilayered defense which is initiated sequentially during plant–attacker interactions (Chisholm et al. 2006; Jones and Dangl 2006; Nishimura and Dangl 2010).

The defensive response of the plant upon pathogen infection is the outcome of highly coordinated sequential changes at the cellular level, during which the plant hormones play important roles (Ton et al. 2009). SA, jasmonic acid (JA), and ethylene (ET) are the primary defense hormones, and their importance in plant innate immunity is well documented, particularly in the model plant *Arabidopsis thaliana*

(Grant and Jones 2009; Robert-Seilaniantz et al. 2011). The SA signaling pathway is mainly linked to resistance to biotrophic pathogens that feed on living host tissues and often associated with hypersensitive response (HR; Durrant and Dong 2004; Vlot et al. 2009), while the JA and ET signaling pathways are predominantly associated with resistance to necrotrophic pathogens that promote host cell death at early stages of infection (Glazebrook 2004). It is well known that there has an mutual antagonism between SA-mediated and JA/ET-mediated defense signaling pathways (Kunkel and Brooks 2002). More recently, other plant hormones, including the stress hormone ABA and the developmental hormones, such as cytokinins (CKs), auxin, brassinosteroids (BRs), and gibberellins (GAs), have also been demonstrated to be important regulators during plant–microbe interactions (Mauch-Mani and Mauch 2005; Robert-Seilaniantz et al. 2007; Grant and Jones 2009). Interestingly, numerous studies have suggested that these hormones participated in disease responses by interaction antagonistically or synergistically with the SA–JA–ET backbone of the plant immune signaling network (Verhage et al. 2010; Pieterse et al. 2012; Robert-Seilaniantz et al. 2007). Thus, it is interesting to reveal how pathogens disturb and evade plant defense responses by manipulation of hormone signaling and how hosts counteract this manipulation (Grant and Jones 2009). In this chapter, we focus on recent literature dealing with the roles of ABA in plant immunity.

20.3 Roles of ABA in Plant–Pathogen Interaction

Beside of its prominent roles in plant response to abiotic stress, the plant hormone ABA has also been demonstrated to function as a crucial regulator in plant biotic defense responses. Accumulated studies have shown that ABA can exert a positive or negative influence on the host defense against almost all types of attackers, such as bacterial, fungal, and oomycete pathogens, as well as herbivore and virus, and its efficacy is dependent on the specific plant–attacker combination (Ton et al. 2009; Cao et al. 2011). Further more, studies have demonstrated that ABA has variable roles throughout different phases of plant defense (Ton et al. 2009). Below, we will first talk about the role of ABA in defense against bacterial, fungal, and oomycete pathogens.

20.4 Expression Pattern of ABA Pathway Core Components Under Pathogen Challenge

The public availability of micro-array datasets allows researchers to obtain some clues as to certain gene's function toward particular stress conditions. Recently, Chan summarized the transcriptional profiling of *Arabidopsis* ABA pathway core components under different pathogen infection based on both publicly available micro-array data and their qRT-PCR results. Their results showed that most ABA pathway core components (from ABA synthesis, metabolism, transport, to signal transduction) were responded to pathogen infection. For example, the expression of three key ABA biosynthesis enzymes, namely *NCED3*, *ABA3/LOS5* and *AAO3*, was all induced by *Pseudomonas syringae pv. Tomato DC3000*, *Pseudomonas syringae pv. maculicola ES4326,* and *Botrytis Cinerea* infection; similarly, the expression of two ABA catabolism genes, *CYP707A1* and *CYP707A4*, was also slightly induced in a long-term manner by the infection of the three pathogens (Chan 2012). Thus, homeostasis of endogenous ABA is important for the plant to properly respond to pathogen infection. Recently, PYR/PYLs, a family of novel START domain proteins, were identified as ABA receptors (Ma et al. 2009; Park et al. 2009). The ABA-bound receptors are able to inactivate the PP2Cs and disrupt or decrease the physical interaction between the PP2Cs and the SnRK2 s and finally transduce the signal to downstream targets (Fujii et al. 2009). Chan's results also showed that the expression levels of most *PYR/PYL*s were inhibited while the expression levels of most *PP2C*s were induced by pathogen infection (Chan 2012). Therefore, there has an increased ratio of PP2Cs:PYR/PYLs upon pathogen infection which may be required for activation of the downstream ABA signal pathway. Consistent with the these findings, one recent research has demonstrated that *Arabidopsis* plants impaired in ABA biosynthesis, such as *aba1-6*, or in ABA signaling, like the quadruple *pyr/pyl* mutant (*pyr1pyl1pyl2pyl4*), showed enhanced resistance to the fungus *Plectosphaerella cucumerina* while the *hab1-1abi1-2abi2-2* mutant impaired in three phosphatases that negatively regulate ABA signaling, are more susceptible to this fungus (Sánchez-Vallet et al. 2012). In tomato, the expression of ABA pathway core components was also affected in *B.cinerea*-infected fruits (Blanco-Ulate et al. 2013). Based on both the expression pattern and function of ABA pathway core components under pathogen infection, we can deduce that ABA may play important roles during plant–pathogen interactions.

20.5 ABA and Stomatal Innate Immunity

Plants use innate physical and biochemical barriers to protect themselves from a variety of pathogens. Thus, to successfully infect a host plant, pathogens need to evolve mechanisms that allow them to circumvent plant mechanical barriers, such as cell walls and waxy epidermal cuticles. Some fungal pathogens penetrate plant tissue using mechanical force or by secreting cuticle- and cell-wall-degrading enzymes (Mendgen et al. 1996; van Kan J.A. 2006). But unlike fungal pathogens, bacteria pathogens cannot directly penetrate the leaf epidermis and instead enter leaf tissues using preexisting openings, such as stomata, hydathodes, nectarthodes, and wounds (Melotto et al. 2008). Among them, stomata represent a major route for many phytopathogens to enter the plant tissues (Zeng et al. 2010). In order to prevent pathogen entry into host tissue through the stomata, plants can actively close the stomata by activating stomatal innate immunity, which usually occurs

1 h after exposure to pathogenic and non-pathogenic bacteria (Melotto et al. 2006, 2008; Schulze-Lefert and Robatzek 2006; Zeng et al. 2010; Faulkner and Robatzek 2012). Furthermore, perception of several pathogen- or microbe-associated molecular patterns (PAMPs or MAMPs), such as the flagellin derivative flg22 and lipopolysaccharides, can rapidly induce stomatal closure and then activate stomatal innate immunity (Melotto et al. 2006; Zeng et al. 2010). Interestingly, besides of its prominent role in regulating stomatal closure upon drought stresses (Schroeder et al. 2001), several studies have also demonstrated the involvement of ABA in stomatal innate immunity. For example, the *Arabidopsis* mutants defective in ABA synthesis (*aba3-1*) or ABA signaling (*ost1-2*) are unable to perceive two PAMPs, flg22 and lipopolysaccharide, and fail to induce stomatal closure. Thus, it is evident that ABA can positively prevent pathogen infection by functioning in PAMP-induced stomatal closure (Melotto et al. 2006). The ABA-controlled PAMP-induced stomatal closure required several key steps, including ABA synthesis, NO production, and the *OST1* kinase (Melotto et al. 2006). Similarly, SA can also induce stomatal closure, and the SA-deficient mutants *eds5-1*, *eds16-1/sid2,* and NahG transgenic plants, or SA signaling mutants, such as *npr1*, show impaired ability in closing the stomata in response to the *Pst DC3000* or PAMPs, indicating the important role of SA in stomatal innate immunity (Melotto et al. 2006; Zeng et al. 2010). Further studies demonstrated that the SA signaling regulator NPR1 acts downstream of SA, but upstream of ABA in stomatal immunity (Zeng and He 2010). Recently, it was reported that the rhizobacteria *Bacillus subtilis* FB17-mediated stomatal closure involved both SA and ABA signaling (Kumar et al. 2012). However, the mechanism of possible cross talk between SA and ABA during stomatal innate immunity is still unclear.

In turn, pathogens have evolved a variety of virulence factors to counteract stomatal innate immunity and finally colonize the host tissues through reopening the stomata. *Pst DC3000* can reopen stomata by producing the effector molecule coronatine (Melotto et al. 2006; Zeng and He 2010). Recent study demonstrated that *Pst DC3000* can also produce another effector HopM1 to reduce both bacterial PAMP flg22 and the fungal PAMP chitin-induced stomatal closure (Lozano-Durán et al. 2013). Similarly, *Xanthomonas campestris pv. campestrisis* (*Xcc*) can also overcome stomatal innate immunity through a DSF cell-to-cell signal-regulated virulence factor which has the ability to revert bacteria, lipopolysaccharide, or ABA-induced stomatal closure (Gudesblat et al. 2009). The bacterial citrus pathogen, *X. axonopodis pv. citri*, can produce a plant natriuretic peptide (PNP) to antagonize ABA-dependent stomatal closure and finally create a favorable host environment for its own survival (Gottig et al. 2008). Some fungal pathogens also use virulence factors, such as fusicoccin and oxalate, to counteract stomatal closure (Turner et al. 1969; Guimaraes et al. 2004). Hence, based on these reports, we can deduced that ABA functions in stomatal innate immunity by controlling stomatal closure upon pathogen infection while some successful pathogens have evolved virulence factors to counteract stomatal innate immunity. The ABA-mediated stomatal innate immunity constitutes an important part of the plant pre-invasive penetration resistance (Ton et al. 2009).

20.6 ABA and Callose Deposition

Upon the recognition of a successful invasion, the host plant would activate post-invasive resistance responses, such as rapid callose deposition and generation of reactive oxygen species (ROS), during which ABA involved. Compared with its positive roles in stomatal innate immunity, the role of ABA in post-invasive resistance responses is controversial and seems to vary among different plant–pathogen interactions.

Accumulating studies have demonstrated that both ABA and callose can affect the consequence of many plant–pathogen interactions, and ABA can both negatively and positively affect the deposition of callose which is dependent on the plant–pathogen combination (Flors et al. 2001). However, the exact molecular mechanism behind the modulation of callose by ABA upon pathogen attack is unclear.

In plant–bacteria interactions, several studies have demonstrated that ABA generally functions as a negative regulator in the post-invasive defense by suppressing bacteria-induced callose formation. When challenged with *Pst DC3000*, ABA hyper-responsive *abi2* mutants strongly decreased callose deposition accompanied by enhanced susceptibility, while ABA-insensitive *abi1-1* and *abi2-1* mutants deposit augmented levels of callose which coincided with enhanced resistance (de Torres-Zabala et al. 2007; Goritschnig et al. 2008). Furthermore, pre-treatment with ABA also strongly reduced callose deposition in subsequently *Pst DC3000* inoculated plants (de Torres-Zabala et al. 2007).

Contrarily, in some plant–fungal–oomycetes interactions, ABA can exert an positive role by stimulating callose deposition or by modulating the priming of callose deposition (Ton and Mauch-Mani 2004; Flors et al. 2001). Treatment with exogenous ABA can mimic the effect of β-aminobutyric acid (BABA) and result in priming for pathogen-inducible callose deposition, coinciding with enhanced resistance against *P. cucumerina* and *Alternaria brassicicola*. It was also demonstrated that both the intact ABA signaling and callose formation are crucial for BABA-induced resistance against *P. cucumerina* and *A. brassicicola* in *Arabidopsis*, since the ABA-deficient mutant *aba1-5*, the ABA-insensitive mutant *abi4-1* and the callose-deficient mutant *pmr4-1* failed to express the BABA-induced resistance. The reduction or absence of primed callose deposition in BABA-treated *aba1/ibs3* and *abi4* plants upon infection by *Hyaloperonospora parasitica* and *P. cucumerina*, respectively, further provides a causal link between ABA signaling and callose deposition (Ton and Mauch-Mani 2004; Ton et al. 2005). Thus, these studies clearly showed that ABA functions in the BABA-induced augmentation of pathogen-induced callose deposition. However, it should be noted that besides of its role in callose deposition, ABA-dependent resistance can also to be exerted in a callose-independent manner, such as activation of JA biosynthesis, as reflected in *Arabidopsis–Pythium irregulare* interaction (Adie et al. 2007).

20.7 ABA and ROS

Reactive oxygen species (ROS) burst is one of the most important plant defense responses to pathogens, whose efficacy in plant defense remains controversial (Elad 1992; Mehdy 1994; Asselbergh et al. 2007). Several studies implied that there may exist an correlation between ABA level or signaling and ROS generation. Studies using the ABA-deficient *sitiens* tomato mutant revealed that the timely and localized hyper-induction of hydrogen peroxide (H_2O_2, a major ROS) and subsequent HR-mediated defense response in the epidermal cell wall can effectively arrest the necrotrophic pathogens *B. cinerea* and *Erwinia chrysanthemi* (Asselbergh et al. 2007, 2008). Consistently, treatment with exogenous ABA can restore the pathogen susceptibility of *sitiens* and also suppress H_2O_2 accumulation in this mutant (Asselbergh et al. 2007). Thus, ABA deficiency in *sitiens* contributes to the generation of H_2O_2 which may result in the epidermal ROS/HR-mediated defense responses. Recently, Seifi et al. further showed that both epidermal HR-mediated response and mesophyllic GS1/GABA-shunt-mediated anti-cell death mechanism constitute the multifaceted resistance mechanism in *sitiens* against *B. cinerea* (Seifi et al. 2013).

On the other hand, ROS are important messengers in ABA-mediated stress responses. Studies have demonstrated that ROS play important roles in both MAMP- and ABA-induced stomatal closure. ABA and flg22 activate the plasma membrane-resident NADPH oxidases to trigger H_2O_2 production (Kwak et al. 2003; Mersmann et al. 2010). The NADPH oxidase *RESPIRATORY BURST OXIDASE HOMOLOGUE D* and *F* (*RBOHD* and *RBOHF*) function redundantly in ABA-induced stomatal closure, and the *atrbohD/F* double mutations severely impair this process (Kwak et al. 2003). However, the MAMP-triggered ROS production is mainly dependent on the *RBOHD*, whose mutation abolished the flg22- and elf18-dependent stomatal closure (Mersmann et al. 2010; Macho et al. 2012). In addition, the chloroplastic enzyme ASPARTATE OXIDASE (AO), which catalyzes the first irreversible step in the de novo biosynthesis of NAD, is also required for PAMP-induced RBOHD-dependent ROS burst and stomatal closure; however, it does not impact RBOHF-mediated ABA-induced ROS burst or ABA-induced stomatal closure (Macho et al. 2012). Thus, the ABA and MAMP-triggered production of ROS is achieved partially by sharing some common components.

20.8 Interplay Between ABA and SA–JA–ET During Plant–Pathogen Interaction

Based on the above observations, impairment in ABA synthesis or signal resulted in altered responses of host plants to various pathogen attack, indicating the involvement of ABA in plant biotic stresses. Besides of the above-mentioned effect of ABA in callose deposition and ROS generation, ABA can also affect the expression of defense-related genes. Accumulating evidences have demonstrated that ABA can participate in plant biotic stress responses by connecting to the well-documented SA and JA/ET-mediated signaling pathways. Compared to the predominantly antagonistic relation between ABA and SA, the interaction between ABA and JA/ET can be both synergistic and antagonistic.

20.9 ABA and SA

Numerous studies have provided evidences that ABA can negatively regulate SA-mediated pathogen responses. The production of ABA was enhanced by *Pst DC3000* infection, implying that ABA may increase susceptibility of host plant to this bacterium (Truman et al. 2006; de Torres-Zabala et al. 2009). Coincidently, both exogenous ABA application and addition of exogenous ABA by over-expression of ABA biosynthesis-related *NCED* genes result in increased growth of bacterial pathogens, while mutations in *Arabidopsis* ABA synthesis mutants, such as *aba3-1* and *aao3*, confer enhanced resistance to *Pst DC3000* infection (Mohr and Cahill 2003; Fan et al. 2009; de Torres-Zabala et al. 2009). Similarly, exogenous ABA application renders rice hypersusceptible to the infection by both the rice leaf blight bacteria *Xanthomonas oryzae pv oryzae* (*Xoo*) and the blast fungus *Magnaporthe grisea*, and further analysis demonstrated that ABA promotes susceptibility to the two pathogens by suppressing SA-regulated defenses (Jiang et al. 2010; Xu et al. 2013). Furthermore, ABA suppresses the induction of SAR by inhibition of SA-dependent gene expression (Yasuda et al. 2008). Studies also showed that ABA can suppress SA accumulation. For example, exogenous ABA application can reduce SA content in *P. syringae pv. Tomato* (*Pst 1065*)-infected leaves (Mohr and Cahill 2007). This finding was echoed by a subsequent study where the increased resistance in the ABA-deficient *aao3* mutant was accompanied by a higher basal and induced SA level, coincident with the enhanced basal and induced expression level of the SA biosynthetic gene *ICS1*, implying that ABA can suppress SA biosynthesis through downregulation of *ICS1* (de Torres-Zabala et al. 2009). In ABA-deficient *sitiens* tomato mutants, it was showed that the enhanced resistance to *B. cinerea* was correlated with the elevated expression of the SA biosynthesis gene *PHENYLALANINE AMMONIA LYASE* (*PAL*) and increased sensitivity to BTH-induced *PR1* expression, indicating that ABA can promote susceptibility to *B. cinerea* in tomato by negatively modulating the SA-dependent defense pathway (Audenaert et al. 2002). Together, these results demonstrated that ABA appears to suppress the SA-dependent signaling mechanisms.

Interestingly, SA can both enhance and suppress the ABA-mediated signaling pathway. For example, it was showed that SA can positively regulate ABA accumulation—at least in the latter stages of infection, as reflected by the reduced accumulation of endogenous ABA in the *Pst DC3000*-infected *sid2-1* mutants (de Torres-Zabala et al. 2009), while the activation of the SA-mediated SAR can inhibit the expression of ABA biosynthesis-related and ABA responsive genes, in which the NPR1 protein or signaling downstream of NPR1 may participate (Yasuda et al. 2008). Collectively, we can deduce that the bidirectional interaction between ABA and SA can occur at multiple points of their signaling pathways with their effect of cross talk prominently antagonistic.

20.10 ABA and JA/ET

Numerous studies have also suggested that there is a complex cross talk between ABA and JA/ET-mediated signaling pathways. As we know that the JA signal pathway can be divided into two antagonistic branches, namely the ET co-regulated ethylene response factor (ERF)-branch and the ABA co-regulated MYC-branch. The ERF-branch is mainly responsible for the resistance to necrotrophic pathogens and is controlled by the AP2/ERF domain transcription factors ERF1 and ORA59, leading to activation of the downstream marker gene *PDF1.2* (Penninckx et al. 1998; Lorenzo et al. 2004; Pré et al. 2008),while the MYC-branch is activated during plant–herbivore interactions and is regulated by the basic helix-loop-helix leucine zipper proteins MYC2/3/4, resulting in the transcription of *VSP1* and *VSP2* marker genes (Anderson et al. 2004; Thaler and Bostock 2004; Lorenzo and Solano 2005; Fernández-Calvo et al. 2011; Niu et al. 2011). Interestingly, ABA can confer synergistic effects on the MYC-branch while antagonistic effects on the ERF-branch, as demonstrated by the opposite effects of ABA on two branches' master genes, with enhanced expression for *MYC2* and *VSP2* but suppressed expression for *ERF1*, *ORA59*, and *PDF1.2* (Anderson et al. 2004; Kazan and Manners 2013). Consistent with these findings, it was reported that ABA-deficient mutants showed increased susceptibility to herbivory (Thaler and Bostock 2004; Bodenhausen and Reymond 2007) and enhanced resistance to necrotrophic pathogens (Anderson et al. 2004; Sánchez-Valletetal et al. 2012). Studies also demonstrated that ABA can function as an essential signal to promote JA biosynthesis and then lead to the activation of defense responses against the damping-off oomycete *P. irregulare* (Adie et al. 2007). Additionally, there exist bidirectional antagonism between ABA and ET signaling. For example, the ethylene insensitive mutant *ein2* had elevated ABA levels (Ghassemian et al. 2000), while increased ABA levels may result in reduced ethylene levels as demonstrated by the *OsMPK5* over-expression plants or exogenous ABA treatment in rice (Xiong and Yang 2003; Yang 2007). Together, ABA can both suppress and enhance the JA/ET-mediated defense responses whose efficacy is determined by the combination between host plants and pathogens.

Furthermore, there may have complex cross talk between ABA, JA, and SA during the interaction between *A. brassicicola* and susceptible *Brassica juncea* or resistant *Sinapis alba*. The susceptible *B. Juncea* triggered an SA-mediated biotrophic mode of defense response upon challenge with *A. Brassicicola*, while the resistant *S. Alba* initiated enhanced ABA response to counteract SA response and synchronously restore the necrotrophic mode of resistance by enhancing JA biosynthesis (Mazumder et al. 2013). Thus, ABA can function in plant disease resistance by enhancing one defense pathway but inhibiting another defense pathway.

Besides of the above-mentioned regulation of ABA during plant pathogen defense, such as stomatal innate immunity, callose deposition, ROS, and cross talk with SA/JA/ET, ABA can affect plant defense responses by modulating the cell wall composition (Sánchez-Vallet et al. 2012) and also functions in induced systemic resistance (ISR) that is elicited by below-ground rhizobacteria in *Arabidopsis* (Van der Ent et al. 2009; Kumar et al. 2012).

20.11 Roles of ABA in Plant–Herbivore Interaction

In order to efficiently counteract the attack by herbivorous insects, plants have evolved surveillance systems aimed to recognize attacking harmful insects and respond with proper defense mechanisms, including the "early warning" response, wound-induced resistance (WIR), and herbivore-associated molecular pattern (HAMP)-induced immunity (HTI) (Erb et al. 2012). Following the recognition of an herbivorous insect, plants use an complex signaling cascades to reprogram their phenotype. Although the recognition of pathogens or herbivores can be very specific, plants may use a "common downstream signaling machinery" to activate defense responses (Katagiri et al. 2010). Similar to their important roles in plant response to pathogen infection, phytohormones are also involved in herbivore defense, with the JA pathway playing a dominant role in host resistance. In addition to this core hormonal signal, other plant hormones, including ABA, are also been shown to participate in herbivore defense, and more and more evidence demonstrated that different multiple hormone signaling pathways can cross-talk or interact to translate initial perception events into appropriate responses that increase survival and/or reproduction of plants under herbivore attack (Erb et al. 2012). Below, we will focus on the role of ABA in herbivore defense.

Several studies have showed that herbivore attack or treatment with herbivorous oral secretions (OS) increases the ABA level in the host plant (Erb et al. 2009; Schäfer et al. 2011; Erb et al. 2011; Tooker and De Moraes 2011). Analysis using of mutants with altered ABA biosynthesis or signaling also supports participation of ABA in herbivore defense. In most cases, ABA-deficient mutants were reported to be more susceptible to herbivory (Thaler and Bostock 2004; Bodenhausen and Reymond 2007). Transcription profiling analysis also revealed a modulating influence of ABA on herbivore-induced gene expression (Bodenhausen and Reymond 2007). Interestingly, recent studies have shown that *ABI4* may integrate redox signaling pathway to promote resistance to aphid feeding, while the *aba2* and *abi1* mutant plants showed opposite aphid resistance to those of *abi4* (Kerchev et al. 2013), suggesting that the ABI4-mediated redox regulation may have a greater effect than ABA on aphid resistance in *Arabidopsis*. Another study showed that mutation in *aba2* gene compromised *Pieris rapae*-induced resistance and

P.rapae-induced expression of *VSP1*, a marker defense gene during plant–herbivore interaction (Vos et al. 2013). These studies implied that ABA may play important roles in herbivore defense. Further studies demonstrated that the ABA-mediated defense response against herbivore attack is tightly interconnected with the JA pathway and that the modulation of ABA synthesis or signaling in response to herbivore attack can modulate JA-driven defense responses. For example, during plant–herbivore interactions, ABA and JA can synergistically regulate the transcription of the MYC2 transcription factor, which is activated upon feeding by herbivorous insects and finally regulated the expression of downstream marker genes, such as VSP1 and VSP2 (Anderson et al. 2004; Thaler and Bostock 2004; Lorenzo and Solano 2005; Fernández-Calvo et al. 2011; Niu et al. 2011). Furthermore, ABA functions as a crucial regulator of *P. Rapae*-induced resistance in systemic tissues by activating primed JA-regulated defense responses in *Arabidopsis* (Vos et al. 2013). Research in *Nicotiana attenuata* also demonstrated that the ABA signaling was required for OS-triggered defense responses partly by enhancing the accumulation of the JA and its associated responses (Dinh et al. 2013). In *Dionaea muscipula*, the ABA signaling pathway was used to protect the carnivore from untimely prey catching during periods of drought, while the jasmonate signaling pathway plays essential roles in signaling systemically the presence of a predator and also subsequently eliciting secretions (Escalante-Pérez et al. 2011). Thus, ABA and JA can synergistically or separately function in plant responses to insect attack.

20.12 Roles of ABA in Plant–Virus Interaction

The interaction between plants and viral pathogens reflects a sophisticated coevolution of recognition, defense, and counter-defense mechanisms. Because of the limited genetic information encoded by the virus genome, plant viruses must depend entirely on host cells to replicate their genome and produce infectious progeny, during which they should use a variety of strategies to suppress or bypass host defense. In plants, these strategies involve enhancing infection by manipulating host resources, such as the formation of replication complexes (Hills et al. 1987), enlargement of the plasmodesma (PD) size exclusion limit (Wolf et al. 1989; Waigmann et al. 1994), evolution of viral suppressors of RNA silencing (VSRs) to counteract antiviral silencing (Burgyán and Havelda 2011), interference with regulation of the plant cell cycle (Gutierrez 2000; Lai et al. 2008), and usage of host components for its own replication (Cui et al. 2007). In turn, plants have developed diverse defense mechanisms to fight viral infection, such as HR and SAR responses, RNA silencing, hormone-mediated signaling pathways, as well as regulation of metabolism (Durrant and Dong 2004; Seo et al. 2004; Herbers et al. 1996; Chen et al. 2010; Ding and Voinnet 2007; Ding 2010). Numerous reports have demonstrated that several plant hormones, such as SA, ET, and ABA, are involved in plant basal defense responses in plant–virus interactions. Here, we will mainly talk about the role of ABA in plant antivirus defense.

So far, there has relatively less information related to the ABA–virus interaction when compared with the interactions between ABA and non-virus pathogens, such as bacteria, fungi, and oomycetes. Several researches have demonstrated that infection with plant virus can increase the ABA levels in host plants which indicates the involvement of ABA in plant antivirus defense. For example, infection with banana bunchy top virus (BBTV) or turnip mosaic virus (TuMV) can increase the accumulation of ABA in *banana* and non-heading Chinese cabbage, respectively (Zhang et al. 1997; Wang et al. 2011). Compared with both negative and positive roles of ABA in plant defense against non-viral pathogens, previous studies mainly support a positive correlation between ABA levels and antiviral resistance. In tobacco, exogenous application of ABA increases resistance to tobacco mosaic virus (TMV) infection, and TMV infection also increases ABA concentrations (Whenham et al. 1986). Iriti and Faoro also demonstrated that exogenous application of ABA induces a significant resistance to *tobacco necrosis virus* (Iriti and Faoro 2008). Our recent results also suggest that ABA has a positive role in tobacco mosaic virus (TMV)-cg infection. Exogenous application of ABA greatly inhibited the accumulation of TMV-cg RNA in systemically infected leaves, and studies using both ABA-deficient and ABA-insensitive mutants showed that the transportation of TMV-cg in systemically infected leaves was faster in these ABA-related mutants (Chen et al. 2013). However, as to the bamboo mosaic virus (BaMV), the ABA pathway may have multifaceted effects on this virus accumulation in host plants. In this study, although exogenous application of ABA enhances resistance to *BaMV* infection, mutations in ABA biosynthesis genes, including *NCED3*, *ABA2,* and *AAO3*, differentially affected BaMV accumulation. Their results demonstrated that *ABA2* functions in *BaMV* defense by dividing the ABA pathway into two parts, with upstream genes (such *NCED3*) required for BaMV accumulation and downstream genes (such as *AAO3*, *ABI1*, *ABI3*, and *ABI4*) required for plant resistance (Alazem et al. 2014). Thus, their finding is contrary to ours related to the role of *ABA2* in plant antivirus defense. This strengthens the probability that the role of ABA2 in plant virus defense seems to vary among different plant–virus interactions. Furthermore, ABA also can exert its positive effect on virus infection by downregulating the transcriptional level of *β-1,3-glucanase* genes that can degrade the b-1,3-glucan callose, forming a physical barrier to viral spread through plasmodesmata (Rezzonico et al. 1998).

Despite the above observations, our knowledge about the role of ABA in antiviral defense is still limited and needs further investigation. As we know that each plant virus encodes an average of 4–10 proteins required to coordinate the complex biochemical and intermolecular interactions that facilitate infection. Really, several studies have demonstrated that virus-encoded proteins, such as cucumber mosaic virus-encoded 2b and beet severe curly top virus C2, can physically interact with Argonaute1 and SAMDC1, respectively, to attenuate host antiviral RNA silencing or interferes with DNA methylation-mediated gene silencing in *Arabidopsis* (Zhang et al. 2006, 2011). Thus, it would be interesting to further investigate how and where these viral proteins interact with host factors and how various plant virus use these interactions to modify the hormone-mediated signaling pathways, especially the ABA pathway, to disturb host defense and finally promote infection and vice versa.

20.13 ABA-Mediated Cross Talk Between Biotic and Abiotic Stress

In nature, plants always simultaneously suffered from biotic and abiotic stresses that cross-talk with each other. In order to successfully adapt to such complex environments, plants have developed diverse signaling pathways for combating and tolerating them. Numerous evidences have accumulated supporting the notion that different signaling pathways are interconnected to constitute the complicated networks that lead to various plant responses (López-Pérez et al. 2009; Knight and Knight 2001; Chen and Zhu 2004). Among them, the ABA signaling pathway is well known for its role in abiotic stresses, especially in drought, and has also emerged as an important regulator in plant immunity. Thus, ABA may play important roles in mediating the cross talk between biotic and abiotic stresses. Expression profile analysis demonstrated that both abiotic and biotic stress treatments can affect the expression levels of ABA signaling pathway core components, indicating that abiotic and biotic stress responses shared ABA signal pathway in *Arabidopsis* (Chan 2012). Indeed, numerous studies demonstrated that mutations in ABA signaling pathway core components cause altered responses to both abiotic and biotic stresses. Further studies demonstrated that ABA participated in this process possibly by antagonistically or synergistically interacting with the prominent defense phytohormones SA, JA, and ET, implying that pre-treatment with abiotic stresses may affect the plant's response to subsequent pathogen or herbivore attack and vice versa (Mauch-Mani and Mauch 2005; Robert-Seilaniantz et al. 2007; Yasuda et al. 2008; Anderson et al. 2004). Increasing evidences have showed that abiotic stresses, such as high temperature and humidity as well as drought and salinity stress, can enhance the susceptibility of plant to biotic stresses (Mohr and Cahill 2003; Koga et al. 2004; Moeder and Yoshioka 2009). For example, drought stresses have been shown to enhance the susceptibility of host plants to various pathogens, such as *Arabidipsis* to avirulent *Pst 1065*, bean plants (*Phaseolus vulgaris*) to the charcoal rot causal fungus *Macrophomina phaseolina*, as well as vine (*Parthenocissus quinquefolia*) to the xylem-limited bacteria *Xylella fastidiosa* (Mohr and Cahill 2003; Mayek-Perez et al. 2002; McElrone et al. 2001). Although abiotic stress responses generally decreased the disease resistance, in some cases, pre-exposure to abiotic stress can also enhance the resistance to certain pathogens. Examples includes the enhanced resistance of tomato against *B.cinerea* (necrotroph) and *Oidium neolycopersici* (biotroph) after pre-exposure to drought stress and *Arabidopsis* against virulent bacterial pathogen *Pst DC3000* after pre-treatment with submergence (Achuo et al. 2006; Hsu et al. 2013). Thus, abiotic stresses can both positively or negatively affect the resistance of host plant to pathogen or herbivore attack. Abiotic stresses, such as drought and submergence, can affect the ABA metabolism, signal transduction, and catabolism (Saika et al. 2007; Xiong et al. 2002; Chan et al. 2012), indicating that the altered responses of host plants to pathogen or herbivore attack after pre-exposure with certain abiotic stress are at least partially dependent on the ABA signaling pathway. Based on the above observations, the role of ABA in plant biotic stresses should not be generalized to one particular pathogen lifestyle (biotroph, hemibiotroph, or necrotroph), and the exact role that conferred by ABA may correlate with the specific virulence strategies adopted by individual pathogen species. In turn, since the cross talk between ABA and SA or JA/ET may be bidirectional, and indeed, accumulated studies have demonstrated that the SA and JA/ET are also involved in various plant abiotic stresses (Miura and Tada 2014; Fujita et al. 2006; Ballaré et al. 2011), it would be reasonable to deduce that biotic stresses may also affect the response of plants to abiotic stresses (Yasuda et al. 2008; Mosher et al. 2010; Moeder et al. 2010). Thus, ABA functions as a crucial regulator during the adaption of plant to the variable environment by mediating the cross talk between biotic and abiotic stresses.

Furthermore, besides of the ABA pathway core components, more and more proteins, such as various transcription factors (such as WRKYs, MYBs, NACs) and histone deacetylases, were all demonstrated to be involved in both ABA signaling and biotic stresses (Rushton et al. 2010; Chen et al. 2012; Seo and Park 2010; Jensen et al. 2008; Chen and Wu 2010; Kim et al. 2008); thus, these proteins may also participate in the cross talk between biotic and abiotic stresses through the coordination with the ABA pathway core components. Further molecular studies of such proteins will add new insights into the ABA-regulated cross talk between biotic and abiotic stress responses.

20.14 Conclusions and Perspective

As presented in this chapter, the abiotic hormone ABA has been emerged as an critical regulator in plant biotic stresses. Compared with the positive roles of ABA in stomata innate immunity, it has both positive and negative effects on callose deposition and ROS generation which may be determined by the time and location of infection or the combination of host plant and attackers. Generally, ABA functions in plant biotic stresses by cross talk antagonistically or synergistically with the SA–JA–ET backbone of the plant immune signaling network. However, the underlying mechanisms that control the cross talk remained largely unknown. We supposed that the cross talk may be conducted through direct interactions of one hormone pathway's core component with another pathway's core component or direct binding of a given DNA binding protein in one hormone pathway to its target gene that functions in another pathway. Future research on this fine-tuning of signaling pathways will add our knowledge to better understand the complex cross talk between ABA and SA–JA–ET-mediated stress responses. Furthermore, since there exists an evident trade-off between abiotic stress tolerance and biotic defense, we should consider more when attempting to manipulate the ABA signaling for modern agriculture.

Acknowledgments The work in the authors' laboratory was supported by the Natural Science Foundation of China (31200915), the West Light Foundation of CAS, and Yong Innovation Promotion Association of the Chinese Academy of Sciences. We apologize to the colleagues whose work could not be cited here because of space restrictions.

References

Achuo EA, Prinsen E, Hofte M. Influence of drought, salt stress and abscisic acid on the resistance of tomato to *Botrytis cinerea* and *Oidium neolycopersici*. Plant Pathol. 2006;55:178–86.

Adie BA, Pérez-Pérez J, Pérez-Pérez MM, Godoy M, Sánchez-Serrano JJ, Schmelz EA, Solano R. ABA is an essential signal for plant resistance to pathogens affecting JA biosynthesis and the activation of defenses in *Arabidopsis*. Plant Cell. 2007;19:1665–81.

Alazem M, Lin KY, Lin NS. The abscisic acid pathway has multifaceted effects on the accumulation of Bamboo mosaic virus. Mol Plant Microbe Interact. 2014;27:177–89.

Anderson JP, Badruzsaufari E, Schenk PM, Manners JM, Desmond OJ, Ehlert C, Maclean DJ, Ebert PR, Kazan K. Antagonistic interaction between abscisic acid and jasmonate-ethylene signaling pathways modulates defense gene expression and disease resistance in *Arabidopsis*. Plant Cell. 2004;16:3460–79.

Asselbergh B, Curvers K, Franca SC, Audenaert K, Vuylsteke M, Van Breusegem F, Höfte M. Resistance to Botrytis cinerea in sitiens, an abscisic acid-deficient tomato mutant, involves timely production of hydrogen peroxide and cell wall modifications in the epidermis. Plant Physiol. 2007;144:1863–77.

Asselbergh B, Achuo EA, Höfte M, Van Gijsegem F. Abscisic acid-deficiency leads to rapid activation of tomato defense responses upon infection with *Erwinia chrysanthemi*. Mol Plant Pathol. 2008;9:11–24.

Audenaert K, De Meyer GB, Hofte MM. Abscisic acid determines basal susceptibility of tomato to *Botrytis cinerea* and suppresses salicylic acid-dependent signaling mechanisms. Plant Physiol. 2002;128:491–501.

Ballaré CL. Jasmonate-induced defenses: a tale of intelligence, collaborators and rascals. Trends Plant Sci. 2011;16:249–57.

Blanco-Ulate B, Vincenti E, Powell AL, Cantu D. Tomato transcriptome and mutant analyses suggest a role for plant stress hormones in the interaction between fruit and *Botrytis cinerea*. Front Plant Sci. 2013;4:142.

Bodenhausen N, Reymond P. Signaling pathways controlling induced resistance to insect herbivores in *Arabidopsis*. Mol Plant–Microbe Interact. 2007;20:1406–20.

Burgyán J, Havelda Z. Viral suppressors of RNA silencing. Trends Plant Sci. 2011;16:265–72.

Cao FY, Yoshioka K, Desveaux D. The roles of ABA in plant–pathogen interactions. J Plant Res. 2011;124:489–99.

Chan Z. Expression profiling of ABA pathway transcripts indicates crosstalk between abiotic and biotic stress responses in *Arabidopsis*. Genomics. 2012;100:110–5.

Chen WJ, Zhu T. Networks of transcription factors with roles in environmental stress response. Trends Plant Sci. 2004;9:591–6.

Chen LT, Wu K. Role of histone deacetylases HDA6 and HDA19 in ABA and abiotic stress response. Plant Signal Behav. 2010;5:1318–20.

Chen H, Zhang Z, Teng K, Lai J, Zhang Y, Huang Y, Li Y, Liang L, Wang Y, Chu C, Guo H, Xie Q. Up-regulation of LSB1/GDU3 impacts geminivirus infection by activating the salicylic acid pathway. Plant J. 2010;62:12–23.

Chen L, Song Y, Li S, Zhang L, Zou C, Yu D. The role of WRKY transcription factors in plant abiotic stresses. Biochim Biophys Acta. 2012;1819:120–8.

Chen L, Zhang L, Li D, Wang F, Yu D. WRKY8 transcription factor functions in the TMV-cg defense response by mediating both abscisic acid and ethylene signaling in *Arabidopsis*. Proc Natl Acad Sci USA. 2013;110:E1963–71.

Chisholm ST, Coaker G, Day B, Staskawicz BJ. Host-microbe interactions: shaping the evolution of the plant immune response. Cell. 2006;124:803–14.

Cui X, Fan B, Scholz J, Chen Z. Roles of *Arabidopsis* cyclin-dependent kinase C complexes in cauliflower mosaic virus infection, plant growth, and development. Plant Cell. 2007;19:1388–402.

de Torres-Zabala M, Truman W, Bennett MH, Lafforguel G, Mansfield JW, Egea PR, Bogre L, Grant M. *Pseudomonas syringae* pv. tomato hijacks the *Arabidopsis* abscisic acid signaling pathway to cause disease. EMBO J. 2007;26:1434–43.

de Torres-Zabala M, Bennett MH, Truman W, Grant M. Antagonism between salicylic and abscisic acid reflects early host-pathogen conflict and moulds plant defence responses. Plant J. 2009;59:375–86.

Ding SW. RNA-based antiviral immunity. Nat Rev Immunol. 2010;10:632–44.

Ding SW, Voinnet O. Antiviral immunity directed by small RNAs. Cell. 2007;130:413–26.

Dinh ST, Baldwin IT, Galis I. The HERBIVORE ELICITOR-REGULATED1 gene enhances abscisic acid levels and defenses against herbivores in *Nicotiana attenuata* plants. Plant Physiol. 2013;162:2106–24.

Durrant WE, Dong X. Systemic acquired resistance. Annu Rev Phytopathol. 2004;42:185–209.

Elad Y. The use of antioxidants (free radical scavengers) to control grey mould (*Botrytis cinerea*) and white mould (*Sclerotinia sclerotiorum*) in various crops. Plant Pathol. 1992;41:417–26.

Erb M, Flors V, Karlen D, de Lange E, Planchamp C, D'Alessandro M, Turlings TC, Ton J. Signal signature of aboveground-induced resistance upon belowground herbivory in maize. Plant J. 2009;59:292–302.

Erb M, Köllner TG, Degenhardt J, Zwahlen C, Hibbard BE, Turlings TC. The role of abscisic acid and water stress in root herbivore-induced leaf resistance. New Phytol. 2011;189:308–20.

Erb M, Meldau S, Howe GA. Role of phytohormones in insect–specific plant reactions. Trends Plant Sci. 2012;17:250–9.

Escalante-Pérez M, Krol E, Stange A, Geiger D, Al-Rasheid KA, Hause B, Neher E, Hedrich R. A special pair of phytohormones controls excitability, slow closure, and external stomach formation in the *Venus flytrap*. Proc Natl Acad Sci USA. 2011;108:15492–7.

Fan J, Hill L, Crooks C, Doerner P, Lamb C. Abscisic acid has a key role in modulating diverse plant–pathogen interactions. Plant Physiol. 2009;150:1750–61.

Faulkner C, Robatzek S. Plants and pathogens : putting infection strategies and defense mechanisms on the map. Curr Opin Plant Biol. 2012;15:699–707.

Fernández-Calvo P, Chini A, Fernández-Barbero G, Chico JM, Gimenez-Ibanez S, Geerinck J, Eeckhout D, Schweizer F, Godoy M, Franco-Zorrilla JM, Pauwels L, Witters E, Puga MI, Paz-Ares J, Goossens A, Reymond P, De Jaeger G, Solano R. The *Arabidopsis* bHLH transcription factors MYC3 and MYC4 are targets of JAZ repressors and act additively with MYC2 in the activation of jasmonate responses. Plant Cell. 2011;23:701–15.

Flors V, Ton J, Jakab G, Mauch-Mani B. Abscisic acid and callose: team players in defence against pathogens? J Phytopathol. 2001;153:377–83.

Fujii H, Chinnusamy V, Rodrigues A, Rubio S, Antoni R, Park SY, Cutler SR, Sheen J, Rodriguez PL, Zhu JK. In vitro reconstitution of an abscisic acid signalling pathway. Nature. 2009;462:660–4.

Fujita M, Fujita Y, Noutoshi Y, Takahashi F, Narusaka Y, Yamaguchi-Shinogaki K, Shinozaki K. Crosstalk between abiotic and biotic stress responses: a current view from the points of convergence in the stress signaling networks. Curr Opin Plant Biol. 2006;9:436–42.

Ghassemian M, Nambara E, Cutler S, Kawaide H, Kamiya Y, McCourt P. Regulation of abscisic acid signaling by the ethylene response pathway in *Arabidopsis*. Plant Cell. 2000;12:1117–26.

Glazebrook J. Contrasting mechanisms of defense against biotrophic and necrotrophic pathogens. Annu Rev Phytopathol. 2004;43:205–27.

Goritschnig S, Weihmann T, Zhang Y, Fobert P, McCourt P, Li X. A novel role for protein farnesylation in plant innate immunity. Plant Physiol. 2008;148:348–57.

Gottig N, Garavaglia BS, Daurelio LD, Valentine A, Gehring C, Orellano EG, Ottado J. *Xanthomonas axonopodis* pv. citri uses a plant natriuretic peptide-like protein to modify host homeostasis. Proc Natl Acad Sci USA. 2008;105:18631–6.

Grant MR, Jones JDG. Hormone (dis)harmony moulds plant health and disease. Science. 2009;324:750–2.
Gudesblat GE, Torres PS, Vojnov AA. *Xanthomonas campestris* overcomes *Arabidopsis* stomatal innate immunity through a DSF cell-to-cell signal-regulated virulence factor. Plant Physiol. 2009;149:1017–27.
Guimaraes RL, Stotz HU. Oxalate production by *Sclerotinia sclerotiorum* deregulates guard cells during infection. Plant Physiol. 2004;136:3703–11.
Gutierrez C. DNA replication and cell cycle in plants: learning from geminiviruses. EMBO J. 2000;19:792–9.
Herbers K, Meuwly P, Frommer WB, Metraux JP, Sonnewald U. Systemic acquired resistance mediated by the ectopic expression of invertase: possible hexose sensing in the secretory pathway. Plant Cell. 1996;8:793–803.
Hills GJ, Plaskitt KA, Young ND, Dunigan DD, Watts JW, Wilson TM, Zaitlin M. Immunogold localization of the intracellular sites of structural and nonstructural tobacco mosaic virus proteins. Virology. 1987;161:488–96.
Hsu FC, Chou MY, Chou SJ, Li YR, Peng HP, Shih MC. Submergence confers immunity mediated by the WRKY22 transcription factor in *Arabidopsis*. Plant Cell. 2013;25:2699–713.
Iriti M, Faoro F. Abscisic aicd is involved in chitosan-induced resistance to tobacco necrosis virus (TNV). Plant Physiol Bioch. 2008;46:1106–11.
Jensen MK, Hagedorn PH, de Torres-Zabala M, Grant MR, Rung JH, Collinge DB, Lyngkjaer MF. Transcriptional regulation by an NAC (NAM–ATAF1,2–CUC2) transcription factor attenuates ABA signalling for efficient basal defence towards *Blumeria graminis* f.sp. hordei in *Arabidopsis*. Plant J. 2008;56:867–80.
Jiang CJ, Shimono M, Sugano S, Kojima M, Yazawa K, Yoshida R, Inoue H, Hayashi N, Sakakibara H, Takatsuki H. Abscisic acid interacts antagonistically with salicylic acid signaling pathway in rice–*Magnaporthe grisea* interaction. Mol Plant Microbe Interact. 2010;23:791–8.
Jones JDG, Dangl JL. The plant immune system. Nature. 2006;444:323–9.
Katagiri F, Tsuda K. Understanding the plant immune system. Mol Plant Microbe Interact. 2010;23:1531–6.
Kazan K, Manners JM. MYC2: the master in action. Mol Plant. 2013;6:686–703.
Kerchev P, Karpinska B, Morris J, Hussain A, Verrall S, Hedley P, Fenton B, Foyer CH, Hancock R. Vitamin C and the abscisic acid-insensitive 4 (ABI4) transcription factor are important determinants of aphid resistance in *Arabidopsis*. Antioxid Redox Signal. 2013;18:2091–105.
Kim KC, Lai Z, Fan B, Chen Z. *Arabidopsis* WRKY38 and WRKY62 transcription factors interact with histone deacetylase 19 in basal defense. Plant Cell. 2008;20:2357–71.
Knight H, Knight MR. Abiotic stress signalling pathways: specificity and cross-talk. Trends Plant Sci. 2001;6:262–7.
Koga H, Dohi K, Mori M. Abscisic acid and low temperatures suppress the whole plant-specific resistance reaction of rice plants to the infection of *Magnaporthe grisea*. Physiol Mol Plant Path. 2004;65:3–9.
Kumar AS, Lakshmanan V, Caplan JL, Powell D, Czymmek KJ, Levia DF, Bais HP. Rhizobacteria *Bacillus subtilis* restricts foliar pathogen entry through stomata. Plant J. 2012;72:694–706.
Kunkel BN, Brooks DM. Cross talk between signaling pathways in pathogen defense. Curr Opin Plant Biol. 2002;5:325–31.
Kwak JM, Mori IC, Pei ZM, Leonhardt N, Torres MA, Dangl JL, Bloom RE, Bodde S, Jones JD, Schroeder JI. NADPH oxidase AtrbohD and AtrbohF genes function in ROS dependent ABA signaling in *Arabidopsis*. EMBO J. 2003;22:2623–33.
Lai J, Chen H, Teng K, Zhao Q, Zhang Z, Li Y, Liang L, Xia R, Wu Y, Guo H, Xie Q. RKP, a RING finger E3 ligase induced by BSCTV C4 protein, affects geminivirus infection by regulation of the plant cell cycle. Plant J. 2008;57:905–17.
López-Pérez L, Martínez-Ballesta Mdel C, Maurel C, Carvajal M. Changes in plasma membrane lipids, aquaporins and proton pump of broccoli roots, as an adaptation mechanism to salinity. Phytochemistry. 2009;70:492–500.

Lorenzo O, Chico JM, Sánchez-Serrano JJ, Solano R. JASMONATE-INSENSITIVE1 encodes a MYC transcription factor essential to discriminate between different jasmonate-regulated defense responses in *Arabidopsis*. Plant Cell. 2004;16:1938–50.

Lorenzo O, Solano R. Molecular players regulating the jasmonate signaling network. Curr Opin Plant Biol. 2005;8:532–40.

Lozano-Durán R, Bourdais G, He SY, Robatzek S. The bacterial effector HopM1 suppresses PAMP-triggered oxidative burst and stomatal immunity. New Phytol. 2013;202:259.

Ma Y, Szostkiewicz I, Korte A, Moes D, Yang Y, Christmann A, Grill E. Regulators of PP2C phosphatase activity function as abscisic acid sensors. Science. 2009;324:1064–8.

Macho AP, Boutrot F, Rathjen JP, Zipfel C. ASPARTATE OXIDASE plays an important role in *Arabidopsis* stomatal immunity. Plant Physiol. 2012;159:1845–56.

Mauch-Mani B, Mauch F. The role of abscisic acid and plant–pathogen interactions. Curr Opin Plant Biol. 2005;8:409–14.

Mayek-Perez N, Garcia-Espinosa R, Lopez-Castaneda C, Acosta-Gallegos J, Simpson J. Water relations, histopathology and growth of common bean (*Phaseolus vulgaris* L.) during pathogenesis of *Macrophomina phaseolina* under drought stress. Physiol Mol Plant Path. 2002;60:185–95.

Mazumder M, Das S, Saha U, Chatterjee M, Bannerjee K, Basu D. Salicylic acid-mediated establishment of the compatibility between *Alternaria brassicicola* and *Brassica juncea* is mitigated by abscisic acid in *Sinapis alba*. Plant Physiol Biochem. 2013;70:43–51.

McElrone AJ, Sherald JL, Forseth IN. Effects of water stress on symptomatology and growth of *Parthenocissus quinquefolia* by *Xylella fastiosa*. Plant Dis. 2001;85:1160–4.

Mehdy MC. Active oxygen species in plant defense against pathogens. Plant Physiol. 1994;105:467–72.

Melotto M, Underwood W, Koczan J, Nomura K, He SY. Plant stomata function in innate immunity against bacterial invasion. Cell. 2006;126:969–80.

Melotto M, Underwood W, He SY. Role of stomata in plant innate immunity and foliar bacterial diseases. Annu Rev Phytopathol. 2008;46:101–22.

Mendgen K, Hahn M, Deising H. Morphogenesis and mechanisms of penetration by plant pathogenic fungi. Annu Rev Phytopathol. 1996;34:367–86.

Mersmann S, Bourdais G, Rietz S, Robatzek S. Ethylene signaling regulates accumulation of the FLS2 receptor and is required for the oxidative burst contributing to plant immunity. Plant Physiol. 2010;154:391–400.

Miura K, Tada Y. Regulation of water, salinity, and cold stress responses by salicylic acid. Front Plant Sci. 2014;23(5):4.

Moeder W, Yoshioka K. Environmental sensitivity in pathogen resistant *Arabidopsis* mutants. In: Yoshioka K, Shinozaki K, editors. Signal crosstalk in plant stress responses. Iowa: Wiley; 2009. p. 113–35.

Moeder W, Ung H, Mosher S, Yoshioka K. SA-ABA antagonism in defense responses. Plant Signal Behav. 2010;5:1231–3.

Mohr PG, Cahill DM. Abscisic acid influences the susceptibility of *Arabidopsis* thaliana to *Pseudomonas syringae* pv. Tomato and *Peronospora parasitica*. Func Plant Biol. 2003;30:461–9.

Mohr PG, Cahill DM. Suppression by ABA of salicylic acid and lignin accumulation and the expression of multiple genes, in *Arabidopsis* infected with *Pseudomonas syringae* pv Tomato. Funct Integr Genom. 2007;7:181–91.

Mosher S, Moeder W, Nishimura N, Jikumaru Y, Joo SH, Urquhart W, Klessig DF, Kim SK, Nambara E, Yoshioka K. The lesion-mimic mutant cpr22 shows alterations in abscisic acid signaling and abscisic acid insensitivity in a salicylic aciddependent manner. Plant Physiol. 2010;152:1901–13.

Nishimura MT, Dangl JL. *Arabidopsis* and the plant immune system. Plant J. 2010;61:1053–66.

Niu Y, Figueroa P, Browse J. Characterization of JAZ-interacting bHLH transcription factors that regulate jasmonate responses in *Arabidopsis*. J Exp Bot. 2011;62:2143–54.

Park SY, Fung P, Nishimura N, Jensen DR, Fujii H, Zhao Y, Lumba S, Santiago J, Rodrigues A, Chow TF, Alfred SE, Bonetta D, Finkelstein R, Provart NJ, Desveaux D, Rodriguez

PL, McCourt P, Zhu JK, Schroeder JI, Volkman BF, Cutler SR. Abscisic acid inhibits type 2C protein phosphatases via the PYR/PYL family of START proteins. Science. 2009;324:1068–71.

Penninckx IA, Thomma BP, Buchala A, Métraux JP, Broekaert WF. Concomitant activation of jasmonate and ethylene response pathways is required for induction of a plant defensin gene in *Arabidopsis*. Plant Cell. 1998;10:2103–14.

Pieterse CM, Van der Does D, Zamioudis C, Leon-Reyes A, Van Wees SC. Hormonal modulation of plant immunity. Annu Rev Cell Dev Biol. 2012;28:489–521.

Pré M, Atallah M, Champion A, DeVos M, Pieterse CMJ, Memelink J. The AP2/ERF domain transcription factor ORA59 integrates jasmonic acid and ethylene signals in plant defense. Plant Physiol. 2008;147:1347–57.

Rezzonico E, Flury N, Meins F, Beffa R. Transcriptional down-regulation by abscisic acid of pathogenesis-related beta-1,3-glucanase genes in tobacco cell cultures. Plant Physiol. 1998;117:585–92.

Robert-Seilaniantz A, Navarro L, Bari R, Jones JD. Pathological hormone imbalances. Curr Opin Plant Biol. 2007;10:372–9.

Robert-Seilaniantz A, Grant MR, Jones JDG. Hormone crosstalk in plant disease and defense: more than just jasmonate-salicylate antagonism. Annu Rev Phytopathol. 2011;49:26.21–26.27.

Rushton PJ, Somssich IE, Ringler P, Shen QJ. WRKY transcription factors. Trends Plant Sci. 2010;15:247–58.

Saika H, Okamoto M, Miyoshi K, Kushiro T, Shinoda S, Jikumaru Y, Fujimoto M, Arikawa T, Takahashi H, Ando M, Arimura S, Miyao A, Hirochika H, Kamiya Y, Tsutsumi N, Nambara E, Nakazono M. Ethylene promotes submergence-induced expression of OsABA8ox1, a gene that encodes ABA 80-hydroxylase in rice. Plant Cell Physiol. 2007;48:287–98.

Sánchez-Vallet A, López G, Ramos B, Delgado-Cerezo M, Riviere M, Llorente F, Fernández PV, Miedes E, Estevez JM, Grant M, Molina A. Disruption of abscisic acid signaling constitutively activates *Arabidopsis* resistance to the necrotrophic fungus *Plectosphaerella cucumerina*. Plant Physiol. 2012;160:2109–24.

Schäfer M, Fischer C, Meldau S, Seebald E, Oelmüller R, Baldwin IT. Lipase activity in insect oral secretions mediates defense responses in *Arabidopsis*. Plant Physiol. 2011;156:1520–34.

Schroeder JI, Kwak JM, Allen GJ. Guard cell abscisic acid signaling and engineering drought hardiness in plants. Nature. 2001;410:317–30.

Schulze-Lefert P, Robatzek S. Plant pathogen strick guard cells into opening the gates. Cell. 2006;126:831–4.

Seifi HS, Curvers K, De Vleesschauwer D, Delaere I, Aziz A, Höfte M. Concurrent overactivation of the cytosolic glutamine synthetase and the GABA shunt in the ABA-deficient sitiens mutant of tomato leads to resistance against *Botrytis cinerea*. New Phytol. 2013;199:490–504.

Seo PJ, Park CM. MYB96-mediated abscisic acid signals induce pathogen resistance response by promoting salicylic acid biosynthesis in *Arabidopsis*. New Phytol. 2010;186:471–83.

Seo YS, Gepts P, Gilbertson RL. Genetics of resistance to the geminivirus, bean dwarf mosaic virus, and the role of the hypersensitive response in common bean. Theor Appl Genet. 2004;108:786–93.

Thaler JS, Bostock RM. Interactions between abscisic-acid-mediated responses and plant resistance to pathogens and insects. Ecology. 2004;85:48–58.

Ton J, Mauch-Mani B. Beta-amino-butyric acid-induced resistance against necrotrophic pathogens is based on ABA-dependent priming for callose. Plant J. 2004;38:119–30.

Ton J, Jakab G, Toquin V, Flors V, Iavicoli A, Maeder MN, Metraux JP, Mauch-Mani B. Dissecting the b-aminobutyric acid induced priming phenomenon in *Arabidopsis*. Plant Cell. 2005;17:987–99.

Ton J, Flors V, Mauch-Mani B. The multifaceted role of ABA in disease resistance. Trends Plant Sci. 2009;14:310–7.

Tooker JF, De Moraes CM. Feeding by a gall-inducing caterpillar species alters levels of indole-3-acetic and abscisic acid in *Solidago altissima* (Asteraceae) stems. Arthropod Plant Interact. 2011;5:115–24.

Truman W, de Zabala MT, Grant M. Type III effectors orchestrate a complex interplay between transcriptional networks to modify basal defence responses during pathogenesis and resistance. Plant J. 2006;46:14–33.

Turner NC, Graniti A. Fusicoccin: a fungal toxin that opens stomata. Nature. 1969;223:1070–1.

Van der Ent S, Van Hulten M, Pozo MJ, Czechowski T, Udvardi MK, Pieterse CM, Ton J. Priming of plant innate immunity by rhizobacteria and β-amino butyric acid: differences and similarities in regulation. New Phytol. 2009;183:419–31.

van Kan JA. Licensed to kill: the lifestyle of a necrotrophic plant pathogen. Trends Plant Sci. 2006;11:247–53.

Verhage A, van Wees SC, Pieterse CM. Plant immunity: it's the hormones talking, but what do they say? Plant Physiol. 2010;154:536–40.

Vlot AC, Dempsey DA, Klessig DF. Salicylic acid, a multifaceted hormone to combat disease. Annu Rev Phytopathol. 2009;47:177–206.

Vos IA, Verhage A, Schuurink RC, Watt LG, Pieterse CM, Van Wees SC. On set of herbivore-induced resistance in systemic tissue primed for jasmonate-dependent defenses is activated by abscisic acid. Front Plant Sci. 2013;4:539.

Wang SM, Hou XL, Ying L, Cao XW, Zhang S, Wang F. Effects of turnip mosaic virus (TuMV) on endogenous hormones and transcriptional level of related genes in infected non-heading Chinese cabbage. J Nanjing Agric Univ. 2011;5:13–9.

Waigmann E, Lucas WJ, Citovsky V, Zambryski PC. Direct functional assay for tobacco mosaic virus cell-to-cell movement protein and identification of a domain involved in increasing plasmodesmal permeability. Proc Natl Acad Sci USA. 1994;91:1433–7.

Whenham RJ, Fraser RSS, Brown LP, Payne JA. Tobacco mosaic virus-induced increase in abscisic acid concentration in tobacco leaves: intracellular location in light and dark green areas, and relationship to symptom development. Planta. 1986;168:592–8.

Wolf S, Deom CM, Beachy RN, Lucas WJ. Movement protein of tobacco mosaic virus modifies plasmodesmatal size exclusion limit. Science. 1989;246:377–9.

Xiong L, Schumaker KS, Zhu JK. Cell signaling during cold, drought, and salt stress. Plant Cell. 2002;14(suppl.):S165–S183.

Xiong LZ, Yang YN. Disease resistance and abiotic stress tolerance in rice are inversely modulated by an abscisic acid-inducible mitogen-activated protein kinase. Plant Cell. 2003;15:745–59.

Xu J, Audenaert K, Hofte M, De Vleesschauwer D. Abscisic acid promotes susceptibility to the rice leaf blight pathogen *Xanthomonas oryzae* pv oryzae by suppressing salicylic acid-mediated defenses. PLoS ONE. 2013;27:e67413.

Yang Y. Towards understanding of signal perception and transduction in rice blast resistance. The 4th International Rice Blast Conference, Changsha, China;2007.

Yasuda M, Ishikawa A, Jikumaru Y, Umezawa T, Asami T, Maruyama-Nakashita A, Kudo T, Shinozaki K, Yoshida S, Nakashita H. Antagonistic interaction between systemic acquired resistance and the abscisic acid-mediated abiotic stress responses in *Arabidopsis*. Plant Cell. 2008;20:1678–92.

Zeng W, He SY. A prominent role of the flagellin receptor FLAGELLIN-SENSING2 in mediating stomatal response to *Pseudomonas syringae* pv. tomato DC3000 in *Arabidopsis*. Plant Physiol. 2010;3:1188–98.

Zeng W, Melotto M, He SY. Plant stomata: a checkpoint of host immunity and pathogen virulence. Curr Opin Bio technol. 2010;21:599–603.

Zhang H, Zhu X, Liu H. Effect of banana bunchy top virus (BBTV) on endogenous hormone of banana plant. Acta Phytopathol Sin. 1997;27:79–83.

Zhang X, Yuan YR, Pei Y, Lin SS, Tuschl T, Patel DJ, Chua NH. Cucumber mosaic virus-encoded 2b suppressor inhibits *Arabidopsis* Argonaute1 cleavage activity to counter plant defense. Genes Dev. 2006;20:3255–68.

Zhang Z, Chen H, Huang X, Xia R, Zhao Q, Lai J, Teng K, Li Y, Liang L, Du Q, Zhou X, Guo H, Xie Q. BSCTV C2 attenuates the degradation of SAMDC1 to suppress DNA methylation-mediated gene silencing in *Arabidopsis*. Plant Cell. 2011;23:273–88.

Chapter 21
Principles and Practice of ABA Analysis

Zhigang Huang, Mohammed Humayun Kabir, Yuan Xiao, Qing Liu, Jianhua Tong and Langtao Xiao

Abstract The development of fast, sensitive, and accurate analytical methods for abscisic acid (ABA) is of increasing importance for elucidation of its metabolism, transport, and molecular regulatory mechanisms in plants. In the past several decades, significant technical advances including sample preparation and detection techniques for ABA determination have been achieved. This chapter reviews the principles and practice in the extraction, purification, and analysis of ABA. Special attention is paid to the recent advances in the solid-phase extraction (SPE) and chromatographic methods coupled with mass spectrometry (MS), especially with tandem MS. A sum-up protocol based on analysis practice and service of several years in the authors' laboratory is also presented.

Keywords Extraction · Purification · ABA analysis

21.1 Introduction

Quantitative analysis of abscisic acid (ABA) is very important for in-depth study of its metabolism, transport, and molecular regulatory mechanisms in plants. Therefore, great efforts have been invested in ABA assay methodology. ABA is a sesquiterpene with an α,β-unsaturated ketone in the ring and a conjugated diene side-chain. Consequently, it has both a high extinction coefficient in the ultraviolet region and strong electron-capturing properties (Quarrie et al. 1988). Since the first report of ABA isolation and characterization (Ohkuma et al. 1963), different types of methods including bioassays, immunoassays, and chromatographic methods,

Z. Huang · M.H. Kabir · Y. Xiao · Q. Liu · J. Tong · L. Xiao (✉)
Hunan Provincial Key Laboratory of Phytohormones and Growth Development, Hunan Provincial Key Laboratory for Crop Germplasm Innovation and Utilization, Hunan Agricultural University, Changsha 410128, Hunan, China
e-mail: langtaoxiao@163.com

D.-P. Zhang (ed.), *Abscisic Acid: Metabolism, Transport and Signaling*,
DOI 10.1007/978-94-017-9424-4_21

have been described to analyze ABA in plants, and significant technical advances on quantifying ABA are achieved (Weiler 1979; Blintsov and Gusakovskaya 2006; Forcat et al. 2008; Hou et al. 2008; Su et al. 2013). Because of the good separation and qualitative abilities, chromatographic methods coupled with mass spectrometry (MS), especially with tandem MS, have become the most powerful tools for ABA analysis from complex plant samples.

Accurate quantitative and qualitative analysis of ABA is difficult because of its instability and low physiological concentrations in plants (in the ng/g fresh weight range) (Weiler 1979). One of the challenges is to enrich these low-level ABA from crude plant extracts in which large amounts of interfering substances are present. In addition, the limited quantity of plant tissue available for ABA quantification is another challenge. Consequently, the analytic methods must be extremely selective and sensitive (Fu et al. 2011). Even the advanced tandem MS systems are used, many random errors that bring about inaccurate and imprecise analysis of ABA are still unavoidable. This chapter reviews the principles and practice of ABA analysis by focusing on the recent advances.

21.2 Sample Preparation

The determination of trace ABA in complex biological samples often requires extensive sample preparation techniques prior to analysis, whether quantitative or qualitative. Sample preparation is thus of great importance in the rapid separation and ultra-sensitive analysis of trace ABA, and more sample preparation techniques, including sampling, extracting, purifying, and concentrating methods, have been developed to analyze ABA (Bai et al. 2010; Fu et al. 2011). For the analysis of ABA, sample preparation generally comprises several procedures, including sampling, freeze-drying, comminuting, homogenization, exhaustive extraction from the matrix, and subsequent removal of co-extracted interferences by several cleanup steps prior to instrumental analysis. Such a preliminary sample preparation is often coupled with ABA loss, but the loss can be estimated and corrected by recovery analysis of the sample preparation processes using isotope-labeled ABA.

21.2.1 Sample Extraction

The concentrations and distribution of phytohormones including ABA in plant tissues are affected by temperature, watering regime, air humidity, light intensity, and many other environmental conditions (Peleg and Blumwald 2011; Alcázar and Parker 2011; Qaderi et al. 2012; Susawaengsup et al. 2011). Therefore, it is important to freeze the collected fresh tissues in liquid nitrogen immediately after detaching the tissues from plants, and then, the tissues are ground into fine powder with mortar or tissue lyser. No matter what method is selected, the homogenizing process of tissue must be kept at low temperature to minimize the ABA loss caused by enzymatic metabolism and chemical degradation.

The extracting method should be based on the physicochemical properties of ABA, and the ideal method would provide high extraction efficiency. Solvent extraction is still the most widely used method for ABA extraction. Plant material extracted in distilled water has also been reported the same extraction efficiency as organic solvents for ABA in several plant tissues (Loveys and van Dijk 1988; Quarrie et al. 1988; Vernieri et al. 1989). However, most of the reports show that the polarity of the extraction solvent should match closely that of the target compound, and thereby, the ratio of organic solvent to water has been defined according to the polarity of ABA (Fu et al. 2011). Many procedures for extraction have been developed and diverse solvents such as chloroform, methanol, ethanol, acetone, propanol, ethylacetate, and acetic acid have been used (Barkawi et al. 2010; Vine et al. 1987; Ross et al. 2004; Vilaró et al. 2006; Novák et al. 2008; Engelberth et al. 2003). In the organic solvents, methanol has become the preferred solvent for its small molecular size and low molecular weight (Fu et al. 2011). In our practice of ABA analysis, 80 % methanol is applied for ABA extraction with good results (Li et al. 2012). The modified Bieleski's solvent (methanol/formic acid/water 15:1:4) is a good choice for the simultaneous extraction of multiple classes of phytohormones, including auxins, CTKs, ABA, JAs, and a number of related metabolites, because it blocks the enzymatic degradation of phytohormones without extracting large quantities of lipids (Kojima et al. 2009; Izumi et al. 2009; Giannarelli et al. 2010).

21.2.2 *Sample Purification*

Determination of ABA can be hindered by hundreds of other abundant primary and secondary metabolites co-extracted by organic solvents, the purification and enrichment of ABA is crucial for the final assay. An increasing number of methods and techniques have been adapted for plant sample purification by considering both the physicochemical properties of ABA and the method types of analysis.

21.2.2.1 Liquid–Liquid Extraction

Liquid–liquid extraction (LLE) is based on the solubility difference of target analytes between two liquid phases. The most classical extraction is performed in separating funnels to extract analytes from an aqueous sample solution into a nonpolar or less polar organic solvent. Many organic solvents, such as hexane, diethyl ether, butanol, ethyl acetate, dichloromethane/isopropanol, and methylene chloride/1-propanol, have been used to purify phytohormones including ABA (Schmelz et al. 2003; Duffield and Netting 2001; Blake et al. 2002; Durgbanshi et al. 2005; Pacáková et al. 1997; Liu et al. 2002; Xie et al. 2011). However, traditional LLE method is time- and solvent consuming because it involves too many steps. Therefore, it has been used decreasingly in phytohormonal analysis (Fu et al. 2011).

Recently, liquid–liquid microextraction (LLME) in two application modes including hollow fiber-based liquid–liquid–liquid microextraction (HF-LLLME) and dispersive liquid–liquid microextraction (DLLME) have been applied for the further purification of phytohormones (Wu and Hu 2009; Lu et al. 2010; Gupta et al. 2011). HF-LLLME has been used for the simultaneous extraction and enrichment of four acidic phytohormones (IAA, ABA, SA, and JA) from natural coconut juice samples. This method provided good enrichment factors (48- to 243-fold) and high recoveries (88.3–119.1 %) for the target phytohormones, along with excellent sample cleanup capability (Wu and Hu 2009). DLLME has been applied for the simultaneous extraction and enrichment of major classes of phytohormones and plant growth regulators (GA, ABA, SA, IAA, IBA, indole-3-propionic acid (IPA), 1-naphthylacetic acid (NAA), and kinetin riboside (KR)) from crude extract, and exhibited several merits including high enrichment efficiency and reduced sample preparation time (Lu et al. 2010). However, both $CHCl_3$ (extraction solvent) and acetone (disperser solvent) are toxic, volatile, and flammable, thus safe and environmentally friendly solvents are desired in the LLME purification of phytohormones. Although LLME is a fast and low solvent-consuming method, the complicated organic solvent selection criteria and experimental design still limit its application (Fu et al. 2011).

21.2.2.2 Solid-Phase Extraction

Solid-phase extraction (SPE) using cartridges and disc-shaped devices is a widely used sample preparation technique for isolation, enrichment, clean-up, and medium exchange. The target compounds are separated from other compounds dissolved or suspended in the same liquid mixture by SPE according to different physicochemical properties of the compounds on the basis of diverse interactions, including adsorption, hydrogen bonding, polar and nonpolar interactions, cation, anion exchange or size exclusion (Poole 2003). Compared to the LLE method, SPE benefits from low intrinsic cost, shorter processing time, low solvent consumption, and simpler processing procedure, thus it affords the possibility of high-throughput device for sample preparation.

Many kinds of solid particulate medium based on inorganic oxides, low-specificity compound (chemically bonded, porous polymer, and carbon), and group-selective (ion exchange, mixed-mode, macrocyclic, restricted access, immunoaffinity, and molecularly imprinted polymer) materials can be used as sorbents of SPE (Poole 2003), and several commercial SPE columns, such as Sep-Pak C_{18}, Oasis HLB, Oasis MCX, and Oasis MAX, are available for the further purification of phytohormones including ABA from crude plant extracts by using different separation modes (Dobrev et al. 2005; Liu et al. 2010; Fu et al. 2011). Among the sorbents, spherical particles of hydrophobic silica bonded with octadecyl chain (C_{18}) are the most widely used for the sample pre-purification. When the SPE column is conditioned with weak acid solution and the plant extracts' pH is adjusted with weaker acid, the acidic ABA will remain the neutral form and can be trapped on the C_{18} column after sample loading. Then after, ABA is eluted with the corresponding acidic aqueous solution of methanol or ethanol (Fu et al. 2011).

With the progress of analysis techniques, high-performance liquid chromatography (HPLC) has also been used for sample pre-purification. HPLC can be regarded as a special solid-phase extraction technique and can enrich the target hormones efficiently and remove interfering compounds significantly. Fractionation by HPLC is more accurate and reliable than that by other methods. However, HPLC has disadvantages such as longer running time, more organic solvents consumption and higher cost compared to ordinary SPE methods.

Solid-phase microextraction (SPME) is a more developed SPE technique which allows simultaneous extraction and enrichment of analytes from different sample matrices (Prosen and Zupancic-Kralj 1999; Ouyang and Pawliszyn 2006). Since SPME was introduced in the early 1990s, it has been widely applied in combination with GC/HPLC to the sampling and analysis of food, aroma, forensic, environmental, and pharmaceutical samples (Kataoka et al. 2000; Ouyang and Pawliszyn 2006). SPME was also used for the purification of phytohormones and plant growth regulators including ABA, IAA, indole-3-butyric acid (IBA), and NAA in plant samples. SPME was capable of selective extraction of phytohormones with high recovery and was simpler and faster than other procedures for phytohormonal extraction (Liu et al. 2007). As a sampling and sample preparation method, it eliminates the need for solvents and combines sampling, isolation and enrichment in one step, and is suitable for plant samples of small quantities. However, application of SPME is still limited because of the commercial availabilities of fiber coatings and special fiber desorption chamber to couple with the GC/HPLC systems.

21.2.2.3 Immunoaffinity Purification

Immunoaffinity purification methods are based on antibody–antigen (Ab–Ag) recognition, and this specific interaction can provide extremely selective enrichment of the sample and thus greatly enhance the sensitivity of the analysis (Tarkowski et al. 2009). Several papers have reported on the immunoaffinity chromatography (IAC) purification and immunoaffinity gel (IAG) purification of ABA, the results suggest that the method is useful for quantifying ABA in plant material and represents the advantage of a short-time sample preparation with a high accuracy and capability. (Vaňková et al. 1998; Hauserová et al. 2005; Hradecká et al. 2007; López-Carbonell et al. 2009).

Immunoaffinity purification methods purify analytes according to structural similarities, so the setup for a suitable affinity system requires a highly specific anti-ABA antibody. However, as small organic molecules, ABA and its metabolites are haptens, which caused problems for specific recognition by anti-ABA antibodies because many phytohormonal analogs, metabolites, and other structurally similar compounds present in plant sample. The antibody preparation step is the key factor for establishment of immunoaffinity purification methods for ABA. To improve the durability and reusability of immunoaffinity purification methods, a pre-cleanup process with C_{18}-based SPE (Hradecká et al. 2007; Pěnčík et al. 2009; Novák et al. 2008; Simerský et al. 2009), mixed-mode SPE (Liang et al. 2012), or semi-preparative HPLC (Du et al. 2010) was often necessary in the purification of extracts from real plant tissues. Because of their higher selectivity but lower throughput than conventional SPE, immunoaffinity purification methods still have good potential for purification of trace ABA in plant samples of small quantities.

21.2.3 Techniques for Analysis

In plant tissues, phytohormones including ABA are present at very low levels against a background of a wide range of more abundant primary and secondary metabolites. Despite the application of complex purification steps to separate the analyte from crude plant extracts, there are still a large number of interfering metabolites in the samples for final assay, highly selective and sensitive analytical methods for ABA analysis are still essential. For the determination and quantification of ABA, many analytical methods, such as bioassay, immunoassay, biosensor, chromatography, and chromatography/mass spectrometry, have been adopted. Many advanced analytical techniques, such as single-cell capillary electrophoresis, nanoelectrospray, and quantum dots, are increasingly being explored in the development of ABA analytical methods. In this section, two types of commonly used methods including immuno-based methods and chromatographic methods are summarized.

21.2.3.1 Immuno-based Methods

Immuno-based methods including radioimmunoassay (RIA), enzyme-linked immunosorbent assay (ELISA) and immunosensor are specific detection methods based on the specific antibody–antigen (Ab–Ag) binding property. The specificity of the Ab–Ag interaction is the key factor affecting quantification results. However, due to the low-affinity and strong cross-reactions of polyclonal antibodies (pAbs) in the presence of high levels of ABA precursors, ABA catabolites or conjugates, the application of immunoassay techniques for the quantitative analysis of ABA were limited initially. In 1980s, the introduction of monoclonal antibodies (mAbs) has provided attractive alternatives to the traditional immunoassays of ABA by reducing the need for purification and speeding up analysis for a large number of samples (Mertens et al. 1983; Weiler 1984; Harris and Dugger 1986, Harris et al. 1988, Leroux et al. 1985).

Radioactively labeled Ag (Ag*) to bind with Ab is used to form the specific Ag*–Ab complex in RIA, following which the analytes are quantified by measurement of radioactivity. In the assay, special security measures should be taken in the assay because radioisotopes are used (Fu et al. 2011). RIA is a highly sensitive method, which can detect target molecules at nmol or pmol lever (Tarkowski et al. 2009). After mAbs instead of pAbs were used, one of the major objectives in developing RIA methods was to reduce the purification procedures prior to quantification, thus some RIA systems without complex purification steps for ABA analysis were developed (Weiler 1979). Because RIA is relatively quick, it is still used in ABA assay today especially when a large number of small samples to be processed (Su et al. 2013; Thiagarajan et al. 2013).

Ag can be labeled with easily assayed enzyme instead of radioisotope in ELISA. Accordingly, ELISA combines the specificity of Ag–Ab reaction with the sensitivity of enzymic assays and provides a sensitive method with lower detection limit for ABA analysis. Weiler (1982) developed a solid-phase ELISA for abscisic acid, and abscisic acid conjugates at fmol level (detection limit: 50–60 fmol). Only little procedural effort was required for the assay which can be completed within 6 h. After that, several ELISA assays including direct ELISA and indirect ELISA for free and conjugated ABA have been reported (Ross et al. 1987; Norman et al. 1988; Cahill and Ward 1989; Zhang et al. 1991). Two solid-phase competitive ELISA systems were also developed for detecting free and bound ABA (Blintsov and Gusakovskaya 2006). Currently, ELISA kits are commercially available from biotechnology companies and research institutions. Compared with RIA, ELISA is less expensive and easier to set up despite the common problem of cross-reaction with structurally similar compounds; moreover, the hazards of RIA associated with the radioactive waste disposal can be excluded (Tarkowski et al. 2009).

Immunosensors are affinity ligand-based solid-state biosensor devices in which the immunochemical reaction is coupled to a transducer. The fundamental basis of all immunosensors is also the specificity of the molecular recognition of Ag and Ab to form a stable complex. In contrast to traditional immunoassays of RIA and ELISA, modern transducer technology enables the label-free detection and quantification of the immune complex (Luppa et al. 2001). Electrochemical immunosensor, one of the commonly used immunosensors, combines the sensitivity of electroanalytical methods with the inherent bioselectivity of the biological component (e.g., antibody or aptamer). These immunosensors presented good specificity and high sensitivity and have been successfully applied to ABA analysis (Li et al. 2008, 2010; Wang et al. 2009). In addition, immobilization protocols, biosensor regeneration, signal amplification, equipment miniaturization, and antibody (or aptamer) properties are considered in the development and applications of these immunosensors. For example, to develop an amperometric immunosensor for ABA, an approximate 10-nm-thick gold layer was first sputtered uniformly onto the electrode surface, and then gold nanoparticles were chemically grown directly on the gold layer for antibody adsorption by immersing the electrode into H_2AuCl_4 solution (Wang et al. 2009), or an anti-ABA antibody was adsorbed directly on a porous nanogold film (Li et al. 2008).

Although immuno-based methods are excellent for the detection of ABA because of their high selectivity and sensitivity, the specificity of antibodies restricts the applications of the method in the simultaneous detection of multiple classes of phytohormones. On the other hand, the immuno-based methods are less dependent on large equipment and are more amenable to miniaturization, they are more portable and suitable for in situ screening of target hormones in plant tissues (Bai et al. 2010). Compared with the existing chromatographic methods, though the established immuno-based methods are not as sensitive as LC-MS/MS, even LC-MS combined with solid-phase extraction, they offer the advantages of fast detection, simplicity in operation because no complex purification steps are required prior to analysis, thus are frequently used today.

21.2.3.2 Chromatographic Methods

Gas chromatography (GC) was the earliest chromatographic method used for quantitative analysis of ABA (Lenton et al. 1968; Davis et al. 1968). As ABA is not a volatile compound, it needs to be derivatized to increase the volatility and also to improve the thermal stability prior to GC-based analysis (Fu et al. 2011). After the application of mass spectrometry (MS) detector, GC with MS (GC-MS) was more reliable and specific for identification and quantification of ABA (Li et al. 1992; Müller et al. 2002). Although GC-based methods provide high resolution and low limit of detection, they are generally not used currently to analyze ABA because they are labor-intensive and costly, more and more studies turn to use alternative techniques such as HPLC or LC-MS.

HPLC, which enables rapid, high-resolution purification of ABA from extracts, can provide relatively reliable identification and quantification of ABA by UV absorbance without derivatization when using a UV detector. Therefore, HPLC coupled with UV detector was a popular tool used for ABA analysis in a long period of time. However, it is inadequate for the accurate measurement of ABA only based on the retention time and UV or fluorescence spectroscopy because many components may be present in one peak with a complicated matrix. Owing to the introduction of MS detector, HPLC combined with MS detector (LC-MS), especially with tandem MS detectors (LC-MS/MS), overcomes the disadvantages of HPLC with UV detector and become a more powerful and more commonly applied method for ABA determination. Ultra-performance liquid chromatography (UPLC), which can achieve higher resolution, higher sensitivity and more rapid separation than traditional HPLC, has also been employed to combine with tandem MS detectors (UPLC-MS/MS) for ABA analysis in recent years (Müller and Munné-Bosch 2011; Fu et al. 2012; Balcke et al. 2012; Niu et al. 2013).

MS is an analytical technique that measures the mass-to-charge ratio (*m/z*) of charged particles. The analytes are ionized in the ion source, and then formed an ion beam to pass through the ion guide system, followed by analysis with a mass analyzer based on the *m/z* ratio (Fu et al. 2011). Electrospray ionization (ESI) is the most common ion source used in MS systems because of its fairly high sensitivity and low background (Tarkowski et al. 2009). Selective ion monitoring (SIM) mode, in which only specific characteristic ions of the target compound are measured, provides higher sensitivity and selectivity than monitoring all ions simultaneously, a data acquisition mode named full scan. A better alternative is the multiple reactions monitoring (MRM) mode with higher selectivity based on the tandem mass spectrometry (MS/MS) technique, which can be performed either in time or space, corresponding to triple quadrupole mass spectrometry (QQQ MS) and ion trap mass spectrometry (IT MS), respectively. In MRM, a precursor mass ion is selected in the first-stage quadrupole and then fragmented in the collision cell to yield diagnostic product ions filtered by the third-stage quadrupole. A signal is detected only when the selected precursor ion passes the first-stage quadrupole and the selected product ion passes the third-stage quadrupole; thus each ionized compound gives a distinct precursor-to-product ion transition in the MRM mode, which is diagnostic for the presence of a particular compound in an extract (Fu et al. 2011).

Although chromatographic methods have high separation efficiency, direct loading of crude extracts onto chromatographic columns would cause separation efficiency deteriorating, irreparable column damage and fouling, and MS signal suppressing, and thus, purification and enrichment of the crude extracts is necessary. Overall, advances of highly sensitive MS techniques have greatly reduced the amount of plant material required and overcome the low detectability of ABA. In recent years, LC-MS/MS becomes the most accurate technique for ABA analysis (López-Carbonell and Jáuregui 2005; Fletcher and Mader 2007; Pan et al. 2008, 2010; Ma et al. 2008; Izumi et al. 2009; Balcke et al. 2012; Yu et al. 2013).

Since the metabolomics is of more and more importance in the mechanism and cross-talk research of ABA (Forcat et al. 2008; Kanno et al. 2010); Pan et al. (2008) developed a rapid and sensitive method for simultaneous quantification of multiple classes of phytohormones including ABA and some related metabolites in 50–100 mg of fresh *Arabidopsis* leaves without purification or derivatization. After being frozen in liquid nitrogen, the leaves were ground into powder, and 500 μL of 1-propanol/H_2O/concentrated HCl (2:1:0.002, vol/vol/vol) with internal standards were added, followed by agitation for 30 min at 4 °C. One milliliter of CH_2Cl_2 was added, followed by agitation for another 30 min and then centrifugation at 13,000 g for 5 min. After centrifugation, two phases were formed and plant debris was in the middle of two layers. The lower layer was concentrated and re-solubilized in 0.3 mL of MeOH and then 25 μL was injected to column for analysis. The mixtures of standard compounds were separated by reversed-phase HPLC and analyzed by tandem mass spectrometry (RP-HPLC/ESI–MS/MS) in the MRM mode. The identities of phytohormones in the crude plant extracts were confirmed by analysis of product ion fragments obtained by the hybrid triple quadrupole/linear ion trap mass spectrometry, operating in the information dependent acquisition (IDA) mode. A variety of synthetic, isotopically labeled, or modified compounds were selected as internal standard. The extraction efficiencies of the phytohormones in this study ranged from 85 to 98 % in one round of extraction. By this method, one person can easily analyze 80 samples in 6 h.

Forcat et al. (2008) also described an efficient method for the rapid quantitative determination of ABA. Plant material was harvested into liquid nitrogen and freeze dried, then 10 mg of powdered tissue was weighed into a new 2-mL microfuge tube and extracted with 400 μL of 10 % methanol containing 1 % acetic acid to which internal standards had been added in a bead beater with 3-mm tungsten beads for 2 min, placed on ice for 30 min then centrifuged at 13,000 g for 10 min at 4 °C. The supernatant was carefully removed and the pellet re-extracted with 400 μL of 10 % methanol containing 1 % acetic acid. Following a further 30 min incubation on ice, the extract was centrifuged and the supernatants pooled. The two extractions resulted in 90–95 % recovery of the targeted analytes. The extracts (50 μL) were analyzed by HPLC-electrospray ionization/MS-MS using an Agilent 1100 HPLC coupled to an Applied Biosystems Q-TRAP 2000 in the MRM mode. The method requires minimal 10 mg freeze dried tissue and is highly reproducible and can accurately measure ABA across the expected physiological dynamic range. Moreover, it compares well with other methods that have more complex extraction methods. The use of freeze dried material promotes ease of handling and automation, and it is more convenient for scaling up extraction, especially dealing with multiple samples during a time course analysis.

A simple and fast protocol for ABA analysis developed by our group is as follows: 50–200 mg of fresh plant tissue was well ground with a small glass pestle in a 2-mL vial. Following the addition of 1 mL of 80 % methanol, homogenates were well mixed in an ultrasonic bath and then kept at 4 °C overnight. After being centrifuged at 15,200 g for 10 min, the supernatant was collected and then vacuumed to dryness in a Jouan RCT-60 concentrator. Dried extract was dissolved in 200 μL of sodium phosphate solution (0.1 mol/L, pH 7.8) and later passed through a Waters Sep-Pak C_{18} cartridge. The cartridge was eluted with 1500 μL of 80 % methanol and the eluate was vacuumed to dryness again. After being dissolved in 50 μL of 10 % methanol and injected 5 μL into the Shimadzu LC-MS/MS 8030 system in which an Acquity UPLC BEH column (2.1 mm I.D. × 100 mm, 1.7 μm) was used. The column temperature was set at 40 °C. Elution of the samples was carried out with 0.02 % aqueous acetic acid (solvent A) and acetonitrile (solvent B), and a gradient mode [(min/%/%) for 0/90/10, 3/20/80; 5.0/20/80, 6.0/90/10] at a flow rate of 0.25 mL/min. MS/MS conditions were as follows: collision energy −18.0 eV, m/z 181/134.1. The mass spectrometer was set to MRM mode using ESI in negative ion mode, with a nebulizing gas flow at 3L/min, a drying gas flow at 15L/min, a desolvation temperature at 250 °C, a heat block temperature at 480 °C. $^2H_6$-ABA was used as an internal standard. For ABA, the ionization conditions (pre-bias voltages of 19 V for quadrupole 1 and 28 V for quadrupole 3; collision energy of 10 eV; m/z of 263/153.2) were employed. While for $^2H_6$-ABA, the ionization conditions (pre-bias voltages of 20 V for quadrupole 1 and 15 V for quadrupole 3; collision energy of 11 eV; m/z of 269/159.2) were employed. The ABA concentrations were calculated according to calibration curve which created by internal standard of deuterium labeled ABA. The method is capable to analyze pg level ABA in about 100 mg of fresh plant samples.

21.3 Summary

Since the isolation and characterization of ABA have been reported (Ohkuma et al. 1963), many analytical systems for the precise and accurate identification and quantitative determination of trace ABA has been developed for the better understanding of the molecular mechanisms of ABA. Based on the literature reviewed in this paper, a sum-up protocol for ABA analysis is given in Fig. 21.1.

All of the current analytical techniques require prior preparation of highly purified extracts which are achieved by one or more steps. The pretreatment steps should be minimized to curb matrix effect of interfering substances. Universal and selective sample preparation techniques, including LLE, SPE, IAE, and MIP extraction, have been successfully applied to separate and enrich ABA from plant tissues. SPE benefits from low intrinsic cost, shorter processing time, low solvent consumption, and simpler processing procedure, and becomes the most popular method for fast and high-throughput purification of samples.

Immuno-based analytical methods depended on the antigen–antibody specific interaction have high sensitivity and specificity for purification and detection of ABA, but cross-reactions cause many problems. Biosensor, the real-time quantification approach, is somehow restricted by its limited reproducibility and short service life. The capability of combination with GC or LC broadens detection with mass spectrometry. Compared with other assays, the MS methods, which exhibits more advantages including high sensitivity, high selectivity, high-throughput and excellent accuracy, have been proved to be a reliable tool for ABA analysis. In particular, GC or LC coupled with tandem mass spectrometry not only significantly improves the sensitivity but also provides structural information based on

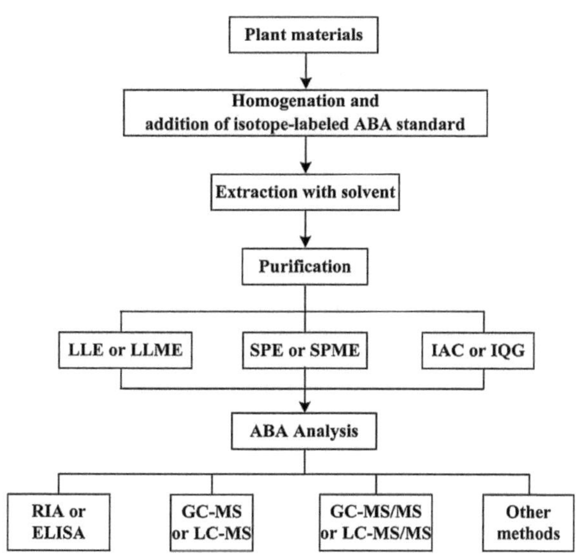

Fig. 21.1 A typical sum-up protocol for ABA analysis. See detailed description in the text

fragmentation patterns. LC-MS/MS has become the most powerful tool in ABA analysis because ABA can be analyzed directly without derivatization, which is needed by the GC-MS or GC-MS/MS. In addition, the metabolomics potential of LC-MS/MS in simultaneous analysis of ABA-related conjugates and metabolites, multi-phytohormonal profiles is very promising. However, the high cost of instruments and the consumption of stable isotope-labeled internal standards also hinder its wider application (Fu et al. 2011). Additionally, although the amount of plant material needed per sample has been greatly decreased for LC-MS/MS, sensitivity improvement for future methods to ultra-high or single-cell level is still desired; since in many research cases, the amount of available sample is quite limited.

Currently, the established procedures for ABA analysis involve the detaching of tissues from the plants, which may induce some changes in phytohormonal level. Also, the samples are usually homogenized in the extraction solvent with the quantified results showing only the average ABA level in the whole tissue (Bai et al. 2010). However, in-depth investigation of the molecular mechanism has uncovered the temporal and spatial patterns of phytohormones signaling in plants. To date, there have no established methods to follow the in situ and real-time changes in endogenous ABA concentration, the introduction of some new analytical techniques, such as the microminiaturization of biosensor probes, sequence-specific DNA biosensors and in vivo imaging may contribute to its realization.

Acknowledgments The work in the authors laboratory was financially supported by Natural Science Foundation of China (Grants 90817101, 91117006 and 91317312), Scientific Research Fund of Hunan Provincial Education Department (12K060, 12K061) and Program for Science and Technology Innovative Research Team in Higher Educational Institutions of Hunan Province.

References

Alcázar R, Parker JE. The impact of temperature on balancing immune responsiveness and growth in *Arabidopsis*. Trends Plant Sci. 2011;16(12):666–75.

Bai Y, Du F, Liu H. Determination strategies of phytohormones: recent advances. Anal Methods. 2010;2(12):1867–73.

Balcke GU, Handrick V, Bergau N, Fichtner M, Henning A, Stellmach H, Tissier A, Hause B, Frolov A. An UPLC-MS/MS method for highly sensitive high-throughput analysis of phytohormones in plant tissues. Plant Methods. 2012;8(1):47.

Barkawi LS, Tam Y-Y, Tillman JA, Normanly J, Cohen JD. A high-throughput method for the quantitative analysis of auxins. Nat Protoc. 2010;5(10):1609–18.

Blake PS, Taylor JM, Finch-Savage WE. Identification of abscisic acid, indole-3-acetic acid, jasmonic acid, indole-3-acetonitrile, methyl jasmonate and gibberellins in developing, dormant and stratified seeds of ash (*Fraxinus excelsior*). Plant Growth Regul. 2002;37(2):119–25.

Blintsov A, Gusakovskaya M. Immunoenzyme method for differential assay of free and bound ABA. Russ J Plant Physiol. 2006;53(3):407–12.

Cahill D, Ward E. An indirect enzyme-linked immunosorbent assay for measurement of abscisic acid in soybean inoculated with *Phytophthora megasperma* f. sp. *glycinea*. Phytopathology. 1989;79(11):1238–42.

Davis LA, Heinz D, Addicott F. Gas-liquid chromatography of trimethylsilyl derivatives of abscisic acid and other plant hormones. Plant Physiol. 1968;43(9):1389–94.

Dobrev P, Havlíček L, Vágner M, Malbeck J, Kamínek M. Purification and determination of plant hormones auxin and abscisic acid using solid phase extraction and two-dimensional high performance liquid chromatography. J Chromatogr A. 1075;1:159–66.

Du F, Bai Y, Bai Y, Liu H. Quantitative detection of trace systemins in Solanaceous plants by immunoaffinity purification combined with liquid chromatography/electrospray quadrupole time-of-flight mass spectrometry. Anal Chem. 2010;82(22):9374–83.

Duffield P, Netting A. Methods for the quantitation of abscisic acid and its precursors from plant tissues. Anal Biochem. 2001;289(2):251–9.

Durgbanshi A, Arbona V, Pozo O, Miersch O, Sancho JV, Gomez-Cadenas A. Simultaneous determination of multiple phytohormones in plant extracts by liquid chromatography-electrospray tandem mass spectrometry. J Agric Food Chem. 2005;53(22):8437–42.

Engelberth J, Schmelz EA, Alborn HT, Cardoza YJ, Huang J, Tumlinson JH. Simultaneous quantification of jasmonic acid and salicylic acid in plants by vapor-phase extraction and gas chromatography-chemical ionization-mass spectrometry. Anal Biochem. 2003;312(2):242–50.

Fletcher AT, Mader JC. Hormone profiling by LC-QToF-MS/MS in dormant macadamia integrifolia: correlations with abnormal vertical growth. J Plant Growth Regul. 2007;26(4):351–61.

Forcat S, Bennett MH, Mansfield JW, Grant MR. A rapid and robust method for simultaneously measuring changes in the phytohormones ABA, JA and SA in plants following biotic and abiotic stress. Plant Methods. 2008;4(1):16.

Fu J, Chu J, Sun X, Wang J, Yan C. Simple, rapid, and simultaneous assay of multiple carboxyl containing phytohormones in wounded tomatoes by UPLC-MS/MS using single SPE purification and isotope dilution. Anal Sci. 2012;28(11):1081–7.

Fu J, Sun X, Wang J, Chu J, Yan C. Progress in quantitative analysis of plant hormones. Chin Sci Bull. 2011;56(4–5):355–66.

Giannarelli S, Muscatello B, Bogani P, Spiriti M, Buiatti M, Fuoco R. Comparative determination of some phytohormones in wild-type and genetically modified plants by gas chromatography–mass spectrometry and high-performance liquid chromatography–tandem mass spectrometry. Anal Biochem. 2010;398(1):60–8.

Gupta V, Kumar M, Brahmbhatt H, Reddy C, Seth A, Jha B. Simultaneous determination of different endogenetic plant growth regulators in common green seaweeds using dispersive liquid–liquid microextraction method. Plant Physiol Biochem. 2011;49(11):1259–63.

Harris MJ, Dugger WM. The occurrence of abscisic acid and abscisyl-β-D-glucopyranoside in developing and mature citrus fruit as determined by enzyme immunoassay. Plant Physiol. 1986;82(2):339–45.

Harris MJ, Outlaw WH, Mertens R, Weiler EW. Water-stress-induced changes in the abscisic acid content of guard cells and other cells of *Vicia faba* L. leaves as determined by enzyme-amplified immunoassay. Proc Nat Acad Sci. 1988;85(8):2584–8.

Hauserová E, Swaczynová J, Doležal K, Lenobel R, Popa I, Hajdúch M, Vydra D, Fuksová K, Strnad M. Batch immunoextraction method for efficient purification of aromatic cytokinins. J Chromatogr A. 2005;1100(1):116–25.

Hou S, Zhu J, Ding M, Lv G. Simultaneous determination of gibberellic acid, indole-3-acetic acid and abscisic acid in wheat extracts by solid-phase extraction and liquid chromatography–electrospray tandem mass spectrometry. Talanta. 2008;76(4):798–802.

Hradecká V, Novák O, Havlíček L, Strnad M. Immunoaffinity chromatography of abscisic acid combined with electrospray liquid chromatography–mass spectrometry. J Chromatogr B. 2007;847(2):162–73.

Izumi Y, Okazawa A, Bamba T, Kobayashi A, Fukusaki E. Development of a method for comprehensive and quantitative analysis of plant hormones by highly sensitive nanoflow liquid chromatography–electrospray ionization-ion trap mass spectrometry. Anal Chim Acta. 2009;648(2):215–25.

Kanno Y, Jikumaru Y, Hanada A, Nambara E, Abrams SR, Kamiya Y, Seo M. Comprehensive hormone profiling in developing *Arabidopsis* seeds: examination of the site of ABA biosynthesis, ABA transport and hormone interactions. Plant Cell Physiol. 2010;51(12):1988–2001.

Kataoka H, Lord HL, Pawliszyn J. Applications of solid-phase microextraction in food analysis. J Chromatogr A. 2000;880(1):35–62.

Kojima M, Kamada-Nobusada T, Komatsu H, Takei K, Kuroha T, Mizutani M, Ashikari M, Ueguchi-Tanaka M, Matsuoka M, Suzuki K. Highly sensitive and high-throughput analysis of plant hormones using MS-probe modification and liquid chromatography–tandem mass spectrometry: an application for hormone profiling in *Oryza sativa*. Plant Cell Physiol. 2009;50(7):1201–14.

Lenton J, Bowen M, Saunders P. Detection of abscisic acid in the xylem sap of Willow (*Salix viminalis* L.) by Gas–Liquid Chromatography; 1968.

Leroux B, Maldiney R, Miginiac E, Sossountzov L, Sotta B. Comparative quantitation of abscisic acid in plant extracts by gas-liquid chromatography and an enzyme-linked immunosorbent assay using the avidin-biotin system. Planta. 1985;166(4):524–9.

Li C, Liu Z, Zhang Q, Wang R, Xiao L, Ma H, Chong K, Xu Y. SKP1 is involved in abscisic acid signalling to regulate seed germination, stomatal opening and root growth in *Arabidopsis thaliana*. Plant Cell Environ. 2012;35(5):952–65.

Li Q, Wang RZ, Huang ZG, Li HS, Xiao LT. A novel impedance immunosensor based on O-phenylenediamine modified gold electrode to analyze abscisic acid. Chin Chem Lett. 2010;21(4):472–5.

Li X, La Motte CE, Stewart CR, Cloud NP, Wear-Bagnall S, Jiang C-Z. Determination of IAA and ABA in the same plant sample by a widely applicable method using GC-MS with selected ion monitoring. J Plant Growth Regul. 1992;11(1):55–65.

Li Y, Xia K, Wang R, Jiang J, Xiao L. An impedance immunosensor for the detection of the phytohormone abscisic acid. Anal Bioanal Chem. 2008;391(8):2869–74.

Liang Y, Zhu X, Zhao M, Liu H. Sensitive quantification of isoprenoid cytokinins in plants by selective immunoaffinity purification and high performance liquid chromatography–quadrupole-time of flight mass spectrometry. Methods. 2012;56(2):174–9.

Liu B, Zhong X, Lu Y. Analysis of plant hormones in tobacco flowers by micellar electrokinetic capillary chromatography coupled with on-line large volume sample stacking. J Chromatogr A. 2002;945(1):257–65.

Liu H, Li Y, Luan T, Lan C, Shu W. Simultaneous determination of phytohormones in plant extracts using SPME and HPLC. Chromatographia. 2007;66(7–8):515–20.

Liu X, Yang Y, Lin W, Tong J, Huang Z, Xiao L. Determination of both jasmonic acid and methyl jasmonate in plant samples by liquid chromatography tandem mass spectrometry. Chin Sci Bull. 2010;55(21):2231–5.

López-Carbonell M, Gabasa M, Jáuregui O. Enhanced determination of abscisic acid (ABA) and abscisic acid glucose ester (ABA-GE) in *Cistus albidus* plants by liquid chromatography–mass spectrometry in tandem mode. Plant Physiol Biochem. 2009;47(4):256–61.

López-Carbonell M, Jáuregui O. A rapid method for analysis of abscisic acid (ABA) in crude extracts of water stressed *Arabidopsis thaliana* plants by liquid chromatography–mass spectrometry in tandem mode. Plant Physiol Biochem. 2005;43(4):407–11.

Loveys B, Van Dijk H. Improved extraction of abscisic acid from plant tissue. Funct Plant Biol. 1988;15(3):421–7.

Lu Q, Chen L, Lu M, Chen G, Zhang L. Extraction and analysis of auxins in plants using dispersive liquid–liquid microextraction followed by high-performance liquid chromatography with fluorescence detection. J Agric Food Chem. 2010;58(5):2763–70.

Luppa PB, Sokoll LJ, Chan DW. Immunosensors—principles and applications to clinical chemistry. Clin Chim Acta. 2001;314(1):1–26.

Ma Z, Ge L, Lee AS, Yong JWH, Tan SN, Ong ES. Simultaneous analysis of different classes of phytohormones in coconut (*Cocos nucifera* L.) water using high-performance liquid chromatography and liquid chromatography–tandem mass spectrometry after solid-phase extraction. Anal Chim Acta. 2008;610(2):274–81.

Mertens R, Deus-Neumann B, Weiler E. Monoclonal antibodies for the detection and quantitation of the endogenous plant growth regulator, abscisic acid. FEBS Lett. 1983;160(1):269–72.

Müller A, Düchting P, Weiler EW. A multiplex GC-MS/MS technique for the sensitive and quantitative single-run analysis of acidic phytohormones and related compounds, and its application to *Arabidopsis thaliana*. Planta. 2002;216(1):44–56.

Müller M, Munné-Bosch S. Rapid and sensitive hormonal profiling of complex plant samples by liquid chromatography coupled to electrospray ionization tandem mass spectrometry. Plant Methods. 2011;7(1):37.

Niu Q, Zong Y, Qian M, Yang F, Teng Y. Simultaneous quantitative determination of major plant hormones in pear flowers and fruit by UPLC/ESI–MS/MS. Analytical Methods; 2013.

Norman SM, Poling SM, Maier VP. An indirect enzyme-linked immunosorbent assay for (+)-abscisic acid in *Citrus*, *Ricinus*, and *Xanthium* leaves. J Agric Food Chem. 1988;36(1):225–31.

Novák O, Hauserová E, Amakorová P, Doležal K, Strnad M. Cytokinin profiling in plant tissues using ultra-performance liquid chromatography–electrospray tandem mass spectrometry. Phytochemistry. 2008;69(11):2214–24.

Ohkuma K, Lyon J, Addicott F, Smith O. Abscisin II, an abscission-accelerating substance from young cotton fruit. Science. 1963;142(3599):1592–3.

Ouyang G, Pawliszyn J. SPME in environmental analysis. Anal Bioanal Chem. 2006;386(4):1059–73.

Pacáková V, Štulík K, Vlasáková V, Březinová A. Capillary electrophoresis of cytokinins and cytokinin ribosides. J Chromatogr A. 1997;764(2):331–5.

Pan X, Welti R, Wang X. Simultaneous quantification of major phytohormones and related compounds in crude plant extracts by liquid chromatography–electrospray tandem mass spectrometry. Phytochemistry. 2008;69(8):1773–81.

Pan X, Welti R, Wang X. Quantitative analysis of major plant hormones in crude plant extracts by high-performance liquid chromatography–mass spectrometry. Nat Protoc. 2010;5(6):986–92.

Peleg Z, Blumwald E. Hormone balance and abiotic stress tolerance in crop plants. Curr Opin Plant Biol. 2011;14(3):290–5.

Pěnčík A, Rolčík J, Novák O, Magnus V, Barták P, Buchtík R, Salopek-Sondi B, Strnad M. Isolation of novel indole-3-acetic acid conjugates by immunoaffinity extraction. Talanta. 2009;80(2):651–5.

Poole CF. New trends in solid-phase extraction. TrAC Trends Anal Chem. 2003;22(6):362–73.

Prosen H, Zupančič-Kralj L. Solid-phase microextraction. TrAC Trends Anal Chem. 1999;18(4):272–82.

Qaderi MM, Kurepin LV, Reid DM. Effects of temperature and watering regime on growth, gas exchange and abscisic acid content of canola (*Brassica napus*) seedlings. Environ Exp Bot. 2012;75:107–13.

Quarrie S, Whitford P, Appleford N, Wang T, Cook S, Henson I, Loveys B. A monoclonal antibody to (S)-abscisic acid: its characterisation and use in a radioimmunoassay for measuring abscisic acid in crude extracts of cereal and lupin leaves. Planta. 1988;173(3):330–9.

Ross AR, Ambrose SJ, Cutler AJ, Allan Feurtado J, Kermode AR, Nelson K, Zhou R, Abrams SR. Determination of endogenous and supplied deuterated abscisic acid in plant tissues by high-performance liquid chromatography-electrospray ionization tandem mass spectrometry with multiple reaction monitoring. Anal Biochem. 2004;329(2):324–33.

Ross GS, Elder PA, McWha JA, Pearce D, Pharis RP. The development of an indirect enzyme linked immunoassay for abscisic acid. Plant Physiol. 1987;85(1):46–50.

Schmelz EA, Engelberth J, Alborn HT, O'Donnell P, Sammons M, Toshima H, Tumlinson JH. Simultaneous analysis of phytohormones, phytotoxins, and volatile organic compounds in plants. Proc Nat Acad Sci. 2003;100(18):10552–7.

Simerský R, Novák O, Morris DA, Pouzar V, Strnad M. Identification and quantification of several mammalian steroid hormones in plants by UPLC-MS/MS. J Plant Growth Regul. 2009;28(2):125–36.

Su YH, Su YX, Liu YG, Zhang XS. Abscisic acid is required for somatic embryo initiation through mediating spatial auxin response in *Arabidopsis*. Plant Growth Regul. 2013;69(2):167–76.

Susawaengsup C, Rayanakorn M, Wongpornchai S, Wangkarn S. Investigation of plant hormone level changes in shoot tips of longan (*Dimocarpus longan* Lour) treated with potassium chlorate by liquid chromatography–electrospray ionization mass spectrometry. Talanta. 2011;85(2):897–905.

Tarkowski P, Ge L, Yong JWH, Tan SN. Analytical methods for cytokinins. TrAC Trends Anal Chem. 2009;28(3):323–35.

Thiagarajan A, Lada R, Pepin S, Forney C, Desjardins Y, Dorais M. Temperature and photoperiod influence postharvest needle abscission of selected Balsam Fir (*Abies balsamea* L. (Mill.)) genotypes by modulating ABA levels. J Plant Growth Regul. 2013;32(4):843–51.

Vaňková Rr, Gaudinová A, Süssenbeková H, Dobrev P, Strnad M, Holík J, Lenfeld J. Comparison of oriented and random antibody immobilization in immunoaffinity chromatography of cytokinins. J Chromatogr A. 1998;811(1):77–84.

Vernieri P, Perata P, Armellini D, Bugnoli M, Presentini R, Lorenzi R, Ceccarelli N, Alpi A, Tognoni F. Solid phase radioimmunoassay for the quantitation of abscisic acid in plant crude extracts using a new monoclonal antibody. J Plant Physiol. 1989;134(4):441–6.

Vilaró F, Canela-Xandri A, Canela R. Quantification of abscisic acid in grapevine leaf (*Vitis vinifera*) by isotope-dilution liquid chromatography–mass spectrometry. Anal Bioanal Chem. 2006;386(2):306–12.

Vine JH, Noiton D, Plummer JA, Baleriola-Lucas C, Mullins MG. Simultaneous quantitation of indole 3-acetic acid and abscisic acid in small samples of plant tissue by gas chromatography/mass spectrometry/selected ion monitoring. Plant Physiol. 1987;85(2):419–22.

Wang R, Li Y, Li Q, Shen G, Xiao L. A novel amperometric immunosensor for phytohormone abscisic acid based on in situ chemical reductive growth of gold nanoparticles on glassy carbon electrode. Anal Lett. 2009;42(17):2893–904.

Weiler EW. Immunoassay of plant growth regulators. Annual review of plant physiology. 1984;35(1):85–95.

Weiler EW. Radioimmunoassay for the determination of free and conjugated abscisic acid. Planta. 1979;144(3):255–63.

Weiler EW. An enzyme-immunoassay for cis-(+)-abscisic acid. Physiol Plant. 1982;54(4):510–4.

Wu Y, Hu B. Simultaneous determination of several phytohormones in natural coconut juice by hollow fiber-based liquid–liquid–liquid microextraction-high performance liquid chromatography. J Chromatogr A. 2009;1216(45):7657–63.

Xie W, Han C, Zheng Z, Chen X, Qian Y, Ding H, Shi L, Lv C. Determination of Gibberellin A3 residue in fruit samples by liquid chromatography–tandem mass spectrometry. Food Chem. 2011;127(2):890–2.

Yu J, Meng Q, Liu W, Lu Y, Ren X. Analysis of acidic endogenous phytohormones in grapes by using online solid-phase extraction coupled with LC-MS/MS. J Chromatogr Sci. 2013. doi: 10.1093/chromsci/bmt162..

Zhang N, Xu Y, Zhou X. The development of an indirect enzyme linked immunosorbent assay for abscisic acid. Journal of Nanjing Agricultural University. 1991;14(3):21–4.

Chapter 22
Improvement of Stress Tolerance in Crops by Genetic Manipulation of ABA Metabolism, Signaling, and Regulation

Hao Du and Lizhong Xiong

Abstract Abiotic stresses such as extremes in water, temperature, and salinity are major limiting factors for the productivity of most crops. The plant response and adaptation to these stresses are controlled by numerous small-effect loci and regulated by hundreds of genes controlling various developmental and physiological processes. The elucidation of the abscisic acid (ABA) functional mechanisms in mediating the response and adaptation to stresses as described in the previous chapters has provided many opportunities for the development of new crop varieties with enhanced stress resistance. In this chapter, we mainly focus on the recent studies of ABA-related genes (involved in metabolism, signaling, and regulation) that have been well characterized and are known to affect stress resistance and especially those genes which have been engineered in crops for stress resistance improvement. Although there is evidence that many ABA-related genes have a biologically significant effect in improving stress resistance, a large gap remains to generate crops with significantly improved stress resistance in the field. A significant challenge is to unveil the complex mechanisms of stress resistance in crops by more intensive and integrative studies to find the key functional components involved in the ABA-mediated biological processes as tools for engineering and breeding stress-resistant crops.

Keywords Stress tolerance · Genetic manipulation · ABA-related genes

22.1 Introduction

Climate change, the distribution of contagious water, and soil salinization are global problems relevant to all of the living organisms on the earth and are caused mainly by the increasing world population. These extreme environmental

H. Du · L. Xiong (✉)
National Key Laboratory of Crop Genetic Improvement and National Center of Plant Gene Research (Wuhan), Huazhong Agricultural University, Wuhan 430070, China
e-mail: lizhongx@mail.hzau.edu.cn

D.-P. Zhang (ed.), *Abscisic Acid: Metabolism, Transport and Signaling*,
DOI 10.1007/978-94-017-9424-4_22

factors are known as abiotic stresses, and they affect plant growth and development and in particular the productivity of field crops worldwide. Thus, the production of increased quantities of food with limited land and adverse environmental conditions are immense challenges to the sustainable development of agriculture (Mittler 2006; Borlaug 2007). These goals might be attained by conventional plant breeding approaches, and in fact, using the traditional methods of crossing and selecting progeny, breeders have produced new varieties with improved stress resistance (Witcombe et al. 2008). Meanwhile, increasing evidence suggests that modern plant biotechnologies have much greater potential to make substantial stress resistance improvements in major crops (Zurbriggen et al. 2010).

Abscisic acid (ABA) is recognized as a stress hormone that coordinates the complex networks of stress responses. Under drought or salt stress conditions, the endogenous plant ABA level can rise to about 40-fold, triggering the closure of stomata, and the accumulation of dehydrins and late embryogenesis abundant proteins for osmotic adjustment (Verslues et al. 2006). Cold or heat stress also increases endogenous ABA levels in plants but to a much lesser extent. Many ABA-deficient rice mutants showed enhanced water loss from stomata under drought stress conditions and displayed drought-sensitive phenotypes (Du et al. 2013). Due to the physiological importance of ABA in drought responses, the genetic basis of drought-induced ABA accumulation was investigated in rice (Quarrie et al. 1997) and maize (Tuberosa et al. 1998; Landi et al. 2005).

ABA signaling also plays critical roles in stress response pathways by producing various yet specific outputs such as stomatal closure, changes in the root system, and an increase in the osmotic adjustment ability. Recently, one of the breakthroughs in ABA biology was the discovery of a core ABA signaling cascade consisting of the ABA receptor PYR/PYL/RCAR, negative regulator PP2C protein phosphatases, and positive regulator SnRK2 protein kinases which can activate transcription factors to regulate the expression of downstream genes (Described in the previous chapters). Numerous genes involved in ABA signaling and regulation have been tested by transgenic approaches for their effects in influencing stress tolerance through proper regulation of ABA responsiveness, water loss, and stress-related gene expression (Hu and Xiong 2014; Nakashima and Yamaguchi-Shinozaki 2013), some of which may be promising for improving crop performance under stress conditions.

As a "stress hormone," ABA has been well studied from production to signaling pathways, especially as a key regulator under drought stress conditions. The above chapters on ABA focused on metabolism, transportation, signaling and regulation, and responses to particular stresses, mainly in the model plant Arabidopsis. This chapter focuses on the recent studies of ABA-related genes that have been well characterized in physiological, molecular, or biochemical processes involved in abiotic stress tolerance. For ease of reference, the ABA-related genes that have been tested in crops for improving stress resistance are listed in Table 22.1, and these genes are described in different categories: ABA metabolism, ABA signaling,

Table 22.1 ABA-related genes having been tested for stress resistance in major crops

Genes	Protein function	Gene origin	Testing crop	Engineering method	Testing condition	References
OsMAPK5	MAPK	*O. sativa*	*O. sativa*	OE, *CaMV35SP*	DT, ST, CT; VS	(Xiong and Yang 2003)
OsbZIP23	bZIP	*O. sativa*	*O. sativa*	OE, Maize *Ubi1P*	DT, ST; VS/RS	(Xiang et al. 2008)
OsbZIP46CA1	bZIP	*O. sativa*	*O. sativa*	OE, Maize *Ubi1P*	HO/DT; VS/RS	(Tang et al. 2012)
OsbZIP72	bZIP	*O. sativa*	*O. sativa*	OE, *CaMV35SP*	DT; VS	(Lu et al. 2009)
ABF3	bZIP	*A. thaliana*	*O. sativa*	OE, Maize *Ubi1P*	DT; VS	(Oh et al. 2005)
SlAREB1	bZIP	*S. lycopersicum*	*S. lycopersicum*	OE, *CaMV35SP*	DT; VS	(Orellana et al. 2010)
SNAC1	NAC	*O. sativa*	*O. sativa*	OE, *CaMV35SP*	DT, ST; VS/RS	(Hu et al. 2006)
SNAC2	NAC	*O. sativa*	*O. sativa*	OE, *CaMV35SP*	CT; VS	(Hu et al. 2008)
OsWRKY45-2	WRKY	*O. sativa*	*O. sativa*	RNAi	ST; VS	(Tao et al. 2011)
OsWRKY13	WRKY	*O. sativa*	*O. sativa*	RNAi	DT; VS	(Xiao et al. 2013)
ARAG1	DREB	*O. sativa*	*O. sativa*	OE, *CaMV35SP*	DT; VS	(Zhao et al. 2010)
JERF3	ERF	*Lycopersicon esculentum*	*Nicotiana tabacum*	OE, *CaMV35SP*	DT, CT; VS	(Wu et al. 2008)
JERF3	ERF	*Lycopersicon esculentum*	*O. sativa*	OE, *CaMV35SP*	DT; HO; VS	(Zhang et al. 2010)
TSRF1	ERF	*Lycopersicon esculentum*	*O. sativa*	OE, *CaMV35SP*	DT; HO; VS	(Quan et al. 2010)
SUB1A	ERF	*O. sativa*	*O. sativa*	OE, *CaMV35SP*	DT; VS	(Fukao et al. 2011)
AmMYB1	MYB	*Avicennia marina*	*Nicotiana tabacum*	OE, *CaMV35SP*	ST; VS	(Ganesan et al. 2012)
OsMYB2	MYB	*O. sativa*	*O. sativa*	OE, Maize *Ubi1P*	DT, ST; VS	(Yang et al. 2012)
OsMYB3R-2	MYB	*O. sativa*	*O. sativa*	OE, Maize *Ubi1P*	CT; VS	(Ma et al. 2009a)

(continued)

Table 22.1 (continued)

Genes	Protein function	Gene origin	Testing crop	Engineering method	Testing condition	References
ZFP245	C2H2 zinc finger	*O. sativa*	*O. sativa*	OE, *CaMV35SP*	DT, CT; VS	(Huang et al. 2009)
ZFP179	C2H2 zinc finger	*O. sativa*	*O. sativa*	OE, *CaMV35SP*	ST; VS	(Sun et al. 2010)
OsBIRF1	RING-H2 finger	*O. sativa*	*Nicotiana tabacum*	OE, *CaMV35SP*	DT; VS	(Liu et al. 2008)
SDIR1	E3 ubiquitin ligase	*A. thaliana*	*Nicotiana tabacum*	OE, *CaMV35SP*	DT; VS	(Zhang et al. 2008b)
SDIR1	E3 ubiquitin ligase	*A. thaliana*	*O. sativa*	OE, Maize *Ubi1P*	DT; VS	(Zhang et al. 2008b)
OsSDIR1	E3 ubiquitin ligase	*O. sativa*	*O. sativa*	OE, Maize Ubi1P	DT; VS	(Gao et al. 2011b)
OsDSG1	E3 ubiquitin ligase	*O. sativa*	*O. sativa*	KM	DT, ST; VS	(Park et al. 2010)
SQS1	Farnesyltransferase/ squalene synthase	*O. sativa*	*O. sativa*	RNAi	DT; VS/RS	(Manavalan et al. 2012)
ERA1	Farnesyltransferase	*A. thaliana*	*B. napus*	OE, RD29A promoter	DT; RS	(Wang et al. 2009)
DSM2	Carotene hydroxylase	*O. sativa*	*O. sativa*	OE, *CaMV35SP*	DT; VS/RS	(Du et al. 2010)
LOS5	Molybdenum cofactor sulfurase	*A. thaliana*	*O. sativa*	OE, rice *Actin1P/LEA3-1P*	DT; RS	(Xiao et al. 2009)
LOS5	Molybdenum cofactor sulfurase	*A. thaliana*	*G. max*	OE, super promoter (enhancer and promoter)	DT; VS/RS	(Li et al. 2013)
LOS5	Molybdenum cofactor sulfurase	*A. thaliana*	*G. hirsutum*	OE, super promoter	DT; VS	(Yue et al. 2012)
OsSDIR1	E3 ubiquitin ligase	*O. sativa*	*O. sativa*	OE, Maize *Ubi1P*	DT; VS	(Gao et al. 2011b)
PSH3	Carotene cis–trans isomerase	*O. sativa*	*O. sativa*	KM	CT; VS	(Du et al. 2013)
PDS	Phytoene desaturase	*O. sativa*	*O. sativa*	RNAi	CT; VS	(Du et al. 2013)
TaABA8′OH1	Cytochrome P450 8′-hydroxylase	wheat	*O. sativa*	Tapetum-specific promoter	CT; VS	(Ji et al. 2011)

(continued)

Table 22.1 (continued)

Genes	Protein function	Gene origin	Testing crop	Engineering method	Testing condition	References
OsGH3-2	IAA-amido synthetase	*O. sativa*	*O. sativa*	OE, Maize *Ubi1P*	CT; VS	(Du et al. 2012)
TLD1/GH3.13	IAA-amido synthetase	*O. sativa*	*O. sativa*	OE, Maize *Ubi1P*	DT; VS	(Zhang et al. 2009)

Notes Testing conditions: *DT* drought tolerance, *ST* salt tolerance, *CT* cold tolerance, *HO* high osmotic medium supplied with PEG or mannitol, *VS* vegetative or seedling stage, *RS* reproductive or flowering stage. Abbreviations: *OE* over-expression, *RI* RNAi repression, *KM* knockout mutant regulation of ABA signaling, and downstream functional proteins. A comprehensive review on this information will be useful to the communities for the future efforts in developing stress-resistant crops by exploring the knowledge of ABA biology.

22.2 Enhancing Stress Resistance by Manipulating ABA Metabolism

The endogenous ABA level is determined by ABA biosynthesis, catabolism, and the release of ABA from ABA–glucose conjugates (Nambara and Marion-Poll 2005; Lee et al. 2006). Therefore, the identification of all of the components affecting ABA metabolism is essential for a complete understanding of the action of the hormone, which in turn helps in manipulating ABA levels for stress resistance. The mutants that were unable to produce carotenoid precursors for endogenous ABA synthesis exhibited pre-harvest seed sprouting and/or wilted leaves in maize, rice, and tomato, and such mutants often showed drought-sensitive phenotypes (Schwartz et al. 1997; Burbidge et al. 1999; Fang et al. 2008). Some genes controlling the early steps of ABA biosynthesis have been used in genetic engineering for improving plant drought tolerance. For example, constitutive over-expression of a bacterial β-carotene hydroxygenase (BCH) gene *crtZ* in tobacco led to increased zeaxanthin (an ABA precursor) synthesis and enhanced UV tolerance (Gotz et al. 2002), providing an association between the xanthophyll cycle and stress tolerance in plants. Similarly, over-expression of a rice *DSM2* gene which encodes BCH and which converts β-carotene to zeaxanthin resulted in significantly improved drought resistance of the transgenic rice at both the seedling and reproductive stages (Du et al. 2010). *NCED* (9-cis-epoxycarotenoid dioxygenase) is a drought-inducible gene encoding a rate-limiting enzyme for ABA biosynthesis. Over-expression of *NCED* could increase endogenous ABA levels, trigger stomatal closure, and lead to improved drought tolerance in transgenic tomato (Thompson et al. 2000) and tobacco (Qin and Zeevaart 2002). The over-expression of the *NCED* gene by a constitutive promoter resulted in increased ABA accumulation and water-use efficiency (WUE), but with apparently little impact on long-term biomass production. Positive effects on drought stress under field conditions were also observed in transgenic rice over-expressing *AtNCED2* (Xiao et al. 2009). *LOS5/ABA3* encodes a molybdenum cofactor (MoCo) sulfurase which functions in the final step of ABA biosynthesis in Arabidopsis. When *LOS5/ABA3* was over-expressed in rice, transgenic rice showed improved spikelet fertility and grain yield under drought stress conditions in the field (Xiao et al. 2009). *LOS5* also had a positive effect on improving drought tolerance when it was over-expressed in soybean, cotton, and maize, and the phenotype was associated with ABA accumulation and the expression of stress-related genes via ABA signaling, which sequentially caused a set of stress-related physiological and biochemical responses (Yue et al. 2012; Li et al. 2013; Lu et al. 2013). Genetic association mapping identified allelic variation in *ZmAO3* which controls the biosynthesis of ABA, and the variation was associated significantly with the endogenous ABA levels in the silks of water-stressed plants (Setter et al. 2011). Fleury et al. (2010) found that the "drought tolerant" wheat genotype RAC875 produced fewer tillers, showed moderate osmotic adjustment, increased ABA content, and reduced stomatal conductance, providing a link between drought resistance and allelic variation of ABA biosynthesis in natural crop varieties.

ABA biosynthesis can also be affected by changes in the levels of other endogenous phytohormones. For example, *TLD1* is a rice *GH3-13* gene which encodes indole-3-acetic acid (IAA)-amido synthetase, and it is dramatically induced by drought stress. A gain-of-function mutant, *tld1-D*, showed increased ABA biosynthesis and expression of late embryogenesis abundant (LEA) protein genes and enhanced drought tolerance (Zhang et al. 2009). However, over-expressing *OsGH3-2*, which also encodes an IAA-amido synthetase, led to significantly reduced ABA levels but increased resistance to cold stress (Du et al. 2012).

22.3 Improving Stress Tolerance by Manipulating ABA Signaling

A core ABA signaling model has been proposed which includes the binding of ABA to the receptors PYR/PYL/RCAR which inhibit the type 2C protein phosphatases (PP2C) of group A, resulting in the activation of SNF1-related type 2 protein kinases (SnRK2s), which subsequently activate various ion channels and ABA-dependent gene expression by phosphorylating bZIP transcription factors (Ma et al. 2009b; Raghavendra et al. 2010). The group A PP2C are negative regulators of ABA signaling. The elimination of group A PP2C is sufficient to ensure moss survival to full desiccation, implying that group A PP2Cs emerged in land plants to confer desiccation tolerance, possibly facilitating the propagation of plants on land (Komatsu et al. 2013). The maize ZmSAPK8/ZmOST1 sequence is highly homologous to that of the AtOST1 and is required for drought and salt tolerance responses, and this kinase can phosphorylate ZmSNAC1 together with ZmOST1 to function as a positive regulator of water deficit signaling in guard cells (Vilela et al. 2013). However, whether manipulation of *ZmOST1* can enhance drought tolerance in maize remains to be investigated. The SnRK2 subfamily consists of ten members that were designated as SAPK1 to SAPK10 (osmotic stress/ABA-activated protein kinases) in rice. All of the SAPKs were shown to be activated by hyperosmotic stress (Kobayashi et al. 2004). The SAPK8, SAPK9, and SAPK10 kinases were also activated by ABA and were able to phosphorylate the rice bZIP transcription factor TRAB1, suggesting that the regulation of the bZIP transcription factors by SnRK2 kinases is conserved among plant species (Kobayashi et al. 2005). In vitro phosphorylation assays demonstrated that the rice OsbZIP46 protein could be phosphorylated by

SAPK2 and SAPK6, suggesting that OsbZIP46 may be activated by posttranslational phosphorylation modification, which is related to the increased tolerance to drought and osmotic stresses in the *OsbZIP46CA1* over-expression rice (Tang et al. 2012). To date, despite the critical roles of ABA signaling in stress responses, very limited effort has been spent on exploring the ABA signaling components for improving stress resistance.

22.4 Improving Stress Tolerance by Engineering ABA-Related Transcription Factors

22.4.1 AREB/bZIPs

Transcription factors are involved in regulating or amplifying ABA signaling. Among them, ABA-responsive element-binding proteins/factors (AREB/ABF) (or bZIP transcription factors) play major roles in this aspect. These transcription factors contain a basic region which binds DNA and a leucine zipper dimerization motif to create an amphipathic helix and function in ABA-dependent regulatory systems in response to stress (Kang et al. 2002). A large number of studies reported that many bZIPs mediate the stress response by modulating ABA signaling intensively in model plants. Therefore, numerous efforts on enhancing stress tolerance by modulating bZIP transcription factors have been attempted in many crops. When an Arabidopsis bZIP gene *ABF3* was over-expressed in rice under the control of the maize ubiquitin promoter, the transgenic rice seedlings showed delayed leaf rolling and wilting after drought stress treatment (Oh et al. 2005). In rice, the bZIP family is comprised of approximately 100 members, and at least 33 OsbZIP genes are responsive to drought stress (Nijhawan et al. 2008). Constitutive over-expression of *OsbZIP23* in rice resulted in enhanced sensitivity to ABA and significantly improved tolerance to drought and high-salinity stresses (Xiang et al. 2008). Transgenic rice over-expressing *OsbZIP72* showed hypersensitivity to ABA, elevated levels of expression of ABA-responsive genes such as *LEAs*, and enhanced drought tolerance (Lu et al. 2009). Recently, OsbZIP46 in the constitutively active form was tested in rice, and over-expression of *OsbZIP46* resulted in improved drought tolerance (Tang et al. 2012). Meanwhile, some ABA-responsive bZIP genes in other crops have been identified or functionally verified by transgenic studies in model plants. For example, over-expression of the maize bZIP transcription factor *ZmbZIP72* enhanced the expression of ABA-inducible genes, such as *RD29B*, *RAB18*, and *HIS1-3*, and improved the drought and salt tolerance of transgenic *Arabidopsis* plants (Ying et al. 2012). Transgenic Arabidopsis over-expressing soybean *GmbZIP132* showed an increase in salt tolerance during germination (Liao et al. 2008a), and over-expression of *GmbZIP44*, *GmbZIP62*, and *GmbZIP78* resulted in increased tolerance to salt and freezing stresses (Liao et al. 2008b). Of special note is that a bZIP gene *GmbZIP1* from soybean confers stress tolerance not only to drought, but also to salt and cold stresses (Gao et al. 2011a).

Improvement of multi-stress tolerance was reported for transgenic tomato over-expressing *SlAREB1*, which showed increased tolerance to salt and water stress (Orellana et al. 2010). These reports suggest that the roles of ABA-dependent bZIP transcription factors in the modulation of stress tolerance are conserved in plants, and some of the bZIP genes are promising candidates for genetic engineering for developing stress-resistant crops in the future.

22.4.2 WRKYs

WRKY transcription factors, originally isolated from plants containing one or two conserved WRKY domains, are key regulators of many processes including the responses to biotic and abiotic stresses (Zou et al. 2004; Eulgem and Somssich 2007). Recent studies on the diverse ABA signaling pathways suggest that WRKY transcription factors may also play important roles in regulating plant responses to ABA. The magnesium-protoporphyrin IX chelatase H subunit (CHLH/ABAR) functions as a receptor of ABA in Arabidopsis (Shen et al. 2006). ABAR spans the chloroplast envelope, and the cytosolic C terminus of ABAR interacts with a group of WRKY transcription factors (WRKY40, WRKY18, and WRKY60) which function as negative regulators of ABA signaling in seed germination and post-germination growth (Shang et al. 2010). WRKY40 binds in vivo to W box-containing fragments of the promoters of many ABA-dependent genes such as *ABI4*, *ABI5*, *ABF4*, and *RAB18*, which places these WRKYs upstream of other known ABA-responsive transcription factors (Shang et al. 2010).

In crops, a few WRKYs were reported for their involvement in ABA and stress response regulation. Tao et al. (2011) reported that two alleles of *OsWRKY45* play different roles in ABA signaling and salt stress adaptation in rice. *OsWRKY45-1*-over-expressing lines showed reduced ABA sensitivity, but no obvious differences in response to salt stress. In contrast, *OsWRKY45-2*-over-expressing lines showed increased ABA sensitivity and reduced salt stress tolerance. In addition, *OsWRKY45-1* and *OsWRKY45-2* transgenic plants showed differential expression of a set of ABA- and abiotic stress-responsive genes, although they showed similar responses to cold and drought stresses. The different roles of the two alleles in ABA signaling and salt stress may be due to their transcriptional mediation of different signaling pathways (Tao et al. 2011). Over-expression of another WRKY gene *OsWRKY11* caused significant desiccation tolerance and induction of genes encoding raffinose synthase and galactinol synthase (Wu et al. 2009). Recently, OsWRKY13 was reported to selectively bind to the DNA segments harboring known *cis*-elements in the promoters of *SNAC1* and *OsWRKY45-1* in vivo during abiotic stresses, thus directly suppressing the transcription of *SNAC1* and *WRKY45-1* in vascular tissue and guard cells (Xiao et al. 2013). *OsWRKY13*-suppressing plants showed an increased *SNAC1* transcript level and enhanced tolerances to drought stress (Xiao et al. 2013). These results together suggest that some members of the WRKY family are also involved in the regulation ABA signaling and are potential candidate genes for improving stress resistance, although different WRKYs may have different roles in the regulation of stress responses.

22.4.3 Other ABA-Related Transcription Factors

Dehydration-responsive element-binding (DREB) transcription factors specifically interact with the dehydration-responsive element/C-repeat (DRE/CRT) *cis*-acting element and control the expression of many stress-inducible genes in plants (Yamaguchi-Shinozaki and Shinozaki 1994). Previous studies often presumed that DREBs act in the regulation of ABA-independent pathways. However, several studies have suggested that some DREBs may also function in an ABA-dependent manner for regulating stress tolerance. For example, *OsDREB1F* was induced by ABA, and over-expression of *OsDREB1F* in Arabidopsis resulted in the upregulation of *RD29B* and *RAB18*, which are typical ABA-dependent stress-induced genes, suggesting that activation of stress-responsive genes by OsDREB1F may be achieved via both ABA-dependent and ABA-independent pathways (Wang et al. 2008). *ARAG1*, another ABA-responsive *DREB* gene, plays a role in seed germination and drought tolerance of rice, and an *ARAG1*-knockdown line was hypersensitive to ABA during seed germination and seedling growth (Zhao et al. 2010).

NAC (NAM, ATAF, and CUC) is a plant-specific transcription factor family with diverse roles in development and stress regulation (Nakashima et al. 2012). *SNAC1* is an ABA and stress-responsive NAC gene conferring drought resistance in rice (Hu et al. 2006). *SNAC1*-over-expressing transgenic rice plants were hypersensitive to ABA application during seedling growth, and the over-expression lines exhibited 22–34 % higher seed setting in the field than the negative control under severe drought stress conditions imposed at the reproductive stage (Hu et al. 2006). Over-expression of another rice stress-responsive NAC gene *SNAC2* resulted in significantly increased sensitivity to ABA, and the transgenic rice showed enhanced cold tolerance (Hu et al. 2008). These results suggest that some NAC transcription factors also function as important transcriptional activators in ABA-dependent gene expression and may be useful in stress tolerance improvement in crops.

Ethylene response factors (ERFs) are plant transcriptional regulators mediating the progression of plant development and stress response (Gutterson and Reuber 2004). The tomato ERF protein TSRF1 is able to interact with a GCC box-like sequence containing the core sequence of the ZmABI4-binding-CE1-like element and regulates the ABA response, and over-expression of *TSRF1* in tobacco enhanced ABA sensitivity during germination, cotyledon expansion, and root elongation (Zhang et al. 2008a). In addition, over-expression of *TSRF1* in rice led to improved drought and osmotic tolerance (Quan et al. 2010). Another tomato ERF transcription activator, TERF1, was proposed to act as a linker between the ethylene and osmotic signaling pathways, and over-expression of TERF1 in tobacco enhanced drought tolerance and ABA sensitivity during seedling development

(Huang et al. 2004; Zhang et al. 2005). In rice, ABA suppresses the expression of *SUB1A* (also encoding an ERF protein), and over-expression of *SUB1A* resulted in an ABA-hypersensitive phenotype, thereby activating stress-inducible gene expression and improving drought and submergence resistance (Fukao et al. 2011).

Some members of the MYB transcription factor family also regulate ABA-related stress responses in plants. Transgenic tobacco constitutively expressing *AmMYB1*, a single-repeat MYB isolated from the salt-tolerant mangrove tree *Avicennia marina*, showed reduced sensitivity to ABA but enhanced tolerance to salinity stress, suggesting that AmMYB1 may function as a regulator of ABA and stress signaling (Ganesan et al. 2012). In rice, the expression of *OsMYB3R-2* was induced by cold, drought, and salt stress. Transgenic Arabidopsis plants over-expressing *OsMYB3R-2* showed increased tolerance to cold, drought, salt, and ABA treatments (Dai et al. 2007). A subsequent study demonstrated that *OsMYB3R-2* may play an important role in the cold stress signaling pathway modulated by the cell cycle and a putative DREB/CBF pathway in rice (Ma et al. 2009a). In addition, over-expression of the R2R3-type MYB gene *OsMYB2* in rice enhanced salt and dehydration tolerance, and the transgenic plants were more sensitive to ABA (Yang et al. 2012).

A few zinc finger transcription factors are also involved in the regulation of ABA and stress responses. Transgenic rice over-expressing a zinc finger gene *ZFP245* showed increased sensitivity to exogenous ABA and tolerance to cold and drought stresses by regulating proline levels and reactive oxygen species-scavenging activities (Huang et al. 2009). Over-expression of another zinc finger gene *ZFP179* in rice also resulted in hypersensitivity to exogenous ABA and increased salt tolerance at the seedling stage, and it was suggested that *ZFP179* was involved in both the ABA-dependent and ABA-independent pathways (Sun et al. 2010).

Nuclear factor Y (NF-Y) is a heterotrimeric transcription factor complex composed of the NF-YA, NF-YB, and NF-YC proteins. Over-expression of *GmNFYA3*, encoding the NF-YA subunit of the NF-Y complex in soybeans, resulted in reduced leaf water loss and enhanced drought tolerance in Arabidopsis (Ni et al. 2013). In addition, the transcript levels of ABA biosynthesis (*ABA1*, *ABA2*), ABA signaling (*ABI1*, *ABI2*), and stress-responsive genes (*RD29A* and *CBF3*) were generally higher in *GmNFYA3*-over-expression plants than in the controls, suggesting that *GmNFYA3* functions in the positive modulation of drought stress tolerance, partially through the regulation of ABA-related pathways (Ni et al. 2013).

22.5 Manipulating Posttranscriptional Regulators of ABA Signaling for Improving Stress Tolerance

Increasing evidence suggests that ABA signaling may also be regulated at the posttranscriptional or posttranslational levels. SDIR1 is a RING finger E3 ligase which positively regulates stress-responsive ABA signaling in Arabidopsis, and over-expression of *SDIR1* leads to enhanced ABA-induced stomatal closure and drought tolerance (Zhang et al. 2007). Ectopic expression of SDIR1 in rice and tobacco plants also led to improved drought tolerance (Zhang et al. 2008b). Transgenic rice over-expressing an ortholog of SDIR1 and OsSDIR1 showed strong drought tolerance compared to control plants (Gao et al. 2011b). Drought-induced expression of *ZmRFP1*, a putative ortholog of SDIR1 in maize, was impaired in the ABA-deficient mutant *viviparous14*, indicating that *ZmRFP1* responds to drought stress in an ABA-dependent way (Xia et al. 2012). OsDSG1, which is most similar to Arabidopsis AIP2, has E3 ubiquitin ligase activity and can target ABI3, and the *osdsg1* mutant plants that exhibited significantly increased expression of ABA signaling and responsive genes were more tolerant to salt and drought stresses (Park et al. 2010). In addition, transgenic tobacco constitutively over-expressing *OsBIRF1*, which encodes a rice RING-H2 finger protein, exhibited reduced ABA sensitivity and increased drought tolerance (Liu et al. 2008).

Small interfering RNAs (siRNAs) may also be involved in the regulation of ABA-related stress responses. In rice, *siR441* and *siR446* are positive regulators of ABA signaling and tolerance to abiotic stress, possibly by regulating *MAIF1* expression (Yan et al. 2011a). The expression of *MAIF1*, encoding an F-box domain protein mainly localized in the plasma membrane and nucleus, is induced rapidly and strongly by ABA and abiotic stresses, and over-expression of *MAIF1* reduced ABA sensitivity and abiotic stress tolerance in rice (Yan et al. 2011b).

Mitogen-activated protein kinase (MAPK) cascades play an important role in mediating ABA and stress responses in eukaryotic organisms. In rice, basal-level *OsMAPK5* can be activated by ABA very quickly before the accumulation of its mRNA and protein, and *OsMAPK5*-over-expression rice exhibited increased OsMAPK5 kinase activity and increased tolerance to drought, salt, and cold stresses (Xiong and Yang 2003).

22.6 Improving Stress Resistance by Manipulating Other ABA-Related Functional Genes

The output of ABA signaling and/or its regulation includes expression changes or the activation of numerous downstream functional proteins which translate the original signal input into diverse physiological responses to the stresses (see reviews by Nambara and Marion-Poll 2005; Boursiac et al. 2013; Fukao and Xiong 2013; Hu and Xiong 2014). Here, we only introduce some of the ABA-related downstream genes which were seldom reviewed previously.

ABA accumulation in the roots of stressed plants can stimulate root growth and increase root hydraulic conductivity (Zhu et al. 2005; Lian et al. 2006). In this process, aquaporins, or water channel proteins which are members of major intrinsic proteins (MIPs) and translocate water across cell membranes, have been demonstrated for their roles in stress-induced physiological processes (Ishibashi et al. 2000; Javot and Maurel 2002). Plasma membrane intrinsic proteins (PIPs) have also been investigated for their induction under water deficit conditions (Malz and

Sauter 1999). ABA application can increase hydraulic conductance, thus promoting water movement from roots to leaves (Hose et al. 2000). The early induction of *ZmPIP2-4* may also be mediated by ABA, implying that ABA may transiently induce the water conductivity of roots and the water permeability of cortex cells in maize (Zhu et al. 2005). The expression of the water channel protein gene *RWC3* was increased in upland rice at the early stages of treatment with 20 % polyethylene glycol (PEG) 6000, whereas no significant expression change was detected in lowland rice, indicating that *RWC3* probably plays a role in drought avoidance in rice (Lian et al. 2004). Treatment with ABA also enhanced the expression of several PIP genes in upland rice (Lian et al. 2006), and transgenic Arabidopsis over-expressing *OsPIP2-2* showed enhanced tolerance to salt and drought stresses (Guo et al. 2006).

The functions of some protein-modification enzymes are also involved in ABA responses. Squalene synthase (SQS) is a farnesyl-diphosphate farnesyltransferase which catalyses the first reaction of the branch of the isoprenoid metabolic pathway committed specifically to sterol biosynthesis (Tansey and Shechter 2001). It has been demonstrated that RNAi-mediated disruption of a rice SQS gene resulted in reduced sensitivity to ABA and improved drought tolerance at both the vegetative and reproductive stages (Manavalan et al. 2012). Transgenic canola over-expressing a farnesyltransferase gene *ERA1* driven by a drought and ABA-inducible *rd29A* promoter showed increased ABA sensitivity, a significant decrease in stomatal conductance, and increased drought tolerance at mid-flowering under field conditions (Wang et al. 2005).

22.7 Conclusions and Perspectives

It is apparent that ABA, as an important signal molecule, can mediate signal transduction involved in the responses to multiple abiotic stresses. The core ABA signal perception components which involve PYR/PYL/RCAR, PPC2, and SnRK2 in Arabidopsis are especially important in understanding the molecular basis of a regulatory network controlling ABA-dependent stress responses (Ma et al. 2009b; Nishimura et al. 2009). A recent study showed that a large number of ABA biosynthesis and signaling components are conserved in plants (Hauser et al. 2011). These findings will help us to discover similar ABA-regulated pathways and networks in crops and to discover potential candidate genes for genetic improvement of stress tolerance in crops. However, when compared to Arabidopsis, the amount of data on ABA-regulated genes and their functions in crop species is scattered. Further intensive studies on ABA biology in economically important crops need to be carried out to gain better insights into the mechanisms of stress responses.

It should also be noted that there are negative side effects of ABA on normal growth and yield potential. For example, over-expression of *NCED* by a constitutive promoter in tomato had negative effects (increased seed dormancy), while transgenic tomato over-expressing *LeNCED1* by a Rubisco rbcS3C promoter, which is light-responsive and expressed in a circadian manner, showed much higher levels of ABA accumulation in leaves and sap, but no negative effects on seed dormancy. However, these plants exhibited a set of severe abnormal symptoms resulting from the consequences of long-term and high levels of ABA accumulation (Tung et al. 2008). Therefore, the moderate and timely increase in ABA biosynthesis may be critical for genetic improvement of stress resistance in crops. In addition, stress-inducible promoters may be useful for engineering of ABA-related genes. For example, transgenic rice over-expressing the trehalose-6-phosphate synthase/phosphatase gene by a drought-inducible promoter showed increased amounts of trehalose and drought tolerance without obvious negative effects on plant growth or grain yield (Garg et al. 2002). In the future, the application of specific genes on stress resistance needs to be optimized by using stress-inducible and/or tissue-specific promoters.

Even though many ABA-related genes have proven to be effective in enhancing stress resistance by transgenic approaches, the effects of these genes were mostly shown based on greenhouse experiments with plants grown in small pots or only at the seedling stages. Due to the complex nature of stress in the field, those genes proven to be effective in the greenhouse have to be further evaluated in the field before being adopted in breeding programs.

The last challenge is how to tackle the different abiotic stresses that simultaneously occur under natural field conditions. In the future, it is essential to examine the effects of ABA-related genes for stress resistance under multiple stress combination conditions. Meanwhile, the nature of different crops in stress adaptation may further complicate the stress resistance effect for the same gene(s). Therefore, intensive knowledge on ABA biology obtained from model plants needs to be integrated into specific crops according to their inherent nature of stress response and adaptation to specific stresses.

Acknowledgments The work from the authors' laboratory cited in this review was supported by grants from the National Program of China for Transgenic Research (2011ZX08009-003-002, 2011ZX08001-003), the National Program for Basic Research of China (2012CB114305), and the National Program on High Technology Development (2012AA10A303).

References

Borlaug N. Feeding a hungry world. Science. 2007;318:359.

Boursiac Y, Leran S, Corratge-Faillie C, Gojon A, Krouk G, Lacombe B. ABA transport and transporters. Trends Plant Sci. 2013;18:325–33.

Burbidge A, Grieve TM, Jackson A, Thompson A, McCarty DR, Taylor IB. Characterization of the ABA-deficient tomato mutant notabilis and its relationship with maize Vp14. Plant J. 1999;17:427–31.

Dai X, Xu Y, Ma Q, Xu W, Wang T, Xue Y, Chong K. Overexpression of an R1R2R3 MYB gene, OsMYB3R-2, increases tolerance to freezing, drought, and salt stress in transgenic Arabidopsis. Plant Physiol. 2007;143:1739–51.

Du H, Wang N, Cui F, Li X, Xiao J, Xiong L. Characterization of the β-carotene hydroxylase gene DSM2 conferring drought and oxidative stress resistance by increasing xanthophylls and abscisic acid synthesis in rice. Plant Physiol. 2010;154:1304–18.

Du H, Wu N, Fu J, Wang S, Li X, Xiao J, Xiong L. A GH3 family member, OsGH3-2, modulates auxin and abscisic acid levels and differentially affects drought and cold tolerance in rice. J Exp Bot. 2012;63:6467–80.

Du H, Wu N, Chang Y, Li X, Xiao J, Xiong L. Carotenoid deficiency impairs ABA and IAA biosynthesis and differentially affects drought and cold tolerance in rice. Plant Mol Biol. 2013;83:475–88.

Eulgem T, Somssich IE. Networks of WRKY transcription factors in defense signaling. Curr Opin Plant Biol. 2007;10:366–71.

Fang J, Chai C, Qian Q, Li C, Tang J, Sun L, Huang Z, Guo X, Sun C, Liu M, Zhang Y, Lu Q, Wang Y, Lu C, Han B, Chen F, Cheng Z, Chu C. Mutations of genes in synthesis of the carotenoid precursors of ABA lead to pre-harvest sprouting and photo-oxidation in rice. Plant J. 2008;54:177–89.

Fleury D, Jefferies S, Kuchel H, Langridge P. Genetic and genomic tools to improve drought tolerance in wheat. J Exp Bot. 2010;61:3211–22.

Fukao T, Xiong L. Genetic mechanisms conferring adaptation to submergence and drought in rice: simple or complex? Curr Opin Plant Biol. 2013;16:196–204.

Fukao T, Yeung E, Bailey-Serres J. The submergence tolerance regulator SUB1A mediates crosstalk between submergence and drought tolerance in rice. Plant Cell. 2011;23:412–27.

Ganesan G, Sankararamasubramanian HM, Harikrishnan M, Ganpudi A, Parida A. A MYB transcription factor from the grey mangrove is induced by stress and confers NaCl tolerance in tobacco. J Exp Bot. 2012;63:4549–61.

Gao SQ, Chen M, Xu ZS, Zhao CP, Li L, Xu HJ, Tang YM, Zhao X, Ma YZ. The soybean GmbZIP1 transcription factor enhances multiple abiotic stress tolerances in transgenic plants. Plant Mol Biol. 2011a;75:537–53.

Gao T, Wu Y, Zhang Y, Liu L, Ning Y, Wang D, Tong H, Chen S, Chu C, Xie Q. OsSDIR1 overexpression greatly improves drought tolerance in transgenic rice. Plant Mol Biol. 2011b;76:145–56.

Garg AK, Kim JK, Owens TG, Ranwala AP, Choi YD, Kochian LV, Wu RJ. Trehalose accumulation in rice plants confers high tolerance levels to different abiotic stresses. Proc Natl Acad Sci USA. 2002;99:15898–903.

Gotz T, Sandmann G, Romer S. Expression of a bacterial carotene hydroxylase gene (crtZ) enhances UV tolerance in tobacco. Plant Mol Biol. 2002;50:129–42.

Guo L, Wang ZY, Lin H, Cui WE, Chen J, Liu M, Chen ZL, Qu LJ, Gu H. Expression and functional analysis of the rice plasma-membrane intrinsic protein gene family. Cell Res. 2006;16:277–86.

Gutterson N, Reuber TL. Regulation of disease resistance pathways by AP2/ERF transcription factors. Curr Opin Plant Biol. 2004;7:465–71.

Hauser F, Waadt R, Schroeder JI. Evolution of abscisic acid synthesis and signaling mechanisms. Curr Biol. 2011;21:R346–55.

Hose E, Steudle E, Hartung W. Abscisic acid and hydraulic conductivity of maize roots: a study using cell- and root-pressure probes. Planta. 2000;211:874–82.

Hu H, Xiong L. Genetic engineering and breeding of drought-resistant crops. Annu Rev Plant Biol. 2014;65:715–41.

Hu H, Dai M, Yao J, Xiao B, Li X, Zhang Q, Xiong L. Overexpressing a NAM, ATAF, and CUC (NAC) transcription factor enhances drought resistance and salt tolerance in rice. Proc Natl Acad Sci USA. 2006;103:12987–92.

Hu H, You J, Fang Y, Zhu X, Qi Z, Xiong L. Characterization of transcription factor gene SNAC2 conferring cold and salt tolerance in rice. Plant Mol Biol. 2008;67:169–81.

Huang Z, Zhang Z, Zhang X, Zhang H, Huang D, Huang R. Tomato TERF1 modulates ethylene response and enhances osmotic stress tolerance by activating expression of downstream genes. FEBS Lett. 2004;573:110–6.

Huang J, Sun SJ, Xu DQ, Yang X, Bao YM, Wang ZF, Tang HJ, Zhang H. Increased tolerance of rice to cold, drought and oxidative stresses mediated by the overexpression of a gene that encodes the zinc finger protein ZFP245. Biochem Biophys Res Commun. 2009;389:556–61.

Ishibashi K, Kuwahara M, Sasaki S. Molecular biology of aquaporins. Rev Physiol Biochem Pharmacol. 2000;141:1–32.

Javot H, Maurel C. The role of aquaporins in root water uptake. Ann Bot. 2002;90:301–13.

Kang JY, Choi HI, Im MY, Kim SY. Arabidopsis basic leucine zipper proteins that mediate stress-responsive abscisic acid signaling. Plant Cell. 2002;14:343–57.

Ji XM, Dong BD, Shiran B, Talbot MJ, Edlington JE, Hughes T, White RG, Gubler F, Dolferus R. Control of abscisic acid catabolism and abscisic acid homeostasis is important for reproductive stage stress tolerance in cereals. Plant Physiol. 2011;156:647–62.

Kobayashi Y, Yamamoto S, Minami H, Kagaya Y, Hattori T. Differential activation of the rice sucrose nonfermenting1-related protein kinase2 family by hyperosmotic stress and abscisic acid. Plant Cell. 2004;16:1163–77.

Kobayashi Y, Murata M, Minami H, Yamamoto S, Kagaya Y, Hobo T, Yamamoto A, Hattori T. Abscisic acid-activated SNRK2 protein kinases function in the gene-regulation pathway of ABA signal transduction by phosphorylating ABA response element-binding factors. Plant J. 2005;44:939–49.

Komatsu K, Suzuki N, Kuwamura M, Nishikawa Y, Nakatani M, Ohtawa H, Takezawa D, Seki M, Tanaka M, Taji T, Hayashi T, Sakata Y. Group A PP2Cs evolved in land plants as key regulators of intrinsic desiccation tolerance. Nat Commun. 2013;4:2219.

Landi P, Sanguineti MC, Salvi S, Giuliani S, Bellotti M, Maccaferri M, Conti S, Tuberosa R. Validation and characterization of a major QTL affecting leaf ABA concentration in maize. Mol Breed. 2005;15:291–303.

Lee KH, Piao HL, Kim HY, Choi SM, Jiang F, Hartung W, Hwang I, Kwak JM, Lee IJ, Hwang I. Activation of glucosidase via stress-induced polymerization rapidly increases active pools of abscisic acid. Cell. 2006;126:1109–20.

Li Y, Zhang J, Hao L, Hua J, Duan L, Zhang M, Li Z. Expression of an Arabidopsis molybdenum cofactor sulphurase gene in soybean enhances drought tolerance and increases yield under field conditions. Plant Biotechnol J. 2013;11:747–58.

Lian HL, Yu X, Ye Q, Ding X, Kitagawa Y, Kwak SS, Su WA, Tang ZC. The role of aquaporin RWC3 in drought avoidance in rice. Plant Cell Physiol. 2004;45:481–9.

Lian HL, Yu X, Lane D, Sun WN, Tang ZC, Su WA. Upland rice and lowland rice exhibited different PIP expression under water deficit and ABA treatment. Cell Res. 2006;16:651–60.

Liao Y, Zhang JS, Chen SY, Zhang WK. Role of soybean GmbZIP132 under abscisic acid and salt stresses. J Integr Plant Biol. 2008a;50:221–30.

Liao Y, Zou HF, Wei W, Hao YJ, Tian AG, Huang J, Liu YF, Zhang JS, Chen SY. Soybean GmbZIP44, GmbZIP62 and GmbZIP78 genes function as negative regulator of ABA signaling and confer salt and freezing tolerance in transgenic Arabidopsis. Planta. 2008b;228:225–40.

Liu H, Zhang H, Yang Y, Li G, Yang Y, Wang X, Basnayake BM, Li D, Song F. Functional analysis reveals pleiotropic effects of rice RING-H2 finger protein gene OsBIRF1 on regulation of growth and defense responses against abiotic and biotic stresses. Plant Mol Biol. 2008;68:17–30.

Lu G, Gao C, Zheng X, Han B. Identification of OsbZIP72 as a positive regulator of ABA response and drought tolerance in rice. Planta. 2009;229:605–15.

Lu Y, Li Y, Zhang J, Xiao Y, Yue Y, Duan L, Zhang M, Li Z. Overexpression of Arabidopsis molybdenum cofactor sulfurase gene confers drought tolerance in maize (*Zea mays* L.). PLoS ONE. 2013;8:e52126.

Ma Q, Dai X, Xu Y, Guo J, Liu Y, Chen N, Xiao J, Zhang D, Xu Z, Zhang X, Chong K. Enhanced tolerance to chilling stress in OsMYB3R-2 transgenic rice is mediated by alteration in cell cycle and ectopic expression of stress genes. Plant Physiol. 2009a;150:244–56.

Ma Y, Szostkiewicz I, Korte A, Moes D, Yang Y, Christmann A, Grill E. Regulators of PP2C phosphatase activity function as abscisic acid sensors. Science. 2009b;324:1064–8.

Malz S, Sauter M. Expression of two PIP genes in rapidly growing internodes of rice is not primarily controlled by meristem activity or cell expansion. Plant Mol Biol. 1999;40:985–95.

Manavalan LP, Chen X, Clarke J, Salmeron J, Nguyen HT. RNAi-mediated disruption of squalene synthase improves drought tolerance and yield in rice. J Exp Bot. 2012;63:163–75.
Mittler R. Abiotic stress, the field environment and stress combination. Trends Plant Sci. 2006;11:15–9.
Nakashima K, Yamaguchi-Shinozaki K. ABA signaling in stress-response and seed development. Plant Cell Rep. 2013;32:959–70.
Nakashima K, Takasaki H, Mizoi J, Shinozaki K, Yamaguchi-Shinozaki K. NAC transcription factors in plant abiotic stress responses. Biochim Biophys Acta. 2012;1819:97–103.
Nambara E, Marion-Poll A. Abscisic acid biosynthesis and catabolism. Annu Rev Plant Biol. 2005;56:165–85.
Ni Z, Hu Z, Jiang Q, Zhang H. GmNFYA3, a target gene of miR169, is a positive regulator of plant tolerance to drought stress. Plant Mol Biol. 2013;82:113–29.
Nijhawan A, Jain M, Tyagi AK, Khurana JP. Genomic survey and gene expression analysis of the basic leucine zipper transcription factor family in rice. Plant Physiol. 2008;146:333–50.
Nishimura N, Hitomi K, Arvai AS, Rambo RP, Hitomi C, Cutler SR, Schroeder JI, Getzoff ED. Structural mechanism of abscisic acid binding and signaling by dimeric PYR1. Science. 2009;326:1373–9.
Oh S-J, Song SI, Kim YS, Jang H-J, Kim SY, Kim M, Kim Y-K, Nahm BH, Kim J-K. Arabidopsis CBF3/DREB1A and ABF3 in transgenic rice increased tolerance to abiotic stress without stunting growth. Plant Physiol. 2005;138:341–51.
Orellana S, Yanez M, Espinoza A, Verdugo I, Gonzalez E, Ruiz-Lara S, Casaretto JA. The transcription factor SlAREB1 confers drought, salt stress tolerance and regulates biotic and abiotic stress-related genes in tomato. Plant Cell Environ. 2010;33:2191–208.
Park GG, Park JJ, Yoon J, Yu SN, An G. A RING finger E3 ligase gene, *Oryza sativa* delayed seed germination 1 (OsDSG1), controls seed germination and stress responses in rice. Plant Mol Biol. 2010;74:467–78.
Qin X, Zeevaart JA. Overexpression of a 9-cis-epoxycarotenoid dioxygenase gene in *Nicotiana plumbaginifolia* increases abscisic acid and phaseic acid levels and enhances drought tolerance. Plant Physiol. 2002;128:544–51.
Quan R, Hu S, Zhang Z, Zhang H, Zhang Z, Huang R. Overexpression of an ERF transcription factor TSRF1 improves rice drought tolerance. Plant Biotechnol J. 2010;8:476–88.
Quarrie SA, Laurie DA, Zhu J, Lebreton C, Semikhodskii A, Steed A, Witsenboer H, Calestani C. QTL analysis to study the association between leaf size and abscisic acid accumulation in droughted rice leaves and comparisons across cereals. Plant Mol Biol. 1997;35:155–65.
Raghavendra AS, Gonugunta VK, Christmann A, Grill E. ABA perception and signalling. Trends Plant Sci. 2010;15:395–401.
Schwartz SH, Tan BC, Gage DA, Zeevaart JA, McCarty DR. Specific oxidative cleavage of carotenoids by VP14 of maize. Science. 1997;276:1872–4.
Setter TL, Yan J, Warburton M, Ribaut JM, Xu Y, Sawkins M, Buckler ES, Zhang Z, Gore MA. Genetic association mapping identifies single nucleotide polymorphisms in genes that affect abscisic acid levels in maize floral tissues during drought. J Exp Bot. 2011;62:701–16.
Shang Y, Yan L, Liu ZQ, Cao Z, Mei C, Xin Q, Wu FQ, Wang XF, Du SY, Jiang T, Zhang XF, Zhao R, Sun HL, Liu R, Yu YT, Zhang DP. The Mg-chelatase H subunit of Arabidopsis antagonizes a group of WRKY transcription repressors to relieve ABA-responsive genes of inhibition. Plant Cell. 2010;22:1909–35.
Shen YY, Wang XF, Wu FQ, Du SY, Cao Z, Shang Y, Wang XL, Peng CC, Yu XC, Zhu SY, Fan RC, Xu YH, Zhang DP. The Mg-chelatase H subunit is an abscisic acid receptor. Nature. 2006;443:823–6.
Sun SJ, Guo SQ, Yang X, Bao YM, Tang HJ, Sun H, Huang J, Zhang HS. Functional analysis of a novel Cys2/His2-type zinc finger protein involved in salt tolerance in rice. J Exp Bot. 2010;61:2807–18.
Tang N, Zhang H, Li X, Xiao J, Xiong L. Constitutive activation of transcription factor OsbZIP46 improves drought tolerance in rice. Plant Physiol. 2012;158:1755–68.

Tansey TR, Shechter I. Squalene synthase: structure and regulation. Prog Nucleic Acid Res Mol Biol. 2001;65:157–95.

Tao Z, Kou Y, Liu H, Li X, Xiao J, Wang S. OsWRKY45 alleles play different roles in abscisic acid signalling and salt stress tolerance but similar roles in drought and cold tolerance in rice. J Exp Bot. 2011;62:4863–74.

Thompson AJ, Jackson AC, Symonds RC, Mulholland BJ, Dadswell AR, Blake PS, Burbidge A, Taylor IB. Ectopic expression of a tomato 9-cis-epoxycarotenoid dioxygenase gene causes over-production of abscisic acid. Plant J. 2000;23:363–74.

Tuberosa R, Sanguineti MC, Landi P, Salvi S, Casarini E, Conti S. RFLP mapping of quantitative trait loci controlling abscisic acid concentration in leaves of drought-stressed maize (*Zea mays* L.). Theor Appl Genet. 1998;97:744–55.

Tung SA, Smeeton R, White CA, Black CR, Taylor IB, Hilton HW, Thompson AJ. Over-expression of LeNCED1 in tomato (*Solanum lycopersicum* L.) with the rbcS3C promoter allows recovery of lines that accumulate very high levels of abscisic acid and exhibit severe phenotypes. Plant Cell Environ. 2008;31:968–81.

Verslues PE, Agarwal M, Katiyar-Agarwal S, Zhu J, Zhu JK. Methods and concepts in quantifying resistance to drought, salt and freezing, abiotic stresses that affect plant water status. Plant J. 2006;45:523–39.

Vilela B, Moreno-Cortes A, Rabissi A, Leung J, Pages M, Lumbreras V. The maize OST1 kinase homolog phosphorylates and regulates the maize SNAC1-type transcription factor. PLoS ONE. 2013;8:e58105.

Wang Y, Ying J, Kuzma M, Chalifoux M, Sample A, McArthur C, Uchacz T, Sarvas C, Wan J, Dennis DT, McCourt P, Huang Y. Molecular tailoring of farnesylation for plant drought tolerance and yield protection. Plant J. 2005;43:413–24.

Wang Y, Beaith M, Chalifoux M, Ying J, Uchacz T, Sarvas C, Griffiths R, Kuzma M, Wan J, Huang Y. Shoot-specific down-regulation of protein farnesyltransferase (α-subunit) for yield protection against drought in canola. Mol Plant 2009;2:191–200.

Wang Q, Guan Y, Wu Y, Chen H, Chen F, Chu C. Overexpression of a rice OsDREB1F gene increases salt, drought, and low temperature tolerance in both Arabidopsis and rice. Plant Mol Biol. 2008;67:589–602.

Witcombe JR, Hollington PA, Howarth CJ, Reader S, Steele KA. Breeding for abiotic stresses for sustainable agriculture. Philos Trans R Soc Lond B Biol Sci. 2008;363:703–16.

Wu L, Zhang Z, Zhang H, Wang XC, Huang R. Transcriptional modulation of ethylene response factor protein JERF3 in the oxidative stress response enhances tolerance of tobacco seedlings to salt, drought, and freezing. Plant Physiol. 2008;148:1953–63.

Wu X, Shiroto Y, Kishitani S, Ito Y, Toriyama K. Enhanced heat and drought tolerance in transgenic rice seedlings overexpressing OsWRKY11 under the control of HSP101 promoter. Plant Cell Rep. 2009;28:21–30.

Xia Z, Liu Q, Wu J, Ding J. ZmRFP1, the putative ortholog of SDIR1, encodes a RING-H2 E3 ubiquitin ligase and responds to drought stress in an ABA-dependent manner in maize. Gene. 2012;495:146–53.

Xiang Y, Tang N, Du H, Ye H, Xiong L. Characterization of OsbZIP23 as a key player of the basic leucine zipper transcription factor family for conferring abscisic acid sensitivity and salinity and drought tolerance in rice. Plant Physiol. 2008;148:1938–52.

Xiao B-Z, Chen X, Xiang C-B, Tang N, Zhang Q-F, Xiong L-Z. Evaluation of seven function-known candidate genes for their effects on improving drought resistance of transgenic rice under field conditions. Mol Plant. 2009;2:73–83.

Xiao J, Cheng H, Li X, Xiao J, Xu C, Wang S. Rice WRKY13 regulates crosstalk between abiotic and biotic stress signaling pathways by selective binding to different cis-elements. Plant Physiol. 2013;163:1868–82.

Xiong L, Yang Y. Disease resistance and abiotic stress tolerance in rice are inversely modulated by an abscisic acid-inducible mitogen-activated protein kinase. Plant Cell. 2003;15:745–59.

Yamaguchi-Shinozaki K, Shinozaki K. A novel cis-acting element in an Arabidopsis gene is involved in responsiveness to drought, low-temperature, or high-salt stress. Plant Cell. 1994;6:251–64.

Yan Y, Zhang Y, Yang K, Sun Z, Fu Y, Chen X, Fang R. Small RNAs from MITE-derived stem-loop precursors regulate abscisic acid signaling and abiotic stress responses in rice. Plant J. 2011a;65:820–8.

Yan YS, Chen XY, Yang K, Sun ZX, Fu YP, Zhang YM, Fang RX. Overexpression of an F-box protein gene reduces abiotic stress tolerance and promotes root growth in rice. Mol Plant. 2011b;4:190–7.

Yang A, Dai X, Zhang WH. A R2R3-type MYB gene, OsMYB2, is involved in salt, cold, and dehydration tolerance in rice. J Exp Bot. 2012;63:2541–56.

Ying S, Zhang DF, Fu J, Shi YS, Song YC, Wang TY, Li Y. Cloning and characterization of a maize bZIP transcription factor, ZmbZIP72, confers drought and salt tolerance in transgenic Arabidopsis. Planta. 2012;235:253–66.

Yue Y, Zhang M, Zhang J, Tian X, Duan L, Li Z. Overexpression of the AtLOS5 gene increased abscisic acid level and drought tolerance in transgenic cotton. J Exp Bot. 2012;63:3741–8.

Zhang X, Zhang Z, Chen J, Chen Q, Wang XC, Huang R. Expressing TERF1 in tobacco enhances drought tolerance and abscisic acid sensitivity during seedling development. Planta. 2005;222:494–501.

Zhang Y, Yang C, Li Y, Zheng N, Chen H, Zhao Q, Gao T, Guo H, Xie Q. SDIR1 is a RING finger E3 ligase that positively regulates stress-responsive abscisic acid signaling in Arabidopsis. Plant Cell. 2007;19:1912–29.

Zhang H, Yang Y, Zhang Z, Chen J, Wang XC, Huang R. Expression of the ethylene response factor gene TSRF1 enhances abscisic acid responses during seedling development in tobacco. Planta. 2008a;228:777–87.

Zhang H, Liu W, Wan L, Li F, Dai L, Li D, Zhang Z, Huang R. Functional analyses of ethylene response factor JERF3 with the aim of improving tolerance to drought and osmotic stress in transgenic rice. Transgenic Res. 2010;19:809–18.

Zhang YY, Li Y, Gao T, Zhu H, Wang DJ, Zhang HW, Ning YS, Liu LJ, Wu YR, Chu CC, Guo HS, Xie Q. Arabidopsis SDIR1 enhances drought tolerance in crop plants. Biosci Biotechnol Biochem. 2008b;72:2251–4.

Zhang SW, Li CH, Cao J, Zhang YC, Zhang SQ, Xia YF, Sun DY, Sun Y. Altered architecture and enhanced drought tolerance in rice via the down-regulation of indole-3-acetic acid by TLD1/OsGH3.13 activation. Plant Physiol. 2009;151:1889–901.

Zhao L, Hu Y, Chong K, Wang T. ARAG1, an ABA-responsive DREB gene, plays a role in seed germination and drought tolerance of rice. Ann Bot. 2010;105:401–9.

Zhu C, Schraut D, Hartung W, Schaffner AR. Differential responses of maize MIP genes to salt stress and ABA. J Exp Bot. 2005;56:2971–81.

Zou X, Seemann JR, Neuman D, Shen QJ. A WRKY gene from creosote bush encodes an activator of the abscisic acid signaling pathway. J Biol Chem. 2004;279:55770–9.

Zurbriggen MD, Hajirezaei MR, Carrillo N. Engineering the future. Development of transgenic plants with enhanced tolerance to adverse environments. Biotechnol Genet Eng Rev. 2010;27:33–56.

MIX
Papier aus verantwortungsvollen Quellen
Paper from responsible sources
FSC® C105338
www.fsc.org

If you have any concerns about our products,
you can contact us on
ProductSafety@springernature.com

In case Publisher is established outside the EU,
the EU authorized representative is:
Springer Nature Customer Service Center GmbH
Europaplatz 3, 69115 Heidelberg, Germany

Printed by Libri Plureos GmbH
in Hamburg, Germany